Differential Equations on the Internet

The Boston University
Ordinary Differential Equations Project
http://math.bu.edu/odes

DIFFERENTIAL EQUATIONS

Fourth Edition
DIFFERENTIAL EQUATIONS

Paul Blanchard

Robert L. Devaney

Glen R. Hall

Boston University

BROOKS/COLE
CENGAGE Learning™

Australia • Brazil • Japan • Korea • Mexico • Singapore • Spain • United Kingdom • United States

BROOKS/COLE
CENGAGE Learning

Differential Equations, Fourth Edition
Paul Blanchard, Robert L. Devaney,
Glen R. Hall

Publisher: Richard Stratton

Senior Sponsoring Editor: Molly Taylor

Assistant Editor: Shaylin Walsh

Editorial Assistant: Alexander Gontar

Associate Media Editor: Andrew Coppola

Senior Marketing Manager:
Jennifer Pursley Jones

Marketing Coordinator: Michael Ledesma

Marketing Communications Manager:
Maryanne Payumo

Content Project Manager: Susan Miscio

Senior Art Director: Jill Ort

Print Buyer: Diane Gibbons

Permissions Editor: Mandy Groszko

Production Service and Compositor:
MPS Limited, a Macmillan Company

Cover Designer: RHDG | Riezebos Holzbaur

Cover Image: Photolibrary

For product information and technology assistance, contact us at
Cengage Learning Customer & Sales Support, 1-800-354-9706
For permission to use material from this text or product,
submit all requests online at **www.cengage.com/permissions**.
Further permissions questions can be emailed to
permissionrequest@cengage.com.

Library of Congress Control Number: 2011921135

ISBN-13: 978-1-133-10903-7

ISBN-10: 1-133-10903-9

Brooks/Cole
20 Channel Center Street
Boston, MA 02210
USA

Cengage Learning is a leading provider of customized learning solutions
with office locations around the globe, including Singapore, the United
Kingdom, Australia, Mexico, Brazil and Japan. Locate your local office at
international.cengage.com/region

Cengage Learning products are represented in Canada by
Nelson Education, Ltd.

For your course and learning solutions, visit **www.cengage.com**.

Purchase any of our products at your local college store or at our
preferred online store **www.cengagebrain.com**.

Printed in the United States of America
7 8 9 10 11 12 13   23 22 21 20 19

ABOUT THE AUTHORS

Paul Blanchard

Paul Blanchard grew up in Sutton, Massachusetts, spent his undergraduate years at Brown University, and received his Ph.D. from Yale University. He has taught college mathematics for more than thirty years, mostly at Boston University. In 2001, he won the Northeast Section of the Mathematical Association of America's Award for Distinguished Teaching of Mathematics, and in 2011, the conference *Differential Equations Across the Collegiate Curriculum* was held to celebrate his 60th birthday. His main area of mathematical research is complex analytic dynamical systems and the related point sets—Julia sets and the Mandelbrot set. For many of the last fifteen years, his efforts have focused on modernizing the traditional differential equations course. When he becomes exhausted fixing the errors made by his two coauthors, he usually closes up his coffee shop and heads for the golf course with his caddy, Glen Hall.

Robert L. Devaney

Robert L. Devaney was raised in Methuen, Massachusetts. He received his undergraduate degree from Holy Cross College and his Ph.D. from the University of California, Berkeley. Since 1980 he has taught at Boston University where he is the Feld Family Professor of Teaching Excellence in the College of Arts and Sciences. His main area of research is complex dynamical systems, and he has lectured extensively throughout the world on this topic. In 1996 he received the Deborah and Franklin Tepper Halmo Award for Distinguished University Teaching from the Mathematical Association of America. When he gets sick of arguing with his coauthors over which topics to include in the differential equations course, he either turns up the volume of his opera recordings, or heads for waters off New England for a long distance sail.

Glen R. Hall

Glen R. Hall spent his youth in Denver, Colorado, but he never learned to ski. His undergraduate degree comes from Carleton College in Minnesota and his Ph.D. comes from the University of Minnesota. His current research interests are in the field of dynamical systems, particularly celestial mechanics. For his research he has been awarded both NSF Postdoctoral and Sloan Foundation Fellowships. He once bicycled 148 miles in a single day but is now happy to bike 10 miles to campus.

PREFACE

The study of differential equations is a beautiful application of the ideas and techniques of calculus to our everyday lives. Indeed, it could be said that calculus was developed mainly so that the fundamental principles that govern many phenomena could be expressed in the language of differential equations. Unfortunately, it was difficult to convey the beauty of the subject in the traditional first course on differential equations because the number of equations that can be treated by analytic techniques is very limited. Consequently, the course tended to focus on technique rather than on concept.

At Boston University, we decided to revise our course, and we wrote this book to support our efforts. We now approach our course with several goals in mind. First, the traditional emphasis on specialized tricks and techniques for solving differential equations is no longer appropriate given the technology (laptops, ipads, smart phones, ...) that we carry around with us everywhere. Second, many of the most important differential equations are nonlinear, and numerical and qualitative techniques are more effective than analytic techniques in this setting. Finally, the differential equations course is one of the few undergraduate courses where we can give our students a glimpse of the nature of contemporary mathematical research.

The Qualitative, Numeric, and Analytic Approaches

Accordingly, this book is very different from the old-fashioned "cookbook" differential equations text. We have eliminated many of the specialized techniques for deriving formulas for solutions, and we have replaced them with topics that focus on the formulation of differential equations and the interpretation of their solutions. To obtain an understanding of the solutions, we generally attack a given equation from three different points of view.

One major approach we use is qualitative. We expect students to be able to visualize differential equations and their solutions in many geometric ways. For example, we readily use slope fields, graphs of solutions, vector fields, and solution curves in the phase plane as tools to gain a better understanding of solutions. We also ask students to become adept at moving among these geometric representations and more traditional analytic representations.

Since differential equations are easily studied using a computer, we also emphasize numerical techniques. DETools, the software that accompanies this book, provides students with ample computational tools to investigate the behavior of solutions of differential equations both numerically and graphically. Even if we can find an explicit formula for a solution, we often work with the equation both numerically and qualitatively to understand the geometry and the long-term behavior of solutions. When we can find explicit solutions easily, we do the calculations. But we always examine the resulting formulas using qualitative and numerical points of view as well.

How This Book is Different

There are several specific ways in which this book differs from other books at this level. First, we incorporate modeling throughout. We expect students to understand the meaning of the variables and parameters in a differential equation and to be able to interpret this meaning in terms of a particular model. Certain models reappear often as running themes and are drawn from a variety of disciplines so that students with various backgrounds will find something familiar.

We also advocate a dynamical systems point of view. That is, we are always concerned with the long-term behavior of solutions, and using all of the appropriate approaches outlined above, we ask students to predict this long-term behavior. In addition, we emphasize the role of parameters in many of our examples, and we specifically address the manner in which the behavior of solutions changes as these parameters vary.

It is our philosophy that using a computer is as natural and necessary to the study of differential equations as is the use of paper and pencil. DETools should make the inclusion of technology in the course as easy as possible. This suite of computer programs illustrates the basic concepts of differential equations. Three of these programs are solvers which allow the student to compute and graph numerical solutions of both first-order equations and systems of differential equations. The other 26 tools are demonstrations that allow students and teachers to investigate in detail specific topics covered in the text. A number of exercises in the text refer directly to these tools. DETools is available through CengageBrain.com.

As most texts do, we begin with a chapter on first-order equations. However, the only analytic technique we use to find closed-form solutions is separation of variables until we discuss linear equations at the end of the chapter. Instead, we emphasize the meaning of a differential equation and its solutions in terms of its slope field and the graphs of its solutions. If the differential equation is autonomous, we also discuss its phase line. This discussion of the phase line serves as an elementary introduction to the idea of a phase plane, which plays a fundamental role in subsequent chapters.

We then move directly from first-order equations to systems of first-order differential equations. Rather than consider second-order equations separately, we convert these equations to first-order systems. When these equations are viewed as systems, we are able to use qualitative and numerical techniques more readily. Of course, we then use the information about these systems gleaned from these techniques to recover information about the solutions of the original equation.

We also begin the treatment of systems with a general approach. We do not immediately restrict our attention to linear systems. Qualitative and numerical techniques work just as easily on nonlinear systems, and one can proceed a long way toward understanding solutions without resorting to algebraic techniques. However, qualitative ideas do not tell the whole story, and we are led naturally to the idea of linearization. With this background in the fundamental geometric and qualitative concepts, we then discuss linear systems in detail. Not only do we emphasize the formula for the general solution of a linear system, but also the geometry of its solution curves and its relationship to the eigenvalues and eigenvectors of the system.

While our study of systems requires the minimal use of some linear algebra, it is definitely not a prerequisite. Because we deal primarily with two-dimensional systems, we easily develop all of the necessary algebraic techniques as we proceed. In the process, we give considerable insight into the geometry of eigenvectors and eigenvalues.

These topics form the core of our approach. However, there are many additional topics that one would like to cover in the course. Consequently, we have included discussions of forced second-order equations, nonlinear systems, Laplace transforms, numerical methods, and discrete dynamical systems. In Appendix A, we even have a short discussion of Riccati and Bernoulli equations, and Appendix B is an ultra-lite treatment of power series methods. In Appendix B we take the point of view that power series are an algebraic way of finding approximate solutions much like numerical methods. Occasional surprises, such as Hermite and Legendre polynomials, are icing on the cake. Although some of these topics are quite traditional, we always present them in a manner that is consistent with the philosophy developed in the first half of the text.

At the end of each chapter, we have included several "labs." Doing detailed numerical experimentation and writing reports has been our most successful modification of our course at Boston University. Good labs are tough to write and to grade, but we feel that the benefit to students is extraordinary.

Changes in the Fourth Edition

This revision has been our most extensive since we published the first edition in 1998. In Chapter 1, the table of contents remains the same. However, many new exercises have been added, and they often introduce models that are new to the text. For example, the theta model for the spiking of a neuron appears in the exercise sets of Sections 1.3, 1.4, 1.6, and 1.7. The concept of a time constant is introduced in Section 1.1 and discussed in the context of a blinking light in Section 1.3. The velocity of a freefalling skydiver is discussed in three exercise sets. In Section 1.1, we discuss terminal velocity to illustrate the concept of long-term behavior. In Section 1.2, we find the general solution of the velocity equation using the method of of separation of varibles, and in Section 1.4, we study these solutions numerically using Euler's method.

Chapter 2 has undergone a complete overhaul. We added a section (Section 2.7) on the SIR model. We include this topic for two reasons. First, many of our students had first-hand experience with the H1N1 pandemic in 2009–2010. Second, many users of the preliminary edition liked the fact that we discussed nullclines in Chapter 2. Section 2.7 provides some phase plane analysis without going into the detail that is found in in our section on nullclines in Chapter 5.

Chapter 2 now has eight sections rather than five. Sections 2.1 and 2.2 are essentially unchanged. Section 2.3 is a short section in which the damped harmonic oscillator is introduced. This model is so important that it deserves a section of its own rather than being buried at the end of a section as it was in previous editions. The remaining analytic techniques that we presented in the previous editions can now be found in Section 2.4. The Existence and Uniqueness Theorem for systems along with its consequences has its own section (Section 2.6), and the consequences of uniqueness are discussed in more detail. The presentations of Euler's method for systems and Lorenz's chaotic system are essentially unchanged.

This material is presented in smaller sections to give the instructor more flexibility to pick and choose topics from Chapter 2. Only Sections 2.1 and 2.2 are absolute prerequisities for what follows. Chapter 2 has always been the most difficult one to teach, and now instructors can cover as many (or as few) sections from Chapter 2 as they see fit.

Pathways Through This Book

There are a number of possible tracks that instructors can follow in using this book. Chapters 1–3 form the core (with the possible exception of Sections 2.8 and 3.8, which cover systems in three dimensions). Most of the later chapters assume familiarity with this material. Certain sections such as Section 1.7 (bifurcations), Section 1.9 (integrating factors), and Sections 2.4–2.7 can be skipped if some care is taken in choosing material from subsequent sections. However, the material on phase lines and phase planes, qualitative analysis, and solutions of linear systems is central.

A typical track for an engineering-oriented course would follow Chapters 1–3 (perhaps skipping Sections 1.7, 1.9, 2.4, 2.6, 2.7, 2.8, and 3.8). Appendix A (changing variables) can be covered at the end of Chapter 1. These chapters will take roughly two-thirds of a semester. The final third of the course might cover Sections 4.1–4.3 (forced, second-order linear equations and resonance), Section 5.1 (linearization of nonlinear systems), and Chapter 6 (Laplace transforms). Chapters 4 and 5 are independent of each other and can be covered in either order. In particular, Section 5.1 on linearization of nonlinear systems near equilibrium points forms an excellent capstone for the material on linear systems in Chapter 3. Appendix B (power series) goes well after Chapter 4.

Incidentally, it is possible to cover Sections 6.1 and 6.2 (Laplace transforms for first-order equations) immediately after Chapter 1. As we have learned from our colleagues in the College of Engineering at Boston University, some engineering programs teach a circuit theory course that uses the Laplace transform early in the course. Consequently, Sections 6.1 and 6.2 are written so that the differential equations course and such a circuits course could proceed in parallel. However, if possible, we recommend waiting to cover Chapter 6 entirely until after the material in Sections 4.1–4.3 has been discussed.

Instructors can substitute material on discrete dynamics (Chapter 8) for Laplace transforms. A course for students with a strong background in physics might involve more of Chapter 5, including a treatment of systems that are Hamiltonian (Section 5.3) and gradient (Section 5.4). A course geared toward applied mathematics might include a more detailed discussion of numerical methods (Chapter 7).

Our Website and Ancillaries

Readers and instructors are invited to make extensive use of our web site

$$\text{http://math.bu.edu/odes}$$

At this site we have posted an on-line instructor's guide that includes discussions of how we use the text. We have sample syllabi contributed by users at various institutions as well as information about workshops and seminars dealing with the teaching of differential equations. We also maintain a list of errata. Solution Builder, available to instructors who have adopted the text for class use, creates customized, secure PDF copies of solutions matched exactly to the the exercises assigned for class. The web site for Solution Builder is

http://www.cengage.com/solutionbuilder

The *Student Solutions Manual* contains the solutions to all odd-numbered exercises.

The Boston University Differential Equations Project

This book is a product of the now complete National Science Foundation Boston University Differential Equations Project (NSF Grant DUE-9352833) sponsored by the National Science Foundation and Boston University. The goal of that project was to re-think the traditional, sophomore-level differential equations course. We are especially thankful for that support.

Paul Blanchard
Robert L. Devaney
Glen R. Hall
Boston University

ACKNOWLEDGMENTS

As we move from the third to the fourth edition, the number of people we are privileged to thank continues to grow. For this edition, we owe particular thanks to one individual, **Brian Persaud**. Over the course of almost two years, he was there when we needed him. He basically did all of the production work that was done at Boston University. He researched countless topics and gave us valuable feedback at every step in the process. Thanks, Brian. As with his predecessors, Sam Kaplan on the preliminary edition, Gareth Roberts on the first edition, Melissa Vellela and J. Doug Wright on the second edition, Sebastian Marotta and Dan Look on the third edition, Brian has left his mark on the text in many positive ways.

We also had help with the graphics programming from Nick Benes. Nick is currently serving in Afghanistan, but we have the light on in anticipation of his safe return. When we needed some last-minute accuracy checking, Dan Cuzzocreo was the man. We also thank Mark Kramer for giving us copies of his notes on the theta model for the spiking of a neuron.

With the exception of a few professionally drawn figures, this book was entirely produced at Boston University's Department of Mathematics and Statistics using LaTeX 2_ε. We owe a great debt to Alex Kasman for serving as our TeX wizard. He saved us countless hours during the formating of the text and the production of the figures.

Much of the production work, solutions to exercises, accuracy checking and rendering of pictures was done by former Boston University graduate students Bill Basener, Lee DeVille, and Stephanie Jones during the production of the first edition. They spent many long days and nights in an alternately too-hot-or-too-cold windowless computer lab to bring this book to completion. We still rely on much of the work done by Adrian Iovita, Kinya Ono, Adrian Vajiac, and Núria Fagella during the production of the preliminary edition.

Many other individuals at Boston University have made important contributions. In particular, our teaching assistants Duff Campbell, Michael Hayes, Eileen Lee, and Clara Bodelon had to put up with the headaches associated with our experimentation. Duff also advised us during the development of our power series appendix for the third edition, and he continues to provide valuable feedback whenever he teaches differential equations using this book.

We received support from many of our colleagues at Boston University and at other institutions. It was a special pleasure for us to work closely with colleagues in the College of Engineering—Michael Ruane (who often coordinates the circuits course), Moe Wasserman (who permitted one of the authors to audit his course), and John Baillieul (a member of our advisory board). We also thank Donna Molinek, Davidson College; Carolyn Narasimhan, DePaul University; and James Walsh, Oberlin College; for organizing workshops for faculty on their campuses.

As we mentioned in the Preface, this book would not exist if our project had not received support from the National Science Foundation's Division of Undergraduate Education, and we would like to thank the program directors at NSF/DUE for their enthusiastic support. We would also like to thank the members of the advisory board—John Baillieul, Morton Brown, John Franks, Deborah Hughes Hallett, Philip Holmes, and Nancy Kopell. All contributed their scarce time during workshops and trips to Boston University.

The `DETools` software package was originally designed and implemented by Hu Hohn. This software truly enhances the teaching and learning of differential equations. As you will see while using these programs, Hu has the ability to visualize differential equations in a way that only an artist (as he is) can. It is always a pleasure to work with him. We are also grateful to John Cantwell, Jean Marie McDill, Steven Strogatz, and Beverly West for their many helpful discussions about incorporating technology in the differential equations course. Their pathbreaking work with in-class and hands-on software demonstrations has been an inspiration, both to us and to our students. It has had a major impact on the way we teach and think about this subject.

The current version of the software was ported to Java by the folks at Artmedialab International LLC. In particular we thank Lee Stayton for managing the conversion and Dusan A. Koljensic for designing the new interface to the tools.

We were pleased that so many of our colleagues outside of Boston University were willing to help us with this project. Bill Krohn gave us valuable advice regarding our exposition of Laplace transforms, and Bruce Elenbogen did a thorough reading of early drafts of the beginning chapters. Preliminary drafts of our original notes were class tested in a number of different settings by Gregory Buck, Scott Sutherland, Kathleen Alligood, Diego Benardete, Jack Dockery, Mako Haruta, Jim Henle, Ed Packel, and Ben Pollina.

We have been pleased with the reception given to this text. We particularly wish to thank the many people who have generously shared their ideas and suggestions. Also, thanks to all who caught the typo we made in the third edition. It is corrected in this edition.

Thoughtful and insightful reviews have also been a tremendous help in the evolution of this text from preliminary to fourth edition, and we thank all those who took the time to give us valuable feedback. In particular, reviewers for this edition were Josefina Alvarez, New Mexico State University; David Dudley, Scottsdale Community College; Faiz Al-Rubaee, University of North Florida; and Richard Penn, Montgomery College.

The final months of production coincided with Paul Blanchard's sabbatical during Fall 2010. From October through December, he traveled extensively, and he was the guest of many organizations and individuals. While in Europe, he lived at the Stefan Banach International Mathematical Center in Warsaw, Poland. He visited the University of Barcelona, and he stayed at the infamous Math Houses at the University of Warwick in England. He was also the guest of Beth and Scott Sutherland on Long Island, and he spent a very productive week writing at the Battles house in Asheville, North Carolina. Sean McLaughlin was kind enough to arrange that visit.

Scott Sutherland also provided some last-minute computer expertise to help bring the production of this edition to its conclusion.

We thank Charu Khanna and her colleagues at MPS Limited and all of our Cengage Learning team, particularly Susan Miscio, Content Project Manager; Suzanne St. Clair, Production Manager; Stacy Green, Development Editor; Jennifer Jones,

Senior Marketing Manager; Dan Seibert, Assistant Editor; Shaylin Walsh, Assistant Editor; Andrew Coppola, Associate Media Editor and Alexander Gontar, Editorial Assistant. RHDG: Reizebos Holzbaur designed another great cover that involves the Zakim Bridge in Boston. A very special thanks goes to our Senior Sponsoring Editor (and BU grad), Molly Taylor. For a number of minor reasons and a couple of major ones, the timing of this edition did not go as planned. We wish to thank everyone at Cengage for their patience and understanding.

Finally, as any author knows, writing a book requires significant sacrifices from one's family. Extra special thanks goes to Lori, Kathy, and Dottie.

G.R.H., R.L.D., P.B.

A NOTE TO THE STUDENT

This book might be different from most of your previous mathematics texts. If you thumb through it, you will see that there are few "boxed" formulas, no margin notes, and very few n-step procedures. We wrote the book this way because we think that you are now at a point in your education where you should be learning to identify and work effectively with the mathematics inherent in everyday life. As you pursue your careers, no one is going to ask you to do all of the odd exercises at the end of some employee manual. They are going to give you some problem with mathematical components that might be difficult to identify and ask you to do your best with it. One of our goals in this book is to start preparing you for this type of work by avoiding artificial algorithmic exercises.

Our intention is that you will read this book as you would any other text, then work on the exercises, rereading sections and examples as necessary. Even though there are no template examples, you will find the discussions full of examples. Since one of our main goals is to demonstrate how differential equations are used to model physical systems, we often start with the description of a physical system, build a model, and then study the model to make conclusions and predictions about the original system. Many of the exercises ask you to produce or modify a model of a physical system, analyze it, and explain your conclusions. This is hard stuff, and you will need to practice. Since the days when you could make a living plugging and chugging through computations are over (computers do that now), you will need to learn these skills, and we hope that this book helps you develop them.

Another way in which this book may differ from your previous texts is that we expect you to make judicious use of a computer as you work the exercises and labs. Included in the software that accompanies this text are solvers that let you compute solutions of differential equations and graph the results. We encourage you to start playing with them immediately. As far as we know, a computer cannot think for you (yet), but it can provide you with numerical evidence that is essentially impossible for you to get in any other way. One of our goals is to give you practice as a sophisticated consumer of computer cycles as well as a skeptic of computer results.

In this book, solutions of differential equations involve motion of a system over time—the changes in population over time, the motion of a pendulum, and so forth. It is therefore important that you be able to visualize how certain systems evolve as time passes. A static text book is not the ideal mechanism to illustrate motion. Accordingly, we have included a number of demonstrations in the DETools software that show the actual motion associated with differential equations. These demos are fun to use, and we encourage you to refer to them early and often.

Incidentally, one of the authors is known to have made a mistake or two in his life that the other two authors have overlooked. So we maintain a very short list of errata at our web site http://math.bu.edu/odes. Please check this page if you think that something you have read is not quite right.

Finally, you should know that the authors take the study of differential equations very seriously. However, we don't take ourselves very seriously (and we certainly don't take the other two authors seriously). We have tried to express both the beauty of the mathematics and some of the fun we have doing mathematics. If you think the jokes are old or stupid, you're probably right.

All of us who worked on this book have learned something about differential equations along the way, and we hope that we are able to communicate our appreciation for the subject's beauty and range of application. We would enjoy hearing your comments. Feel free to send us e-mail at odes@math.bu.edu. We sometimes get busy and cannot always respond, but we do read and appreciate your feedback.

We had fun writing this book. We hope you have fun reading it.

G.R.H., P.B., R.L.D.

CONTENTS

1 FIRST-ORDER DIFFERENTIAL EQUATIONS   1

1.1 Modeling via Differential Equations   2

1.2 Analytic Technique: Separation of Variables   21

1.3 Qualitative Technique: Slope Fields   36

1.4 Numerical Technique: Euler's Method   52

1.5 Existence and Uniqueness of Solutions   63

1.6 Equilibria and the Phase Line   74

1.7 Bifurcations   94

1.8 Linear Equations   110

1.9 Integrating Factors for Linear Equations   124

Review Exercises for Chapter 1   136

Labs for Chapter 1   142

2 FIRST-ORDER SYSTEMS   149

2.1 Modeling via Systems   150

2.2 The Geometry of Systems   166

2.3 The Damped Harmonic Oscillator   183

2.4 Additional Analytic Methods for Special Systems   189

2.5 Euler's Method for Systems   196

2.6 Existence and Uniqueness for Systems   204

2.7 The SIR Model of an Epidemic   209

2.8 The Lorenz Equations   217

Review Exercises for Chapter 2   224

Labs for Chapter 2   229

xvii

3 LINEAR SYSTEMS   239

3.1 Properties of Linear Systems and the Linearity Principle   240

3.2 Straight-Line Solutions   264

3.3 Phase Portraits for Linear Systems with Real Eigenvalues   280

3.4 Complex Eigenvalues   296

3.5 Special Cases: Repeated and Zero Eigenvalues   315

3.6 Second-Order Linear Equations   330

3.7 The Trace-Determinant Plane   347

3.8 Linear Systems in Three Dimensions   360

Review Exercises for Chapter 3   376

Labs for Chapter 3   381

4 FORCING AND RESONANCE   387

4.1 Forced Harmonic Oscillators   388

4.2 Sinusoidal Forcing   403

4.3 Undamped Forcing and Resonance   415

4.4 Amplitude and Phase of the Steady State   427

4.5 The Tacoma Narrows Bridge   439

Review Exercises for Chapter 4   449

Labs for Chapter 4   452

5 NONLINEAR SYSTEMS   457

5.1 Equilibrium Point Analysis   458

5.2 Qualitative Analysis   477

5.3 Hamiltonian Systems   490

5.4 Dissipative Systems   508

5.5 Nonlinear Systems in Three Dimensions   530

5.6 Periodic Forcing of Nonlinear Systems and Chaos   538

Review Exercises for Chapter 5   555

Labs for Chapter 5   558

6 LAPLACE TRANSFORMS   565

6.1 Laplace Transforms   566

6.2 Discontinuous Functions   578

6.3 Second-Order Equations   587

6.4 Delta Functions and Impulse Forcing   601

6.5 Convolutions   609

6.6 The Qualitative Theory of Laplace Transforms   618

Table of Laplace Transforms   626

Review Exercises for Chapter 6   627

Labs for Chapter 6   630

7 NUMERICAL METHODS   633

7.1 Numerical Error in Euler's Method   634

7.2 Improving Euler's Method   647

7.3 The Runge-Kutta Method   655

7.4 The Effects of Finite Arithmetic   665

Review Exercises for Chapter 7   670

Labs for Chapter 7   671

8 DISCRETE DYNAMICAL SYSTEMS   675

8.1 The Discrete Logistic Equation   676

8.2 Fixed Points and Periodic Points   689

8.3 Bifurcations   698

8.4 Chaos   707

8.5 Chaos in the Lorenz System   715

Review Exercises for Chapter 8   721

Labs for Chapter 8   723

APPENDICES   729

A Changing Variables   730

B Power Series: The Ultimate Guess   742

C Complex Numbers and Euler's Formula   750

HINTS AND ANSWERS   755

INDEX   825

1 FIRST-ORDER DIFFERENTIAL EQUATIONS

This book is about how to predict the future. To do so, all we have is a knowledge of how things are and an understanding of the rules that govern the changes that will occur. From calculus we know that change is measured by the derivative. Using the derivative to describe how a quantity changes is what the subject of differential equations is all about.

Turning the rules that govern the evolution of a quantity into a differential equation is called modeling, and in this book we study many models. Our goal is to use the differential equation to predict the future value of the quantity being modeled.

There are three basic types of techniques for making these predictions. Analytical techniques involve finding formulas for the future values of the quantity. Qualitative techniques involve obtaining a rough sketch of the graph of the quantity as a function of time as well as a description of its long-term behavior. Numerical techniques involve doing arithmetic (or having a computer do arithmetic) that yields approximations of the future values of the quantity. We introduce and use all three of these approaches in this chapter.

1.1   MODELING VIA DIFFERENTIAL EQUATIONS

The hardest part of using mathematics to study an application is the translation from real life into mathematical formalism. This translation is usually difficult because it involves the conversion of imprecise assumptions into precise formulas. There is no way to avoid it. Modeling is difficult, and the best way to get good at it is the same way you get to play Carnegie Hall—practice, practice, practice.

What Is a Model?

It is important to remember that mathematical models are like other types of models. The goal is not to produce an exact copy of the "real" object but rather to give a representation of some aspect of the real thing. For example, a portrait of a person, a store mannequin, and a pig can all be models of a human being. None is a perfect copy of a human, but each has certain aspects in common with a human. The painting gives a description of what a particular person looks like; the mannequin wears clothes as a person does; and the pig is alive. Which of the three models is "best" depends on how we use the model—to remember old friends, to buy clothes, or to study biology.

We study mathematical models of systems that evolve over time, but they often depend on other variables as well. In fact, real-world systems can be notoriously complicated—the population of rabbits in Wyoming depends on the number of coyotes, the number of bobcats, the number of mountain lions, the number of mice (alternative food for the predators), farming practices, the weather, any number of rabbit diseases, etc. We can make a model of the rabbit population simple enough to understand only by making simplifying assumptions and lumping together effects that may or may not belong together.

Once we've built the model, we should compare predictions of the model with data from the system. If the model and the system agree, then we gain confidence that the assumptions we made in creating the model are reasonable, and we can use the model to make predictions. If the system and the model disagree, then we must study and improve our assumptions. In either case we learn more about the system by comparing it to the model.

The types of predictions that are reasonable depend on our assumptions. If our model is based on precise rules such as Newton's laws of motion or the rules of compound interest, then we can use the model to make very accurate quantitative predictions. If the assumptions are less precise or if the model is a simplified version of the system, then precise quantitative predictions would be silly. In this case we would use the model to make qualitative predictions such as "the population of rabbits in Wyoming will increase ...." The dividing line between qualitative and quantitative prediction is itself imprecise, but we will see that it is frequently better and easier to make qualitative use of even the most precise models.

Some hints for model building
The basic steps in creating the model are:

Step 1 Clearly state the assumptions on which the model will be based. These assumptions should describe the relationships among the quantities to be studied.

Step 2 Completely describe the variables and parameters to be used in the model—"you can't tell the players without a score card."

Step 3 Use the assumptions formulated in Step 1 to derive equations relating the quantities in Step 2.

Step 1 is the "science" step. In Step 1, we describe how we think the physical system works or, at least, what the most important aspects of the system are. In some cases these assumptions are fairly speculative, as, for example, "rabbits don't mind being overcrowded." In other cases the assumptions are quite precise and well accepted, such as "force is equal to the product of mass and acceleration." The quality of the assumptions determines the validity of the model and the situations to which the model is relevant. For example, some population models apply only to small populations in large environments, whereas others consider limited space and resources. Most important, we must avoid "hidden assumptions" that make the model seem mysterious or magical.

Step 2 is where we name the quantities to be studied and, if necessary, describe the units and scales involved. Leaving this step out is like deciding you will speak your own language without telling anyone what the words mean.

The quantities in our models fall into three basic categories: the **independent variable**, the **dependent variables**, and the **parameters**. In this book the independent variable is almost always time. Time is "independent" of any other quantity in the model. On the other hand, the dependent variables are quantities that are functions of the independent variable. For example, if we say that "position is a function of time," we mean that position is a variable that depends on time. We can vaguely state the goal of a model expressed in terms of a differential equation as "Describe the behavior of the dependent variable as the independent variable changes." For example, we may ask whether the dependent variable increases or decreases, or whether it oscillates or tends to a limit.

Parameters are quantities that do not change with time (or with the independent variable) but that can be adjusted (by natural causes or by a scientist running the experiment). For example, if we are studying the motion of a rocket, the initial mass of the rocket is a parameter. If we are studying the amount of ozone in the upper atmosphere, then the rate of release of fluorocarbons from refrigerators is a parameter. Determining how the behavior of the dependent variables changes as we adjust the parameters can be the most important aspect of the study of a model.

In Step 3 we create the equations. Most of the models we consider in this book are expressed mathematically as differential equations. In other words, we expect to find derivatives in our equations. Look for phrases such as "rate of change of ..." or "rate of increase of ...," since rate of change is synonymous with derivative. Of course, also watch for "velocity" (derivative of position) and "acceleration" (derivative of velocity) in models from physics. The word *is* means "equals" and indicates where the equality lies. The phrase "A is proportional to B" means $A = kB$, where k is a proportionality constant (often a parameter in the model).

When we formulate a model, we follow the advice of Albert Einstein: "Make everything as simple as possible, but not simpler." In this case, *we make the algebra as simple as possible.* For example, when modeling the velocity v of a cat falling from a tall building, we could assume:

- Air resistance increases as the cat's velocity increases.*

This assumption says that air resistance provides a force that counteracts the force of gravity and that this force increases as the velocity v of the cat increases. We could choose kv or kv^2 for the air resistance term, where k is the friction coefficient, a parameter. Both expressions increase as v increases, so they satisfy the assumption. However, we most likely would try kv first because it is the simplest expression that satisfies the assumption. In fact, it turns out that kv yields a good model for falling bodies with low densities such as snowflakes, but kv^2 is a more appropriate model for dense objects such as raindrops (see Exercise 12).

Now we turn to a series of models of population growth based on various assumptions about the species involved. Our goal here is to study how to go from a set of assumptions to a model that involves differential equations. These examples are not state-of-the-art models from population ecology, but they are good ones to consider initially. We also begin to describe the analytic, qualitative, and numerical techniques that we use to make predictions based on these models. Our approach is meant to be illustrative only; we discuss these mathematical techniques in much more detail throughout the entire book.

Figure 1.1
Well prepared cat.

Unlimited Population Growth

An elementary model of population growth is based on the assumption that

- The rate of growth of the population is proportional to the size of the population.

Note that the rate of change of a population depends on only the size of the population and nothing else. In particular, limitations of space or resources are ignored. This assumption is reasonable for small populations in large environments—for example, the first few spots of mold on a piece of bread or the first European settlers in the United States.

---

*In 1987, veterinarians at Manhattan's Animal Medical Center conducted a study of cats that had fallen from high-rise buildings ("High-Rise syndrome in cats" by W. O. Whitney and C. J. Mehlhaff, in *Journal of the American Veterinary Medical Association*, Vol. 191, No. 11, 1987, pp. 1399–1403). They found that 90% of the cats that they treated survived. More than one-half suffered serious injuries, and more than one-third required life-saving treatments. However, slightly under one-third did not require any treatment at all.

Counterintuitively, this study found that cats that fell from heights of 7 to 32 stories were less likely to die than cats that fell from 2 to 6 stories. One might assume that falling from a greater distance gives the cat more time to adopt a Rocky-the-flying-squirrel pose.

Of course, this study suffers from one obvious design flaw. That is, data was collected only from cats that were brought into clinics for veterinary care. It is unknown how many cats died on impact.

Because the assumption is so simple, we expect the model to be simple as well. The quantities involved are

$$t = \text{time (independent variable)},$$

$$P = \text{population (dependent variable), and}$$

$$k = \text{proportionality constant (parameter) between the rate}$$
$$\text{of growth of the population and the size of the population.}$$

The parameter k is often called the "growth-rate coefficient."

The units for these quantities depend on the application. If we are modeling the growth of mold on bread, then t might be measured in days and $P(t)$ might be either the area of bread covered by the mold or the weight of the mold. If we are talking about the European population of the United States, then t probably should be measured in years and $P(t)$ in millions of people. In this case we could let $t = 0$ correspond to any time we wanted. The year 1790 (the year of the first census) is a convenient choice.

Now let's express our assumption using this notation. The rate of growth of the population P is the derivative dP/dt. Being proportional to the population is expressed as the product, kP, of the population P and the proportionality constant k. Hence our assumption is expressed as the differential equation

$$\frac{dP}{dt} = kP.$$

In other words, the rate of change of P is proportional to P. Note that, since the units associated to both sides of the equation much agree, we see that the units associated to the growth-rate coefficient k are 1/time.

This equation is our first example of a differential equation. Associated with it are a number of adjectives that describe the type of differential equation that we are considering. In particular, it is a **first-order** equation because it contains only first derivatives of the dependent variable, and it is an **ordinary differential equation** because it does not contain partial derivatives. In this book we deal only with ordinary differential equations.

We have written this differential equation using the dP/dt Leibniz notation—the notation that we tend to use. However, there are many other ways to express the same differential equation. In particular, we could also write this equation as $P' = kP$ or as $\dot{P} = kP$. The "dot" notation is often used when the independent variable is time t.

What does the model predict?

More important than the adjectives or how the equation is written is what the equation tells us about the situation being modeled. Since $dP/dt = kP$ for some constant k, $dP/dt = 0$ if $P = 0$. Thus the constant function $P(t) = 0$ is a solution of the differential equation. This special type of solution is called an **equilibrium solution** because it is constant forever. In terms of the population model, it corresponds to a species that is nonexistent.

If $P(t_0) \neq 0$ at some time t_0, then

$$\frac{dP}{dt} = k\,P(t_0) \neq 0.$$

at t_0. As a consequence, the population is not constant. If $k > 0$ and $P(t_0) > 0$, we have

$$\frac{dP}{dt} = kP(t_0) > 0,$$

at time $t = t_0$ and the population is increasing (as one would expect). As t increases, $P(t)$ becomes larger, so dP/dt becomes larger. In turn, $P(t)$ increases even faster. That is, the rate of growth increases as the population increases. We therefore expect that the graph of the function $P(t)$ might look like Figure 1.2.

The value of $P(t)$ at $t = 0$ is called an **initial condition**. If we start with a different initial condition we get a different function $P(t)$ as is indicated in Figure 1.3. If $P(0)$ is negative (remembering $k > 0$), we then have $dP/dt < 0$ for $t = 0$, so $P(t)$ is initially decreasing. As t increases, $P(t)$ becomes more negative. The graphs below the t-axis are mirror images of the graphs above the t-axis, although they are not physically meaningful because a negative population doesn't make much sense.

Our analysis of the way in which $P(t)$ increases as t increases is called a **qualitative analysis** of the differential equation. If all we care about is whether the model predicts "population explosions," then we can answer "yes, as long as $P(0) > 0$."

Analytic solutions of the differential equation

If, on the other hand, we know the exact value P_0 of $P(0)$ and we want to predict the value of $P(10)$ or $P(100)$, then we need more precise information about function $P(t)$.

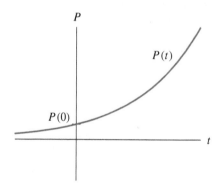

Figure 1.2
The graph of a function that satisfies the differential equation

$$\frac{dP}{dt} = kP.$$

Its initial value at $t = 0$ is $P(0)$.

Figure 1.3
The graphs of several different functions that satisfy the differential equation

$$\frac{dP}{dt} = kP.$$

Each has a different value at $t = 0$.

The pair of equations

$$\frac{dP}{dt} = kP, \quad P(0) = P_0,$$

is called an **initial-value problem**. A **solution** to the initial-value problem is a function $P(t)$ that satisfies both equations. That is,

$$\frac{dP}{dt} = kP \text{ for all } t \quad \text{and} \quad P(0) = P_0.$$

Consequently, to find a solution to this differential equation we must find a function $P(t)$ whose derivative is the product of k with $P(t)$. One (not very subtle) way to find such a function is to guess. In this case, it is relatively easy to guess the right form for $P(t)$ because we know that the derivative of an exponential function is essentially itself. (We can eliminate this guesswork by using the method of separation of variables, which we describe in the next section. But for now, let's just try the exponential and see where that leads us.) After a couple of tries with various forms of the exponential, we see that

$$P(t) = e^{kt}$$

is a function whose derivative, $dP/dt = ke^{kt}$, is the product of k with $P(t)$. But there are other possible solutions, since $P(t) = ce^{kt}$ (where c is a constant) yields $dP/dt = c(ke^{kt}) = k(ce^{kt}) = kP(t)$. Thus $dP/dt = kP$ for all t for any value of the constant c.

We have infinitely many solutions to the differential equation, one for each value of c. To determine which of these solutions is the correct one for the situation at hand, we use the given initial condition. We have

$$P_0 = P(0) = c \cdot e^{k \cdot 0} = c \cdot e^0 = c \cdot 1 = c.$$

Consequently, we should choose $c = P_0$, so a solution to the initial-value problem is

$$P(t) = P_0 e^{kt}.$$

We have obtained an actual formula for our solution, not just a qualitative picture of its graph.

The function $P(t)$ is called the solution to the initial-value problem as well as a **particular solution** of the differential equation. The collection of functions $P(t) = ce^{kt}$ is called the **general solution** of the differential equation because we can use it to find the particular solution corresponding to any initial-value problem. Figure 1.3 consists of the graphs of exponential functions of the form $P(t) = ce^{kt}$ with various values of the constant c, that is, with different initial values. In other words, it is a picture of the general solution to the differential equation.

The U.S. Population

As an example of how this model can be used, consider the U.S. census figures since 1790 given in Table 1.1.

Let's see how well the unlimited growth model fits this data. We measure time in years and the population $P(t)$ in millions of people. We also let $t = 0$ be the year 1790, so the initial condition is $P(0) = 3.9$. The corresponding initial-value problem

$$\frac{dP}{dt} = kP, \quad P(0) = 3.9,$$

has $P(t) = 3.9e^{kt}$ as a solution. We cannot use this model to make predictions yet because we don't know the value of k. However, we are assuming that k is a constant, so we can use the initial condition along with the population in the year 1800 to estimate k. If we set

$$5.3 = P(10) = 3.9e^{k \cdot 10},$$

then we have

$$e^{k \cdot 10} = \frac{5.3}{3.9}$$

$$10k = \ln\left(\frac{5.3}{3.9}\right)$$

$$k \approx 0.03067.$$

Table 1.1
U.S. census figures, in millions of people (see www.census.gov)

Year	t	Actual	$P(t) = 3.9e^{0.03067t}$	Year	t	Actual	$P(t) = 3.9e^{0.03067t}$
1790	0	3.9	3.9	1930	140	123	286
1800	10	5.3	5.3	1940	150	132	388
1810	20	7.2	7.2	1950	160	151	528
1820	30	9.6	9.8	1960	170	179	717
1830	40	13	13	1970	180	203	975
1840	50	17	18	1980	190	227	1,320
1850	60	23	25	1990	200	249	1,800
1860	70	31	33	2000	210	281	2,450
1870	80	39	45	2010	220		3,320
1880	90	50	62	2020	230		4,520
1890	100	63	84	2030	240		6,140
1900	110	76	114	2040	250		8,340
1910	120	91	155	2050	260		11,300
1920	130	106	210				

Thus our model predicts that the United States population is given by

$$P(t) = 3.9e^{0.03067t}.$$

As we see from Figure 1.4, this model of $P(t)$ does a decent job of predicting the population until roughly 1860, but after 1860 the prediction is much too large. (Table 1.1 includes a comparison of the predicted values to the actual data.)

Our model is fairly good provided the population is relatively small. However, as time goes by, the model predicts that the population will continue to grow without any limits, which obviously cannot happen in the real world. Consequently, if we want a model that is accurate over a large time scale, we should account for the fact that populations exist in a finite amount of space and with limited resources.

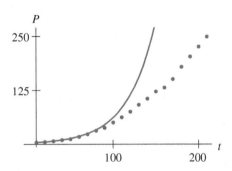

Figure 1.4
The dots represent actual census data and the solid line is the solution of the exponential growth model

$$\frac{dP}{dt} = 0.03067P.$$

Time t is measured in years since the year 1790.

Limited Resources and the Logistic Population Model

To adjust the exponential growth population model to account for a limited environment and limited resources, we add the assumptions:

- If the population is small, the rate of growth of the population is proportional to its size.
- If the population is too large to be supported by its environment and resources, the population will decrease. That is, the rate of growth is negative.

For this model, we again use

$$t = \text{time (independent variable)},$$
$$P = \text{population (dependent variable)},$$
$$k = \text{growth-rate coefficient for small}$$
$$\text{populations (parameter)}.$$

However, our assumption about limited resources introduces another quantity, the size of the population that corresponds to being "too large." This quantity is a second parameter, denoted by N, that we call the carrying capacity of the environment. In terms of the carrying capacity, we are assuming that $P(t)$ is increasing if $P(t) < N$. However, if $P(t) > N$, we assume that $P(t)$ is decreasing.

Using this notation, we can restate our assumptions as:

1. $\dfrac{dP}{dt} \approx kP$ if P is small.

2. If $P > N$, $\dfrac{dP}{dt} < 0$.

We also want the model to be "algebraically simple," or at least as simple as possible, so we try to modify the exponential model as little as possible. For instance, we might look for an expression of the form

$$\frac{dP}{dt} = k \cdot (\text{something}) \cdot P.$$

We want the "something" factor to be close to 1 if P is small, but if $P > N$ we want "something" to be negative. The simplest expression that has these properties is the function

$$(\text{something}) = \left(1 - \frac{P}{N}\right).$$

Note that this expression equals 1 if $P = 0$, and it is negative if $P > N$. Thus our model is

$$\frac{dP}{dt} = k\left(1 - \frac{P}{N}\right)P.$$

This is called the **logistic population model** with growth rate k and carrying capacity N. It is another first-order differential equation. This equation is said to be **nonlinear** because its right-hand side is not a linear function of P as it was in the exponential growth model.

Qualitative analysis of the logistic model

Although the logistic differential equation is just slightly more complicated than the exponential growth model, there is no way that we can just guess solutions. The method of separation of variables discussed in the next section produces a formula for the solution of this particular differential equation. But for now, we rely solely on qualitative methods to see what this model predicts over the long term.

First, let

$$f(P) = k\left(1 - \frac{P}{N}\right)P$$

denote the right-hand side of the differential equation. In other words, the differential equation can be written as

$$\frac{dP}{dt} = f(P) = k\left(1 - \frac{P}{N}\right)P.$$

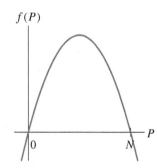

Figure 1.5
Graph of the right-hand side $f(P) = k\,(1 - P/N)\,P$ of the logistic differential equation.

We can derive qualitative information about the solutions to the differential equation from a knowledge of where dP/dt is zero, where it is positive, and where it is negative.

If we sketch the graph of the quadratic function f (see Figure 1.5), we see that it crosses the P-axis at exactly two points, $P = 0$ and $P = N$. In either case we have

$dP/dt = 0$. Since the derivative of P vanishes for all t, the population remains constant if $P = 0$ or $P = N$. That is, the constant functions $P(t) = 0$ and $P(t) = N$ are solutions of the differential equation. These two constant solutions make perfect sense: If the population is zero, the population remains zero indefinitely; if the population is exactly at the carrying capacity, it neither increases nor decreases. As before, we say that $P = 0$ and $P = N$ are *equilibria*. The constant functions $P(t) = 0$ and $P(t) = N$ are called *equilibrium solutions* (see Figure 1.6).

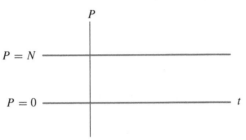

Figure 1.6
The equilibrium solutions of the logistic differential equation

$$\frac{dP}{dt} = k\left(1 - \frac{P}{N}\right)P.$$

The long-term behavior of the population is very different for other values of the population. If the initial population lies between 0 and N, then we have $f(P) > 0$. In this case the rate of growth $dP/dt = f(P)$ is positive, and consequently the population $P(t)$ is increasing. As long as $P(t)$ lies between 0 and N, the population continues to increase. However, as the population approaches the carrying capacity N, $dP/dt = f(P)$ approaches zero, so we expect that the population might level off as it approaches N (see Figure 1.7).

If $P(0) > N$, then $dP/dt = f(P) < 0$, and the population is decreasing. As above, when the population approaches the carrying capacity N, dP/dt approaches zero, and we again expect the population to level off at N.

Finally, if $P(0) < 0$ (which does not make much sense in terms of populations), we also have $dP/dt = f(P) < 0$. Again we see that $P(t)$ decreases, but this time it does not level off at any particular value since dP/dt becomes more and more negative as $P(t)$ decreases.

Thus, just from a knowledge of the graph of f, we can sketch a number of different solutions with different initial conditions, all on the same axes. The only information that we need is the fact that $P = 0$ and $P = N$ are equilibrium solutions, $P(t)$ increases if $0 < P < N$ and $P(t)$ decreases if $P > N$ or $P < 0$. Of course the exact values of $P(t)$ at any given time t depend on the values of $P(0)$, k, and N (see Figure 1.8).

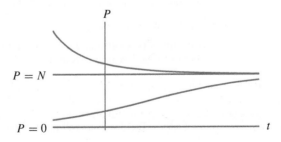

Figure 1.7
Solutions of the logistic differential equation

$$\frac{dP}{dt} = k\left(1 - \frac{P}{N}\right)P$$

approaching the equilibrium solution $P = N$.

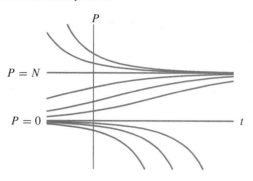

Figure 1.8

Solutions of the logistic differential equation

$$\frac{dP}{dt} = k\left(1 - \frac{P}{N}\right)P$$

approaching the equilibrium solution $P = N$ and moving away from the equilibrium solution $P = 0$.

Predator-Prey Systems

No species lives in isolation, and the interactions among species give some of the most interesting models to study. We conclude this section by introducing a simple predator-prey system of differential equations where one species "eats" another. The most obvious difference between the model here and previous models is that we have *two* quantities that depend on time. Thus our model has two dependent variables that are both functions of time. Since both predator and prey begin with "p," we call the prey "rabbits" and the predators "foxes," and we denote the prey by R and the predators by F. The assumptions for our model are:

- If no foxes are present, the rabbits reproduce at a rate proportional to their population, and they are not affected by overcrowding.
- The foxes eat the rabbits, and the rate at which the rabbits are eaten is proportional to the rate at which the foxes and rabbits interact.
- Without rabbits to eat, the fox population declines at a rate proportional to itself.
- The rate at which foxes are born is proportional to the number of rabbits eaten by foxes which, by the second assumption, is proportional to the rate at which the foxes and rabbits interact.*

To formulate this model in mathematical terms, we need four parameters in addition to our independent variable t and our two dependent variables F and R. The parameters are

$\alpha =$ growth-rate coefficient of rabbits,

$\beta =$ constant of proportionality that measures the number of rabbit-fox interactions in which the rabbit is eaten,

$\gamma =$ death-rate coefficient of foxes,

$\delta =$ constant of proportionality that measures the benefit to the fox population of an eaten rabbit.

When we formulate our model, we follow the convention that α, β, γ, and δ are all positive.

*Actually, foxes rarely eat rabbits. They focus on smaller prey, mostly mice and especially grasshoppers.

Our first and third assumptions above are similar to the assumption in the unlimited growth model discussed earlier in this section. Consequently, they give terms of the form αR in the equation for dR/dt and $-\gamma F$ (since the fox population declines) in the equation for dF/dt.

The rate at which the rabbits are eaten is proportional to the rate at which the foxes and rabbits interact, so we need a term that models the rate of interaction of the two populations. We want a term that increases if either R or F increases, but it should vanish if either $R = 0$ or $F = 0$. A simple term that incorporates these assumptions is RF. Thus we model the effects of rabbit-fox interactions on dR/dt by a term of the form $-\beta RF$. The fourth assumption gives a similar term in the equation for dF/dt. In this case, eating rabbits helps the foxes, so we add a term of the form δRF.

Given these assumptions, we obtain the model

$$\frac{dR}{dt} = \alpha R - \beta RF$$
$$\frac{dF}{dt} = -\gamma F + \delta RF.$$

Considered together, this pair of equations is called a **first-order system** (only first derivatives, but more than one dependent variable) of ordinary differential equations. The system is said to be *coupled* because the rates of change of R and F depend on both R and F.

It is important to note the signs of the terms in this system. Because $\beta > 0$, the term "$-\beta RF$" is nonpositive, so an increase in the number of foxes decreases the growth rate of the rabbit population. Also, since $\delta > 0$, the term "δRF" is nonnegative. Consequently, an increase in the number of rabbits increases the growth rate of the fox population.

Although this model may seem relatively simpleminded, it has been the basis of some interesting ecological studies. In particular, Volterra and D'Ancona successfully used the model to explain the increase in the population of sharks in the Mediterranean during World War I when the fishing of "prey" species decreased. The model can also be used as the basis for studying the effects of pesticides on the populations of predator and prey insects.

A *solution* to this system of equations is, unlike our previous models, a pair of functions, $R(t)$ and $F(t)$, that describe the populations of rabbits and foxes as functions of time. Since the system is coupled, we cannot simply determine one of these functions first and then the other. Rather, we must solve both differential equations simultaneously. Unfortunately, for most values of the parameters, it is impossible to determine explicit formulas for $R(t)$ and $F(t)$. These functions cannot be expressed in terms of known functions such as polynomials, sines, cosines, exponentials, and the like. However, as we will see in Chapter 2, these solutions do exist, although we have no hope of ever finding them exactly. Since analytic methods for solving this system are destined to fail, we must use either qualitative or numerical methods to study $R(t)$ and $F(t)$.

The Analytic, Qualitative, and Numerical Approaches

Our discussion of the three population models in this section illustrates three different approaches to the study of the solutions of differential equations. The **analytic** approach searches for explicit formulas that describe the behavior of the solutions. Here we saw that exponential functions give us explicit solutions to the exponential growth model. Unfortunately, a large number of important equations cannot be handled with the analytic approach; there simply is no way to find an exact formula that describes the situation. We are therefore forced to turn to alternative methods.

One particularly powerful method of describing the behavior of solutions is the **qualitative** approach. This method involves using geometry to give an overview of the behavior of the model, just as we did with the logistic population growth model. We do not use this method to give precise values of the solution at specific times, but we are often able to use this method to determine the long-term behavior of the solutions. Frequently, this is just the kind of information we need.

The third approach to solving differential equations is **numerical**. The computer approximates the solution we seek. Although we did not illustrate any numerical techniques in this section, we will soon see that numerical approximation techniques are a powerful tool for giving us intuition regarding the solutions we desire.

All three of the methods we use have certain advantages, and all have drawbacks. Sometimes certain methods are useful while others are not. One of our main tasks as we study the solutions to differential equations will be to determine which method or combination of methods works in each specific case. In the next three sections, we elaborate on these three techniques.

EXERCISES FOR SECTION 1.1

In Exercises 1 and 2, find the equilibrium solutions of the differential equation specified.

1. $\dfrac{dy}{dt} = \dfrac{y + 3}{1 - y}$

2. $\dfrac{dy}{dt} = \dfrac{(t^2 - 1)(y^2 - 2)}{y^2 - 4}$

3. Consider the population model

$$\frac{dP}{dt} = 0.4P \left(1 - \frac{P}{230}\right),$$

where $P(t)$ is the population at time t.

(a) For what values of P is the population in equilibrium?

(b) For what values of P is the population increasing?

(c) For what values of P is the population decreasing?

4. Consider the population model

$$\frac{dP}{dt} = 0.3 \left(1 - \frac{P}{200}\right)\left(\frac{P}{50} - 1\right) P,$$

where $P(t)$ is the population at time t.

(a) For what values of P is the population in equilibrium?

(b) For what values of P is the population increasing?

(c) For what values of P is the population decreasing?

5. Consider the differential equation

$$\frac{dy}{dt} = y^3 - y^2 - 12y.$$

(a) For what values of y is $y(t)$ in equilibrium?

(b) For what values of y is $y(t)$ increasing?

(c) For what values of y is $y(t)$ decreasing?

In Exercises 6–10, we consider the phenomenon of radioactive decay which, from experimentation, we know behaves according to the law:

The rate at which a quantity of a radioactive isotope decays is proportional to the amount of the isotope present. The proportionality constant depends only on which radioactive isotope is used.

6. Model radioactive decay using the notation

$$t = \text{time (independent variable)},$$
$$r(t) = \text{amount of particular radioactive isotope}$$
$$\text{present at time } t \text{ (dependent variable)},$$
$$-\lambda = \text{decay rate (parameter)}.$$

Note that the minus sign is used so that $\lambda > 0$.

(a) Using this notation, write a model for the decay of a particular radioactive isotope.

(b) If the amount of the isotope present at $t = 0$ is r_0, state the corresponding initial-value problem for the model in part (a).

7. The **half-life** of a radioactive isotope is the amount of time it takes for a quantity of radioactive material to decay to one-half of its original amount.

(a) The half-life of Carbon 14 (C-14) is 5230 years. Determine the decay-rate parameter λ for C-14.

(b) The half-life of Iodine 131 (I-131) is 8 days. Determine the decay-rate parameter for I-131.

(c) What are the units of the decay-rate parameters in parts (a) and (b)?

(d) To determine the half-life of an isotope, we could start with 1000 atoms of the isotope and measure the amount of time it takes 500 of them to decay, or we could start with 10,000 atoms of the isotope and measure the amount of time it takes 5000 of them to decay. Will we get the same answer? Why?

8. Carbon dating is a method of determining the time elapsed since the death of organic material. The assumptions implicit in carbon dating are that

- Carbon 14 (C-14) makes up a constant proportion of the carbon that living matter ingests on a regular basis, and

- once the matter dies, the C-14 present decays, but no new carbon is added to the matter.

Hence, by measuring the amount of C-14 still in the organic matter and comparing it to the amount of C-14 typically found in living matter, a "time since death" can be approximated. Using the decay-rate parameter you computed in Exercise 7, determine the time since death if

(a) 88% of the original C-14 is still in the material.

(b) 12% of the original C-14 is still in the material.

(c) 2% of the original C-14 is still in the material.

(d) 98% of the original C-14 is still in the material.

Remark: There has been speculation that the amount of C-14 available to living creatures has not been exactly constant over long periods (thousands of years). This makes accurate dates much trickier to determine.

9. Engineers and scientists often measure the rate of decay of an exponentially decaying quantity using its **time constant**. The time constant τ is the amount of time that an exponentially decaying quantity takes to decay by a factor of $1/e$. Because $1/e$ is approximately 0.368, τ is the amount of time that the quantity takes to decay to approximately 36.8% of its original amount.

(a) How are the time constant τ and the decay rate λ related?

(b) Express the time constant in terms of the half-life.

(c) What are the time constants for Carbon 14 and Iodine 131?

(d) Given an exponentially decaying quantity $r(t)$ with initial value $r_0 = r(0)$, show that its time constant is the time at which the tangent line to the graph of $r(t)/r_0$ at $(0, 1)$ crosses the t-axis. [*Hint*: Start by sketching the graph of $r(t)/r_0$ and the line tangent to the graph at $(0, 1)$.]

(e) It is often said that an exponentially decaying quantity reaches its steady state in five time constants, that is, at $t = 5\tau$. Explain why this statement is not literally true but is correct for all practical purposes.

10. The radioactive isotope I-131 is used in the treatment of hyperthyroidism. When administered to a patient, I-131 accumulates in the thyroid gland, where it decays and kills part of that gland.

    (a) Suppose that it takes 72 hours to ship I-131 from the producer to the hospital. What percentage of the original amount shipped actually arrives at the hospital? (See Exercise 7.)

    (b) If the I-131 is stored at the hospital for an additional 48 hours before it is used, how much of the original amount shipped from the producer is left when it is used?

    (c) How long will it take for the I-131 to decay *completely* so that the remnants can be thrown away without special precautions?

11. MacQuarie Island is a small island about half-way between Antarctica and New Zealand. Between 2000 and 2006, the population of rabbits on the island rose from 4,000 to 130,000. Model the growth in the rabbit population $R(t)$ at time t using an exponential growth model

$$\frac{dR}{dt} = kR,$$

where $t = 0$ corresponds to the year 2000. What is an appropriate value for the growth-rate parameter k, and what does this model predict for the population in the year 2010. (For more information on why the population of rabbits exploded, see Review Exercise 22 in Chapter 2.)

12. The velocity v of a freefalling skydiver is well modeled by the differential equation

$$m\frac{dv}{dt} = mg - kv^2,$$

where m is the mass of the skydiver, g is the gravitational constant, and k is the drag coefficient determined by the position of the diver during the dive. (Note that the constants m, g, and k are positive.)

    (a) Perform a qualitative analysis of this model.

    (b) Calculate the terminal velocity of the skydiver. Express your answer in terms of m, g, and k.

Exercises 13–15 consider an elementary model of the learning process: Although human learning is an extremely complicated process, it is possible to build models of certain simple types of memorization. For example, consider a person presented with a list to be studied. The subject is given periodic quizzes to determine exactly how much of the list has been memorized. (The lists are usually things like nonsense syllables, randomly generated three-digit numbers, or entries from tables of integrals.) If we let $L(t)$ be the fraction of the list learned at time t, where $L = 0$ corresponds to knowing nothing and $L = 1$ corresponds to knowing the entire list, then we can form a simple model of this type of learning based on the assumption:

- The rate dL/dt is proportional to the fraction of the list left to be learned.

Since $L = 1$ corresponds to knowing the entire list, the model is

$$\frac{dL}{dt} = k(1 - L),$$

where k is the constant of proportionality.

13. For what value of L, $0 \leq L \leq 1$, does learning occur most rapidly?

14. Suppose two students memorize lists according to the model

$$\frac{dL}{dt} = 2(1 - L).$$

   (a) If one of the students knows one-half of the list at time $t = 0$ and the other knows none of the list, which student is learning more rapidly at this instant?

   (b) Will the student who starts out knowing none of the list ever catch up to the student who starts out knowing one-half of the list?

15. Consider the following two differential equations that model two students' rates of memorizing a poem. Aly's rate is proportional to the amount to be learned with proportionality constant $k = 2$. Beth's rate is proportional to the square of the amount to be learned with proportionality constant 3. The corresponding differential equations are

$$\frac{dL_A}{dt} = 2(1 - L_A) \quad \text{and} \quad \frac{dL_B}{dt} = 3(1 - L_B)^2,$$

where $L_A(t)$ and $L_B(t)$ are the fractions of the poem learned at time t by Aly and Beth, respectively.

   (a) Which student has a faster rate of learning at $t = 0$ if they both start memorizing together having never seen the poem before?

   (b) Which student has a faster rate of learning at $t = 0$ if they both start memorizing together having already learned one-half of the poem?

   (c) Which student has a faster rate of learning at $t = 0$ if they both start memorizing together having already learned one-third of the poem?

16. The expenditure on education in the U.S. is given in the following table. (Amounts are expressed in millions of 2001 constant dollars.)

Year	Expenditure	Year	Expenditure	Year	Expenditure
1900	5,669	1940	39,559	1980	380,165
1910	10,081	1950	67,048	1990	535,417
1920	12,110	1960	114,700	2000	714,064
1930	30,700	1970	322,935		

   (a) Let $s(t) = s_0 e^{kt}$ be an exponential function. Show that the graph of $\ln s(t)$ as a function of t is a line. What is its slope and vertical intercept?

(b) Is spending on education in the U.S. rising exponentially fast? If so, what is the growth-rate coefficient? [*Hint*: Use your solution to part (a).]

17. Suppose a species of fish in a particular lake has a population that is modeled by the logistic population model with growth rate k, carrying capacity N, and time t measured in years. Adjust the model to account for each of the following situations.

   (a) One hundred fish are harvested each year.

   (b) One-third of the fish population is harvested annually.

   (c) The number of fish harvested each year is proportional to the square root of the number of fish in the lake.

18. Suppose that the growth-rate parameter $k = 0.3$ and the carrying capacity $N = 2500$ in the logistic population model of Exercise 17. Suppose $P(0) = 2500$.

   (a) If 100 fish are harvested each year, what does the model predict for the long-term behavior of the fish population? In other words, what does a qualitative analysis of the model yield?

   (b) If one-third of the fish are harvested each year, what does the model predict for the long-term behavior of the fish population?

19. The rhinoceros is now extremely rare. Suppose enough game preserve land is set aside so that there is sufficient room for many more rhinoceros territories than there are rhinoceroses. Consequently, there will be no danger of overcrowding. However, if the population is too small, fertile adults have difficulty finding each other when it is time to mate. Write a differential equation that models the rhinoceros population based on these assumptions. (Note that there is more than one reasonable model that fits these assumptions.)

20. While it is difficult to imagine a time before cell phones, such a time did exist. The table below gives the number (in millions) of cell phone subscriptions in the United States from the U.S. census (see www.census.gov).

Year	Subscriptions	Year	Subscriptions	Year	Subscriptions
1985	0.34	1993	16	2001	128
1986	0.68	1994	24	2002	141
1987	1.23	1995	34	2003	159
1988	2.1	1996	44	2004	182
1989	3.5	1997	55	2005	208
1990	5.3	1998	69	2006	233
1991	7.6	1999	86	2007	250
1992	11	2000	110	2008	263

Let $s(t)$ be the number of cell phone subscriptions at time t, measured in years since 1989. The **relative growth rate** of $s(t)$ is its growth rate divided by the number of subscriptions. In other words, the relative growth rate is

$$\frac{1}{s(t)}\frac{ds}{dt},$$

and it is often expressed as a percentage.

(a) Estimate the relative growth rate of $s(t)$ at $t = 1$. That is, estimate the relative rate for the year 1990. Express this growth rate as a percentage. [*Hint*: The best estimate involves the number of cell phones at 1989 and 1991.]

(b) In general, if a quantity grows exponentially, how does its relative growth rate change?

(c) Also estimate the relative growth rates of $s(t)$ for the years 1991–2007.

(d) How long after 1989 was the number of subscriptions growing exponentially?

(e) In general, if a quantity grows according to a logistic model, how does its relative growth rate change?

(f) Using your results in part (c), calculate the carrying capacity for this model. [*Hint*: There is more than one way to do this calculation.]

21. For the following predator-prey systems, identify which dependent variable, x or y, is the prey population and which is the predator population. Is the growth of the prey limited by any factors other than the number of predators? Do the predators have sources of food other than the prey? (Assume that the parameters α, β, γ, δ, and N are all positive.)

(a) $\dfrac{dx}{dt} = -\alpha x + \beta xy$

$\dfrac{dy}{dt} = \gamma y - \delta xy$

(b) $\dfrac{dx}{dt} = \alpha x - \alpha\dfrac{x^2}{N} - \beta xy$

$\dfrac{dy}{dt} = \gamma y + \delta xy$

22. In the following predator-prey population models, x represents the prey, and y represents the predators.

(i) $\dfrac{dx}{dt} = 5x - 3xy$

$\dfrac{dy}{dt} = -2y + \frac{1}{2}xy$

(ii) $\dfrac{dx}{dt} = x - 8xy$

$\dfrac{dy}{dt} = -2y + 6xy$

(a) In which system does the prey reproduce more quickly when there are no predators (when $y = 0$) and equal numbers of prey?

(b) In which system are the predators more successful at catching prey? In other words, if the number of predators and prey are equal for the two systems, in which system do the predators have a greater effect on the rate of change of the prey?

(c) Which system requires more prey for the predators to achieve a given growth rate (assuming identical numbers of predators in both cases)?

23. The following systems are models of the populations of pairs of species that either *compete* for resources (an increase in one species decreases the growth rate of the other) or *cooperate* (an increase in one species increases the growth rate of the other). For each system, identify the variables (independent and dependent) and the parameters (carrying capacity, measures of interaction between species, etc.) Do the species compete or cooperate? (Assume all parameters are positive.)

(a)
$$\frac{dx}{dt} = \alpha x - \alpha \frac{x^2}{N} + \beta xy$$
$$\frac{dy}{dt} = \gamma y + \delta xy$$

(b)
$$\frac{dx}{dt} = -\gamma x - \delta xy$$
$$\frac{dy}{dt} = \alpha y - \beta xy$$

1.2 ANALYTIC TECHNIQUE: SEPARATION OF VARIABLES

What Is a Differential Equation and What Is a Solution?

A first-order differential equation is an equation for an unknown function in terms of its derivative. As we saw in Section 1.1, there are three types of "variables" in differential equations—the independent variable (almost always t for time in our examples), one or more dependent variables (which are functions of the independent variable), and the parameters. This terminology is standard but a bit confusing. The dependent variable is actually a function, so technically it should be called the dependent function.

The standard form for a first-order differential equation is

$$\frac{dy}{dt} = f(t, y).$$

Here the right-hand side typically depends on both the dependent and independent variables, although we often encounter cases where either t or y is missing.

A **solution** of the differential equation is a function of the independent variable that, when substituted into the equation as the dependent variable, satisfies the equation for all values of the independent variable. That is, a function $y(t)$ is a solution if it satisfies $dy/dt = y'(t) = f(t, y(t))$. This terminology doesn't tell us how to find solutions, but it does tell us how to check whether a candidate function is or is not a solution. For example, consider the simple differential equation

$$\frac{dy}{dt} = y.$$

We can easily check that the function $y_1(t) = 3e^t$ is a solution, whereas $y_2(t) = \sin t$ is not a solution. The function $y_1(t)$ is a solution because

$$\frac{dy_1}{dt} = \frac{d(3e^t)}{dt} = 3e^t = y_1 \quad \text{for all } t.$$

On the other hand, $y_2(t)$ is not a solution since

$$\frac{dy_2}{dt} = \frac{d(\sin t)}{dt} = \cos t,$$

and certainly the function $\cos t$ is not the same function as $y_2(t) = \sin t$.

Checking that a given function is a solution to a given equation

If we look at a more complicated equation such as

$$\frac{dy}{dt} = \frac{y^2 - 1}{t^2 + 2t},$$

then we have considerably more trouble finding a solution. On the other hand, if some-body hands us a function $y(t)$, then we know how to check whether or not it is a solution.

For example, suppose we meet three differential equations textbook authors—say Paul, Bob, and Glen—at our local espresso bar, and we ask them to find solutions of this differential equation. After a few minutes of furious calculation, Paul says that

$$y_1(t) = 1 + t$$

is a solution. Glen then says that

$$y_2(t) = 1 + 2t$$

is a solution. After several more minutes, Bob says that

$$y_3(t) = 1$$

is a solution. Which of these functions is a solution? Let's see who is right by substi-tuting each function into the differential equation.

First we test Paul's function. We compute the left-hand side by differentiating $y_1(t)$. We have

$$\frac{dy_1}{dt} = \frac{d(1 + t)}{dt} = 1.$$

Substituting $y_1(t)$ into the right-hand side, we find

$$\frac{(y_1(t))^2 - 1}{t^2 + 2t} = \frac{(1 + t)^2 - 1}{t^2 + 2t} = \frac{t^2 + 2t}{t^2 + 2t} = 1.$$

The left-hand side and the right-hand side of the differential equation are identical, so Paul is correct.

To check Glen's function, we again compute the derivative

$$\frac{dy_2}{dt} = \frac{d(1 + 2t)}{dt} = 2.$$

With $y_2(t)$, the right-hand side of the differential equation is

$$\frac{(y_2(t))^2 - 1}{t^2 + 2t} = \frac{(1 + 2t)^2 - 1}{t^2 + 2t} = \frac{4t^2 + 4t}{t^2 + 2t} = \frac{4(t + 1)}{t + 2}.$$

The left-hand side of the differential equation does not equal the right-hand side for all t since the right-hand side is not the constant function 2. Glen's function is *not* a solution.

Finally, we check Bob's function the same way. The left-hand side is

$$\frac{dy_3}{dt} = \frac{d(1)}{dt} = 0$$

because $y_3(t) = 1$ is a constant. The right-hand side is

$$\frac{y_3(t)^2 - 1}{t^2 + 2t} = \frac{1 - 1}{t^2 + 2t} = 0.$$

Both the left-hand side and the right-hand side of the differential equation approaches zero for all t. Hence, Bob's function *is* a solution of the differential equation.

The lessons we learn from this example are that a differential equation may have solutions that look very different from each other algebraically and that (of course) not every function is a solution. Given a function, we can test to see whether it is a solution by just substituting it into the differential equation and checking to see whether the left-hand side is identical to the right-hand side. This is a very nice aspect of differential equations: *We can always check our answers.* So we should never be wrong.

Initial-Value Problems and the General Solution

When we encounter differential equations in practice, they often come with **initial conditions**. We seek a solution of the given equation that assumes a given value at a particular time. A differential equation along with an initial condition is called an **initial-value problem**. Thus the usual form of an initial-value problem is

$$\frac{dy}{dt} = f(t, y), \quad y(t_0) = y_0.$$

Here we are looking for a function $y(t)$ that is a solution of the differential equation *and* assumes the value y_0 at time t_0. Often, the particular time in question is $t = 0$ (hence the name *initial condition*), but any other time could be specified.

For example,

$$\frac{dy}{dt} = 12t^3 - 2\sin t, \quad y(0) = 3,$$

is an initial-value problem. To solve this problem, note that the right-hand side of the differential equation depends only on t, not on y. We are looking for a function whose derivative is $12t^3 - 2\sin t$. This is a typical antidifferentiation problem from calculus, so all we need to do is to integrate this expression. We find

$$\int (12t^3 - 2\sin t)\, dt = 3t^4 + 2\cos t + c,$$

where c is a constant of integration. Thus the solution must be of the form

$$y(t) = 3t^4 + 2\cos t + c.$$

We now use the initial condition $y(0) = 3$ to determine c by

$$3 = y(0) = 3 \cdot 0^4 + 2\cos 0 + c = 0 + 2 \cdot 1 + c = 2 + c.$$

Thus $c = 1$, and the solution to this initial-value problem is $y(t) = 3t^4 + 2\cos t + 1$.
    The expression

$$y(t) = 3t^4 + 2\cos t + c$$

is called the **general solution** of the differential equation because we can use it to solve
any initial-value problem whatsoever. For example, if the initial condition is $y(0) = \pi$,
then we choose $c = \pi - 2$ to solve the initial-value problem $dy/dt = 12t^3 - 2\sin t$,
$y(0) = \pi$.

Separable Equations

Now that we know how to check that a given function is a solution to a differential
equation, the question is: How can we get our hands on a solution in the first place?
Unfortunately, it is rarely the case that we can find explicit solutions of a differential equation. Many differential equations have solutions that cannot be expressed in
terms of known functions such as polynomials, exponentials, or trigonometric functions. However, there are a few special types of differential equations for which we can
derive explicit solutions, and in this section we discuss one of these types of differential
equations.
    The typical first-order differential equation is given in the form

$$\frac{dy}{dt} = f(t, y).$$

The right-hand side of this equation generally involves both the independent variable t
and the dependent variable y (although there are many important examples where either
the t or the y is missing). A differential equation is called **separable** if the function
$f(t, y)$ can be written as the product of two functions: one that depends on t alone and
another that depends only on y. That is, a differential equation is separable if it can be
written in the form

$$\frac{dy}{dt} = g(t)h(y).$$

For example, the differential equation

$$\frac{dy}{dt} = yt$$

is clearly separable, and the equation

$$\frac{dy}{dt} = y + t$$

is not. We might have to do a little work to see that an equation is separable. For
instance,

$$\frac{dy}{dt} = \frac{t + 1}{ty + t}$$

is separable since we can rewrite the equation as

$$\frac{dy}{dt} = \frac{(t+1)}{t(y+1)} = \left(\frac{t+1}{t}\right)\left(\frac{1}{y+1}\right).$$

Two important types of separable equations occur if either t or y is missing from the right-hand side of the equation. The differential equation

$$\frac{dy}{dt} = g(t)$$

is separable since we may regard the right-hand side as $g(t) \cdot 1$, where we consider 1 as a (very simple) function of y. Similarly,

$$\frac{dy}{dt} = h(y)$$

is also separable. This last type of differential equation is said to be **autonomous**. Many of the most important first-order differential equations that arise in applications (including all of our models in the previous section) are autonomous. For example, the right-hand side of the logistic equation

$$\frac{dP}{dt} = kP\left(1 - \frac{P}{N}\right)$$

depends on the dependent variable P alone, so this equation is autonomous.

How to solve separable differential equations

To find explicit solutions of separable differential equations, we use a technique familiar from calculus. To illustrate the method, consider the differential equation

$$\frac{dy}{dt} = \frac{t}{y^2}.$$

There is a temptation to solve this equation by simply integrating both sides of the equation with respect to t. This yields

$$\int \frac{dy}{dt}\, dt = \int \frac{t}{y^2}\, dt,$$

and, consequently,

$$y(t) = \int \frac{t}{y^2}\, dt.$$

Now we are stuck. We can't evaluate the integral on the right-hand side because we don't know the function $y(t)$. In fact, that is precisely the function we wish to find. We have simply replaced the differential equation with an *integral equation*.

We need to do something to this equation *before* we try to integrate. Returning to the original differential equation

$$\frac{dy}{dt} = \frac{t}{y^2},$$

we first do some "informal" algebra and rewrite this equation in the form

$$y^2 \, dy = t \, dt.$$

That is, we multiply both sides by $y^2 \, dt$. Of course, it makes no sense to split up dy/dt by multiplying by dt. However, this should remind you of the technique of integration known as u-substitution in calculus. We will soon see that substitution is exactly what we are doing here.

We now integrate both sides: the left with respect to y and the right with respect to t. We have

$$\int y^2 \, dy = \int t \, dt,$$

which yields

$$\frac{y^3}{3} = \frac{t^2}{2} + c.$$

Technically there is a constant of integration on both sides of this equation, but we can lump them together as a single constant c on the right. We may rewrite this expression as

$$y(t) = \left(\frac{3t^2}{2} + 3c \right)^{1/3};$$

and since c is an arbitrary constant, we may write this even more compactly as

$$y(t) = \left(\frac{3t^2}{2} + k \right)^{1/3},$$

where k is an arbitrary constant. As usual, we can check that this expression really is a solution of the differential equation, so despite the questionable separation we just performed, we do obtain infinitely many solutions.

Note that this process yields many solutions of the differential equation. Each choice of the constant k gives a different solution.

What is really going on in our informal algebra

If you read the previous example closely, you probably became nervous at one point. Treating dt as a variable is a tip-off that something a little more complicated is actually going on. Here is the real story.

We began with a separable equation

$$\frac{dy}{dt} = g(t)h(y),$$

and then rewrote it as

$$\frac{1}{h(y)} \frac{dy}{dt} = g(t).$$

This equation actually has a function of t on each side of the equals sign because y is a function of t. So we really should write it as

$$\frac{1}{h(y(t))} \frac{dy}{dt} = g(t).$$

In this form, we can integrate both sides with respect to t to get

$$\int \frac{1}{h(y(t))} \frac{dy}{dt} \, dt = \int g(t) \, dt.$$

Now for the important step: We make a "u-substitution" just as in calculus by replacing the function $y(t)$ by the new variable, say y. (In this case, the substitution is actually a y-substitution.) Of course, we must also replace the expression $(dy/dt) \, dt$ by dy. The method of substitution from calculus tells us that

$$\int \frac{1}{h(y(t))} \frac{dy}{dt} \, dt = \int \frac{1}{h(y)} \, dy,$$

and therefore we can combine the last two equations to obtain

$$\int \frac{1}{h(y)} \, dy = \int g(t) \, dt.$$

Hence, we can integrate the left-hand side with respect to y and the right-hand side with respect to t.

Separating variables and multiplying both sides of the differential equation by dt is simply a notational convention that helps us remember the method. It is justified by the argument above.

Missing Solutions

If it is possible to separate variables in a differential equation, it appears that solving the equation reduces to a matter of computing several integrals. This is true, but there are some hidden pitfalls, as the following example shows. Consider the differential equation

$$\frac{dy}{dt} = y^2.$$

This is an autonomous and hence separable equation, and its solution looks straightforward. If we separate and integrate as usual, we obtain

$$\int \frac{dy}{y^2} = \int dt$$

$$-\frac{1}{y} = t + c$$

$$y(t) = -\frac{1}{t + c}.$$

We are tempted to say that this expression

$$y(t) = -\frac{1}{t + c}$$

is the general solution. However, we cannot solve all initial-value problems with solutions of this form. In fact, we have $y(0) = -1/c$, so we cannot use this expression to solve the initial-value problem $y(0) = 0$.

What's wrong? Note that the right-hand side of the differential equation vanishes if $y = 0$. So the constant function $y(t) = 0$ is a solution to this differential equation. In other words, in addition to those solutions that we derived using the method of separation of variables, this differential equation possesses the equilibrium solution $y(t) = 0$ for all t, and it is this equilibrium solution that satisfies the initial-value problem $y(0) = 0$. Even though it is "missing" from the family of solutions that we obtain by separating variables, it is a solution that we need if we want to solve every initial-value problem for this differential equation. Thus the general solution consists of functions of the form $y(t) = -1/(t + c)$ together with the equilibrium solution $y(t) = 0$.

Getting Stuck

As another example, consider the differential equation

$$\frac{dy}{dt} = \frac{y}{1 + y^2}.$$

As before, this equation is autonomous. So we first separate variables to obtain

$$\left(\frac{1 + y^2}{y}\right) dy = dt.$$

Then we integrate

$$\int \left(\frac{1}{y} + y\right) dy = \int dt,$$

which yields

$$\ln |y| + \frac{y^2}{2} = t + c.$$

But now we are stuck; there is no way to solve the equation

$$\ln |y| + \frac{y^2}{2} = t + c$$

for y alone. Thus we cannot generate an *explicit* formula for y. We do, however, have an **implicit form** for the solution which, for many purposes, is perfectly acceptable.

Even though we don't obtain explicit solutions by separating variables for this equation, we can find one explicit solution. The right-hand side is zero if $y = 0$. Thus the constant function $y(t) = 0$ for all t is an equilibrium solution. Note that this equilibrium solution does not appear in the implicit solution we derived from the method of separation of variables.

There is another problem that arises with this method. It is often impossible to perform the necessary integrations. For example, the differential equation

$$\frac{dy}{dt} = \sec(y^2)$$

is autonomous. Separating variables and integrating we get

$$\int \frac{1}{\sec(y^2)} \, dy = \int dt.$$

In other words,

$$\int \cos(y^2) \, dy = \int dt.$$

The integral on the left-hand side is difficult, to say the least. (In fact, there is a special function that was defined just to give us a name for this integral.) The lesson is that, even for autonomous equations

$$\frac{dy}{dt} = f(y),$$

carrying out the required algebra or integration is frequently impossible. We will not be able to rely solely on analytic tools and explicit solutions when studying differential equations, even if we can separate variables.

A Savings Model

Suppose we deposit $5000 in a savings account with interest accruing at the rate of 2% compounded continuously. If we let $A(t)$ denote the amount of money in the account at time t, then the differential equation for A is

$$\frac{dA}{dt} = 0.02A.$$

As we saw in the previous section, the general solution to this equation is the exponential function

$$A(t) = ce^{0.02t},$$

where $c = A(0)$. Thus $A(t) = 5000e^{0.02t}$ is our particular solution.

Assuming interest rates never change, after 10 years we will have

$$A(10) = 5000e^{0.2} \approx 6107$$

dollars in this account. That is a nice little nest egg, so we decide we should have some fun in life. We decide to withdraw $500 (mad money) from the account each year in a continuous way beginning in year 10. How long will this money last? Will we ever go broke?

The differential equation for $A(t)$ must change, but only beginning in year 10. For $0 \leq t \leq 10$, our previous model works fine. However, for $t > 10$, the differential equation becomes

$$\frac{dA}{dt} = 0.02A - 500.$$

Thus we really have a differential equation of the form

$$\frac{dA}{dt} = \begin{cases} 0.02A & \text{for } t < 10; \\ 0.02A - 500 & \text{for } t > 10, \end{cases}$$

whose right-hand side consists of two pieces.

To solve this two-part equation, we solve the first part and determine $A(10)$. We just did that and obtained $A(10) \approx 6107$. Then we solve the second equation using $A(10) \approx 6107$ as the initial value. This equation is also separable, and we have

$$\int \frac{dA}{0.02A - 500} = \int dt.$$

We calculate this integral using substitution and the natural logarithm function. Let $u = 0.02A - 500$. Then $du = 0.02 \, dA$, or $50 \, du = dA$ since $0.02 = 1/50$. We obtain

$$\int \frac{50 \, du}{u} = t + c_1$$

$$50 \ln |u| = t + c_1$$

$$50 \ln |0.02A - 500| = t + c_1,$$

for some constant c_1.

At $t = 10$, we know that $A \approx 6107$. Thus at $t = 10$,

$$\frac{dA}{dt} = 0.02A - 500 \approx -377.9 < 0.$$

In other words, we are withdrawing at a rate that exceeds the rate at which we are earning interest. Since dA/dt at $t = 10$ is negative, A will decrease and $0.02A - 500$ remains negative for all $t > 10$. If $0.02A - 500 < 0$, then

$$|0.02A - 500| = -(0.02A - 500) = 500 - 0.02A.$$

Consequently, we have

$$50 \ln(500 - 0.02A) = t + c_1.$$

Since dividing by 50 is the same as multiplying 0.02, we get

$$\ln(500 - 0.02A) = 0.02(t + c_1).$$

We exponentiate and obtain

$$500 - 0.02A = e^{0.02(t+c_1)}$$
$$= c_2 e^{0.02t}$$

where $c_2 = e^{0.02c_1}$. Solving for A, we have

$$A = \frac{500 - c_2 e^{0.02t}}{0.02}$$
$$= 50\left(500 - c_2 e^{0.02t}\right) = 25000 - c_3 e^{0.02t},$$

where $c_3 = 50c_2$. (Although we have been careful to spell out the relationships among the constants c_1, c_2, and c_3, we need only remember that c_3 is a constant that is determined from the initial condition.)

Now we use the initial condition to determine c_3. We know that

$$6107 \approx A(10) = 25000 - c_3 e^{0.02(10)} \approx 25000 - c_3(1.2214).$$

Solving for c_3, we obtain $c_3 \approx 15468$. Our solution for $t \geq 10$ is

$$A(t) \approx 25000 - 15468 e^{0.02t}.$$

We see that

$$A(11) \approx 5726$$
$$A(12) \approx 5336$$

and so forth. Our account is being depleted, but not by that much. In fact, we can find out just how long the good times will last by asking when our money will run out. In other words, we solve the equation $A(t) = 0$ for t. We have

$$0 = 25000 - 15468 e^{0.02t},$$

which yields

$$t = 50\ln\left(\frac{25000}{15468}\right) \approx 24.01.$$

After letting the $5000 accumulate interest for ten years, we can withdraw $500 per year for more than twenty years.

A Mixing Problem

The name *mixing problem* refers to a large collection of different problems where two or more substances are mixed together at various rates. Examples range from the mixing of pollutants in a lake to the mixing of chemicals in a vat to the diffusion of cigar smoke in the air in a room to the blending of spices in a serving of curry.

Mixing in a vat

Consider a large vat containing sugar water that is to be made into soft drinks (see Figure 1.9). Suppose:

- The vat contains 100 gallons of liquid. Moreover, the amount flowing in is the same as the amount flowing out, so there are always 100 gallons in the vat.
- The vat is kept well mixed, so the sugar concentration is uniform throughout the vat.
- Sugar water containing 5 tablespoons of sugar per gallon enters the vat through pipe A at a rate of 2 gallons per minute.
- Sugar water containing 10 tablespoons of sugar per gallon enters the vat through pipe B at a rate of 1 gallon per minute.
- Sugar water leaves the vat through pipe C at a rate of 3 gallons per minute.

Figure 1.9
Mixing vat.

To make the model, we let t be time measured in minutes (the independent variable). For the dependent variable, we have two choices. We could choose either the total amount of sugar, $S(t)$, in the vat at time t measured in tablespoons, or $C(t)$, the concentration of sugar in the vat at time t measured in tablespoons per gallon. We develop the model for S, leaving the model for C as an exercise for the reader.

Using the total sugar $S(t)$ in the vat as the dependent variable, the rate of change of S is the difference between the amount of sugar being added and the amount of sugar being removed. The sugar entering the vat comes from pipes A and B and can be easily computed by multiplying the number of gallons per minute of sugar mixture entering the vat by the amount of sugar per gallon. The amount of sugar leaving the vat through pipe C at any given moment depends on the concentration of sugar in the vat at that moment. The concentration is given by $S/100$, so the sugar leaving the vat is the product of the number of gallons leaving per minute (3 gallons per minute) and the concentration ($S/100$). The model is

$$\frac{dS}{dt} = \underbrace{2 \cdot 5}_{\substack{\text{sugar in} \\ \text{from pipe A}}} + \underbrace{1 \cdot 10}_{\substack{\text{sugar in} \\ \text{from pipe B}}} - \underbrace{3 \cdot \frac{S}{100}}_{\substack{\text{sugar out} \\ \text{from pipe C}}}.$$

That is,

$$\frac{dS}{dt} = 20 - \frac{3S}{100} = \frac{2000 - 3S}{100}.$$

To solve this equation analytically, we separate and integrate. We find

$$\frac{dS}{2000 - 3S} = \frac{dt}{100}$$

$$\frac{\ln|2000 - 3S|}{-3} = \frac{t}{100} + c_1$$

$$\ln|2000 - 3S| = -\frac{3t}{100} - 3c_1$$

$$\ln|2000 - 3S| = -0.03t + c_2,$$

where $c_2 = -3c_1$. Exponentiating we obtain

$$|2000 - 3S| = e^{(-0.03t + c_2)} = c_3 e^{-0.03t},$$

where $c_3 = e^{c_2}$. Note that this means that c_3 is a positive constant. Now we must be careful. Removing the absolute value signs yields

$$2000 - 3S = \pm c_3 e^{-0.03t},$$

where we choose the plus sign if $S(t) < 2000/3$ and the minus sign if $S(t) > 2000/3$. Therefore we may write this equation more simply as

$$2000 - 3S = c_4 e^{-0.03t},$$

where c_4 is an arbitrary constant (positive, negative, or zero). Solving for S yields the general solution

$$S(t) = ce^{-0.03t} + \frac{2000}{3},$$

where $c = -c_4/3$ is an arbitrary constant. We can determine the precise value of c if we know the exact amount of sugar that is initially in the vat. Note that, if $c = 0$, the solution is simply $S(t) = 2000/3$, an equilibrium solution.

EXERCISES FOR SECTION 1.2

1. Bob, Glen, and Paul are once again sitting around enjoying their nice, cold glasses of iced cappucino when one of their students asks them to come up with solutions to the differential equation

$$\frac{dy}{dt} = \frac{y+1}{t+1}.$$

After much discussion, Bob says $y(t) = t$, Glen says $y(t) = 2t + 1$, and Paul says $y(t) = t^2 - 2$.

   (a) Who is right?

   (b) What solution should they have seen right away?

2. Make up a differential equation of the form

$$\frac{dy}{dt} = 2y - t + g(y)$$

that has the function $y(t) = e^{2t}$ as a solution.

3. Make up a differential equation of the form $dy/dt = f(t, y)$ that has $y(t) = e^{t^3}$ as a solution. (Try to come up with one whose right-hand side $f(t, y)$ depends explicitly on both t and y.)

4. In Section 1.1, we guessed solutions to the exponential growth model $dP/dt = kP$, where k is a constant (see page 6). Using the fact that this equation is separable, derive these solutions by separating variables.

In Exercises 5–24, find the general solution of the differential equation specified. (You may not be able to reach the ideal answer of an equation with only the dependent variable on the left and only the independent variable on the right, but get as far as you can.)

5. $\dfrac{dy}{dt} = (ty)^2$

6. $\dfrac{dy}{dt} = t^4 y$

7. $\dfrac{dy}{dt} = 2y + 1$

8. $\dfrac{dy}{dt} = 2 - y$

9. $\dfrac{dy}{dt} = e^{-y}$

10. $\dfrac{dx}{dt} = 1 + x^2$

11. $\dfrac{dy}{dt} = 2ty^2 + 3y^2$

12. $\dfrac{dy}{dt} = \dfrac{t}{y}$

13. $\dfrac{dy}{dt} = \dfrac{t}{t^2 y + y}$

14. $\dfrac{dy}{dt} = t\sqrt[3]{y}$

15. $\dfrac{dy}{dt} = \dfrac{1}{2y + 1}$

16. $\dfrac{dy}{dt} = \dfrac{2y + 1}{t}$

17. $\dfrac{dy}{dt} = y(1 - y)$

18. $\dfrac{dy}{dt} = \dfrac{4t}{1 + 3y^2}$

19. $\dfrac{dv}{dt} = t^2 v - 2 - 2v + t^2$

20. $\dfrac{dy}{dt} = \dfrac{1}{ty + t + y + 1}$

21. $\dfrac{dy}{dt} = \dfrac{e^t y}{1 + y^2}$

22. $\dfrac{dy}{dt} = y^2 - 4$

23. $\dfrac{dw}{dt} = \dfrac{w}{t}$

24. $\dfrac{dy}{dx} = \sec y$

In Exercises 25–38, solve the given initial-value problem.

25. $\dfrac{dx}{dt} = -xt, \quad x(0) = 1/\sqrt{\pi}$

26. $\dfrac{dy}{dt} = ty, \quad y(0) = 3$

27. $\dfrac{dy}{dt} = -y^2, \quad y(0) = 1/2$

28. $\dfrac{dy}{dt} = t^2 y^3, \quad y(0) = -1$

29. $\dfrac{dy}{dt} = -y^2, \quad y(0) = 0$

30. $\dfrac{dy}{dt} = \dfrac{t}{y - t^2 y}, \quad y(0) = 4$

31. $\dfrac{dy}{dt} = 2y + 1, \quad y(0) = 3$

32. $\dfrac{dy}{dt} = ty^2 + 2y^2, \quad y(0) = 1$

33. $\dfrac{dx}{dt} = \dfrac{t^2}{x + t^3 x}, \quad x(0) = -2$

34. $\dfrac{dy}{dt} = \dfrac{1 - y^2}{y}, \quad y(0) = -2$

35. $\dfrac{dy}{dt} = (y^2 + 1)t, \quad y(0) = 1$

36. $\dfrac{dy}{dt} = \dfrac{1}{2y + 3}, \quad y(0) = 1$

37. $\dfrac{dy}{dt} = 2ty^2 + 3t^2 y^2, \quad y(1) = -1$

38. $\dfrac{dy}{dt} = \dfrac{y^2 + 5}{y}, \quad y(0) = -2$

39. A 5-gallon bucket is full of pure water. Suppose we begin dumping salt into the bucket at a rate of 1/4 pounds per minute. Also, we open the spigot so that 1/2 gallons per minute leaves the bucket, and we add pure water to keep the bucket full. If the salt water solution is always well mixed, what is the amount of salt in the bucket after

(a) 1 minute?   **(b)**   10 minutes?   **(c)**   60 minutes?

(d) 1000 minutes?   **(e)**   a very, very long time?

40. Consider the following very simple model of blood cholesterol levels based on the fact that cholesterol is manufactured by the body for use in the construction of cell walls and is absorbed from foods containing cholesterol: Let $C(t)$ be the amount (in milligrams per deciliter) of cholesterol in the blood of a particular person at time t (in days). Then

$$\frac{dC}{dt} = k_1(N - C) + k_2 E,$$

where

N = the person's natural cholesterol level,

k_1 = production parameter,

E = daily rate at which cholesterol is eaten, and

k_2 = absorption parameter.

(a) Suppose $N = 200$, $k_1 = 0.1$, $k_2 = 0.1$, $E = 400$, and $C(0) = 150$. What will the person's cholesterol level be after 2 days on this diet?

(b) With the initial conditions as above, what will the person's cholesterol level be after 5 days on this diet?

(c) What will the person's cholesterol level be after a long time on this diet?

(d) High levels of cholesterol in the blood are known to be a risk factor for heart disease. Suppose that, after a long time on the high cholesterol diet described above, the person goes on a very low cholesterol diet, so E changes to $E = 100$. (The initial cholesterol level at the starting time of this diet is the result of part (c).) What will the person's cholesterol level be after 1 day on the new diet, after 5 days on the new diet, and after a very long time on the new diet?

(e) Suppose the person stays on the high cholesterol diet but takes drugs that block some of the uptake of cholesterol from food, so k_2 changes to $k_2 = 0.075$. With the cholesterol level from part (c), what will the person's cholesterol level be after 1 day, after 5 days, and after a very long time?

41. A cup of hot chocolate is initially 170° F and is left in a room with an ambient temperature of 70° F. Suppose that at time $t = 0$ it is cooling at a rate of 20° per minute.

(a) Assume that Newton's law of cooling applies: The rate of cooling is proportional to the difference between the current temperature and the ambient temperature. Write an initial-value problem that models the temperature of the hot chocolate.

(b) How long does it take the hot chocolate to cool to a temperature of 110° F?

42. Suppose you are having a dinner party for a large group of people, and you decide to make 2 gallons of chili. The recipe calls for 2 teaspoons of hot sauce per gallon, but you misread the instructions and put in 2 tablespoons of hot sauce per gallon. (Since each tablespoon is 3 teaspoons, you have put in 6 teaspoons per gallon, which is a total of 12 teaspoons of hot sauce in the chili.) You don't want to throw the chili out because there isn't much else to eat (and some people like hot chili), so you serve the chili anyway. However, as each person takes some chili, you fill up the pot with beans and tomatoes without hot sauce until the concentration of hot sauce agrees with the recipe. Suppose the guests take 1 cup of chili per minute from the pot (there are 16 cups in a gallon), how long will it take to get the chili back to the recipe's concentration of hot sauce? How many cups of chili will have been taken from the pot?

43. In Exercise 12 of Section 1.1, we saw that the velocity v of a freefalling skydiver is well modeled by the differential equation

$$m\frac{dv}{dt} = mg - kv^2,$$

where m is the mass of the skydiver, g is the gravitational constant, and k is the drag coefficient determined by the position of the driver during the dive.

(a) Find the general solution of this differential equation.

(b) Confirm your answer to Exercise 12 of Section 1.1 by calculating the limit of $v(t)$ as $t \to \infty$.

1.3 QUALITATIVE TECHNIQUE: SLOPE FIELDS

Finding an analytic expression (in other words, finding a formula) for a solution to a differential equation is often a useful way to describe a solution of a differential equation. However, there are other ways to describe solutions, and these alternative representations are frequently easier to understand and use. In this section we focus on geometric techniques for representing solutions, and we develop a method for visualizing the graphs of the solutions to the differential equation

$$\frac{dy}{dt} = f(t, y).$$

The Geometry of $dy/dt = f(t, y)$

If the function $y(t)$ is a solution of the equation $dy/dt = f(t, y)$ and if its graph passes through the point (t_1, y_1) where $y_1 = y(t_1)$, then the differential equation says that the derivative dy/dt at $t = t_1$ is given by the number $f(t_1, y_1)$. Geometrically, this equality of dy/dt at $t = t_1$ with $f(t_1, y_1)$ means that the slope of the tangent line to the graph of $y(t)$ at the point (t_1, y_1) is $f(t_1, y_1)$ (see Figure 1.10). Note that there is nothing special about the point (t_1, y_1) other than the fact that it is a point on the graph

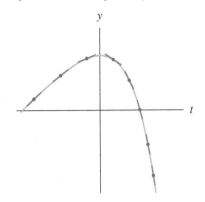

Figure 1.10
Slope of the tangent at the point (t_1, y_1)
is given by the value of $f(t_1, y_1)$.

Figure 1.11
If $y = y(t)$ is a solution, then the slope
of any tangent must equal $f(t, y)$.

of the solution $y(t)$. The equality of dy/dt and $f(t, y)$ must hold for all t for which $y(t)$ satisfies the differential equation. In other words, the values of the right-hand side of the differential equation yield the slopes of the tangents at all points on the graph of $y(t)$ (see Figure 1.11).

Slope Fields

This simple geometric observation leads to our main device for the visualization of the solutions to a first-order differential equation

$$\frac{dy}{dt} = f(t, y).$$

If we are given the function $f(t, y)$, we obtain a rough idea of the graphs of the solutions to the differential equation by sketching its corresponding **slope field**. We make this sketch by selecting points in the ty-plane and computing the numbers $f(t, y)$ at these points. At each point (t, y) selected, we use $f(t, y)$ to draw a minitangent line whose slope is $f(t, y)$ (see Figure 1.12). These minitangent lines are also called slope marks. Once we have a lot of slope marks, we can visualize the graphs of the solutions. For example, consider the differential equation

$$\frac{dy}{dt} = y - t.$$

In other words, the right-hand side of the differential equation is given by the function $f(t, y) = y - t$. To get some practice with the idea of a slope field, we sketch its slope field by hand at a small number of points. Then we discuss a computer-generated version of this slope field.

Generating slope fields by hand is tedious, so we consider only the nine points in the ty-plane. For example, at the point $(t, y) = (1, -1)$, we have $f(t, y) = f(1, -1) = -1 - 1 = -2$. Therefore we sketch a "small" line segment with slope -2 centered at

slope of minitangent
line is $f(t, y)$

Figure 1.12
The slope of the minitangent
at the point (t, y) is
determined by the right-hand
side $f(t, y)$ of the
differential equation.

Nancy Kopell (1942–) received her doctorate in mathematics at the University of California, Berkeley, where she wrote her thesis under the direction of Stephen Smale. She is one of the leading figures in the world in the use of differential equations to model natural phenomena. Kopell has employed techniques similar to those that we study in this book to tackle such diverse problems as spontaneous pattern formation in chemical systems and the networks of neurons that govern rhythmic motion in animals and other oscillations in the central nervous system.

For her work, she has received numerous awards, including a MacArthur Fellowship "genius grant" in 1990. In 1996, she was elected to the National Academy of Sciences. She is currently professor of mathematics and founding director of the Center for BioDynamics (and the authors' colleague) at Boston University.

the point $(1, -1)$ (see Figure 1.13). To sketch the slope field for all nine points, we use the function $f(t, y)$ to compute the appropriate slopes. The results are summarized in Table 1.2. Once we have these values, we use them to give a sparse sketch of the slope field for this equation (see Figure 1.13).

Sketching slope fields is best done using a computer. Figure 1.14 is a sketch of the slope field for this equation over the region $-3 \leq t \leq 3$ and $-3 \leq y \leq 3$ in the ty-plane. We calculated values of the function $f(t, y)$ over 25×25 points (625 points) in that region.

A glance at this slope field suggests that the graph of one solution is a line passing through the points $(-1, 0)$ and $(0, 1)$. Solutions corresponding to initial conditions that are below this line seem to increase until they reach an absolute maximum. Solutions corresponding to initial conditions that are above the line seem to increase more and more rapidly.

In fact, in Section 1.8 we will learn an analytic technique for finding solutions of this equation. We will see that the general solution consists of the family of functions

$$y(t) = t + 1 + ce^t,$$

where c is an arbitrary constant. (At this point it is important to emphasize that, even though we have not studied the technique that gives us these solutions, we can still

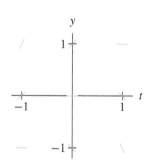

Figure 1.13
A "sparse" slope field generated from Table 1.2.

Table 1.2
Selected slopes corresponding to the differential equation $dy/dt = y - t$

(t, y)	$f(t, y)$	(t, y)	$f(t, y)$	(t, y)	$f(t, y)$
$(-1, 1)$	2	$(0, 1)$	1	$(1, 1)$	0
$(-1, 0)$	1	$(0, 0)$	0	$(1, 0)$	-1
$(-1, -1)$	0	$(0, -1)$	-1	$(1, -1)$	-2

check to see whether these functions are indeed solutions. If $y(t) = t + 1 + ce^t$, then $dy/dt = 1 + ce^t$. Also $f(t, y) = y - t = (t + 1 + ce^t) - t = 1 + ce^t$. Hence all of these functions are solutions.)

In Figure 1.15 we sketch the graphs of these functions with $c = -2, -1, 0, 1, 2, 3$. Note that each of these graphs is tangent to the slope field. Also note that, if $c = 0$, the graph is a line whose slope is 1. It goes through the points $(-1, 0)$ and $(0, 1)$.

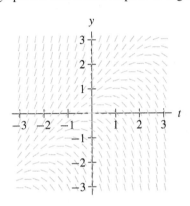

Figure 1.14
A computer-generated version of the slope field for $dy/dt = y - t$.

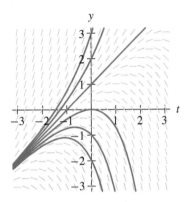

Figure 1.15
The graphs of six solutions to $dy/dt = y - t$ superimposed on its slope field.

Important Special Cases

From an analytic point of view, differential equations of the forms

$$\frac{dy}{dt} = f(t) \quad \text{and} \quad \frac{dy}{dt} = f(y)$$

are somewhat easier to consider than more complicated equations because they are separable. The geometry of their slope fields is equally special.

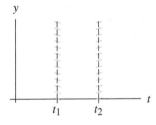

Figure 1.16
If the right-hand side of the differential equation is a function of t alone, that is,

$$\frac{dy}{dt} = f(t),$$

then the slope marks in the slope field are determined solely by their t-coordinate

Slope fields for $dy/dt = f(t)$

If the right-hand side of the differential equation in question is solely a function of t, or in other words, if $dy/dt = f(t)$, the slope at any point is the same as the slope of any other point with the same t-coordinate (see Figure 1.16).

Geometrically, this implies that all of the slope marks on each vertical line are parallel. Whenever a slope field has this geometric property for all vertical lines throughout the domain in question, we know that the corresponding differential equation is really an equation of the form

$$\frac{dy}{dt} = f(t).$$

(Note that finding solutions to this type of differential equation is the same thing as finding an antiderivative of $f(t)$ in calculus.)

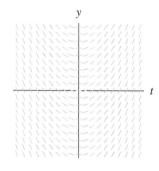

Figure 1.17
The slope field for

$$\frac{dy}{dt} = 2t.$$

Note the parallel slopes along vertical lines.

For example, consider the slope field shown in Figure 1.17. We generated this slope field from the equation

$$\frac{dy}{dt} = 2t,$$

and from calculus we know that

$$y(t) = \int 2t \, dt = t^2 + c,$$

where c is the constant of integration. Hence the general solution of the differential equation consists of functions of the form

$$y(t) = t^2 + c.$$

In Figure 1.18 we have superimposed graphs of such solutions on this field. Note that all of these graphs simply differ by a vertical translation. If one graph is tangent to the slope field, we can get infinitely many graphs—all tangent to the slope field—by translating the original graph either up or down.

Slope fields for autonomous equations

In the case of an autonomous differential equation

$$\frac{dy}{dt} = f(y),$$

the right-hand side of the equation does not depend on the independent variable t. The slope field in this case is also somewhat special. Here, the slopes that correspond to two different points with the same y-coordinate are equal. That is, $f(t_1, y) = f(t_2, y) = f(y)$ since the right-hand side of the differential equation depends only on y. In other words, the slope field of an autonomous equation is parallel along each horizontal line (see Figure 1.19).

For example, the slope field for the autonomous equation

$$\frac{dy}{dt} = 4y(1 - y)$$

is given in Figure 1.20. Note that, along each horizontal line, the slope marks are parallel. In fact, if $0 < y < 1$, then dy/dt is positive, and the tangents suggest that a solution

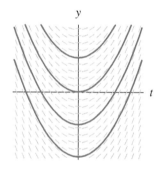

Figure 1.18
Graphs of solutions to

$$\frac{dy}{dt} = 2t$$

superimposed on its slope field.

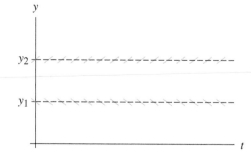

Figure 1.19
If the right-hand side of the differential equation is a function of y alone, that is, if

$$\frac{dy}{dt} = f(y),$$

then the slope marks in the slope field are determined solely by their y-coordinate.

with $0 < y < 1$ is increasing. On the other hand, if $y < 0$ or if $y > 1$, then dy/dt is negative and any solution with either $y < 0$ or $y > 1$ is decreasing.

We have equilibrium solutions at $y = 0$ and at $y = 1$ since the right-hand side of the differential equation vanishes along these lines. The slope field is horizontal all along these two horizontal lines, and therefore we know that these lines are the graphs of solutions. Solutions whose graphs are between these two lines are increasing. Solutions that are above the line $y = 1$ or that are below the line $y = 0$ are decreasing (see Figure 1.21).

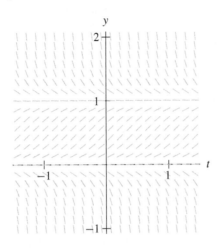

Figure 1.20
The slope field for $dy/dt = 4y(1 - y)$.

Figure 1.21
The graphs of five solutions superimposed on the slope field for $dy/dt = 4y(1 - y)$.

The fact that autonomous equations produce slope fields that are parallel along horizontal lines indicates that we can get infinitely many solutions from one solution simply by translating the graph of the given solution left or right (see Figure 1.22). We will make extensive use of this simple geometric observation about the solutions to autonomous equations in Section 1.6.

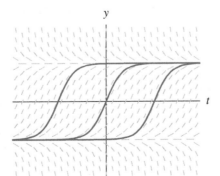

Figure 1.22
The graphs of three solutions to an autonomous equation, that is, an equation of the form

$$\frac{dy}{dt} = f(y).$$

Note that each graph is a horizontal translate of the others.

Analytic versus Qualitative Analysis

For the autonomous equation

$$\frac{dy}{dt} = 4y(1 - y),$$

we could have used the analytic techniques of the previous section to find explicit formulas for the solutions. In fact, we can perform all of the required integrations to determine the general solution (see Exercise 17 on page 34). However, these integrations are complicated, and the formulas that result are by no means easy to interpret. This points out the power of geometric and qualitative methods for solving differential equations. With very little work, we gain a lot of insight into the behavior of solutions. Although we cannot use qualitative methods to answer specific questions, such as what the exact value of the solution is at any given time, we can use these methods to understand the long-term behavior of a solution.

These ideas are especially important if the differential equation in question cannot be handled by analytic techniques. As an example, consider the differential equation

$$\frac{dy}{dt} = e^{y^2/10} \sin^2 y.$$

This equation is autonomous and hence separable. To solve this equation analytically, we must evaluate the integrals

$$\int \frac{dy}{e^{y^2/10} \sin^2 y} = \int dt.$$

However, the integral on the left-hand side cannot be evaluated so easily. Thus we resort to qualitative methods. The right-hand side of this differential equation is positive except if $y = n\pi$ for any integer n. These special lines correspond to equilibrium solutions of the equation. Between these equilibria, solutions must always increase. From the slope field, we expect that their graphs either lie on one of the horizontal lines $y = n\pi$ or increase from one of these lines to the next higher as $t \to \infty$ (see Figure 1.23). Hence we can predict the long-term behavior of the solutions even though we cannot explicitly solve the equation.

Figure 1.23
The slope field and graphs of solutions for the differential equation

$$\frac{dy}{dt} = e^{y^2/10} \sin^2 y.$$

The lines $y = n\pi$ are the graphs of the equilibrium solutions, and between these lines, all solutions are increasing.

Although the computer pictures of solutions of this differential equation are convincing, some subtle questions remain. For example, how do we *really* know that these pictures are correct? In particular, for $dy/dt = e^{y^2/10} \sin^2 y$, how do we know that the graphs of solutions do not cross the horizontal lines that are the graphs of the equilibrium solutions (see Figure 1.23)? Such a solution could not cross these lines at a nonzero angle since we know that the tangent line to the solution must be horizontal. But what prevents certain solutions from crossing these lines tangentially and then continuing to increase?

For the differential equation

$$\frac{dy}{dt} = 4y(1-y)$$

we can eliminate these questions because we can evaluate all of the integrals and check the accuracy of the pictures using analytic techniques. But using analytic techniques to check our qualitative analysis does not work if we cannot find explicit solutions. Besides, having to resort to analytic techniques to check the qualitative results defeats the purpose of using these methods in the first place. In Section 1.5 we discuss powerful theorems that answer many of these questions without undue effort.

The Mixing Problem Revisited

Recall that in the previous section (page 32) we found precise analytic solutions for the differential equation

$$\frac{dS}{dt} = \frac{2000 - 3S}{100},$$

where S describes the amount of sugar in a vat at time t. We found that the general solution of this equation was

$$S(t) = ce^{-0.03t} + \frac{2000}{3},$$

where c is an arbitrary constant.

Using the slope field of this equation, we can easily derive a qualitative description of these solutions. In Figure 1.24, we display the slope field and graphs of selected solutions. Note that, as expected, the slope field is horizontal if $S = 2000/3$, the equilibrium solution. Slopes are positive if $S < 2000/3$ and negative if $S > 2000/3$. So

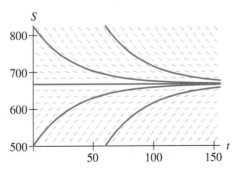

Figure 1.24
The slope field and graphs of a few solutions of

$$\frac{dS}{dt} = \frac{2000 - 3S}{100}.$$

The horizontal line is the graph of the equilibrium solution $S(t) = 2000/3$ for all t. Solutions below the equilibrium value are increasing, and solutions above that value are decreasing.

we expect solutions to tend toward the equilibrium solution as t increases. This qualitative analysis indicates that, no matter what the initial amount of sugar, the amount of sugar in the vat tends to 2000/3 as $t \to \infty$. Of course, we obtain the same information by taking the limit of the general solution as $t \to \infty$, but it is nice to see the same result in a geometric setting. Furthermore, in other examples, taking such a limit may not be as easy as in this case, but qualitative methods may still be used to determine the long-term behavior of the solutions.

An RC Circuit

Figure 1.25
Circuit diagram with resistor, capacitor, and voltage source.

The simple electric circuit pictured in Figure 1.25 contains a capacitor, a resistor, and a voltage source. The behavior of the resistor is specified by a positive parameter R (the "resistance"), and the behavior of the capacitor is specified by a positive parameter C (the "capacitance").* The input voltage across the voltage source at time t is denoted by $V(t)$. This voltage source could be a constant source such as a battery, or it could be a source that varies with time such as alternating current. In any case, we consider $V(t)$ to be a function that is specified by the circuit designer. In other words, it is part of the design of the circuit.

The quantities that specify the behavior of the circuit at a particular time t are the current $i(t)$ and the voltage across the capacitor $v_c(t)$. In this example we are interested in the voltage $v_c(t)$ across the capacitor. From the theory of electric circuits, we know that $v_c(t)$ satisfies the differential equation

$$RC\frac{dv_c}{dt} + v_c = V(t).$$

If we rewrite this in our standard form $dv_c/dt = f(t, v_c)$, we have

$$\frac{dv_c}{dt} = \frac{V(t) - v_c}{RC}.$$

We use slope fields to visualize solutions for four different types of voltage sources $V(t)$. (If you don't know anything about electric circuits, don't worry; Paul, Bob, and Glen don't either. In examples like this, all we need to do is accept the differential equation and "go with it.")

Zero input
If $V(t) = 0$ for all t, the equation becomes

$$\frac{dv_c}{dt} = \frac{-v_c}{RC}.$$

A slope field for a particular choice of R and C is given in Figure 1.26. We see clearly that all solutions "decay" toward $v_c = 0$ as t increases. If there is no voltage source,

---

*The usual units are ohms for resistance and farads for capacitance. In this section and in Section 1.4, we chose values of of R and C so that the numbers in the examples work out nicely. A 1 farad capacitor would be extremely large.

the voltage across the capacitor $v_c(t)$ decays to zero. This prediction for the voltage agrees with what we obtain analytically since the general solution of this equation is $v_c(t) = v_0 e^{-t/RC}$, where v_0 is the initial voltage across the capacitor. (Note that this equation is essentially the same as the exponential growth model that we studied in Section 1.1, and consequently we can solve it analytically by either guessing the correct form of a solution or by separating variables—see Exercise 20.)

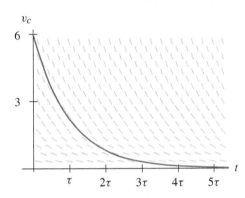

Figure 1.26
Slope field for

$$\frac{dv_c}{dt} = -\frac{v_c}{RC}$$

with $R = 0.5$ and $C = 1$, and the graph of the solution with initial value $v_c(0) = 6$. The time constant τ for this equation exponentially decaying solution is $\tau = 0.5$. (For more information about time constants, see Exercise 9 in Section 1.1.)

Constant nonzero voltage source

Suppose $V(t)$ is a nonzero constant K for all t. The equation for voltage across the capacitor becomes

$$\frac{dv_c}{dt} = \frac{K - v_c}{RC}.$$

This equation is autonomous with one equilibrium solution at $v_c = K$. The slope field for this equation shows that all solutions tend toward this equilibrium as t increases (see Figure 1.27). Given any initial voltage $v_c(0)$ across the capacitor, the voltage $v_c(t)$ tends to the value $v = K$ as time increases.

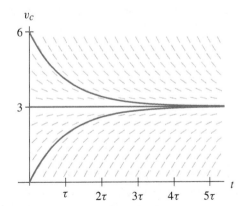

Figure 1.27
Slope field for

$$\frac{dv_c}{dt} = \frac{K - v_c}{RC}$$

for $R = 0.5$, $C = 1$, and $K = 3$, and the graphs of three solutions with different initial conditions. The time constant for this equation is the same as the time constant for the equation in Figure 1.26.

We could find a formula for the general solution by separating variables and integrating, but we leave this as an exercise (see Exercise 21).

On-off voltage source

Suppose $V(t) = K > 0$ for $0 \leq t < 3$, but at $t = 3$, this voltage is "turned off," perhaps by someone flicking a light switch. Then $V(t) = 0$ for $t > 3$. Our differential equation is

$$\frac{dv_c}{dt} = \frac{V(t) - v_c}{RC} = \begin{cases} \dfrac{K - v_c}{RC} & \text{for } 0 \leq t < 3; \\[2mm] \dfrac{-v_c}{RC} & \text{for } t > 3. \end{cases}$$

The right-hand side is given by two different formulas depending on the value of t. We can see this discontinuity in the slope field for this equation (see Figure 1.28). It resembles Figures 1.26 and 1.27 pasted together along the vertical line $t = 3$. Since the differential equation is not defined at $t = 3$, we must add an additional assumption to our model. We assume that the voltage $v_c(t)$ is a continuous function at $t = 3$.

The particular solution with the initial condition $v_c(0) = K$ is constant for $t < 3$, but for $t > 3$ it decays exponentially. Solutions with $v_c(0) \neq K$ move toward K for $t < 3$, but then decay toward zero for $t > 3$. We could find formulas for the solutions by first calculating $v_c(t)$ for $t \leq 3$, and then using the value $v_c(3)$ to solve the equation for $t > 3$ (see Section 1.2). We again leave this derivation as an exercise (see Exercise 22).

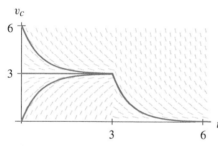

Figure 1.28
Slope field for

$$\frac{dv_c}{dt} = \frac{V(t) - v_c}{RC}$$

for $V(t)$, which "turns off" at $t = 3$ for $R = 0.5$, $C = 1$, and $K = 3$, along with graphs of three solutions with different initial conditions.

A flashing light

Figure 1.29
Circuit diagram for a flashing light.

The circuit in Figure 1.25 can be modified to produce a flashing light such as those that are used on cell phone towers and flashing road hazard signs (see Figure 1.29). The switch periodically opens and closes. It is open for an interval of time T_O. (The letter o stands for open. The constant T_O is *not* an initial value.)   After the switch is open for the time interval T_O, the switch closes and remains closed for a different (and shorter) interval T_C, where the letter c stands for closed. When the switch is open, the capacitor is charging according to the equation

$$\frac{dv_c}{dt} = \frac{V(t) - v_c}{RC},$$

where $V(t)$ is the voltage source. If $V(t)$ is a constant K and v_0 is the initial value $v_c(0)$, then $v_c(t)$ satisfies the initial-value problem

$$\frac{dv_c}{dt} = \frac{K - v_c}{RC}, \quad v_c(0) = v_0.$$

The voltage v_c satisfies this equation for $0 \leq t \leq T_O$.

At time T_O, the switch closes and the light turns on. While the lamp is lit, it acts as a resistor in parallel with the other resistor. Let R_L be the resistance that is due solely to the lamp, then it can be shown that the differential equation that governs v_c over the time interval $T_0 < t < T_0 + T_C$ is

$$\frac{dv_c}{dt} = \frac{K}{RC} - \left(\frac{R + R_L}{RR_LC}\right) v_c.$$

Note that increased resistance due to the lamp causes v_c to decrease faster than it increased when the switch was open.

The light switch remains closed over the interval $T_0 < t < T_0 + T_C$, and we pick T_C so that $v_c(T_0 + T_C) = v_0$. In other words, we pick T_C so that the voltage v_c is periodic with period $T_0 + T_C$ (see Figure 1.30). For this example, the slope field is discontinuous along infinitely many vertical lines, that is, the lines $t = T_0, t = T_0 + T_C,$ $t = 2T_0 + T_C, t = 2(T_0 + T_C), \ldots$ (see Figure 1.30).

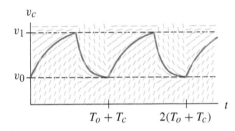

Figure 1.30
The graph of the solution and its corresponding slope field for the flashing light example in the case where $R = 0.5$, $R_L = 0.25$, $C = 1$, and $K = 2.5$. In this case, we have made $T_0 = 1$ and $T_C = 0.7$, so the solution is periodic with period 1.7.

Combining Qualitative with Quantitative Results

When only knowledge of the qualitative behavior of the solution is required, sketches of solutions obtained from slope fields can sometimes suffice. In other applications it is necessary to know the exact value (or almost exact value) of the solution with a given initial condition. In these situations analytic and/or numerical methods can't be avoided. But even then, it is nice to have graphs of the solutions.

EXERCISES FOR SECTION 1.3

In Exercises 1–6, sketch the slope fields for the differential equation as follows:

(a) Pick a few points (t, y) with both $-2 \leq t \leq 2$ and $-2 \leq y \leq 2$ and plot the associated slope marks without the use of technology.

(b) Use HPGSolver to check these individual slope marks.

(c) Make a more detailed drawing of the slope field and then use HPGSolver to confirm your answer.

For more details about HPGSolver and other programs that are part of the DETools package, see the description of DETools inside the front cover of this book.

1. $\dfrac{dy}{dt} = t^2 + t$ **2.** $\dfrac{dy}{dt} = t^2 + 1$ **3.** $\dfrac{dy}{dt} = 1 - 2y$

4. $\dfrac{dy}{dt} = 4y^2$ **5.** $\dfrac{dy}{dt} = 2y(1 - y)$ **6.** $\dfrac{dy}{dt} = y + t + 1$

In Exercises 7–10, a differential equation and its associated slope field are given. For each equation,

(a) sketch a number of different solutions on the slope field, and

(b) describe briefly the behavior of the solution with $y(0) = 1/2$ as t increases.

You should first answer these exercises without using any technology, and then you should confirm your answer using HPGSolver.

7. $\dfrac{dy}{dt} = 3y(1 - y)$

8. $\dfrac{dy}{dt} = 2y - t$

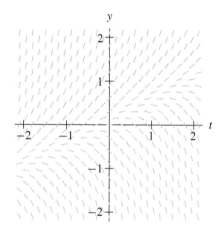

9. $\dfrac{dy}{dt} = \left(y + \tfrac{1}{2}\right)(y + t)$

10. $\dfrac{dy}{dt} = (t + 1)y$

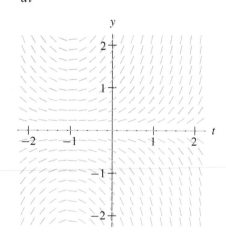

11. Suppose we know that the function $f(t, y)$ is continuous and that $f(t, 3) = -1$ for all t.

   (a) What does this information tell us about the slope field for the differential equation $dy/dt = f(t, y)$?

   (b) What can we conclude about solutions $y(t)$ of $dy/dt = f(t, y)$? For example, if $y(0) < 3$, can $y(t) \to \infty$ as t increases?

12. Suppose the constant function $y(t) = 2$ for all t is a solution of the differential equation

$$\frac{dy}{dt} = f(t, y).$$

   (a) What does this tell you about the function $f(t, y)$?

   (b) What does this tell you about the slope field? In other words, how much of the slope field can you sketch using this information?

   (c) What does this tell you about solutions with initial conditions $y(0) \neq 2$?

13. Suppose we know that the graph to the right is the graph of the right-hand side $f(t)$ of the differential equation

$$\frac{dy}{dt} = f(t).$$

$f(t)$

Give a rough sketch of the slope field that corresponds to this differential equation.

14. Suppose we know that the graph to the right is the graph of the right-hand side $f(y)$ of the differential equation

$$\frac{dy}{dt} = f(y).$$

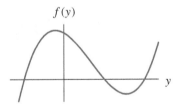

$f(y)$

Give a rough sketch of the slope field that corresponds to this differential equation.

15. Consider the autonomous differential equation

$$\frac{dS}{dt} = S^3 - 2S^2 + S.$$

   (a) Make a rough sketch of the slope field without using any technology.

   (b) Using this drawing, sketch the graphs of the solutions $S(t)$ with the initial conditions $S(0) = 1/2$, $S(1) = 1/2$, $S(0) = 1$, $S(0) = 3/2$, and $S(0) = -1/2$.

   (c) Confirm your answer using HPGSolver.

16. Eight differential equations and four slope fields are given below. Determine the equation that corresponds to each slope field and state briefly how you know your choice is correct. You should do this exercise without using technology.

(i) $\dfrac{dy}{dt} = y^2 + y$ (ii) $\dfrac{dy}{dt} = y^2 - y$ (iii) $\dfrac{dy}{dt} = y^3 + y^2$ (iv) $\dfrac{dy}{dt} = 2 - t^2$

(v) $\dfrac{dy}{dt} = ty + ty^2$ (vi) $\dfrac{dy}{dt} = t^2 + t^2 y$ (vii) $\dfrac{dy}{dt} = t + ty$ (viii) $\dfrac{dy}{dt} = t^2 - 2$

(a)

(b)

(c)

(d)

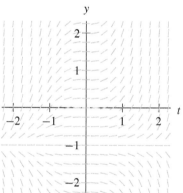

17. Suppose we know that the graph below is the graph of a solution to $dy/dt = f(t)$.

(a) How much of the slope field can you sketch from this information? [*Hint:* Note that the differential equation depends only on t.]

(b) What can you say about the solution with $y(0) = 2$? (For example, can you sketch the graph of this solution?)

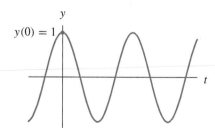

$y(0) = 1$

18. Suppose we know that the graph below is the graph of a solution to $dy/dt = f(y)$.

(a) How much of the slope field can you sketch from this information? [*Hint*: Note that the equation is autonomous.]

(b) What can you say about the solution with $y(0) = 2$? Sketch this solution.

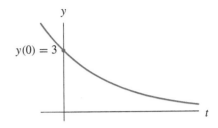

19. The spiking of a neuron can be modeled* by the differential equation

$$\frac{d\theta}{dt} = 1 - \cos\theta + (1 + \cos\theta)I(t),$$

where $I(t)$ is the input. Often the input function $I(t)$ is a constant I. When θ is an odd multiple of π, the neuron spikes.

(a) Using HPGSolver, sketch three slope fields, one for each of the following values of I: $I_1 = -0.1$, $I_2 = 0.0$, and $I_3 = 0.1$.

(b) Calculate the equilbrium solutions for each of these three values.

(c) Using the slope field, describe the long-term behavior of the solutions in each of the three cases.

20. By separating variables, find the general solution of the differential equation

$$\frac{dv_c}{dt} = -\frac{v_c}{RC},$$

where R and C are constants. Then check your answer by substituting it back into the differential equation.

21. By separating variables, find the general solution of the differential equation

$$\frac{dv_c}{dt} = \frac{K - v_c}{RC},$$

where R, C, and K are constants. Then check your answer by substituting it back into the differential equation.

22. By separating variables, find the solution of the initial-value problem

$$\frac{dv_c}{dt} = \frac{V(t) - v_c}{RC}, \quad v_c(0) = 6$$

where $R = 0.5$, $C = 1.0$, and $V(t)$ is the function that is constantly 3 for $t < 3$ and 0 for $t > 3$. Then check your answer by substituting it back into the differential equation. [*Hint*: Do this exercise in two steps. First, solve the equation for $t \leq 3$. Then use the value $v_c(3)$ to state another initial-value problem.]

---

*This model is often referred to as the theta model, but it is also called the Ermentrout-Kopell canonical model. See "Parabolic bursting in an excitable system coupled with a slow oscillation" by G. B. Ermentrout and N. Kopell, in *SIAM J. Applied Math*, Vol. 44, 1984, pp. 1133–1149.

1.4 NUMERICAL TECHNIQUE: EULER'S METHOD

The geometric concept of a slope field as discussed in the previous section is closely related to a fundamental numerical method for approximating solutions to a differential equation. Given an initial-value problem

$$\frac{dy}{dt} = f(t, y), \quad y(t_0) = y_0,$$

we can get a rough idea of the graph of its solution by first sketching the slope field in the ty-plane and then, starting at the initial value (t_0, y_0), sketching the solution by drawing a graph that is tangent to the slope field at each point along the graph. In this section we describe a numerical procedure that automates this idea. Using a computer or a calculator, we obtain numbers and graphs that approximate solutions to initial-value problems.

Numerical methods provide quantitative information about solutions even if we cannot find their formulas. There is also the advantage that most of the work can be done by machine. The disadvantage is that we obtain only approximations, not precise solutions. If we remain aware of this fact and are prudent, numerical methods become powerful tools for the study of differential equations. It is not uncommon to turn to numerical methods even when it is possible to find formulas for solutions. (Most of the graphs of solutions of differential equations in this text were drawn using numerical approximations even when formulas were available.)

The numerical technique that we discuss in this section is called *Euler's method*. A more detailed discussion of the accuracy of Euler's method as well as other numerical methods is given in Chapter 7.

Stepping along the Slope Field

To describe Euler's method, we begin with the initial-value problem

$$\frac{dy}{dt} = f(t, y), \quad y(t_0) = y_0.$$

Since we are given $f(t, y)$, we can plot its slope field in the ty-plane. The idea of the method is to start at the point (t_0, y_0) in the slope field and take tiny steps dictated by the tangents in the slope field.

We begin by choosing a (small) **step size** Δt. The slope of the approximate solution is updated every Δt units of t. In other words, for each step, we move Δt units along the t-axis. The size of Δt determines the accuracy of the approximate solution as well as the number of computations that are necessary to obtain the approximation.

Starting at (t_0, y_0), our first step is to the point (t_1, y_1) where $t_1 = t_0 + \Delta t$ and (t_1, y_1) is the point on the line through (t_0, y_0) with slope given by the slope field at (t_0, y_0) (see Figure 1.31). At (t_1, y_1) we repeat the procedure. Taking a step whose size along the t-axis is Δt and whose direction is determined by the slope field at (t_1, y_1), we reach the new point (t_2, y_2). The new time is given by $t_2 = t_1 + \Delta t$ and (t_2, y_2) is on the line segment that starts at (t_1, y_1) and has slope $f(t_1, y_1)$. Continuing, we use the slope field at the point (t_k, y_k) to determine the next point (t_{k+1}, y_{k+1}). The sequence of values $y_0, y_1, y_2, \ldots$ serves as an approximation to the solution at the times $t_0, t_1, t_2, \ldots$. Geometrically, we think of the method as producing a sequence of tiny line segments connecting (t_k, y_k) to (t_{k+1}, y_{k+1}) (see Figure 1.32). Basically, we are stitching together little pieces of the slope field to form a graph that approximates our solution curve.

This method uses tangent line segments, given by the slope field, to approximate the graph of the solution. Consequently, at each stage we make a slight error (see Figure 1.32). Hopefully, if the step size is sufficiently small, these errors do not get out of hand as we continue to step, and the resulting graph is close to the desired solution.

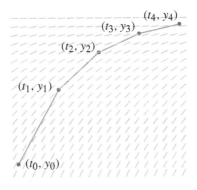

Figure 1.31
Stepping along the slope field.

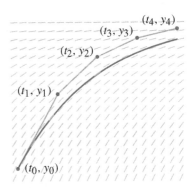

Figure 1.32
The graph of a solution and its approximation obtained using Euler's method.

Euler's Method

To put Euler's method into practice, we need a formula for determining (t_{k+1}, y_{k+1}) from (t_k, y_k). Finding t_{k+1} is easy. We specify the step size Δt at the outset, so

$$t_{k+1} = t_k + \Delta t.$$

To obtain y_{k+1} from (t_k, y_k), we use the differential equation. We know that the slope of the solution to the equation $dy/dt = f(t, y)$ at the point (t_k, y_k) is $f(t_k, y_k)$, and Euler's method uses this slope to determine y_{k+1}. In fact, the method determines the point (t_{k+1}, y_{k+1}) by assuming that it lies on the line through (t_k, y_k) with slope $f(t_k, y_k)$ (see Figure 1.33).

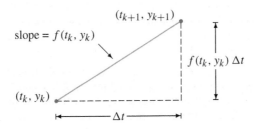

Figure 1.33
Euler's method uses the slope at the point (t_k, y_k) to approximate the solution for $t_k \leq t \leq t_{k+1}$.

Now we can use our basic knowledge of slopes to determine y_{k+1}. The formula for the slope of a line gives

$$\frac{y_{k+1} - y_k}{t_{k+1} - t_k} = f(t_k, y_k).$$

Since $t_{k+1} = t_k + \Delta t$, the denominator $t_{k+1} - t_k$ is just Δt, and therefore we have

$$\frac{y_{k+1} - y_k}{\Delta t} = f(t_k, y_k)$$

$$y_{k+1} - y_k = f(t_k, y_k)\,\Delta t$$

$$y_{k+1} = y_k + f(t_k, y_k)\,\Delta t.$$

This is the formula for Euler's method (see Figures 1.33 and 1.34).

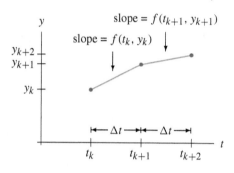

Figure 1.34
Two successive steps of Euler's method. Note that the slope used in the $k+1$st step is $f(t_k, y_k)$, and this slope determines y_{k+1} by the formula

$$y_{k+1} = y_k + f(t_k, y_k)\Delta t.$$

The slope used at the $k+2$nd step is $f(t_{k+1}, y_{k+1})$, and y_{k+2} is determined similarly.

Euler's method for $\dfrac{dy}{dt} = f(t, y)$

Given the initial condition $y(t_0) = y_0$ and the step size Δt, compute the point (t_{k+1}, y_{k+1}) from the preceding point (t_k, y_k) as follows:

1. Use the differential equation to compute the slope $f(t_k, y_k)$.

2. Calculate the next point (t_{k+1}, y_{k+1}) using the formulas

$$t_{k+1} = t_k + \Delta t$$

and

$$y_{k+1} = y_k + f(t_k, y_k)\Delta t.$$

Approximating an Autonomous Equation

To illustrate Euler's method, we first use it to approximate the solution to a differential equation whose solution we already know. In this way, we are able to compare the approximation we obtain to the known solution. Consequently, we are able to gain some insight into the effectiveness of the method in addition to seeing how it is implemented.

Consider the initial-value problem

$$\frac{dy}{dt} = 2y - 1, \quad y(0) = 1.$$

This equation is separable, and by separating and integrating we obtain the solution

$$y(t) = \frac{e^{2t} + 1}{2}.$$

In this example, $f(t, y) = 2y - 1$, so Euler's method is given by

$$y_{k+1} = y_k + (2y_k - 1)\Delta t.$$

To illustrate the method, we start with a relatively large step size of $\Delta t = 0.1$ and approximate the solution over the interval $0 \leq t \leq 1$. In order to approximate the solution over an interval whose length is 1 with a step size of 0.1, we must compute ten iterations of the method. The initial condition $y(0) = 1$ provides the initial value $y_0 = 1$. Given $\Delta t = 0.1$, we have $t_1 = t_0 + 0.1 = 0 + 0.1 = 0.1$. We compute the y-coordinate for the first step by

$$y_1 = y_0 + (2y_0 - 1)\Delta t = 1 + (1)\,0.1 = 1.1.$$

Thus the first point (t_1, y_1) on the graph of the approximate solution is $(0.1, 1.1)$.

To compute the y-coordinate y_2 for the second step, we now use y_1 rather than y_0. That is,

$$y_2 = y_1 + (2y_1 - 1)\Delta t = 1.1 + (1.2)\,0.1 = 1.22,$$

and the second point for our approximate solution is $(t_2, y_2) = (0.2, 1.22)$.

Continuing this procedure, we obtain the results given in Table 1.3. After ten steps, we obtain the approximation of $y(1)$ by $y_{10} = 3.596$. (Different machines use different algorithms for rounding numbers, so you may get slightly different results on

Table 1.3
Euler's method (to three decimal places) for $dy/dt = 2y - 1$, $y(0) = 1$ with $\Delta t = 0.1$

k	t_k	y_k	$f(t_k, y_k)$
0	0	1	1
1	0.1	1.100	1.20
2	0.2	1.220	1.44
3	0.3	1.364	1.73
4	0.4	1.537	2.07
5	0.5	1.744	2.49
6	0.6	1.993	2.98
7	0.7	2.292	3.58
8	0.8	2.650	4.30
9	0.9	3.080	5.16
10	1.0	3.596	

your computer or calculator. Keep this fact in mind whenever you compare the numerical results presented in this book with the results of your calculation.) Since we know that

$$y(1) = \frac{e^2 + 1}{2} \approx 4.195,$$

the approximation y_{10} is off by slightly less than 0.6. This is not a very good approximation, but we'll soon see how to avoid this (usually). The reason for the error can be seen by looking at the graph of the solution and its approximation. The slope field for this differential equation always lies below the graph (see Figure 1.35), so we expect our approximation to come up short.

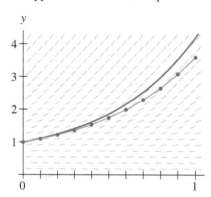

Figure 1.35
The graph of the solution to

$$\frac{dy}{dt} = 2y - 1$$

with $y(0) = 1$ and the approximation produced by Euler's method with $\Delta t = 0.1$.

Using a smaller step size usually reduces the error, but more computations must be done to approximate the solution over the same interval. For example, if we halve the step size in this example ($\Delta t = 0.05$), then we must calculate twice as many steps, since $t_1 = 0.05$, $t_2 = 0.1, \ldots, t_{20} = 1.0$. Again we start with $(t_0, y_0) = (0, 1)$ as specified by the initial condition. However, with $\Delta t = 0.05$, we obtain

$$y_1 = y_0 + (2y_0 - 1)\Delta t = 1 + (1)\,0.05 = 1.05.$$

This step yields the point $(t_1, y_1) = (0.05, 1.05)$ on the graph of our approximate solution. For the next step, we compute

$$y_2 = y_1 + (2y_1 - 1)\Delta t = 1.05 + (1.1)\,0.05 = 1.105.$$

Now we have the point $(t_2, y_2) = (1.1, 1.105)$. This type of calculation gets tedious fairly quickly, but luckily calculations such as these are perfect for a computer or a calculator. For $\Delta t = 0.05$, the results of Euler's method are given in Table 1.4.

If we carefully compare the final results of our two computations, we see that, with $\Delta t = 0.1$, we approximate $y(1) \approx 4.195$ with $y_{10} = 3.596$. With $\Delta t = 0.05$, we approximate $y(1)$ with $y_{20} = 3.864$. The error in the first approximation is slightly less than 0.6, whereas the error in the second approximation is 0.331. Roughly speaking we halve the error by halving the step size. This type of improvement is typical of Euler's method. (We will be much more precise about how the error in Euler's method is related to the step size in Chapter 7.)

Table 1.4

Euler's method (to three decimal places) for $dy/dt = 2y - 1$, $y(0) = 1$ with $\Delta t = 0.05$

k	t_k	y_k	$f(t_k, y_k)$
0	0	1	1
1	0.05	1.050	1.100
2	0.10	1.105	1.210
3	0.15	1.166	1.331
$\vdots$	$\vdots$	$\vdots$	$\vdots$
19	0.95	3.558	6.116
20	1.00	3.864	

With the even smaller step size of $\Delta t = 0.01$, we must do much more work since we need 100 steps to go from $t = 0$ to $t = 1$. However, in the end, we obtain a much better approximation to the solution (see Table 1.5).

This example illustrates the typical trade-off that occurs with numerical methods. There are always decisions to be made such as the choice of the step size Δt. Lowering Δt often results in a better approximation—at the expense of more computation.

Table 1.5

Euler's method (to four decimal places) for $dy/dt = 2y - 1$, $y(0) = 1$ with $\Delta t = 0.01$

k	t_k	y_k	$f(t_k, y_k)$
0	0	1	1
1	0.01	1.0100	1.0200
2	0.02	1.0202	1.0404
3	0.03	1.0306	1.0612
$\vdots$	$\vdots$	$\vdots$	$\vdots$
98	0.98	3.9817	6.9633
99	0.99	4.0513	7.1026
100	1.00	4.1223	

A Nonautonomous Example

Note that it is the value $f(t_k, y_k)$ of the right-hand side of the differential equation at (t_k, y_k) that determines the next point (t_{k+1}, y_{k+1}). The last example was an autonomous differential equation, so the right-hand side $f(t_k, y_k)$ depended only on y_k. However, if the differential equation is nonautonomous, the value of t_k also plays a role in the computations.

To illustrate Euler's method applied to a nonautonomous equation, we consider the initial-value problem

$$\frac{dy}{dt} = -2ty^2, \quad y(0) = 1.$$

This differential equation is also separable, and we can separate variables to obtain the solution

$$y(t) = \frac{1}{1+t^2}.$$

We use Euler's method to approximate this solution over the interval $0 \le t \le 2$. The value of the solution at $t = 2$ is $y(2) = 1/5$. Again, it is interesting to see how close we come to this value with various choices of Δt. The formula for Euler's method is

$$y_{k+1} = y_k + f(t_k, y_k)\, \Delta t = y_k - (2t_k y_k^2)\Delta t$$

with $t_0 = 0$ and $y_0 = 1$. We begin by approximating the solution from $t = 0$ to $t = 2$ using just four steps. This involves so few computations that we can perform the arithmetic "by hand." To cover an interval of length 2 in four steps, we must use $\Delta t = 2/4 = 1/2$. The entire calculation is displayed in Table 1.6.

Table 1.6
Euler's method for $dy/dt = -2ty^2$, $y(0) = 1$ with $\Delta t = 1/2$

k	t_k	y_k	$f(t_k, y_k)$
0	0	1	0
1	1/2	1	−1
2	1	1/2	
	−1/2		
3	3/2	1/4	−3/16
4	2	5/32	

Note that we end up approximating the exact value $y(2) = 1/5 = 0.2$ by $y_4 = 5/32 = 0.15625$. Figure 1.36 shows the graph of the solution as compared to the results of Euler's method over this interval.

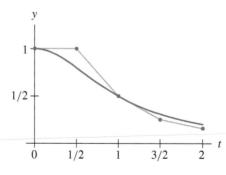

Figure 1.36
The graph of the solution to the initial-value problem

$$\frac{dy}{dt} = -2ty^2, \quad y(0) = 1,$$

and the approximation produced by Euler's method with $\Delta t = 1/2$.

As before, choosing smaller values of Δt yields better approximations. For example, if $\Delta t = 0.1$, the Euler approximation that gives the exact value $y(2) = 0.2$ is

Table 1.7
Euler's method (to four decimal places) for
$dy/dt = -2ty^2$, $y(0) = 1$ with $\Delta t = 0.1$

k	t_k	y_k
0	0	1
1	0.1	1.0000
2	0.2	0.9800
3	0.3	0.9416
⋮	⋮	⋮
19	1.9	0.2101
20	2.0	0.1933

Table 1.8
Euler's method (to six decimal places) for
$dy/dt = -2ty^2$, $y(0) = 1$ with $\Delta t = 0.001$

k	t_k	y_k
0	0	1
1	0.001	1.000000
2	0.002	0.999998
3	0.003	0.999994
⋮	⋮	⋮
1999	1.999	0.200097
2000	2	0.199937

$y_{20} = 0.1933$. If $\Delta t = 0.001$, we need to compute 2000 steps, but the approximation improves to $y_{2000} = 0.199937$ (see Tables 1.7 and 1.8).

Note that the convergence of the approximation to the actual value is slow. We computed 2000 steps and obtained an answer that is only accurate to three decimal places. In Chapter 7, we present more complicated algorithms for numerical approximation of solutions. Although the algorithms are more complicated from a conceptual point of view, they obtain better accuracy with less computation.

An RC Circuit with Periodic Input

R

Figure 1.37
Circuit diagram with resistor, capacitor, and voltage source.

Recall from Section 1.3 that the voltage v_c across the capacitor in the simple circuit shown in Figure 1.37 is given by the differential equation

$$\frac{dv_c}{dt} = \frac{V(t) - v_c}{RC}$$

where R is the resistance, C is the capacitance, and $V(t)$ is the source or input voltage. We have seen how we can use slope fields to give a qualitative sketch of solutions. Using Euler's method we can also obtain numerical approximations of the solutions.

Suppose we consider a circuit where $R = 0.5$ and $C = 1$ (see the footnote on page 44 in Section 1.3 for a comment regarding our choice of units in these examples). Then the differential equation is

$$\frac{dv_c}{dt} = \frac{V(t) - v_c}{0.5} = 2(V(t) - v_c).$$

Figure 1.38
Graph of $V(t) = \sin(2\pi t)$, the input voltage.

To understand how the voltage v_c varies if the voltage source $V(t)$ is periodic in time, we consider the case where $V(t) = \sin(2\pi t)$. Consequently, the voltage oscillates between -1 and 1 once each unit of time (see Figure 1.38). The differential equation is now

$$\frac{dv_c}{dt} = -2v_c + 2\sin(2\pi t).$$

From the slope field for this equation (see Figure 1.39), we might predict that the solutions oscillate. Using Euler's method applied to this equation for several different initial conditions, we see that the solutions do indeed oscillate. In addition, we see that they also approach each other and collect around a single solution (see Figure 1.40). This uniformity of long-term behavior is not so easily predicted from the slope field alone.

Figure 1.39
Slope field for
$dv_C/dt = -2v_C + 2\sin(2\pi t)$.

Figure 1.40
Graphs of approximate solutions to
$dv_C/dt = -2v_C + 2\sin(2\pi t)$ obtained
using Euler's method.

Errors in Numerical Methods

By its very nature, any numerical approximation scheme is inaccurate. For instance, in each step of Euler's method, we almost always make an error of some sort. These errors can accumulate and sometimes lead to disastrously wrong approximations. As an example, consider the differential equation

$$\frac{dy}{dt} = e^t \sin y.$$

There are equilibrium solutions for this equation if $\sin y = 0$. In other words, any constant function of the form $y(t) = n\pi$ for any integer n is a solution.

Using the initial value $y(0) = 5$ and a step size $\Delta t = 0.1$, Euler's method yields the approximation graphed in Figure 1.41. It seems that something must be wrong. At first, the solution tends toward the equilibrium solution $y(t) = \pi$, but then just before $t = 5$ something strange happens. The graph of the approximation jumps dramatically. If we lower Δt to 0.05, we still find erratic behavior, although t is slightly greater than 5 before this happens (see Figure 1.42).

The difficulty arises in Euler's method for this equation because of the term e^t on the right-hand side. It becomes very large as t increases, and consequently slopes in the slope field are quite large for large t. Even a very small step in the t-direction throws us far from the actual solution.

This problem is typical of the use of numerics in the study of differential equations. Numerical methods, when they work, work beautifully. But they sometimes fail. We must always be aware of this possibility and be ready with an alternate approach.

Figure 1.41
Euler's method applied to

$$\frac{dy}{dt} = e^t \sin y$$

with $\Delta t = 0.1$

Figure 1.42
Euler's method applied to

$$\frac{dy}{dt} = e^t \sin y$$

with $\Delta t = 0.05$.

In the next section we present theoretical results that help identify when numerical approximations have gone awry.

The Big Three

We have now introduced examples of all three of the fundamental methods for attacking differential equations—the analytic, the numeric, and the qualitative approaches. Which method is the best depends both on the differential equation in question and on what we want to know about the solutions. Often all three methods "work," but a great deal of labor can be saved if we think first about which method gives the most direct route to the information we need.

EXERCISES FOR SECTION 1.4

In Exercises 1–4, use `EulersMethod` to perform Euler's method with the given step size Δt on the given initial-value problem over the time interval specified. Your answer should include a table of the approximate values of the dependent variable. It should also include a sketch of the graph of the approximate solution.

1. $\dfrac{dy}{dt} = 2y + 1, \quad y(0) = 3, \quad 0 \le t \le 2, \quad \Delta t = 0.5$

2. $\dfrac{dy}{dt} = t - y^2, \quad y(0) = 1, \quad 0 \le t \le 1, \quad \Delta t = 0.25$

3. $\dfrac{dy}{dt} = y^2 - 4t, \quad y(0) = 0.5, \quad 0 \le t \le 2, \quad \Delta t = 0.25$

4. $\dfrac{dy}{dt} = \sin y, \quad y(0) = 1, \quad 0 \le t \le 3, \quad \Delta t = 0.5$

In Exercises 5–10, use Euler's method with the given step size Δt to approximate the solution to the given initial-value problem over the time interval specified. Your answer should include a table of the approximate values of the dependent variable. It should also include a sketch of the graph of the approximate solution.

5. $\dfrac{dw}{dt} = (3 - w)(w + 1)$, $\quad w(0) = 4$, $\quad 0 \le t \le 5$, $\quad \Delta t = 1.0$

6. $\dfrac{dw}{dt} = (3 - w)(w + 1)$, $\quad w(0) = 0$, $\quad 0 \le t \le 5$, $\quad \Delta t = 0.5$

7. $\dfrac{dy}{dt} = e^{2/y}$, $\quad y(0) = 2$, $\quad 0 \le t \le 2$, $\quad \Delta t = 0.5$

8. $\dfrac{dy}{dt} = e^{2/y}$, $\quad y(1) = 2$, $\quad 1 \le t \le 3$, $\quad \Delta t = 0.5$

9. $\dfrac{dy}{dt} = y^2 - y^3$, $\quad y(0) = 0.2$, $\quad 0 \le t \le 10$, $\quad \Delta t = 0.1$

10. $\dfrac{dy}{dt} = 2y^3 + t^2$, $\quad y(0) = -0.5$, $\quad -2 \le t \le 2$, $\quad \Delta t = 0.1$

[*Hint*: Euler's method also works with a negative Δt.]

11. Do a qualitative analysis of the solution of the initial-value problem in Exercise 6 and compare your conclusions with your results in Exercise 6. What's wrong with the approximate solution given by Euler's method?

12. As we saw in Exercise 12 of Section 1.1, the velocity v of a freefalling skydiver is well modeled by the differential equation

$$m\dfrac{dv}{dt} = mg - kv^2,$$

where m is the mass of the skydiver, $g = 9.8$ m/s^2 is the gravitational constant, and k is the drag coefficient determined by the position of the diver during the dive. Consider a diver of mass $m = 54$ kg (120 lb) with a drag coefficient of 0.18 kg/m. Use Euler's method to determine how long it will take the diver to reach 95% of her terminal velocity after she jumps from the plane. [*Hint*: Use the formula for terminal velocity that was derived in Exercise 12 of Section 1.1.]

13. Compare your answers to Exercises 7 and 8, and explain your observations.

14. Compare your answers to Exercises 5 and 6. Is Euler's method doing a good job in this case? What would you do to avoid the difficulties that arise in this case?

15. Consider the initial-value problem $dy/dt = \sqrt{y}$, $y(0) = 1$. Using Euler's method, compute three different approximate solutions corresponding to $\Delta t = 1.0, 0.5$, and 0.25 over the interval $0 \le t \le 4$. Graph all three solutions. What predictions do you make about the actual solution to the initial-value problem?

16. Consider the initial-value problem $dy/dt = 2 - y$, $y(0) = 1$. Using Euler's method, compute three different approximate solutions corresponding to $\Delta t = 1.0, 0.5$, and 0.25 over the interval $0 \le t \le 4$. Graph all three solutions. What predictions do you make about the actual solution to the initial-value problem? How do the graphs of these approximate solutions relate to the graph of the actual solution? Why?

17. As we saw in Exercise 19 of Section 1.3, the spiking of a neuron can be modeled by the differential equation $d\theta/dt = 1 - \cos\theta + (1 + \cos\theta)I(t)$, where $I(t)$ is the input. Assume that $I(t)$ is constantly equal to -0.1. Using Euler's method with $\Delta t = 0.1$, graph the solution that solves the initial value $\theta(0) = 1.0$ over the interval $0 \le t \le 5$. When does the neuron spike?

In Exercises 18–21, we consider the RC circuit equation $dv_c/dt = (V(t) - v_c)/(RC)$ that is discussed on page 59. Suppose $V(t) = 2\cos 3t$ (the voltage source $V(t)$ is oscillating periodically). If $R = 4$ and $C = 0.5$, use Euler's method to compute values of the solutions with the given initial conditions over the interval $0 \le t \le 10$.

18. $v_c(0) = 2$ **19.** $v_c(0) = 1$ **20.** $v_c(0) = -1$ **21.** $v_c(0) = -2$

1.5 EXISTENCE AND UNIQUENESS OF SOLUTIONS

What Does It Mean to Say Solutions Exist?

We have seen analytic, qualitative, and numerical techniques for studying solutions of differential equations. One problem we have not considered is: How do we know there are solutions? Although this may seem to be a subtle and abstract question, it is also a question of great importance. If solutions to the differential equation do not exist, then there is no use trying to find or approximate them. More important, if a differential equation is supposed to model a physical system but the solutions of the differential equation do not exist, then we should have serious doubts about the validity of the model.

To get an idea of what is meant by the existence of solutions, consider the algebraic equation

$$2x^5 - 10x + 5 = 0.$$

A solution to this equation is a value of x for which the left-hand side is zero. In other words, it is a root of the fifth-degree polynomial $2x^5 - 10x + 5$. We can easily compute that the value of $2x^5 - 10x + 5$ is -3 if $x = 1$ and 13 if $x = -1$. Since polynomials are continuous, there must therefore be a value of x between -1 and 1 for which the left-hand side is zero.

So we have established the existence of at least one solution of this equation between -1 and 1. We did not construct this value of x or approximate it (other than to say it is between -1 and 1). Unfortunately, there is no "quadratic formula" for finding roots of fifth-degree polynomials, so there is no way to write down the exact values of the solutions of this equation. But this does not make us any less sure of the existence of this solution. The point here is that we can discuss the existence of solutions without having to compute them.

It is also possible that there is more than one solution between -1 and 1. In other words, the solution may not be unique.

In the same way, if we are given an initial-value problem

$$\frac{dy}{dt} = f(t, y), \quad y(0) = y_0,$$

we can ask whether there is a solution. This is a different question than asking what the solution is or what its graph looks like. We can say there is a solution without having any knowledge of a formula for the solution, just as we can say that the algebraic equation above has a solution between -1 and 1 without knowing its exact or even approximate value.

Existence

Luckily, the question of existence of solutions for differential equations has been extensively studied and some very good results have been established. For our purposes, we will use the standard existence theorem.

EXISTENCE THEOREM   Suppose $f(t, y)$ is a continuous function in a rectangle of the form $\{(t, y) \mid a < t < b, c < y < d\}$ in the ty-plane. If (t_0, y_0) is a point in this rectangle, then there exists an $\epsilon > 0$ and a function $y(t)$ defined for $t_0 - \epsilon < t < t_0 + \epsilon$ that solves the initial-value problem

$$\frac{dy}{dt} = f(t, y), \quad y(t_0) = y_0. \quad \blacksquare$$

This theorem says that as long as the function on the right-hand side of the differential equation is reasonable, solutions exist. (It does not rule out the possibility that solutions exist even if $f(t, y)$ is not a nice function, but it doesn't guarantee it either.) This is reassuring. When we are studying the solutions of a reasonable initial-value problem, there is something there to study.

Extendability

Given an initial-value problem $dy/dt = f(t, y)$, $y(t_0) = y_0$, the Existence Theorem guarantees that there is a solution. If you read the theorem very closely (with a lawyer's eye for loopholes), you will see that the solution may have a very small domain of definition. The theorem says that there exists an $\epsilon > 0$ and that the solution has a domain that includes the open interval $(t_0 - \epsilon, t_0 + \epsilon)$. The ϵ may be very, very small, so although the theorem guarantees that a solution exists, it may be defined for only a very short interval of time.

Unfortunately, this is a serious but necessary restriction. Consider the initial-value problem

$$\frac{dy}{dt} = 1 + y^2, \quad y(0) = 0.$$

The slopes in the slope field for this equation increase in steepness very rapidly as y increases (see Figure 1.43). Hence, dy/dt increases more and more rapidly as $y(t)$

increases. There is a danger that solutions "blow up" (tend to infinity very quickly) as t increases. By looking at solutions sketched by the slope field, we can't really tell if the solutions blow up in finite time or if they stay finite for all time, so we try analytic methods.

This is an autonomous equation, so we can separate variables and integrate as usual. We have

$$\int \frac{1}{1+y^2}\,dy = \int dt.$$

Integration yields

$$\arctan y = t + c,$$

where c is an arbitrary constant. Therefore

$$y(t) = \tan(t + c),$$

which is the general solution of the differential equation. Using the initial value

$$0 = y(0) = \tan(0 + c),$$

we find $c = 0$ (or $c = n\pi$ for any integer n). Thus, the particular solution is $y(t) = \tan t$, and the domain of definition for this particular solution is $-\pi/2 < t < \pi/2$.

As we see from Figure 1.44, our fears were well founded. The graph of this particular solution has vertical asymptotes at $t = \pm\pi/2$. As t approaches $\pi/2$ from the left and $-\pi/2$ from the right, the solution blows up. If this differential equation were a model of a physical system, then we would expect the system to break as t approaches $\pi/2$.

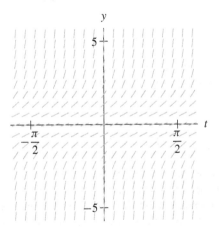

Figure 1.43
The slope field for the equation $dy/dt = 1 + y^2$. Note that the slopes are quite large if y is either moderately positive or moderately negative. In either case, the solutions increase rapidly.

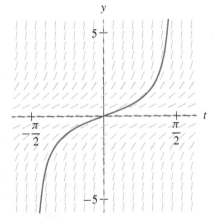

Figure 1.44
The graph of the solution $y(t) = \tan t$ with initial condition $y(0) = 0$ along with the slope field for $dy/dt = 1 + y^2$. As t approaches $\pi/2$ from the left, $y(t) = \tan t \to \infty$. As t approaches $-\pi/2$ from the right, $y(t) = \tan t \to -\infty$.

Solutions that blow up (or down) in finite time is a common phenomenon. Many relatively simple-looking differential equations have solutions that tend to infinity in finite time, and we should always be alert to this possibility.

Uniqueness

When dealing with initial-value problems of the form

$$\frac{dy}{dt} = f(t, y), \quad y(t_0) = y_0,$$

we have always said "consider *the* solution." By the Existence Theorem we know there is a solution, but how do we know there is only one? Why don't we have to say "consider *a* solution" instead of "consider *the* solution?" In other words, how do we know the solution is unique?

Knowing that the solution to an initial-value problem is unique is very valuable from both theoretical and practical standpoints. If solutions weren't unique, then we would have to worry about all possible solutions, even when we were doing numerical or qualitative work. Different solutions could give completely different predictions for how the system would work. Fortunately, there is a very good theorem that guarantees that solutions of initial-value problems are unique.

UNIQUENESS THEOREM   Suppose $f(t, y)$ and $\partial f / \partial y$ are continuous functions in a rectangle of the form $\{(t, y) \mid a < t < b, \ c < y < d\}$ in the ty-plane. If (t_0, y_0) is a point in this rectangle and if $y_1(t)$ and $y_2(t)$ are two functions that solve the initial-value problem

$$\frac{dy}{dt} = f(t, y), \quad y(t_0) = y_0$$

for all t in the interval $t_0 - \epsilon < t < t_0 + \epsilon$ (where ϵ is some positive number), then

$$y_1(t) = y_2(t)$$

for $t_0 - \epsilon < t < t_0 + \epsilon$. That is, the solution to the initial-value problem is *unique*.   ∎

Before giving applications of the Uniqueness Theorem we should emphasize that both the Existence and the Uniqueness Theorems have *hypotheses*—conditions that must hold before we can use these theorems. Before we say that the solution of an initial-value problem

$$\frac{dy}{dt} = f(t, y), \quad y(t_0) = y_0$$

exists and is unique, we must check that $f(t, y)$ satisfies the necessary hypotheses.

Often we lump these two theorems together (using the more restrictive hypotheses of the Uniqueness Theorem) and refer to the combination as the **Existence and Uniqueness Theorem**.

Lack of Uniqueness

It is pretty difficult to construct an example of a sensible differential equation that does not have solutions. However, it is not so hard to find examples where $f(t, y)$ is a decent function but where uniqueness fails. (Of course, in these examples, either $f(t, y)$ or $\partial f/\partial y$ is not continuous.)

For example, consider the differential equation

$$\frac{dy}{dt} = 3y^{2/3}.$$

The right-hand side is a continuous function on the entire ty-plane. Unfortunately, the partial derivative of $y^{2/3}$ with respect to y fails to exist if $y = 0$, so the Uniqueness Theorem does not tell us anything about the number of solutions to an initial-value problem of the form $y(t_0) = 0$.

Let's apply the qualitative and analytic techniques that we have already discussed. First, if we look for equilibrium solutions, we see that the function $y(t) = 0$ for all t is a solution. Second, we note that this equation is separable, so we separate variables and obtain

$$\int y^{-2/3}\, dy = \int 3\, dt.$$

Integrating, we obtain the solutions

$$y(t) = (t + c)^3$$

where c is an arbitrary constant.

Now consider the initial-value problem

$$\frac{dy}{dt} = 3y^{2/3}, \quad y(0) = 0.$$

One solution is the equilibrium solution $y_1(t) = 0$ for all t. However, a second solution is obtained by setting $c = 0$ after we separate variables. We have $y_2(t) = t^3$. Consequently, we have two solutions, $y_1(t) = 0$ and $y_2(t) = t^3$, to the same initial-value problem (see Figure 1.45).

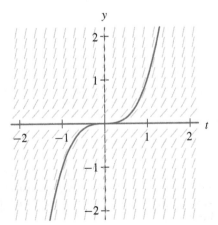

Figure 1.45
The slope field and the graphs of two solutions to the initial-value problem

$$\frac{dy}{dt} = 3y^{2/3}, \quad y(0) = 0.$$

This differential equation does not satisfy the hypothesis of the Uniqueness Theorem if $y = 0$. Note that we have two different solutions whose graphs intersect at $(0, 0)$.

Applications of the Uniqueness Theorem

The Uniqueness Theorem says that two solutions to the same initial-value problem are identical. This result is reassuring, but it may not sound useful in a practical sense. Here we discuss a few examples to illustrate why this theorem is, in fact, very useful.

Suppose $y_1(t)$ and $y_2(t)$ are both solutions of a differential equation

$$\frac{dy}{dt} = f(t, y),$$

where $f(t, y)$ satisfies the hypotheses of the Uniqueness Theorem. If for some t_0 we have $y_1(t_0) = y_2(t_0)$, then both of these functions are solutions of the same initial-value problem

$$\frac{dy}{dt} = f(t, y), \quad y(t_0) = y_1(t_0) = y_2(t_0).$$

The Uniqueness Theorem guarantees that $y_1(t) = y_2(t)$, at least for all t for which both solutions are defined. We can paraphrase the Uniqueness Theorem as:

> "If two solutions are ever in the same place at the same time, then they are the same function."

This form of the Uniqueness Theorem is very valuable, as the following examples show.

Role of equilibrium solutions

Consider the initial-value problem

$$\frac{dy}{dt} = \frac{(y^2 - 4)(\sin^2 y^3 + \cos y - 2)}{2}, \quad y(0) = \frac{1}{2}.$$

Finding the explicit solution to this equation is not easy because, even though the equation is autonomous and hence separable, the integrals involved are very difficult (try them). On the other hand, if $y = 2$, the right-hand side of the equation vanishes. Thus the constant function $y_1(t) = 2$ is an equilibrium solution for this equation.

Suppose $y_2(t)$ is the solution to the differential equation that satisfies the initial condition $y_2(0) = 1/2$. The Uniqueness Theorem implies that $y_2(t) < 2$ for all t since the graph of $y_2(t)$ cannot touch the line $y = 2$, which is the graph of the constant solution $y_1(t)$ (see Figure 1.46).

Figure 1.46
The slope field and the graphs of two solutions of

$$\frac{dy}{dt} = \frac{(y^2 - 4)(\sin^2 y^3 + \cos y - 2)}{2}.$$

Although it looks as if these two graphs agree for $t > 2$, the Uniqueness Theorem tells us that there is always a little space between them.

This observation is not a lot of information about the solution of the initial-value problem with $y(0) = 1/2$. On the other hand, we didn't have to do a lot of work to get this information. Identifying $y_1(t) = 2$ as a solution is pretty easy, and the rest follows from the Uniqueness Theorem. By doing a little bit of work, we get some information. If all we care about is how large the solution of the original initial-value problem can possibly become, then the fact that it is bounded above by $y = 2$ may suffice. If we need more detailed information, we must look more carefully at the equation.

Comparing solutions

We can also use this technique to obtain information about solutions by comparing them to "known" solutions. For example, consider the differential equation

$$\frac{dy}{dt} = \frac{(1+t)^2}{(1+y)^2}.$$

It is easy to check that $y_1(t) = t$ is a solution to the differential equation with the initial condition $y_1(0) = 0$. If $y_2(t)$ is the solution satisfying the initial condition $y(0) = -0.1$, then $y_2(0) < y_1(0)$, so $y_2(t) < y_1(t)$ for all t. Thus $y_2(t) < t$ for all t (see Figure 1.47). Again, this is only a little bit of information about the solution of the initial-value problem, but then we only did a little work.

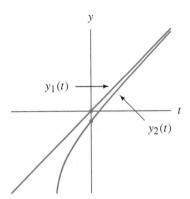

y

$y_1(t)$

$y_2(t)$

t

Figure 1.47
The graphs of two solutions $y_1(t)$ and $y_2(t)$ of

$$\frac{dy}{dt} = \frac{(1+t)^2}{(1+y)^2}.$$

The graph of the solution $y_1(t)$ that satisfies the initial condition $y_1(0) = 0$ is the diagonal line, and the graph of the solution that satisfies the initial condition $y_2(0) = -0.1$ must lie below the line.

Uniqueness and qualitative analysis

In some cases we can use the Uniqueness Theorem and some qualitative information to give more exact information about solutions. For example, consider the differential equation

$$\frac{dy}{dt} = (y - 2)(y + 1).$$

The right-hand side of this autonomous equation is the function $f(y) = (y-2)(y+1)$. Note that $f(2) = f(-1) = 0$. Thus $y = 2$ and $y = -1$ are equilibrium solutions (see the slope field in Figure 1.48). By the Existence and Uniqueness Theorem, any solution $y(t)$ with an initial condition $y(0)$ that satisfies $-1 < y(0) < 2$ must also satisfy $-1 < y(t) < 2$ for all t.

In this case we can say even more about these solutions. For example, consider the solution with the initial condition $y(0) = 0.5$. Not only do we know that $-1 < y(t) < 2$ for all t, but because this equation is autonomous, the sign of dy/dt depends only on the value of y. For $-1 < y < 2$, $dy/dt = f(y) < 0$. Hence the solution $y(t)$ with the initial condition $y(0) = 0.5$ satisfies $dy/dt < 0$ for all t. Consequently this solution is decreasing for all t.

Since the solution is decreasing for all t and since it always remains above $y = -1$, we might guess that $y(t) \to -1$ as $t \to \infty$. In fact this is precisely what happens. If $y(t)$ were to limit to any value y_0 larger than -1 as $t \to \infty$, then when t is very large, $y(t)$ must be close to y_0. But $f(y_0)$ is negative because $-1 < y_0 < 2$. So when $y(t)$ is close to y_0, we have dy/dt close to $f(y_0)$, which is negative, so the solution must continue to decrease past y_0. That is, solutions of this differential equation can be asymptotic only to the equilibrium solutions.

We can sketch the solution of this initial-value problem. For all t the graph is between the lines $y = -1$ and $y = 2$, and for all t it decreases (see Figure 1.49).

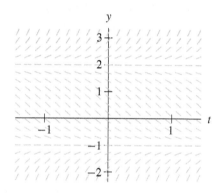

Figure 1.48
The slope field for $dy/dt = (y - 2)(y + 1)$.

Figure 1.49
Graphs of the equilibrium solutions and the solution with initial condition $y(0) = 0.5$ for $dy/dt = (y - 2)(y + 1)$.

Uniqueness and Numerical Approximation

As the preceding examples show, the Uniqueness Theorem gives us qualitative information concerning the behavior of solutions. We can use this information to check the behavior of numerical approximations of solutions. If numerical approximations of solutions violate the Uniqueness Theorem, then we are certain that something is wrong.

The graph of the Euler approximation to the solution of the initial-value problem

$$\frac{dy}{dt} = e^t \sin y, \quad y(0) = 5,$$

with $\Delta t = 0.05$ is shown in Figure 1.50. As noted in Section 1.4, the behavior seems erratic, and hence we are suspicious.

Figure 1.50
Euler's method applied to

$$\frac{dy}{dt} = e^t \sin y$$

with $\Delta t = 0.05$. The graph of the approximation behaves as expected for $t < 5$, but for t slightly larger than 5, the approximation is no longer valid.

We can easily check that the constant function $y(t) = n\pi$ is a solution for any integer n and, hence by the Uniqueness Theorem, each solution is trapped between $y = n\pi$ and $y = (n + 1)\pi$ for some integer n. The approximations in Figure 1.50 violate this requirement. This confirms our suspicions that the numerical results in this case are not to be believed.

This equation is unusual because of the e^t term on the right-hand side. When t is large, the slopes of solutions become gigantic and hence Euler's method overshoots the true solution for even a very small step size.

EXERCISES FOR SECTION 1.5

In Exercises 1–4, we refer to a function f, but we do not provide its formula. However, we do assume that f satisfies the hypotheses of the Uniqueness Theorem in the entire ty-plane, and we do provide various solutions to the given differential equation. Finally, we specify an initial condition. Using the Uniqueness Theorem, what can you conclude about the solution to the equation with the given initial condition?

1. $\dfrac{dy}{dt} = f(t, y)$

   $y_1(t) = 3$ for all t is a solution,

   initial condition $y(0) = 1$

2. $\dfrac{dy}{dt} = f(y)$

   $y_1(t) = 4$ for all t is a solution,

   $y_2(t) = 2$ for all t is a solution,

   $y_3(t) = 0$ for all t is a solution,

   initial condition $y(0) = 1$

3. $\dfrac{dy}{dt} = f(t, y)$

   $y_1(t) = t + 2$ for all t is a solution,

   $y_2(t) = -t^2$ for all t is a solution,

   initial condition $y(0) = 1$

4. $\dfrac{dy}{dt} = f(t, y)$

   $y_1(t) = -1$ for all t is a solution,

   $y_2(t) = 1 + t^2$ for all t is a solution,

   initial condition $y(0) = 0$

In Exercises 5–8, an initial condition for the differential equation

$$\frac{dy}{dt} = y(y-1)(y-3)$$

is given. What does the Existence and Uniqueness Theorem say about the corresponding solution?

5. $y(0) = 4$ **6.** $y(0) = 0$ **7.** $y(0) = 2$ **8.** $y(0) = -1$

9. **(a)** Show that $y_1(t) = t^2$ and $y_2(t) = t^2 + 1$ are solutions to

$$\frac{dy}{dt} = -y^2 + y + 2yt^2 + 2t - t^2 - t^4.$$

(b) Show that if $y(t)$ is a solution to the differential equation in part (a) and if $0 < y(0) < 1$, then $t^2 < y(t) < t^2 + 1$ for all t.

(c) Illustrate your answer using HPGSolver.

10. Consider the differential equation $dy/dt = 2\sqrt{|y|}$.

(a) Show that the function $y(t) = 0$ for all t is an equilibrium solution.

(b) Find all solutions. [*Hint*: Consider the cases $y > 0$ and $y < 0$ separately. Then you need to define the solutions using language like "$y(t) = \ldots$ when $t \leq 0$ and $y(t) = \ldots$ when $t > 0$."]

(c) Why doesn't this differential equation contradict the Uniqueness Theorem?

(d) What does HPGSolver do with this equation?

11. Consider the differential equation

$$\frac{dy}{dt} = \frac{y}{t^2}.$$

(a) Show that the constant function $y_1(t) = 0$ is a solution.

(b) Show that there are infinitely many other functions that satisfy the differential equation, that agree with this solution when $t \leq 0$, but that are nonzero when $t > 0$. [*Hint*: You need to define these functions using language like "$y(t) = \ldots$ when $t \leq 0$ and $y(t) = \ldots$ when $t > 0$."]

(c) Why doesn't this example contradict the Uniqueness Theorem?

12. **(a)** Show that

$$y_1(t) = \frac{1}{t-1} \quad \text{and} \quad y_2(t) = \frac{1}{t-2}$$

are solutions of $dy/dt = -y^2$.

(b) What can you say about solutions of $dy/dt = -y^2$ for which the initial condition $y(0)$ satisfies the inequality $-1 < y(0) < -1/2$? [*Hint*: You could find the general solution, but what information can you get from your answer to part (a) alone?]

In Exercises 13–16, an initial-value problem is given.

(a) Find a formula for the solution.

(b) State the domain of definition of the solution.

(c) Describe what happens to the solution as it approachs the limits of its domain of definition. Why can't the solution be extended for more time?

13. $\dfrac{dy}{dt} = y^3, \quad y(0) = 1$

14. $\dfrac{dy}{dt} = \dfrac{1}{(y+1)(t-2)}, \quad y(0) = 0$

15. $\dfrac{dy}{dt} = \dfrac{1}{(y+2)^2}, \quad y(0) = 1$

16. $\dfrac{dy}{dt} = \dfrac{t}{y-2}, \quad y(-1) = 0$

17. Consider a differential equation of the form $dy/dt = f(y)$, an autonomous equation, and assume that the function $f(y)$ is continuously differentiable.

(a) Suppose $y_1(t)$ is a solution and $y_1(t)$ has a local maximum at $t = t_0$. Let $y_0 = y_1(t_0)$. Show that $f(y_0) = 0$.

(b) Use the information of part (a) to sketch the slope field along the line $y = y_0$ in the ty-plane.

(c) Show that the constant function $y_2(t) = y_0$ is a solution (in other words, $y_2(t)$ is an equilibrium solution).

(d) Show that $y_1(t) = y_0$ for all t.

(e) Show that if a solution of $dy/dt = f(y)$ has a local minimum, then it is a constant function; that is, it also corresponds to an equilibrium solution.

18. We have emphasized that the Uniqueness Theorem does not apply to every differential equation. There are hypotheses that must be verified before we can apply the theorem. However, there is a temptation to think that, since models of "real-world" problems must obviously have solutions, we don't need to worry about the hypotheses of the Uniqueness Theorem when we are working with differential equations modeling the physical world. The following model illustrates the flaw in this assumption.

   Suppose we wish to study the formation of raindrops in the atmosphere. We make the reasonable assumption that raindrops are approximately spherical. We also assume that the rate of growth of the volume of a raindrop is proportional to its surface area.

   Let $r(t)$ be the radius of the raindrop at time t, $s(t)$ be its surface area at time t, and $v(t)$ be its volume at time t. From three-dimensional geometry, we know that

$$s = 4\pi r^2 \qquad \text{and} \qquad v = \tfrac{4}{3}\pi r^3.$$

(a) Show that the differential equation that models the volume of the raindrop under these assumptions is

$$\frac{dv}{dt} = kv^{2/3},$$

where k is a proportionality constant.

(b) Why doesn't this equation satisfy the hypotheses of the Uniqueness Theorem?

(c) Give a physical interpretation of the fact that solutions to this equation with the initial condition $v(0) = 0$ are not unique. Does this model say anything about the way raindrops begin to form?

1.6 EQUILIBRIA AND THE PHASE LINE

Given a differential equation

$$\frac{dy}{dt} = f(t, y),$$

we can get an idea of how solutions behave by drawing slope fields and sketching their graphs or by using Euler's method and computing approximate solutions. Sometimes we can even derive explicit formulas for solutions and plot the results. All of these techniques require quite a bit of work, either numerical (computation of slopes or Euler's method) or analytic (integration).

In this section we consider differential equations where the right-hand side is independent of t. Such equations are said to be **autonomous differential equations**. The word *autonomous* means "self-governing," and roughly speaking, an autonomous system is self-governing because it evolves according to differential equations that are determined entirely by the values of the dependent variables. For autonomous differential equations, there are qualitative techniques that help us sketch the graphs of the solutions with less arithmetic than with other methods.

Autonomous Equations

Autonomous equations are differential equations of the form

$$\frac{dy}{dt} = f(y).$$

In other words, the rate of change of the dependent variable can be expressed as a function of the dependent variable alone. Autonomous equations appear frequently as models for two reasons. First, many physical systems work the same way at any time. For example, a spring compressed the same amount at 10:00 AM and at 3:00 PM provides the same force. Second, for many systems, the time dependence "averages out" over the time scales being considered. For example, if we are studying how wolves and field mice interact, we might find that wolves eat many more field mice during the day than they do at night. However, if we are interested in how the wolf and mouse populations behave over a period of years or decades, then we can average the number of mice eaten by each wolf per week. We ignore the daily fluctuations.

We have already noticed that autonomous equations have slope fields that have a special form (see page 40 in Section 1.3). Because the right-hand side of the equation does not depend on t, the slope marks are parallel along horizontal lines in the ty-plane. That is, for an autonomous equation, two points with the same y-coordinate but different t-coordinates have the same slope marks (see Figure 1.51).

Figure 1.51
Slope field for the autonomous differential equation

$$\frac{dy}{dt} = (y - 2)(y + 1).$$

The slopes are parallel along horizontal lines. Note that the slope field indicates that there are two equilibrium solutions, $y_1(t) = -1$ for all t and $y_2(t) = 2$ for all t. Also, solutions with initial values that lie between -1 and 2 are decreasing and defined for all time.

Hence there is a great deal of redundancy in the slope field of an autonomous equation. If we know the slope field along a single vertical line $t = t_0$, then we know the slope field in the entire ty-plane. So instead of drawing the entire slope field, we should be able to draw just one line containing the same information. This line is called the **phase line** for the autonomous equation.

Metaphor of the rope

Suppose you are given an autonomous differential equation

$$\frac{dy}{dt} = f(y).$$

Think of a rope hanging vertically and stretching infinitely far up and infinitely far down. The dependent variable y tells you a position on the rope (the rope is the y-axis). The function $f(y)$ gives a number for each position on the rope. Suppose the number $f(y)$ is actually printed on the rope at height y for every value of y. For example, at the height $y = 2.17$, the value $f(2.17)$ is printed on the rope.

Suppose that you are placed on the rope at height y_0 at time $t = 0$ and given the following instructions: Read the number that is printed on the rope and climb up or down the rope with velocity equal to that number. Climb up the rope if the number is positive or down the rope if the number is negative. (A large positive number means you climb up very quickly, whereas a negative number near zero means you climb down slowly.) As you move, continue to read the numbers on the rope and adjust your velocity so that it always agrees with the number printed on the rope.

If you follow this rather bizarre set of instructions, you will generate a function $y(t)$ that gives your position on the rope at time t. Your position at time $t = 0$ is $y(0) = y_0$ because that is where you were placed initially. The velocity of your motion dy/dt at time t will be given by the number on the rope, so $dy/dt = f(y(t))$ for all t. Hence, your position function $y(t)$ is a solution to the initial-value problem

$$\frac{dy}{dt} = f(y), \quad y(0) = y_0.$$

The phase line is a picture of this rope. Because it is tedious to record the numerical values of all the velocities, we only mark the phase line with the numbers where the velocity is zero and indicate the sign of the velocity on the intervals in between. The phase line provides qualitative information about the solutions.

Phase Line of a Logistic Equation

For example, consider the differential equation

$$\frac{dy}{dt} = (1 - y)y.$$

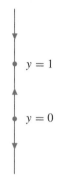

Figure 1.52
Phase line for
$dy/dt = (1 - y)y$.

The right-hand side of the differential equation is $f(y) = (1 - y)y$. In this case, $f(y) = 0$ precisely when $y = 0$ and $y = 1$. Therefore the constant function $y_1(t) = 0$ for all t and $y_2(t) = 1$ for all t are equilibrium solutions for this equation. We call the points $y = 0$ and $y = 1$ on the y-axis **equilibrium points**. Also note that $f(y)$ is positive if $0 < y < 1$, whereas $f(y)$ is negative if $y < 0$ or $y > 1$. We can draw the phase line (or "rope") by placing dots at the equilibrium points $y = 0$ and $y = 1$. For $0 < y < 1$, we put arrows pointing up because $f(y) > 0$ means you climb up; and for $y < 0$ or $y > 1$, we put arrows pointing down because $f(y) < 0$ means you climb down (see Figure 1.52).

If we compare the phase line to the slope field, we see that the phase line contains all the information about the equilibrium solutions and whether the solutions are increasing or decreasing. Information about the *speed* of increase or decrease of solutions is lost (see Figure 1.53), But we can give rough sketches of the graphs of solutions using the phase line alone. These sketches will not be quite as accurate as the sketches from the slope field, but they will contain all the information about the behavior of solutions as t gets large (see Figure 1.54).

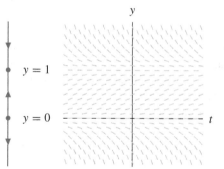

Figure 1.53
Phase line and slope field of
$dy/dt = (1 - y)y$.

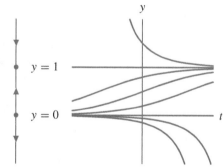

Figure 1.54
Phase line and sketches of the graphs of
solutions for $dy/dt = (1 - y)y$.

How to Draw Phase Lines

We can give a more precise definition of the phase line by giving the steps required to draw it. For the autonomous equation $dy/dt = f(y)$:

- Draw the y-line.
- Find the equilibrium points (the numbers such that $f(y) = 0$), and mark them on the line.

- Find the intervals of y-values for which $f(y) > 0$, and draw arrows pointing up in these intervals.
- Find the intervals of y-values for which $f(y) < 0$, and draw arrows pointing down in these intervals.

We sketch several examples of phase lines in Figure 1.55. When looking at the phase line, you should remember the metaphor of the rope and think of solutions of the differential equation "dynamically"—people climbing up and down the rope as time increases.

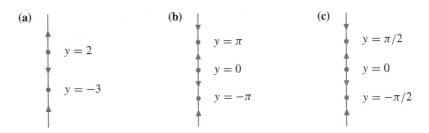

Figure 1.55
Phase lines for **(a)** $dy/dt = (y - 2)(y + 3)$,   **(b)** $dy/dt = \sin y$, and
(c) $dy/dt = y \cos y$.

How to Use Phase Lines to Sketch Solutions

Figure 1.56
Phase line for
$dw/dt = (2 - w) \sin w$.

We can obtain rough sketches of the graphs of solutions directly from the phase lines, provided we are careful in interpreting these sketches. The sort of information that phase lines are very good at predicting is the limiting behavior of solutions as t increases or decreases.

Consider the equation

$$\frac{dw}{dt} = (2 - w) \sin w.$$

The phase line for this differential equation is given in Figure 1.56. Note that the equilibrium points are $w = 2$ and $w = k\pi$ for any integer k. Suppose we want to sketch the graph of the solution $w(t)$ with the initial value $w(0) = 0.4$. Because $w = 0$ and $w = 2$ are equilibrium points of this equation and $0 < 0.4 < 2$, we know from the Existence and Uniqueness Theorem that $0 < w(t) < 2$ for all t. Moreover, because $(2 - w) \sin w > 0$ for $0 < w < 2$, the solution is always increasing. Because the velocity of the solution is small only when $(2 - w) \sin w$ is close to zero and because this happens only near equilibrium points, we know that the solution $w(t)$ increases toward $w = 2$ as $t \to \infty$ (see Section 1.5).

Similarly, if we run the clock backward, the solution $w(t)$ decreases. It always remains above $w = 0$ and cannot stop, since $0 < w < 2$. Thus as $t \to -\infty$, the solution tends toward $w = 0$. We can draw a qualitative picture of the graph of the solution with the initial condition $w(0) = 0.4$ (see Figure 1.57).

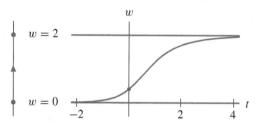

Figure 1.57
Graph of the solution to the initial-value problem

$$\frac{dw}{dt} = (2 - w)\sin w, \quad w(0) = 0.4.$$

Likewise, we can sketch other solutions in the tw-plane from the information on the phase line. The equilibrium solutions are easy to find and draw because they are marked on the phase line. The intervals on the phase line with upward-pointing arrows correspond to increasing solutions, and those with downward-pointing arrows correspond to decreasing solutions. Graphs of the solutions do not cross by the Uniqueness Theorem. In particular, they cannot cross the graphs of the equilibrium solutions. Also, solutions must continue to increase or decrease until they come close to an equilibrium solution. Hence we can sketch many solutions with different initial conditions quite easily. The only information that we do not have is how quickly the solutions increase or decrease (see Figure 1.58).

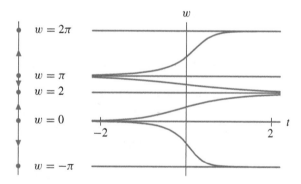

Figure 1.58
Graphs of many solutions to $dw/dt = (2 - w)\sin w$.

These observations lead to some general statements that can be made for all solutions of autonomous equations. Suppose $y(t)$ is a solution to an autonomous equation

$$\frac{dy}{dt} = f(y),$$

where $f(y)$ is continuously differentiable for all y.

- If $f(y(0)) = 0$, then $y(0)$ is an equilibrium point and $y(t) = y(0)$ for all t.
- If $f(y(0)) > 0$, then $y(t)$ is increasing for all t and either $y(t) \to \infty$ as t increases or $y(t)$ tends to the first equilibrium point larger than $y(0)$.
- If $f(y(0)) < 0$, then $y(t)$ is decreasing for all t and either $y(t) \to -\infty$ as t increases or $y(t)$ tends to the first equilibrium point smaller than $y(0)$.

Similar results hold as t decreases (as time runs backward). If $f(y(0)) > 0$, then $y(t)$ either tends (in negative time) to $-\infty$ or to the next smaller equilibrium point. If $f(y(0)) < 0$, then $y(t)$ either tends (in negative time) to $+\infty$ or the next larger equilibrium point.

An example with three equilibrium points

For example, consider the differential equation

$$\frac{dP}{dt} = \left(1 - \frac{P}{20}\right)^3 \left(\frac{P}{5} - 1\right) P^7.$$

If the initial condition is given by $P(0) = 8$, what happens as t becomes very large? First we draw the phase line for this equation. Let

$$f(P) = \left(1 - \frac{P}{20}\right)^3 \left(\frac{P}{5} - 1\right) P^7.$$

We find the equilibrium points by solving $f(P) = 0$. Thus $P = 0$, $P = 5$, and $P = 20$ are the equilibrium points.

If $0 < P < 5$, $f(P)$ is negative; if $P < 0$ or $5 < P < 20$, $f(P)$ is positive; and if $P > 20$, $f(P)$ is negative. We can place the arrows on the phase line appropriately (see Figure 1.59). Note that we only have to check the value of $f(P)$ at one point in each of these intervals to determine the sign of $f(P)$ in the entire interval.

The solution $P(t)$ with initial condition $P(0) = 8$ is in the region between the equilibrium points $P = 5$ and $P = 20$, so $5 < P(t) < 20$ for all t. The arrows point up in this interval, so $P(t)$ is increasing for all t. As $t \to \infty$, $P(t)$ tends toward the equilibrium point $P = 20$.

As $t \to -\infty$, the solution with initial condition $P(0) = 8$ decreases toward the next smaller equilibrium point, which is $P = 5$. Hence $P(t)$ is always greater than $P = 5$. If we compute the solution $P(t)$ numerically, we see that it increases from $P(0) = 8$ to close to $P = 20$ very quickly (see Figure 1.60). From the phase line alone, we cannot tell how quickly the solution increases.

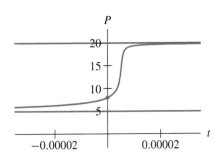

Figure 1.59
Phase line for $dP/dt = f(P) = (1 - P/20)^3 ((P/5) - 1) P^7$.

Figure 1.60
Graph of the solution to the initial-value problem $dP/dt = (1 - P/20)^3 ((P/5) - 1) P^7$, $P(0) = 8$.

Warning: Not All Solutions Exist for All Time

Suppose y_0 is an equilibrium point for the equation $dy/dt = f(y)$. Then $f(y_0) = 0$. We are assuming $f(y)$ is continuous, so if solutions are close to y_0, the value of f is small. Thus solutions move slowly when they are close to equilibrium points. A solution that approaches an equilibrium point as t increases (or decreases) moves more and more slowly as it approaches the equilibrium point. By the Existence and Uniqueness Theorem, a solution that approaches an equilibrium point never actually gets there. It is *asymptotic* to the equilibrium point, and the graph of the solution in the ty-plane has a horizontal asymptote.

On the other hand, unbounded solutions often speed up as they move. For example, the equation

$$\frac{dy}{dt} = (1 + y)^2$$

has one equilibrium point at $y = -1$ and $dy/dt > 0$ for all other values of y (see Figure 1.61).

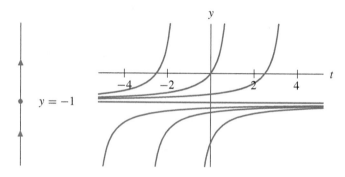

Figure 1.61
Phase line for $dy/dt = (1 + y)^2$ and graphs of solutions that are unbounded in finite time.

The phase line indicates that solutions with initial conditions that are greater than -1 increase for all t and tend to $+\infty$ as t increases.

If we separate variables and compute the explicit form of the solution, we can determine that these solutions actually blow up in finite time. In fact, the explicit form of any nonconstant solution is given by

$$y(t) = -1 - \frac{1}{t + c}$$

for some constant c. Since we are assuming that $y(0) > -1$, we must have

$$y(0) = -1 - \frac{1}{c} > -1,$$

which implies that $c < 0$. Therefore these solutions are defined only for $t < -c$, and they tend to ∞ as $t \rightarrow -c$ from below (see Figure 1.61). We cannot tell if solutions blow up in finite time like this simply by looking at the phase line.

The solutions with initial conditions $y(0) < -1$ are asymptotic to the equilibrium point $y = -1$ as t increases, so they are defined for all $t > 0$. However, these solutions tend to $-\infty$ in finite time as t decreases.

Another dangerous example is

$$\frac{dy}{dt} = \frac{1}{1-y}.$$

If $y > 1$, dy/dt is negative, and if $y < 1$, dy/dt is positive. If $y = 1$, dy/dt does not exist. The phase line has a hole in it. There is no standard way to denote such points on the phase line, but we will use a small empty circle to mark them (see Figure 1.62).

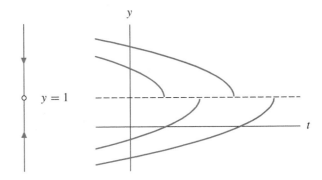

Figure 1.62
Phase line for $dy/dt = 1/(1 - y)$. Note that dy/dt is not defined for $y = 1$. Also, the graphs of solutions reach the "hole" at $y = 1$ in finite time.

All solutions tend toward $y = 1$ as t increases. Because the value of dy/dt is large if y is close to 1, solutions speed up as they get close to $y = 1$, and solutions reach $y = 1$ in a finite amount of time. Once a solution reaches $y = 1$, it cannot be continued because it has left the domain of definition of the differential equation. It has fallen into a hole in the phase line.

Drawing Phase Lines from Qualitative Information Alone

To draw the phase line for the differential equation $dy/dt = f(y)$, we need to know the location of the equilibrium points and the intervals over which the solutions are increasing or decreasing. That is, we need to know the points where $f(y) = 0$, the intervals where $f(y) > 0$, and the intervals where $f(y) < 0$. Consequently, we can draw the phase line for the differential equation with only qualitative information about the function $f(y)$.

For example, suppose we do not know a formula for $f(y)$, but we do have its graph (see Figure 1.63). From the graph we can determine the values of y for which $f(y) = 0$ and decide on which intervals $f(y) > 0$ and $f(y) < 0$. With this information we can draw the phase line (see Figure 1.64). From the phase line we can then get qualitative sketches of solutions (see Figure 1.65). Thus we can go from qualitative information about $f(y)$ to graphs of solutions of the differential equation $dy/dt = f(y)$ without ever writing down a formula. For models where the information available is completely qualitative, this approach is very appropriate.

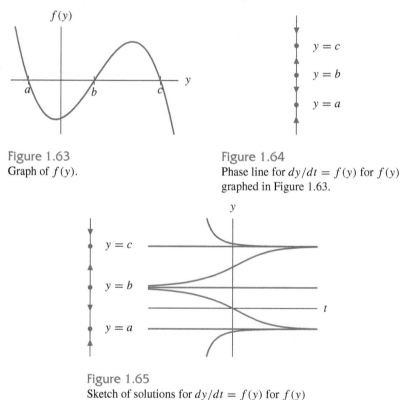

Figure 1.63
Graph of $f(y)$.

Figure 1.64
Phase line for $dy/dt = f(y)$ for $f(y)$ graphed in Figure 1.63.

Figure 1.65
Sketch of solutions for $dy/dt = f(y)$ for $f(y)$ graphed in Figure 1.63.

The Role of Equilibrium Points

If $f(y)$ is continuously differentiable for all y, we have already determined that every solution to the autonomous equation $dy/dt = f(y)$ either tends to $+\infty$ or $-\infty$ as t increases (perhaps becoming infinite in finite time) or tends asymptotically to an equilibrium point as t increases. Hence the equilibrium points are extremely important in understanding the long-term behavior of solutions.

Also we have seen that, when drawing a phase line, we need to find the equilibrium points, the intervals on which $f(y)$ is positive, and the intervals on which $f(y)$ is negative. If f is continuous, it can switch from positive to negative only at points y_0

for which $f(y_0) = 0$, that is, at equilibrium points. Hence the equilibrium points also play a crucial role in sketching the phase line.

In fact, the equilibrium points are the key to understanding the entire phase line. For example, suppose we have an autonomous differential equation $dy/dt = g(y)$ where $g(y)$ is continuous for all y. Suppose all we know about this differential equation is that it has exactly two equilibrium points, at $y = 2$ and $y = 7$, and that the phase line near $y = 2$ and $y = 7$ is as shown on the left-hand side of Figure 1.66. We can use this information to sketch the entire phase line. We know that the sign of $g(y)$ can change only at an equilibrium point. Hence the sign of $g(y)$ does not change for $2 < y < 7$, for $y < 2$, or for $y > 7$. Thus if we know the direction of the arrows anywhere in these intervals (say near the equilibrium points), then we know the directions on the entire phase line (see Figure 1.66). Consequently if we understand the equilibrium points for an autonomous differential equation, we should be able to understand (at least qualitatively) any solution of the equation.

$y = 7$ $y = 7$

$y = 2$ $y = 2$

Figure 1.66
On the left we have two pieces of the phase line, one piece for each of the two equilibrium points $y = 2$ and $y = 7$. On the right we construct the entire phase line of $dy/dt = g(y)$ from these individual pieces.

Stephen Smale (1930–) is one of the founders of modern-day dynamical systems theory. In the mid-1960s, Smale began to advocate taking a more qualitative approach to the study of differential equations, as we do in this book. Using this approach, he was among the first mathematicians to encounter and analyze a "chaotic" dynamical system. Since this discovery, scientists have found that many important physical systems exhibit chaos.

Smale's research has spanned many disciplines, including economics, theoretical computer science, mathematical biology, as well as many subareas of mathematics. In 1966 he was awarded the Fields Medal, the equivalent of the Nobel Prize in mathematics. He is currently Professor Emeritus at the University of California, Berkeley.

Smale is an avid collector of rare minerals. His father gave him his first specimen in 1968, and ever since, he has traveled to many exotic locations to add to his collection. He now owns more than 1000 world-class specimens. Pictures of some of the minerals in his collection are available at http://math.berkeley.edu/ smale/crystals.html.

Classification of Equilibrium Points

Given their significance, it is useful to name the different types of equilibrium points and to classify them according to the behavior of nearby solutions. Consider an equilibrium point $y = y_0$, as shown in Figure 1.67. For y slightly less than y_0, the arrows point up, and for y slightly larger than y_0, the arrows point down. A solution with initial condition close to y_0 is asymptotic to y_0 as $t \to \infty$.

We say an equilibrium point y_0 is a **sink** if any solution with initial condition sufficiently close to y_0 is asymptotic to y_0 as t increases. (The name *sink* is supposed to bring to mind a kitchen sink with the equilibrium point as the drain. If water starts close enough to the drain, it will run toward it.)

Another possible phase line near an equilibrium point y_0 is shown in Figure 1.68. Here, the arrows point up for values of y just above y_0 and down for values of y just below y_0. A solution that has an initial value near y_0 tends away from y_0 as t increases. If time is run backward, solutions that start near y_0 tend toward y_0.

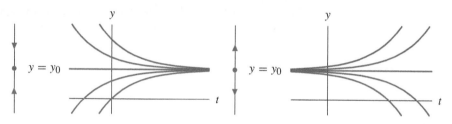

Figure 1.67
Phase line at a sink and graphs of solutions near a sink.

Figure 1.68
Phase line at a source and graphs of solutions near a source.

We say an equilibrium point y_0 is a **source** if all solutions that start sufficiently close to y_0 tend toward y_0 as t decreases. This means that all solutions that start close to y_0 (but not at y_0) will tend away from y_0 as t increases. So a source is a sink if time is run backward. (The name *source* is supposed to help you picture solutions flowing out of or away from a point.)

Sinks and sources are the two major types of equilibrium points. Every equilibrium point that is neither a source nor a sink is called a **node**. Two possible phase line pictures near nodes are shown in Figure 1.69.

Figure 1.69
Examples of node equilibrium points and graphs of nearby solutions.

Given a differential equation, we can classify the equilibrium points as sinks, sources, or nodes from the phase line. For example, consider

$$\frac{dy}{dt} = y^2 + y - 6 = (y + 3)(y - 2).$$

The equilibrium points are $y = -3$ and $y = 2$. Also $dy/dt < 0$ for $-3 < y < 2$, and $dy/dt > 0$ for $y < -3$ and $y > 2$. Given this information, we can draw the phase line, and from the phase line we see that $y = -3$ is a sink and $y = 2$ is a source (see Figure 1.70).

Suppose we are given a differential equation $dw/dt = g(w)$, where the right-hand side $g(w)$ is specified in terms of a graph rather than in terms of a formula. Then we can still sketch the phase line. For example, suppose that $g(w)$ is the function graphed in Figure 1.71. The corresponding differential equation has three equilibrium points, $w = -0.5$, $w = 1$, and $w = 2.5$; and $g(w) > 0$ if $w < -0.5$, $1 < w < 2.5$, and $w > 2.5$. For $-0.5 < w < 1$, $g(w) < 0$. Using this information, we can draw the phase line (see Figure 1.72) and classify the equilibrium points. The point $w = -0.5$ is a sink, the point $w = 1$ is a source, and the point $w = 2.5$ is a node.

Figure 1.70
Phase line for
$dy/dt = y^2 + y - 6$.

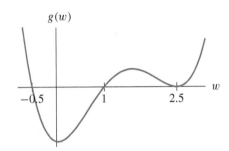

Figure 1.71
Graph of $g(w)$.

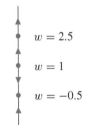

Figure 1.72
Phase line for $dw/dt = g(w)$
for $g(w)$, as displayed in
Figure 1.71.

Identifying the type of an equilibrium point and "linearization"

From the previous examples we know that we can determine the phase line and classify the equilibrium points for an autonomous differential equation $dy/dt = f(y)$ from the graph of $f(y)$ alone. Since the classification of an equilibrium point depends only on the phase line near the equilibrium point, then we should be able to determine the type of an equilibrium point y_0 from the graph of $f(y)$ near y_0.

If y_0 is a sink, then the arrows on the phase line just below y_0 point up and the arrows just above y_0 point down. Hence $f(y)$ must be positive for y just smaller than y_0 and negative for y just larger than y_0 (see Figure 1.73). So f must be decreasing for y near y_0. Conversely, if $f(y_0) = 0$ and f is decreasing for all y near y_0, then $f(y)$ is positive just to the left of y_0 and negative just to the right of y_0. Hence, y_0 is a sink. Similarly, the equilibrium point y_0 is a source if and only if f is increasing for all y near y_0 (see Figure 1.74).

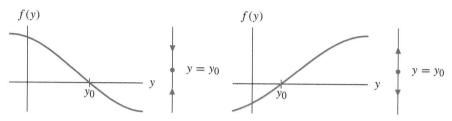

Figure 1.73
Phase line near a sink at $y = y_0$ for $dy/dt = f(y)$ and graph of $f(y)$ near $y = y_0$.

Figure 1.74
Phase line near a source at $y = y_0$ for $dy/dt = f(y)$ and graph of $f(y)$ near $y = y_0$.

From calculus we have a powerful tool for telling whether a function is increasing or decreasing at a particular point—the derivative. Using the derivative of $f(y)$ combined with the geometric observations above, we can give criteria that specify the type of the equilibrium point.

LINEARIZATION THEOREM   Suppose y_0 is an equilibrium point of the differential equation $dy/dt = f(y)$ where f is a continuously differentiable function. Then,

- if $f'(y_0) < 0$, then y_0 is a sink;
- if $f'(y_0) > 0$, then y_0 is a source; or
- if $f'(y_0) = 0$, then we need additional information to determine the type of y_0.   ■

This theorem follows immediately from the discussion prior to its statement once we recall that if $f'(y_0) < 0$, then f is decreasing near y_0, and if $f'(y_0) > 0$, then f is increasing near y_0. This analysis and these conclusions are an example of **linearization**, a technique that we will often find useful. The derivative $f'(y_0)$ tells us the behavior of the best linear approximation to f near y_0. If we replace f with its best linear approximation, then the differential equation we obtain is very close to the original differential equation for y near y_0.

We cannot make any conclusion about the classification of y_0 if $f'(y_0) = 0$, because all three possibilities can occur (see Figure 1.75).

As another example, consider the differential equation

$$\frac{dy}{dt} = h(y) = y(\cos(y^5 + 2y) - 27\pi y^4).$$

What does the phase line look like near $y = 0$? Drawing the phase line for this equation would be a very complicated affair. We would have to find the equilibrium points and determine the sign of $h(y)$. On the other hand, it is easy to see that $y = 0$ is an equilibrium point because $h(0) = 0$. We compute

$$h'(y) = (\cos(y^5 + 2y) - 27\pi y^4) + y\frac{d}{dy}(\cos(y^5 + 2y) - 27\pi y^4).$$

Thus $h'(0) = (\cos(0) - 0) + 0 = 1$. By the Linearization Theorem, we conclude that $y = 0$ is a source. Solutions that start sufficiently close to $y = 0$ move away from

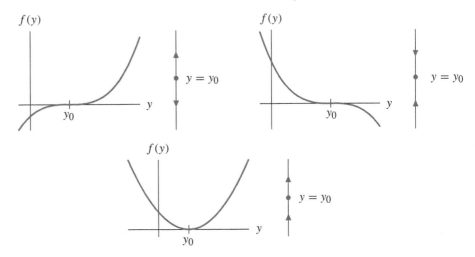

Figure 1.75
Graphs of various functions f along with the corresponding phase lines for the differential equation $dy/dt = f(y)$. In all cases, $y_0 = 0$ is an equilibrium point and $f'(y_0) = 0$.

$y = 0$ as t increases. Of course, there is the dangerous loophole clause "sufficiently close." Initial conditions might have to be very, very close to $y = 0$ for the above to apply. Again we did a little work and got a little information. To get more information, we would need to study the function $h(y)$ more carefully.

Modified Logistic Model

As an application of these ideas, we use the techniques of this section to discuss a modification of the logistic population model we introduced in Section 1.1.

The pine squirrel is a small mammal native to the Rocky Mountains. These squirrels are very territorial, so if their population is large, their rate of growth decreases and may even become negative. On the other hand, if the population is too small, fertile adults run the risk of not being able to find suitable mates, so again the rate of growth is negative.*

The model
We can restate these assumptions succinctly:

- If the population is too big, the rate of growth is negative.
- If the population is too small, the rate of growth is negative.

So the population grows only if it is between "too big" and "too small." Also, it is reasonable to assume that, if the population is zero, it will stay zero. Thus we also assume:

- If the population is zero, the growth rate is zero. (Compare these assumptions with those of the logistic population model of Section 1.1.)

---

*This phenomenon is called the Allee effect. See Allee, W. C., *Animal Aggregations: A Study in General Sociology*, University of Chicago Press, 1931.

We let

$$t = \text{time (independent variable)},$$
$$S(t) = \text{population of squirrels at time } t \text{ (dependent variable)},$$
$$k = \text{growth-rate coefficient (parameter)},$$
$$N = \text{carrying capacity (parameter), and}$$
$$M = \text{``sparsity'' constant (parameter)}.$$

The carrying capacity N indicates what population is "too big," and the sparsity parameter M indicates what population is "too small."

Now we want a model of the form $dS/dt = g(S)$ that conforms to the assumptions. We can think of the assumptions as determining the shape of the graph of $g(S)$, in particular where $g(S)$ is positive and where it is negative. Note that $dS/dt = g(S) < 0$ if $S > N$ because the population decreases if it is too big. Also $g(S) < 0$ when $S < M$ because the population decreases if it is too small. Finally, $g(S) > 0$ when $M < S < N$ and $g(0) = 0$. That is, we want $g(S)$ to have a graph shaped like Figure 1.76. The graph of g for $S < 0$ does not matter because a negative number of squirrels (anti-squirrels?) is meaningless.

The logistic model would give "correct" behavior for populations near the carrying capacity, but for small populations (below the "sparsity" level M), the solutions of the logistic model do not agree with the assumptions. Hence we will need to modify the logistic model to include the behavior of small populations and to include the parameter M. We make a model of the form

$$\frac{dS}{dt} = g(S) = kS\left(1 - \frac{S}{N}\right) \text{(something)}.$$

The "something" term must be positive if $S > M$ and negative if $S < M$. The simplest choice that satisfies these conditions is

$$\text{(something)} = \left(\frac{S}{M} - 1\right).$$

Hence our model is

$$\frac{dS}{dt} = kS\left(1 - \frac{S}{N}\right)\left(\frac{S}{M} - 1\right).$$

This is the logistic model with the extra term

$$\left(\frac{S}{M} - 1\right).$$

We call it the modified logistic population model. (Other models might also be called the modified logistic, but modified in a different way.)

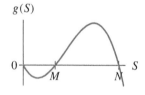

$g(S)$

0

M

N

S

Figure 1.76
Graph of $g(S)$.

Analysis of the model

To analyze solutions of this differential equation, we could use analytic techniques, since the equation is separable. However, qualitative techniques provide a lot of information about the solutions with a lot less work. The differential equation is

$$\frac{dS}{dt} = g(S) = kS\left(1 - \frac{S}{N}\right)\left(\frac{S}{M} - 1\right),$$

with $0 < M < N$ and $k > 0$. There are three equilibrium points—$S = 0$, $S = M$, and $S = N$. If $0 < S < M$, we have $g(S) < 0$, so solutions with initial conditions between 0 and M decrease. Similarly, if $S > N$, $g(S) < 0$, solutions with initial conditions larger than N also decrease. For $M < S < N$, we have $g(S) > 0$. Consequently, solutions with initial conditions between M and N increase. Thus we conclude that the equilibria at 0 and N are sinks, and the equilibrium point at M is a source. The phase line and graphs of typical solutions are shown in Figure 1.77.

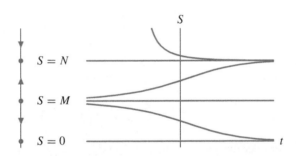

Figure 1.77
Solutions of the modified logistic equation

$$\frac{dS}{dt} = k\left(1 - \frac{S}{N}\right)\left(\frac{S}{M} - 1\right)S,$$

with various initial conditions.

EXERCISES FOR SECTION 1.6

In Exercises 1–12, sketch the phase lines for the given differential equation. Identify the equilibrium points as sinks, sources, or nodes.

1. $\dfrac{dy}{dt} = 3y(y - 2)$ **2.** $\dfrac{dy}{dt} = y^2 - 4y - 12$ **3.** $\dfrac{dy}{dt} = \cos y$

4. $\dfrac{dw}{dt} = w \cos w$ **5.** $\dfrac{dw}{dt} = (1 - w)\sin w$ **6.** $\dfrac{dy}{dt} = \dfrac{1}{y - 2}$

7. $\dfrac{dv}{dt} = -v^2 - 2v - 2$ **8.** $\dfrac{dw}{dt} = 3w^3 - 12w^2$ **9.** $\dfrac{dy}{dt} = 1 + \cos y$

10. $\dfrac{dy}{dt} = \tan y$ **11.** $\dfrac{dy}{dt} = y \ln |y|$ **12.** $\dfrac{dw}{dt} = (w^2 - 2)\arctan w$

In Exercises 13–21, a differential equation and various initial conditions are specified. Sketch the graphs of the solutions satisfying these initial conditions. For each exercise, put all your graphs on one pair of axes.

13. Equation from Exercise 1; $y(0) = 1$, $y(-2) = -1$, $y(0) = 3$, $y(0) = 2$.

14. Equation from Exercise 2; $y(0) = 1$, $y(1) = 0$, $y(0) = 6$, $y(0) = 5$.

15. Equation from Exercise 3; $y(0) = 0$, $y(-1) = 1$, $y(0) = -\pi/2$, $y(0) = \pi$.

16. Equation from Exercise 4; $w(0) = 0$, $w(3) = 1$, $w(0) = 2$, $w(0) = -1$.

17. Equation from Exercise 5; $w(0) = -3/2$, $w(0) = 1$, $w(0) = 2$, $w(0) = 3$.

18. Equation from Exercise 6; $y(0) = 0$, $y(1) = 3$, $y(0) = 2$ (trick question).

19. Equation from Exercise 7; $v(0) = 0$, $v(1) = 1$, $v(0) = 1$.

20. Equation from Exercise 8; $w(0) = -1$, $w(0) = 0$, $w(0) = 3$, $w(1) = 3$.

21. Equation from Exercise 9; $y(0) = -\pi$, $y(0) = 0$, $y(0) = \pi$, $y(0) = 2\pi$.

In Exercises 22–27, describe the long-term behavior of the solution to the differential equation

$$\frac{dy}{dt} = y^2 - 4y + 2$$

with the given initial condition.

22. $y(0) = -1$ **23.** $y(0) = 2$ **24.** $y(0) = -2$

25. $y(0) = -4$ **26.** $y(0) = 4$ **27.** $y(3) = 1$

28. Consider the autonomous equation $dy/dt = f(y)$ where $f(y)$ is continuously differentiable, and suppose we know that $f(-1) = f(2) = 0$.

    (a) Describe all the possible behaviors of the solution $y(t)$ that satisfies the initial condition $y(0) = 1$.

    (b) Suppose also that $f(y) > 0$ for $-1 < y < 2$. Describe all the possible behaviors of the solution $y(t)$ that satisfies the initial condition $y(0) = 1$.

In Exercises 29–32, the graph of a function $f(y)$ is given. Sketch the phase line for the autonomous differential equation $dy/dt = f(y)$.

29.

30.

31.

32.

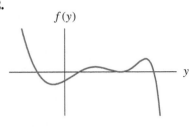

In Exercises 33–36, a phase line for an autonomous equation $dy/dt = f(y)$ is shown. Make a rough sketch of the graph of the corresponding function $f(y)$. (Assume $y = 0$ is in the middle of the segment shown in each case.)

33. **34.** **35.** **36.**

37. Eight differential equations and four phase lines are given below. Determine the equation that corresponds to each phase line and state briefly how you know your choice is correct.

(i) $\dfrac{dy}{dt} = y \cos\left(\frac{\pi}{2} y\right)$ (ii) $\dfrac{dy}{dt} = y - y^2$ (iii) $\dfrac{dy}{dt} = |y| \sin\left(\frac{\pi}{2} y\right)$ (iv) $\dfrac{dy}{dt} = y^3 - y^2$

(v) $\dfrac{dy}{dt} = \cos\left(\frac{\pi}{2} y\right)$ (vi) $\dfrac{dy}{dt} = y^2 - y$ (vii) $\dfrac{dy}{dt} = y \sin\left(\frac{\pi}{2} y\right)$ (viii) $\dfrac{dy}{dt} = y^2 - y^3$

(a) **(b)** **(c)** **(d)**

38. Let $f(y)$ be a continuous function.

  (a) Suppose that $f(-10) > 0$ and $f(10) < 0$. Show that there is an equilibrium point for $dy/dt = f(y)$ between $y = -10$ and $y = 10$.

  (b) Suppose that $f(-10) > 0$, that $f(10) < 0$, and that there are finitely many equilibrium points between $y = -10$ and $y = 10$. If $y = 1$ is a source, show that $dy/dt = f(y)$ must have at least two sinks between $y = -10$ and $y = 10$. (Can you say where they are located?)

39. Suppose you wish to model a population with a differential equation of the form $dP/dt = f(P)$, where $P(t)$ is the population at time t. Experiments have been performed on the population that give the following information:

  • The only equilibrium points in the population are $P = 0$, $P = 10$, and $P = 50$.
  • If the population is 100, the population decreases.
  • If the population is 25, the population increases.

  (a) Sketch the possible phase lines for this system for $P > 0$ (there are two).

  (b) Give a rough sketch of the corresponding functions $f(P)$ for each of your phase lines.

  (c) Give a formula for functions $f(P)$ whose graph agrees (qualitatively) with the rough sketches in part (b) for each of your phase lines.

40. Consider the Ermentrout-Kopell model for the spiking of a neuron

$$\frac{d\theta}{dt} = 1 - \cos\theta + (1 + \cos\theta)I(t)$$

introduced in Exercise 19 of Section 1.4. Let the input function $I(t)$ be the function that is constantly $-1/3$.

  (a) Determine the equilibrium points for this input.

  (b) Classify these equilibria.

41. Use PhaseLines to describe the phase line for the differential equation

$$\frac{dy}{dt} = y^2 + a$$

for various values of the parameter a.

  (a) For which values of a is the phase line qualitatively the same?

  (b) At which value(s) of a does the phase line undergo a qualitative change?

42. Use PhaseLines to describe the phase line for the differential equation

$$\frac{dy}{dt} = ay - y^3$$

for various values of the parameter a.

  (a) For which values of a is the phase line qualitatively the same?

  (b) At which value(s) of a does the phase line undergo a qualitative change?

43. Suppose $dy/dt = f(y)$ has an equilibrium point at $y = y_0$ and

(a) $f'(y_0) = 0$, $f''(y_0) = 0$, and $f'''(y_0) > 0$: Is y_0 a source, a sink, or a node?

(b) $f'(y_0) = 0$, $f''(y_0) = 0$, and $f'''(y_0) < 0$: Is y_0 a source, a sink, or a node?

(c) $f'(y_0) = 0$ and $f''(y_0) > 0$: Is y_0 a source, a sink, or a node?

44. **(a)** Sketch the phase line for the differential equation

$$\frac{dy}{dt} = \frac{1}{(y-2)(y+1)},$$

and discuss the behavior of the solution with initial condition $y(0) = 1/2$.

(b) Apply analytic techniques to the initial-value problem

$$\frac{dy}{dt} = \frac{1}{(y-2)(y+1)}, \quad y(0) = \frac{1}{2},$$

and compare your results with your discussion in part (a).

The proper scheduling of city bus and train systems is a difficult problem, which the City of Boston seems to ignore. It is not uncommon in Boston to wait a long time for the trolley, only to have several trolleys arrive simultaneously. In Exercises 45–48, we study a very simple model of the behavior of trolley cars.

Consider two trolley cars on the same track moving toward downtown Boston. Let $x(t)$ denote the amount of time between the two cars at time t. That is, if the first car arrives at a particular stop at time t, then the other car will arrive at the stop $x(t)$ time units later. We assume that the first car runs at a constant average speed (not a bad assumption for a car running before rush hour). We wish to model how $x(t)$ changes as t increases.

We first assume that, if no passengers are waiting for the second train, then it has an average speed greater than the first train and hence will catch up to the first train. Thus the time between trains $x(t)$ will decrease at a constant rate if no people are waiting for the second train. However, the speed of the second train decreases if there are passengers to pick up. We assume that the speed of the second train decreases at a rate proportional to the number of passengers it picks up and that the passengers arrive at the stops at a constant rate. Hence the number of passengers waiting for the second train is proportional to the time between trains.

45. Let $x(t)$ be the amount of time between two consecutive trolley cars as described above. We claim that a reasonable model for $x(t)$ is

$$\frac{dx}{dt} = \beta x - \alpha.$$

Which term represents the rate of decrease of the time between the trains if no people are waiting, and which term represents the effect of the people waiting for the second train? (Justify your answer.) Should the parameters α and β be positive or negative?

46. For the model in Exercise 45:

    (a) Find the equilibrium points.
    (b) Classify the equilibrium points (source, sink, or node).
    (c) Sketch the phase line.
    (d) Sketch the graphs of solutions.
    (e) Find the formula for the general solution.

47. Use the model in Exercise 45 to predict what happens to $x(t)$ as t increases. Include the effect of the initial value $x(0)$. Is it possible for the trains to run at regular intervals? Given that there are always slight variations in the number of passengers waiting at each stop, is it likely that a regular interval can be maintained? Write two brief reports (of one or two paragraphs):

    (a) The first report is addressed to other students in the class (hence you may use technical language we use in class).
    (b) The second report is addressed to the Mayor of Boston.

48. Assuming the model for $x(t)$ from Exercise 45, what happens if trolley cars leave the station at fixed intervals? Can you use the model to predict what will happen for a whole sequence of trains? Will it help to increase the number of trains so that they leave the station more frequently?

1.7 BIFURCATIONS

Equations with Parameters

In many of our models, a common feature is the presence of **parameters** along with the other variables involved. Parameters are quantities that do not depend on time (the independent variable) but that assume different values depending on the specifics of the application at hand. For instance, the exponential growth model for population

$$\frac{dP}{dt} = kP$$

contains the parameter k, the constant of proportionality for the growth rate dP/dt versus the total population P. One of the underlying assumptions of this model is that the growth rate dP/dt is a constant multiple of the total population. However, when we apply this model to different species, we expect to use different values for the constant of proportionality. For example, the value of k that we would use for rabbits would be significantly larger than the value for humans.

How the behavior of solutions changes as the parameters vary is a particularly important aspect of the study of differential equations. For some models, we must study the behavior of solutions for all parameter values in a certain range. As an example, consider a model for the motion of a bridge over time. In this case, the number of cars on the bridge may affect how the bridge reacts to wind, and a model for the motion of the bridge might contain a parameter for the total mass of the cars on the bridge. In that case, we would want to know the behavior of various solutions of the model for a variety of different values of the mass.

In many models we know only approximate values for the parameters. However, in order for the model to be useful to us, we must know the effect of slight variations in the values of the parameters on the behavior of the solutions. Also there may be effects that we have not included in our model that make the parameters vary in unexpected ways. In many complicated physical systems, the long-term effect of these intentional or unintentional adjustments in the parameters can be very dramatic.

In this section we study how solutions of a differential equation change as a parameter is varied. We study autonomous equations with one parameter. We find that a small change in the parameter usually results in only a small change in the nature of the solutions. However, occasionally a small change in the parameter can lead to a drastic change in the long-term behavior of solutions. Such a change is called a **bifurcation**. We say that a differential equation that depends on a parameter *bifurcates* if there is a qualitative change in the behavior of solutions as the parameter changes.

Notation for differential equations depending on a parameter

An example of an autonomous differential equation that depends on a parameter is

$$\frac{dy}{dt} = y^2 - 2y + \mu.$$

The parameter is μ. The independent variable is t and the dependent variable is y, as usual. Note that this equation really represents infinitely many different equations, one for each value of μ. We think of the value of μ as a constant in each equation, but different values of μ yield different differential equations, each with a different set of solutions. Because of their different roles in the differential equation, we use a notation that distinguishes the dependence of the right-hand side on y and μ. We let

$$f_\mu(y) = y^2 - 2y + \mu.$$

The parameter μ appears in the subscript, and the dependent variable y is the argument of the function f_μ. If we want to specify a particular value of μ, say $\mu = 3$, then we write

$$f_3(y) = y^2 - 2y + 3.$$

With $\mu = 3$, we obtain the corresponding differential equation

$$\frac{dy}{dt} = f_3(y) = y^2 - 2y + 3.$$

We use this notation in general. A function of the dependent variable y, which also depends on a parameter μ, is denoted by $f_\mu(y)$. The corresponding differential equation with dependent variable y and parameter μ is

$$\frac{dy}{dt} = f_\mu(y).$$

Since such a differential equation really refers to a collection of different equations, one for each value of μ, we call such an equation a **one-parameter family** of differential equations.

A One-Parameter Family with One Bifurcation

Let's consider the one-parameter family

$$\frac{dy}{dt} = f_\mu(y) = y^2 - 2y + \mu$$

more closely. For each value of μ we have an autonomous differential equation, and we can draw its phase line and analyze it using the techniques of the previous section. We begin our study of this family by studying the differential equations obtained from particular choices of μ. Since we do not yet know the most interesting values of μ, we just pick integer values, say $\mu = -4$, $\mu = -2$, $\mu = 0$, $\mu = 2$, and $\mu = 4$, for starters. (In general, μ *need not be an integer*, but we might as well begin our analysis with integer values of μ.) For each μ, we have an autonomous differential equation and its phase line. For example, for $\mu = -2$, the equation is

$$\frac{dy}{dt} = f_{-2}(y) = y^2 - 2y - 2.$$

This differential equation has equilibrium points at values of y for which

$$f_{-2}(y) = y^2 - 2y - 2 = 0.$$

The equilibrium points are $y = 1 - \sqrt{3}$ and $y = 1 + \sqrt{3}$. Between the equilibrium points, the function f_{-2} is negative, and above and below the equilibrium points, f_{-2} is positive. Hence $y = 1 - \sqrt{3}$ is a sink and $y = 1 + \sqrt{3}$ is a source. With this information we can draw the phase line. For the other values of μ we follow a similar procedure and draw the phase lines. All these phase lines are shown in Figure 1.78.

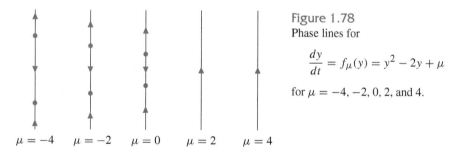

Figure 1.78
Phase lines for

$$\frac{dy}{dt} = f_\mu(y) = y^2 - 2y + \mu$$

for $\mu = -4, -2, 0, 2,$ and 4.

$\mu = -4$ $\mu = -2$ $\mu = 0$ $\mu = 2$ $\mu = 4$

Each of the phase lines is somewhat different from the others. However, the basic description of the phase lines for $\mu = -4$, $\mu = -2$, and $\mu = 0$ is the same: There are exactly two equilibrium points; the smaller one is a sink and the larger one is a source. Although the exact position of these equilibrium points changes as μ increases, their relative position and type do not change. Solutions of these equations with large initial values blow up in finite time as t increases and tend to an equilibrium point as t decreases. Solutions with very negative initial conditions tend to an equilibrium point as t increases and to $-\infty$ as t decreases. Solutions with initial values between the equilibrium points tend to the smaller equilibrium point as t increases and to the larger equilibrium point as t decreases (see Figure 1.79).

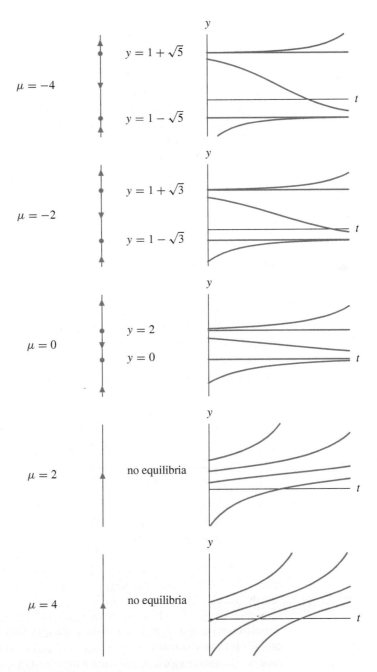

Figure 1.79
Phase lines and sketches of solutions for $dy/dt = f_\mu(y) = y^2 - 2y + \mu$ for $\mu = -4, -2, 0, 2, 4$.

Figure 1.80
Graphs of
$f_\mu(y) = y^2 - 2y + \mu$
for $\mu = -4, -2, 0, 2,$ and 4.

If $\mu = 2$ and $\mu = 4$, we have something very different. There are no equilibrium points. All solutions tend to $+\infty$ as t increases and to $-\infty$ as t decreases. Because there is a significant change in the nature of the solutions, we say that a bifurcation has occurred somewhere between $\mu = 0$ and $\mu = 2$.

To investigate the nature of this bifurcation, we draw the graphs of f_μ for the μ-values above (see Figure 1.80). For $\mu = -4, -2,$ and 0, $f_\mu(y)$ has 2 roots, but for $\mu = 2$ and 4, the graph of $f_\mu(y)$ does not cross the y-axis. Somewhere between $\mu = 0$ and $\mu = 2$ the graph of $f_\mu(y)$ must be tangent to the y-axis.

The roots of the quadratic equation

$$y^2 - 2y + \mu = 0$$

are $y = 1 \pm \sqrt{1 - \mu}$. If $\mu < 1$, this quadratic has two real roots; if $\mu = 1$, it has only one root; and if $\mu > 1$, it has no real roots. The corresponding differential equations have two equilibrium points if $\mu < 1$, one equilibrium point if $\mu = 1$, and no equilibrium points if $\mu > 1$. Hence the qualitative nature of the phase lines changes when $\mu = 1$. We say that a bifurcation occurs at $\mu = 1$ and that $\mu = 1$ is a **bifurcation value**.

The graph of $f_1(y)$ and the phase line for $dy/dt = f_1(y)$ are shown in Figures 1.81 and 1.82. The phase line has one equilibrium point (which is a node), and everywhere else solutions increase. The fact that the bifurcation occurs at the parameter value for which the equilibrium point is a node is no coincidence. In fact, this entire bifurcation scenario is quite common.

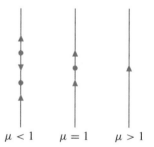

Figure 1.81
Graphs of $f_\mu(y) = y^2 - 2y + \mu$ for μ slightly less than 1, equal to 1, and slightly greater than 1.

Figure 1.82
Corresponding phase lines for $dy/dt = f_\mu(y) = y^2 - 2y + \mu$.

The Bifurcation Diagram

An extremely helpful way to understand the qualitative behavior of solutions is through the **bifurcation diagram**. This diagram is a picture (in the μy-plane) of the phase lines near a bifurcation value. It highlights the changes that the phase lines undergo as the parameter passes through this value.

To plot the bifurcation diagram, we plot the parameter values along the horizontal axis. For each μ-value (not just integers), we draw the phase line corresponding to μ

on the vertical line through μ. We think of the bifurcation diagram as a movie: As our eye scans the picture from left to right, we see the phase lines evolve through the bifurcation. Figure 1.83 shows the bifurcation diagram for $f_\mu(y) = y^2 - 2y + \mu$.

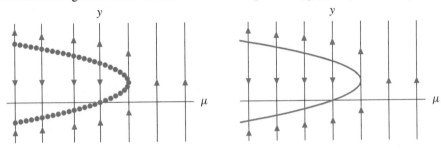

Figure 1.83
Bifurcation diagram for the differential equation $dy/dt = f_\mu(y) = y^2 - 2y + \mu$. The horizontal axis is the μ-value and the vertical lines are the phase lines for the differential equations with the corresponding μ-values.

A bifurcation from one to three equilibria

Let's look now at another one-parameter family of differential equations

$$\frac{dy}{dt} = g_\alpha(y) = y^3 - \alpha y = y(y^2 - \alpha).$$

In this equation, α is the parameter. There are three equilibria if $\alpha > 0$ ($y = 0, \pm\sqrt{\alpha}$), but there is only one equilibrium point ($y = 0$) if $\alpha \leq 0$. Therefore a bifurcation occurs when $\alpha = 0$. To understand this bifurcation, we plot the bifurcation diagram.

First, if $\alpha < 0$, the term $y^2 - \alpha$ is always positive. Thus $g_\alpha(y) = y(y^2 - \alpha)$ has the same sign as y. Solutions tend to ∞ if $y(0) > 0$ and to $-\infty$ if $y(0) < 0$. If $\alpha > 0$, the situation is different. The graph of $g_\alpha(y)$ shows that $g_\alpha(y) > 0$ in the intervals $\sqrt{\alpha} < y < \infty$ and $-\sqrt{\alpha} < y < 0$ (see Figure 1.84). Thus solutions increase in these intervals. In the other intervals, $g_\alpha(y) < 0$, so solutions decrease. The bifurcation diagram is depicted in Figure 1.85.

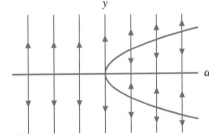

Figure 1.84
Graphs of $g_\alpha(y)$ for $\alpha > 0$, $\alpha = 0$, and $\alpha < 0$. Note that for $\alpha \leq 0$ the graph crosses the y-axis once, whereas if $\alpha > 0$, the graph crosses the y-axis three times.

Figure 1.85
Bifurcation diagram for the one-parameter family $dy/dt = g_\alpha(y) = y^3 - \alpha y$.

Bifurcations of Equilibrium Points

Throughout the rest of this section, we assume that all the one-parameter families of differential equations that we consider depend smoothly on the parameter. That is, for the one-parameter family

$$\frac{dy}{dt} = f_\mu(y),$$

the partial derivatives of $f_\mu(y)$ with respect to y and μ exist and are continuous. So changing μ a little changes the graph of $f_\mu(y)$ only slightly.

When bifurcations do not happen

The most important fact about bifurcations is that they usually do not happen. A small change in the parameter usually leads to only a small change in the behavior of solutions. This is very reassuring. For example, suppose we have a one-parameter family

$$\frac{dy}{dt} = f_\mu(y),$$

and the differential equation for $\mu = \mu_0$ has an equilibrium point at $y = y_0$. Also suppose that $f'_{\mu_0}(y_0) < 0$, so the equilibrium point is a sink. We sketch the phase line and the graph of $f_{\mu_0}(y)$ near $y = y_0$ in Figure 1.86.

Now if we change μ just a little bit, say from μ_0 to μ_1, then the graph of $f_{\mu_1}(y)$ is very close to the graph of $f_{\mu_0}(y)$ (see Figure 1.87). So the graph of $f_{\mu_1}(y)$ is strictly decreasing near y_0, passing through the horizontal axis near $y = y_0$. The corresponding differential equation

$$\frac{dy}{dt} = f_{\mu_1}(y)$$

has a sink at some point $y = y_1$ very near y_0.

We can make this more precise: If y_0 is a sink for a differential equation

$$\frac{dy}{dt} = f_{\mu_0}(y)$$

Figure 1.86
Graph of $f_{\mu_0}(y)$ near the sink y_0 and the phase line for the differential equation $dy/dt = f_{\mu_0}(y)$ near y_0.

Figure 1.87
Graphs of $f_{\mu_0}(y)$ and $f_{\mu_1}(y)$ for μ_1 close to μ_0. Note that $f_{\mu_1}(y)$ decreases across the y-axis at $y = y_1$ near y_0, so $dy/dt = f_{\mu_1}(y)$ has a sink at $y = y_1$.

with $f'_{\mu_0}(y_0) < 0$, then for all μ_1 sufficiently close to μ_0, the differential equation

$$\frac{dy}{dt} = f_{\mu_1}(y)$$

has a sink at a point $y = y_1$ very near y_0 (and no other equilibrium points near y_0). A similar statement holds if y_0 is a source and $f'_{\mu_0}(y_0) > 0$. These are the situations in which we can say for sure that no bifurcation occurs, at least not near y_0.

With these observations in mind, we see that bifurcations occur only if the above conditions do not hold. Consequently, given a one-parameter family of differential equations

$$\frac{dy}{dt} = f_\mu(y),$$

we look for values $\mu = \mu_0$ and $y = y_0$ for which $f_{\mu_0}(y_0) = 0$ and $f'_{\mu_0}(y_0) = 0$.

Determining bifurcation values

Consider the one-parameter family of differential equations given by

$$\frac{dy}{dt} = f_\mu(y) = y(1 - y)^2 + \mu.$$

If $\mu = 0$, the equilibrium points are $y = 0$ and $y = 1$. Also $f'_0(0) = 1$. Hence $y = 0$ is a source for the differential equation $dy/dt = f_0(y)$. Thus for all μ sufficiently close to zero, the differential equation $dy/dt = f_\mu(y)$ has a source near $y = 0$.

On the other hand, for the equilibrium point $y = 1$, $f'_0(1) = 0$. The Linearization Theorem from Section 1.6 says nothing about what happens in this case. To see what is going on, we sketch the graph of $f_\mu(y)$ for several μ-values near $\mu = 0$ (see Figure 1.88). If $\mu = 0$, the graph of f_μ is tangent to the horizontal axis at $y = 1$. Since $f_0(y) > 0$ for all $y > 0$ except $y = 1$, it follows that the equilibrium point at $y = 1$ is a node for this parameter value. Changing μ moves the graph of $f_\mu(y)$ up (if μ is positive) or down (if μ is negative). If we make μ slightly positive, $f_\mu(y)$ does not touch the horizontal axis near $y = 1$. So the equilibrium point at $y = 1$ for $\mu = 0$ disappears. A bifurcation occurs at $\mu = 0$. For μ slightly negative, the corresponding differential equation has two equilibrium points near $y = 1$. Since f_μ is decreasing at one of these equilibria and increasing at the other, one of these equilibria is a source and the other is a sink.

$f_\mu(y)$

y

1

Figure 1.88
Graphs of

$$f_\mu(y) = y(1 - y)^2 + \mu$$

for μ slightly greater than zero, μ equal to zero, and μ slightly less than zero.

There is a second bifurcation in this one-parameter family. To see this, note what happens as μ decreases. There is a value of μ for which the graph of $f_\mu(y)$ again has a tangency with the horizontal axis (see Figure 1.89). For larger μ-values, the graph crosses the horizontal axis three times, but for lower μ-values, the graph crosses only once. Thus a second bifurcation occurs at this μ-value.

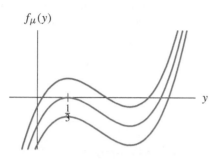

$f_\mu(y)$

y

Figure 1.89
Graphs of

$$f_\mu(y) = y(1-y)^2 + \mu$$

for μ slightly greater than $-4/27$, for μ equal to $-4/27$, and for μ slightly less than $-4/27$.

To find this bifurcation value exactly, we must find the μ-values for which the graph of f_μ is tangent to the horizontal axis. That is, we must find the μ-values for which, at some equilibrium point y, we have $f'_\mu(y) = 0$. Since

$$f'_\mu(y) = (1-y)^2 - 2y(1-y) = (1-y)(1-3y),$$

it follows that the graph of $f_\mu(y)$ is horizontal at the two points $y = 1$ and $y = 1/3$. We know that the graph of $f_0(y)$ is tangent to the horizontal axis $y = 1$, so let's look at $y = 1/3$. We have $f_\mu(1/3) = \mu + 4/27$, so the graph is also tangent to the horizontal axis if $\mu = -4/27$. This is our second bifurcation value. Using analogous arguments to those above, we find that f_μ has three equilibria for $-4/27 < \mu < 0$ and only one equilibrium point when $\mu < -4/27$. The bifurcation diagram summarizes all this information in one picture (see Figure 1.90).

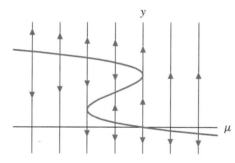

y

μ

Figure 1.90
Bifurcation diagram for

$$\frac{dy}{dt} = f_\mu(y) = y(1-y)^2 + \mu.$$

Note the two bifurcation values of μ, $\mu = -4/27$ and $\mu = 0$.

Sustainability

When harvesting a natural resource, it is important to control the amount harvested so that the resource is not completely depleted. To accomplish this, we must study the particular species involved and pay close attention to the possible changes that may occur if the harvesting level is increased.

Suppose we model the population $P(t)$ of a particular species of fish with a logistic model

$$\frac{dP}{dt} = kP\left(1 - \frac{P}{N}\right),$$

where k is the growth-rate parameter and N is the carrying capacity of the habitat. Suppose that fishing removes a certain constant number C (for catch) of fish per season from the population. Then a modification of the model that takes fishing into account is

$$\frac{dP}{dt} = kP\left(1 - \frac{P}{N}\right) - C.$$

How does the population of fish vary as C is increased?

This model has three parameters, k, N, and C; but we are concerned only with what happens if C is varied. Therefore we think of k and N as fixed constants determined by the type of fish and their habitat. Our predictions involve the values of k and N. For example, if $C = 0$, we know from Section 1.1 that all positive initial conditions yield solutions that tend toward the equilibrium point $P = N$. So if fishing is prohibited, we expect the population to be close to $P = N$.

Let

$$f_C(P) = kP\left(1 - \frac{P}{N}\right) - C.$$

As C increases, the graph of $f_C(P)$ slides down (see Figure 1.91). The points where $f_C(P)$ crosses the P-axis tend toward each other. In other words, the equilibrium points for the corresponding differential equations slide together.

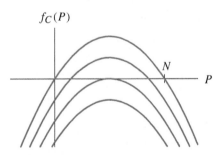

$f_C(P)$

N

P

Figure 1.91
Graphs of

$$f_C(P) = kP\left(1 - \frac{P}{N}\right) - C$$

for several values of C. Note that, as C increases, the graph of $f_C(P)$ slides down the vertical axis.

We can compute the equilibrium points by solving $f_C(P) = 0$. We have

$$kP\left(1 - \frac{P}{N}\right) - C = 0,$$

which yields

$$-kP^2 + kNP - CN = 0.$$

This quadratic equation has solutions

$$P = \frac{N}{2} \pm \sqrt{\frac{N^2}{4} - \frac{CN}{k}}.$$

As long as the term under the square root (the discriminant of the quadratic) is positive, the function crosses the horizontal axis twice and the corresponding differential equation has two equilibrium points—a source and a sink. Thus, for small values of C, the phase line has two equilibrium points (see Figure 1.91).

If

$$\frac{N^2}{4} - \frac{CN}{k} < 0,$$

then the graph of $f_C(P)$ does not cross the P-axis and the corresponding differential equation has no equilibrium points. Thus, if

$$\frac{N^2}{4} < \frac{CN}{k}$$

or equivalently if

$$C > \frac{kN}{4},$$

then there are no equilibria. For these values of C, the function $f_C(P)$ is negative for all values of P and the solutions of the corresponding differential equation tend toward $-\infty$. Since negative populations do not make any sense, we say that the species has become extinct when the population reaches zero.

With this information, we can sketch the bifurcation diagram for this system (see Figure 1.92). A bifurcation occurs as we increase C. The bifurcation value for the parameter C is $kN/4$ because, at this value, the graph of $f_C(P)$ is tangent to the P-axis. The corresponding differential equation has a node at $P = N/2$. If C is slightly less than $kN/4$, the corresponding differential equation has two equilibrium points, a source and a sink, near $P = N/2$. If C is slightly greater than $kN/4$, the corresponding differential equation has no equilibrium points (see Figure 1.92).

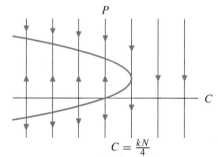

$$C = \frac{kN}{4}$$

Figure 1.92
Bifurcation diagram for

$$\frac{dP}{dt} = f_C(P) = kP\left(1 - \frac{P}{N}\right) - C.$$

Note that if $C < kN/4$, the phase line has two equilibrium points, whereas if $C > kN/4$, the phase line has no equilibrium points and all solutions decrease.

It is interesting to consider what happens to the fish population as the parameter C is slowly increased. If $C = 0$, the population tends to the sink at $P = N$. Then, if there is a relatively small amount of fishing, the fish population is close to $P = N$. That is, if C is slightly positive, the sink for $C = 0$ at $P = N$ moves to the slightly smaller value

$$P = \frac{N}{2} + \sqrt{\frac{N^2}{4} - \frac{CN}{k}}.$$

For somewhat larger values of C, the value of the sink continues to decrease, and the fish population adjusts to stay close to this sink. We observe a gradual decrease in the fish population. When C is close to $kN/4$, the fish population is close to the sink for the corresponding differential equation, which is close to $P = N/2$. If C increases just a little more so that $C > kN/4$, then the corresponding differential equation has no equilibrium points and all solutions decrease. If C is slightly larger than $kN/4$, $f_C(P)$ is slightly negative near $P = N/2$, so the population decreases slowly at first. As P decreases, $f_C(P)$ becomes more negative and the rate of decrease of P accelerates. The population reaches zero in a finite amount of time, and the fish species becomes extinct.

So as the number of fish removed by fishing increases gradually, we initially expect a gradual decline in the fish population. This decline continues until the fishing parameter C reaches the bifurcation value $C = kN/4$. At this point, if we allow even slightly more fishing, the fish population decreases slowly at first and then collapses, and the fish become extinct in the area. This is a pretty frightening scenario. The fact that a little fishing causes only a small population decline over the long term does not necessarily imply that a little more fishing causes only a little more population decline. Once the bifurcation value is passed, the fish population tends to zero.

This model is a very simple one, and as such it should not be taken too seriously. The lesson to be learned is that, if this sort of behavior can be observed in simple models, we would expect that the same (and even more surprising behavior) occurs in more complicated models and in the actual populations. To properly manage resources, we need to have accurate models and to be aware of possible bifurcations.

Mary Lou Zeeman (1961–) grew up in England, learning about bifurcations and catastrophe theory from her father, Sir Christopher Zeeman (1925–). She has applied methods of dynamical systems to population interactions, disease dynamics, neuroscience, cell networks, and hormone surges in the menstrual cycle. She enjoys collaborating with scientists: interweaving experiment and data collection with mathematical modeling.

Zeeman is also involved in several interdisciplinary initiatives focused on the health of the planet. In 2008, she helped found the Institute for Computational Sustainability based at Cornell University. In 2010, she and her colleagues founded the NSF-funded Mathematics and Climate Research Network, to identify and attack mathematical challenges underlying climate modeling. She has taught at MIT and the University of Texas at San Antonio. She is currently Wells Johnson Professor of Mathematics at Bowdoin College.

EXERCISES FOR SECTION 1.7

In Exercises 1–6, locate the bifurcation values for the one-parameter family and draw the phase lines for values of the parameter slightly smaller than, slightly larger than, and at the bifurcation values.

1. $\dfrac{dy}{dt} = y^2 + a$

2. $\dfrac{dy}{dt} = y^2 + 3y + a$

3. $\dfrac{dy}{dt} = y^2 - ay + 1$

4. $\dfrac{dy}{dt} = y^3 + \alpha y^2$

5. $\dfrac{dy}{dt} = (y^2 - \alpha)(y^2 - 4)$

6. $\dfrac{dy}{dt} = \alpha - |y|$

In Exercises 7–10, locate the bifurcation values of α for the one-parameter family and describe the bifurcation that takes place at each such value.

7. $\dfrac{dy}{dt} = y^4 + \alpha y^2$

8. $\dfrac{dy}{dt} = y^6 - 2y^3 + \alpha$

9. $\dfrac{dy}{dt} = \sin y + \alpha$

10. $\dfrac{dy}{dt} = e^{-y^2} + \alpha$

11. The graph to the right is the graph of a function $f(y)$. Describe the bifurcations that occur in the one-parameter family

$$\frac{dy}{dt} = f(y) + \alpha.$$

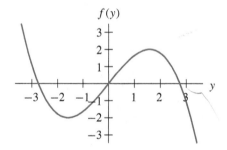

12. The graph to the right is the graph of a function $g(y)$. Describe the bifurcations that occur in the one-parameter family

$$\frac{dy}{dt} = g(y) + \alpha y.$$

[*Hint*: Note that the equilibria of this function occur at values of y where $g(y) = -\alpha y$.]

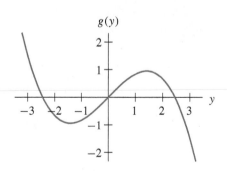

13. Six one-parameter families of differential equations depending on the parameter A and four bifurcation diagrams are given below. Determine the one-parameter family that corresponds to each bifurcation diagram, and state briefly how you know your choice is correct.

(i) $\dfrac{dy}{dt} = Ay - y^2$ (ii) $\dfrac{dy}{dt} = A + y^2$ (iii) $\dfrac{dy}{dt} = Ay - y^3$

(iv) $\dfrac{dy}{dt} = A - y^2$ (v) $\dfrac{dy}{dt} = y^2 - A$ (vi) $\dfrac{dy}{dt} = Ay + y^2$

(a)

(b)

(c)

(d)

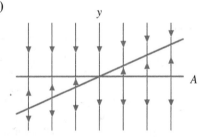

14. Consider the Ermentrout-Kopell model for the spiking of a neuron

$$\frac{d\theta}{dt} = 1 - \cos\theta + (1 + \cos\theta)I(t)$$

introduced in Exercise 19 of Section 1.3. Suppose that the input function $I(t)$ is a constant function, that is, $I(t) = I$ where I is a constant. Describe the bifurcations that occur as the parameter I varies.

15. Sketch the graph of a function $f(y)$ such that the one-parameter family of differential equations $dy/dt = f(y) + \alpha$ satisfies all of the following properties:

- For all $\alpha \le -3$, the differential equation has exactly two equilibria.
- For all $\alpha \ge 3$, the equation has no equilibria.
- For $\alpha = 0$, the equation has exactly four equilibria.

[There are many possible functions $f(y)$ that satisfy these conditions. Sketch just one graph.]

16. Sketch the graph of a function $g(y)$ such that the one-parameter family of differential equations $dy/dt = g(y) + \alpha$ satisfies all of the following properties:

- For all $\alpha \leq -4$, the differential equation has one sink and no other equilibria.
- For all $\alpha \geq 4$, the equation has one sink and no other equilibria.
- For $\alpha = 0$, the differential equation has exactly six equilibria.

[There are many possible functions $g(y)$ that satisfy these conditions. Sketch just one graph.]

17. Is it possible to find a continuous function $f(y)$ such that the one-parameter family of differential equations $dy/dt = f(y) + \alpha$ satisfies both of the following statements?

- For $\alpha = 0$, the differential equation has exactly one equilibrium point and that equilibrium is a sink.
- For $\alpha = 1$, the equation has exactly one equilibrium point and that equilibrium is a source.

If so, sketch the graph of one such $f(y)$. If not, why not?

18. Consider an exponential growth model with harvesting

$$\frac{dP}{dt} = kP - C,$$

where P is the population, $k > 0$ is the growth-rate parameter, and $C \geq 0$ is the harvest rate.

(a) Does a bifurcation occur as the parameter C varies?

(b) Describe the long-term behavior of the population $P(t)$ if $P(0) > 0$.

19. Consider the population model

$$\frac{dP}{dt} = 2P - \frac{P^2}{50}$$

for a species of fish in a lake. Suppose it is decided that fishing will be allowed, but it is unclear how many fishing licenses should be issued. Suppose the average catch of a fisherman with a license is 3 fish per year (these are hard fish to catch).

(a) What is the largest number of licenses that can be issued if the fish are to have a chance to survive in the lake?

(b) Suppose the number of fishing licenses in part (a) is issued. What will happen to the fish population—that is, how does the behavior of the population depend on the initial population?

(c) The simple population model above can be thought of as a model of an ideal fish population that is not subject to many of the environmental problems of an actual lake. For the actual fish population, there will be occasional changes in the population that were not considered when this model was constructed. For example, if the water level increases due to a heavy rainstorm, a few extra fish might be able to swim down a usually dry stream bed to reach the lake, or the extra water might wash toxic waste into the lake, killing a few fish. Given the possibility of unexpected perturbations of the population not included in the model, what do you think will happen to the actual fish population if we allow fishing at the level determined in part (b)?

20. Consider our model

$$\frac{dS}{dt} = f(S) = kS\left(1 - \frac{S}{N}\right)\left(\frac{S}{M} - 1\right)$$

of a fox squirrel population from the previous section. Suppose that the parameters M and k remain relatively constant over the long term but as more people move into the area, the parameter N (the carrying capacity) decreases.

(a) Assuming that $M \leq N$, sketch the graph of the function $f(S)$ for fixed values of k and M and several values of N.

(b) At what value of N does a bifurcation occur?

(c) How does the population of fox squirrels behave if the parameter N slowly and continuously decreases toward the bifurcation value?

21. For the differential equation that models fish populations with harvesting,

$$\frac{dP}{dt} = f_C(P) = kP\left(1 - \frac{P}{N}\right) - C,$$

we saw that if $C > kN/4$ the fish population will become extinct. If the fish population falls to near zero because the fishing level C is slightly greater than $kN/4$, why must fishing be banned completely in order for the population to recover? That is, if a level of fishing just above $C = kN/4$ causes a collapse of the population, why can't the population be restored by reducing the fishing level to just below $C = kN/4$?

22. (a) Use PhaseLines to investigate the bifurcation diagram for the differential equation

$$\frac{dy}{dt} = ay - y^2,$$

where a is a parameter. Describe the different types of phase lines that occur.

(b) What are the bifurcation values for the one-parameter family in part (a)?

(c) Use PhaseLines to investigate the bifurcation diagram for the differential equation

$$\frac{dy}{dt} = r + ay - y^2,$$

where r is a positive parameter. How does the bifurcation diagram change from the $r = 0$ case (see part (a))?

(d) Suppose r is negative in the equation in part (c). How does the bifurcation diagram change?

23. **(a)** Use PhaseLines to investigate the bifurcation diagram for the differential equation

$$\frac{dy}{dt} = ay - y^3,$$

where a is a parameter. Describe the different types of phase lines that occur.

(b) What are the bifurcation values for the one-parameter family in part (a)?

(c) Use PhaseLines to investigate the bifurcation diagram for the differential equation

$$\frac{dy}{dt} = r + ay - y^3,$$

where r is a positive parameter. How does the bifurcation diagram change from the $r = 0$ case (see part (a))?

(d) Suppose r is negative in the equation in part (c). How does the bifurcation diagram change?

1.8 LINEAR EQUATIONS

In Section 1.2 we developed an analytic method for finding explicit solutions to separable differential equations. Although many interesting problems lead to separable equations, most differential equations are not separable. The qualitative and numerical techniques we developed in Sections 1.3–1.6 apply to a much wider range of problems. It would be nice if we could also extend our analytic methods by developing ways to find explicit solutions of equations that are not separable.

Unfortunately, there is no general technique for computing explicit solutions that works for every differential equation. Although we know from the Existence Theorem that every reasonable differential equation has solutions, we have no guarantee that these solutions are made up of familiar functions such as polynomials, sines, cosines, and exponentials. In fact, they usually are not. Over the centuries, mathematicians have dealt with this dilemma by developing numerous specialized techniques for various types of differential equations. Today these techniques are available to us as one-line commands in sophisticated computer packages such as Maple and *Mathematica*. Nevertheless, you should be familiar with a few of the standard analytic techniques that apply to the most commonly encountered types of equations. In this section and Section 1.9, we develop two of the standard techniques for solving the most important type of differential equation—the *linear* differential equation.

Linear Differential Equations

A first-order differential equation is **linear** if it can be written in the form

$$\frac{dy}{dt} = a(t)y + b(t),$$

where $a(t)$ and $b(t)$ are arbitrary functions of t. Examples of linear equations include

$$\frac{dy}{dt} = t^2y + \cos t,$$

where $a(t) = t^2$ and $b(t) = \cos t$, and

$$\frac{dy}{dt} = \frac{e^{4\sin t}}{t^3 + 7t}y + 23t^3 - 7t^2 + 3,$$

where $a(t) = e^{4\sin t}/(t^3 + 7t)$ and $b(t) = 23t^3 - 7t^2 + 3$.

Sometimes it is necessary to do a little algebra in order to see that an equation is linear. For example, the differential equation

$$\frac{dy}{dt} - 3y = ty + 2$$

can be rewritten as

$$\frac{dy}{dt} = (t + 3)y + 2.$$

In this form we see that the equation is linear with $a(t) = t + 3$ and $b(t) = 2$.

Some differential equations fit into several categories. For example, the equation

$$\frac{dy}{dt} = 2y + 8$$

is linear with $a(t) = 2$ and $b(t) = 8$. (Both $a(t)$ and $b(t)$ are constant functions of t.) It is also autonomous and consequently separable.

The term *linear* refers to the fact that the dependent variable y appears in the equation only to the first power. The differential equation

$$\frac{dy}{dt} = y^2$$

is not linear because y^2 cannot be rewritten in the form $a(t)y + b(t)$, no matter how $a(t)$ and $b(t)$ are chosen.

Of course, there is nothing magical about the names of the variables. The equation

$$\frac{dP}{dt} = e^{2t}P - \sin t$$

is linear with $a(t) = e^{2t}$ and $b(t) = -\sin t$. Also,

$$\frac{dw}{dt} = (\sin t)w$$

is both linear ($a(t) = \sin t$ and $b(t) = 0$) and separable. However,

$$\frac{dz}{dt} = t \sin z$$

is not linear but is separable.

Additional terminology for linear equations

Linear differential equations come in two flavors. If $b(t) = 0$ for all t, then the equation is said to be **homogeneous** or *unforced*. Otherwise it is **nonhomogeneous** or *forced*. For example,

$$\frac{dy}{dt} = (\sin 2t)y$$

is homogeneous, and

$$\frac{dy}{dt} = y + \sin 2t$$

is nonhomogeneous.

A first-order linear differential equation is a *constant-coefficient* equation if $a(t)$ is a constant. In other words, the linear equation is a constant-coefficient equation if it has the form

$$\frac{dy}{dt} = \lambda y + b(t),$$

where λ is a constant.

Linearity Principles

Linear differential equations are important for many reasons. They are used to model a wide range of phenomena such as the decay of radioactive elements, the cooling of a cup of coffee, and the mixing of chemicals in a solution. In fact, when we start the modeling process, we almost always try a linear model first. Not only do we want to keep the model as simple as possible, but we also want to exploit the fact that the solutions to a linear equation are all related in a simple way. Given one or two nontrivial solutions, we get the rest by using the appropriate linearity principle.

The homogeneous case

There are two linearity principles, one for homogeneous equations and a different one for nonhomogeneous equations. We begin with the homogeneous case.

LINEARITY PRINCIPLE   If $y_h(t)$ is a solution of the homogeneous linear equation

$$\frac{dy}{dt} = a(t)y,$$

then any constant multiple of $y_h(t)$ is also a solution. That is, $ky_h(t)$ is a solution for any constant k.   ■

We verify this theorem simply by checking that $ky_h(t)$ satisfies the differential equation. In other words, if $y_h(t)$ is a solution, then

$$\frac{dy_h}{dt} = a(t)y_h$$

for all t. If k is a constant, then

$$\frac{d(ky_h)}{dt} = k\frac{dy_h}{dt}$$

$$= ka(t)y_h$$

$$= a(t)(ky_h).$$

We conclude that $ky_h(t)$ is also a solution to $dy/dt = a(t)y$.

This theorem is not very surprising. A homogeneous linear equation

$$\frac{dy}{dt} = a(t)y$$

is separable. Separating variables yields

$$\int \frac{1}{y}\, dy = \int a(t)\, dt,$$

and if we integrate the left-hand side, we get $\ln|y| + c = \int a(t)\, dt$, where c is a constant of integration. Exponentiating both sides, removing the absolute value sign, and rewriting the constant produces

$$y(t) = ke^{\int a(t)\, dt},$$

where k is an arbitrary constant. In this form, we can see that the nonzero solutions are constant multiples of each other. (Note that the equilibrium solution $y(t) = 0$ for all t is a solution to *every* homogeneous equation.)

For example, consider the homogeneous equation

$$\frac{dy}{dt} = (\cos t)y.$$

All solutions are constant multiples of

$$y(t) = e^{\int \cos t\, dt} = e^{\sin t}.$$

In other words, the general solution of this equation is $y(t) = ke^{\sin t}$, where k is an arbitrary constant (see Figure 1.93).

Figure 1.93

The slope field and graphs of various solutions to

$$\frac{dy}{dt} = (\cos t)y.$$

Note that the solutions are constant multiples of one another.

Remember that the Linearity Principle applies only to homogeneous linear equations. For example, it is easy to check that $y_1(t) = 1/(1 - t)$ is a solution of the nonlinear equation

$$\frac{dy}{dt} = y^2$$

and that $y_2(t) = 2y_1(t) = 2/(1 - t)$ is *not* a solution (see Exercise 17).

The nonhomogeneous case

Although the Linearity Principle does not hold for a nonhomogeneous linear equation (see Exercises 18 and 34), there is a nice relationship between its solutions and the solutions to its *associated homogeneous equation.*

EXTENDED LINEARITY PRINCIPLE   Consider the nonhomogeneous equation

$$\frac{dy}{dt} = a(t)y + b(t)$$

and its associated homogeneous equation

$$\frac{dy}{dt} = a(t)y.$$

1. If $y_h(t)$ is *any* solution of the homogeneous equation and $y_p(t)$ is *any* solution of the nonhomogeneous equation ("p" stands for particular), then $y_h(t) + y_p(t)$ is also a solution of the nonhomogeneous equation.

2. Suppose $y_p(t)$ and $y_q(t)$ are two solutions of the nonhomogeneous equation. Then $y_p(t) - y_q(t)$ is a solution of the associated homogeneous equation.

Therefore, if $y_h(t)$ is nonzero, $ky_h(t) + y_p(t)$ is the general solution of the nonhomogeneous equation.   ■

If $ky_h(t)$ is the general solution of the homogeneous equation, then the first half of the Extended Linearity Principle says that

$$ky_h(t) + y_p(t)$$

is a solution of the nonhomogeneous equation for *any* value of the constant k. The second half of the Extended Linearity Principle says that *any* solution $y_q(t)$ of the nonhomogeneous equation can be written as

$$ky_h(t) + y_p(t)$$

for some value of k. Therefore, $ky_h(t) + y_p(t)$ is the general solution of the nonhomogeneous equation. We often summarize this observation by saying that

"The general solution of the nonhomogeneous equation is the sum of the general solution of the homogeneous equation and one solution of the nonhomogeneous equation."

For example, consider the nonhomogeneous equation

$$\frac{dy}{dt} = (\cos t)y + \tfrac{1}{5}(1 - t\cos t).$$

We have already seen that the general solution to its associated homogeneous equation $dy/dt = (\cos t)y$ is $y(t) = ke^{\sin t}$, where k is an arbitrary constant. It is also easy to verify that $y_p(t) = t/5$ is a solution to the nonhomogeneous equation (see Exercise 32). Once we have the particular solution $y_p(t) = t/5$, the Extended Linearity Principle tells us that the general solution of the nonhomogeneous equation is

$$y(t) = \frac{t}{5} + ke^{\sin t},$$

where k is an arbitrary constant (see Figure 1.94).

We can verify the Extended Linearity Principle by substituting the functions into the differential equation just as we did when we verified the Linearity Principle earlier in this section (see Exercise 33).

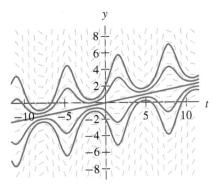

Figure 1.94
The slope field and graphs of various solutions to

$$\frac{dy}{dt} = (\cos t)y + \tfrac{1}{5}(1 - t\cos t).$$

We obtain these graphs by taking the graphs in Figure 1.93 and adding them to the graph of $y = t/5$.

Solving Linear Equations

We now have a three-step procedure for solving linear equations. First, we find the general solution of the homogeneous equation, separating variables if necessary. Then we find one "particular" solution of the nonhomogeneous equation. Finally, we obtain the general solution of the nonhomogeneous equation by adding the general solution of the homogeneous equation to the particular solution of the nonhomogeneous equation.

In theory, we could solve any linear differential equation using this procedure. In practice, however, this technique is used only for special linear equations such as constant-coefficient equations. The limitation is caused by the fact that the second step requires that we produce a particular solution of the nonhomogeneous equation. If $a(t)$ is not a constant, this step can be quite difficult. If $a(t)$ is a constant, then we can sometimes succeed using a time-honored mathematical technique. We guess.

The lucky guess

For example, consider the nonhomogeneous linear equation

$$\frac{dy}{dt} = -2y + e^t.$$

The associated homogeneous equation is $dy/dt = -2y$, and its general solution is $y(t) = ke^{-2t}$. (You could solve this homogeneous equation by separating variables, but its general solution should be second nature by now. See page 6.)

The hardest part of guessing a solution to the nonhomogeneous equation is deciding what to guess, and this task is made easier if we rewrite the equation so that all terms that involve y are on the left-hand side. In other words, we rewrite the equation in question as

$$\frac{dy}{dt} + 2y = e^t.$$

Now we need to guess a function $y_p(t)$ such that, if we insert $y_p(t)$ into the left-hand side of the equation, out pops e^t on the right-hand side. We probably should not guess sines or cosines for $y_p(t)$ because the left-hand side would still involve trigonometric functions after the computation. Similarly, polynomials would not work. What we need to guess is an exponential function. Guessing $y_p(t) = e^t$ seems to be a natural choice because its derivative is also e^t. Unfortunately, when we compute

$$\frac{dy_p}{dt} + 2y_p,$$

we get $e^t + 2e^t$, which does not equal e^t. Close, but no cigar.

This guess $y_p(t) = e^t$ almost worked. We were only off by the constant factor of 3. Perhaps we should guess a constant multiple of e^t, and in fact, perhaps we should let the differential equation tell us what the constant should be. In other words, we should replace the guess $y_p(t) = e^t$ with the guess $y_p(t) = \alpha e^t$, where α is a constant to be named later. This method is called the Method of the Undetermined Coefficient: We must determine the coefficient α so that $y_p(t) = \alpha e^t$ is a solution of the nonhomogeneous equation.

Starting with this more flexible guess of $y_p(t) = \alpha e^t$, we check to see if it works. We substitute $y_p(t)$ into $dy/dt + 2y$ and obtain

$$\frac{dy_p}{dt} + 2y_p = \alpha e^t + 2\alpha e^t$$
$$= 3\alpha e^t.$$

In order for $y_p(t)$ to be a solution, $3\alpha e^t$ must equal e^t. That is, $3\alpha = 1$, which implies $\alpha = 1/3$. Therefore, the guess of $y_p(t) = e^t/3$ is a solution, and the general solution of $dy/dt = -2y + e^t$ is

$$y(t) = ke^{-2t} + \tfrac{1}{3}e^t,$$

where k is an arbitrary constant.

Another lucky guess

In the previous example, we guessed $y_p(t) = \alpha e^t$ because the equation was

$$\frac{dy}{dt} + 2y = b(t),$$

where $b(t)$ was an exponential involving e^t. Now let's consider a nonhomogeneous equation where $b(t)$ is a trigonometric function. For example,

$$\frac{dy}{dt} + 2y = \cos 3t.$$

Then the general solution of the homogeneous equation is still $y(t) = ke^{-2t}$. However, guessing an exponential will not work for this equation. This time we try

$$y_p(t) = \alpha \cos 3t + \beta \sin 3t.$$

Note that the simpler guesses of $y_p(t) = \alpha \cos 3t$ and $y_p(t) = \alpha \sin 3t$ are destined to fail because we end up with both sines and cosines when we compute $dy/dt + 2y$ (see Exercise 13).

To determine α and β, we substitute $y_p(t)$ into $dy/dt + 2y$ and obtain

$$\frac{dy_p}{dt} + 2y_p = \frac{d(\alpha \cos 3t + \beta \sin 3t)}{dt} + 2(\alpha \cos 3t + \beta \sin 3t)$$

$$= -3\alpha \sin 3t + 3\beta \cos 3t + 2\alpha \cos 3t + 2\beta \sin 3t$$

$$= (-3\alpha + 2\beta) \sin 3t + (2\alpha + 3\beta) \cos 3t.$$

In order for $y_p(t)$ to be a solution, we must find α and β so that

$$(-3\alpha + 2\beta) \sin 3t + (2\alpha + 3\beta) \cos 3t = \cos 3t$$

for all t. To accomplish this, we solve the simultaneous algebraic equations

$$\begin{cases} -3\alpha + 2\beta = 0 \\ 2\alpha + 3\beta = 1 \end{cases}$$

for α and β. We obtain $\alpha = 2/13$ and $\beta = 3/13$. So

$$y_p(t) = \tfrac{2}{13} \cos 3t + \tfrac{3}{13} \sin 3t$$

is a solution of the nonhomogeneous equation.

Therefore, the general solution of $dy/dt + 2y = \cos 3t$ is

$$y(t) = ke^{-2t} + \tfrac{2}{13} \cos 3t + \tfrac{3}{13} \sin 3t,$$

where k is an arbitrary constant. Solutions for several different initial conditions are shown in Figure 1.95.

Figure 1.95
Graphs of several solutions of

$$\frac{dy}{dt} + 2y = \cos 3t.$$

Note that all of these graphs tend to merge relatively quickly.

How lucky do you need to be?

After a little practice, you will find that there really isn't much luck involved. If $b(t)$ is made up of nice functions (sines, cosines, exponentials, ...), you guess a particular solution made up of the same types of functions. If you make an inappropriate guess (for example, forgetting the $\beta \sin 3t$ term in the second example), then it will be impossible to find choices of the constants that make the guess a solution. If that happens, simply refine the original guess based on what you learned from the previous computation.

Also, you should be careful to avoid a common mistake. Throughout this process, it is important to remember that the undetermined constants are treated as constants during the differentiation step. Do not force a guess of the wrong form to work by turning α (or any other undetermined constant) into a nonconstant function $\alpha(t)$ during the last step in the computation.

Qualitative Analysis

The previous example gives a great deal of insight into the qualitative behavior of solutions of many nonhomogeneous, linear differential equations. Note that the general solution of the associated homogeneous equation, ke^{-2t}, tends to zero quickly. Consequently, every solution is eventually close to the particular solution

$$y_p(t) = \tfrac{2}{13} \cos 3t + \tfrac{3}{13} \sin 3t.$$

We see this clearly in Figure 1.95, where solutions with different initial conditions tend toward the same periodic function. (This periodic solution is called a **steady-state solution** because every solution tends toward it in the long term. Note that this steady-state solution oscillates in a periodic fashion unlike an equilibrium solution that remains constant for all time.)

We could have predicted some of this behavior without computation. If we look at the slope field for this equation (see Figure 1.96), we see that for $y > 1/2$, the slopes are negative, and for $y < -1/2$, the slopes are positive. Graphs of solutions with initial conditions that are outside the interval $-1/2 \le y \le 1/2$ eventually enter the strip of the ty-plane determined by the inequalities $-1/2 \le y \le 1/2$. The detailed behavior of solutions near the t-axis is harder to see from the slope field. However, it is clear that solutions oscillate in some manner.

Figure 1.96
Slope field of

$$\frac{dy}{dt} = -2y + \cos 3t.$$

Note that, if $y \geq 1$, then $dy/dt \leq -1$. Similarly, if $y \leq -1$, the $dy/dt \geq 1$. Hence, any solution that enters the strip $-1 \leq y \leq 1$ remains in that strip as $t \to \infty$.

Looking again at the general solution

$$y(t) = ke^{-2t} + \tfrac{2}{13} \cos 3t + \tfrac{3}{13} \sin 3t,$$

we see that the long-term behavior of the solution is an oscillation with period $2\pi/3$ (see Figure 1.95). Note that this period is the same as the period of $\cos 3t$. However, the amplitude and the phase (that is, the locations of the maxima and minima) for the solution are not exactly the same as the amplitude and phase of $\cos 3t$. (We study the amplitude and phase of solutions to linear equations in detail in Chapter 4.)

These same ideas hold for any nonhomogeneous equation of the form

$$\frac{dy}{dt} = \lambda y + b(t)$$

as long as λ is negative. As before, the homogeneous equation associated with this equation is

$$\frac{dy}{dt} = \lambda y,$$

whose general solution is $ke^{\lambda t}$. If $\lambda < 0$, these functions tend to zero exponentially fast. If one solution of the nonhomogeneous equation is $y_p(t)$, then the general solution of the nonhomogeneous equation is

$$y(t) = ke^{\lambda t} + y_p(t),$$

and we see that all solutions are close to $y_p(t)$ for large t. In other words, the solution of the homogeneous part of the equation tends to zero, and all solutions merge toward $y_p(t)$ over the long term.

The fact that all solutions converge over time definitely relies on the fact that λ is negative. If $\lambda \geq 0$, very different behavior is possible (see Exercises 25–28).

Second Guessing

Sometimes, our first guess may not work no matter how reasonable it is. If this happens, we simply guess again.

Consider the equation

$$\frac{dy}{dt} = -2y + 3e^{-2t}.$$

To compute the general solution, we first note that the general solution of the homogeneous equation is $y(t) = ke^{-2t}$. To find a particular solution of the nonhomogeneous equation we rewrite the equation as

$$\frac{dy}{dt} + 2y = 3e^{-2t}$$

and guess $y_p(t) = \alpha e^{-2t}$, with α as the undetermined coefficient. Substituting this guess into $dy/dt + 2y$, we get

$$\frac{dy_p}{dt} + 2y_p = \frac{d(\alpha e^{-2t})}{dt} + 2\alpha e^{-2t}$$

$$= -2\alpha e^{-2t} + 2\alpha e^{-2t}$$

$$= 0.$$

This is upsetting. No matter how we pick the coefficient α, we always get zero when we substitute $y_p(t)$ into $dy/dt + 2y$. None of the solutions of the nonhomogeneous equation are of the form $y_p(t) = \alpha e^{-2t}$. We failed because our guess, αe^{-2t}, is a solution of the associated homogeneous equation. When we substitute $y_p(t) = \alpha e^{-2t}$ into $dy/dt + 2y$, we are guaranteed to get zero.

Our guess must contain a factor of e^{-2t} to have any hope of being a solution. Unfortunately, there is a wide variety of possible choices. We need a second guess for $y_p(t)$ that contains an e^{-2t} term, is not a solution of the homogeneous equation, and is as simple as possible. Guesses of the form $\alpha e^{-2t} \sin t$ or αe^{bt} are clearly destined to fail. We need a guess whose derivative has one term that is just like itself and another term that involves e^{-2t}. The Product Rule suggests a product of t and our first guess, so we try

$$y_p(t) = \alpha t e^{-2t},$$

where α is our undetermined coefficient. The derivative of $y_p(t)$ is $\alpha(1 - 2t)e^{-2t}$, and substituting this derivative into $dy/dt + 2y$, we obtain

$$\frac{dy_p}{dt} + 2y_p = \alpha(1 - 2t)e^{-2t} + 2\alpha t e^{-2t}$$

$$= \alpha e^{-2t}.$$

Since we want $dy/dt + 2y$ to be $3e^{-2t}$, the guess $y_p(t) = \alpha t e^{-2t}$ is a solution if $\alpha = 3$. (This calculation illustrates why multiplying our first guess by t is a good idea.) The general solution to this nonhomogeneous equation is

$$y(t) = ke^{-2t} + 3te^{-2t},$$

where k is an arbitrary constant.

Rule of thumb for second guessing

The last example indicates what is so unsatisfying about guessing techniques. How did we know to make the second guess a product of t and our first guess? The answer is that we have either seen a similar problem before or we can figure out at least the form of the guess by another technique. Methods for arriving at the second guess with less guesswork but more computation are given in Exercise 23 of Section 1.9, in Chapter 6, and in Exercises 17 and 18 of Appendix B.

EXERCISES FOR SECTION 1.8

In Exercises 1–6, find the general solution of the equation specified.

1. $\dfrac{dy}{dt} = -4y + 9e^{-t}$

2. $\dfrac{dy}{dt} = -4y + 3e^{-t}$

3. $\dfrac{dy}{dt} = -3y + 4\cos 2t$

4. $\dfrac{dy}{dt} = 2y + \sin 2t$

5. $\dfrac{dy}{dt} = 3y - 4e^{3t}$

6. $\dfrac{dy}{dt} = \dfrac{y}{2} + 4e^{t/2}$

In Exercises 7–12, solve the given initial-value problem.

7. $\dfrac{dy}{dt} + 2y = e^{t/3}, \quad y(0) = 1$

8. $\dfrac{dy}{dt} - 2y = 3e^{-2t}, \quad y(0) = 10$

9. $\dfrac{dy}{dt} + y = \cos 2t, \quad y(0) = 5$

10. $\dfrac{dy}{dt} + 3y = \cos 2t, \quad y(0) = -1$

11. $\dfrac{dy}{dt} - 2y = 7e^{2t}, \quad y(0) = 3$

12. $\dfrac{dy}{dt} - 2y = 7e^{2t}, \quad y(0) = 3$

13. Consider the nonhomogeneous linear equation

$$\frac{dy}{dt} + 2y = \cos 3t.$$

To find a particular solution, it is pretty clear that our guess must contain a cosine function, but it is not so clear that the guess must also contain a sine function.

    (a) Guess $y_p(t) = \alpha \cos 3t$ and substitute this guess into the equation. Is there a value of α such that $y_p(t)$ is a solution?

    (b) Write a brief paragraph explaining why the proper guess for a particular solution is $y_p(t) = \alpha \cos 3t + \beta \sin 3t$.

14. Consider the nonhomogeneous linear equation

$$\frac{dy}{dt} = \lambda y + \cos 2t.$$

To find its general solution, we add the general solution of the associated homogeneous equation and a particular solution $y_p(t)$ of the nonhomogeneous equation. Briefly explain why it does not matter which solution of the nonhomogeneous equation we use for $y_p(t)$.

15. The graph to the right is the graph of a solution of a homogeneous linear equation $dy/dt = a(t)y$. Give rough sketches of the graphs of the solutions to this equation that satisfy the initial conditions $y(0) = 0$, $y(0) = 2$, $y(0) = 3$, $y(0) = -1$, and $y(0) = -2.5$.

16. The two graphs to the right are graphs of solutions of a nonhomogeneous linear equation $dy/dt = a(t)y + b(t)$. Give rough sketches of the graphs of the solutions to this equation that satisfy the initial conditions $y(0) = 2$, $y(0) = 3.5$, $y(0) = -1$, and $y(0) = -2$.

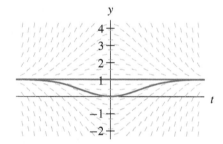

17. Consider the nonlinear differential equation $dy/dt = y^2$.
    (a) Show that $y_1(t) = 1/(1 - t)$ is a solution.
    (b) Show that $y_2(t) = 2/(1 - t)$ is not a solution.
    (c) Why don't these two facts contradict the Linearity Principle?

18. Consider the nonhomogeneous linear equation $dy/dt = -y + 2$.
    (a) Compute an equilibrium solution for this equation.
    (b) Verify that $y(t) = 2 - e^{-t}$ is a solution for this equation.
    (c) Using your results in parts (a) and (b) and the Uniqueness Theorem, explain why the Linearity Principle does not hold for this equation.

19. Consider a nonhomogeneous linear equation of the form

$$\frac{dy}{dt} + a(t)y = b_1(t) + b_2(t),$$

that is, $b(t)$ is written as a sum of two functions. Suppose that $y_h(t)$ is a solution of the associated homogeneous equation $dy/dt + a(t)y = 0$, that $y_1(t)$ is a solution of the equation $dy/dt + a(t)y = b_1(t)$, and that $y_2(t)$ is a solution of the equation $dy/dt + a(t)y = b_2(t)$. Show that $y_h(t) + y_1(t) + y_2(t)$ is a solution of the original nonhomogeneous equation.

20. Consider the nonhomogeneous linear equation

$$\frac{dy}{dt} + 2y = 3t^2 + 2t - 1.$$

In order to find the general solution, we must guess a particular solution $y_p(t)$. Since the right-hand side is a quadratic polynomial, it is reasonable to guess a quadratic for $y_p(t)$, so let

$$y_p(t) = at^2 + bt + c,$$

where a, b, and c are constants. Determine values for these constants so that $y_p(t)$ is a solution.

In Exercises 21–24, find the general solution and the solution that satisfies the initial condition $y(0) = 0$.

21. $\dfrac{dy}{dt} + 2y = t^2 + 2t + 1 + e^{4t}$

22. $\dfrac{dy}{dt} + y = t^3 + \sin 3t$

23. $\dfrac{dy}{dt} - 3y = 2t - e^{4t}$

24. $\dfrac{dy}{dt} + y = \cos 2t + 3 \sin 2t + e^{-t}$

In Exercises 25–28, give a brief qualitative description of the behavior of solutions. Note that we only give partial information about the functions in the differential equation, so your description must allow for various possibilities. Be sure to deal with initial conditions of different sizes and to discuss the long-term behavior of solutions.

25. $\dfrac{dy}{dt} + 2y = b(t)$, where $-1 < b(t) < 2$ for all t.

26. $\dfrac{dy}{dt} - 2y = b(t)$, where $-1 < b(t) < 2$ for all t.

27. $\dfrac{dy}{dt} + y = b(t)$, where $b(t) \to 3$ as $t \to \infty$.

28. $\dfrac{dy}{dt} + ay = \cos 3t + b$, where a and b are positive constants.

29. A person initially places $1,000 in a savings account that pays interest at the rate of 1.1% per year compounded continuously. Suppose the person arranges for $20 per week to be deposited automatically into the savings account.

   (a) Write a differential equation for $P(t)$, the amount on deposit after t years (assume that "weekly deposits" is close enough to "continuous deposits" so that we may model the balance with a differential equation.)

   (b) Find the amount on deposit after 5 years.

30. A student has saved $70,000 for her college tuition. When she starts college, she invests the money in a savings account that pays 1.5% interest per year, compounded continuously. Suppose her college tuition is $30,000 per year and she arranges with the college that the money will be deducted from her savings account in small payments. In other words, we assume that she is paying continuously. How long will she be able to stay in school before she runs out of money?

31. A college professor contributes $5,000 per year into her retirement fund by making many small deposits throughout the year. The fund grows at a rate of 7% per year compounded continuously. After 30 years, she retires and begins withdrawing from her fund at a rate of $3000 per month. If she does not make any deposits after retirement, how long will the money last? [*Hint*: Solve this in two steps, before retirement and after retirement.]

32. Verify that the function $y(t) = t/5$ satisfies the nonhomogeneous linear equation

$$\frac{dy}{dt} = (\cos t)y + \tfrac{1}{5}(1 - t \cos t).$$

33. In this exercise, we verify the Extended Linearity Principle for the nonhomogeneous equation

$$\frac{dy}{dt} = a(t)y + b(t).$$

(a) Let $y_h(t)$ be a solution of the associated homogeneous equation and let $y_p(t)$ be any solution of the nonhomogeneous equation. Show that $y_h(t) + y_p(t)$ satisfies the nonhomogeneous equation by calculating $d(y_h + y_p)/dt$.

(b) Assume that $y_p(t)$ and $y_q(t)$ are solutions to the nonhomogeneous equation. Show that $y_p(t) - y_q(t)$ is a solution to the associated homogeneous equation by computing $d(y_p - y_q)/dt$.

34. Suppose that every constant multiple of a solution is also a solution for a first-order differential equation $dy/dt = f(t, y)$, where $f(t, y)$ is continuous on the entire ty-plane. What can be said about the differential equation?

1.9 INTEGRATING FACTORS FOR LINEAR EQUATIONS

In Section 1.8 we described a guessing technique for solving certain first-order nonhomogeneous linear differential equations. In this section we develop a different analytic method for solving these equations. It is more general than the technique of the previous section, so it can be applied successfully to more equations. It also avoids "guessing." Unfortunately, this method involves the calculation of an integral, which may be a problem as we will see. It is also not as amenable to qualitative analysis. At the end of this section, we discuss the pros and cons of both methods.

Integrating Factors

Given a nonhomogeneous linear differential equation

$$\frac{dy}{dt} = a(t)y + b(t),$$

how can we go about finding the general solution? There is a clever trick that turns an equation of this form into a differential equation that can be solved by integration. As with many techniques in mathematics, the cleverness of this trick might leave you with that "how could I ever think of something like this?" feeling. The thing to remember is that differential equations have been around for more than 300 years. Given three centuries, it is not so surprising that mathematicians were able to discover and refine a slick way to treat these equations.

The idea behind the method

We begin by rewriting the nonhomogeneous equation as

$$\frac{dy}{dt} + g(t)y = b(t),$$

where $g(t) = -a(t)$. We use this form and change the notation for two reasons. The form of the left-hand side of the equation suggests this method, and replacing $-a(t)$ by $g(t)$ avoids a number of annoying minus signs in the calculations.

After staring at this equation for a while (a couple of decades or so), we notice that, with sufficiently poor eyesight, the left-hand side looks somewhat like what we get when we differentiate using the Product Rule. That is, the Product Rule says that the derivative of the product of $y(t)$ and a function $\mu(t)$ is

$$\frac{d(\mu(t)\,y(t))}{dt} = \mu(t)\frac{dy}{dt} + \frac{d\mu}{dt}y(t).$$

Note that one term on the right-hand side has dy/dt in it and the other term has y in it just like the left-hand side of our nonhomogeneous linear equation.

Here's the clever part. Multiply both sides of the original differential equation by an (as yet unspecified) function $\mu(t)$. We obtain the new differential equation

$$\mu(t)\frac{dy}{dt} + \mu(t)g(t)y = \mu(t)b(t)$$

whose left-hand side looks even more like the derivative of a product of two functions. For the moment, let's *assume* that we have a function $\mu(t)$ so that the left-hand side actually *is* the derivative of the product $\mu(t)\,y(t)$. That is, suppose we have found a function $\mu(t)$ that satisfies

$$\frac{d(\mu(t)\,y(t))}{dt} = \mu(t)\frac{dy}{dt} + \mu(t)g(t)y.$$

Then the new differential equation is just

$$\frac{d(\mu(t)\,y(t))}{dt} = \mu(t)b(t).$$

How does this help? We can integrate both sides of this equation with respect to t to obtain

$$\mu(t)\,y(t) = \int \mu(t)\,b(t)\,dt,$$

and consequently,

$$y(t) = \frac{1}{\mu(t)} \int \mu(t)\,b(t)\,dt.$$

That is, assuming we have such a $\mu(t)$ and can evaluate $\int \mu(t)\,b(t)\,dt$, we can compute our solution $y(t)$.

Finding the integrating factor

This derivation of $y(t)$ is based on one pretty big assumption. How can we find a function $\mu(t)$ such that

$$\frac{d(\mu(t)\,y(t))}{dt} = \mu(t)\frac{dy}{dt} + \mu(t)g(t)y(t)$$

in the first place?

Applying the Product Rule to the left-hand side, we see that the desired $\mu(t)$ must satisfy

$$\mu(t)\frac{dy}{dt} + \frac{d\mu}{dt}y(t) = \mu(t)\frac{dy}{dt} + \mu(t)g(t)y(t).$$

Canceling the $\mu(t)(dy/dt)$ term on both sides leaves

$$\frac{d\mu}{dt}y(t) = \mu(t)\,g(t)\,y(t).$$

So, if we find a function $\mu(t)$ that satisfies the equation

$$\frac{d\mu}{dt} = \mu(t)\,g(t),$$

we get our desired $\mu(t)$. However, this last equation is just $d\mu/dt = g(t)\mu$, which is a homogeneous linear differential equation, and we already know that

$$\mu(t) = e^{\int g(t)\,dt}.$$

(See page 113 for the derivation of this solution.)

Given this formula for $\mu(t)$, we now see that this strategy is going to work. The function $\mu(t)$ is called an **integrating factor** for the original nonhomogeneous equation because we can solve the equation by integration if we multiply it by the factor $\mu(t)$. In other words, whenever we want to determine an explicit solution to

$$\frac{dy}{dt} + g(t)y = b(t),$$

we first compute the integrating factor $\mu(t)$. Then we solve the equation by multiplying both sides by $\mu(t)$ and integrating. Note that, when we calculate $\mu(t)$, there is an arbitrary constant of integration in the exponent. Since we only need one integrating factor $\mu(t)$ to solve the equation, we choose the constant to be whatever is most convenient. That choice is usually zero.

To see this method at work, let's look at some examples. The method looks very general. However, because there are two integrals to calculate, we may get stuck before we obtain an explicit solution.

Complete success

Consider the nonhomogeneous linear equation

$$\frac{dy}{dt} + \frac{2}{t}y = t - 1.$$

First we compute the integrating factor

$$\mu(t) = e^{\int g(t)\,dt} = e^{\int (2/t)\,dt} = e^{2\ln t} = e^{\ln(t^2)} = t^2.$$

Remember that the idea behind this method is to multiply both sides of the differential equation by $\mu(t)$ so that the left-hand side of the new equation is the result of the Product Rule. In this case, multiplying by $\mu(t) = t^2$ yields

$$t^2\frac{dy}{dt} + 2ty = t^2(t - 1).$$

Note that the left-hand side is the derivative of the product of t^2 and $y(t)$. In other words, this equation is the same as

$$\frac{d}{dt}(t^2 y) = t^3 - t^2.$$

Integrating both sides with respect to t yields

$$t^2 y = \frac{t^4}{4} - \frac{t^3}{3} + k,$$

where k is an arbitrary constant. The general solution is

$$y(t) = \frac{t^2}{4} - \frac{t}{3} + \frac{k}{t^2}.$$

Of course, we can check that these functions satisfy the differential equation by substituting them back into the equation.

It is important to note the role of constants of integration in this example. When we calculated $\mu(t) = t^2$, we ignored the constant because we only need one integrating factor. However, after we multiplied both sides of the original equation by $\mu(t)$ and integrated, it was important to include the constant of integration on the right-hand side. If we had omitted that constant, we would have computed just one solution to the nonhomogeneous equation rather than the general solution.

This example is also a good illustration of the Extended Linearity Principle. Note that k/t^2 is the general solution of the associated homogeneous equation

$$\frac{dy}{dt} = -\frac{2}{t}y,$$

and

$$y(t) = \frac{t^2}{4} - \frac{t}{3}$$

is one solution of the nonhomogeneous equation.

Problems with the integration

The previous example was chosen carefully. Another linear equation which does not look any more difficult is

$$\frac{dy}{dt} = t^2 y + t - 1.$$

We rewrite the differential equation as

$$\frac{dy}{dt} - t^2 y = t - 1$$

and compute the integrating factor

$$\mu(t) = e^{\int -t^2\, dt} = e^{-t^3/3}.$$

Next we multiply both sides by $\mu(t)$ and obtain

$$e^{-t^3/3}\frac{dy}{dt} - t^2 e^{-t^3/3}\, y = e^{-t^3/3}(t - 1).$$

Note that the left-hand side is the derivative of the product of $e^{-t^3/3}$ and $y(t)$, so we have

$$\frac{d}{dt}\left(e^{-t^3/3}\, y\right) = e^{-t^3/3}(t - 1).$$

Integrating both sides yields

$$e^{-t^3/3}\, y = \int e^{-t^3/3}(t - 1)\, dt$$

but then we are stuck. It turns out that the integral on the right-hand side of this equation is not expressible in terms of the familiar functions (sin, cos, ln, and so on), so we cannot obtain explicit formulas for the solutions.

This example indicates what can go wrong with techniques that involve the calculation of explicit integrals. Even reasonable-looking functions can quickly lead to complicated integrating factors and integrals. On the other hand, we can express the solution in terms of integrals with respect to t, and although many integrals are impossible to calculate explicitly, many others are possible. Indeed, as we have mentioned before, there are a number of computer programs that are quite good at calculating the indefinite integrals involved in this technique.

Mixing Problems Revisited

In Section 1.2 we considered a model of the concentration of a substance in solution. Typically in these problems we have a container in which there is a certain amount of fluid (such as water or air) to which a contaminant is added at some rate. The fluid is kept well mixed at all times. If the total volume of fluid is kept fixed, then the resulting differential equation for the amount of contaminant is autonomous and can be solved either by separating variables or by the Extended Linearity Principle along with a guessing technique. If the total volume of fluid changes with time, then the differential equation is nonautonomous and must be solved using an integrating factor.

A polluted pond

Consider a pond that has an initial volume of 10,000 cubic meters. Suppose that at time $t = 0$, the water in the pond is clean and that the pond has two streams flowing into it, stream A and stream B, and one stream flowing out, stream C (see Figure 1.97). Suppose 500 cubic meters per day of water flow into the pond from stream A, 750 cubic meters per day flow into the pond from stream B, and 1250 cubic meters flow out of the pond via stream C.

At time $t = 0$, the water flowing into the pond from stream A becomes contaminated with road salt at a concentration of 5 kilograms per 1000 cubic meters. Suppose the water in the pond is well mixed so the concentration of salt at any given time is constant. To make matters worse, suppose also that at time $t = 0$ someone begins dumping trash into the pond at a rate of 50 cubic meters per day. The trash settles to the bottom of the pond, reducing the volume by 50 cubic meters per day. To adjust for the incoming trash, the rate that water flows out via stream C increases to 1300 cubic meters per day and the banks of the pond do not overflow.

The description looks very much like the mixing problems we have already considered (where "pond" replaces "vat" and "stream" replaces "pipe"). The new element here is that the total volume is not constant. Because of the dumping of trash, the volume decreases by 50 cubic meters per day.

If we let $S(t)$ be the amount of salt (in kilograms) in the pond at time t, then dS/dt is the difference between the rate that salt enters the pond and the rate that salt

Figure 1.97
Schematic of the pond with three streams.

leaves the pond. Salt enters the pond from stream A only. The rate at which it enters is the product of its concentration in the water and the rate at which the water flows in through stream A. Since the concentration is 5 kilograms per 1000 cubic meters and the rate that water flows into the pond from stream A is 500 cubic meters per day, the rate at which salt enters the pond is $(500)(5/1000) = 5/2$ kilograms per day. The rate at which the salt leaves the pond via stream C is the product of its concentration *in the pond* and the rate at which water flows out of the pond. The rate at which water flows out is 1300 cubic meters per day. To determine the concentration, we note that it is the quotient of the amount S of salt in the pond by the volume V. Because the volume is initially 10,000 cubic meters and it decreases by 50 cubic meters per day, we know that $V(t) = 10,000 - 50t$. Hence, the concentration is $S/(10,000 - 50t)$, and the rate at which salt flows out of the pond is

$$1300 \left(\frac{S}{10,000 - 50t} \right),$$

which simplifies to $26S/(200 - t)$. Therefore, the differential equation that models the amount of salt in the pond is

$$\frac{dS}{dt} = \frac{5}{2} - \frac{26S}{200 - t}.$$

This model is valid only as long as there is water in the pond—that is, as long as the volume $V(t) = 10,000 - 50t$ is positive. So the differential equation is valid for $0 \leq t < 200$. Because the water is clean at time $t = 0$, the initial condition is $S(0) = 0$.

Since this equation is nonautonomous, we solve this initial-value problem using an integrating factor. Rewriting the differential equation as

$$\frac{dS}{dt} + \left(\frac{26}{200 - t} \right) S = \frac{5}{2}$$

indicates that the integrating factor is

$$\mu(t) = e^{\int \frac{26}{200-t} \, dt} = e^{-26 \ln(200-t)} = e^{\ln((200-t)^{-26})} = (200 - t)^{-26}.$$

Multiplying both sides by $\mu(t)$ gives

$$(200 - t)^{-26} \frac{dS}{dt} + 26(200 - t)^{-27} S = \frac{5}{2}(200 - t)^{-26}.$$

By the Product Rule, this equation is the same as the differential equation

$$\frac{d}{dt} \left((200 - t)^{-26} S \right) = \frac{5}{2}(200 - t)^{-26}.$$

Integrating both sides yields

$$(200 - t)^{-26} S = \frac{5}{2} \int (200 - t)^{-26} \, dt$$

$$= \frac{5}{2} \frac{(200 - t)^{-25}}{25} + c,$$

where c is an arbitrary constant. Solving for S, we obtain the general solution

$$S = \frac{200 - t}{10} + c(200 - t)^{26}.$$

Using the initial condition $S(0) = 0$, we find that $c = -20/200^{26}$ and the particular solution for the initial-value problem is

$$S = \frac{200 - t}{10} - 20\left(\frac{200 - t}{200}\right)^{26}.$$

This is an unusual-looking expression because of the large number 200^{26}. However, the graph reveals that its behavior is not at all unusual (see Figure 1.98). The amount of salt in the pond rises fairly quickly, reaching a maximum close to $S = 20$ at $t \approx 25$. After that time, the amount of salt decreases almost linearly, reaching zero at $t = 200$.

The behavior of this solution is quite reasonable if we recall that the pond starts out containing no salt and that eventually it is completely filled with trash. (It contains no salt or water at time $t = 200$.) As we mentioned above, the concentration of salt in the pond water is given by $C(t) = S(t)/V(t) = S(t)/(10{,}000 - 50t)$. Graphing $C(t)$, we see that it increases asymptotically toward 0.002 kilograms per cubic meter even as the water level decreases (see Figure 1.99).

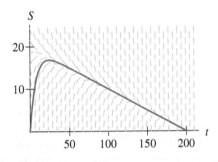

Figure 1.98
Graph of the solution of
$dS/dt = 5/2 - 26S/(200 - t)$, with
$S(0) = 0$.

Figure 1.99
Graph of concentration of salt versus time for the solution graphed in Figure 1.98.

Comparing the Methods of Solution for Linear Equations

There is an old saying that goes

"If the only tool you have is a hammer, then every problem looks like a nail."

If you know only one method for solving linear differential equations, then you certainly save time thinking about which method to use when confronted with such an equation. However, we have two and each method has its advantages and disadvantages.

Which method should you use for a given linear differential equation? Trying to guess a solution to the nonhomogeneous equation that we just solved would be a nightmare. Hence, the method of integrating factors is the only reasonable choice for that equation.

On the other hand, consider a linear equation such as

$$\frac{dv}{dt} + 0.4v = 3\cos 2t,$$

which is typical for the voltage over a capacitor in an RC circuit with a periodic voltage source (see Section 1.4). The integrating factor for this equation is $\mu(t) = e^{0.4t}$. Therefore, the integral you must compute is

$$\int e^{0.4t} (3\cos 2t)\, dt.$$

This integral can certainly be done by hand using integration by parts but it would take some effort.

If you use a guessing technique, you would guess a particular solution of the form

$$v_p(t) = \alpha \cos 2t + \beta \sin 2t$$

and solve for α and β. The computation requires some algebra but not much calculus (see Exercise 21).

So which method is better for this equation? Both end up with the same general solution but the guessing method is arguably faster. One advantage of the guessing method is that it exploits the Extended Linearity Principle and we see the qualitative behavior of the solutions more directly. We know the general solution of the homogeneous equation is $ke^{-0.4t}$, which tends to zero in an exponential fashion, and over the long term, all solutions converge to the periodic solution $v_p(t)$ (see Figure 1.100).

In theory, the method of integrating factors works more generally but the integrals involved might be difficult or impossible to do. The guessing technique described in the previous section avoids the integration but is only practical for certain linear equations such as constant-coefficient equations with relatively simple functions $b(t)$.

Most importantly, you need to understand what it means to be a linear equation and the implications of the Linearity and Extended Linearity Principles. It is also important that you remember the clever idea behind the development of integrating factors. Each of the methods teaches us something about linear differential equations.

Figure 1.100
Graphs of various solutions of

$$\frac{dv}{dt} + 0.4v = 3\cos 2t.$$

Note that all solutions converge to the solution

$$v_p(t) = \tfrac{15}{52}\cos 2t + \tfrac{75}{52}\sin 2t$$

over the long term (see Exercise 21).

EXERCISES FOR SECTION 1.9

In Exercises 1–6, find the general solution of the differential equation specified.

1. $\dfrac{dy}{dt} = -\dfrac{y}{t} + 2$

2. $\dfrac{dy}{dt} = \dfrac{3}{t}y + t^5$

3. $\dfrac{dy}{dt} = -\dfrac{y}{1+t} + t^2$

4. $\dfrac{dy}{dt} = -2ty + 4e^{-t^2}$

5. $\dfrac{dy}{dt} - \dfrac{2t}{1+t^2}y = 3$

6. $\dfrac{dy}{dt} - \dfrac{2}{t}y = t^3 e^t$

In Exercises 7–12, solve the given initial-value problem.

7. $\dfrac{dy}{dt} = -\dfrac{y}{1+t} + 2, \quad y(0) = 3$

8. $\dfrac{dy}{dt} = \dfrac{1}{t+1}y + 4t^2 + 4t, \quad y(1) = 10$

9. $\dfrac{dy}{dt} = -\dfrac{y}{t} + 2, \quad y(1) = 3$

10. $\dfrac{dy}{dt} = -2ty + 4e^{-t^2}, \quad y(0) = 3$

11. $\dfrac{dy}{dt} - \dfrac{2y}{t} = 2t^2, \quad y(-2) = 4$

12. $\dfrac{dy}{dt} - \dfrac{3}{t}y = 2t^3 e^{2t}, \quad y(1) = 0$

In Exercises 13–18, the differential equation is linear, and in theory, we can find its general solution using the method of integrating factors. However, since this method involves computing two integrals, in practice it is frequently impossible to reach a formula for the solution that is free of integrals. For these exercises, determine the general solution to the equation and express it with as few integrals as possible.

13. $\dfrac{dy}{dt} = (\sin t)y + 4$

14. $\dfrac{dy}{dt} = t^2 y + 4$

15. $\dfrac{dy}{dt} = \dfrac{y}{t^2} + 4\cos t$

16. $\dfrac{dy}{dt} = y + 4\cos t^2$

17. $\dfrac{dy}{dt} = -\dfrac{y}{e^{t^2}} + \cos t$

18. $\dfrac{dy}{dt} = \dfrac{y}{\sqrt{t^3 - 3}} + t$

19. For what value(s) of the parameter a is it possible to find explicit formulas (without integrals) for the solutions to

$$\frac{dy}{dt} = aty + 4e^{-t^2}?$$

20. For what value(s) of the parameter r is it possible to find explicit formulas (without integrals) for the solutions to

$$\frac{dy}{dt} = t^r y + 4?$$

21. Consider the nonhomogeneous equation

$$\frac{dv}{dt} + 0.4v = 3\cos 2t.$$

(a) Find the general solution using the method of integrating factors.

(b) Find the general solution using the guessing technique from Section 1.8.

Comment on which method was easier for you.

22. In this exercise, we explore the connections between the method of integrating factors discussed in this section and the Extended Linearity Principle. Consider the nonhomogeneous linear equation

$$\frac{dy}{dt} = a(t)y + b(t),$$

where $a(t)$ and $b(t)$ are continuous for all t.

(a) Let

$$\mu(t) = e^{-\int_0^t a(\tau)\,d\tau}.$$

Show that $\mu(t)$ is an integrating factor for the nonhomogeneous equation.

(b) Show that $1/\mu(t)$ is a solution to the associated homogeneous equation.

(c) Show that

$$y_p(t) = \frac{1}{\mu(t)} \int_0^t \mu(\tau)\,b(\tau)\,d\tau$$

is a solution to the nonhomogeneous equation.

(d) Use the Extended Linearity Principle to find the general solution of the nonhomogeneous equation.

(e) Compare your result in part (d) to the formula

$$y(t) = \frac{1}{\mu(t)} \int \mu(t)\,b(t)\,dt$$

for the general solution that we obtained on page 126.

(f) Illustrate the calculations that you did in this exercise for the example

$$\frac{dy}{dt} = -2ty + 4e^{-t^2}.$$

23. Consider the nonhomogeneous equation

$$\frac{dy}{dt} + 2y = 3e^{-2t}.$$

In Section 1.8, we saw that the guess $y_p(t) = \alpha e^{-2t}$ does not produce a solution because it is a solution to the associated homogeneous equation. We then guessed $y_p(t) = \alpha t e^{-2t}$. Use the method of integrating factors to explain why this guess is a good idea.

24. A 30-gallon tank initially contains 15 gallons of salt water containing 6 pounds of salt. Suppose salt water containing 1 pound of salt per gallon is pumped into the top of the tank at the rate of 2 gallons per minute, while a well-mixed solution leaves the bottom of the tank at a rate of 1 gallon per minute. How much salt is in the tank when the tank is full?

25. A 400-gallon tank initially contains 200 gallons of water containing 2 parts per billion by weight of dioxin, an extremely potent carcinogen. Suppose water containing 5 parts per billion of dioxin flows into the top of the tank at a rate of 4 gallons per minute. The water in the tank is kept well mixed, and 2 gallons per minute are removed from the bottom of the tank. How much dioxin is in the tank when the tank is full?

26. A 100-gallon tank initially contains 100 gallons of sugar water at a concentration of 0.25 pounds of sugar per gallon. Suppose that sugar is added to the tank at a rate of p pounds per minute, that sugar water is removed at a rate of 1 gallon per minute, and that the water in the tank is kept well mixed.

   (a) What value of p should we pick so that, when 5 gallons of sugar solution is left in the tank, the concentration is 0.5 pounds of sugar per gallon?

   (b) Is it possible to choose p so that the last drop of water out of the bucket has a concentration of 0.75 pounds of sugar per gallon?

27. Suppose a 50-gallon tank contains a volume V_0 of clean water at time $t = 0$. At time $t = 0$, we begin dumping 2 gallons per minute of salt solution containing 0.25 pounds of salt per gallon into the tank. Also at time $t = 0$, we begin removing 1 gallon per minute of salt water from the tank. As usual, suppose the water in the tank is well mixed so that the salt concentration at any given time is constant throughout the tank.

   (a) Set up the initial-value problem for the amount of salt in the tank. [*Hint*: The initial value of V_0 will appear in the differential equation.]

   (b) What is your equation if $V_0 = 0$ (the tank is initially empty)? Comment on the validity of the model in this situation. What will be the amount of salt in the tank at time t for this situation?

REVIEW EXERCISES FOR CHAPTER 1

Short answer exercises: Exercises 1–10 focus on the basic ideas, definitions, and vocabulary of this chapter. Their answers are short (a single sentence or drawing), and you should be able to do them with little or no computation. However, they vary in difficulty, so think carefully before you answer.

1. Give an example of a first-order differential equation that has the function $y(t) = 2t + 3$ as a solution.

2. What is the general solution of the differential equation $dy/dt = 3y$?

3. Find all equilibrium solutions for the differential equation $dy/dt = t^2(t^2 + 1)$.

4. Find one solution of the differential equation $dy/dt = -|\sin^5 y|$.

5. Find all of the equilibrium solutions for the differential equation

$$\frac{dy}{dt} = \frac{(t^2 - 4)(1 + y)e^y}{(t - 1)(3 - y)}.$$

6. Sketch the phase line for the autonomous equation $dy/dt = \sin^2 y$.

7. Give an example of a first-order differential equation that is autonomous, separable, linear, and homogeneous.

8. Give an example of a first-order, autonomous, linear, nonhomogeneous differential equation that has the equilibrium solution $y(t) = 2$ for all t.

9. Suppose the phase line to the right is the phase line for the autonomous differential equation $dy/dt = f(y)$. What can you say about the graph of $f(y)$?

$y = 0$

10. What are the bifurcation values of the one-parameter family of differential equations $dy/dt = a + 4$?

True-false: For Exercises 11–20, determine if the statement is true or false. If it is true, explain why. If it is false, provide a counterexample or an explanation.

11. The function $y(t) = -e^{-t}$ is a solution to the differential equation $dy/dt = |y|$.

12. Every separable differential equation is autonomous.

13. Every autonomous differential equation is separable.

14. Every linear differential equation is separable.

15. Every separable differential equation is a homogeneous linear equation.

16. Every homogeneous linear differential equation is separable.

17. The solution of $dy/dt = (y - 3)(\sin y \sin t + \cos t + 1)$ with $y(0) = 4$ satisfies $y(t) > 3$ for all t.

18. Suppose that $f(y)$ is a continuous function for all y. The phase line for $dy/dt = f(y)$ must have the same number of sources as sinks.

19. Suppose that $f(y)$ is continuously differentiable for all y. Exactly one solution of $dy/dt = f(y)$ tends to ∞ as t increases.

20. Every solution of $dy/dt = y + e^{-t}$ tends to $+\infty$ or $-\infty$ as $t \to \infty$.

In Exercises 21–29,

(a) specify if the given equation is autonomous, linear and homogeneous, linear and nonhomogeneous, and/or separable, and

(b) find its general solution.

21. $\dfrac{dy}{dt} = 3 - 2y$

22. $\dfrac{dy}{dt} = ty$

23. $\dfrac{dy}{dt} = 3y + e^{7t}$

24. $\dfrac{dy}{dt} = \dfrac{ty}{1+t^2}$

25. $\dfrac{dy}{dt} = -5y + \sin 3t$

26. $\dfrac{dy}{dt} = t + \dfrac{2y}{1+t}$

27. $\dfrac{dy}{dt} = 3 + y^2$

28. $\dfrac{dy}{dt} = 2y - y^2$

29. $\dfrac{dy}{dt} = -3y + e^{-2t} + t^2$

In Exercises 30–39,

(a) specify if the given equation is autonomous, linear and homogeneous, linear and nonhomogeneous, and/or separable, and

(b) solve the initial-value problem.

30. $\dfrac{dx}{dt} = -2tx, \quad x(0) = e$

31. $\dfrac{dy}{dt} = 2y + \cos 4t, \quad y(0) = 1$

32. $\dfrac{dy}{dt} = 3y + 2e^{3t}, \quad y(0) = -1$

33. $\dfrac{dy}{dt} = t^2 y^3 + y^3, \quad y(0) = -1/2$

34. $\dfrac{dy}{dt} + 5y = 3e^{-5t}, \quad y(0) = -2$

35. $\dfrac{dy}{dt} = 2ty + 3te^{t^2}, \quad y(0) = 1$

36. $\dfrac{dy}{dt} = \dfrac{(t+1)^2}{(y+1)^2}, \quad y(0) = 0$

37. $\dfrac{dy}{dt} = 2ty^2 + 3t^2 y^2, \quad y(1) = -1$

38. $\dfrac{dy}{dt} = 1 - y^2, \quad y(0) = 1$

39. $\dfrac{dy}{dt} = \dfrac{t^2}{y + t^3 y}, \quad y(0) = -2$

40. Consider the initial-value problem $dy/dt = y^2 - 2y + 1, \quad y(0) = 2$.

(a) Using Euler's method with $\Delta t = 0.5$, graph an approximate solution over the interval $0 \le t \le 2$.

(b) What happens when you try to repeat part (a) with $\Delta t = 0.05$?

(c) Solve this initial-value problem by separating variables, and use the result to explain your observations in parts (a) and (b).

41. Consider the autonomous differential equation $dy/dt = f(y)$ where the graph of $f(y)$ is given below.

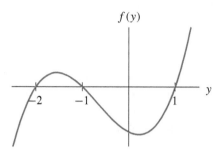

$f(y)$

(a) Give a rough sketch of the slope field that corresponds to this equation.

(b) Give a rough sketch of the graph of the solution to $dy/dt = f(y)$ that satisfies the initial condition $y(0) = 0$.

42. Consider the autonomous differential equation $dy/dt = f(y)$ where the graph of $f(y)$ is given below.

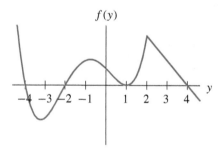

$f(y)$

(a) Sketch the phase line for this equation and identify the equilibrium points as sinks, sources, or nodes.

(b) Give a rough sketch of the slope field that corresponds to this equation.

(c) Give rough sketches of the graphs of the solutions that satisfy the initial conditions $y(0) = -3$, $y(0) = 0$, $y(0) = 1$, and $y(0) = 2$.

43. The slope field to the right is the field for the differential equation

$$\frac{dy}{dt} = (y - 2)(y + 1 - \cos t).$$

Describe the long-term behavior of solutions with various initial values at $t = 0$. Then confirm your answer with HPGSolver.

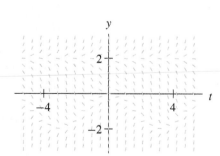

44. The slope field to the right is the
field for the differential equation

$$\frac{dy}{dt} = (y - 1)(y - 2)(y - e^{t/2}).$$

Describe the long-term behavior of
solutions with various initial values
at $t = 0$. Then confirm your answer
with HPGSolver.

45. Consider the differential equation

$$\frac{dy}{dt} = t^2 y + 1 + y + t^2.$$

(a) Find its general solution by separating variables.
(b) Note that this equation is also a nonhomogeneous linear equation. Find the general solution of its associated homogeneous equation.
(c) Calculate the equilibrium solutions of the nonhomogeneous equation.
(d) Using the Extended Linearity Principle, find the general solution of the nonhomogeneous equation. Compare your result to the one you obtained in part (a).

46. Consider the differential equation

$$\frac{dy}{dt} = \frac{2y + 1}{t}.$$

(a) Compute its general solution by separating variables.
(b) What happens to these solutions as $t \to 0$?
(c) Why doesn't this example violate the Uniqueness Theorem?

47. Consider the initial-value problem $dy/dt = 3 - y^2$, $y(0) = 0$.
(a) Using Euler's method with $\Delta t = 0.5$, plot the graph of an approximate solution over the interval $0 \le t \le 2$.
(b) Sketch the phase line for this differential equation.
(c) What does the phase line tell you about the approximate values that you computed in part (a)?

48. A cup of soup is initially 150°. Suppose that it cools to 140° in 1 minute in a room with an ambient temperature of 70°.
(a) Assume that Newton's law of cooling applies: The rate of cooling is proportional to the difference between the current temperature and the ambient temperature. Write an initial-value problem that models the temperature of the soup.
(b) How long does it take the soup to cool to a temperature of 100°?

49. Eight differential equations and four slope fields are given below. Determine the equation that corresponds to each slope field and state briefly how you know your choice is correct. You should do this exercise without using technology.

(i) $\dfrac{dy}{dt} = t - 1$ (ii) $\dfrac{dy}{dt} = 1 - y^2$ (iii) $\dfrac{dy}{dt} = y - t^2$ (iv) $\dfrac{dy}{dt} = 1 - t$

(v) $\dfrac{dy}{dt} = 1 - y$ (vi) $\dfrac{dy}{dt} = y + t^2$ (vii) $\dfrac{dy}{dt} = ty - t$ (viii) $\dfrac{dy}{dt} = y^2 - 1$

(a)   **(b)**

(c)   **(d)**

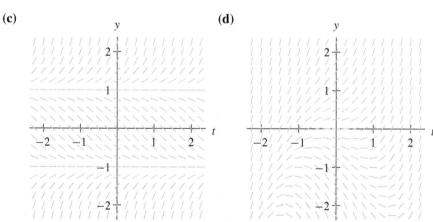

50. Beth initially deposits $400 in a savings account that pays interest at the rate of 1.1% per year compounded continuously. She also arranges for $20 per week to be deposited automatically into the account.

   (a) Assume that weekly deposits are close enough to continuous deposits so that we can reasonably approximate her balance using a differential equation. Write an initial-value problem for her balance over time.

   (b) Approximate Beth's balance after 4 years by solving the initial-value problem in part (a).

51. Consider the linear differential equation

$$a\frac{dy}{dt} + y = b,$$

where a and b are positive constants.

(a) Sketch the phase line associated with this equation.

(b) Describe the long-term behavior of all solutions.

(c) How many different methods do you know to calculate its general solution?

(d) Using your favorite method, calculate the general solution.

(e) Using your least favorite method, calculate the general solution.

(f) Using your answer in parts (d) and (e), confirm your answer to part (b).

52. Consider the differential equation $dy/dt = -2ty^2$.

(a) Calculate its general solution.

(b) Find all values of y_0 such that the solution to the initial-value problem

$$\frac{dy}{dt} = -2ty^2, \quad y(-1) = y_0,$$

does not blow up (or down) in finite time. In other words, find all y_0 such that the solution is defined for all real t.

53. The air in a small rectangular room 20 ft by 5 ft by 10 ft is 3% carbon monoxide. Starting at $t = 0$, air containing 1% carbon monoxide is blown into the room at the rate of 100 ft^3 per hour and well mixed air flows out through a vent at the same rate.

(a) Write an initial-value problem for the amount of carbon monoxide in the room over time.

(b) Sketch the phase line corresponding to the initial-value problem in part (a), and determine how much carbon monoxide will be in the room over the long term.

(c) When will the air in the room be 2% carbon monoxide?

54. A 1000-gallon tank initially contains a mixture of 450 gallons of cola and 50 gallons of cherry syrup. Cola is added at the rate of 8 gallons per minute, and cherry syrup is added at the rate of 2 gallons per minute. At the same time, a well mixed solution of cherry cola is withdrawn at the rate of 5 gallons per minute. What percentage of the mixture is cherry syrup when the tank is full?

LAB 1.1 Rate of Memorization Model

Human learning is, to say the least, an extremely complicated process. The biology and chemistry of learning is far from understood. While simple models of learning cannot hope to encompass this complexity, they can illuminate limited aspects of the learning process. In this lab we study a simple model of the process of memorization of lists (lists of nonsense syllables or entries from tables of integrals).

The model is based on the assumption that the rate of learning is proportional to the amount left to be learned. We let $L(t)$ be the fraction of the list already committed to memory at time t. So $L = 0$ corresponds to knowing none of the list, and $L = 1$ corresponds to knowing the entire list. The differential equation is

$$\frac{dL}{dt} = k(1 - L).$$

Different people take different amounts of time to memorize a list. According to the model this means that each person has his or her own personal value of k. The value of k for a given individual must be determined by experiment.

Carry out the following steps:

1. Four lists of three-digit numbers are given in Table 1.9, and additional lists can be generated by a random number generator on a computer. Collect the data necessary to determine your personal k value as follows:

    (a) Spend one minute studying one of the lists of numbers in table Table 1.9. (Measure the time carefully. A friend can help.)

    (b) Quiz yourself on how many of the numbers you have memorized by writing down as many of the numbers as you remember in their correct order. (You may skip over numbers you don't remember and obtain "credit" for numbers you remember later in the list.) Put your quiz aside to be graded later.

    (c) Spend another minute studying the same list.

    (d) Quiz yourself again.

    Repeat the process ten times (or until you have learned the entire list). Grade your quizzes (a correct answer is having a correct number in its correct position in the list). Compile your data in a graph with t, the amount of time spent studying, on the horizontal axis, and L, the fraction of the list learned, on the vertical axis.

2. Use this data to approximate your personal k-value and compare your data with the predictions of the model. You may use numeric or analytic methods, but be sure to carefully explain your work. Estimate how long it would take you to learn a list of 50 and 100 three-digit numbers.

3. Repeat the process in Part 1 on two of the other lists and compute your k-value on these lists. Is your personal k-value really constant, or does it improve with practice? If k does improve with practice, how would you modify the model to include this?

Table 1.9
Four lists of random three-digit numbers

	List 1	List 2	List 3	List 4
1	457	167	733	240
2	938	603	297	897
3	363	980	184	935
4	246	326	784	105
5	219	189	277	679
6	538	846	274	011
7	790	040	516	020
8	895	891	051	013
9	073	519	925	144
10	951	306	102	209
11	777	424	826	419
12	300	559	937	191
13	048	911	182	551
14	918	439	951	282
15	524	140	643	587
16	203	155	434	609
17	719	847	921	391
18	518	245	820	364
19	130	752	017	733
20	874	552	389	735

Your report: In your report, you should give your data in Parts 1 and 3 neatly and clearly. Your answer to the questions in Parts 2 and 3 should be in the form of short essays. You should include hand- or computer-drawn graphs of your data and solutions of the model *as appropriate*. (Remember that one carefully chosen picture can be worth a thousand words, but a thousand pictures aren't worth anything.)

LAB 1.2   Growth of a Population of Mold

In the text, we modeled the U.S. population using both an exponential growth model and a logistic growth model. The assumptions we used to create the models are easy to state. For the exponential model we assumed only that the growth of the population is proportional to the size of the population. For the logistic model we added the assumption that the ratio of the population to the growth rate decreases as the population increases. In this lab we apply these same principles to model the colonization of a piece of bread by mold.

Place a piece of mold-free bread in a plastic bag with a small amount of water and leave the bread in a warm place. Each day, record the area of the bread that is covered with mold. (One way to do this is to trace the grid from a piece of graph paper onto a clear piece of plastic. Hold the plastic over the bread and count the number of squares that are mostly covered by mold.)

Warning: It takes at least two weeks to accumulate a reasonable amount of data. Some types of bread seem to be resistant to mold growth, and the bread just dries out. If the mold grows, then after about a week the bread will look pretty disgusting. Take precautions to make sure your assignment isn't thrown out. In your report, address the following questions:

1. Model the growth of mold using an exponential growth model. How accurately does the model fit the data? Be sure to explain carefully how you obtained the value for the growth-rate parameter.

2. Model the growth of mold using a logistic growth model. How accurately does the model fit the data? Be sure to explain carefully how you obtained the value of the growth-rate parameter and carrying capacity.

3. Discuss the models for mold growth population. Were there any surprises? Does it matter that we are measuring the area covered by the mold rather than the total weight of the mold? To what extent would you believe predictions of future mold populations based on these models?

Your report: You should include in your report the details of the type of bread used, where it was kept, and how and how often the mold was measured. Your analysis of the models may include qualitative, numerical, and analytic arguments, and graphs of data and solutions of your models as appropriate. (Remember that a well-chosen picture can be worth a thousand words, but a thousand pictures aren't worth anything.) **Do not hand in the piece of bread.**

LAB 1.3 Logistic Population Models with Harvesting

In this lab, we consider logistic models of population growth that have been modified to include terms that account for "harvesting." In particular, you should imagine a fish population subject to various degrees and types of fishing. The differential equation models are given below. (Your instructor will indicate the values of the parameters k, N, a_1, and a_2 you should use. Several possible choices are listed in Table 1.10.) In your report, you should include a discussion of the meaning of each variable and parameter and an explanation of why the equation is written the way it is.

We have discussed three general approaches that can be employed to study a differential equation: Numerical techniques yield graphs of approximate solutions, geometric/qualitative techniques provide predictions of the long-term behavior of the solution and in special cases analytic techniques provide explicit formulas for the solution. In your report, you should employ as many of these techniques as is appropriate to help understand the models, and you should consider the following equations:

1. (Logistic growth with constant harvesting) The equation

$$\frac{dp}{dt} = kp\left(1 - \frac{p}{N}\right) - a$$

represents a logistic model of population growth with constant harvesting at a rate a. For $a = a_1$, what will happen to the fish population for various initial conditions? (Note: This equation is autonomous, so you can take advantage of the special techniques that are available for autonomous equations.)

2. (Logistic growth with periodic harvesting) The equation

$$\frac{dp}{dt} = kp\left(1 - \frac{p}{N}\right) - a(1 + \sin bt)$$

is a nonautonomous equation that considers periodic harvesting. What do the parameters a and b represent? Let $b = 1$. If $a = a_1$, what will happen to the fish population for various initial conditions?

3. Consider the same equation as in Part 2 above, but let $a = a_2$. What will happen to the fish population for various initial conditions with this value of a?

Your report: In your report you should address these three questions, one at a time, in the form of a short essay. Begin Questions 1 and 2 with a description of the meaning of each of the variables and parameters and an explanation of why the differential equation is the way it is. You should include pictures and graphs of data and of solutions of your models *as appropriate*. (Remember that one carefully chosen picture can be worth a thousand words, but a thousand pictures aren't worth anything.)

Table 1.10
Possible choices for the parameters

Choice	k	N	a_1	a_2
1	0.25	4	0.16	0.25
2	0.50	2	0.21	0.25
3	0.20	5	0.21	0.25
4	0.20	5	0.16	0.25
5	0.25	4	0.09	0.25
6	0.20	5	0.09	0.25
7	0.50	2	0.16	0.25
8	0.20	5	0.24	0.25
9	0.25	4	0.21	0.25
10	0.50	2	0.09	0.25

LAB 1.4 Exponential and Logistic Population Models

In the text, we modeled the U.S. population over the last 210 years using both an exponential growth model and a logistic growth model. For this lab project, we ask that you model the population growth of a particular state. Population data for several states are given in Table 1.11. (Your instructor will assign the state(s) you should consider.)

We have also discussed three general approaches that can be employed to study a differential equation: numerical techniques yield graphs of approximate solutions, geometric/qualitative techniques provide predictions of the long-term behavior of the solution, and in special cases analytic techniques provide explicit formulas for the solution. In your report, you should use as many of these techniques as is appropriate to help understand the models.

Table 1.11
Population (in thousands) of selected states (see www.census.gov)

Year	Massachusetts	New York	North Carolina	Alabama	Florida	California	Montana	Hawaii
1790	379	340	394					
1800	423	589	478	1				
1810	472	959	556	9				
1820	523	1373	639	128				
1830	610	1919	738	310	35			
1840	738	2429	753	591	54			
1850	995	3097	869	772	87	93		
1860	1231	3881	993	964	140	380		
1870	1457	4383	1071	997	188	560	21	
1880	1783	5083	1400	1263	269	865	39	
1890	2239	6003	1618	1513	391	1213	143	
1900	2805	7269	1893	1829	529	1485	243	154
1910	3366	9114	2206	2138	753	2378	376	192
1920	3852	10385	2559	2348	968	3427	549	226
1930	4250	12588	3170	2646	1468	5677	538	368
1940	4317	13479	3572	2833	1897	6907	559	423
1950	4691	14830	4062	3062	2771	10586	591	500
1960	5149	16782	4556	3267	4952	15717	675	633
1970	5689	18241	5084	3444	6791	19971	694	770
1980	5737	17558	5880	3894	9747	23668	787	965
1990	6016	17990	6628	4041	12938	29760	799	1108
2000	6349	18976	8049	4447	15982	33871	902	1212

Your report should address the following items:

1. Using an exponential growth model, determine as accurate a prediction as possible for the population of your state in the year 2010. How much does your prediction differ from the prediction that comes from linear extrapolation using the populations in 1990 and 2000? To what extent do solutions of your model agree with the historical data?

2. Produce a logistic growth model for the population of your state. What is the carrying capacity for your model? Using Euler's method, predict the population in the years 2010 and 2050. Using analytic techniques, obtain a formula for the population function $P(t)$ that satisfies your model. To what extent do solutions of your model agree with the historical data?

3. Comment on how much confidence you have in your predictions of the future populations. Discuss which model, exponential or logistic growth, is better for your data and why (and if neither is very good, suggest alternatives).

Your report: The body of your report should address all three items, one at a time, in the form of a short essay. For each model, you must choose specific values for certain parameters (the growth-rate parameter and the carrying capacity). Be sure to give a complete justification of why you made the choices that you did. You should include pictures and graphs of data and of solutions of your models *as appropriate*. (Remember that one carefully chosen picture can be worth a thousand words, but a thousand pictures aren't worth anything.)

LAB 1.5 Modeling Oil Production

There are two things that are clear about crude oil. One is that we use a lot of it. The world consumption of crude oil is approximately 80 million barrels per day, and world consumption grew by 3.4% in 2004.*

The other is that the earth's oil reserves are finite. The processes that created the crude oil that we use today are fairly well understood. There may be significant deposits of crude oil yet to be discovered, but it is a limited resource.

Governments, economists, and scientists argue endlessly about almost every other aspect of oil production. Exactly how much oil is left in the earth and what fraction of that oil can or will ever be removed is difficult to estimate and has significant financial ramifications. Substantial disagreement on oil policy is not surprising.

Predictions of the decline in production are notoriously difficult, and it is easy to find examples of such predictions that ended up being absurdly wrong.♣On the other hand, sometimes predictions of decline in production are accurate. In *Hubbert's Peak,*♡ Kenneth Deffeyes recounts the work of geologist M. King Hubbert. Hubbert fit a logistic model, precisely like those in this chapter, to the production data for crude oil in the

---

*See *New Scientist*, 21 May, 2005, page 7.

♣See, for example, http://www.econlib.org/library/Enc/NaturalResources.html.

♡Deffeyes, K. S., *Hubbert's Peak*, Princeton University Press, Princeton and Oxford, 2001.

United States. Using production data up to the mid 1950s along with approximations of the total amount of recoverable crude oil, Hubbert predicted that production would peak in the U.S. in the 1970s. He was right.

In this lab we model the U.S. and world crude oil production using a logistic model, where the carrying capacity represents the total possible recoverable crude oil. Your report should address the following items:

1. Find parameter values for a logistic differential equation that fit the crude oil production data for the U.S. (see Table 1.12).*

2. Predicting both the growth rate and the total amount of recoverable crude oil from the data is difficult. Model the crude oil production of the U.S. assuming the total amount of recoverable crude oil in the U.S. is 200 billion barrels. (This assumption includes what has already been recovered and serves the role of the carrying capacity in the logistic model.)

3. Repeat Part 2 replacing 200 billion barrels with 300 billion barrels.

4. Model the world crude oil production based on estimates of total recoverable crude oil (past and future) of 2.1 trillion barrels and of 3 trillion barrels. (Both of these estimates are commonly used. They are based on differing assumptions concerning what it means for crude oil to be "recoverable.") When do the models predict that the rate of production of oil reaches its maximum?

5. The decline in production of crude oil will certainly result in an increase in price of oil products. This price increase will provide more funds for crude oil production, perhaps slowing the rate of decline. Describe how this price increase might affect the predictions of your model for world oil production and how you might modify your model to reflect these assumptions.

Your report: Present your models one at a time. Discuss how well they fit the data and how sensitive this fit is to small changes in the parameters.

Table 1.12

Oil production per five year periods in billions of barrels

Year	U.S. Oil	World Oil	Year	U.S. Oil	World Oil
1920–24	2.9	4.3	1965–69	15.8	65.4
1925–29	4.2	6.2	1970–74	17.0	93.9
1930–34	4.3	7.0	1975–79	15.3	107
1935–39	5.8	9.6	1980–84	15.8	101
1940–44	7.5	11.3	1985–89	15.2	104
1945–49	9.2	15.2	1990–94	12.9	110
1950–54	11.2	22.4	1995–99	11.5	118
1955–59	12.7	31.9	2000–04	10.4	126
1960–64	13.4	44.6	2005–08	7.4	107

*Data from *Twentieth Century Petroleum Statistics*, 1984, by DeGolyer and MacNaughton and www.eia.doe.gov.

2 FIRST-ORDER SYSTEMS

Few phenomena are completely described by a single number. For example, the size of a population of rabbits can be represented using one number, but to know its rate of change, we should consider other quantities such as the size of predator populations and the availability of food.

In this chapter we begin the study of systems of differential equations—systems of equations that involve more than one dependent variable. As with first-order equations, the techniques for studying these systems fall into three general categories: analytic, qualitative, and numeric. Only special types of systems such as linear systems can be solved using analytic methods, so we focus primarily on qualitative and numerical methods in this chapter.

We continue to study models that involve differential equations by discussing some that have more than one dependent variable. One particularly important example is the harmonic oscillator, which has numerous applications in many branches of science such as mechanics, electronics, and physics.

2.1   MODELING VIA SYSTEMS

In this section we discuss models of two very different phenomena—the evolution of the two populations in a predator-prey system, and the motion of a mass-spring system. Initially these models seem quite different, but from the correct point of view, they possess a number of similarities.

The Predator-Prey System Revisited

We begin our study of systems of differential equations by considering two versions of the predator-prey model discussed briefly in Section 1.1. Recall that $R(t)$ denotes the population (in thousands, or millions, or whatever) of prey present at time t and that $F(t)$ denotes the population of predators. We assume that both $R(t)$ and $F(t)$ are nonnegative. One system of differential equations that might govern the changes in the population of these two species is

$$\frac{dR}{dt} = 2R - 1.2RF$$
$$\frac{dF}{dt} = -F + 0.9RF.$$

The $2R$ term in the equation for dR/dt represents exponential growth of the prey in the absence of predators, and the $-1.2RF$ term corresponds to the negative effect on the prey of predator-prey interaction. The $-F$ term in dF/dt corresponds to the assumption that the predators die off if there are no prey to eat, and the $0.9RF$ term corresponds to the positive effect on the predators of predator-prey interaction.

The coefficients 2, -1.2, -1, and 0.9 depend on the species involved. Similar systems with different coefficients are considered in the exercises. (We choose these values of the parameters in this example solely for convenience.)*

The presence of the RF terms in these equations makes this system difficult to solve. It is impossible to derive explicit formulas for the general solution, but there are some initial conditions that do yield simple solutions. For instance, suppose that both $R = 0$ and $F = 0$. Then the right-hand sides of both equations vanish ($dR/dt = dF/dt = 0$) for all t, and consequently the pair of constant functions $R(t) = 0$ and $F(t) = 0$ form a solution to the system. By analogy to first-order equations, we call such a pair of constant functions an **equilibrium solution** to the system. This equilibrium solution makes perfect sense: If both the predator and prey populations vanish, we certainly do not expect the populations to grow at any later time.

We can also look for other values of R and F that correspond to constant

---

*For more details on the development, use, and limitations of this system as a model of predator-prey interactions in the wild, we refer the reader to the excellent discussions in J.P. Dempster, *Animal Population Ecology* (New York: Academic Press, 1975) and M. Braun, *Differential Equations and Their Applications* (New York: Springer-Verlag, 1993).

solutions. We rewrite the system as

$$\frac{dR}{dt} = (2 - 1.2F)R$$

$$\frac{dF}{dt} = (-1 + 0.9R)F$$

and note that both equations vanish if $R = 1/0.9 \approx 1.11$ and $F = 2/1.2 \approx 1.67$. Thus the pair of constant functions $R(t) \approx 1.11$ and $F(t) \approx 1.67$ together form another equilibrium solution. This solution says that, if the prey population is 1.11 and the predator population is 1.67, the system is in perfect balance. There are just enough prey to support a constant predator population of 1.67, and similarly there are neither too many predators (which would cause the population of prey to fall) nor too few (in which case the number of prey would rise). Each species' birth rate is exactly equal to its death rate, and these populations are maintained indefinitely. The system is in *equilibrium*.

For certain initial conditions, we can use the techniques that we have already developed for first-order equations to study systems. For example, if $R = 0$, the first equation in this system vanishes. Therefore the constant function $R(t) = 0$ satisfies this differential equation no matter what initial condition we choose for F. In this case the second differential equation reduces to

$$\frac{dF}{dt} = -F,$$

which we recognize as the exponential decay model for the predator population—a familiar and very simple differential equation. From this equation we know that the population of predators tends exponentially to zero. This entire scenario for $R = 0$ is reasonable because, if there are no prey at some time, then there never will be any prey no matter how many predators there are. Moreover, without a food supply, the predators must die out.

In similar fashion, note that the equation for dF/dt vanishes if $F = 0$, and the equation for dR/dt reduces to

$$\frac{dR}{dt} = 2R,$$

which is an exponential growth model. As we saw in Section 1.1, any nonzero prey population grows without bound under these assumptions. Again, these conclusions make sense because there are no predators to control the growth of the prey population. On the other hand, we could make the more realistic assumption that the prey population obeys a logistic growth law. Our second example in this section incorporates this additional assumption.

$R(t)$- and $F(t)$-graphs

In order to understand all solutions of this predator-prey system

$$\frac{dR}{dt} = 2R - 1.2RF$$
$$\frac{dF}{dt} = -F + 0.9RF,$$

it is important to note that the rate of change of either population depends both on R and on F. Hence we need two numbers, an initial value R_0 of R and an initial value F_0 for F, to determine the manner in which these populations evolve over time. In other words, an **initial condition** which determines a solution to this system of equations is a pair of numbers, R_0 and F_0, which are then used to determine the initial values of dR/dt and dF/dt. This initial condition yields a **solution** of the system which consists of two functions $R(t)$ and $F(t)$ that, taken together, satisfy the system of equations.

For the study of solutions to systems of differential equations, there is good news and bad news. The bad news is that for many systems there are few analytic techniques that yield formulas for the solutions. The good news is that there are numerical and qualitative methods that give us a good understanding of the solutions even if we cannot find analytic representations for them. For example, if we specify the initial conditions $R_0 = 1$ and $F_0 = 0.5$, we can use a numerical method akin to Euler's method to obtain approximate values for the corresponding solutions $R(t)$ and $F(t)$. (We will develop this method in Section 2.5.)

In Figures 2.1 and 2.2 we graph the solutions $R(t)$ and $F(t)$ that correspond to the initial condition $R_0 = 1$ and $F_0 = 0.5$, and we see that both $R(t)$ and $F(t)$ rise and fall in a *periodic* fashion.

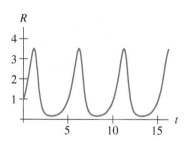

Figure 2.1
The $R(t)$-graph if $R_0 = 1$ and $F_0 = 0.5$.

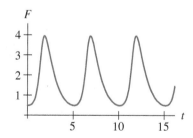

Figure 2.2
The $F(t)$-graph if $R_0 = 1$ and $F_0 = 0.5$.

In Figure 2.3 we graph both $R(t)$ and $F(t)$ on the same set of axes. Although this graph is somewhat misleading because there are really two scales on the vertical axis—one corresponding to the units of $R(t)$ and the other corresponding to the units of $F(t)$, it does provide information that is hard to read from the individual $R(t)$- and $F(t)$-graphs. For example, for this particular solution we see that the increases in the predator population lag the increases in the prey population and that the predator population continues to increase for a short amount of time after the prey population starts

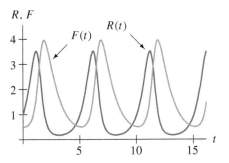

Figure 2.3
The $R(t)$- and $F(t)$-graphs given by the initial condition $R_0 = 1$ and $F_0 = 0.5$. Note that there are really two scales on the vertical axis—one corresponding to the units of R and the other corresponding to the units of F. Note also that both $R(t)$ and $F(t)$ repeat with the same period.

to decline. Perhaps the most important observation that we can make from this graph is that both $R(t)$ and $F(t)$ seem to repeat with the same period (roughly five time units). Although we could reach the same conclusion by closely studying Figures 2.1 and 2.2, this fact is much easier to observe if both the $R(t)$- and $F(t)$-graphs are drawn on the same pair of axes.

The phase portrait for this system

There is another way to graph the solution of the system that corresponds to the initial condition $(R_0, F_0) = (1, 0.5)$. Given $R(t)$, $F(t)$, and a value of t, we can form the pair $(R(t), F(t))$ and think of it as a point in the RF-plane. In other words, the co-ordinates of the point are the values of the two populations at time t. As t varies, the pair $(R(t), F(t))$ sweeps out a curve in the RF-plane. This curve is the **solution curve** determined by the original initial condition. The coordinates of each point on the curve are the prey and predator populations at the associated time t, and the point (R_0, F_0) that corresponds to the initial condition for the solution is often referred to as the **initial point** of this solution curve.

It is often helpful to view a solution curve for a system of differential equations not merely as a set of points in the plane but, rather in a more dynamic fashion, as a point following a curve that is determined by the solution to the differential equation. In Figure 2.4 we show the solution curve corresponding to the solution with initial

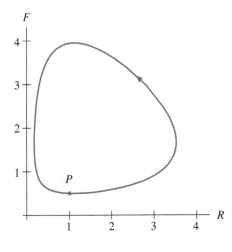

Figure 2.4
The solution curve for the predator-prey system

$$\frac{dR}{dt} = 2R - 1.2RF$$

$$\frac{dF}{dt} = -F + 0.9RF,$$

corresponding to the initial condition $P = (R_0, F_0) = (1, 0.5)$.

conditions $R_0 = 1$ and $F_0 = 0.5$ in the RF-plane. This curve starts at the point $P = (1, 0.5)$. As t increases, the corresponding point on the curve moves to the right. This motion implies that $R(t)$ is increasing but that $F(t)$ initially stays relatively constant. Near $R = 3$, the solution curve turns significantly upward. Thus the predator population $F(t)$ starts increasing significantly. As $F(t)$ nears $F = 2$, the curve starts heading to the left. Thus $R(t)$ has reached a maximum and is starting to decrease. As t increases, the values of $R(t)$ and $F(t)$ change as indicated by the shape of the solution curve. Eventually the solution curve returns to its starting point P and begins its cycle again.

The RF-plane is called the **phase plane**, and it is analogous to the phase line for an autonomous first-order differential equation. Just as the phase line has a point for each value of the dependent variable but does not explicitly show the corresponding value of time, the phase plane has a point for each ordered pair (R, F) of populations. The dependence of a solution on the independent variable t can only be imagined as a point moving along the solution curve as t evolves.

We can plot many solutions curves on the phase plane simultaneously. In Figure 2.5 we see the complete **phase portrait** for our predator-prey system. Of course, since negative populations do not make sense for this model, we restrict our attention to the first quadrant of the RF-plane.

Equilibrium solutions are solutions that are constant, and consequently they produce solution curves $(R(t), F(t))$, where $R(t)$ and $F(t)$ never vary. In other words, the solution curves that correspond to equilibrium solutions are really just points, and we refer to them as **equilibrium points**. Just as with the phase line, the equilibrium points in the phase plane are especially important parts of the phase portrait, and therefore we usually mark them with large dots. (Note the dots at the equilibrium points at $(0, 0)$ and $(1.11, 1.67)$ in Figure 2.5.)

In this predator-prey system, all other solutions for which $R_0 > 0$ and $F_0 > 0$ yield solution curves that loop around the equilibrium point $(1.11, 1.67)$ in a counterclockwise fashion. Ultimately, they return to their initial points, and hence this model predicts that except for the equilibrium solution, both $R(t)$ and $F(t)$ rise and fall in a periodic fashion.

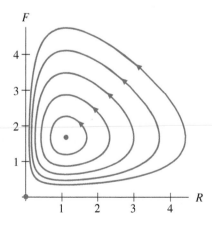

Figure 2.5
The phase plane for the predator-prey system

$$\frac{dR}{dt} = 2R - 1.2RF$$

$$\frac{dF}{dt} = -F + 0.9RF.$$

Note the two equilibruim points. The other solution curves correspond to solutions that are periodic.

A Modified Predator-Prey Model

We now consider a modification of the predator-prey model in which we assume that in the absence of predators, the prey population obeys a logistic rather than an exponential growth model. One such model for this situation is the system

$$\frac{dR}{dt} = 2R\left(1 - \frac{R}{2}\right) - 1.2RF$$
$$\frac{dF}{dt} = -F + 0.9RF.$$

In this system, when the predators are not present (that is, $F = 0$), the prey population obeys a logistic growth model with carrying capacity 2. Once again, with the use of numerical methods we see that the behavior of solution curves (and therefore the predictions made by this model) are quite different from those made by the previous predator-prey model.

First, let's find the equilibrium solutions for this system. Recall that these solutions occur at points (R, F) where the right-hand sides of both of the differential equations vanish. As before, $(R, F) = (0, 0)$ is one equilibrium solution. There are two other equilibria—$(R, F) = (2, 0)$ and $(R, F) = (10/9, 20/27) \approx (1.11, 0.74)$ (see Exercise 12 in Section 2.2).

As in our first predator-prey model, if there are no prey present, the predator population declines exponentially. In other words, if $R = 0$, then $dR/dt = 0$ for all t, so $R(t) = 0$. Then the equation for dF/dt reduces to the familiar exponential decay model

$$\frac{dF}{dt} = -F.$$

In the absence of predators the situation is somewhat different. If $F = 0$, we have $dF/dt = 0$ for all t, and the equation for R simplifies to the familiar logistic model

$$\frac{dR}{dt} = 2R\left(1 - \frac{R}{2}\right).$$

From this equation we see that the growth coefficient for low populations of prey is 2 and the carrying capacity is 2. Thus, if $F = 0$, we expect any nonzero initial population of prey to approach 2 eventually.

When both R and F are nonzero, the evolution of the two populations is more complicated. In Figure 2.6 we plot three solution curves for $t \geq 0$. Note that, in all cases, the solutions tend to the equilibrium point A, which has coordinates $(R, F) = (1.11, 0.74)$. Once we have the solution curve that corresponds to a given initial condition, we know what the model predicts for the solution that satisfies this initial condition. For example, we see that the initial condition B in Figure 2.6 corresponds to an overabundance of both predators and prey. Following the solution curve we see that the predator population initially rises while the prey population declines. However, once the supply of prey is sufficiently low, the predator population declines and eventually approaches the equilibrium value $F = 0.74$. On the other hand, the prey population

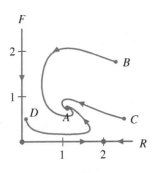

Figure 2.6
The three equilibrium points and three solution curves for the logistic predator-prey model.

eventually recovers, and this population also tends to stabilize at the equilibrium value $R = 1.11$. This evolution of $R(t)$ and $F(t)$ is exactly what we see if we plot the corresponding $R(t)$- and $F(t)$-graphs (see Figure 2.7).

Figure 2.7
The $R(t)$- and $F(t)$-graphs for the solution curve B in Figure 2.6.

The other two solution curves that are shown in Figure 2.6 can be interpreted in a similar fashion (see Figures 2.8 and 2.9). Note that the graphs of both $F(t)$ and $R(t)$ tend to the equilibrium values $R = 1.11$ and $F = 0.74$. We can predict this from the solution curves in the phase plane (see Figure 2.6).

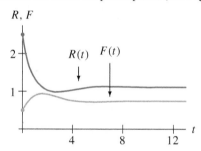

Figure 2.8
The $R(t)$- and $F(t)$-graphs for the solution curve C in Figure 2.6.

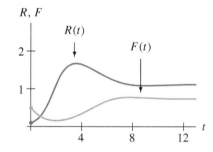

Figure 2.9
The $R(t)$- and $F(t)$-graphs for the solution curve D in Figure 2.6.

The Motion of a Mass Attached to a Spring

At first glance, the standard model of the motion of an undamped mass-spring system seems quite different from the population models that we have just discussed, but there are some important similarities in the corresponding mathematical models.

Consider a mass that is attached to a spring and that slides on a frictionless table (see Figure 2.10). We wish to understand its horizontal motion when the spring is

Figure 2.10
A mass-spring system.

stretched (or compressed) and then released. In order to keep the model as simple as possible, we assume that the only force acting on the mass is the force of the spring. In particular, we ignore air resistance and other forces that would dampen the motion of the mass.

There are two key quantities in this model—a quantity that measures the displacement of the mass from its natural rest position and the restoring force on the mass caused by the spring. We wish to determine the position of the mass as a function of time, so we let $y(t)$ denote the position of the mass at time t. It is convenient to let $y = 0$ represent the rest position of the mass (see Figure 2.11). At the rest position the spring is neither stretched nor compressed, and it exerts no force on the mass. We adopt the convention that $y(t) < 0$ if the spring is compressed and $y(t) > 0$ if the spring is stretched using whatever units are convenient (see Figures 2.11–2.13).

The main idea from physics needed to derive the differential equation that models this motion is Newton's second law,

$$\text{Force } F = \text{mass} \times \text{acceleration}.$$

Since the displacement is $y(t)$, the acceleration is d^2y/dt^2. If we let m denote the mass, Newton's law becomes

$$F = m\frac{d^2y}{dt^2}.$$

To complete the model we must specify an expression for the force that the spring exerts on the mass. We use **Hooke's law of springs** as our model for the restoring force F_s of the spring:

The restoring force exerted by a spring is linearly proportional to the spring's displacement from its rest position and is directed toward the rest position.

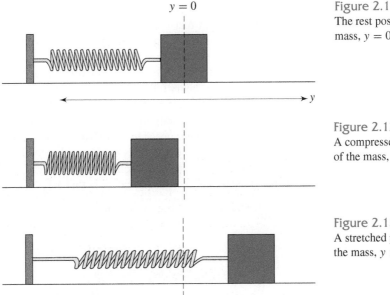

$y = 0$

Figure 2.11
The rest position of the mass, $y = 0$.

Figure 2.12
A compressed position of the mass, $y < 0$.

Figure 2.13
A stretched position of the mass, $y > 0$.

Therefore we have

$$F_s = -ky,$$

where $k > 0$ is a constant of proportionality called the **spring constant**—a parameter we can adjust by changing springs. Combining this expression for the force with Newton's law, we obtain the differential equation

$$F_s = -ky = m\frac{d^2y}{dt^2},$$

which models the motion of the mass. It is traditional to rewrite this equation in the form

$$\frac{d^2y}{dt^2} + \frac{k}{m}y = 0.$$

This equation is the differential equation for what is often called a simple (or undamped) **harmonic oscillator**. Since the equation contains the second derivative of the dependent variable y, it is a **second-order** differential equation. The coefficients m and k are parameters that are determined by the particular mass and spring involved.

From a notational point of view, this second-order equation seems to have little in common with the first-order predator-prey systems that we discussed earlier in this section. In particular, the equation contains only a single dependent variable, and it involves a second derivative rather than two first derivatives.

However, once we attempt to use this second-order equation to describe the motion of a particular mass-spring system, the similarities start to emerge. For example, suppose that we want to describe the motion of the mass. What do we need for initial conditions? Certainly we need an initial condition y_0 that corresponds to the initial displacement of the mass, but does y_0 alone determine the subsequent motion of the mass? The answer is no because we cannot ignore the initial velocity v_0 of the mass. For example, the motion that results from extending the mass-spring system by 1 foot and releasing it is different than the motion that results from extending the system by 1 foot and then pushing with an initial velocity of 1 foot/second. There is a theory of existence and uniqueness for solutions to this equation just as with first-order equations (see Section 2.5), and this theory tells us that we need two numbers, y_0 and v_0, to determine the motion of the simple harmonic oscillator.

Now that the velocity of the motion has been identified as a key part of the overall picture, we are only one step away from completing the analogy between first-order systems such as the predator-prey system and second-order equations such as the equation for the simple harmonic oscillator. If we let $v(t)$ denote the velocity of the mass at time t, then we know from calculus that $v = dy/dt$. Therefore, the acceleration d^2y/dt^2 is the derivative dv/dt of the velocity, and we can rewrite our second-order equation

$$\frac{d^2y}{dt^2} = -\frac{k}{m}y$$

as

$$\frac{dv}{dt} = -\frac{k}{m}y.$$

In other words, we can rewrite the second-order equation as the first-order system

$$\frac{dy}{dt} = v$$
$$\frac{dv}{dt} = -\frac{k}{m}y.$$

This technique of reducing the order of the system by increasing the number of dependent variables gives us two ways of representing the same model for the motion of the mass. Each representation has its advantages and disadvantages. The representation of the mass-spring system as a second-order equation involving one variable is more convenient for certain analytic techniques, whereas the representation as a first-order system is much better for numerical and qualitative analysis.

An initial-value problem

To demonstrate the connections between these two points of view, we consider a very specific initial-value problem. Suppose m and k are fixed so that $k/m = 1$. Then the second-order equation simplifies to

$$\frac{d^2y}{dt^2} = -y.$$

In other words, the second derivative of $y(t)$ is $-y(t)$. Two such functions, sine and cosine, come to mind immediately. As we will see in Chapter 3, there are many other functions that also satisfy this differential equation, but for the purposes of this discussion, we focus on the initial-value problem $(y(0), v(0)) = (y_0, v_0) = (1, 0)$. In this case the function $y(t) = \cos t$ satisfies this initial condition since $y(0) = \cos 0 = 1$ and $y'(0) = -\sin 0 = 0$.

If we convert this second-order equation to a first-order system where $v = dy/dt$, we obtain

$$\frac{dy}{dt} = v$$
$$\frac{dv}{dt} = -y.$$

In this context the same initial condition yields a solution that consists of the pair of functions $y(t) = \cos t$ and $v(t) = -\sin t$. Their $y(t)$- and $v(t)$-graphs are shown in Figure 2.14.

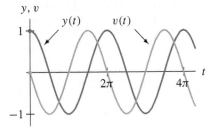

Figure 2.14
Graphs of the solutions $y(t)$ and $v(t)$ for the initial-value problem

$$\frac{d^2y}{dt^2} + y = 0, \quad y(0) = 1, \quad v(0) = 0.$$

Figure 2.15
The solution curve in the yv-phase plane for the initial-value problem

$$\frac{d^2y}{dt^2} + y = 0,$$

with the initial-condition $(y(0), v(0)) = (1, 0)$ is a circle of radius 1 centered at the origin.

In the yv-phase plane the corresponding solution curve is

$$(y(t), v(t)) = (\cos t, -\sin t).$$

With the help of a little trigonometry we see that

$$y^2 + v^2 = (\cos t)^2 + (-\sin t)^2 = 1,$$

and therefore this curve sweeps out the unit circle centered at the origin. Due to the minus sign in $v(t) = -\sin t$, the unit circle is swept out in a clockwise direction (as indicated by the arrowhead on the circle in Figure 2.15).

Either the periodic $y(t)$- and $v(t)$-graphs (Figure 2.14) or the parameterization of the unit circle in the yv-plane (Figure 2.15) indicates that the solution is periodic, with $y(t)$ and $v(t)$ alternately increasing and decreasing, repeating the same cycle again and again. The mass oscillates back and forth across its rest position, $y = 0$, forever. Of course, this phenomenon is possible only because we have neglected damping.

Taken together, Figures 2.14 and 2.15 give a complete picture of the solution. It would be nice if we could make one picture that included all of the information in both Figures 2.14 and 2.15. Such a picture must be three-dimensional since three important variables—t, y, and v—are involved. Due to the fact that we are so familiar with the functions that arise in this example, we can be successful for this equation (see Figure 2.16). Note that Figure 2.14 comes from the projections of Figure 2.16 into both the ty- and tv-planes and that Figure 2.15 is the projection of Figure 2.16 into the yv-phase plane.

Drawing these types of three-dimensional figures requires considerable graphical skill even when the solution $y(t)$ is the very familiar cosine function. In addition, interpreting these pictures requires an even greater skill in visualization. We therefore generally avoid graphs that involve all three variables at once. We restrict our attention to the graphs of the solutions, the $y(t)$- and $v(t)$-graphs, and the solution curve in the yv-phase plane.

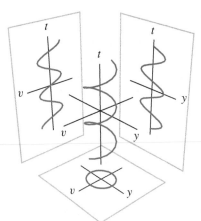

Figure 2.16
The graph of a solution of

$$\frac{d^2y}{dt^2} + y = 0$$

in tyv-space and its projections onto the ty-, tv-, and yv-coordinate planes. Note that the $y(t)$-graph is the graph of $\cos t$, the $v(t)$-graph is the graph of $-\sin t$, and the solution curve in the yv-phase plane is the unit circle.

The Study of Systems of Differential Equations

In Chapter 1 we learned that there are three basic ways to understand the solutions of a differential equation—with the use of analytic, geometric (or qualitative), and numeric techniques. In the subsequent sections of this chapter, we will concentrate on analogous approaches for systems and second-order equations. In the next section we introduce vector notation in order to provide a geometric approach. In Sections 2.3 and 2.4, we discuss analytic techniques that we can use to find explicit formulas for solutions in somewhat specialized situations, and in Section 2.5 we generalize Euler's method to systems of differential equations.

EXERCISES FOR SECTION 2.1

Exercises 1–6 refer to the following systems of equations:

(i)
$$\frac{dx}{dt} = 10x\left(1 - \frac{x}{10}\right) - 20xy$$
$$\frac{dy}{dt} = -5y + \frac{xy}{20}$$

(ii)
$$\frac{dx}{dt} = 0.3x - \frac{xy}{100}$$
$$\frac{dy}{dt} = 15y\left(1 - \frac{y}{15}\right) + 25xy.$$

1. In one of these systems, the prey are very large animals and the predators are very small animals, such as elephants and mosquitoes. Thus it takes many predators to eat one prey, but each prey eaten is a tremendous benefit for the predator population. The other system has very large predators and very small prey. Determine which system is which and provide a justification for your answer.

2. Find all equilibrium points for the two systems. Explain the significance of these points in terms of the predator and prey populations.

3. Suppose that the predators are extinct at time $t_0 = 0$. For each system, verify that the predators remain extinct for all time.

4. For each system, describe the behavior of the prey population if the predators are extinct. (Sketch the phase line for the prey population assuming that the predators are extinct, and sketch the graphs of the prey population as a function of time for several solutions. Then interpret these graphs for the prey population.)

5. For each system, suppose that the prey are extinct at time $t_0 = 0$. Verify that the prey remain extinct for all time.

6. For each system, describe the behavior of the predator population if the prey are extinct. (Sketch the phase line for the predator population assuming that the prey are extinct, and sketch the graphs of the predator population as a function of time for several solutions. Then interpret these graphs for the predator population.)

7. Consider the predator-prey system

$$\frac{dR}{dt} = 2\left(1 - \frac{R}{3}\right)R - RF$$

$$\frac{dF}{dt} = -2F + 4RF.$$

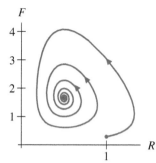

The figure to the right shows a computer-generated plot of a solution curve for this system in the RF-plane.

(a) Describe the fate of the prey (R) and predator (F) populations based on this image.

(b) Confirm your answer using HPGSystemSolver.

8. Consider the predator-prey system

$$\frac{dR}{dt} = 2R\left(1 - \frac{R}{2.5}\right) - 1.5RF$$

$$\frac{dF}{dt} = -F + 0.8RF$$

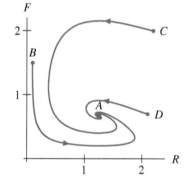

and the solution curves in the phase plane on the right.

(a) Sketch the $R(t)$- and $F(t)$-graphs for the solutions with initial points A, B, C, and D.

(b) Interpret each solution curve in terms of the behavior of the populations over time.

(c) Confirm your answer using HPGSystemSolver.

Exercises 9–14 refer to the predator-prey and the modified predator-prey systems discussed in the text (repeated here for convenience):

(i)
$$\frac{dR}{dt} = 2R - 1.2RF$$
$$\frac{dF}{dt} = -F + 0.9RF$$

(ii)
$$\frac{dR}{dt} = 2R\left(1 - \frac{R}{2}\right) - 1.2RF$$
$$\frac{dF}{dt} = -F + 0.9RF.$$

9. How would you modify these systems to include the effect of hunting of the prey at a rate of α units of prey per unit of time?

10. How would you modify these systems to include the effect of hunting of the predators at a rate proportional to the number of predators?

11. Suppose the predators discover a second, unlimited source of food, but they still prefer to eat prey when they can catch them. How would you modify these systems to include this assumption?

12. Suppose the predators found a second food source that is limited in supply. How would you modify these systems to include this fact?

13. Suppose predators migrate to an area if there are five times as many prey as predators in that area (that is, if $R > 5F$), and they move away if there are fewer than five times as many prey as predators. How would you modify these systems to take this into account?

14. Suppose prey move out of an area at a rate proportional to the number of predators in the area. How would you modify these systems to take this into account?

15. Consider the two systems of differential equations

(i)
$$\frac{dx}{dt} = 0.3x - 0.1xy$$
$$\frac{dy}{dt} = -0.1y + 2xy$$

(ii)
$$\frac{dx}{dt} = 0.3x - 3xy$$
$$\frac{dy}{dt} = -2y + 0.1xy.$$

One of these systems refers to a predator-prey system with very lethargic predators—those who seldom catch prey but who can live for a long time on a single prey (for example, boa constrictors). The other system refers to a very active predator that requires many prey to stay healthy (such as a small cat). The prey in each case is the same. Identify which system is which and justify your answer.

16. Consider the system of predator-prey equations

$$\frac{dR}{dt} = 2\left(1 - \frac{R}{3}\right)R - RF$$
$$\frac{dF}{dt} = -16F + 4RF.$$

The figure below shows a computer-generated plot of a solution curve for this system in the RF-plane.

(a) What can you say about the fate of the rabbit R and fox F populations based on this image?

(b) Confirm your answer using HPGSystemSolver.

17. Pesticides that kill all insect species are not only bad for the environment, but they can also be inefficient at controlling pest species. Suppose a pest insect species in a particular field has population $R(t)$ at time t, and suppose its primary predator is another insect species with population $F(t)$ at time t. Suppose the populations of these species are accurately modeled by the system

$$\frac{dR}{dt} = 2R - 1.2RF$$

$$\frac{dF}{dt} = -F + 0.9RF$$

studied in this section. Finally, suppose that at time $t = 0$ a pesticide is applied to the field that reduces both the pest and predator populations to very small but nonzero numbers.

   (a) Using Figures 2.3 and 2.5, predict what will happen as t increases to the population of the pest species.

   (b) Write a short essay, in nontechnical language, warning of the possibility of the "paradoxical" effect that pesticide application can have on pest populations.

18. Some predator species seldom capture healthy adult prey, eating only injured or weak prey. Because weak prey consume resources but are not as successful at reproduction, the harsh reality is that their removal from the population increases prey population. Discuss how you would modify a predator-prey system to model this sort of interaction.

19. Consider the initial-value problem

$$\frac{d^2y}{dt^2} + y = 0$$

with $y(0) = 0$ and $y'(0) = v(0) = 1$.

   (a) Show that the function $y(t) = \sin t$ is a solution to this initial-value problem.

   (b) Plot the solution curve corresponding to this solution in the yv-plane.

   (c) In what ways is this solution curve the same as the one shown in Figure 2.15?

   (d) How is this curve different from the one shown in Figure 2.15?

20. Consider the equation

$$\frac{d^2y}{dt^2} + \frac{k}{m}y = 0$$

for the motion of a simple harmonic oscillator.

   (a) Consider the function $y(t) = \cos \beta t$. Under what conditions on β is $y(t)$ a solution?

   (b) What initial condition ($t = 0$) in the yv-plane corresponds to this solution?

   (c) In terms of k and m, what is the period of this solution?

   (d) Sketch the solution curve (in the yv-plane) associated to this solution. [*Hint:* Consider the quantity $y^2 + (v/\beta)^2$.]

21. A mass weighing 12 pounds stretches a spring 3 inches. What is the spring constant for this spring?

22. A mass weighing 4 pounds stretches a spring 4 inches.

  (a) Formulate an initial-value problem that corresponds to the motion of this un-damped mass-spring system if the mass is extended 1 foot from its rest position and released (with no initial velocity).

  (b) Using the result of Exercise 20, find the solution of this initial-value problem.

23. Do the springs in an "extra firm" mattress have a large spring constant or a small spring constant?

24. Consider a vertical mass-spring system as shown in the figure below.

Before the mass is placed on the end of the spring, the spring has a natural length. After the mass is placed on the end of the spring, the system has a new equilibrium position, which corresponds to the position where the force on the mass due to gravity is equal to the force on the mass due to the spring.

  (a) Assuming that the only forces acting on the mass are the force due to gravity and the force of the spring, formulate two different (but related) second-order differential equations that describe the motion of the mass. For one equation, let the position $y_1(t)$ be measured from the point at the end of the spring when it hangs without the mass attached. For the other equation, let $y_2(t)$ be measured from the equilibrium position once the mass is attached to the spring.

  (b) Rewrite these two second-order equations as first-order systems and calculate their equilibrium points. Interpret your results in terms of the mass-spring system.

  (c) Given a solution $y_1(t)$ to one system, how can you produce a solution $y_2(t)$ to the second system?

  (d) Which choice of coordinate system, y_1 or y_2, do you prefer? Why?

Exercises 25–30 refer to a situation in which models similar to the predator-prey population models arise. Suppose A and B represent two substances that can combine to form a new substance C (chemists would write A + B → C). Suppose we have a container with a solution containing low concentrations of substances A and B, and A and B molecules react only when they happen to come close to each other. If $a(t)$ and $b(t)$ represent the amount of A and B in the solution, respectively, then the chance that a molecule of A is close to a molecule of B at time t is proportional to the product $a(t) \cdot b(t)$. Hence the rate of reaction of A and B to form C is proportional to ab. Suppose C precipitates out of the solution as soon as it is formed, and the solution is always kept well mixed.

25. Write a system of differential equations that models the evolution of $a(t)$ and $b(t)$. Be sure to identify and describe any parameters you introduce.

26. Describe an experiment you could perform to determine an approximate value for the parameter(s) in the system you developed in Exercise 25. Include the calculations you would perform using the data from your experiment to determine the parameter(s).

27. Suppose substances A and B are added to the solution at constant (perhaps unequal) rates. How would you modify your system to include this assumption?

28. Suppose A and B are being added to the solution at constant (perhaps unequal) rates, and, in addition to the A + B → C reaction, a reaction A + A → D also can occur when two molecules of A are close and substance D precipitates out of the solution. How would you modify your system of equations to include these assumptions?

29. Suppose A and B are being added to the solutions at constant (perhaps unequal) rates, and, in addition to the A + B → C reaction, a reaction B + B → A can also occur when two molecules of B are close. How would you modify your system of equations to include these assumptions?

30. Suppose A and B are being added to the solution at constant (perhaps unequal) rates, and, in addition to the A + B → C reaction, a reaction A + 2B → D can occur when two B and one A molecules are close. Suppose substance D precipitates out of the solution. How would you modify your system of equations to include these assumptions?

2.2 THE GEOMETRY OF SYSTEMS

In Section 2.1 we displayed $R(t)$- and $F(t)$-graphs of solutions to two different predator-prey systems, but we did not describe how we generated these graphs. We will ultimately answer this question in Section 2.5 in which we generalize Euler's method to produce numerical approximations to solutions, but we must begin the explanation by introducing some vector notation. This notation provides a convenient shorthand for writing systems of differential equations, but it is also important for a more fundamental reason. Using vectors, we build a geometric representation of a system of differential equations. As we saw when we used slope fields in Chapter 1, having a geometric representation of a differential equation gives us a convenient way to understand its solutions.

The Predator-Prey Vector Field

Recall that the predator-prey system

$$\frac{dR}{dt} = 2R - 1.2RF$$

$$\frac{dF}{dt} = -F + 0.9RF$$

models the evolution of two populations, R and F, over time. In the previous section we studied two different (but related) ways to visualize this evolution. We can plot the graphs of $R(t)$ and $F(t)$ as functions of t, or we can plot the solution curve $(R(t), F(t))$ in the RF-plane. Although we can think of $(R(t), F(t))$ as simply a combination of the two scalar-valued functions $R(t)$ and $F(t)$, there are advantages if we take a different approach. We consider the pair $(R(t), F(t))$ as a vector-valued function in the RF-plane.

For each t, we let $\mathbf{P}(t)$ denote the vector

$$\mathbf{P}(t) = \begin{pmatrix} R(t) \\ F(t) \end{pmatrix}.$$

Then the vector-valued function $\mathbf{P}(t)$ corresponds to the solution curve $(R(t), F(t))$ in the RF-plane.

To compute the derivative of the vector-valued function $\mathbf{P}(t)$, we compute the derivatives of each component. That is,

$$\frac{d\mathbf{P}}{dt} = \begin{pmatrix} \dfrac{dR}{dt} \\ \dfrac{dF}{dt} \end{pmatrix}.$$

Using this notation, we can rewrite the predator-prey system as the single vector equation

$$\frac{d\mathbf{P}}{dt} = \begin{pmatrix} \dfrac{dR}{dt} \\ \dfrac{dF}{dt} \end{pmatrix} = \begin{pmatrix} 2R - 1.2RF \\ -F + 0.9RF \end{pmatrix}.$$

So far we have only introduced more notation. We have converted our first-order system consisting of two scalar equations into a single vector equation involving vectors with two components.

The advantages of the vector notation start to become evident once we consider the right-hand side of this system as a *vector field*. The right-hand side of the predator-prey system is a function that assigns a vector to each point in the RF-plane. If we denote this function using the vector **V**, we have

$$\mathbf{V}\left(\begin{array}{c} R \\ F \end{array}\right) = \left(\begin{array}{c} 2R - 1.2RF \\ -F + 0.9RF \end{array}\right).$$

For example, at the point $(R, F) = (2, 1)$,

$$\mathbf{V}\left(\begin{array}{c} 2 \\ 1 \end{array}\right) = \left(\begin{array}{c} 2(2) - 1.2(2)(1) \\ -(1) + 0.9(2)(1) \end{array}\right) = \left(\begin{array}{c} 1.6 \\ 0.8 \end{array}\right).$$

To save paper, we will sometimes write vectors vertically (as "column" vectors) and at other times horizontally (as "row" vectors). The vertical notation is more consistent with how we have written systems up to now, whereas the horizontal notation is easier on trees. In any case, we always write vectors in boldface type to distinguish them from scalars. Written as a row vector, the predator-prey vector field is expressed as

$$\mathbf{V}(R, F) = (2R - 1.2RF, -F + 0.9RF),$$

and $\mathbf{V}(2, 1) = (1.6, 0.8)$.

In the previous computation there was nothing special about the point $(R, F) = (2, 1)$. Similarly, we have $\mathbf{V}(1, 1) = (0.8, -0.1)$, $\mathbf{V}(0.5, 2.2) = (-0.32, -1.21)$, and so forth. The function $\mathbf{V}(R, F)$ can be evaluated at any point in the RF-plane.

The use of vectors enables us to simplify the notation considerably. We can now write the predator-prey system very economically as

$$\frac{d\mathbf{P}}{dt} = \mathbf{V}(\mathbf{P}).$$

The vector notation is much more than just a way to save ink. It also gives us a new way to think about and to visualize systems of differential equations. We can sketch the vector field **V** by attaching the vector $\mathbf{V}(\mathbf{P})$ to the corresponding point **P** in the plane. Computing $\mathbf{V}(\mathbf{P})$ for many different values of **P** and carefully sketching these vectors in the plane is tedious work for a human, but it is just the sort of job that computers and calculators are good at. A few vectors in the predator-prey vector field **V** are shown in Figure 2.17. In general we visualize this vector field as a "field" of arrows, one based at each point in the RF-plane.

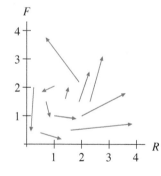

Figure 2.17

Selected vectors in the vector field $\mathbf{V}(R, F)$.

The Vector Field for a Simple Harmonic Oscillator

In Section 2.1 we modeled the motion of an undamped mass-spring system by a second-order differential equation of the form

$$\frac{d^2y}{dt^2} + \frac{k}{m}y = 0,$$

where k is the spring constant and m is the mass. We also saw that this mass-spring system can be written as the first-order system

$$\frac{dy}{dt} = v$$

$$\frac{dv}{dt} = -\frac{k}{m}y,$$

where $v = dy/dt$ is the velocity of the mass. In the special case where $k/m = 1$, we obtained the especially nice system

$$\frac{dy}{dt} = v$$

$$\frac{dv}{dt} = -y.$$

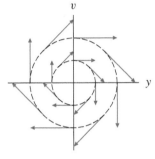

Figure 2.18
Selected vectors in the vector field $\mathbf{F}(y, v) = (v, -y)$

One reason this system is so nice is that its vector field $\mathbf{F}(y, v) = (v, -y)$ in the yv-plane is relatively easy to understand. After plotting a few vectors in the vector field, it is natural to wonder if all of the vectors are tangent to circles centered at the origin and in fact, they are (see Figure 2.18 and Exercise 20).

   Although computers can take the tedium out of the process of plotting vector fields, there is one aspect of vector fields that make them much harder to plot than slope fields. By definition, the vectors in a vector field have various lengths as determined by the system of equations. Some of the vectors can be quite short while others can be quite long. Therefore if we plot a vector field by evaluating it over a regular grid in the plane, we often get overlapping vectors. For example, Figure 2.19 is a plot of the vector field $\mathbf{F}(y, v) = (v, -y)$ for the simple harmonic oscillator. We don't need to take many points before we end up with a picture that is basically useless.

   To avoid the confusion of overlapping vectors in our pictures of vector fields, we often scale the vectors so they all have the same (short) length. The resulting picture is called the **direction field** associated to the original vector field. Figure 2.20 is a plot of the direction field associated to the vector field $\mathbf{F}(y, v) = (v, -y)$ for a simple harmonic oscillator.

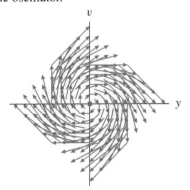

Figure 2.19
Vector field for $\mathbf{F}(y, v) = (v, -y)$.

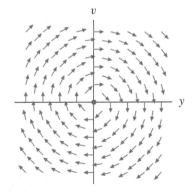

Figure 2.20
Direction field for $\mathbf{F}(y, v) = (v, -y)$.

While the direction field gives a picture that is much easier to visualize than the vector field, there is some loss of information. The lengths of a vector in the vector field give the speed of the solution as it passes through the associated point in the plane. In the direction field all information about the speed of the solution is lost. Because of the artistic advantages of using the direction field, we are almost always willing to live with this loss.

Examples of Systems and Vector Fields

In general, for a system with two dependent variables of the form

$$\frac{dx}{dt} = f(x, y)$$
$$\frac{dy}{dt} = g(x, y),$$

we introduce the vector $\mathbf{Y}(t) = (x(t), y(t))$ and the **vector field**

$$\mathbf{F}(\mathbf{Y}) = \mathbf{F}(x, y) = (f(x, y), g(x, y)).$$

With this notation the system of two equations may be written in the compact form

$$\frac{d\mathbf{Y}}{dt} = \begin{pmatrix} \dfrac{dx}{dt} \\ \dfrac{dy}{dt} \end{pmatrix} = \begin{pmatrix} f(x, y) \\ g(x, y) \end{pmatrix} = \mathbf{F}(\mathbf{Y}),$$

or even more economically as

$$\frac{d\mathbf{Y}}{dt} = \mathbf{F}(\mathbf{Y}).$$

Elementary (but important) examples
The system

$$\frac{dx}{dt} = x$$
$$\frac{dy}{dt} = y$$

yields the vector field $\mathbf{F}(x, y) = (x, y)$, and the vectors in the vector field always point directly away from the origin (see Figure 2.21). On the other hand, the system

$$\frac{dx}{dt} = -x$$
$$\frac{dy}{dt} = -y$$

yields the vector field $\mathbf{G}(x, y) = (-x, -y)$, and the vectors in the vector field always point toward the origin (see Figure 2.22).

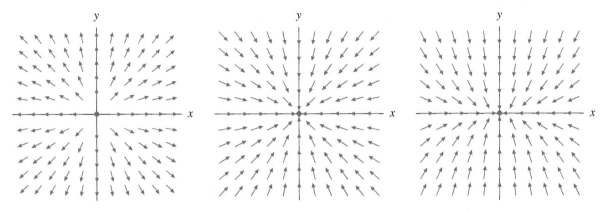

Figure 2.21
Direction field for $\mathbf{F}(x, y) = (x, y)$.

Figure 2.22
Direction field for $\mathbf{G}(x, y) = (-x, -y)$.

Figure 2.23
Direction field for $\mathbf{H}(x, y) = (-x, -2y)$.

The system

$$\frac{dx}{dt} = -x$$

$$\frac{dy}{dt} = -2y$$

also yields a vector field $\mathbf{H}(x, y) = (-x, -2y)$ which (more or less) points toward the origin (see Figure 2.23). We will soon see that the trained eye can distinguish important differences between the vector field $\mathbf{G}(x, y)$ in Figure 2.22 and the vector field $\mathbf{H}(x, y)$ in Figure 2.23.

The Geometry of Solutions

We can think of the picture of a vector field or a direction field as a picture of a system of differential equations, and we can use this picture to sketch solution curves of the system. To be more precise, let's consider a system of the form

$$\frac{dx}{dt} = f(x, y)$$

$$\frac{dy}{dt} = g(x, y).$$

As we have seen, this system yields the vector field $\mathbf{F}(x, y) = (f(x, y), g(x, y))$. Letting $\mathbf{Y}(t) = (x(t), y(t))$, the system can be written in terms of the vector equation

$$\frac{d\mathbf{Y}}{dt} = \mathbf{F}(\mathbf{Y}).$$

Interpreting this vector equation geometrically is the key to a geometric understanding of this system of differential equations. If we think of a solution $\mathbf{Y}(t) = (x(t), y(t))$ as a parameterization of a curve in the xy-plane, then $d\mathbf{Y}/dt$ yields the tangent vectors of the curve. Therefore the equation $d\mathbf{Y}/dt = \mathbf{F}(\mathbf{Y})$ says that the tangent vectors for the solution curves are given by the vectors in the vector field.

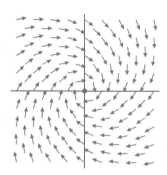

Figure 2.24
A direction field that spirals about the origin.

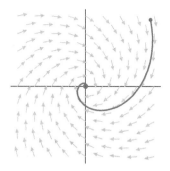

Figure 2.25
A solution curve corresponding to the initial condition indicated.

One consequence of this geometric interpretation is that we can go directly from a vector field **F** (or its direction field) to a sketch of the solution curves of the equation $d\mathbf{Y}/dt = \mathbf{F}(\mathbf{Y})$ without ever knowing a formula for **F** (see Figures 2.24 and 2.25).

Metaphor of the parking lot

To help visualize solution curves of a system from this point of view, imagine an infinite, perfectly flat parking lot. At each point in the lot, an arrow is painted on the pavement. These arrows come from the vector field $\mathbf{F}(\mathbf{Y})$. As you drive through the parking lot, your instructions are to look out your window at the ground and drive so that your velocity vector always agrees with the arrow on the ground. (Imagine you are a professional driver in a closed parking lot.) You steer so that your car goes in the direction given by the direction of the arrow, and you go as fast as the length of the vector indicates. As you move, the arrow outside your window changes, so you must adjust the speed and direction of the car accordingly. The path you follow is the solution curve associated to a solution of the system. In fact, as you will soon see, you can use exactly this idea to sketch solution curves of a system using only this interpretation of the vector field. (*Do not text while you do these exercises.*)

A solution curve of the harmonic oscillator

For example, in Section 2.1 we saw that the functions $y(t) = \cos t$ and $v(t) = -\sin t$ satisfy the simple harmonic oscillator system

$$\frac{dy}{dt} = v$$

$$\frac{dv}{dt} = -y.$$

Since $y^2 + v^2 = 1$, we know that the vector-valued function

$$\mathbf{Y}(t) = (y(t), v(t)) = (\cos t, -\sin t)$$

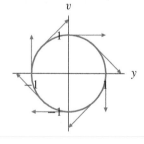

Figure 2.26
The unit circle in the yv-plane is a solution curve for the harmonic oscillator system

sweeps out the unit circle centered at the origin of the yv-plane in a clockwise fashion. As we see in Figure 2.26, the velocity vectors for this motion agree precisely with the vectors in the vector field $\mathbf{F}(y, v) = (v, -y)$.

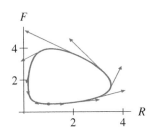

Figure 2.27

A solution curve for a predator-prey system

In Section 2.1 we plotted the solution curve to the system

$$\frac{dR}{dt} = 2R - 1.2RF$$

$$\frac{dF}{dt} = -F + 0.9RF$$

corresponding to the initial condition $(R_0, F_0) = (1, 0.5)$. In Figure 2.27 we see the relationship between the solution curve and the vectors in the vector field from the predator-prey system.

Equilibrium Solutions

Just as there are special points—equilibrium points—on the phase line, there are distinguished points in the phase plane of systems of the form

$$\frac{dx}{dt} = f(x, y)$$

$$\frac{dy}{dt} = g(x, y).$$

These points also correspond to constant solutions.

DEFINITION   The point $\mathbf{Y}_0$ is an **equilibrium point** for the system $d\mathbf{Y}/dt = \mathbf{F}(\mathbf{Y})$ if $\mathbf{F}(\mathbf{Y}_0) = \mathbf{0}$. The constant function $\mathbf{Y}(t) = \mathbf{Y}_0$ is an **equilibrium solution**.   ∎

Equilibrium points are simply points at which the right-hand side of the system vanishes. If $\mathbf{Y}_0$ is an equilibrium point, then the constant function

$$\mathbf{Y}(t) = \mathbf{Y}_0 \quad \text{for } all \ t$$

is a solution of the system. To verify this claim, note that the constant function has $d\mathbf{Y}/dt = (0, 0)$ for all t. On the other hand, $\mathbf{F}(\mathbf{Y}(t)) = \mathbf{F}(\mathbf{Y}_0) = (0, 0)$ at an equilibrium point. Hence equilibrium points in the vector field correspond to constant solutions.

Computation of equilibrium points
The system

$$\frac{dx}{dt} = 3x + y$$

$$\frac{dy}{dt} = x - y$$

has only one equilibrium point, the origin $(0, 0)$. To see why, we simultaneously solve the two equations

$$\begin{cases} 3x + y = 0 \\ x - y = 0, \end{cases}$$

which are given by the right-hand side of the system. (Add the first equation to the second to see that $x = 0$, then use either equation to conclude that $y = 0$.) If we look at the vector field for this system, we see that the vectors near the origin are relatively short (see Figure 2.28). Thus solution curves move slowly as they pass near the origin. Although all nonzero vectors in the direction field are the same length by definition, we can still tell that there must be an equilibrium point at the origin because the directions of the vectors in the direction field change radically near the origin $(0, 0)$ (see Figure 2.29).

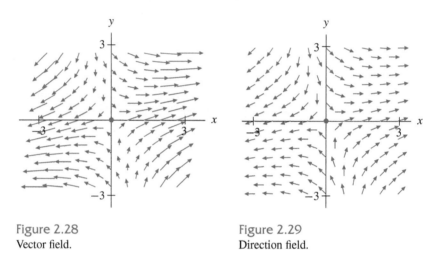

Figure 2.28
Vector field.

Figure 2.29
Direction field.

As a solution passes near an equilibrium point, both dx/dt and dy/dt are close to zero. Therefore, the $x(t)$- and $y(t)$-graphs are nearly flat over the corresponding time interval (see Figure 2.30).

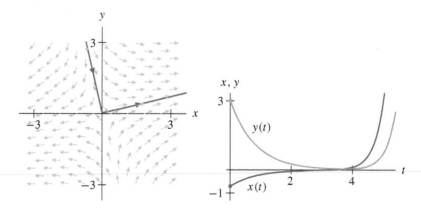

Figure 2.30
As a solution curve travels near an equilibrium point, the $x(t)$- and $y(t)$-graphs are nearly flat.

A Population Model for Two Competing Species

To illustrate all of the concepts introduced in this section, we conclude with an analysis of the system

$$\frac{dx}{dt} = 2x\left(1 - \frac{x}{2}\right) - xy$$

$$\frac{dy}{dt} = 3y\left(1 - \frac{y}{3}\right) - 2xy.$$

We think of x and y as representing the populations of two species that compete for the same resource. Note that, left on its own, each species evolves according to a logistic population growth model. The interaction of the two species is modeled by the xy-terms in both equations. For example, the effect of the population y on the rate of change of x is determined by the term $-xy$ in the dx/dt equation. This term is negative since we are assuming that the two species compete for resources. Similarly, the term $-2xy$ determines the effect of the x population on the rate of change of y. Since x and y represent populations, we focus our attention on the solutions whose initial conditions lie in the first quadrant.

Finding the equilibrium points

First, we find the equilibrium points by setting the right-hand sides of the differential equations to zero and solving for x and y in the resulting system of equations

$$\begin{cases} 2x\left(1 - \frac{x}{2}\right) - xy = 0 \\ 3y\left(1 - \frac{y}{3}\right) - 2xy = 0. \end{cases}$$

These equations can be rewritten in the form

$$\begin{cases} x(2 - x - y) = 0 \\ y(3 - y - 2x) = 0. \end{cases}$$

The first equation is satisfied if $x = 0$ or if $2 - x - y = 0$, and the second equation is satisfied if either $y = 0$ or $3 - y - 2x = 0$.

Suppose first that $x = 0$. Then the equation $y = 0$ yields an equilibrium point at the origin, and the equation $3 - y - 2x = 0$ yields an equilibrium point at $(0, 3)$.

Now suppose that $2 - x - y = 0$. Then the equation $y = 0$ yields an equilibrium point at $(2, 0)$, and the equation $3 - y - 2x = 0$ yields an equilibrium point at $(1, 1)$. (Solve the equations $2 - x - y = 0$ and $3 - y - 2x = 0$ simultaneously.) Hence the equilibrium points are $(0, 0)$, $(0, 3)$, $(2, 0)$, and $(1, 1)$.

Sketching the phase portrait

Next, we use the direction field to sketch solution curves. To get a good sketch of the phase portrait, we must choose enough solutions to see all the different types of solution curves, but not so many curves that the picture gets messy (see Figure 2.31). It is advisable to make the sketch with the aid of a computer or calculator, and in Section 2.5

we will generalize Euler's method to numerically approximate solution curves. Note that the phase portrait for this competing species model suggests that for most initial conditions, one or the other species dies out and the surviving population stabilizes.

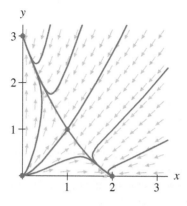

Figure 2.31
Direction field and phase portrait for the competing species model

$$\frac{dx}{dt} = 2x\left(1 - \frac{x}{2}\right) - xy$$
$$\frac{dy}{dt} = 3y\left(1 - \frac{y}{3}\right) - 2xy.$$

Note that this phase portrait suggests that for most initial conditions, one or the other species dies out and the surviving population stabilizes.

Just as we did in Chapter 1 when we started sketching slope fields and graphs of solutions, we should pause and wonder if sketches such as this one represent the true behavior of the solutions. For example, how do we know that distinct solution curves in the phase plane do not cross or even touch? As in Chapter 1 the answer follows from a powerful theorem regarding the uniqueness of solutions. We will study this theorem in Section 2.5, but in the meantime, you should assume that, if the differential equations are sufficiently nice, then distinct solution curves will not cross or even touch.

$x(t)$- and $y(t)$-graphs

As we saw in Section 2.1, the phase portrait is just one way of visualizing the solutions of a system of differential equations. Not all information about a particular solution can be seen by studying its solution curve in the phase plane. In particular, when we look at a solution curve in the phase plane, we do not see the time variable, so we do not know how fast the solution traverses the curve. One way to get information about the time variable is to watch a computer sketch the solution curve in real time. Another way is to view its $x(t)$- and $y(t)$-graphs.

In Figure 2.32, we see the $x(t)$- and $y(t)$-graphs for two solutions of the competing species model. For the initial condition corresponding to the graph on the left, the x population does not die out until at least $t = 15$, but for the initial condition corresponding to the graph on the right, the y population is essentially extinct after $t = 8$.

Even though solution curves and $x(t)$- and $y(t)$-graphs display different information about solutions, it is important to be able to connect the two different representations. The two solution curves that come from these particular initial conditions are shown in Figure 2.33. From the solution curve corresponding to the initial condition on the right, we can conclude that the solution approaches the equilibrium point $(2, 0)$. In particular, for this initial condition the y population becomes extinct. The solution for the left initial condition approaches the equilibrium point $(0, 3)$, so the x population

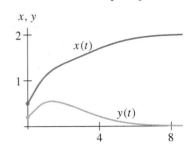

Figure 2.32
The $x(t)$- and $y(t)$-graphs for two solutions with nearby initial conditions.
Note that these graphs illustrate distinctly different long-term behaviors.

becomes extinct. We observed the same long-term behavior when we plotted the $x(t)$-
and $y(t)$-graphs (see Figure 2.32).

In the phase plane we also note that the solution curve for the initial condition on
the left crosses the line $y = x$. In other words, from the solution curve in the phase
plane, we can see from the phase portrait that there is one time t at which the two populations are equal. However, to determine that particular time, we must consult the
corresponding $x(t)$- and $y(t)$-graphs. Similarly, for the other initial condition, we know
that the x population is always larger than the y population.

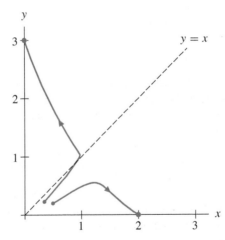

Figure 2.33
Two solution curves for solutions to the system

$$\frac{dx}{dt} = 2x \left(1 - \frac{x}{2} \right) - xy$$

$$\frac{dy}{dt} = 3y \left(1 - \frac{y}{3} \right) - 2xy.$$

These curves correspond to solutions with
nearby initial conditions. The long-term
behavior of these two solutions is also
illustrated in Figure 2.32.

Qualitative Thinking

In all the systems considered so far, the independent variable has not appeared on the
right-hand side. As we said in Section 1.6, systems with this property are said to be **autonomous**. The word *autonomous* means self-governing, and an autonomous system is
self-governing because it evolves according to differential equations that are determined

entirely by the values of the dependent variables. An important geometric consequence of this observation is the fact that the vector field associated with an autonomous system depends *only* on the dependent variables and not explicitly on the value of the independent variable. Therefore, we do not need to consider the independent variable when we sketch the vector field, the direction field, the solution curves, or the phase portrait.

Although we will continue to focus on autonomous systems for the remainder of this chapter and throughout Chapter 3, many important systems are nonautonomous. We will first encounter nonautonomous systems in Chapter 4. In the sections of this chapter that follow this one, we complement the geometric approach introduced here with analytic and numerical approaches.

EXERCISES FOR SECTION 2.2

In Exercises 1–6:

(a) Determine the vector field associated with the first-order system specified.

(b) Sketch enough vectors in the vector field to get a sense of its geometric structure. (You should do this part of the exercise without the use of technology.)

(c) Use HPGSystemSolver to sketch the associated direction field.

(d) Make a rough sketch of the phase portrait of the system and confirm your answer using HPGSystemSolver.

(e) Briefly describe the behavior of the solutions.

1.
$$\frac{dx}{dt} = 1$$
$$\frac{dy}{dt} = 0$$

2.
$$\frac{dx}{dt} = x$$
$$\frac{dy}{dt} = 1$$

3.
$$\frac{dy}{dt} = -v$$
$$\frac{dv}{dt} = y$$

4.
$$\frac{du}{dt} = u - 1$$
$$\frac{dv}{dt} = v - 1$$

5.
$$\frac{dx}{dt} = x$$
$$\frac{dy}{dt} = -y$$

6.
$$\frac{dx}{dt} = x$$
$$\frac{dy}{dt} = 2y$$

7. Convert the second-order differential equation

$$\frac{d^2y}{dt^2} - y = 0$$

into a first-order system in terms of y and v, where $v = dy/dt$.

(a) Determine the vector field associated with the first-order system.

(b) Sketch enough vectors in the vector field to get a sense of its geometric structure. (You should do this part of the exercise without the use of technology.)

(c) Use HPGSystemSolver to sketch the associated direction field.

(d) Make a rough sketch of the phase portrait of the system and confirm your answer using HPGSystemSolver.

(e) Briefly describe the behavior of the solutions.

8. Convert the second-order differential equation

$$\frac{d^2y}{dt^2} + 2y = 0$$

into a first-order system in terms of y and v, where $v = dy/dt$.

(a) Determine the vector field associated with the first-order system.

(b) Sketch enough vectors in the vector field to get a sense of its geometric structure. (You should do this part of the exercise without the use of technology.)

(c) Use HPGSystemSolver to sketch the associated direction field.

(d) Make a rough sketch of the phase portrait of the system and confirm your answer using HPGSystemSolver.

(e) Briefly describe the behavior of the solutions.

9. Consider the system

$$\frac{dx}{dt} = x + 2y$$

$$\frac{dy}{dt} = -y$$

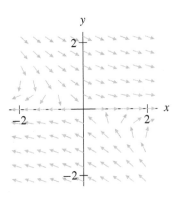

and its corresponding direction field.

(a) Sketch a number of different solution curves on the phase plane.

(b) Describe the behavior of the solution that satisfies the initial condition $(x_0, y_0) = (-2, 2)$.

10. Consider the system

$$\frac{dx}{dt} = -2x + y$$

$$\frac{dy}{dt} = -2y$$

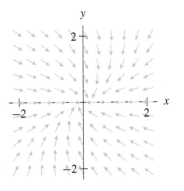

and its corresponding direction field.

(a) Sketch a number of different solution curves on the phase plane.

(b) Describe the behavior of the solution that satisfies the initial condition $(x_0, y_0) = (0, 2)$.

11. Eight systems of differential equations and four direction fields are given below. Determine the system that corresponds to each direction field and state briefly how you know your choice is correct. You should do this exercise without using technology.

(i)
$$\frac{dx}{dt} = -x$$
$$\frac{dy}{dt} = y - 1$$

(ii)
$$\frac{dx}{dt} = x^2 - 1$$
$$\frac{dy}{dt} = y$$

(iii)
$$\frac{dx}{dt} = x + 2y$$
$$\frac{dy}{dt} = -y$$

(iv)
$$\frac{dx}{dt} = 2x$$
$$\frac{dy}{dt} = y$$

(v)
$$\frac{dx}{dt} = x$$
$$\frac{dy}{dt} = 2y$$

(vi)
$$\frac{dx}{dt} = x - 1$$
$$\frac{dy}{dt} = -y$$

(vii)
$$\frac{dx}{dt} = x^2 - 1$$
$$\frac{dy}{dt} = -y$$

(viii)
$$\frac{dx}{dt} = x - 2y$$
$$\frac{dy}{dt} = -y$$

(a)

(b)

(c)

(d)

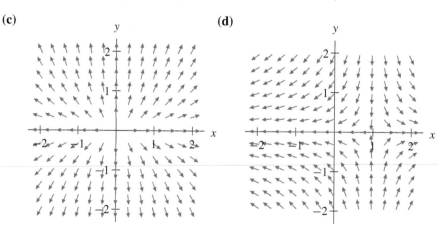

12. Consider the modified predator-prey system

$$\frac{dR}{dt} = 2R\left(1 - \frac{R}{2}\right) - 1.2RF$$

$$\frac{dF}{dt} = -F + 0.9RF$$

discussed in Section 2.1. Find all equilibrium solutions.

In Exercises 13–18,

  (a) find the equilibrium points of the system,

  (b) using `HPGSystemSolver`, sketch the direction field and phase portrait of the system, and

  (c) briefly describe the behavior of typical solutions.

13. $\dfrac{dx}{dt} = 4x - 7y + 2$

    $\dfrac{dy}{dt} = 3x + 6y - 1$

14. $\dfrac{dR}{dt} = 4R - 7F - 1$

    $\dfrac{dF}{dt} = 3R + 6F - 12$

15. $\dfrac{dz}{dt} = \cos w$

    $\dfrac{dw}{dt} = -z + w$

16. $\dfrac{dx}{dt} = y$

    $\dfrac{dy}{dt} = x - x^3 - y$

17. $\dfrac{dx}{dt} = y$

    $\dfrac{dy}{dt} = -\cos x - y$

18. $\dfrac{dx}{dt} = y(x^2 + y^2 - 1)$

    $\dfrac{dy}{dt} = -x(x^2 + y^2 - 1)$

19. Convert the second-order differential equation

$$\frac{d^2x}{dt^2} + 2\frac{dx}{dt} - 3x + x^3 = 0$$

into a first-order system in terms of x and v, where $v = dx/dt$.

  (a) Determine the vector field associated with the first-order system.

  (b) Find all equilibrium points.

  (c) Use `HPGSystemSolver` to sketch the associated direction field.

  (d) Use the direction field to make a rough sketch of the phase portrait of the system and confirm your answer using `HPGSystemSolver`.

  (e) Briefly describe the behavior of the solutions.

20. Show that all vectors in the vector field $\mathbf{F}(y, v) = (v, -y)$ are tangent to circles centered at the origin (see Figure 2.18). [*Hint:* You can verify this fact using slopes or the dot product of two vectors.]

21. Consider the four solution curves in the phase portrait and the four pairs of $x(t)$- and $y(t)$-graphs shown below.

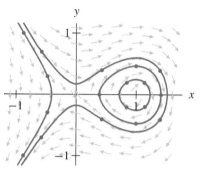

Match each solution curve with its corresponding pair of $x(t)$- and $y(t)$-graphs. Then on the t-axis mark the t-values that correspond to the distinguished points along the curve.

(a)

(b)

(c)

(d)

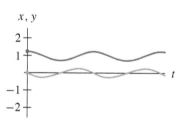

22. Use the DETools program GraphingSolutionsQuiz to practice sketching the $x(t)$- and $y(t)$-graphs associated to a given solution curve in the xy-phase plane.

In Exercises 23–26, a solution curve in the xy-plane and an initial condition on that curve are specified. Sketch the $x(t)$- and $y(t)$-graphs for the solution.

23.

24.

25.

26.

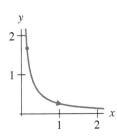

27. The following graphs are the $x(t)$- and the $y(t)$-graphs for a solution curve in the xy-phase plane. Sketch that curve and indicate the direction that the solution travels as time increases.

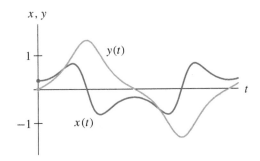

2.3 THE DAMPED HARMONIC OSCILLATOR

In this section we describe an analytic technique that applies to one of the most important models in this book—the *damped harmonic oscillator*. This second-order differential equation is used to model phenomena as varied as mass-spring systems, RLC circuits in electric circuit theory, and the blood glucose regulatory system in humans.

For example, consider the suspension in an automobile. It smooths out the ride on a bumpy road and helps keep the tires in contact with the surface of the road. We are mainly concerned with the spring and shock absorber in the suspension (see Figures 2.34 and 2.35). The springs absorb the forces caused by bumps in the road and keep the tires in contact with the road. The shock absorber consists of a piston that moves through a reservoir of oil. The oil slows the movement of the piston and the spring (see Figure 2.36). Consequently, it absorbs the force caused by the bump.

We start with the simple harmonic oscillator that we discussed in Section 2.1. As before, we let $y(t)$ denote the position of the mass as measured from the rest position of the spring. The undamped harmonic oscillator equation is

$$m\frac{d^2y}{dt^2} = -ky,$$

where m is the mass and k is the spring constant.

Figure 2.34
Location of the shock absorber.

Figure 2.35
Automobile suspension.

Figure 2.36
Shock absorber.

In Sections 2.1 and 2.2, we saw that this equation has solutions that involve sine and cosine functions. Such solutions oscillate forever with constant amplitude, and therefore they correspond to perpetual motion. To make the model more realistic, we include some form of friction or damping. A damping force slows the motion, dissipating energy from the system. A realistic model including air resistance and frictional forces is very complicated because friction is a surprisingly subtle phenomenon.* In this model, we lump together all of the damping forces and assume that the strength of this force is proportional to the velocity. Thus the damping force is given by

$$-b\left(\frac{dy}{dt}\right),$$

where $b > 0$ is called the **damping coefficient**. The minus sign indicates that the damping pushes against the direction of motion, always reducing the speed. The parameter b can be adjusted by adjusting the viscosity of the medium through which the mass moves (for example, by putting the whole mechanism in the bathtub).

To obtain the new model, we use Newton's second law and get

$$-ky - b\frac{dy}{dt} = m\frac{d^2y}{dt^2},$$

which is typically written as

$$m\frac{d^2y}{dt^2} + b\frac{dy}{dt} + ky = 0.$$

This second-order differential equation is called the **damped harmonic oscillator**. To simplify the notation, we often let $p = b/m$ and $q = k/m$, and rewrite the equation as

$$\frac{d^2y}{dt^2} + p\frac{dy}{dt} + qy = 0.$$

*See Jacqueline Krim, "Friction at the Atomic Scale," *Scientific American*, Vol. 275, No. 4, Oct. 1996 for an interesting discussion of friction.

We can convert this second-order equation into a system by letting v denote the velocity. Then $v = dy/dt$, and we have

$$\frac{dy}{dt} = v$$

$$\frac{dv}{dt} = -qy - pv.$$

Guessing Solutions

To get an idea of the behavior of solutions of the damped harmonic oscillator, it would be nice to have some explicit solutions, and we can use another time-honored *guess-and-test* method to obtain some. The idea behind this method is to make a reasonable guess of the form of the solution and then to substitute this guess into the differential equation. The hope is that we can obtain a solution by adjusting the guess.

Consider the equation

$$\frac{d^2y}{dt^2} + 3\frac{dy}{dt} + 2y = 0.$$

A solution $y(t)$ is a function whose second derivative can be expressed in terms of y, dy/dt, and constants. The most familiar function whose derivative is almost exactly itself is the exponential function, so we guess that there is a solution of the form $y(t) = e^{st}$ for some choice of the constant s. (In engineering courses, the variable s is typically used. In Chapter 3, we sometimes use the letter s, and sometimes we use the greek letter λ. The reason for this schizophrenic behavior will become clear at that time.) To determine which (if any) choices of s yield solutions, we substitute $y(t) = e^{st}$ into the left-hand side of the differential equation and obtain

$$\frac{d^2y}{dt^2} + 3\frac{dy}{dt} + 2y = \frac{d^2(e^{st})}{dt^2} + 3\frac{d(e^{st})}{dt} + 2(e^{st})$$

$$= s^2 e^{st} + 3s e^{st} + 2e^{st}$$

$$= (s^2 + 3s + 2)e^{st}.$$

In order for $y(t) = e^{st}$ to be a solution, this expression must equal the right-hand side of the differential equation for all t. In other words, we must have

$$(s^2 + 3s + 2)e^{st} = 0$$

for all t. Now, $e^{st} \neq 0$ for all t, so we must choose s so that

$$s^2 + 3s + 2 = 0.$$

This equation is satisfied only if $s = -1$ or $s = -2$. Hence this process yields two solutions, $y_1(t) = e^{-t}$ and $y_2(t) = e^{-2t}$, of this equation.

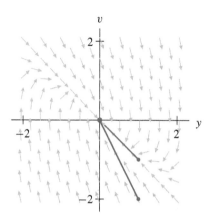

Figure 2.37
The two solution curves that correspond to the
solutions $y_1(t) = e^{-t}$ and $y_2(t) = e^{-2t}$. Both
curves lie on lines in the yv-plane.

These solutions can be converted into solutions of the system by letting

$$v_1 = \frac{dy_1}{dt} = -e^{-t} \quad \text{and} \quad v_2 = \frac{dy_2}{dt} = -2e^{-2t}.$$

So, $\mathbf{Y}_1(t) = (y_1(t), v_1(t)) = (e^{-t}, -e^{-t})$ and $\mathbf{Y}_2(t) = (y_2(t), v_2(t)) = (e^{-2t}, -2e^{-2t})$
are solutions of the associated system.

The solution curves and the $y(t)$- and $v(t)$-graphs for these two solutions are
given in Figures 2.37–2.39. The direction field indicates that all solutions tend to the
origin. This is no surprise because the damping reduces the speed. The two particular solutions we have computed are special because the solution curves lie on lines in
the phase plane. From the direction field we can see that most solution curves are not
straight lines.

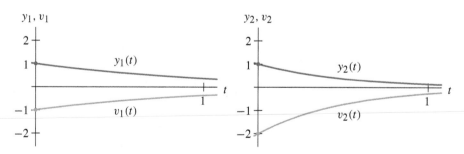

Figure 2.38
The $y(t)$- and $v(t)$-graphs for the solution
$y_1(t) = e^{-t}$.

Figure 2.39
The $y(t)$- and $v(t)$-graphs for the solution
$y_2(t) = e^{-2t}$.

How General is this Method?

This guess-and-test method leads to many questions. For example, what initial-value problems can we solve using this approach? What happens if the roots of the resulting quadratic equation are complex numbers rather than real numbers? In Chapter 3, we study linear systems (including the damped harmonic oscillator) in great detail. We will see that this method can be generalized so that we can always find an analytic expression for the general solution of any damped harmonic oscillator equation.

EXERCISES FOR SECTION 2.3

In Exercises 1–4, a harmonic oscillator equation for $y(t)$ is given.

(a) Using HPGSystemSolver, sketch the associated direction field.

(b) Using the guess-and-test method described in this section, find two nonzero solutions that are not multiples of one another.

(c) For each solution, plot both its solution curve in the yv-plane and its $y(t)$- and $v(t)$-graphs.

1. $\dfrac{d^2y}{dt^2} + 7\dfrac{dy}{dt} + 10y = 0$ **2.** $\dfrac{d^2y}{dt^2} + 5\dfrac{dy}{dt} + 6y = 0$

3. $\dfrac{d^2y}{dt^2} + 4\dfrac{dy}{dt} + y = 0$ **4.** $\dfrac{d^2y}{dt^2} + 6\dfrac{dy}{dt} + 7y = 0$

In the damped harmonic oscillator, we assume that the coefficients m, b, and k are positive. However, the rationale underlying the guess-and-test method made no such assumption, and the same analytic technique can be used if some or all of the coefficients of the equation are negative. In Exercises 5 and 6, make the same graphs and perform the same calculations as were specified in Exercises 1–4. What is different in this case?

5. $\dfrac{d^2y}{dt^2} + 3\dfrac{dy}{dt} - 10y = 0$ **6.** $\dfrac{d^2y}{dt^2} + \dfrac{dy}{dt} - 2y = 0$

7. Consider any damped harmonic oscillator equation

$$m\dfrac{d^2y}{dt^2} + b\dfrac{dy}{dt} + ky = 0.$$

(a) Show that a *constant* multiple of any solution is another solution.

(b) Illustrate this fact using the equation

$$\dfrac{d^2y}{dt^2} + 3\dfrac{dy}{dt} + 2y = 0$$

discussed in the section.

(c) How many solutions to the equation do you get if you use this observation along with the guess-and-test method described in this section?

8. Consider any damped harmonic oscillator equation

$$m\frac{d^2y}{dt^2} + b\frac{dy}{dt} + ky = 0.$$

(a) Show that the sum of any two solutions is another solution.

(b) Using the result of part (a), solve the initial-value problem

$$\frac{d^2y}{dt^2} + 3\frac{dy}{dt} + 2y = 0, \quad y(0) = 2, \quad v(0) = -3.$$

(c) Using the result of part (a) in Exercise 7 along with the result of part (a) of this exercise, solve the initial-value problem

$$\frac{d^2y}{dt^2} + 3\frac{dy}{dt} + 2y = 0, \quad y(0) = 3, \quad v(0) = -5.$$

(d) How many solutions to the equation

$$\frac{d^2y}{dt^2} + 3\frac{dy}{dt} + 2y = 0$$

do you get if you use the results of Exercise 7 and this exercise along with the guess-and-test method described in this section?

In Exercises 9 and 10, we consider a mass sliding on a frictionless table between two walls that are 1 unit apart and connected to both walls with springs, as shown below.

Let k_1 and k_2 be the spring constants of the left and right spring, respectively, let m be the mass, and let b be the damping coefficient of the medium the spring is sliding through. Suppose L_1 and L_2 are the rest lengths of the left and right springs, respectively.

9. Write a second-order differential equation for the position of the mass at time t. [*Hint*: The first step is to pick an origin, that is, a point where the position is 0. The left-hand wall is a natural choice.]

10. **(a)** Convert the second-order equation of Exercise 9 into a first-order system.

(b) Find the equilibrium point of this system.

(c) Using your result from part (b), pick a new coordinate system and rewrite the system in terms of this new coordinate system.

(d) How does this new system compare to the system for a damped harmonic oscillator?

2.4 ADDITIONAL ANALYTIC METHODS FOR SPECIAL SYSTEMS

When we studied first-order differential equations in Chapter 1, we saw that we could sometimes derive a formula for the general solution if the differential equation had a special form. When that happened, the analytic techniques for computing the solutions were especially adapted to the form of the differential equation.

For systems of differential equations, the special forms for which we can apply analytic techniques to find explicit solutions are few and far between. Because they are rare, these special systems are very valuable. We can use them to develop intuition (even wisdom) that we then use when studying systems for which analytic techniques are unavailable.

The most important class of systems that we can solve explicitly, the linear systems, is studied at length in Chapter 3. In this section we discuss analytic techniques that apply to very special classes of systems. We use the formulas that we obtain to become more familiar with solution curves and $x(t)$- and $y(t)$-graphs.

Checking Solutions

As noted above, finding formulas for a solution of a system can range from difficult to impossible. However, once we have the formulas, checking that they give a solution is not so bad. This observation is important for two reasons. First, we can double-check the (sometimes daunting) algebra we did while calculating the formulas. Second, and more important, many of the "techniques" for solving systems are really just sophisticated guessing schemes. Once we make a guess, we test to see if our guess actually is a solution.

Consider the system

$$\frac{dx}{dt} = -x + y$$

$$\frac{dy}{dt} = -3x - 5y.$$

We can rewrite this system in vector notation as

$$\frac{d\mathbf{Y}}{dt} = \mathbf{F}(\mathbf{Y}),$$

where $\mathbf{Y}(t) = (x(t), y(t))$ and $\mathbf{F}(x, y) = (-x + y, -3x - 5y)$. Now suppose someone says that

$$\mathbf{Y}(t) = (x(t), y(t)) = (e^{-4t} - 3e^{-2t}, -3e^{-4t} + 3e^{-2t})$$

is a solution to this system.

To verify this claim, we compute the derivatives of both $x(t)$ and $y(t)$. We have

$$\frac{dx}{dt} = \frac{d(e^{-4t} - 3e^{-2t})}{dt} = -4e^{-4t} + 6e^{-2t}$$

$$\frac{dy}{dt} = \frac{d(-3e^{-4t} + 3e^{-2t})}{dt} = 12e^{-4t} - 6e^{-2t}.$$

We must also substitute $x(t)$ and $y(t)$ into the right-hand side of the system. We get

$$-x + y = -(e^{-4t} - 3e^{-2t}) + (-3e^{-4t} + 3e^{-2t}) = -4e^{-4t} + 6e^{-2t}$$

and

$$-3x - 5y = -3(e^{-4t} - 3e^{-2t}) - 5(-3e^{-4t} + 3e^{-2t}) = 12e^{-4t} - 6e^{-2t}.$$

Thus dx/dt is equal to $-x + y$ and dy/dt is equal to $-3x - 5y$ for all t. Hence

$$\mathbf{Y}(t) = (x(t), y(t)) = (e^{-4t} - 3e^{-2t}, -3e^{-4t} + 3e^{-2t})$$

is a solution.

Note that $\mathbf{Y}(0) = (-2, 0)$. Consequently we have checked that $\mathbf{Y}(t)$ is a solution of the initial-value problem

$$\frac{d\mathbf{Y}}{dt} = \mathbf{F}(\mathbf{Y}), \quad \mathbf{Y}(0) = (-2, 0).$$

As a second example, consider the system

$$\frac{dx}{dt} = 2x - y$$

$$\frac{dy}{dt} = x - 2y,$$

and suppose we want to see if the function $\mathbf{Y}(t) = (e^{-t}, 3e^{-t})$ is a solution that satisfies the initial condition $\mathbf{Y}(0) = (1, 3)$.

To check that $\mathbf{Y}(t)$ satisfies the initial condition, we evaluate it at $t = 0$. This gives $\mathbf{Y}(0) = (e^{-0}, 3e^{-0}) = (1, 3)$. Next we check to see if the first equation of the system is satisfied. We have

$$\frac{dx}{dt} = \frac{d(e^{-t})}{dt} = -e^{-t}.$$

Substituting $x(t)$ and $y(t)$ into the right-hand side of the equation for dx/dt gives

$$2x - y = 2e^{-t} - 3e^{-t} = -e^{-t}.$$

Thus the first equation holds for all t. Finally, we must check the second equation in the system. We have

$$\frac{dy}{dt} = \frac{d(3e^{-t})}{dt} = -3e^{-t},$$

and

$$x - 2y = e^{-t} - 2(3e^{-t}) = -5e^{-t}.$$

Since the second equation is not satisfied, the function $\mathbf{Y}(t) = (e^{-t}, 3e^{-t})$ is not a solution of the initial-value problem.

The moral of these two examples is very important and often overlooked. Given a formula for a function $\mathbf{Y}(t)$, we can always check to see if that function satisfies the system simply by direct computation. This type of computation is certainly not the most exciting part of the subject, but it is straightforward. We can immediately determine if a given vector-valued function is a solution.

Decoupled Systems

One of the things that makes systems of differential equations so difficult (and so interesting) is that the rate of change of each of the dependent variables often depends on the values of other dependent variables. However, sometimes there is not too much interdependence among the variables, and in that case we can often derive the general solution using techniques from Chapter 1.

A system of differential equations is said to *decouple* if the rate of change of one or more of the dependent variables depends only on its own value.

A completely decoupled example

Consider the system

$$\frac{dx}{dt} = -2x$$

$$\frac{dy}{dt} = -y.$$

Since the equation for dx/dt involves only x and the equation for dy/dt involves only y, we can solve the two equations separately. When this happens, we say the system is **completely decoupled**. The general solution of $dx/dt = -2x$ is $x(t) = k_1 e^{-2t}$, where k_1 is any constant. The general solution of $dy/dt = -y$ is $y(t) = k_2 e^{-t}$, where k_2 is any constant. We can put these together to find the general solution

$$(x(t), y(t)) = (k_1 e^{-2t}, k_2 e^{-t})$$

of the system. This general solution has two undetermined constants, k_1 and k_2. These constants can be adjusted so that any given initial condition can be satisfied. For example, given the initial condition $\mathbf{Y}(0) = (1, 1)$, we let $k_1 = 1$ and $k_2 = 1$ to obtain the solution

$$\mathbf{Y}(t) = \begin{pmatrix} e^{-2t} \\ e^{-t} \end{pmatrix}.$$

In Figure 2.40 we plot this curve along with the direction field associated to the vector field $\mathbf{F}(x, y) = (-2x, -y)$. From the formula for $\mathbf{Y}(t)$, we note that $\mathbf{Y}(t)$ gives a parameterization of the upper half of the curve $x = y^2$ in the plane because

$$(y(t))^2 = (e^{-t})^2 = e^{-2t} = x(t).$$

We only obtain the upper half of this parabola because $y(t) = e^{-t} > 0$ for all t.

The solution curve in the phase plane hides the behavior of our solution with respect to the independent variable t. The solution actually tends exponentially toward the origin. Since we have the formulas for $x(t)$ and $y(t)$, it is not difficult to sketch the $x(t)$- and $y(t)$-graphs (see Figure 2.41).

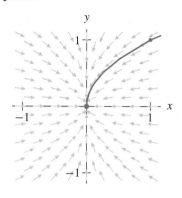

Figure 2.40
The solution curve
$\mathbf{Y}(t) = (e^{-2t}, e^{-t})$.

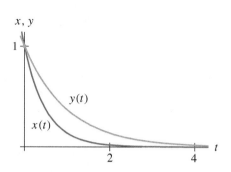

Figure 2.41
The $x(t)$- and $y(t)$-graphs for the solution
$(x(t), y(t)) = (e^{-2t}, e^{-t})$.

A partially decoupled example

Our next example is the system

$$\frac{dx}{dt} = 2x + 3y$$
$$\frac{dy}{dt} = -4y.$$

For this system the rate of change of x depends on both x and y, but the rate of change of y depends only on y. We say that the dependent variable y decouples from the system and the system is **partially decoupled**.

The general solution of the equation for y is $y(t) = k_2 e^{-4t}$, where k_2 is an arbitrary constant. Substituting this expression for y into the equation for x gives

$$\frac{dx}{dt} = 2x + 3k_2 e^{-4t}.$$

This is a first-order linear equation which we can solve using the methods discussed in Sections 1.8 and 1.9. From the Extended Linearity Principle, we know that we need one particular solution $x_p(t)$ of the nonhomogeneous equation as well as the general solution of the associated homogeneous equation. To find $x_p(t)$, we rewrite the equation as

$$\frac{dx}{dt} - 2x = 3k_2 e^{-4t},$$

and guess a solution of the form $x_p(t) = \alpha e^{-4t}$. Substituting this guess into the equation yields

$$-4\alpha e^{-4t} - 2\alpha e^{-4t} = 3k_2 e^{-4t},$$

which simplifies to

$$-6\alpha e^{-4t} = 3k_2 e^{-4t}.$$

Therefore, $x_p(t)$ is a solution if $\alpha = -k_2/2$. The general solution of the associated homogeneous equation is

$$k_1 e^{2t},$$

where k_1 is an arbitrary constant. Combining $x_p(t)$ with the general solution of the homogeneous equation gives the general solution

$$x(t) = k_1 e^{2t} - \tfrac{1}{2} k_2 e^{-4t}$$

of the nonhomogeneous equation.

   Putting this formula for $x(t)$ together with the general solution of the equation for y, we obtain the general solution

$$x(t) = k_1 e^{2t} - \tfrac{1}{2} k_2 e^{-4t}$$
$$y(t) = k_2 e^{-4t}$$

of the partially decoupled system. The constants k_1 and k_2 can be adjusted to obtain any desired initial condition. For example, suppose we have $x(0) = 0$ and $y(0) = 1$. To find the appropriate values of k_1 and k_2, we substitute $t = 0$ into the formula for the general solution and solve. That is,

$$x(0) = 0 = k_1 - \tfrac{1}{2} k_2$$
$$y(0) = 1 = k_2,$$

which gives $k_1 = 1/2$ and $k_2 = 1$. So the solution of the initial-value problem is

$$x(t) = \tfrac{1}{2} e^{2t} - \tfrac{1}{2} e^{-4t}$$
$$y(t) = e^{-4t}.$$

   For the initial condition $(x(0), y(0)) = (-1/2, 1)$, we can follow the same steps as above, obtaining $k_1 = 0$ and $k_2 = 1$. The formula for this solution is

$$x(t) = -\tfrac{1}{2} e^{-4t}$$
$$y(t) = e^{-4t}.$$

Note that $y(t)/x(t) = -2$ for all t and that the solution tends toward the equilibrium point at the origin as t increases and toward infinity as t decreases. Since the ratio y/x is constant, the solution curve lies on a line through the origin in the phase plane (see Figure 2.42). The fact that this system has a solution curve that lies on a line is an artifact of the simple algebra of the equations. This sort of special geometry will be extensively exploited in Chapter 3.

Figure 2.42
Even though the $x(t)$- and $y(t)$-graphs are graphs of exponential functions, the corresponding solution curve lies on a line in the xy-phase plane.

EXERCISES FOR SECTION 2.4

In Exercises 1–4, we consider the system

$$\frac{dx}{dt} = 2x + 2y$$

$$\frac{dy}{dt} = x + 3y.$$

For the given functions $\mathbf{Y}(t) = (x(t), y(t))$, determine if $\mathbf{Y}(t)$ is a solution system.

1. $(x(t), y(t)) = (2e^t, -e^t)$

2. $(x(t), y(t)) = (3e^{2t} + e^t, -e^t + e^{4t})$

3. $(x(t), y(t)) = (2e^t - e^{4t}, -e^t + e^{4t})$

4. $(x(t), y(t)) = (4e^t + e^{4t}, -2e^t + e^{4t})$

In Exercises 5–12, we consider the partially decoupled system

$$\frac{dx}{dt} = 2x + y$$

$$\frac{dy}{dt} = -y.$$

5. Although we can use the method described in this section to derive the general solution to this system, why should we immediately know that $\mathbf{Y}(t) = (x(t), y(t)) = (e^{2t} - e^{-t}, e^{-2t})$ is *not* a solution to the system?

6. Although we can use the method described in this section to derive the general solution to this system, is there an easier way to show that $\mathbf{Y}(t) = (x(t), y(t)) = (4e^{2t} - e^{-t}, 3e^{-t})$ is a solution to the system?

7. Use the method described in this section to derive the general solution to this system.

8. **(a)** Can you choose constants in the general solution obtained in Exercise 7 that yield the function $\mathbf{Y}(t) = (e^{-t}, 3e^{-t})$?

(b) Suppose that the result of Exercise 7 was not immediately available. How could you tell that $\mathbf{Y}(t) = (e^{-t}, 3e^{-t})$ is not a solution?

9. **(a)** Using the result of Exercise 7, determine the solution that satisfies the initial condition $\mathbf{Y}(0) = (x(0), y(0)) = (1, 0)$.

(b) In the xy-phase plane, plot the solution curve associated to this solution.

(c) Plot the corresponding $x(t)$- and $y(t)$-graphs.

10. **(a)** Using the result of Exercise 7, determine the solution that satisfies the initial condition $\mathbf{Y}(0) = (x(0), y(0)) = (-1, 3)$.

(b) In the xy-phase plane, plot the solution curve associated to this solution.

(c) Plot the corresponding $x(t)$- and $y(t)$-graphs.

11. **(a)** Using the result of Exercise 7, determine the solution that satisfies the initial condition $\mathbf{Y}(0) = (x(0), y(0)) = (0, 1)$.

(b) Using HPGSystemSolver, plot the corresponding solution curve in the xy-phase plane and compare the result with the curve that you would have drawn directly from the direction field for the system.

(c) Using only the solution curve, sketch the $x(t)$- and $y(t)$-graphs.

(d) Compare your sketch with the $x(t)$- and $y(t)$-graphs of HPGSystemSolver.

12. **(a)** Using the result of Exercise 7, determine the solution that satisfies the initial condition $\mathbf{Y}(0) = (x(0), y(0)) = (1, -1)$.

(b) Using HPGSystemSolver, plot the corresponding solution curve in the xy-phase plane and compare the result with the curve that you would have drawn directly from the direction field for the system.

(c) Using only the solution curve, sketch the $x(t)$- and $y(t)$-graphs.

(d) Compare your sketch with the $x(t)$- and $y(t)$-graphs of HPGSystemSolver.

13. Consider the partially decoupled system

$$\frac{dx}{dt} = 2x - 8y^2$$

$$\frac{dy}{dt} = -3y.$$

(a) Derive the general solution.

(b) Find the equilibrium points of the system.

(c) Find the solution that satisfies the initial condition $(x_0, y_0) = (0, 1)$.

(d) Use HPGSystemSolver to plot the phase portrait for this system. Identify the solution curve that corresponds to the solution with initial condition $(x_0, y_0) = (0, 1)$.

2.5 EULER'S METHOD FOR SYSTEMS

Many of the examples in this chapter include some type of plot of solutions, either as curves in the phase plane or as $x(t)$- or $y(t)$-graphs. In most cases these plots are provided without any indication of how we obtain them. Occasionally the solutions are line segments or circles or ellipses, and we are able to verify this analytically. But more often the solutions do not lie on familiar curves. For example, consider the predator-prey type system

$$\frac{dx}{dt} = 2x - 1.2xy$$

$$\frac{dy}{dt} = -y + 1.2xy$$

and the solution that satisfies the initial condition $(x(0), y(0)) = (1.75, 1.0)$. Figure 2.43 shows this solution in the phase plane, the xy-plane, and Figure 2.44 contains the corresponding $x(t)$- and $y(t)$-graphs. Figure 2.43 suggests that this solution is a closed curve, but the curve is certainly neither circular nor elliptical. Similarly, the $x(t)$- and $y(t)$-graphs appear to be periodic, although they do not seem to be graphs of any of the standard periodic functions (sine, cosine, etc.). So how did we compute these graphs?

The answer to this question is essentially the same as the answer to the analogous question for first-order equations. We use a dependable numerical technique and the aid of a computer. In this section we define Euler's method for first-order systems. Other numerical methods are discussed in Chapter 7.

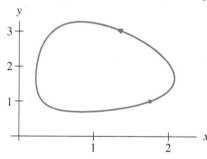

Figure 2.43
A solution curve corresponding to the initial condition $(x_0, y_0) = (1.75, 1.0)$.

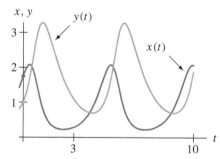

Figure 2.44
The corresponding $x(t)$- and $y(t)$-graphs for the solution curve in Figure 2.43.

Derivation of Euler's Method

Consider the first-order autonomous system

$$\frac{dx}{dt} = f(x, y)$$

$$\frac{dy}{dt} = g(x, y),$$

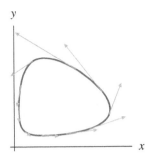

Figure 2.45
A solution curve is a curve that is everywhere tangent to the vector field.

Figure 2.46
The vector at (x_0, y_0) and the point (x_1, y_1) obtained from one step of Euler's method.

along with the initial condition $(x(t_0), y(t_0)) = (x_0, y_0)$. We have seen that we can use vector notation to rewrite this system as

$$\frac{d\mathbf{Y}}{dt} = \mathbf{F}(\mathbf{Y}),$$

where $\mathbf{Y} = (x, y)$, $d\mathbf{Y}/dt = (dx/dt, dy/dt)$, and $\mathbf{F}(\mathbf{Y}) = (f(x, y), g(x, y))$. The vector-valued function $\mathbf{F}$ yields a vector field, and a solution is a curve whose tangent vector at any point on the curve agrees with the vector field (see Figure 2.45). In other words the "velocity" vector for the curve is equal to the vector $\mathbf{F}(x(t), y(t))$.

As we saw in Section 1.4, Euler's method for a first-order equation is based on the idea of approximating the graph of a solution by line segments whose slopes are obtained from the differential equation. Euler's approximation scheme for systems is the same basic idea interpreted in a vector framework.

Given an initial condition (x_0, y_0), how can we use the vector field $\mathbf{F}(x, y)$ to approximate the solution curve? Just as for equations, we first pick a step size Δt. The vector $\mathbf{F}(x_0, y_0)$ is the velocity vector of the solution through (x_0, y_0), so we begin our approximate solution by using $\Delta t\, \mathbf{F}(x_0, y_0)$ to form the first "step." In other words we step from (x_0, y_0) to (x_1, y_1), where the point (x_1, y_1) is given by

$$(x_1, y_1) = (x_0, y_0) + \Delta t\, \mathbf{F}(x_0, y_0)$$

(see Figure 2.46). This corresponds to traveling along a straight line for time Δt with velocity $\mathbf{F}(x_0, y_0)$.

Having calculated a point (x_1, y_1) on the approximate solution curve, we calculate the new velocity vector $\mathbf{F}(x_1, y_1)$. The second step in the approximation is

$$(x_2, y_2) = (x_1, y_1) + \Delta t\, \mathbf{F}(x_1, y_1).$$

We repeat this scheme and obtain an approximate solution curve (see Figure 2.47).

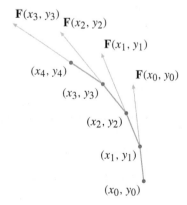

Figure 2.47
The approximate solution curve obtained from four Euler steps.

In practice we choose a step size Δt that is small enough to provide an accurate solution over the given time interval. (See Chapter 7 for a technical discussion of how small is small and how small is too small.)

Euler's Method for Autonomous Systems

Euler's method for systems can be written without the vector notation as follows. Given the system

$$\frac{dx}{dt} = f(x, y)$$

$$\frac{dy}{dt} = g(x, y),$$

the initial condition (x_0, y_0), and the step size Δt, we calculate the Euler approximation by repeating the calculations:

$$m_k = f(x_k, y_k)$$

$$n_k = g(x_k, y_k),$$

$$x_{k+1} = x_k + m_k \Delta t$$

$$y_{k+1} = y_k + n_k \Delta t.$$

Euler's Method Applied to the Van der Pol Equation

For example, consider the second-order differential equation

$$\frac{d^2 x}{dt^2} - (1 - x^2)\frac{dx}{dt} + x = 0.$$

This equation is called the **Van der Pol equation**. To study it numerically, we first convert it into a first-order system by letting $y = dx/dt$. The resulting system is

$$\frac{dx}{dt} = y$$

$$\frac{dy}{dt} = -x + (1 - x^2)y.$$

Suppose we want to find an approximate solution for the initial condition $(x(0), y(0)) = (1, 1)$. We do a few calculations by hand to see how Euler's method works, and then turn to the computer for the repetitive part. The method is best illustrated by doing a calculation with a relatively large step size, although in practice we would never use such a large value for Δt.

Let $\Delta t = 0.25$. Given the initial condition $(x_0, y_0) = (1, 1)$, we compute the vector field

$$\mathbf{F}(x, y) = (y, -x + (1 - x^2)y)$$

at $(1, 1)$. We obtain the vector $\mathbf{F}(1, 1) = (1, -1)$. Thus our first step starts at $(1, 1)$ and

ends at

$$(x_1, y_1) = (x_0, y_0) + \Delta t \, \mathbf{F}(x_0, y_0)$$
$$= (1, 1) + 0.25 \, (1, -1)$$
$$= (1.25, 0.75).$$

In other words, since $\Delta t = 0.25$, we obtain (x_1, y_1) from (x_0, y_0) by stepping one-quarter of the way along the displacement vector $(1, -1)$ (see Figure 2.48).

The next step is obtained by computing the vector field at (x_1, y_1). We have $\mathbf{F}(1.25, 0.75) = (0.75, -1.67)$ (to 2 decimal places). Consequently, our next step starts at $(1.25, 0.75)$ and ends at

$$(x_2, y_2) = (1.25, 0.75) + 0.25 \, (0.75, -1.67)$$
$$= (1.44, 0.33)$$

(see Figure 2.48).

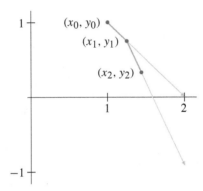

Figure 2.48
Two steps of Euler's method applied to the Van der Pol equation with initial condition $(x_0, y_0) = (1, 1)$ and step size $\Delta t = 0.25$.

James H. Curry (1948–) received his Ph.D. in Mathematics at the University of California at Berkeley in 1976. He spent several years as a Postdoctoral Fellow at MIT and the National Center for Atmospheric Research where he met and worked with E. N. Lorenz. He has also taught at Howard University and the University of Colorado where he currently holds the position of Professor of Applied Mathematics.

Curry's research has focused on qualitative methods in differential equations that model the atmosphere. He has also published extensively on iterative methods for solving nonlinear equations. The methods Curry considers are faster and more advanced than the numerical methods we discuss in this section. We describe some of these more advanced techniques in Chapter 7.

Table 2.1
Ten steps of Euler's method.

i	x_i	y_i	m_i	n_i
0	1	1	1	−1
1	1.25	0.75	0.75	−1.671875
2	1.4375	0.332031	0.332031	−1.791580
3	1.520507	−0.115864	−0.115864	−1.368501
4	1.491542	−0.457989	−0.457989	−0.930644
5	1.377045	−0.690650	−0.690650	−0.758048
6	1.204382	−0.880162	−0.880162	−0.807837
7	0.984342	−1.082121	−1.082121	−1.017965
8	0.713811	−1.336613	−1.336613	−1.369384
9	0.379658	−1.678959	−1.678959	−1.816611
10	−0.040082	−2.133112		

Table 2.1 illustrates the computations necessary to calculate ten steps of Euler's method, starting at the initial condition $(x_0, y_0) = (1, 1)$ with $\Delta t = 0.25$. The resulting approximate solution curve is shown in Figure 2.49.

As we mentioned above, $\Delta t = 0.25$ is much larger than the typical step size, so let's repeat our calculations with $\Delta t = 0.1$. Since we will use a computer to do these calculations, we might as well do more steps, too. Figure 2.50 shows the result of this calculation. In this figure we show both the points obtained in the calculation as well as a graph of an approximate solution curve obtained by joining successive points by line segments. Note that the curve is hardly a "standard" shape and that it is almost a closed curve.

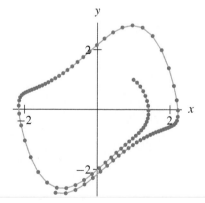

Figure 2.49
Ten steps of Euler's method applied to the Van der Pol equation with initial condition $(x_0, y_0) = (1, 1)$ and step size $\Delta t = 0.25$.

Figure 2.50
One hundred steps of Euler's method applied to the Van der Pol equation with initial condition $(x_0, y_0) = (1, 1)$ and step size $\Delta t = 0.1$.

Table 2.2
Ten steps of Euler's method with $t_0 = 0$.

i	t_i	x_i	y_i	m_i	n_i
0	0	1	1	1	−1
1	0.25	1.25	0.75	0.75	−1.671875
2	0.50	1.4375	0.332031	0.332031	−1.791580
3	0.75	1.520507	−0.115864	−0.115864	−1.368501
4	1.00	1.491542	−0.457989	−0.457989	−0.930644
5	1.25	1.377045	−0.690650	−0.690650	−0.758048
6	1.50	1.204382	−0.880162	−0.880162	−0.807837
7	1.75	0.984342	−1.082121	−1.082121	−1.017965
8	2.00	0.713811	−1.336613	−1.336613	−1.369384
9	2.25	0.379658	−1.678959	−1.678959	−1.816611
10	2.50	−0.040082	−2.133112		

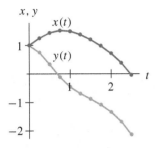

Figure 2.51
The $x(t)$- and $y(t)$-graphs corresponding to the approximate solution curve obtained in Table 2.2.

To show the $x(t)$- and $y(t)$-graphs for this approximate solution, we must include information about the independent variable t in our Euler's method table. If we assume that the initial condition $(x_0, y_0) = (1, 1)$ corresponds to the initial time $t_0 = 0$, we can augment that table by adding the corresponding times (see Table 2.2). Thus we are able to produce $x(t)$- and $y(t)$-graphs of approximate solutions (see Figure 2.51). Figures 2.52 and 2.53 illustrate how the "almost" closed solution curve in the phase plane (the xy-plane) corresponds to the functions $x(t)$ and $y(t)$, which are essentially periodic.

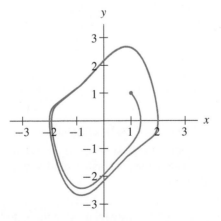

Figure 2.52
The approximate solution curve in the xy-plane.

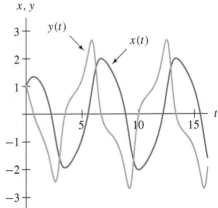

Figure 2.53
The corresponding $x(t)$- and $y(t)$-graphs.

EXERCISES FOR SECTION 2.5

1. For the system

$$\frac{dx}{dt} = -y$$

$$\frac{dy}{dt} = x,$$

the curve $\mathbf{Y}(t) = (\cos t, \sin t)$ is a solution. This solution is periodic. Its initial position is $\mathbf{Y}(0) = (1, 0)$, and it returns to this position when $t = 2\pi$. So $\mathbf{Y}(2\pi) = (1, 0)$ and $\mathbf{Y}(t + 2\pi) = \mathbf{Y}(t)$ for all t.

(a) Check that $\mathbf{Y}(t) = (\cos t, \sin t)$ is a solution.

(b) Use Euler's method with step size 0.5 to approximate this solution, and check how close the approximate solution is to the real solution when $t = 4$, $t = 6$, and $t = 10$.

(c) Use Euler's method with step size 0.1 to approximate this solution, and check how close the approximate solution is to the real solution when $t = 4$, $t = 6$, and $t = 10$.

(d) The points on the solution curve $\mathbf{Y}(t)$ are all 1 unit distance from the origin. Is this true of the approximate solutions? Are they too far from the origin or too close to it? What will happen for other step sizes (that is, will approximate solutions formed with other step sizes be too far or too close to the origin)?

[Use a computer or calculator to perform Euler's method.]

2. For the system

$$\frac{dx}{dt} = 2x$$

$$\frac{dy}{dt} = y,$$

we claim that the curve $\mathbf{Y}(t) = (e^{2t}, 3e^{t})$ is a solution. Its initial position is $\mathbf{Y}(0) = (1, 3)$.

(a) Check that $\mathbf{Y}(t) = (e^{2t}, 3e^{t})$ is a solution.

(b) Use Euler's method with step size $\Delta t = 0.5$ to approximate this solution, and check how close the approximate solution is to the real solution when $t = 2$, $t = 4$, and $t = 6$.

(c) Use Euler's method with step size $\Delta t = 0.1$ to approximate this solution, and check how close the approximate solution is to the real solution when $t = 2$, $t = 4$, and $t = 6$.

(d) Discuss how and why the Euler approximations differ from the solution.

[Use a computer or calculator to perform Euler's method.]

In Exercises 3–6, a system, an initial condition, a step size, and an integer n are given. The direction field for the system is also provided.

(a) Use EulersMethodForSystems to calculate the approximate solution given by Euler's method for the given system with the given initial condition and step size for n steps.

(b) Plot your approximate solution on the direction field. Make sure that your approximate solution is consistent with the direction field.

(c) Using HPGSystemSolver, obtain a more detailed sketch of the phase portrait for the system.

3. $\dfrac{dx}{dt} = y$

$\dfrac{dy}{dt} = -2x - 3y$

$\begin{cases} (x_0, y_0) = (1, 1) \\ \Delta t = 0.25 \\ n = 5 \end{cases}$

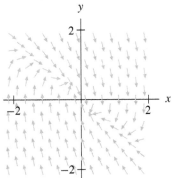

4. $\dfrac{dx}{dt} = y$

$\dfrac{dy}{dt} = -\sin x$

$\begin{cases} (x_0, y_0) = (0, 2) \\ \Delta t = 0.25 \\ n = 8 \end{cases}$

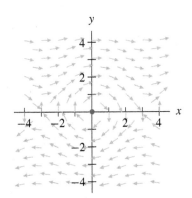

5. $\dfrac{dx}{dt} = y + y^2$

$\dfrac{dy}{dt} = -x + \dfrac{y}{5} - xy + \dfrac{6y^2}{5}$

$\begin{cases} (x_0, y_0) = (1, 1) \\ \Delta t = 0.25 \\ n = 5 \end{cases}$

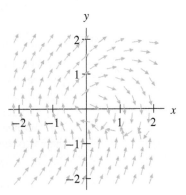

6. $\dfrac{dx}{dt} = y + y^2$

$\dfrac{dy}{dt} = -\dfrac{x}{2} + \dfrac{y}{5} - xy + \dfrac{6y^2}{5}$

$\begin{cases} (x_0, y_0) = (-0.5, 0) \\ \\ \Delta t = 0.25 \\ \\ n = 7 \end{cases}$

7. Using a computer or calculator, apply Euler's method to sketch an approximation to the solution curve for the solution to the initial-value problem

$$2\frac{d^2y}{dt^2} + \frac{dy}{dt} + 4y = 0,$$

where $(y_0, v_0) = (2, 0)$. How does your choice of Δt affect your result?

8. Using a computer or calculator, apply Euler's method to sketch an approximation to the solution curve for the solution to the initial-value problem

$$5\frac{d^2y}{dt^2} + \frac{dy}{dt} + 5y = 0,$$

where $(y_0, v_0) = (0, 1)$. How does your choice of Δt affect your result?

2.6 EXISTENCE AND UNIQUENESS FOR SYSTEMS

Numerical methods, such as Euler's method, give approximations to solutions. Controlling the difference between the numerical approximation and the actual solution is a difficult problem since we usually do not know the actual solution (see Chapter 7). As we saw in Section 1.5, the Existence and Uniqueness Theorem gives us (among other things) qualitative information about solutions, which we can use to check our numerics. The same is true for systems.

The Existence and Uniqueness Theorem

In Chapter 1, we treated existence and uniqueness separately because we wanted to emphasize that uniqueness required a slightly stronger hypothesis than existence. In this section, we focus less on the hypotheses of the theorem and more on its consequences. We consider both autonomous and nonautonomous systems, and we assume that they are continuously differentiable. In other words, if the right-hand side of the system is the vector field

$$\mathbf{F}(t, \mathbf{Y}) = \begin{pmatrix} f(t, x, y) \\ g(t, x, y) \end{pmatrix},$$

then we assume that the six partial derivatives $\partial f/\partial t$, $\partial f/\partial x$, $\partial f/\partial y$, $\partial g/\partial t$, $\partial g/\partial x$, and $\partial g/\partial y$ exist and vary continuously over some open subset of the xy-plane. Under these assumptions, we have both existence and uniqueness of solutions.

EXISTENCE AND UNIQUENESS THEOREM   Let

$$\frac{d\mathbf{Y}}{dt} = \mathbf{F}(t, \mathbf{Y})$$

be a system of differential equations. Suppose that t_0 is an initial time and $\mathbf{Y}_0$ is an initial value. Suppose also that the function $\mathbf{F}$ is continuously differentiable. Then there is an $\epsilon > 0$ and a function $\mathbf{Y}(t)$ defined for $t_0 - \epsilon < t < t_0 + \epsilon$, such that $\mathbf{Y}(t)$ satisfies the initial-value problem

$$\frac{d\mathbf{Y}}{dt} = \mathbf{F}(t, \mathbf{Y}) \quad \text{and} \quad \mathbf{Y}(t_0) = \mathbf{Y}_0.$$

Moreover, for t in this interval, this solution is unique.   ∎

As was the case with first-order equations in Section 1.5, the ϵ is necessary because solutions can blow up in finite time, or they can leave the domain in which the differential equation is defined. For example, consider the system

$$\frac{dx}{dt} = x^2 + 1$$
$$\frac{dy}{dt} = 1$$

with initial condition $(x(0), y(0)) = (0, 0)$. As we saw in Section 1.5, the solution to the initial-value problem

$$\frac{dx}{dt} = x^2 + 1, \quad x(0) = 0,$$

is $x(t) = \tan t$. So the solution to the system is $(x(t), y(t)) = (\tan t, t)$, which blows up as $t \to \pi/2$ from below.

Consequences of Uniqueness for Autonomous Systems

The existence half of the theorem is mostly just reassuring. If we are studying a certain system, then it is nice to know that what we are studying exists. The uniqueness part of the theorem is useful in a much more practical way. Roughly speaking, the Uniqueness Theorem says that two solutions cannot start at the same place at the same time.

For autonomous systems, the vector field does not vary with time, and we have two related and very useful consequences. First, the solution curve for a single solution cannot loop back and intersect itself unless the solution is periodic and the solution curve is a simple, closed curve (see Figure 2.54). In Figure 2.55, we see that nonautonomous systems do not have this property.

Figure 2.54

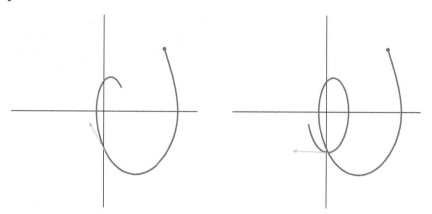

Figure 2.55
A solution curve (left) of a nonautonomous system and its velocity vector as it passes through a point in the phase plane. For this nonautonomous system, the velocity vector points in a different direction when the solution curve passes through the point a second time. Hence, the solution curve crosses itself.

Second, the solution curves for two different solutions cannot intersect unless they sweep out the same curve. For example, consider the second-order equation

$$\frac{d^2y}{dt^2} + y = 0,$$

which is equivalent to the system

$$\frac{dy}{dt} = v$$
$$\frac{dv}{dt} = -y.$$

Note that both $\mathbf{Y}_1(t) = (\cos t, -\sin t)$ and $\mathbf{Y}_2(t) = (\sin t, \cos t)$ are solutions. Note also that $\mathbf{Y}_1(0) = (1, 0)$ and $\mathbf{Y}_2(\pi/2) = (1, 0)$. Hence, the two solution curves intersect, and they must agree. (In this case, both solution curves are the unit circle centered at the origin.)

Informally, these two properties of intersecting solution curves follow from the metaphor of the parking lot as described on page 172. If the system is autonomous, the vector at a given point in the phase plane does not vary with time. Hence, two solution curves that visit the same point at different times must trace out the same curve (see Exercises 6 and 7). We will make this argument more precise at the end of the section.

Consequences for two-dimensional autonomous systems

If a system has exactly two dependent variables (as is the case throughout most of this book), the consequences of uniqueness for solution curves of automous systems limit what can happen in the phase plane given the behavior of certain systems. For example, a periodic solution corresponds to a closed curve in the the phase plane, and any solution with an initial condition that is inside the curve is trapped for all time (see Figure 2.56). Also, a number of solution curves can form a "fence." That is, they can divide the phase plane into separate regions, and solutions with initial conditions on one side of the fence cannot cross over to the other side (see Figure 2.56).

Figure 2.56
A solution curve that is trapped inside a periodic solution.

Figure 2.57
Three solution curves (including an equilibrium point) when taken together divide the phase plane into two regions.

Formal verification of the consequences for autonomous systems

To verify our assertions about the consequences of uniqueness for systems, we suppose that two solution curves intersect at the point $\mathbf{Y}_0$ in the phase plane. In other words, suppose

$$\mathbf{Y}_1(t_1) = \mathbf{Y}_0 = \mathbf{Y}_2(t_2)$$

for two solutions $\mathbf{Y}_1(t)$ and $\mathbf{Y}_2(t)$. Then we consider the function

$$\mathbf{Y}_3(t) = \mathbf{Y}_1(t - (t_2 - t_1)).$$

Both $\mathbf{Y}_1(t)$ and $\mathbf{Y}_3(t)$ sweep out the same solution curve because $\mathbf{Y}_3(t)$ is a time translate of $\mathbf{Y}_1(t)$. On the other hand,

$$\frac{d\mathbf{Y}_3}{dt}(t) = \frac{d\mathbf{Y}_1}{dt}(t - (t_2 - t_1)) \quad \text{(by the Chain Rule)}$$

$$= \mathbf{F}(\mathbf{Y}_1(t - (t_2 - t_1))) \quad \text{(because } \mathbf{Y}_1(t) \text{ is a solution)}$$

$$= \mathbf{F}(\mathbf{Y}_3(t)) \quad \text{(because the system is autonomous)}.$$

Hence, $\mathbf{Y}_3(t)$ is a solution to the system. Moreover, $\mathbf{Y}_3(t_2) = \mathbf{Y}_1(t_1)$, so uniqueness implies that $\mathbf{Y}_2(t) = \mathbf{Y}_3(t)$ for all t. In other words,

$$\mathbf{Y}_2(t) = \mathbf{Y}_1(t - (t_2 - t_1)).$$

(See Exercises 8 and 9.)

EXERCISES FOR SECTION 2.6

1. Consider the system

$$\frac{dx}{dt} = -x + y$$

$$\frac{dy}{dt} = -y.$$

(a) Show that the x-axis consists of three solution curves.

(b) Using HPGSystemSolver, sketch the solution curves for a number of initial conditions above and below the x-axis. Do these curve intersect the x-axis? Do they touch the origin? Justify your assertions.

2. Consider the system

$$\frac{dx}{dt} = y$$

$$\frac{dy}{dt} = -x + (1 - x^2)y.$$

(a) Using HPGSystemSolver, determine (to two decimal places) an initial condition for a periodic solution.

(b) The periodic solution in part (a) produces a closed curve in the phase plane. Describe briefly what happens to all solutions with initial conditions that lie inside this curve.

In Exercises 3–5, consider the system

$$\frac{dx}{dt} = -x + 3y$$

$$\frac{dy}{dt} = -3x - y.$$

3. Verify that $\mathbf{Y}_1(t) = (e^{-t}\sin(3t), e^{-t}\cos(3t))$ is a solution of this system.

4. Verify that $\mathbf{Y}_2(t) = (e^{-(t-1)}\sin(3(t-1)), e^{-(t-1)}\cos(3(t-1)))$ is a solution.

5. Using HPGSystemSolver, sketch the solution curves for $\mathbf{Y}_1(t)$ and $\mathbf{Y}_2(t)$ in the xy-phase plane. Why don't $\mathbf{Y}_1(t)$ and $\mathbf{Y}_2(t)$ contradict the Uniqueness Theorem?

6. Recall the Metaphor of the Parking Lot on page 172. Suppose two people, say Gib and Harry, are both driving cars on the parking lot and both are carefully following the rules prescribed in the metaphor. If they start at time $t = 0$ at different points, will they ever collide? (Neglect the width of their cars.)

7. Consider the two drivers, Gib and Harry, from Exercise 6. Suppose that at time $t = 0$ they start at different points in the parking lot, but at time $t = 1$ Gib drives over the point where Harry started. Will they ever collide? What can you say about their paths?

8. **(a)** Suppose $\mathbf{Y}_1(t)$ is a solution of an autonomous system $d\mathbf{Y}/dt = \mathbf{F}(\mathbf{Y})$. Show that $\mathbf{Y}_2(t) = \mathbf{Y}_1(t + t_0)$ is also a solution for any constant t_0.

   (b) What is the relationship between the solution curves of $\mathbf{Y}_1(t)$ and $\mathbf{Y}_2(t)$?

9. Suppose $\mathbf{Y}_1(t)$ and $\mathbf{Y}_2(t)$ are solutions of an autonomous system $d\mathbf{Y}/dt = \mathbf{F}(\mathbf{Y})$, where $\mathbf{F}(\mathbf{Y})$ satisfies the hypotheses of the Uniqueness Theorem. Suppose also that $\mathbf{Y}_2(1) = \mathbf{Y}_1(0)$. How are $\mathbf{Y}_1(t)$ and $\mathbf{Y}_2(t)$ related?

10. Consider the system

$$\frac{dx}{dt} = 2$$
$$\frac{dy}{dt} = y^2.$$

   (a) Calculate the general solution for the system.

   (b) What solutions go to infinity?

   (c) What solutions blow up in finite time?

11. Consider the system

$$\frac{dx}{dt} = x^2 + y$$
$$\frac{dy}{dt} = x^2 y^2.$$

Show that, for the solution $(x(t), y(t))$ with initial condition $(x(0), y(0)) = (0, 1)$, there is a time t_* such that $x(t) \to \infty$ as $t \to t_*$. In other words the solution blows up in finite time. [*Hint*: Note that $dy/dt \geq 0$ for all x and y.]

2.7 THE SIR MODEL OF AN EPIDEMIC

H1N1 flu, often called "swine flu," caused a worldwide pandemic in 2009. The outbreak began in Mexico early in the year, and eventually the Mexican government closed many public and private facilities in Mexico City in an attempt to restrict the spread of the disease. Nevertheless, the virus spread worldwide. Against the advice of public health officials, some summer camps in the U.S. went so far as to use drugs such as Tamiflu in a prophylactic fashion. More typically, we were encouraged to wash our hands frequently, cough into our sleeves, and stay home during exams.

The pandemic seemed to peak in November of 2009, and by spring of 2010 the number of cases was in rapid decline. The World Health Organization announced the end of the pandemic in August of 2010.*

Modeling an Epidemic

The spread of a contagious disease through a population involves intricate interactions from the level of populations down to the level of individual cells and viruses. How-

*To compare this pandemic to others, see the video "Secrets of the Dead: Killer Flu (1918)" at http://www.pbs.org/wnet/secrets/episodes/preview-of-killer-flu/222/

ever, it is still possible to learn interesting and useful information from relatively simple models. A classical model, introduced by Kermack and McKendrick in 1927* is called the SIR model. In this model, a population is divided into three groups—the susceptible individuals, the infected individuals, and the recovered individuals.

In this model $S(t)$ denotes the fraction of the population that can catch the disease at time t, $I(t)$ denotes the fraction of the population that has the disease and can spread it to the susceptibles, and $R(t)$ denotes the fraction of the population that has recovered from the disease and cannot catch it again. This model is appropriate for the spread of a flu epidemic since once a person has had a particular strain of flu, their immune system prevents them from catching that strain again. Since flu spreads fairly quickly, we can assume that time is measured in days. While most people recover from the flu fairly easily, there is a low mortality rate. Those who do not survive are included in $R(t)$.

We assume everyone in the population is either susceptible, infected, or recovered, that is,

$$S(t) + I(t) + R(t) = 1$$

for all t. In addition, we assume that the disease spreads relatively quickly, so it is reasonable to assume that the only change in the size of these groups is due to the disease.

To set up the model, we make some more specific assumptions. First, we assume that the rate that susceptible people and infected people interact is proportional to both the number of susceptibles and the number of infecteds, that is, proportional to the product of $S(t)$ and $I(t)$. Some fraction of these interactions lead to a susceptible becoming infected. We also assume that the infected individuals recover at a rate that is proportional to the number of infecteds.

Based on these assumptions, our model is

$$\frac{dS}{dt} = -\alpha SI$$
$$\frac{dI}{dt} = \alpha SI - \beta I$$
$$\frac{dR}{dt} = \beta I,$$

where α is the "contagion" parameter and β is the "recovery" parameter. If we know $S(t)$ and $I(t)$, then $R(t) = 1 - (S(t) + I(t))$ (see Exercise 1). Consequently, we need only keep track of $S(t)$ and $I(t)$, and we can consider the planar system

$$\frac{dS}{dt} = -\alpha SI$$
$$\frac{dI}{dt} = \alpha SI - \beta I.$$

The equilibria of this system are the solutions of the simultaneous system of equa-

---

*See "A contribution to the mathematical theory of epidemics" by W. O. Kermack and A. G. McKendrick, Proceedings Royal Society of London A 115, 1927, pp. 700–721.

tions

$$\begin{cases} -\alpha S I = 0 \\ (\alpha S - \beta)I = 0, \end{cases}$$

which is precisely the line $I = 0$. This makes sense since if there are no infecteds, then no one can catch the disease. The next step is to sketch the direction field, and to do so, we must choose values for the parameters. The recovery parameter β gives the rate at which infecteds recover. If we assume that an infected person is contagious for an average of ten days, then roughly 10% of the infecteds recover each day and $\beta = 0.1$.

Choosing α is more difficult since it contains the proportionality constant that measures the likelihood of interaction within the population as well as the likelihood of the disease spreading during an interaction. We test several different values of α starting with $\alpha = 0.2$. We focus on an initial condition $(S(0), I(0)) \approx (1, 0)$ with $I(0) > 0$. It corresponds to a few infecteds existing in a population that is otherwise entirely susceptible. In fact, we use $(S(0), I(0)) = (0.99, 0.01)$, that is, one person in 100 is infected.

Note that the smaller we make $I(0)$, the longer it takes the epidemic to manifest itself. Both $S(t)$ and $I(t)$ change very slowly near the line of equilibrium points along the S-axis.

The solution shows very interesting behavior (see Figure 2.58). The number $I(t)$ of infecteds grows initially. It peaks near $t = 45$ with $I(45) \approx 0.15$. Finally, $I(t) \to 0$ as $t \to \infty$. The number $S(t)$ of susceptibles initially decreases and then almost levels off as $t \to \infty$. However, note that $S(t)$ does not tend to zero as $t \to \infty$. Rather, it tends toward $S \approx 0.2$. In terms of the disease, the model predicts that the percentage of the population that is infected will reach a maximum of approximately 15% after 45 days and then quickly decrease to close to zero after 100 days. The fraction of the population that contracts the disease during the epidemic is approximately 80%. Approximately 20% of the population never gets the disease.

If we try different values of the parameter α, we see that the predictions made by the model vary quantitatively. As α increases, the maximum number of infecteds

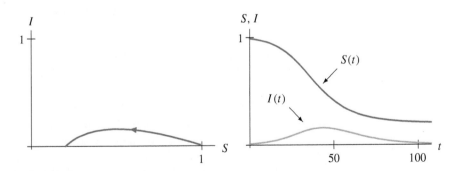

Figure 2.58
The solution curve and $S(t)$- and $I(t)$-graphs for the initial condition
$(S(0), I(0)) = (0.99, 0.01)$ for $\alpha = 0.2$ and $\beta = 0.1$.

increases while the number of susceptibles that avoid the disease decreases. We can use the model to predict the effect of public health measures that alter the values of the parameters α and β (see Figure 2.59).

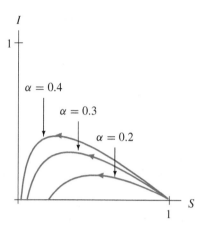

Figure 2.59
Solution curves in the SI-phase plane that correspond to the initial condition $(S(0), I(0)) = (0.99, 0.01)$ with $\beta = 0.1$ and $\alpha = 0.2, 0.3,$ and 0.4. As α increases, the maximum number of infecteds increases while the number of susceptibles that avoid the disease decreases. If you compare this figure with Figure 2.58, the solution curves corresponding to $\alpha = 0.2$ look different. This apparent difference is caused by the fact that the distance between 0 and 1 is the same on both axes in this figure while the distance between 0 and 1 is smaller on the vertical axis than on the horizontal axis in Figure 2.58.

A little phase plane analysis
If we write the SIR system as

$$\frac{dS}{dt} = -\alpha S I$$
$$\frac{dI}{dt} = (\alpha S - \beta) I,$$

we see that $dI/dt = 0$ if $\alpha S - \beta = 0$. In other words, if $S = \beta/\alpha$, the vectors in the vector field are horizontal. To the right of the vertical line in the SI-phase plane, $dI/dt > 0$, and the disease is spreading. To the left of this line, $dI/dt < 0$, and the disease is decreasing. Hence, the value $S = \beta/\alpha$ plays an important role in the evolution of the disease. It is called the **threshold value** of the model. Given values of α and β, if $S(0) > \beta/\alpha$, then an epidemic occurs. If $S(0) < \beta/\alpha$, then there is no epidemic (see Figure 2.60).

The line $S = \beta/\alpha$ along which the vector field is horizontal is one example of what is called a "nullcline." In Section 5.2, we will study nullclines in great detail.

Figure 2.60
The threshold value for $\alpha = 0.2$ and $\beta = 0.1$

An analytic description of the solution curves
Phase plane analysis and numerical solutions for the SIR model give insight into the evolution of a flu epidemic. Because the equations in this system are relatively simple from an algebraic point of view, we can go one step further in describing the solutions curves precisely. The technique we use is one of those ideas that is good to remember. When it works, it gives a great deal of insight into the behavior of solutions.

Because both $S(t)$ and $I(t)$ are nonnegative, we see that $dS/dt < 0$, and $S(t)$ decreases monotonically as $t \to \infty$. As as result, we can view the solution curves as graphs of functions of the variable S. Moreover, we have

$$\frac{dI}{dS} = \frac{dI/dt}{dS/dt}$$

$$= \frac{\alpha SI - \beta I}{-\alpha SI}$$

$$= -1 + \left(\frac{\beta}{\alpha}\right)\frac{1}{S}.$$

This differential equation is one that we can solve by integrating both sides with respect to S. We get

$$I(S) = -S + \frac{\beta}{\alpha}\ln(S) + c,$$

where c is a constant of integration.

When an epidemic starts, there are only a few infected individuals, and almost the entire population is susceptible. That is, $S \approx 1$, and we have

$$0 \approx I(1) = -1 + \frac{\beta}{\alpha}\ln(1) + c = -1 + c.$$

In this case, it makes sense to take $c = 1$. We obtain the function

$$I(S) = -S + \frac{\beta}{\alpha}\ln(S) + 1.$$

The graph of this function $I(S)$ for $\alpha = 0.2$ and $\beta = 0.1$ is almost identical to the solution curve that is shown in Figure 2.58.

For any values of the parameters α and β, we can explicitly compute the maximum value of $I(S)$, that is, the maximum fraction of the population that is ill during the epidemic by doing a maximization problem (see Exercise 4). We can also compute the fraction of the population that completely avoids getting the disease by computing the value of S, $0 < S < 1$, such that $I(S) = 0$. Unfortunately, we cannot solve this equation algebraically for S, but we can understand the behavior of its roots by graphing the function $I(S)$ for various values of α and β (see Exercise 5).

Perhaps the most interesting consequence of the computation of the function $I(S)$ is the fact that $I(S)$ is determined by the ratio of β to α. In other words, if two choices of α and β have the same ratio β/α, then the maximum number of infecteds and the number who escape infection altogether are the same.

Concluding Remark

A dose of reality is in order. We have made a number of simplifying assumptions while setting up the SIR model. The situations where this model gives precise, quantitative predictions are limited to closed environments with simple social dynamics and limited geographical separation (see Exercises 9 and 10). For an epidemic spreading around the world, we must include geographical effects as well as the differences in rates of contact among and within different groups of people. In these cases, the SIR model is the starting point for more involved models.

EXERCISES FOR SECTION 2.7

1. For the SIR-model, show that $S(t) + I(t) + R(t) = 1$ for all t directly from the system of differential equations.

2. In the SIR model, we assume that everyone in the population is susceptible at time $t = 0$ except the very small fraction that is already infected. Suppose that some fraction of the population has received a vaccine, so they cannot catch the disease. The vaccine makes the fraction of the population that is susceptible at time $t = 0$ smaller.

   (a) Using HPGSystemSolver applied to the SIR model with $\alpha = 0.25$ and $\beta = 0.1$, describe the behavior of the solutions with $I(0) = 0.01$ and $S(0) = 0.9, 0.8, 0.7, \ldots$. Pay particular attention to the maximum of $I(t)$, that is, the maximum number of infecteds for each choice of $S(0)$. Also, note the limit of $S(t)$ as $t \to \infty$. (This limit is the fraction of the population that does not catch the disease during the epidemic.)

   (b) If $\alpha = 0.25$ and $\beta = 0.1$, how large a fraction of the population must be vaccinated in order to keep the epidemic from getting started with $I(0) = 0.01$?

3. Vaccines make it possible to prevent epidemics. However, the time it takes to develop a vaccine may make it impossible to vaccinate everyone in a population before a disease arrives.

   (a) For the SIR model, which initial conditions guarantee that $dI/dt < 0$? [Hint: Your answer should be expressed in terms of the parameters α and β.]

   (b) For given values of α and β, what fraction of a population must be vaccinated before a disease arrives in order to prevent an epidemic?

4. In this section we showed that solution curves of the SIR model with $S(0) \approx 1$ and $I(0) \approx 0$ are graphs of the function

$$I(S) = -S + \frac{\beta}{\alpha} \ln(S) + 1.$$

   (Note that the graph depends only on the ratio $\rho = \beta/\alpha$ of the parameters. Different values of the parameters can give the same value of ρ.)

   (a) Determine the maximum value of $I(S)$ in terms of ρ.

   (b) Is the statement "The epidemic cannot get started if $\beta > \alpha$" true or false? Justify your answer.

5. Let ρ denote the ratio β/α of the parameters α and β in the SIR model. Then

$$I(S) = -S + \rho \ln(S) + 1.$$

   (a) Using graphing technology, graph $I(S)$ over the interval $0 < S \leq 1$ for various values of ρ between 0.1 and 1.0.

   (b) Using the graphs that you produced in part (a), graph the solution of $I(S) = 0$ for $0 < S < 1$ as a function of ρ.

(c) What does the graph that you produced in part (b) tell you about the long-term predictions of the SIR model in terms of the ratio ρ?

6. One of the basic assumptions of the SIR model is that individuals who recover from the disease never get it again. However, diseases continually evolve, and new strains can emerge that can infect those who have recovered from the previous strain. In this exercise, we modify the SIR model so that recovereds become susceptible again in a linear rate. We obtain the system of equations

$$\frac{dS}{dt} = -\alpha SI + \gamma R$$

$$\frac{dI}{dt} = \alpha SI - \beta I$$

$$\frac{dR}{dt} = \beta I - \gamma R$$

(a) Show that the sum $S(t) + I(t) + R(t)$ is constant as a function of t for this model.

(b) Derive a system in the two dependent variables S and I using the fact that $R = 1 - (S + I)$.

(c) What are the equilibrium points for this model of the two variables S and I? (*Hint*: Both S and I are nonnegative, and $S(t) + I(t) \leq 1$ for all t.)

(d) Fix $\alpha = 0.3$, $\beta = 0.15$, and $\gamma = 0.05$ and use HPGSystemSolver to sketch the phase portrait. Describe the behavior of solutions.

(e) How does the system change if we fix $\alpha = 0.3$ and $\beta = 0.15$, but vary γ over a small interval surrounding $\gamma = 0.05$?

7. In the movie *I Am Legend*, the infecteds work together to increase the number of infecteds. We can modify the SIR model to include the assumption that zombies actively infect susceptibles by replacing I by $\sqrt{I}$ in the interaction term. (Note that $0 \leq I \leq 1$, so $\sqrt{I} \geq I$.) We obtain the system

$$\frac{dS}{dt} = -\alpha S\sqrt{I}$$

$$\frac{dI}{dt} = \alpha S\sqrt{I} - \beta I.$$

(a) Calculate the equilibrium points of this model.

(b) Find the region of the phase plane where $dI/dt > 0$.

(c) Use $\alpha = 0.2$ and $\beta = 0.1$ and sketch the phase portrait. What does the model predict for the spread of the zombies in this case?

8. Many zombie movies are based on the premise that zombies do not stop infecting new victims until they are destroyed by a susceptible. In addition, the susceptibles destroy as many zombies as they can. We can model the spread of zombies in such a movie by assuming that infecteds (zombies) become recovereds (zombies who can

not infect susceptibles) at a rate proportional to the size of the remaining susceptible population. We obtain the system

$$\frac{dS}{dt} = -\alpha S I$$

$$\frac{dI}{dt} = \alpha S I - \gamma S.$$

(a) Calculate the equilibrium points of this model.

(b) Find the region of the phase plane where $dI/dt > 0$.

(c) Use $\alpha = 0.2$ and $\gamma = 0.1$ and sketch the phase portrait. What does the model predict for the spread of the zombies in this case?

The SIR model is particularly relevant to a homogenous population in an enviroment with little geographic distribution. A famous example of exactly this situation occurred in 1978 at a British boarding school.* A single boy in the school of 763 students contracted the flu and the epidemic spread rapidly, as shown in Table 2.3. (We are assuming that the number of students confined to bed was the same as the number of infected students.)

Table 2.3
The daily count of the number of infected students.

t	Infected	t	Infected	t	Infected
0	1	5	222	10	123
1	3	6	282	11	70
2	7	7	256	12	25
3	25	8	233	13	11
4	72	9	189	14	4

9. Assume that the parameter $\alpha = 1.66$ in the SIR model for the data in Table 2.3.

(a) Using whatever technology that is most convenient, determine an appropriate value of β that matches the data in Table 2.3.

(b) Using the value of β that you computed in part (a), calculate the total number of students who caught the flu during the epidemic.

(c) Interpret the value of β that you computed in part (a) in terms of the length of time that students with the flu remained infected.

10. Using $\alpha = 1.66$ and the value of β that you determined in Exercise 9, how would the progress of the epidemic have changed if 200 students had been vaccinated before the disease started? (Give as precise an answer as possible.)

---

*Anonymous, "Epidemiology: Influenza in a boarding school," *British Medical Journal*, Vol. 4, 1978, p. 587.

2.8 THE LORENZ EQUATIONS

As we have seen, the behavior of solutions of autonomous systems of differential equations can be much more interesting and complicated than solutions of single autonomous equations. For autonomous equations with one dependent variable, the solutions live on a phase line, and their behavior is completely governed by the position and nature of the equilibrium points. Solutions of systems with two dependent variables live in two-dimensional phase planes. A plane has much more "room" than a line, so solutions in a phase plane can do many more interesting things. This includes forming loops (periodic solutions) and approaching and retreating from equilibrium points.

However, there are still definite restrictions on the types of phase portraits that are possible for autonomous systems. As we learned in Section 2.6, the Uniqueness Theorem implies that solution curves must agree entirely if they intersect at all. So, for example, if there is a periodic solution that forms a loop in the phase plane, then solutions with initial conditions inside the loop must stay inside for all time (see Figure 2.61). Also, two or three solutions can fit together to divide the phase plane into distinct regions, and solutions must stay in the same region as their initial condition (see Figure 2.62). Additional details regarding the implications of the Uniqueness Theorem are discussed in Section 2.6.

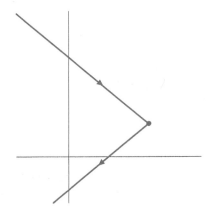

Figure 2.61
Solutions with initial conditions inside a periodic solution must stay inside for all time.

Figure 2.62
Two solutions and an equilibrium point cut the phase plane into regions. Solutions with initial conditions in one region must stay in that region for all time.

If we raise the number of dependent variables to three, the situation becomes much more complicated. A solution of an autonomous system with three dependent variables is a curve in a three-dimensional phase space. The Uniqueness Theorem still applies, so solution curves either agree completely or do not intersect, but in three dimensions this restriction is not nearly so confining as it is in two dimensions. In Figure 2.63 we see examples of curves in three dimensions that do not intersect. These curves can knot and link in very complicated ways.

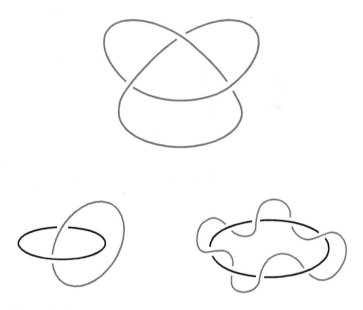

Figure 2.63
A knot and two links in space.

The first to realize the possible complications of three-dimensional systems was the French mathematician Henri Poincaré. In the 1890s, Poincaré, while working on the Newtonian three-body problem for the motion of the planets, realized that systems with three dependent variables can have behavior so complicated that he did not even attempt to draw them. Today we can easily draw numerical approximations of complicated solution curves with a computer. The problem now is to make sense of the pictures. This is an active area of current research in dynamical systems, and the complete story for systems with three dependent variables is still far from being written.

In this section we study a three-dimensional system known as the Lorenz equations. This system was first studied by Ed Lorenz in 1963 in an effort to model the weather. It is important because the vector field is formed by very simple equations, yet solutions are very complicated curves.

The Lorenz System

The behavior of a physical system like the weather on Earth is extremely complicated. To predict the weather, many mathematical models have been developed. The readings from weather stations and satellites are used as initial conditions, and numerical approximations of solutions are used to obtain predictions.

Edward N. Lorenz (1917–2008) began his career as a mathematics graduate student at Harvard but turned his attention to meteorology during World War II. In 1961, using a computer primitive by today's standards, Lorenz attempted to solve a much simplified model for weather prediction. His model seemed to simulate real weather patterns quite well, but it also illustrated something much more important: When Lorenz changed the initial conditions slightly, the resulting weather patterns changed completely after a short time. Lorenz had discovered the fact that simple differential equations can behave "chaotically." We describe additional aspects of this important discovery in Chapters 5 and 8.

The success of long-range weather forecasts (that is, more than five days in the future) is limited. This lack of long-term precision might be due to inaccuracies in the model. It is also possible that the model is accurate but that some property of the equations makes prediction difficult. Consequently it is important to study these models theoretically as well as numerically.

Since the weather is so complicated, it is necessary to start the theoretical study by looking at simplifications. After simplifying, meteorologist Ed Lorenz arrived at the system

$$\frac{dx}{dt} = \sigma(y - x)$$
$$\frac{dy}{dt} = \rho x - y - xz$$
$$\frac{dz}{dt} = -\beta z + xy,$$

where x, y, and z are dependent variables and σ, ρ, and β are parameters.* This system is so much simpler than the one used for modeling the weather that it has nothing to tell us about tomorrow's temperature. However, by studying this system, Lorenz helped start a scientific revolution by making scientists and engineers aware of the field of mathematics now called Chaos Theory.♣

---

*Lorenz's original paper is "Deterministic nonperiodic flow," *Journal of Atmospheric Science*, Vol. 20, 1963, pp. 130–141.

♣*Chaos: Making a New Science* by James Gleick (Viking, 1987) is a popular account of the development of Chaos Theory. Ian Stewart's *Does God Play Dice?* (Wiley-Blackwell, 2nd ed, 2002) is a nice introduction to the mathematics that underlies this theory.

The Vector Field

Lorenz chose to study the system with the parameter values $\sigma = 10$, $\beta = 8/3$, and $\rho = 28$. That is,

$$\frac{dx}{dt} = 10(y - x)$$

$$\frac{dy}{dt} = 28x - y - xz$$

$$\frac{dz}{dt} = -\frac{8}{3}z + xy.$$

The right-hand sides of these equations define a vector field in three-dimensional space

$$\mathbf{F}(x, y, z) = (10(y - x), 28x - y - xz, -\tfrac{8}{3}z + xy).$$

It assigns a three-component vector to each point (x, y, z). Just as in two dimensions, the equilibrium points are the points (x, y, z) where the vector field vanishes, that is, the points (x, y, z) where $\mathbf{F}(x, y, z) = (0, 0, 0)$. We can compute directly that the equilibrium points are $(0, 0, 0)$, $(6\sqrt{2}, 6\sqrt{2}, 27)$, and $(-6\sqrt{2}, -6\sqrt{2}, 27)$ (see Exercise 1).

An initial condition for a solution consists of values for the three coordinates x, y, and z, and consequently, we think of it as a point in the phase space of the system. With the exception of the equilibrium points and solutions with initial conditions on the z-axis, there is little hope of finding formulas for solutions (see Exercise 3). Hence we turn to numerical methods. Euler's method for three-dimensional systems works as it does in two dimensions. The approximate solution is constructed by following the vector field for short time steps. Lorenz began his study of this system by finding numerical approximations of solutions, and we follow in his footsteps.

Numerical Approximation of Solutions

We begin by looking at a numerical approximation of the solution with initial condition $(0, 1, 0)$ (that is, $x(0) = 0$, $y(0) = 1$, $z(0) = 0$—see Figure 2.64). Clearly something interesting is happening. The solution does not seem to have any particular pattern. For example, the x-coordinate jumps from positive to negative values in an unpredictable way.

Although this seemingly random behavior is somewhat unnerving, we see something even more interesting if we compare the behaviors of the solution with initial condition $(0, 1, 0)$ and the solution with initial condition $(0, 1.001, 0)$. The second solution starts very close to the first, with x, y, and z oscillating unpredictably. However, we see that eventually they oscillate in quite different ways (see Figure 2.65). A very small change in the initial condition leads to a big change in the behavior of the solution.

In fact, this strange behavior occurs for almost every solution curve. The functions $x(t)$, $y(t)$, and $z(t)$ oscillate in an unpredictable and unique way, but the solution curve in three dimensions generates a figure that is roughly the same for every initial condition (compare Figures 2.64 and 2.66).

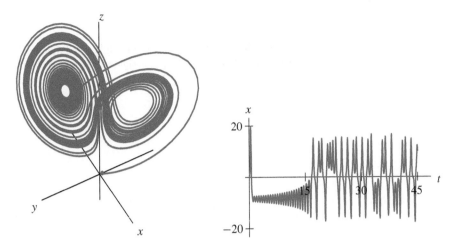

Figure 2.64
The left-hand figure is the solution curve in xyz-space for the initial condition $(0, 1, 0)$ and the right-hand figure is the corresponding $x(t)$-graph.

The solution curves seem to loop around the equilibrium points above the plane $z = 0$ in increasing spirals. Once the radius gets too large, the solution passes close to $(0, 0, 0)$ and then is reinjected toward one of the two equilibria. (It is very instructive to watch an animation of a solution moving in real time through the phase space. In addition, DETools contains the tools LorenzEquations and ButterflyEffect that illustrate the complicated nature of the solutions.) In Chapters 3 and 5, we develop the tools to study the behavior of these solutions near the equilibrium points.

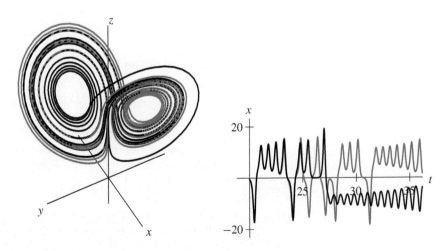

Figure 2.65
The solution curves and $x(t)$-graphs for two solutions with nearby initial conditions.

Figure 2.66
Solution curve for the solution of the Lorenz system with the initial condition $(10, -1, 14)$. There is no special significance to this choice of initial condition. Essentially all initial conditions generate the same figure.

Chaos

The qualitative analysis of this system is a difficult undertaking that must wait until Chapters 5 and 8. However, there is a moral to what we have seen so far that is having an important effect on many different branches of science. The Lorenz system has two important properties. The first is that a small change in initial conditions leads fairly quickly to large differences in the corresponding solutions. If a system as simple as the Lorenz equations can have this property, it is entirely reasonable to think that much more complicated systems (such as the weather) might have it as well. Any small error in the initial conditions leads quickly to a big error in prediction of the solution. This might be why physical systems like the weather are so hard to predict.

The second property of the Lorenz system is that although the details of individual solutions are quite different, the pictures of the solution curves in the three-dimensional phase space look remarkably alike. Many solutions seem to be sweeping out the same "surface" in three dimensions. So the solutions of the Lorenz system still have structure that we can study. We don't have to give up studying the solutions of the Lorenz equations. We just have to ask the right questions.

EXERCISES FOR SECTION 2.8

In Exercises 1–3, we consider the Lorenz system

$$\frac{dx}{dt} = \sigma(y - x)$$

$$\frac{dy}{dt} = \rho x - y - xz$$

$$\frac{dz}{dt} = -\beta z + xy,$$

where x, y, and z are dependent variables and σ, ρ, and β are parameters.

1. Let $\sigma = 10$, $\beta = 8/3$, and $\rho = 28$ in the Lorenz system.
   (a) Verify that $(0, 0, 0)$, $(6\sqrt{2}, 6\sqrt{2}, 27)$, and $(-6\sqrt{2}, -6\sqrt{2}, 27)$ are equilibrium points.
   (b) Verify that these three points are the only equilibrium points.

2. Suppose we fix $\sigma = 10$ and $\beta = 8/3$ in the Lorenz system but leave ρ as a parameter.

    (a) Show that there is only one equilibrium point for the system if $\rho \le 1$.

    (b) Assume that $\rho > 1$ and show that there are three equilibrium points. Express them in terms of ρ.

    (c) What do you conclude about the value $\rho = 1$?

3. For the Lorenz system with $\sigma = 10$, $\rho = 28$, and $\beta = 8/3$,

    (a) verify that if $(x(t), y(t), z(t))$ is a solution with $x(0) = y(0) = 0$, then $x(t) = y(t) = 0$ for all t;

    (b) find the solution with initial condition $(0, 0, 1)$; and

    (c) find the solution with initial condition $(0, 0, z_0)$, where z_0 is any constant, and sketch its solution curve in xyz-phase space.

4. Using `ButterflyEffect`, choose several different pairs of nearby initial conditions. Comment on how long it takes the corresponding solutions to separate.

5. Close to the origin, where x, y, and z are very small, the quadratic terms $-xz$ and $+xy$ will be very, very small. So, near $(x, y, z) = (0, 0, 0)$, we can approximate the Lorenz system with the system

$$\frac{dx}{dt} = 10(y - x)$$

$$\frac{dy}{dt} = 28x - y$$

$$\frac{dz}{dt} = -\frac{8}{3}z.$$

(This is called *linearization* at the origin and will be studied in detail in Chapter 5.) Notice that z does not appear in the equations for dx/dt and dy/dt, and the equation for dz/dt does not contain x or y. That is, the system decouples into a two-dimensional system and a one-dimensional equation.

    (a) Using `HPGSystemSolver`, sketch the direction field and the phase plane for the planar system

$$\frac{dx}{dt} = 10(y - x)$$

$$\frac{dy}{dt} = 28x - y.$$

    (b) Sketch the phase line for the equation

$$\frac{dz}{dt} = -\frac{8}{3}z.$$

    (c) Sketch solutions in the three-dimensional phase space for the system above. (This picture gives the behavior of the Lorenz system *near* $(0, 0, 0)$.)

REVIEW EXERCISES FOR CHAPTER 2

Short answer exercises: Exercises 1–14 focus on the basic ideas, definitions, and vocabulary of this chapter. Their answers are short (a single sentence or drawing), and you should be able to do them with little or no computation. However, they vary in difficulty, so think carefully before you answer.

1. Find one solution of the system $dx/dt = |x| \sin y$ and $dy/dt = |y| \cos x$.

2. Find all equilibrium points of the system $dx/dt = y$ and $dy/dt = e^y + x^2$.

3. Convert the second-order differential equation $d^2y/dt^2 = 1$ to a first-order system.

4. Find the general solution of the system of equations in Exercise 3.

5. Find all equilibrium points of the system $dx/dt = y$ and $dy/dt = \sin(xy)$.

6. How many equilibrium solutions does the system of differential equations $dx/dt = x(x - y)$ and $dy/dt = (x^2 - 4)(y^2 - 9)$ have? What are they?

7. Is the function $(x(t), y(t)) = (e^{-6t}, 2e^{-3t})$ a solution to the system of differential equations $dx/dt = 2x - 2y^2$ and $dy/dt = -3y$? Why?

8. Write the second-order equation and the corresponding first-order system for the mass-spring system with spring constant α, mass β, and damping coefficient γ.

9. Find the general solution of the system $dx/dt = 2x$ and $dy/dt = -3y$.

10. Sketch the $x(t)$- and $y(t)$-graphs corresponding to the solution of the initial-value problem $dx/dt = y^2 - 4$, $dy/dt = x^2 - 2x$, and $(x(0), y(0)) = (0, 2)$.

11. Give an example of a first-order system of differential equations with exactly ten equilibrium points.

12. Suppose that $\mathbf{F}(2, 1) = (3, 2)$. What is the result of one step of Euler's method applied to the initial-value problem $d\mathbf{Y}/dt = \mathbf{F}(\mathbf{Y})$, $\mathbf{Y}(0) = (2, 1)$, with $\Delta t = 0.5$?

13. Sketch the solution curve for the initial-value problem $dx/dt = -x$, $dy/dt = -y$, and $(x(0), y(0)) = (1, 1)$.

14. Suppose that all solutions of the system $dx/dt = f(x, y)$ and $dy/dt = g(x, y)$ tend to an equilibrium point at the origin as t increases. What can you say about solutions of the system $dx/dt = -f(x, y)$ and $dy/dt = -g(x, y)$?

True-false: For Exercises 15–21, determine if the statement is true or false. If it is true, explain why. If it is false, provide a counterexample or an explanation.

15. The function $(x(t), y(t)) = (e^{-6t}, 2e^{-3t})$ is a solution to the system of differential equations $dx/dt = 2x - 2y^2$ and $dy/dt = -3y$.

16. The function $x(t) = 2$ for all t is an equilibrium solution of the system of differential equations $dx/dt = x - 2$ and $dy/dt = -y$.

17. Two different first-order autonomous systems can have the same vector field.

18. Two different first-order autonomous systems can have the same direction field.

19. The function $(x(t), y(t)) = (\sin t, \sin t)$ is a solution of some first-order autonomous system of differential equations.

20. If the function $(x_1(t), y_1(t)) = (\cos t, \sin t)$ is a solution to an autonomous first-order system, then the function $(x_2(t), y_2(t)) = (\cos(t - 1), \sin(t - 1))$ is also a solution.

21. If the function $(x_1(t), y_1(t)) = (\cos t, \sin t)$ is a solution of a first-order autonomous system, then the function $(x_2(t), y_2(t)) = (-\sin t, \cos t)$ is also a solution of the same system.

22. MacQuarie Island is a small island about half-way between Antarctica and New Zealand. As was mentioned in Exercise 11 of Section 1.1, its rabbit population underwent an explosion during the six-year period between 2000 and 2006. Before the year 2000, it was home to approximately 4,000 rabbits. It was also home to 160 feral cats and was an important nesting site for seabirds* The cats, being cats, attacked the nests of the seabirds. To protect the endangered birds, the cats were "eliminated" in 2000. However, the cats ate rabbits as well as seabirds. By 2006, the number of rabbits had grown to about 130,000.

Let $R(t)$ be the rabbit population and $C(t)$ be the cat population where time t is measured in years. Suppose the cat population is well approximated by a logistic model, while the rabbit population is modeled by a modified logistic model. We use

$$\frac{dC}{dt} = C\left(1 - \frac{C}{160}\right)$$

$$\frac{dR}{dt} = R\left(1 - \frac{R}{130,000}\right) - \alpha RC,$$

where the $-\alpha RC$ term measures the negative effects on the rabbits during their interactions with the cats.

(a) What value of α gives an equilibrium point at $C = 160$ and $R = 4000$?

(b) Using the value of α from part (a), calculate the contribution that the term $-\alpha RC$ makes to dR/dt when $C = 160$ and $R = 4000$. Assuming that this value represents the decrease in the rabbit population per year caused by the cats, approximately how many rabbits did each cat eliminate per year (when $C = 160$ and $R = 4000$)?

(c) A plan is being developed to "remove" the rabbits and other rodents. Could the rabbit population be controlled by instituting a constant harvesting parameter? If so, how many rabbits would have to be harvested per year?

---

*See "Rampant rabbits trash World Heritage island" by Rachel Nowak *New Scientist*, January 14, 2009. Available at www.newscientist.com

True-false: Solution curves for several solutions of the system

$$\frac{dx}{dt} = y$$

$$\frac{dy}{dt} = 2x - 3x^2$$

are shown in the figure below. For Exercises 23–28, determine if the statement is true or false for this system. Justify your answer.

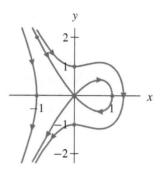

23. The solution curve corresponding to the initial condition $(1, 0)$ includes the point $(0, 0)$.

24. The $x(t)$- and $y(t)$-graphs of the solution with $(x(0), y(0)) = (1/2, 0)$ tend to infinity as t increases.

25. The solution with initial condition $(x(0), y(0)) = (0, 1)$ is the same as the solution with initial condition $(x(0), y(0)) = (0, -1)$.

26. The function $y(t)$ for the solution with initial condition $(x(0), y(0)) = (-1, 2)$ is positive for all $t > 0$.

27. The functions $x(t)$ and $y(t)$ for the solution with initial condition $(x(0), y(0)) = (-1, 0)$ decrease monotonically for all t.

28. The $x(t)$- and $y(t)$-graphs of the solution with initial condition $(x(0), y(0)) = (0, 1)$ each have exactly one critical point.

29. Consider the system

$$\frac{dx}{dt} = \cos 2y$$

$$\frac{dy}{dt} = 2y - x.$$

(a) Find its equilibrium points.

(b) Use HPGSystemSolver to plot its direction field and phase portrait.

(c) Briefly describe the behavior of typical solutions.

30. Consider a decoupled system of the form

$$\frac{dx}{dt} = f(x)$$
$$\frac{dy}{dt} = g(y).$$

What special features does the phase portrait of this system have?

In Exercises 31–34, a solution curve in the xy-plane and an initial condition on that curve are specified. Sketch the $x(t)$- and $y(t)$-graphs for the solution.

31.

32.

33.

34.

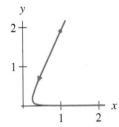

35. Consider the partially decoupled system

$$\frac{dx}{dt} = x + 2y + 1$$
$$\frac{dy}{dt} = 3y.$$

(a) Derive the general solution.
(b) Find the equilibrium points of the system.
(c) Find the solution that satisfies the initial condition $(x_0, y_0) = (-1, 3)$.
(d) Use HPGSystemSolver to plot the phase portrait for this system. Identify the solution curve that corresponds to the solution with initial condition $(x_0, y_0) = (-1, 3)$.

36. Consider the partially decoupled system

$$\frac{dx}{dt} = xy$$
$$\frac{dy}{dt} = y + 1.$$

(a) Derive the general solution.

(b) Find the equilibrium points of the system.

(c) Find the solution that satisfies the initial condition $(x_0, y_0) = (1, 0)$.

(d) Use HPGSystemSolver to plot the phase portrait for this system. Identify the solution curve that corresponds to the solution with initial condition $(x_0, y_0) = (1, 0)$.

37. A simple model of a glider flying along up and down but not left or right ("planar" motion) is given by

$$\frac{d\theta}{dt} = \frac{s^2 - \cos\theta}{s}$$

$$\frac{ds}{dt} = -\sin\theta - Ds^2,$$

where θ represents the angle of the nose of the glider with the horizon, $s > 0$ represents its speed, and $D \geq 0$ is a parameter that represents drag (see the DETools program HMSGlider).

(a) Calculate the equilibrium points for this system.

(b) Give a physical description of the motion of the glider that corresponds to these points.

LAB 2.1 Two Magnets and a Spring

In this lab we consider the motion of a mass that can slide freely along the x-axis. The mass is attached to a spring that has its other end attached to the point $(0, 2)$ on the y-axis. In addition, the mass is made of iron and is attracted to two magnets of equal strength—one located at the point $(-1, -a)$ and the other at $(1, -a)$ (see Figure 2.67).

We assume that the spring obeys Hooke's Law, and the magnets attract the mass with a force proportional to the inverse of the square of the distance of the mass to the magnet (the inverse square law). If we choose the spring constant, mass, strength of the magnets, and units of distance and time appropriately, then we can model the motion of the mass along the x-axis with the equation

$$\frac{d^2x}{dt^2} = -0.3x - \frac{x-1}{((x-1)^2 + a^2)^{3/2}} - \frac{x+1}{((x+1)^2 + a^2)^{3/2}}.$$

(A good exercise for engineering and physics students: Derive this equation and determine the units and choices of spring constant, rest length of the spring, mass, and strength of the magnets involved.)

The goal of this lab is to study this system numerically. Use technology to find equilibria and study the behavior of solutions. Be careful to consider the correct regions of the phase plane at the correct scale so that you can find the important aspects of the system.

In your report, you should address the following items:

1. Consider the system with the parameter value $a = 2.0$. Discuss the behavior of solutions in the phase plane. Relate the phase portrait to the possible motions of the mass along the x-axis.

Figure 2.67
Schematic of a mass sliding on the x-axis attached to a spring and attracted by two magnets.

2. Consider the system with the parameter value $a = 0.5$. Discuss the behavior of solutions in the phase plane. Relate the phase portrait to the possible motions of the mass along the x-axis. Be particularly careful to describe the solutions that separate different types of qualitative behavior.

3. Describe how the system changes as a varies from $a = 0.5$ to $a = 2.0$. That is, describe the bifurcation that occurs.

4. Finally, repeat the analysis in Parts 1–3 with the magnets located at $(\pm 2, -a)$. In other words, use the equation

$$\frac{d^2x}{dt^2} = -0.3x - \frac{x-2}{((x-2)^2 + a^2)^{3/2}} - \frac{x+2}{((x+2)^2 + a^2)^{3/2}}.$$

Note the differences between this system and the previous one and interpret these differences in terms of the possible motions of the mass as it slides along the x-axis.

Your report: Address each of the items above. Pay particular attention to the physical interpretation of the solutions in terms of the possible motions of the mass as it slides along the x-axis. You may include graphs and phase portraits to illustrate your discussion, but pictures alone are not sufficient.

LAB 2.2 Cooperative and Competitive Species

In this chapter we have focused on first-order autonomous systems of differential equations, such as the predator-prey systems described in Section 2.1. In particular, we have seen how such systems can be studied using vector fields and phase plane analysis and how solution curves in the phase plane relate to the $x(t)$- and $y(t)$-graphs of the solutions. In this lab project you will use these concepts and related numerical computations to study the behavior of the solutions to two different systems.

We have discussed predator-prey systems at length. These are systems in which one species benefits while the other species is harmed by the interaction of the two species. In this lab you will study two other types of systems—competitive and cooperative systems. A competitive system is one in which both species are harmed by interaction, for example, cars and pedestrians. A cooperative system is one in which both species benefit from interaction, for example, bees and flowers. Your overall goal is to understand what happens in both systems for all possible nonnegative initial conditions. Several pairs of cooperative and competitive systems are given at the end of this lab. (Your instructor will tell you which pair(s) of systems you should study.) The analytic techniques that are appropriate to analyze these systems have not been discussed so far, so you will employ mostly geometric/qualitative and numeric techniques to establish your conclusions. Since these are population models, you need consider only x and y in the first quadrant ($x \geq 0$ and $y \geq 0$).

Your report should include:

1. A brief discussion of all terms in each system. For example, what does the coefficient to the x term in equation for dx/dt represent? Which system is cooperative and which is competitive?

2. For each system, determine all relevant equilibrium points and analyze the behavior of solutions whose initial conditions satisfy either $x_0 = 0$ or $y_0 = 0$. Determine the curves in the phase plane along which the vector field is either horizontal or vertical. Which way does the vector field point along these curves?

3. For each system, describe all possible population evolution scenarios using the phase portrait as well as $x(t)$- and $y(t)$-graphs. Give special attention to the interpretation of the computer output in terms of the long-term behavior of the populations.

Your report: The text of your report should address the three items above, one at a time, in the form of a short essay. You should include a description of all "hand" computations that you did. You may include a limited number of pictures and graphs. (You should spend some time organizing the qualitative and numerical information since a few well-organized figures are much more useful than a long catalog.)

Systems:
Pair (1):

$$\text{A. } \frac{dx}{dt} = -5x + 2xy \qquad \text{B. } \frac{dx}{dt} = 6x - x^2 - 4xy$$

$$\frac{dy}{dt} = -4y + 3xy \qquad \frac{dy}{dt} = 5y - 2xy - 2y^2$$

Pair (2):

$$\text{A. } \frac{dx}{dt} = -3x + 2xy \qquad \text{B. } \frac{dx}{dt} = 5x - x^2 - 3xy$$

$$\frac{dy}{dt} = -5y + 3xy \qquad \frac{dy}{dt} = 8y - 3xy - 3y^2$$

Pair (3):

$$\text{A. } \frac{dx}{dt} = -4x + 3xy \qquad \text{B. } \frac{dx}{dt} = 5x - 2x^2 - 4xy$$

$$\frac{dy}{dt} = -3y + 2xy \qquad \frac{dy}{dt} = 7y - 4xy - 3y^2$$

Pair (4):

$$\text{A. } \frac{dx}{dt} = -5x + 3xy \qquad \text{B. } \frac{dx}{dt} = 9x - 2x^2 - 4xy$$

$$\frac{dy}{dt} = -3y + 2xy \qquad \frac{dy}{dt} = 8y - 5xy - 3y^2$$

LAB 2.3 The Harmonic Oscillator with Modified Damping

Autonomous second-order differential equations are studied numerically by reducing them to first-order systems with two dependent variables. In this lab you will use the computer to analyze three somewhat related second-order equations. In particular, you will analyze phase planes and $y(t)$- and $v(t)$-graphs to describe the long-term behavior of the solutions.

In Sections 2.1 and 2.3, we discuss the most classic of all second-order equations, the harmonic oscillator. The harmonic oscillator is

$$m\frac{d^2y}{dt^2} + b\frac{dy}{dt} + ky = 0.$$

It is an example of a second-order, homogeneous, linear equation with constant coefficients. In the text we explain how this equation is used to model the motion of a spring. The force due to the spring is assumed to obey Hooke's law (the force is proportional to the amount the spring is compressed or stretched). The force due to damping is assumed to be proportional to the velocity. In your report you should describe the motion of the spring assuming certain values of m, b, and k. (A table of values of the parameters is given below. Your instructor will tell you what values of m, b, and k to consider.) Your report should discuss the following:

1. (Undamped harmonic oscillator) The first equation that you should study is the harmonic oscillator with no damping; that is, $b = 0$ and with $k \neq 0$. Examine solutions using both their graphs and the phase plane. Are the solutions periodic? If so, what does the period seem to be? Describe the behavior of three different solutions that have especially different initial conditions and be specific about the physical interpretation of the different initial conditions. (Analytic methods to answer these questions are discussed in Chapter 3. For now, work numerically.)

2. (Harmonic oscillator with damping) Repeat Part 1 using the equation

$$m\frac{d^2y}{dt^2} + b\frac{dy}{dt} + ky = 0.$$

3. (Harmonic oscillator with nonlinear damping) Repeat Part 1 using the equation

$$m\frac{d^2y}{dt^2} + b\left|\frac{dy}{dt}\right|\frac{dy}{dt} + ky = 0$$

in place of the usual harmonic oscillator equation. (Note that even with the same value of the parameter b, the drag forces in this equation and the equation in Part 2 have the same magnitude only for velocity ± 1. Also, note that the sign of the term

$$\left|\frac{dy}{dt}\right|\frac{dy}{dt}$$

is the same as the sign of dy/dt, hence this damping force is always directed opposite the direction of motion. The difference between this equation and that in Part 2 is the size of the damping for small and large velocities. One of the many examples of situations for which this is a better model than linear damping is the drag on airplane tires from wet snow or slush. Drag from only four inches of slush was enough to cause the 1958 crash during take-off of the plane carrying the Manchester United soccer team. Currently, large airplanes are allowed to take off and land in no more than one-half inch of wet snow or slush.*

4. (Nonlinear second-order equation) Finally, consider a somewhat related second-order equation where the damping coefficient b is replaced by the factor $(y^2 - \alpha)$; that is,

$$m\frac{d^2y}{dt^2} + (y^2 - \alpha)\frac{dy}{dt} + ky = 0.$$

Is it reasonable to interpret this factor as some type of damping? Provide a complete description of the long-term behavior of the solutions. Are the solutions periodic? If so, what does the period seem to be? Explain why this equation is not a good model for something like a mass-spring system. Give an example of some other type of physical or biological phenomenon that could be modeled by this equation.

Your report: Address the questions in each item above in the form of a short essay. Be particularly sure to describe the behavior of the solution and the corresponding behavior of the mass-spring system. You may use the phase planes and graphs of $y(t)$ to illustrate the points you make in your essay. (However, please remember that, although one good illustration may be worth 1000 words, 1000 illustrations are usually worth nothing.)

Table 2.4
Possible choices for the parameters.

Choice	m	k	b	α
1	2	5	2	3
2	3	5	3	3
3	5	5	4	3
4	2	6	3	5
5	3	6	3	5
6	5	6	3	5
7	5	4	4	2
8	5	5	4	2
9	5	6	4	2
10	5	4	4	2

*See Stanley Stewart, *Air Disasters*, Barnes & Noble, 1986.

LAB 2.4   A Mass-Spring System with a Rubber Band

In an idealized mass-spring system, the spring provides a restoring force proportional to its displacement from its rest position (Hooke's Law). This ideal spring is never overstretched or completely compressed. These sorts of assumptions (hopefully) yield models that are simple enough to be tractable but accurate enough to be useful. But we should always be mindful of the underlying assumptions.

For example, we frequently think of a rubber band as a kind of spring. When stretched, it provides a restoring force toward its rest length. However, there are limitations. We cannot stretch the rubber band too far or it will break. More importantly, a compressed rubber band provides no force. We expect the behavior of a system with a spring and a rubber band to be different from one with a spring alone.

In this lab we compare models for a mass-spring system and a mass-spring system with the addition of a rubber band (see Figure 2.68). The rubber band adds extra restoring force when the displacement is positive but adds no force when the displacement is negative.

The mass-spring system depicted in Figure 2.68 is modeled by

$$m\frac{d^2y}{dt^2} + b\frac{dy}{dt} + k_1 y = 10m,$$

where y measures the vertical displacement (with down as positive) in meters. The parameters are the mass m, the damping coefficient b, and the spring constant k_1. The constant $10m$ on the right-hand side of the equation is a rough approximation of the force due to gravity.

To include the rubber band, we add an extra term to the equation above. We assume that the rubber band obeys Hooke's Law when it is stretched but that it exerts no

Figure 2.68
A mass-spring system and a mass-spring system with a rubber band.

force when it is compressed. Let $h(y)$ be the function that is y if y is positive and zero if y is negative, that is,

$$h(y) = \begin{cases} y, & \text{if } y \geq 0; \\ 0, & \text{if } y < 0. \end{cases}$$

Then the term $k_2 h(y)$ models the restoring force of a rubber band with a "spring constant" k_2 when it is stretched and no effect when it is compressed. We obtain

$$m\frac{d^2y}{dt^2} + b\frac{dy}{dt} + k_1 y + k_2 h(y) = 10m,$$

where the parameters m, b, and k_1 are as above. We choose $m = 1$ for simplicity.

Consider the following cases:

1. (Ideal mass-spring system with no rubber band) Choose a value of k_1 such that $12 < k_1 < 13$ and study solutions of the equation

$$\frac{d^2y}{dt^2} + k_1 y = 10.$$

Examine solutions using both their graphs and the phase portrait. Are solutions periodic? If so, approximate the period of the solutions. Be specific about the physical interpretation of the solutions for different initial conditions.

2. (Mass-spring system with damping but no rubber band) In Part 1, $b = 0$. Now repeat your analysis for

$$\frac{d^2y}{dt^2} + b\frac{dy}{dt} + k_1 y = 10$$

using the same value of k_1 as in Part 1 and various values of b. In particular, try $b = 1.0$ and $b = 10.0$. Describe how solutions change as b is adjusted. In fact, there is a particularly important b-value between $b = 1.0$ and $b = 10.0$ that separates the $b = 1.0$ behavior from the $b = 10.0$ behavior. This "bifurcation" value of b is difficult to locate numerically, but try your best.

3. (Mass-spring system with rubber band but no damping) Once again let $b = 0$, but now add the rubber band to the system. That is, consider the equation

$$\frac{d^2y}{dt^2} + k_1 y + k_2 h(y) = 10.$$

Use the same value for k_1 as you used in Part 1 and choose a value of k_2 such that $4.5 < k_2 < 5.0$. Repeat the analysis described in Part 1 for this equation.

4. (Damped mass-spring system with rubber band) We now add damping into the system in Part 3 and obtain

$$\frac{d^2y}{dt^2} + b\frac{dy}{dt} + k_1 y + k_2 h(y) = 10.$$

Repeat the analysis described in Part 2 for this equation.

Your report: Address each of the previous items. You may provide illustrations from the computer, but remember that although a good illustration is worth 1000 words, 1000 illustrations are worth nothing. Make sure you use your conclusions about the solutions of these equations to describe how the mass oscillates.

LAB 2.5   Active Shock Absorbers

Recent advances in materials science have created reliable, long-lasting fluids, MR fluids, which change their properties when subjected to a magnetic field. If an MR fluid is placed in a shock absorber, a change in an applied magnetic field can alter the damping capabilities of the fluid, so the damping coefficient can be adjusted dynamically. These "active" shock absorber systems have found application in such diverse objects as washing machines, prosthetic limbs, and car suspensions.[*]

One of the first applications of this technology is called the Motion Master Ride Management System, an active shock absorber system for truck and school bus seats. Schematically, we can think of a truck seat as being attached to the rest of the truck by a spring and a dashpot (see Figure 2.69). For the perfect ride, we would want the spring to have spring constant $k = 0$ and the dashpot to have damping coefficient $b = 0$. In this case, the seat would float above the truck. For obvious reasons, the seat does have to be connected to the truck, so at least one of the two constants must be nonzero. The springs are chosen so that k is large enough to hold the seat firmly to the truck, and the damping coefficient b is chosen with the comfort of the driver in mind.

If b is very large, the seat is rigidly attached to the truck, which makes the ride very uncomfortable. On the other hand, if b is too small, the seat may "bottom out" when the truck hits a large bump. That is, the spring compresses so much that the seat

Figure 2.69
Schematic of truck and truck seat.

[*]See *Scientific American*, May 2001, p. 28 and http://www.lord.com/mr. For a commercial see http://www.cadillac.com. Look for models with "Magnetic ride control," for example, the STS and SRX models.

violently strikes the base. This response is both dangerous and uncomfortable. In practice, designers compromise between having b small (a smooth ride that has danger from large bumps) and b large (protection from large bumps but a rough ride). Another constant, the mass of a typical driver, must also be considered as a factor in the choice of b.

Active damping allows adjustment of the damping coefficient according to the state of the system. That is, the damping coefficient b can be replaced by a function of y and $v = dy/dt$. As a first step in studying the possibilities in such a system, we consider a modification of the harmonic oscillator of the form

$$m\frac{d^2y}{dt^2} + b(v)\frac{dy}{dt} + ky = 0,$$

where m is the mass of the driver. In this case, the damping coefficient $b(v)$ is assumed to be a function of the velocity v. For this lab, we assume that the units of mass and distance are chosen so that $k = m = 1$, and we study the equation

$$\frac{d^2y}{dt^2} + b(v)\frac{dy}{dt} + y = 0.$$

When the vertical velocity of the seat is near zero, we want small damping so that small bumps are not transmitted to the seat. When the vertical velocity of the seat is large, we want $b(v)$ to be large to protect from "bottoming out" (and "topping out"). These criteria leave considerable freedom in the choice of the function $b(v)$.

In this lab, we consider the behavior of a truck seat for three possible choices of the damping function $b(v)$:

1. Investigate solutions of the equation

$$\frac{d^2y}{dt^2} + b(v)\frac{dy}{dt} + y = 0$$

for $b(v) = v^4$. Describe solutions with a variety of different initial conditions.

2. Repeat Part 1 for $b(v) = 1 - e^{-10v^2}$.

3. Repeat Part 1 for $b(v) = \arctan v$.

4. Suppose you are choosing from among the three possible functions $b(v)$ above for a truck that drives on relatively smooth roads with an occasional large pothole. In this case, y and v are usually small, but occasionally v suddenly becomes large when the truck hits a pothole. Which of the functions $b(v)$ above would you choose to control the damping coefficient? Justify your answer in a paragraph.

Your report: Address all of the items above. Be sure to keep the application in mind when describing the behavior of solutions. Phase portraits and $y(t)$-graphs are useful, but illustrations alone are not enough. In Part 4, address your analysis to an audience having your mathematical background but who have not considered this problem.

3 LINEAR SYSTEMS

In Chapter 2 we focused on qualitative and numerical techniques for studying systems of differential equations. We did so because we can rarely find explicit formulas for solutions of a system with two or more dependent variables. The only exception to this basic truth is the linear system.

In this chapter we show how to use the algebraic and geometric forms of the vector field to produce the general solution of an autonomous linear system. Along the way, we find that understanding the qualitative behavior of a linear system is much easier than finding its general solution. The description of the qualitative behavior of linear systems leads to a classification scheme for these systems, which is particularly useful in applications. We also continue our study of models that yield linear systems, especially the damped harmonic oscillator.

3.1 PROPERTIES OF LINEAR SYSTEMS AND THE LINEARITY PRINCIPLE

In this chapter we investigate the behavior of the simplest types of systems of differential equations—autonomous linear systems. These systems are important both in their own right and as a tool in the study of nonlinear systems. We are able to classify the linear systems by their qualitative behavior and even give formulas for solutions.

Throughout this chapter we use two models repeatedly to illustrate the techniques we develop. One is the harmonic oscillator, the most important of all second-order equations. We derived this model in Sections 2.1 and 2.3. Now, using the techniques of this chapter, we can give a complete description of its solutions for all possible values of the parameters. The other model is an artificial one, which we present to illustrate all possibilities that can arise for planar linear systems. Study our analysis, but don't invest any money based on it.

The Harmonic Oscillator

The harmonic oscillator is a model for (among other things) the motion of a mass attached to a spring. The spring provides a restoring force that obeys Hooke's law, and the only other force considered is that due to damping. Let $y(t)$ be the position of the mass at time t, with $y = 0$ corresponding to the rest position of the spring. Newton's law of motion,

$$\text{force} = \text{mass} \times \text{acceleration},$$

when applied to a mass-spring system, yields the second-order differential equation

$$-ky - b\frac{dy}{dt} = m\frac{d^2y}{dt^2},$$

where m is the mass, k is the spring constant, and b is the damping coefficient. The $-ky$ term on the left-hand side comes from Hooke's law, and the $-b(dy/dt)$ term is the force from damping (see Section 2.3, page 184). This second-order equation is more commonly written as

$$m\frac{d^2y}{dt^2} + b\frac{dy}{dt} + ky = 0.$$

As we did in Sections 2.1 and 2.3, we can convert this equation into a linear system by letting $v = dy/dt$ be the velocity at time t. We obtain

$$\frac{dy}{dt} = v$$
$$\frac{dv}{dt} = -\frac{k}{m}y - \frac{b}{m}v.$$

Note that the derivatives dy/dt and dv/dt depend linearly on y and v. As we will see, the behavior of the solutions depends on the values of the parameters m, k, and b.

Two Cafés

Harmonic oscillators do not display all of the possible behaviors we will encounter in this chapter, so we present the following apocryphal model from microeconomics.

After retiring from writing differential equations textbooks, Paul and Bob both decide to open small cafés near campus. "Paul's High Test Coffee" and "Bob's Gourmet Tea" open on the same block, and Paul and Bob soon become concerned about the effect each café has on the other. Having two beverage shops so close might make their block a more popular destination for students. On the other hand, the cafés might compete for a limited supply of thirsty customers. Paul and Bob argue about this until they are so sick of arguing that they hire their former colleague and famous mathematician Glen to settle the matter. Glen decides to make a differential equations model for the relationship between the profits of the two cafés. Remembering all that he learned from Paul and Bob, Glen starts with as simple a model as possible and suggests the following system.

Let

$$x(t) = \text{daily profit of Paul's café at time } t; \text{ and}$$

$$y(t) = \text{daily profit of Bob's café at time } t.$$

That is, if $x(t) > 0$, then Paul's café is making money, but if $x(t) < 0$, then Paul's café is losing money. Since there is no hard information yet about how the profits of each café affect the change in profits of the other, Glen formulates the simplest possible model that allows each café to affect the other—a linear model. The system is

$$\frac{dx}{dt} = ax + by$$

$$\frac{dy}{dt} = cx + dy,$$

where a, b, c, and d are parameters. The rate of change of Paul's profits depends linearly on both Paul's profits and Bob's profits (and nothing else). The same assumptions apply to Bob's profits. In Chapter 5 we will see that using a model of this form is usually justified as long as both cafés are operating near the break-even point.

We cannot yet use this model to predict future profits because we do not know the values of the parameters a, b, c, and d. However, we can develop a basic understanding of the significance of the signs and magnitudes of the parameters. Consider, for example, the parameter a. It measures the effect of Paul's profits on the rate of change dx/dt of that profit. Suppose, for instance, that a is positive. If Paul is making money, then $x > 0$ and so $ax > 0$. The ax term contributes positively to dx/dt, and so Paul makes more money in that case. In other words, Paul hopes that $a > 0$ if $x > 0$. On the other hand, being profitable ($x > 0$) could conceivably have a negative effect on Paul's profits. (For example, the café might be crowded, and customers might go elsewhere.) In this case profits would decrease, and under this assumption the parameter a should be negative in our model.

The parameter b measures the effect of Bob's profits on the rate of change of Paul's profits. If $b > 0$ and Bob makes money ($y > 0$), then Paul's profits also benefit because the by term contributes positively to dx/dt. On the other hand, if $b < 0$ then, when Bob makes money ($y > 0$), Paul's profits suffer. Conceivably we could interpret $b < 0$ as a measure of Bob's stealing customers from Paul.

Similarly, both Paul's profits and Bob's profits affect the rate of change of Bob's profits, and the parameters c and d have similar interpretations relative to dy/dt. This model assumes that only the profit of the two cafés influences the change in those profits. These assumptions are clearly vast oversimplifications. However, this model does give us a simple situation for which we can interpret the solutions of various linear systems.

In Figure 3.1 we plot the phase portrait for the system

$$\frac{dx}{dt} = ax + by$$

$$\frac{dy}{dt} = cx + dy,$$

assuming that $a = d = 0$, $b = 1$, and $c = -1$, and we get circular solution curves. In Figure 3.2 we consider the case where $a = -1$, $b = 4$, $c = -3$, and $d = -1$. In this case the solution curves spiral toward the origin.

In terms of the model, Figure 3.1 implies that both Paul's and Bob's profits periodically oscillate between making money and losing money. The solution curve in Fig-

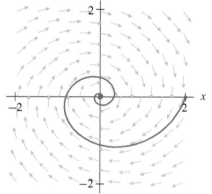

Figure 3.1
The direction field and three solution curves for the system

$$\frac{dx}{dt} = y$$

$$\frac{dy}{dt} = -x.$$

Note that all three curves are circles centered at the origin.

Figure 3.2
The direction field and a solution curve for the system

$$\frac{dx}{dt} = -x + 4y$$

$$\frac{dy}{dt} = -3x - y.$$

This solution curve spirals toward the origin as t increases.

ure 3.2 suggests that the profits oscillate while tending toward the point $(0, 0)$, which is the break-even point for both cafés. The corresponding $x(t)$- and $y(t)$-graphs illustrate these behaviors (see Figures 3.3 and 3.4). As we will see, there are a number of other possible phase portraits for this model, depending on the values of the parameters a, b, c, and d. In this chapter we develop techniques to handle all possibilities.

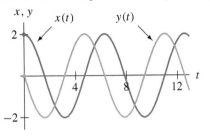

Figure 3.3
The $x(t)$- and $y(t)$-graphs corresponding to the solution curve in Figure 3.1, with initial condition $(x_0, y_0) = (2, 0)$.

Figure 3.4
The $x(t)$- and $y(t)$-graphs corresponding to the solution curve in Figure 3.2 with initial condition $(x_0, y_0) = (2, 0)$.

Linear Systems and Matrix Notation

In this chapter we mainly consider systems of differential equations of the form

$$\frac{dx}{dt} = ax + by$$
$$\frac{dy}{dt} = cx + dy,$$

where a, b, c, and d are constants (which may be 0). Such a system is said to be a **linear system with constant coefficients**. The constants a, b, c, and d are the **coefficients**. Both the harmonic oscillator model and the model of the cafés are, up to changes in the names of the dependent variables and the coefficients, systems of this form.

The most important adjective—linear—refers to the fact that the equations for dx/dt and for dy/dt involve only first powers of the dependent variables. In other words they are linear functions of x and y. Since the coefficients a, b, c, and d are constants, this type of system is also *autonomous*, and therefore we know that distinct solution curves in the phase plane do not touch. These systems have two dependent variables, so we say that they are *planar* or *two-dimensional*. Since "two-dimensional, linear system with constant coefficients" is quite a mouthful, we usually just call these systems *planar linear systems* or even just *linear systems*. Note that these systems are two-variable generalizations of the homogeneous, constant-coefficient, first-order linear equation $dx/dt = ax$, which we discussed in Chapter 1 (see pages 6 and 112).

We can use vector and matrix notation to write this system much more efficiently. Let **A** be the "2-by-2" square matrix (the 2×2 matrix)

$$\mathbf{A} = \begin{pmatrix} a & b \\ c & d \end{pmatrix},$$

and let

$$\mathbf{Y} = \begin{pmatrix} x \\ y \end{pmatrix}$$

denote the column vector of dependent variables. Then the **product** of a 2×2 matrix $\mathbf{A}$ and a column vector $\mathbf{Y}$ is the column vector $\mathbf{AY}$ given by

$$\mathbf{AY} = \begin{pmatrix} a & b \\ c & d \end{pmatrix} \begin{pmatrix} x \\ y \end{pmatrix} = \begin{pmatrix} ax + by \\ cx + dy \end{pmatrix}.$$

For example,

$$\begin{pmatrix} 5 & 2 \\ -1 & 3 \end{pmatrix} \begin{pmatrix} 3 \\ 4 \end{pmatrix} = \begin{pmatrix} 5 \cdot 3 + 2 \cdot 4 \\ -1 \cdot 3 + 3 \cdot 4 \end{pmatrix} = \begin{pmatrix} 23 \\ 9 \end{pmatrix}$$

and

$$\begin{pmatrix} (2 - a) & \pi \\ e & y \end{pmatrix} \begin{pmatrix} y \\ 2v \end{pmatrix} = \begin{pmatrix} (2 - a)y + 2\pi v \\ ey + 2yv \end{pmatrix}.$$

As in Chapter 2, if x and y are dependent variables, then we write

$$\mathbf{Y}(t) = \begin{pmatrix} x(t) \\ y(t) \end{pmatrix} \quad \text{and} \quad \frac{d\mathbf{Y}}{dt} = \begin{pmatrix} \dfrac{dx}{dt} \\ \dfrac{dy}{dt} \end{pmatrix}.$$

Using this matrix notation, we can write the two-dimensional linear system

$$\frac{dx}{dt} = ax + by$$
$$\frac{dy}{dt} = cx + dy$$

as

$$\begin{pmatrix} \dfrac{dx}{dt} \\ \dfrac{dy}{dt} \end{pmatrix} = \begin{pmatrix} ax + by \\ cx + dy \end{pmatrix} = \begin{pmatrix} a & b \\ c & d \end{pmatrix} \begin{pmatrix} x \\ y \end{pmatrix},$$

or more compactly as

$$\frac{d\mathbf{Y}}{dt} = \mathbf{AY},$$

where

$$\mathbf{A} = \begin{pmatrix} a & b \\ c & d \end{pmatrix} \quad \text{and} \quad \mathbf{Y} = \begin{pmatrix} x \\ y \end{pmatrix}.$$

The matrix $\mathbf{A}$ of coefficients of the system is called the **coefficient matrix**.

One advantage of the matrix notation is that it helps us see the similarities between first-order linear systems and first-order linear equations. Working with matrices also gives us some very useful algebraic tools, which we will exploit throughout this chapter.

Vector notation can be extended to include systems with any number n of dependent variables $y_1, y_2, \ldots, y_n$. The (constant coefficient) linear system with n dependent variables is

$$\frac{dy_1}{dt} = a_{11}y_1 + a_{12}y_2 + \cdots + a_{1n}y_n$$

$$\frac{dy_2}{dt} = a_{21}y_1 + a_{22}y_2 + \cdots + a_{2n}y_n$$

$$\vdots \qquad\qquad \vdots$$

$$\frac{dy_n}{dt} = a_{n1}y_1 + a_{n2}y_2 + \cdots + a_{nn}y_n.$$

In this case the coefficients of this system are $a_{11}, a_{12}, \ldots, a_{nn}$. Let

$$\mathbf{Y} = \begin{pmatrix} y_1 \\ y_2 \\ \vdots \\ y_n \end{pmatrix}, \quad \text{so} \quad \frac{d\mathbf{Y}}{dt} = \begin{pmatrix} \dfrac{dy_1}{dt} \\[4pt] \dfrac{dy_2}{dt} \\[4pt] \vdots \\[4pt] \dfrac{dy_n}{dt} \end{pmatrix}.$$

The coefficient matrix is the $n \times n$ matrix

$$\mathbf{A} = \begin{pmatrix} a_{11} & a_{12} & \cdots & a_{1n} \\ a_{21} & a_{22} & \cdots & a_{2n} \\ \vdots & \vdots & \ddots & \vdots \\ a_{n1} & a_{n2} & \cdots & a_{nn} \end{pmatrix},$$

and we have

$$\frac{d\mathbf{Y}}{dt} = \mathbf{AY} = \begin{pmatrix} a_{11} & a_{12} & \cdots & a_{1n} \\ a_{21} & a_{22} & \cdots & a_{2n} \\ \vdots & \vdots & \ddots & \vdots \\ a_{n1} & a_{n2} & \cdots & a_{nn} \end{pmatrix} \begin{pmatrix} y_1 \\ y_2 \\ \vdots \\ y_n \end{pmatrix}$$

$$= \begin{pmatrix} a_{11}y_1 + a_{12}y_2 + \ldots + a_{1n}y_n \\ a_{21}y_1 + a_{22}y_2 + \ldots + a_{2n}y_n \\ \vdots \\ a_{n1}y_1 + a_{n2}y_2 + \ldots + a_{nn}y_n \end{pmatrix}.$$

The number of dependent variables is called the **dimension** of the system, so this system is n**-dimensional**. For example, the three-dimensional system

$$\frac{dx}{dt} = \sqrt{2}\,x + y$$

$$\frac{dy}{dt} = z$$

$$\frac{dz}{dt} = -x - y + 2z$$

can be written as

$$\frac{d\mathbf{Y}}{dt} = \mathbf{AY},$$

where

$$\mathbf{Y} = \begin{pmatrix} x \\ y \\ z \end{pmatrix} \quad \text{and} \quad \mathbf{A} = \begin{pmatrix} \sqrt{2} & 1 & 0 \\ 0 & 0 & 1 \\ -1 & -1 & 2 \end{pmatrix}.$$

In this text we work primarily with planar, or two-dimensional, systems. However, readers familiar with linear algebra will recognize that many of the concepts we discuss carry over to higher dimensional systems with little or no modification.

Linear systems are like other systems of differential equations, only simpler. All of the methods of Chapter 2 apply, and we use these methods to understand the associated vector fields, direction fields, and graphs of solutions. In addition, because linear systems are relatively simple algebraically, it is reasonable to hope that we can "read off" the behavior of solutions just from the coefficients. That is, if we are given the planar linear system

$$\frac{d\mathbf{Y}}{dt} = \mathbf{AY}, \quad \text{where } \mathbf{A} = \begin{pmatrix} a & b \\ c & d \end{pmatrix}$$

is the coefficient matrix, then we would like to understand the system completely if we know the four numbers a, b, c, and d. In fact we might even hope to come up with an explicit formula for the general solution. From the four numbers a, b, c, and d, we are able to give a geometric description of the behavior of the solutions in the xy-plane, describe the $x(t)$- and $y(t)$-graphs of solutions, and even give a formula for the general solution. Hence we are able to produce explicit formulas that solve any initial-value problem.

Equilibrium Points of Linear Systems and the Determinant

We start by looking for the simplest solutions—the equilibrium solutions. Recall that a point $\mathbf{Y}_0 = (x_0, y_0)$ is an equilibrium point of a system if and only if the vector field at $\mathbf{Y}_0$ is the zero vector. Since the vector field of a system at the point $\mathbf{Y}_0$ is given by the right-hand side of the differential equation evaluated at that point and since

$$\frac{d\mathbf{Y}}{dt} = \mathbf{AY}$$

for a linear system, we know that the vector field $\mathbf{F}(\mathbf{Y}_0)$ at $\mathbf{Y}_0$ for a linear system is given by

$$\mathbf{F}(\mathbf{Y}_0) = \mathbf{A}\mathbf{Y}_0.$$

In other words the vector at $\mathbf{Y}_0$ is computed by taking the product of the matrix $\mathbf{A}$ with the vector $\mathbf{Y}_0$. Consequently, the equilibrium points are the points $\mathbf{Y}_0$ such that

$$\mathbf{A}\mathbf{Y}_0 = \begin{pmatrix} 0 \\ 0 \end{pmatrix}.$$

That is,

$$\begin{pmatrix} a & b \\ c & d \end{pmatrix} \begin{pmatrix} x_0 \\ y_0 \end{pmatrix} = \begin{pmatrix} ax_0 + by_0 \\ cx_0 + dy_0 \end{pmatrix} = \begin{pmatrix} 0 \\ 0 \end{pmatrix}.$$

Written in scalar form, this vector equation is a pair of simultaneous linear equations

$$\begin{cases} ax_0 + by_0 = 0 \\ cx_0 + dy_0 = 0. \end{cases}$$

Clearly $(x_0, y_0) = (0, 0)$ is a solution to these equations. Therefore the point $\mathbf{Y}_0 = (0, 0)$ is an equilibrium point, and the constant function

$$\mathbf{Y}(t) = (0, 0) \quad \text{for all } t$$

is a solution to the linear system. This solution is often called the **trivial solution** of the system. (Note that this computation does not depend on the values of the coefficients a, b, c, and d. In other words *every* linear system has an equilibrium point at the origin.)

Any other equilibrium points (x_0, y_0) must also satisfy

$$\begin{cases} ax_0 + by_0 = 0 \\ cx_0 + dy_0 = 0. \end{cases}$$

To find them, assume for the moment that $a \neq 0$. Using the first equation, we get

$$x_0 = -\frac{b}{a} y_0.$$

The second equation then yields

$$c\left(-\frac{b}{a}\right) y_0 + dy_0 = 0,$$

which can be rewritten as

$$(ad - bc)y_0 = 0.$$

Hence either $y_0 = 0$ or $ad - bc = 0$. If $y_0 = 0$, then $x_0 = 0$, and once again we have the trivial solution. Therefore a linear system has nontrivial equilibrium points only if

$ad - bc = 0$. This quantity, $ad - bc$, is a particularly important number associated with the 2×2 matrix $\mathbf{A}$.

DEFINITION   The **determinant** of a 2×2 matrix

$$\mathbf{A} = \begin{pmatrix} a & b \\ c & d \end{pmatrix}$$

is the number $ad - bc$. It is denoted det $\mathbf{A}$.   ∎

With this definition, we are able to summarize the results of the above computation for equilibrium points of linear systems.

THEOREM   If $\mathbf{A}$ is a matrix with det $\mathbf{A} \neq 0$, then the only equilibrium point for the linear system $d\mathbf{Y}/dt = \mathbf{A}\mathbf{Y}$ is the origin.   ∎

The argument above proves this theorem, provided the upper left-hand corner entry, a, of $\mathbf{A}$ is nonzero, but there is nothing special about this entry. By similar steps we can obtain the same result as long as at least one of the entries of A is nonzero (see Exercise 14). If all the entries of $\mathbf{A}$ are zero, then every point in the plane is an equilibrium point.

As an example, suppose

$$\mathbf{A} = \begin{pmatrix} 2 & 1 \\ -4 & 0.3 \end{pmatrix}.$$

Then det $\mathbf{A} = (2)(0.3) - (1)(-4) = 4.6$. Since det $\mathbf{A} \neq 0$, the only equilibrium point for linear system $d\mathbf{Y}/dt = \mathbf{A}\mathbf{Y}$ is the origin, $(0, 0)$.

An Important Property of the Determinant

The determinant is a quantity that pops up repeatedly throughout this chapter. For us its significance usually is whether or not it is zero. If we pick four numbers a, b, c, and d at random, then it is unlikely that the number $ad - bc$ is exactly zero. Thus matrices whose determinant is zero are often called **singular** or **degenerate**.

From the theorem we just discussed, we know that, if a linear system $d\mathbf{Y}/dt = \mathbf{A}\mathbf{Y}$ is **nondegenerate** (det $\mathbf{A} \neq 0$), then it has exactly one equilibrium point, which is $(0, 0)$. In other words an initial condition of $(0, 0)$ corresponds to a solution curve that sits at $(0, 0)$ for all time. Any other initial condition yields a solution that changes with time.

In Section 3.2 we need to use the determinant again, so it is important to understand exactly what we verified when we justified this theorem. For linear systems, equilibria correspond to points $\mathbf{Y}_0$ for which

$$\mathbf{A}\mathbf{Y}_0 = \begin{pmatrix} 0 \\ 0 \end{pmatrix}.$$

Written in terms of scalars, this vector equation is identical to the the simultaneous system of linear equations

$$\begin{cases} ax_0 + by_0 = 0 \\ cx_0 + dy_0 = 0. \end{cases}$$

What we actually verified is that this system of equations has **nontrivial solutions**—solutions other than $(0, 0)$—if and only if $\det \mathbf{A} = 0$.

The Linearity Principle

The solutions of linear systems have special properties that solutions of arbitrary systems do not have. These properties are so useful that we take advantage of them repeatedly. In fact they are exactly the reason that we will be so successful in our analysis of linear systems.

However, a note of caution is in order: It is important to make sure that the system under consideration actually *is* a linear system before you use any of these special properties. This is the equivalent of making sure that the car is in reverse before trying to back out of the garage. If the car is in drive instead of reverse when you start backing out of the garage, there can be dire consequences.

The most important property of linear systems is the **Linearity Principle**.

LINEARITY PRINCIPLE   Suppose $d\mathbf{Y}/dt = \mathbf{AY}$ is a linear system of differential equations.

1. If $\mathbf{Y}(t)$ is a solution of this system and k is any constant, then $k\mathbf{Y}(t)$ is also a solution.

2. If $\mathbf{Y}_1(t)$ and $\mathbf{Y}_2(t)$ are two solutions of this system, then $\mathbf{Y}_1(t) + \mathbf{Y}_2(t)$ is also a solution.   ■

Using the Linearity Principle (also called the **Principle of Superposition**), we can manufacture infinitely many new solutions from any given solution or pair of solutions. Taken together, the two parts of the Linearity Principle imply that, if $\mathbf{Y}_1(t)$ and $\mathbf{Y}_2(t)$ are solutions of the system and if k_1 and k_2 are any constants, then

$$k_1\mathbf{Y}_1(t) + k_2\mathbf{Y}_2(t)$$

is also a solution. A solution of the form $k_1\mathbf{Y}_1(t) + k_2\mathbf{Y}_2(t)$ is called a **linear combination** of the solutions $\mathbf{Y}_1(t)$ and $\mathbf{Y}_2(t)$. Given two solutions, we can produce infinitely many solutions by forming linear combinations of the original two.

For example, consider the partially decoupled linear system

$$\frac{d\mathbf{Y}}{dt} = \begin{pmatrix} 2 & 3 \\ 0 & -4 \end{pmatrix} \mathbf{Y}.$$

In Section 2.4 we found that

$$\mathbf{Y}_1(t) = \begin{pmatrix} e^{2t} \\ 0 \end{pmatrix} \quad \text{and} \quad \mathbf{Y}_2(t) = \begin{pmatrix} -e^{-4t} \\ 2e^{-4t} \end{pmatrix}$$

are solutions to this system (see page 193). We can double-check this by directly calcu-lating both sides, $d\mathbf{Y}/dt$ and $\mathbf{AY}$, of the differential equation. For example, with $\mathbf{Y}_1(t)$ we have

$$\frac{d\mathbf{Y}_1}{dt} = \begin{pmatrix} 2e^{2t} \\ 0 \end{pmatrix} \quad \text{and} \quad \mathbf{AY}_1 = \begin{pmatrix} 2 & 3 \\ 0 & -4 \end{pmatrix} \begin{pmatrix} e^{2t} \\ 0 \end{pmatrix} = \begin{pmatrix} 2e^{2t} \\ 0 \end{pmatrix},$$

so $\mathbf{Y}_1(t)$ is a solution. (You should double-check that $\mathbf{Y}_2(t)$ is a solution. The verifica-tion is good practice with matrix arithmetic.)

The solution curves for $\mathbf{Y}_1(t)$ and $\mathbf{Y}_2(t)$ are shown in Figure 3.5. Note that each one is a line segment in the xy-plane. The solution curve for $\mathbf{Y}_1(t)$ approaches the equi-librium point at the origin as $t \to -\infty$, and the solution curve for $\mathbf{Y}_2(t)$ approaches the equilibrium point at the origin as $t \to \infty$. In the next section we exploit the geometry of solutions such as these to find them using only algebraic techniques.

The Linearity Principle tell us that any function of the form $k_1\mathbf{Y}_1(t) + k_2\mathbf{Y}_2(t)$ is also a solution to this system for any constants k_1 and k_2. To illustrate this fact, we check directly that $\mathbf{Y}_3(t) = -2\mathbf{Y}_1(t) + 5\mathbf{Y}_2(t)$ is a solution. Note that

$$\mathbf{Y}_3(t) = -2\mathbf{Y}_1(t) + 5\mathbf{Y}_2(t) = -2\begin{pmatrix} e^{2t} \\ 0 \end{pmatrix} + 5\begin{pmatrix} -e^{-4t} \\ 2e^{-4t} \end{pmatrix} = \begin{pmatrix} -2e^{2t} - 5e^{-4t} \\ 10e^{-4t} \end{pmatrix},$$

and therefore

$$\frac{d\mathbf{Y}_3}{dt} = \begin{pmatrix} -4e^{2t} + 20e^{-4t} \\ -40e^{-4t} \end{pmatrix}.$$

Also we compute

$$\mathbf{AY}_3 = \begin{pmatrix} 2 & 3 \\ 0 & -4 \end{pmatrix} \begin{pmatrix} -2e^{2t} - 5e^{-4t} \\ 10e^{-4t} \end{pmatrix}$$

$$= \begin{pmatrix} 2(-2e^{2t} - 5e^{-4t}) + 3(10e^{-4t}) \\ -4(10e^{-4t}) \end{pmatrix}$$

$$= \begin{pmatrix} -4e^{2t} + 20e^{-4t} \\ -40e^{-4t} \end{pmatrix}.$$

Since both computations yield the same function, this linear combination of the two solutions $\mathbf{Y}_1(t)$ and $\mathbf{Y}_2(t)$ is also a solution. (In the future we will not bother to double-check the consequences of the Linearity Principle.)

Again we emphasize that the solution curves for the solutions $\mathbf{Y}_1(t)$ and $\mathbf{Y}_2(t)$ possess a very special and useful geometric property. The fact that they form line seg-ments is not typical of solution curves in general. In fact the typical solution curve of this system is not a straight line. For example, as we see in Figure 3.5, the solution curve of the solution $\mathbf{Y}_1(t) + \mathbf{Y}_2(t)$ is not straight.

Figure 3.5
The Linearity Principle implies that the function $\mathbf{Y}_1(t) + \mathbf{Y}_2(t)$ is a solution of the system

$$\frac{d\mathbf{Y}}{dt} = \begin{pmatrix} 2 & 3 \\ 0 & -4 \end{pmatrix} \mathbf{Y}$$

because it is the sum of the two solutions $\mathbf{Y}_1(t)$ and $\mathbf{Y}_2(t)$.

Verification of the Linearity Principle

To show that the Linearity Principle holds in general, we first state the following two algebraic properties of matrix multiplication:

1. If $\mathbf{A}$ is a matrix and $\mathbf{Y}$ is a vector, then

$$\mathbf{A}(k\mathbf{Y}) = k\mathbf{AY}$$

for any constant k.

2. If $\mathbf{A}$ is a matrix and $\mathbf{Y}_1$ and $\mathbf{Y}_2$ are vectors, then

$$\mathbf{A}(\mathbf{Y}_1 + \mathbf{Y}_2) = \mathbf{AY}_1 + \mathbf{AY}_2.$$

We can verify these two facts for 2×2 matrices and 2-dimensional vectors by direct computation. For example, to verify property 2, let

$$\mathbf{A} = \begin{pmatrix} a & b \\ c & d \end{pmatrix}$$

be an arbitrary 2×2 matrix and let

$$\mathbf{Y}_1 = \begin{pmatrix} x_1 \\ y_1 \end{pmatrix} \quad \text{and} \quad \mathbf{Y}_2 = \begin{pmatrix} x_2 \\ y_2 \end{pmatrix}$$

be arbitrary vectors. Then

$$\mathbf{A}(\mathbf{Y}_1 + \mathbf{Y}_2) = \begin{pmatrix} a & b \\ c & d \end{pmatrix} \begin{pmatrix} x_1 + x_2 \\ y_1 + y_2 \end{pmatrix}$$

$$= \begin{pmatrix} a(x_1 + x_2) + b(y_1 + y_2) \\ c(x_1 + x_2) + d(y_1 + y_2) \end{pmatrix}$$

$$= \begin{pmatrix} ax_1 + ax_2 + by_1 + by_2 \\ cx_1 + cx_2 + dy_1 + dy_2 \end{pmatrix}$$

and

$$\mathbf{A}\mathbf{Y}_1 + \mathbf{A}\mathbf{Y}_2 = \begin{pmatrix} a & b \\ c & d \end{pmatrix} \begin{pmatrix} x_1 \\ y_1 \end{pmatrix} + \begin{pmatrix} a & b \\ c & d \end{pmatrix} \begin{pmatrix} x_2 \\ y_2 \end{pmatrix}$$

$$= \begin{pmatrix} ax_1 + by_1 \\ cx_1 + dy_1 \end{pmatrix} + \begin{pmatrix} ax_2 + by_2 \\ cx_2 + dy_2 \end{pmatrix}$$

$$= \begin{pmatrix} ax_1 + ax_2 + by_1 + by_2 \\ cx_1 + cx_2 + dy_1 + dy_2 \end{pmatrix}.$$

Thus property 2 holds. The verification of property 1 is left to the exercises (see Exercise 30).

Given these algebraic properties of matrix multiplication, we can verify the Linearity Principle using the standard rules of differentiation. Suppose $\mathbf{Y}_1(t)$ and $\mathbf{Y}_2(t)$ are solutions to $d\mathbf{Y}/dt = \mathbf{A}\mathbf{Y}$; that is, suppose

$$\frac{d\mathbf{Y}_1}{dt} = \mathbf{A}\mathbf{Y}_1 \quad \text{and} \quad \frac{d\mathbf{Y}_2}{dt} = \mathbf{A}\mathbf{Y}_2 \quad \text{for all } t.$$

For any constant k we have

$$\frac{d(k\mathbf{Y}_1)}{dt} = k\frac{d\mathbf{Y}_1}{dt} = k\mathbf{A}\mathbf{Y}_1 = \mathbf{A}(k\mathbf{Y}_1),$$

so $k\mathbf{Y}_1(t)$ is a solution to the system. Also

$$\frac{d(\mathbf{Y}_1 + \mathbf{Y}_2)}{dt} = \frac{d\mathbf{Y}_1}{dt} + \frac{d\mathbf{Y}_2}{dt} = \mathbf{A}\mathbf{Y}_1 + \mathbf{A}\mathbf{Y}_2 = \mathbf{A}(\mathbf{Y}_1 + \mathbf{Y}_2) \quad \text{for all } t.$$

As a result, $\mathbf{Y}_1(t) + \mathbf{Y}_2(t)$ is also a solution and the Linearity Principle is verified. We can see the advantage of the matrix and vector notation. To write out the above equations showing all the components would be a tedious exercise—in fact, it is a tedious exercise in the exercise set at the end of this section (see Exercise 30).

Solving Initial-Value Problems

From the Linearity Principle we know that, given two solutions $\mathbf{Y}_1(t)$ and $\mathbf{Y}_2(t)$, we can make many more solutions of the form $k_1\mathbf{Y}_1(t) + k_2\mathbf{Y}_2(t)$ for any constants k_1 and k_2. This type of expression is called a *two-parameter family of solutions*, since we have two constants, k_1 and k_2, that we can adjust to obtain various solutions. It is reasonable to ask if these are all of the solutions or, put another way, if each solution is one of this form.

To see how the Linearity Principle is used to solve initial-value problems, we return to the differential equation

$$\frac{d\mathbf{Y}}{dt} = \begin{pmatrix} 2 & 3 \\ 0 & -4 \end{pmatrix} \mathbf{Y}$$

that was discussed earlier in this section. Suppose we want to find the solution $\mathbf{Y}(t)$ of this system with initial value $\mathbf{Y}(0) = (2, -3)$. We already know that

$$\mathbf{Y}_1(t) = \begin{pmatrix} e^{2t} \\ 0 \end{pmatrix} \quad \text{and} \quad \mathbf{Y}_2(t) = \begin{pmatrix} -e^{-4t} \\ 2e^{-4t} \end{pmatrix}$$

are solutions, and by direct evaluation we know that

$$\mathbf{Y}_1(0) = \begin{pmatrix} 1 \\ 0 \end{pmatrix} \quad \text{and} \quad \mathbf{Y}_2(0) = \begin{pmatrix} -1 \\ 2 \end{pmatrix},$$

so neither $\mathbf{Y}_1(t)$ nor $\mathbf{Y}_2(t)$ is the solution to the initial-value problem

$$\frac{d\mathbf{Y}}{dt} = \begin{pmatrix} 2 & 3 \\ 0 & -4 \end{pmatrix} \mathbf{Y}, \quad \mathbf{Y}(0) = \begin{pmatrix} 2 \\ -3 \end{pmatrix}.$$

But the Linearity Principle says that we can form any linear combination of $\mathbf{Y}_1(t)$ and $\mathbf{Y}_2(t)$ and still have a solution. Hence we seek k_1 and k_2 so that

$$k_1 \begin{pmatrix} 1 \\ 0 \end{pmatrix} + k_2 \begin{pmatrix} -1 \\ 2 \end{pmatrix} = \begin{pmatrix} 2 \\ -3 \end{pmatrix}.$$

This vector equation is equivalent to the simultaneous equations

$$\begin{cases} k_1 - k_2 = 2 \\ 2k_2 = -3. \end{cases}$$

The second equation yields $k_2 = -3/2$, and consequently, the first equation yields $k_1 = 1/2$. This computation implies

$$\frac{1}{2}\mathbf{Y}_1(0) - \frac{3}{2}\mathbf{Y}_2(0) = \begin{pmatrix} 2 \\ -3 \end{pmatrix},$$

so we consider the function

$$\mathbf{Y}(t) = \frac{1}{2}\mathbf{Y}_1(t) - \frac{3}{2}\mathbf{Y}_2(t)$$

$$= \frac{1}{2}\begin{pmatrix} e^{2t} \\ 0 \end{pmatrix} - \frac{3}{2}\begin{pmatrix} -e^{-4t} \\ 2e^{-4t} \end{pmatrix}$$

$$= \begin{pmatrix} \frac{1}{2}e^{2t} + \frac{3}{2}e^{-4t} \\ -3e^{-4t} \end{pmatrix}.$$

This function has the correct initial condition, and by the Linearity Principle we know that it must be a solution to the system. The Uniqueness Theorem tells us that this is the only function that solves the initial-value problem (see Section 2.5).

In this example we found the solution to the initial-value problem using the two solutions of the system that were already available plus a little arithmetic (but no calculus). By taking the appropriate linear combination of the two known solutions, we were able to find a solution with the desired initial conditions.

Maybe we were just lucky. Will we always be able to find the appropriate k_1 and k_2 no matter what initial condition we have? To check, suppose we consider the same differential equation with an arbitrary initial condition,

$$\frac{d\mathbf{Y}}{dt} = \begin{pmatrix} 2 & 3 \\ 0 & -4 \end{pmatrix} \mathbf{Y}, \quad \mathbf{Y}(0) = \begin{pmatrix} x_0 \\ y_0 \end{pmatrix},$$

and the two solutions $\mathbf{Y}_1(t)$ and $\mathbf{Y}_2(t)$ with which we started. To solve the initial-value problem, we need to find k_1 and k_2 so that

$$k_1 \mathbf{Y}_1(0) + k_2 \mathbf{Y}_2(0) = \mathbf{Y}(0) = \begin{pmatrix} x_0 \\ y_0 \end{pmatrix}.$$

In other words, given arbitrary x_0 and y_0, can we always find k_1 and k_2 such that

$$k_1 \begin{pmatrix} 1 \\ 0 \end{pmatrix} + k_2 \begin{pmatrix} -1 \\ 2 \end{pmatrix} = \begin{pmatrix} x_0 \\ y_0 \end{pmatrix}?$$

This vector equation is equivalent to the simultaneous system of equations

$$\begin{cases} k_1 - k_2 = x_0 \\ 2k_2 = y_0. \end{cases}$$

Since the second equation is so simple, we can always find k_1 and k_2 given x_0 and y_0. We use the second equation to find k_2 first, and then we find k_1 using this value of k_2 in the first equation.

Since we are able to solve every possible initial-value problem for the system

$$\frac{d\mathbf{Y}}{dt} = \begin{pmatrix} 2 & 3 \\ 0 & -4 \end{pmatrix} \mathbf{Y}$$

using a linear combination of

$$\mathbf{Y}_1(t) = \begin{pmatrix} e^{2t} \\ 0 \end{pmatrix} \quad \text{and} \quad \mathbf{Y}_2(t) = \begin{pmatrix} -e^{-4t} \\ 2e^{-4t} \end{pmatrix},$$

we have found the general solution to this system. It is the two-parameter family

$$\mathbf{Y}(t) = k_1 \mathbf{Y}_1(t) + k_2 \mathbf{Y}_2(t) = k_1 \begin{pmatrix} e^{2t} \\ 0 \end{pmatrix} + k_2 \begin{pmatrix} -e^{-4t} \\ 2e^{-4t} \end{pmatrix}.$$

Using vector addition, this two-parameter family can be written as

$$\mathbf{Y}(t) = \begin{pmatrix} k_1 e^{2t} - k_2 e^{-4t} \\ 2k_2 e^{-4t} \end{pmatrix}.$$

Linear Independence

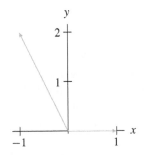

Note that in this example we used the Linearity Principle to produce infinitely many solutions starting with two given solutions $\mathbf{Y}_1(t)$ and $\mathbf{Y}_2(t)$. Then, because we were able to express an arbitrary initial condition as a linear combination of the initial conditions $\mathbf{Y}_1(0)$ and $\mathbf{Y}_2(0)$, we could use $\mathbf{Y}_1(t)$ and $\mathbf{Y}_2(t)$ to form the general solution.

Expressing arbitrary vectors as linear combinations of given vectors is a fundamental topic in linear algebra. In the two-dimensional case the key property that ensures that an arbitrary vector can be written as some linear combination of the given vectors (x_1, y_1) and (x_2, y_2) is that they do *not* lie on the same line through the origin. (Note that in the previous example the initial conditions $(1, 0)$ and $(-1, 2)$ do not lie on the same line through the origin—see Figure 3.6.) We say that the two vectors (x_1, y_1) and (x_2, y_2) are **linearly independent** if they do not lie on the same line through the origin or, equivalently, if neither one is a multiple of the other.

Figure 3.6
The vectors $(1, 0)$ and $(-1, 2)$ are linearly independent.

THEOREM   Suppose (x_1, y_1) and (x_2, y_2) are two linearly independent vectors in the plane. Then given any vector (x_0, y_0), there exist k_1 and k_2 so that

$$k_1 \begin{pmatrix} x_1 \\ y_1 \end{pmatrix} + k_2 \begin{pmatrix} x_2 \\ y_2 \end{pmatrix} = \begin{pmatrix} x_0 \\ y_0 \end{pmatrix}. \quad \blacksquare$$

Two linearly independent vectors can be combined via addition and scalar multiplication to form any other vector in the plane.

Note that the equation

is really a simultaneous system of two linear equations

$$\begin{cases} x_1 k_1 + x_2 k_2 = x_0 \\ y_1 k_1 + y_2 k_2 = y_0 \end{cases}$$

in the two unknowns k_1 and k_2. We are given the x's and the y's, and we must solve for the k's. We can show that there are solutions by writing down formulas for k_1 and k_2 in terms of the x's and y's. As long as the denominators in these formulas are nonzero, the solutions exist. Solving systems of equations of this form involves the same sort of algebra as finding equilibrium points for linear systems. Hence it is not surprising that the determinant plays a role here too (see Exercises 31 and 32.)

The General Solution

The previous example and theorem illustrate the way in which we approach every linear system. Therefore it is worth summarizing the discussion.

THEOREM   Suppose $Y_1(t)$ and $Y_2(t)$ are solutions of the linear system $dY/dt = AY$. If $Y_1(0)$ and $Y_2(0)$ are linearly independent, then for any initial condition $Y(0) = (x_0, y_0)$ we can find constants k_1 and k_2 so that $k_1Y_1(t) + k_2Y_2(t)$ is the solution to the initial-value problem

$$\frac{dY}{dt} = AY, \quad Y(0) = \begin{pmatrix} x_0 \\ y_0 \end{pmatrix}. \quad \blacksquare$$

In this situation we say that the two-parameter family $k_1Y_1(t) + k_2Y_2(t)$, where k_1 and k_2 are arbitrary constants, is the **general solution** of the system.

By the Existence and Uniqueness Theorem for systems, we know that each initial-value problem for a linear system has exactly one solution. Given any two solutions $Y_1(t)$ and $Y_2(t)$ of a linear system with linearly independent initial conditions $Y_1(0)$ and $Y_2(0)$, we can form the general solution of the system by forming the two-parameter family $k_1Y_1(t) + k_2Y_2(t)$. By adjusting the constants k_1 and k_2, we can obtain the solution that satisfies any given initial condition.

This is really excellent progress. We now know that to find all the solutions to a linear system, we need to find only two particular solutions with linearly independent initial positions. Two solutions $Y_1(t)$ and $Y_2(t)$ of a linear system for which $Y_1(0)$ and $Y_2(0)$ are linearly independent are called **linearly independent solutions** of the linear system. (In the exercises we will see that if $Y_1(t)$ and $Y_2(t)$ are solutions of a linear system and the vectors $Y_1(t_0)$ and $Y_2(t_0)$ are linearly independent at any particular t_0, then $Y_1(t)$ and $Y_2(t)$ are linearly independent for all values of t—see Exercise 35.) The next step is to find a general way to come up with two linearly independent solutions $Y_1(t)$ and $Y_2(t)$. Much of the discussion in Sections 3.2 and 3.4 involves techniques which do just that.

An Undamped Harmonic Oscillator

In Section 2.1 we studied the undamped harmonic oscillator given by the second-order differential equation

$$\frac{d^2y}{dt^2} = -y.$$

We guessed that $y_1(t) = \cos t$ is a solution and then checked our guess by verifying that

$$\frac{d^2y_1}{dt^2} + y_1 = \frac{d^2(\cos t)}{dt^2} + \cos t$$

$$= -\cos t + \cos t$$

$$= 0.$$

Similarly, we can check that $y_2(t) = \sin t$ is also a solution. Now that we have the Linearity Principle at our disposal, we can take this discussion one step further.

The second-order equation can be converted to the first-order system

$$\frac{dy}{dt} = v$$
$$\frac{dv}{dt} = -y,$$

which is a linear system. Using vector notation, we write

$$\mathbf{Y}(t) = \begin{pmatrix} y(t) \\ v(t) \end{pmatrix},$$

and the system can be represented as

$$\frac{d\mathbf{Y}}{dt} = \begin{pmatrix} 0 & 1 \\ -1 & 0 \end{pmatrix} \mathbf{Y}.$$

Recall that the second component of the vector-valued function $\mathbf{Y}(t)$ is $v = dy/dt$. We can use the solution $y_1(t)$ to form a vector-valued function

$$\mathbf{Y}_1(t) = \begin{pmatrix} y_1(t) \\ v_1(t) \end{pmatrix} = \begin{pmatrix} \cos t \\ -\sin t \end{pmatrix}.$$

Note that $\mathbf{Y}_1(t)$ is a solution to the system because

$$\frac{d\mathbf{Y}_1}{dt} = \begin{pmatrix} -\sin t \\ -\cos t \end{pmatrix}$$

and

$$\begin{pmatrix} 0 & 1 \\ -1 & 0 \end{pmatrix} \mathbf{Y}_1(t) = \begin{pmatrix} 0 & 1 \\ -1 & 0 \end{pmatrix} \begin{pmatrix} \cos t \\ -\sin t \end{pmatrix} = \begin{pmatrix} -\sin t \\ -\cos t \end{pmatrix}.$$

Similarly, the solution $y_2(t) = \sin t$ to the second-order equation yields

$$\mathbf{Y}_2(t) = \begin{pmatrix} \sin t \\ \cos t \end{pmatrix},$$

which is also a solution to the first-order system. (Double-checking this assertion is good practice with matrix notation.)

We have a first-order linear system with two dependent variables. Therefore we need two linearly independent solutions to obtain the general solution. At $t = 0$,

$$\mathbf{Y}_1(0) = \begin{pmatrix} 1 \\ 0 \end{pmatrix} \quad \text{and} \quad \mathbf{Y}_2(0) = \begin{pmatrix} 0 \\ 1 \end{pmatrix}.$$

That is, $Y_1(0)$ lies on the y-axis and $Y_2(t)$ lies on the v-axis. Consequently, these vectors are linearly independent, and the general solution to the first-order system is

$$Y(t) = k_1 \begin{pmatrix} \cos t \\ -\sin t \end{pmatrix} + k_2 \begin{pmatrix} \sin t \\ \cos t \end{pmatrix}.$$

Recalling that solutions $Y(t)$ to the first-order system are really functions of the form $Y(t) = (y(t), v(t))$ where $y(t)$ is a solution to the original second-order equation, we obtain the general solution to the original second-order equation using the first component of $Y(t)$. The result is

$$y(t) = k_1 \cos t + k_2 \sin t.$$

In Section 3.6 we will discuss a more immediate way to find the general solution of second-order equations such as this one, but it is important to realize that the Linearity Principle also applies to "linear" second-order equations such as the equation for a damped harmonic oscillator.

EXERCISES FOR SECTION 3.1

Recall the model

$$\frac{dx}{dt} = ax + by$$

$$\frac{dy}{dt} = cx + dy,$$

for Paul's and Bob's cafés, where $x(t)$ is Paul's daily profit, $y(t)$ is Bob's daily profit, and a, b, c, and d are parameters governing how the daily profit of each store affects the other. In Exercises 1–4, different choices of the parameters a, b, c, and d are specified. For each exercise write a brief paragraph describing the interaction between the stores, given the specified parameter values. [For example, suppose $a = 1$, $c = -1$, and $b = d = 0$. If Paul's store is making a profit ($x > 0$), then Paul's profit increases more quickly (because $ax > 0$). However, if Paul makes a profit, then Bob's profits suffer (because $cx < 0$). Since $b = d = 0$, Bob's current profits have no impact on his or Paul's future profits.]

1. $a = 1$, $b = -1$, $c = 1$, and $d = -1$ **2.** $a = 2$, $b = -1$, $c = 0$, and $d = 0$
3. $a = 1$, $b = 0$, $c = 2$, and $d = 1$ **4.** $a = -1$, $b = 2$, $c = 2$, and $d = -1$

In Exercises 5–7, rewrite the specified linear system in matrix form.

5. $\dfrac{dx}{dt} = 2x + y$

$\dfrac{dy}{dt} = x + y$

6. $\dfrac{dx}{dt} = 3y$

$\dfrac{dy}{dt} = 3\pi y - 0.3x$

7. $\dfrac{dp}{dt} = 3p - 2q - 7r$

$\dfrac{dq}{dt} = -2p + 6r$

$\dfrac{dr}{dt} = 7.3q + 2r$

In Exercises 8–9, rewrite the specified linear system in component form.

8.
$$\begin{pmatrix} \dfrac{dx}{dt} \\[2mm] \dfrac{dy}{dt} \end{pmatrix} = \begin{pmatrix} -3 & 2\pi \\ 4 & -1 \end{pmatrix} \begin{pmatrix} x \\ y \end{pmatrix}$$

9.
$$\begin{pmatrix} \dfrac{dx}{dt} \\[2mm] \dfrac{dy}{dt} \end{pmatrix} = \begin{pmatrix} 0 & \beta \\ \gamma & -1 \end{pmatrix} \begin{pmatrix} x \\ y \end{pmatrix}$$

For the linear systems given in Exercises 10–13, use `HPGSystemSolver` to sketch the direction field, several solutions, and the $x(t)$- and $y(t)$-graphs for the solution with initial condition $(x, y) = (1, 1)$.

10.
$$\frac{dx}{dt} = 2x + y$$
$$\frac{dy}{dt} = x + y$$

11.
$$\frac{dx}{dt} = x + 2y$$
$$\frac{dy}{dt} = -x - y$$

12.
$$\begin{pmatrix} \dfrac{dx}{dt} \\[2mm] \dfrac{dy}{dt} \end{pmatrix} = \begin{pmatrix} -3 & 2\pi \\ 4 & -1 \end{pmatrix} \begin{pmatrix} x \\ y \end{pmatrix}$$

13.
$$\begin{pmatrix} \dfrac{dx}{dt} \\[2mm] \dfrac{dy}{dt} \end{pmatrix} = \begin{pmatrix} -1 & -11 \\ 6 & 0 \end{pmatrix} \begin{pmatrix} x \\ y \end{pmatrix}$$

14. Let
$$\mathbf{A} = \begin{pmatrix} a & b \\ c & d \end{pmatrix}$$

be a nonsingular matrix ($\det \mathbf{A} \neq 0$).

  (a) Show that, if $a = 0$, then $b \neq 0$ and $c \neq 0$.

  (b) Suppose $a = 0$. Use the result of part (a) to show that the origin is the only equilibrium point.

Along with the verification given in the section, this result shows that, if $\det \mathbf{A} \neq 0$, then the only equilibrium point for the system $d\mathbf{Y}/dt = \mathbf{AY}$ is the origin.

15. Let
$$\mathbf{A} = \begin{pmatrix} a & b \\ c & d \end{pmatrix}$$

be a nonzero matrix. That is, suppose that at least one of its entries is nonzero. Show that, if $\det \mathbf{A} = 0$, then the system $d\mathbf{Y}/dt = \mathbf{AY}$ has an entire line of equilibria. [*Hint*: First consider the case where $a \neq 0$. Show that any point (x_0, y_0) that satisfies $x_0 = (-b/a)y_0$ is an equilibrium point. What if we assume that entries of $\mathbf{A}$ other than a are nonzero?]

16. The general form of a linear, homogeneous, second-order equation with constant coefficients is
$$\frac{d^2y}{dt^2} + p\frac{dy}{dt} + qy = 0.$$

  (a) Write the first-order system for this equation, and write this system in matrix form.

(b) Show that if $q \neq 0$, then the origin is the only equilibrium point of the system.

(c) Show that if $q \neq 0$, then the only solution of the second-order equation with y constant is $y(t) = 0$ for all t.

17. Consider the linear system corresponding to the second-order equation

$$\frac{d^2 y}{dt^2} + p\frac{dy}{dt} + qy = 0.$$

(a) If $q = 0$ and $p \neq 0$, find all the equilibrium points.

(b) If $q = p = 0$, find all the equilibrium points.

18. Convert the second-order equation

$$\frac{d^2 y}{dt^2} = 0$$

into a first-order system using $v = dy/dt$ as usual.

(a) Find the general solution for the dv/dt equation.

(b) Substitute this solution into the dy/dt equation, and find the general solution of the system.

(c) Sketch the phase portrait of the system.

19. Convert the third-order differential equation

$$\frac{d^3 y}{dt^3} + p\frac{d^2 y}{dt^2} + q\frac{dy}{dt} + ry = 0,$$

where p, q, and r are constants, to a three-dimensional linear system written in matrix form.

In Exercises 20–23, we consider the following model of the market for single-family housing in a community. Let $S(t)$ be the number of sellers at time t, and let $B(t)$ be the number of buyers at time t. We assume that there are natural equilibrium levels of buyers and sellers (made up of people who retire, change job locations, or wish to move for family reasons). The equilibrium level of sellers is S_0, and the equilibrium level of buyers is B_0.

However, market forces can entice people to buy or sell under various conditions. For example, if the price of a house is very high, then house owners are tempted to sell their homes. If prices are very low, extra buyers enter the market looking for bargains. We let $b(t) = B(t) - B_0$ denote the deviation of the number of buyers from equilibrium at time t. So if $b(t) > 0$, then there are more buyers than usual, and we say it is a "seller's market." Presumably the competition of the extra buyers for the same number of houses for sale will force the prices up (the law of supply and demand).

Similarly, we let $s(t) = S(t) - S_0$ denote the deviation of the number of sellers from the equilibrium level. If $s(t) > 0$, then there are more sellers on the market than usual; and if the number of buyers is low, there are too many houses on the market and prices decrease, which in turn affects decisions to buy or sell.

We can give a simple model of this situation as follows:

$$\frac{d\mathbf{Y}}{dt} = \mathbf{AY} = \begin{pmatrix} \alpha & \beta \\ \gamma & \delta \end{pmatrix} \begin{pmatrix} b \\ s \end{pmatrix}, \quad \text{where } \mathbf{Y} = \begin{pmatrix} b \\ s \end{pmatrix}.$$

The exact values of the parameters α, β, γ, and δ depend on the economy of a particular community. Nevertheless, if we assume that everybody wants to get a bargain when they are buying a house and to get top dollar when they are selling a house, then we can hope to predict whether the parameters are positive or negative even though we cannot predict their exact values.

Use the information given above to obtain information about the parameters α, β, γ, and δ. Be sure to justify your answers.

20. If there are more than the usual number of buyers competing for houses, we would expect the price of houses to rise, and this increase would make it less likely that new potential buyers will enter the market. What does this say about the parameter α?

21. If there are fewer than the usual number of buyers competing for the houses available for sale, then we would expect the price of houses to decrease. As a result, fewer potential sellers will place their houses on the market. What does this imply about the parameter γ?

22. Consider the effect on house prices if $s > 0$ and the subsequent effect on buyers and sellers. Then determine the sign of the parameter β.

23. Determine the most reasonable sign for the parameter δ.

24. Consider the linear system

$$\frac{d\mathbf{Y}}{dt} = \begin{pmatrix} 2 & 0 \\ 1 & 1 \end{pmatrix} \mathbf{Y}.$$

(a) Show that the two functions

$$\mathbf{Y}_1(t) = \begin{pmatrix} 0 \\ e^t \end{pmatrix} \quad \text{and} \quad \mathbf{Y}_2(t) = \begin{pmatrix} e^{2t} \\ e^{2t} \end{pmatrix}$$

are solutions to the differential equation.

(b) Solve the initial-value problem

$$\frac{d\mathbf{Y}}{dt} = \begin{pmatrix} 2 & 0 \\ 1 & 1 \end{pmatrix} \mathbf{Y}, \quad \mathbf{Y}(0) = \begin{pmatrix} -2 \\ -1 \end{pmatrix}.$$

25. Consider the linear system

$$\frac{d\mathbf{Y}}{dt} = \begin{pmatrix} 1 & -1 \\ 1 & 3 \end{pmatrix} \mathbf{Y}.$$

(a) Show that the function

$$\mathbf{Y}(t) = \begin{pmatrix} te^{2t} \\ -(t+1)e^{2t} \end{pmatrix}$$

is a solution to the differential equation.

(b) Solve the initial-value problem

$$\frac{d\mathbf{Y}}{dt} = \begin{pmatrix} 1 & -1 \\ 1 & 3 \end{pmatrix} \mathbf{Y}, \quad \mathbf{Y}(0) = \begin{pmatrix} 0 \\ 2 \end{pmatrix}.$$

In Exercises 26–29, a coefficient matrix for the linear system

$$\frac{d\mathbf{Y}}{dt} = \mathbf{A}\mathbf{Y}, \quad \text{where } \mathbf{Y}(t) = \begin{pmatrix} x(t) \\ y(t) \end{pmatrix}$$

is specified. Also two functions and an initial value are given. For each system:

(a) Check that the two functions are solutions of the system; if they are not solutions, then stop.

(b) Check that the two solutions are linearly independent; if they are not linearly independent, then stop.

(c) Find the solution to the linear system with the given initial value.

26.
$$\mathbf{A} = \begin{pmatrix} -2 & -1 \\ 2 & -5 \end{pmatrix}$$
Functions: $\mathbf{Y}_1(t) = (e^{-3t}, e^{-3t})$
$\mathbf{Y}_2(t) = (e^{-4t}, 2e^{-4t})$
Initial value: $\mathbf{Y}(0) = (2, 3)$

27.
$$\mathbf{A} = \begin{pmatrix} -2 & -1 \\ 2 & -5 \end{pmatrix}$$
Functions: $\mathbf{Y}_1(t) = (e^{-3t} - 2e^{-4t}, e^{-3t} - 4e^{-4t})$
$\mathbf{Y}_2(t) = (2e^{-3t} + e^{-4t}, 2e^{-3t} + 2e^{-4t})$
Initial value: $\mathbf{Y}(0) = (2, 3)$

28.
$$\mathbf{A} = \begin{pmatrix} -2 & -3 \\ 3 & -2 \end{pmatrix}$$
Functions: $\mathbf{Y}_1(t) = e^{-2t}(\cos 3t, \sin 3t)$
$\mathbf{Y}_2(t) = e^{-2t}(-\sin 3t, \cos 3t)$
Initial value: $\mathbf{Y}(0) = (2, 3)$

29.
$$\mathbf{A} = \begin{pmatrix} 2 & 3 \\ 1 & 0 \end{pmatrix}$$

Functions: $\mathbf{Y}_1(t) = (-e^{-t} + 12e^{3t}, e^{-t} + 4e^{3t})$
$\quad\quad\quad\quad \mathbf{Y}_2(t) = (-e^{-t}, 2e^{-t})$
Initial value: $\mathbf{Y}(0) = (2, 3)$

30. **(a)** Verify property 1, $\mathbf{A}k\mathbf{Y} = k\mathbf{A}\mathbf{Y}$, of matrix multiplication, where $\mathbf{Y}$ is a (two-dimensional) vector, $\mathbf{A}$ is a matrix, and k is a constant.

(b) Using scalar notation, write out and verify the Linearity Principle. (Aren't matrices nice?)

31. Show that the vectors (x_1, y_1) and (x_2, y_2) are linearly dependent—that is, not linearly independent—if any of the following conditions are satisfied.

(a) If $(x_1, y_1) = (0, 0)$.

(b) If $(x_1, y_1) = \lambda(x_2, y_2)$ for some constant λ.

(c) If $x_1y_2 - x_2y_1 = 0$. *Hint*: Assume x_1 is not zero; then $y_2 = x_2y_1/x_1$. But $x_2 = x_2x_1/x_1$, and we can use part b. The other cases are similar. Note that the quantity $x_1y_2 - x_2y_1$ is the determinant of the matrix

$$\begin{pmatrix} x_1 & y_1 \\ x_2 & y_2 \end{pmatrix}.$$

32. Given the vectors (x_1, y_1) and (x_2, y_2), show that they are linearly independent if the quantity $x_1y_2 - x_2y_1$ is nonzero (see part (c) of Exercise 31). [*Hint*: Suppose $x_2 \neq 0$. If (x_1, y_1) and (x_2, y_2) are on the same line through $(0, 0)$, then $(x_1, y_1) = \lambda(x_2, y_2)$ for some λ. But then $\lambda = x_1/x_2$ and $\lambda = y_1/y_2$. What does this say about x_1/x_2 and y_1/y_2? What if $x_2 = 0$?]

33. Suppose that $\mathbf{Y}_1(t) = (-e^{-t}, e^{-t})$ is a solution to some linear system $d\mathbf{Y}/dt = \mathbf{A}\mathbf{Y}$. For which of the following initial conditions can you give the explicit solution of the linear system?

(a) $\mathbf{Y}(0) = (-2, 2)$ **(b)** $\mathbf{Y}(0) = (3, 4)$ **(c)** $\mathbf{Y}(0) = (0, 0)$ **(d)** $\mathbf{Y}(0) = (3, -3)$

34. The Linearity Principle is a fundamental property of systems of the form $d\mathbf{Y}/dt = \mathbf{A}\mathbf{Y}$. However, you should not assume that it is true for systems that are not of this form, no matter how simple. For example, consider the system

$$\frac{dx}{dt} = 1$$
$$\frac{dy}{dt} = x.$$

The following computations show that the Linearity Principle does not hold for this system.

(a) Show that $\mathbf{Y}(t) = (t, t^2/2)$ is a solution to this system.

(b) Show that $2\mathbf{Y}(t)$ is *not* a solution.

An Extended Linearity Principle that applies to systems such as this one is discussed in Chapter 4.

35. Given solutions $\mathbf{Y}_1(t) = (x_1(t), y_1(t))$ and $\mathbf{Y}_2(t) = (x_2(t), y_2(t))$ to the system

$$\frac{d\mathbf{Y}}{dt} = \mathbf{AY}, \quad \text{where } \mathbf{A} = \begin{pmatrix} a & b \\ c & d \end{pmatrix},$$

we define the **Wronskian** of $\mathbf{Y}_1(t)$ and $\mathbf{Y}_2(t)$ to be the (scalar) function

$$W(t) = x_1(t)y_2(t) - x_2(t)y_1(t).$$

(a) Compute dW/dt.

(b) Use the fact that $\mathbf{Y}_1(t)$ and $\mathbf{Y}_2(t)$ are solutions of the linear system to show that

$$\frac{dW}{dt} = (a + d)W(t).$$

(c) Find the general solution of the differential equation $dW/dt = (a + d)W(t)$.

(d) Suppose that $\mathbf{Y}_1(t)$ and $\mathbf{Y}_2(t)$ are solutions to the system $d\mathbf{Y}/dt = \mathbf{AY}$. Verify that if $\mathbf{Y}_1(0)$ and $\mathbf{Y}_2(0)$ are linearly independent, then $\mathbf{Y}_1(t)$ and $\mathbf{Y}_2(t)$ are also linearly independent for every t.

3.2 STRAIGHT-LINE SOLUTIONS

In Section 3.1 we discussed solutions of linear systems without worrying about how we came up with them (the rabbit-out-of-the-hat method). Often we used the time-honored method known as "guess and test." That is, we made a guess, then substituted the guess back into the equation and checked to see if it satisfied the system. However, the guess-and-test method is unsatisfying because it does not give us any understanding of where the formulas came from in the first place. In this section we use the geometry of the vector field to find special solutions of linear systems.

Geometry of Straight-Line Solutions

We begin by reconsidering an example from the previous section. The direction field for the linear system

$$\frac{d\mathbf{Y}}{dt} = \mathbf{AY}, \quad \text{where } \mathbf{A} = \begin{pmatrix} 2 & 3 \\ 0 & -4 \end{pmatrix},$$

is shown in Figure 3.7. Looking at the direction field, we see that there are two special lines through the origin. The first is the x-axis on which the vectors in the direction field all point directly away from the origin. The other special line runs from the second

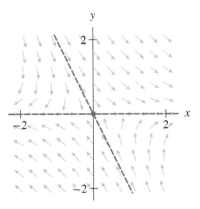

Figure 3.7
The direction field for the linear system

$$\frac{d\mathbf{Y}}{dt} = \begin{pmatrix} 2 & 3 \\ 0 & -4 \end{pmatrix} \mathbf{Y}.$$

There are two special lines through the origin. On the x-axis, the vectors in the direction field all point directly away from the origin. On the distinguished line that runs from the second quadrant to the fourth quadrant, all vectors of the direction field point directly toward the origin.

quadrant to the fourth quadrant. Along this line the vectors of the direction field all point directly toward the origin.

Because solution curves for the system are always tangent to the direction field, a solution that has its initial condition on the positive x-axis moves to the right, directly away from the origin. A solution with an initial condition on the negative x-axis moves to the left, directly away from the origin. Similarly, a solution with an initial condition in the second quadrant on the other special line moves directly toward the origin, and a solution with an initial condition in the fourth quadrant on this line moves directly toward the origin. Thus careful examination of the direction field suggests that there are solutions to this system that lie on straight lines through the origin in the phase plane.

In Section 3.1 we saw that

$$\mathbf{Y}_1(t) = \begin{pmatrix} e^{2t} \\ 0 \end{pmatrix} \quad \text{and} \quad \mathbf{Y}_2(t) = \begin{pmatrix} -e^{-4t} \\ 2e^{-4t} \end{pmatrix}$$

are two linearly independent solutions for the system $d\mathbf{Y}/dt = \mathbf{AY}$. Now let's consider the geometry of these solutions in the phase plane.

To plot the solution curve for $\mathbf{Y}_1(t)$, note that the x-coordinate of $\mathbf{Y}_1(t)$ is e^{2t} and the y-coordinate of $\mathbf{Y}_1(t)$ is always 0. Thus the solution curve lies on the positive x-axis. Moreover, $\mathbf{Y}_1(t) \to \infty$ as $t \to \infty$, and $\mathbf{Y}_1(t)$ tends to the origin as $t \to -\infty$. So $\mathbf{Y}_1(t)$ is a solution that tends directly away from the origin along the x-axis.

For $\mathbf{Y}_2(t)$, it is convenient to rewrite this solution in the form

$$\mathbf{Y}_2(t) = e^{-4t} \begin{pmatrix} -1 \\ 2 \end{pmatrix}.$$

This representation tells us that, as t varies, $\mathbf{Y}_2(t)$ is always a (positive) scalar multiple of the vector $(-1, 2)$. Since positive scalar multiples of a fixed vector always lie on the same ray from the origin, we see that $\mathbf{Y}_2(t)$ parameterizes the ray from $(0, 0)$ with slope -2 in the second quadrant (see Figure 3.7). As $t \to \infty$, $e^{-4t} \to 0$, so this solution tends toward the origin.

We see that the formulas for $\mathbf{Y}_1(t)$ and $\mathbf{Y}_2(t)$ confirm what we guessed by looking at the direction field. There are solutions of this system that lie on two distinguished straight lines in the phase plane.

Straight-line solutions are the simplest solutions (next to equilibrium points) for systems of differential equations. As these solutions move in the xy-plane along straight lines, it is important to remember that the speed at which they move depends on their position on the line. In this example, solutions go to $(0, 0)$ or escape to ∞ at an exponential rate, as can be seen in the $x(t)$- and $y(t)$-graphs for the solutions (see Figures 3.8 and 3.9).

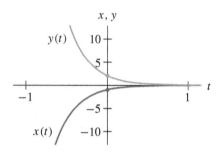

Figure 3.8
The $x(t)$- and $y(t)$-graphs of the straight-line solution

$$\mathbf{Y}_1(t) = \begin{pmatrix} e^{2t} \\ 0 \end{pmatrix}.$$

Figure 3.9
The $x(t)$- and $y(t)$-graphs of the straight-line solution

$$\mathbf{Y}_2(t) = \begin{pmatrix} -e^{-4t} \\ 2e^{-4t} \end{pmatrix}.$$

From the geometry to the algebra of straight-line solutions

Assuming that the system has straight-line solutions (sadly, not all linear systems do), we turn our attention to finding formulas for them. The basic geometric observation is that along a straight-line solution through the origin, the vector field must point either directly toward or directly away from $(0, 0)$ (see Figure 3.7). That is, if $\mathbf{V} = (x, y)$ is on a straight-line solution, then the vector field at (x, y) must point either in the same direction or in exactly the opposite direction as the vector from $(0,0)$ to (x, y).

We now turn this observation into an equation that we can solve to find straight-line solutions. For a linear system of the form $d\mathbf{Y}/dt = \mathbf{AY}$, the vector field at $\mathbf{V} = (x, y)$ is the product $\mathbf{AV}$ of $\mathbf{A}$ and $\mathbf{V}$, which in this example is

$$\begin{pmatrix} 2 & 3 \\ 0 & -4 \end{pmatrix} \begin{pmatrix} x \\ y \end{pmatrix}.$$

Hence we seek vectors $\mathbf{V} = (x, y)$ such that $\mathbf{AV}$ points in the same or in the opposite direction as the vector from $(0, 0)$ to (x, y) or, equivalently, for which there is some number λ such that

$$\mathbf{A} \begin{pmatrix} x \\ y \end{pmatrix} = \lambda \begin{pmatrix} x \\ y \end{pmatrix}.$$

If $\lambda > 0$, then the vector field points in the same direction as (x, y)—away from $(0, 0)$. If $\lambda < 0$, the vector field points in the opposite direction—toward $(0, 0)$.

Using vector notation, this equation can be written more economically as

$$\mathbf{AV} = \lambda \mathbf{V},$$

and it is important to remember that *this equation is the key equation for finding straight-line solutions of the linear system $d\mathbf{Y}/dt = \mathbf{AY}$.*

In our example we seek vectors $\mathbf{V} = (x, y)$ such that $\mathbf{AV} = \lambda \mathbf{V}$, which in coordinates is

$$\begin{pmatrix} 2 & 3 \\ 0 & -4 \end{pmatrix} \begin{pmatrix} x \\ y \end{pmatrix} = \lambda \begin{pmatrix} x \\ y \end{pmatrix}.$$

Multiplying we have

$$\begin{pmatrix} 2x + 3y \\ -4y \end{pmatrix} = \lambda \begin{pmatrix} x \\ y \end{pmatrix},$$

and we can rewrite this equation in the form

$$\begin{pmatrix} 2x + 3y \\ -4y \end{pmatrix} - \lambda \begin{pmatrix} x \\ y \end{pmatrix} = \begin{pmatrix} 0 \\ 0 \end{pmatrix},$$

which is equivalent to the system of simultaneous equations

$$\begin{cases} (2 - \lambda)x + 3y = 0 \\ (-4 - \lambda)y = 0. \end{cases}$$

There is one obvious solution to this system of equations, namely the trivial solution $(x, y) = (0, 0)$. But we already know that the origin is an equilibrium solution of this system, so this solution definitely does not give us a straight-line solution. What we need is a nonzero solution of this system of equations (one where at least one of x or y is nonzero).

To find a nonzero solution, it is important to notice that the simultaneous equations really have three unknowns, x, y, and λ, and in fact we need to determine λ before we can solve for x and y. If we write the simultaneous equations in matrix form, we have

$$\begin{pmatrix} 2 - \lambda & 3 \\ 0 & -4 - \lambda \end{pmatrix} \begin{pmatrix} x \\ y \end{pmatrix} = \begin{pmatrix} 0 \\ 0 \end{pmatrix},$$

and now we recall that we can use the determinant to see if such a system of equations has nontrivial solutions (see Section 3.1, page 249). These equations have nontrivial solutions if and only if

$$\det \begin{pmatrix} 2 - \lambda & 3 \\ 0 & -4 - \lambda \end{pmatrix} = 0.$$

Therefore by computing this determinant, we find that this system has nontrivial solutions if and only if

$$(2 - \lambda)(-4 - \lambda) - (3)(0) = 0.$$

This calculation tells us that we will have nontrivial solutions to our equations only if $\lambda = 2$ or if $\lambda = -4$. All other values of λ do not yield straight-line solutions. (Incidentally, recall that our two straight-line solutions $\mathbf{Y}_1(t)$ and $\mathbf{Y}_2(t)$ involve exponentials of the form e^{2t} and e^{-4t}. In a moment we will see that the appearance of $\lambda = 2$ and $\lambda = -4$ in the exponents is no accident.)

If $\lambda = -4$, then the simultaneous system of equations becomes simply

$$\begin{cases} 6x + 3y = 0 \\ 0 = 0. \end{cases}$$

The second equation always holds, so we need only choose x and y satisfying

$$6x + 3y = 0,$$

which simplifies to

$$y = -2x.$$

There is an entire line of vectors (x, y) that satisfy these equations, and one possible choice is $(x, y) = (-1, 2)$. Note that $(-1, 2)$ is exactly the initial condition for the straight-line solution

$$\mathbf{Y}_2(t) = e^{-4t} \begin{pmatrix} -1 \\ 2 \end{pmatrix}.$$

If $\lambda = 2$, the simultaneous equations become

$$\begin{cases} 3y = 0 \\ -6y = 0, \end{cases}$$

and both equations are satisfied if $y = 0$. Thus any vector of the form $(x, 0)$ with $x \neq 0$ gives a nontrivial solution. That is, anywhere along the x-axis, the vector field points directly away from $(0, 0)$, since $\lambda = 2 > 0$. One vector on this line is $(1, 0)$, which is the initial condition for the straight-line solution

$$\mathbf{Y}_1(t) = e^{2t} \begin{pmatrix} 1 \\ 0 \end{pmatrix}.$$

Eigenvalues and Eigenvectors

We return to this example in a moment, but first we generalize these computations so that we can apply them to any linear system. Consider the general linear system

$$\frac{d\mathbf{Y}}{dt} = \mathbf{AY}.$$

To find straight-line solutions through the origin, we must find nonzero vectors $\mathbf{V} = (x, y)$ such that the vector field at $\mathbf{V}$ points in the same direction as or directly opposite to $\mathbf{V} = (x, y)$. So we seek nonzero vectors $\mathbf{V} = (x, y)$ that satisfy

$$\mathbf{AV} = \lambda \mathbf{V}$$

for some scalar λ. This equation leads to the following definition.

DEFINITION   Given a matrix $\mathbf{A}$, a number λ is called an **eigenvalue** of $\mathbf{A}$ if there is a nonzero vector $\mathbf{V} = (x, y)$ for which

$$\mathbf{AV} = \mathbf{A} \begin{pmatrix} x \\ y \end{pmatrix} = \lambda \begin{pmatrix} x \\ y \end{pmatrix} = \lambda \mathbf{V}.$$

The vector $\mathbf{V}$ is called an **eigenvector** corresponding to the eigenvalue λ.   ∎

The word *eigen* is German for "own" or "self." An eigenvector is a vector where the vector field points in the same or opposite direction as the vector itself.

For example, consider the matrix

$$\mathbf{A} = \begin{pmatrix} 4 & 3 \\ -1 & 0 \end{pmatrix}.$$

The vector $(6, -2)$ is an eigenvector with the eigenvalue 3 because

$$\mathbf{A} \begin{pmatrix} 6 \\ -2 \end{pmatrix} = \begin{pmatrix} 4 & 3 \\ -1 & 0 \end{pmatrix} \begin{pmatrix} 6 \\ -2 \end{pmatrix} = \begin{pmatrix} 18 \\ -6 \end{pmatrix} = 3 \begin{pmatrix} 6 \\ -2 \end{pmatrix}.$$

Also, the vector $(-1, 1)$ is an eigenvector with eigenvalue 1 because

$$\mathbf{A} \begin{pmatrix} -1 \\ 1 \end{pmatrix} = \begin{pmatrix} 4 & 3 \\ -1 & 0 \end{pmatrix} \begin{pmatrix} -1 \\ 1 \end{pmatrix} = \begin{pmatrix} -1 \\ 1 \end{pmatrix} = 1 \begin{pmatrix} -1 \\ 1 \end{pmatrix}.$$

It is important to remember that being an eigenvector is a special property. For a typical matrix, most vectors are not eigenvectors. For example, $(2, 3)$ is not an eigenvector for $\mathbf{A}$ because

$$\mathbf{A} \begin{pmatrix} 2 \\ 3 \end{pmatrix} = \begin{pmatrix} 4 & 3 \\ -1 & 0 \end{pmatrix} \begin{pmatrix} 2 \\ 3 \end{pmatrix} = \begin{pmatrix} 17 \\ -2 \end{pmatrix},$$

and $(17, -2)$ is not a multiple of $(2, 3)$.

Lines of eigenvectors

Given a matrix $\mathbf{A}$, if $\mathbf{V}$ is an eigenvector for eigenvalue λ, then any scalar multiple $k\mathbf{V}$ is also an eigenvector for λ. To verify this, we compute

$$\mathbf{A}(k\mathbf{V}) = k\mathbf{AV} = k(\lambda \mathbf{V}) = \lambda(k\mathbf{V}),$$

where the first equality is a property of matrix multiplication and the second equality uses the fact that $\mathbf{V}$ is an eigenvector. Hence given an eigenvector $\mathbf{V}$ for the eigenvalue λ, the entire line of vectors through $\mathbf{V}$ and the origin are also eigenvectors for λ.

Computation of Eigenvalues

To find straight-line solutions of linear systems, we must find the eigenvalues and eigenvectors of the corresponding coefficient matrix. That is, we need to find the vectors $\mathbf{V} = (x, y)$ such that

$$\mathbf{AV} = \mathbf{A}\begin{pmatrix} x \\ y \end{pmatrix} = \lambda \begin{pmatrix} x \\ y \end{pmatrix} = \lambda\mathbf{V}.$$

If

$$\mathbf{A} = \begin{pmatrix} a & b \\ c & d \end{pmatrix},$$

then we have

$$\begin{pmatrix} a & b \\ c & d \end{pmatrix}\begin{pmatrix} x \\ y \end{pmatrix} = \lambda \begin{pmatrix} x \\ y \end{pmatrix},$$

which is written in components as

$$\begin{cases} ax + by = \lambda x \\ cx + dy = \lambda y. \end{cases}$$

Thus we want nonzero solutions (x, y) to

$$\begin{cases} (a - \lambda)x + \quad by = 0 \\ \quad cx + (d - \lambda)y = 0. \end{cases}$$

From the determinant condition that we derived in Section 3.1 (page 249), we know that this system has nontrivial solutions if and only if

$$\det \begin{pmatrix} a - \lambda & b \\ c & d - \lambda \end{pmatrix} = 0.$$

We encounter this matrix each time we compute eigenvalues and eigenvectors, so we introduce some notation for it. The **identity matrix** is the 2×2 matrix

$$\mathbf{I} = \begin{pmatrix} 1 & 0 \\ 0 & 1 \end{pmatrix}.$$

This matrix is called the identity matrix because $\mathbf{IV} = \mathbf{V}$ for any vector $\mathbf{V}$. Also, $\lambda\mathbf{I}$ represents the matrix

$$\lambda\mathbf{I} = \begin{pmatrix} \lambda & 0 \\ 0 & \lambda \end{pmatrix}.$$

Computing the difference between the matrices $\mathbf{A}$ and $\lambda\mathbf{I}$ by subtracting corresponding entries yields

$$\mathbf{A} - \lambda\mathbf{I} = \begin{pmatrix} a - \lambda & b \\ c & d - \lambda \end{pmatrix}.$$

Thus our determinant condition for a nontrivial solution of the equation $\mathbf{AV} = \lambda\mathbf{V}$ may be written in the compact form

$$\det(\mathbf{A} - \lambda\mathbf{I}) = 0.$$

It is important to remember that the matrix $\mathbf{A} - \lambda\mathbf{I}$ is the matrix $\mathbf{A}$ with λ's subtracted from the upper-left and lower-right entries.

The Characteristic Polynomial

To find the eigenvalues of the matrix $\mathbf{A}$, we must find the values of λ for which

$$\det(\mathbf{A} - \lambda\mathbf{I}) = 0.$$

If we write this equation in terms of the entries of $\mathbf{A}$, we find

$$\det(\mathbf{A} - \lambda\mathbf{I}) = \det\begin{pmatrix} a - \lambda & b \\ c & d - \lambda \end{pmatrix} = (a - \lambda)(d - \lambda) - bc = 0,$$

which expands to the quadratic polynomial

$$\lambda^2 - (a + d)\lambda + (ad - bc) = 0.$$

This polynomial is called the **characteristic polynomial** of the system. Its roots are the eigenvalues of the matrix $\mathbf{A}$.

A quadratic polynomial always has two roots, but these roots need not be real numbers, nor must they be distinct. If the roots of the characteristic polynomial are not real, we say that the matrix $\mathbf{A}$ has **complex eigenvalues**. We will study the behavior of solutions to systems with complex eigenvalues in Section 3.4, and the case of a single root of multiplicity two (a repeated root) is considered in Section 3.5.

Consider the matrix

$$\mathbf{A} = \begin{pmatrix} 2 & 3 \\ 0 & -4 \end{pmatrix}$$

that we discussed earlier in this section. This matrix has the characteristic polynomial

$$\det(\mathbf{A} - \lambda\mathbf{I}) = (2 - \lambda)(-4 - \lambda) - (3)(0) = \lambda^2 + 2\lambda - 8,$$

which has roots $\lambda_1 = 2$ and $\lambda_2 = -4$. As we saw earlier, these numbers are the eigenvalues of this matrix. (This example is somewhat unusual in that it is not necessary to expand the expression $\det(\mathbf{A} - \lambda\mathbf{I}) = (2 - \lambda)(-4 - \lambda) - (3)(0)$ into $\lambda^2 + 2\lambda - 8$ to determine the eigenvalues of $\mathbf{A}$.)

Computation of Eigenvectors

The next step in the process of finding straight-line solutions of a system of differential equations is to find the eigenvectors associated to the eigenvalues. Suppose we are given a matrix

$$\mathbf{A} = \begin{pmatrix} a & b \\ c & d \end{pmatrix}$$

and we know that λ is an eigenvalue. To find a corresponding eigenvector, we must solve the equation $\mathbf{AV} = \lambda\mathbf{V}$ for the vector $\mathbf{V}$. If we write

$$\mathbf{V} = \begin{pmatrix} x \\ y \end{pmatrix},$$

then $\mathbf{AV} = \lambda\mathbf{V}$ becomes a simultaneous system of linear equations in two unknowns, x and y. In fact the equations are

$$\begin{cases} ax + by = \lambda x \\ cx + dy = \lambda y. \end{cases}$$

Since λ is an eigenvalue, we know that there is at least an entire line of eigenvectors (x, y) that satisfy this system of equations. This infinite number of eigenvectors means that the equations are redundant. That is, either the two equations are equivalent, or one of the equations is always satisfied.

For example, suppose we are given the matrix

$$\mathbf{B} = \begin{pmatrix} 2 & 2 \\ 1 & 3 \end{pmatrix}.$$

We find the eigenvalues of $\mathbf{B}$ by finding the roots of the characteristic polynomial

$$\det(\mathbf{B} - \lambda\mathbf{I}) = (2 - \lambda)(3 - \lambda) - (2)(1) = 0,$$

which yields the quadratic equation

$$\lambda^2 - 5\lambda + 4 = 0.$$

The roots of this quadratic polynomial are $\lambda_1 = 4$ and $\lambda_2 = 1$, so 1 and 4 are the eigenvalues of $\mathbf{B}$.

To find an eigenvector $\mathbf{V}_1$ for $\lambda_1 = 4$, we must solve

$$\mathbf{B}\begin{pmatrix} x_1 \\ y_1 \end{pmatrix} = \begin{pmatrix} 2 & 2 \\ 1 & 3 \end{pmatrix}\begin{pmatrix} x_1 \\ y_1 \end{pmatrix} = 4\begin{pmatrix} x_1 \\ y_1 \end{pmatrix}.$$

Rewritten in terms of components, this equation is

$$\begin{cases} 2x_1 + 2y_1 = 4x_1 \\ x_1 + 3y_1 = 4y_1, \end{cases}$$

or, equivalently,

$$\begin{cases} -2x_1 + 2y_1 = 0 \\ x_1 - y_1 = 0. \end{cases}$$

Note that these equations are redundant (multiply both sides of the second by -2 to get the first). So any vector (x_1, y_1) that satisfies the second equation

$$x_1 - y_1 = 0$$

is an eigenvector. This equation specifies the line $y_1 = x_1$ in the plane. Any nonzero vector on this line is an eigenvector of $\mathbf{B}$ corresponding to the eigenvalue $\lambda_1 = 4$. For example, the vectors $(1, 1)$ and $(-\pi, -\pi)$ are two of the infinitely many eigenvectors for $\mathbf{B}$ corresponding to the eigenvalue $\lambda_1 = 4$.

For $\lambda_2 = 1$ we must solve

$$\mathbf{B} \begin{pmatrix} x_2 \\ y_2 \end{pmatrix} = \begin{pmatrix} 2 & 2 \\ 1 & 3 \end{pmatrix} \begin{pmatrix} x_2 \\ y_2 \end{pmatrix} = 1 \begin{pmatrix} x_2 \\ y_2 \end{pmatrix}.$$

In terms of coordinates, this vector equation is the same as the system

$$\begin{cases} 2x_2 + 2y_2 = x_2 \\ x_2 + 3y_2 = y_2 \end{cases}$$

or, equivalently,

$$\begin{cases} x_2 + 2y_2 = 0 \\ x_2 + 2y_2 = 0. \end{cases}$$

Again these equations are redundant. So the eigenvectors corresponding to eigenvalue $\lambda_2 = 1$ are the nonzero vectors (x_2, y_2) that lie on the line $y_2 = -x_2/2$.

Straight-Line Solutions

After all of the algebra of the last few pages, it is time to return to the study of differential equations. To summarize what we have accomplished so far, suppose we are given a linear system of differential equations

$$\frac{d\mathbf{Y}}{dt} = \mathbf{AY}.$$

To find straight-line solutions, we first find the eigenvalues of $\mathbf{A}$ and then their associated eigenvectors. Once we have this information, we have determined the straight-line solutions.

To do this, suppose that λ is an eigenvalue with associated eigenvector $\mathbf{V} = (x, y)$. Then consider the function

$$\mathbf{Y}(t) = e^{\lambda t} \mathbf{V} = \begin{pmatrix} e^{\lambda t} x \\ e^{\lambda t} y \end{pmatrix}.$$

For each t, $\mathbf{Y}(t)$ is a scalar multiple of our eigenvector (x, y), so the curve given by $\mathbf{Y}(t)$ lies on the ray from the origin through (x, y). Moreover, $\mathbf{Y}(t)$ is a solution of the differential equation. We can check this assertion by substituting $\mathbf{Y}(t)$ in the differential equation. We compute

$$\frac{d\mathbf{Y}}{dt} = \frac{d}{dt}\left(\begin{array}{c} e^{\lambda t}x \\ e^{\lambda t}y \end{array}\right) = \left(\begin{array}{c} \lambda e^{\lambda t}x \\ \lambda e^{\lambda t}y \end{array}\right) = \lambda\mathbf{Y}(t).$$

On the other hand, we have

$$\mathbf{AY}(t) = \mathbf{A}e^{\lambda t}\mathbf{V} = e^{\lambda t}\mathbf{AV} = e^{\lambda t}\lambda\mathbf{V} = \lambda e^{\lambda t}\mathbf{V} = \lambda\mathbf{Y}(t)$$

since $\mathbf{V}$ is an eigenvector of $\mathbf{A}$. Comparing the results of these two computations, we see that

$$\frac{d\mathbf{Y}}{dt} = \mathbf{AY},$$

so $\mathbf{Y}(t)$ is indeed a solution.

This is an important observation: We obtain formulas for straight-line solutions using just the eigenvalues and eigenvectors of the matrix $\mathbf{A}$.

Sometimes we can do even better. Suppose we find two real, distinct eigenvalues λ_1 and λ_2 for the system with eigenvectors $\mathbf{V}_1$ and $\mathbf{V}_2$ respectively. Since $\mathbf{V}_1$ and $\mathbf{V}_2$ are eigenvectors for different eigenvalues, they must be linearly independent. That is, any scalar multiple of $\mathbf{V}_1$ is an eigenvector associated to λ_1. Consequently, $\mathbf{V}_2$ does not lie on the line through the origin determined by $\mathbf{V}_1$, and $\mathbf{V}_1$ and $\mathbf{V}_2$ are linearly independent. As a result, the two solutions

$$\mathbf{Y}_1(t) = e^{\lambda_1 t}\mathbf{V}_1 \quad \text{and} \quad \mathbf{Y}_2(t) = e^{\lambda_2 t}\mathbf{V}_2$$

are linearly independent. Therefore, using the Linearity Principle, the general solution of the system is $k_1\mathbf{Y}_1(t) + k_2\mathbf{Y}_2(t) = k_1e^{\lambda_1 t}\mathbf{V}_1 + k_2e^{\lambda_2 t}\mathbf{V}_2$.

THEOREM   Suppose the matrix $\mathbf{A}$ has a real eigenvalue λ with associated eigenvector $\mathbf{V}$. Then the linear system $d\mathbf{Y}/dt = \mathbf{AY}$ has the straight-line solution

$$\mathbf{Y}(t) = e^{\lambda t}\mathbf{V}.$$

Moreover, if λ_1 and λ_2 are distinct, real eigenvalues with eigenvectors $\mathbf{V}_1$ and $\mathbf{V}_2$ respectively, then the solutions $\mathbf{Y}_1(t) = e^{\lambda_1 t}\mathbf{V}_1$ and $\mathbf{Y}_2(t) = e^{\lambda_2 t}\mathbf{V}_2$ are linearly independent and

$$\mathbf{Y}(t) = k_1e^{\lambda_1 t}\mathbf{V}_1 + k_2e^{\lambda_2 t}\mathbf{V}_2$$

is the general solution of the system.   ∎

This is a powerful theorem. It lets us find solutions of linear systems of differential equations using only algebra. All we need to do is to find an eigenvalue and an associated eigenvector. There are no tedious or impossible integrations to perform. (One caveat here is that the eigenvalue must be real; we tackle the case of complex eigenvalues in Section 3.4.)

The theorem also explicitly provides the general solution of certain linear systems, namely those that have two distinct, real eigenvalues. We will treat the possibility that the eigenvalues of $\mathbf{A}$ are real but not distinct in Section 3.5.

Putting Everything Together

Now let's combine the geometry of the direction field with the algebra of this section to produce the general solution of a linear system of differential equations. Consider the linear system

$$\frac{d\mathbf{Y}}{dt} = \mathbf{BY} = \begin{pmatrix} 2 & 2 \\ 1 & 3 \end{pmatrix} \mathbf{Y}.$$

The direction field for this system is depicted in Figure 3.10. There appear to be two distinguished lines of eigenvectors, one cutting diagonally through the first and third quadrants, and another through the second and fourth quadrants. The associated eigenvalues are positive since the direction field points away from the origin.

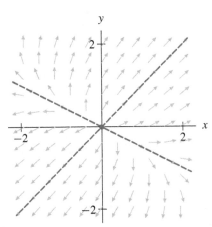

Figure 3.10
The direction field for the system

$$\frac{d\mathbf{Y}}{dt} = \mathbf{BY} = \begin{pmatrix} 2 & 2 \\ 1 & 3 \end{pmatrix} \mathbf{Y}.$$

Note the two distinguished lines of eigenvectors. The one in the first quadrant corresponds to the solution $\mathbf{Y}_1(t) = e^{4t}(1, 1)$ and the one in the second quadrant corresponds to the solution $\mathbf{Y}_2(t) = e^t(-2, 1)$.

To find formulas for corresponding straight-line solutions, we use the eigenvalues and eigenvectors of $\mathbf{B}$, which we computed earlier in the section. The eigenvalues of $\mathbf{B}$ are $\lambda_1 = 4$ and $\lambda_2 = 1$. The eigenvectors $\mathbf{V}_1 = (x_1, y_1)$ associated to λ_1 satisfy the equation $y_1 = x_1$, and the eigenvectors $\mathbf{V}_2 = (x_2, y_2)$ associated to λ_2 satisfy the equation $y_2 = -x_1/2$. In particular, we can use the vectors $\mathbf{V}_1 = (1, 1)$ and $\mathbf{V}_2 = (-2, 1)$ to produce two linearly independent straight-line solutions. The general solution is

$$\mathbf{Y}(t) = k_1 e^{4t} \begin{pmatrix} 1 \\ 1 \end{pmatrix} + k_2 e^t \begin{pmatrix} -2 \\ 1 \end{pmatrix}.$$

Note that there is nothing significant about our choice of $\mathbf{V}_1 = (1, 1)$ and $\mathbf{V}_2 = (-2, 1)$. For $\mathbf{V}_1$ we can use any eigenvector associated to the eigenvalue $\lambda_1 = 4$, and for $\mathbf{V}_2$ we can use any eigenvector associated to the eigenvalue $\lambda_2 = 1$.

A Harmonic Oscillator

Consider the harmonic oscillator with mass $m = 1$, spring constant $k = 10$, and damping coefficient $b = 7$. The second-order equation that models this oscillator is

$$\frac{d^2 y}{dt^2} + 7\frac{dy}{dt} + 10y = 0,$$

and the corresponding system is

$$\frac{d\mathbf{Y}}{dt} = \mathbf{CY}, \quad \text{where } \mathbf{C} = \begin{pmatrix} 0 & 1 \\ -10 & -7 \end{pmatrix} \quad \text{and} \quad \mathbf{Y} = \begin{pmatrix} y \\ v \end{pmatrix}.$$

The phase portrait is shown in Figure 3.11. Note that there appear to be straight-line solutions for this system.

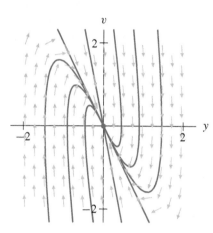

Figure 3.11
Phase portrait for

$$\frac{d\mathbf{Y}}{dt} = \begin{pmatrix} 0 & 1 \\ -10 & -7 \end{pmatrix} \mathbf{Y}.$$

This linear system is obtained from the harmonic oscillator

$$\frac{d^2y}{dt^2} + 7\frac{dy}{dt} + 10y = 0,$$

where $\mathbf{Y} = (y, v)$ and $v = dy/dt$.

The characteristic polynomial for the system is

$$(-\lambda)(-7 - \lambda) + 10 = \lambda^2 + 7\lambda + 10,$$

and the eigenvalues are $\lambda_1 = -5$ and $\lambda_2 = -2$. Note that both of these eigenvalues are negative.

We compute the eigenvector for $\lambda_1 = -5$ by solving $\mathbf{CV}_1 = -5\mathbf{V}_1$. If $\mathbf{V}_1 = (y_1, v_1)$, we have

$$\begin{cases} v_1 = -5y_1 \\ -10y_1 - 7v_1 = -5v_1. \end{cases}$$

If we have done our arithmetic correctly, these two equations are redundant, and the desired eigenvectors must satisfy the equation $v_1 = -5y_1$. (It is a good idea to check the redundancy of these equations. If they are not redundant, then a mistake was made earlier in the computation.) Setting $y_1 = 1$, we obtain the eigenvector $\mathbf{V}_1 = (1, -5)$ corresponding to λ_1. Similarly, we can compute that an eigenvector for $\lambda_2 = -2$ is $\mathbf{V}_2 = (1, -2)$. (Do you notice anything special about these two eigenvectors?)

The general solution for this system is

$$\mathbf{Y}(t) = k_1 e^{-5t} \begin{pmatrix} 1 \\ -5 \end{pmatrix} + k_2 e^{-2t} \begin{pmatrix} 1 \\ -2 \end{pmatrix}.$$

Using this formula, we can find the exact position of the oscillator at any time. More-over, we can also determine qualitative features of the model from these formulas. Each term in the expression for $\mathbf{Y}(t)$ contains an exponential of the form $e^{\lambda t}$ with $\lambda < 0$. Consequently each term tends to 0 as t increases. Note that this is consistent with the directions of the solution curves in the phase portrait (see Figure 3.11), but it is com-forting to see everything fit together so nicely. Since $\mathbf{Y}(t) = (y(t), v(t))$, the general solution of the corresponding second-order equation is the first component of $\mathbf{Y}(t)$, that is,

$$y(t) = k_1 e^{-5t} + k_2 e^{-2t}.$$

One thing that we learned from the eigenvalues that we did not know from the phase portrait alone is the fact that every solution tends to zero at a rate that is at least comparable to the rate at which e^{-2t} tends to 0.

EXERCISES FOR SECTION 3.2

In Exercises 1–10,

(a) compute the eigenvalues;

(b) for each eigenvalue, compute the associated eigenvectors;

(c) using `HPGSystemSolver`, sketch the direction field for the system, and plot the straight-line solutions;

(d) for each eigenvalue, specify a corresponding straight-line solution and plot its $x(t)$- and $y(t)$-graphs; and

(e) if the system has two distinct eigenvalues, compute the general solution.

1. $\dfrac{d\mathbf{Y}}{dt} = \begin{pmatrix} 3 & 2 \\ 0 & -2 \end{pmatrix} \mathbf{Y}$

2. $\dfrac{d\mathbf{Y}}{dt} = \begin{pmatrix} -4 & -2 \\ -1 & -3 \end{pmatrix} \mathbf{Y}$

3. $\begin{pmatrix} \dfrac{dx}{dt} \\ \dfrac{dy}{dt} \end{pmatrix} = \begin{pmatrix} -5 & -2 \\ -1 & -4 \end{pmatrix} \begin{pmatrix} x \\ y \end{pmatrix}$

4. $\begin{pmatrix} \dfrac{dx}{dt} \\ \dfrac{dy}{dt} \end{pmatrix} = \begin{pmatrix} 2 & 1 \\ -1 & 4 \end{pmatrix} \begin{pmatrix} x \\ y \end{pmatrix}$

5. $\dfrac{dx}{dt} = -\dfrac{x}{2}$

$\dfrac{dy}{dt} = x - \dfrac{y}{2}$

6. $\dfrac{dx}{dt} = 5x + 4y$

$\dfrac{dy}{dt} = 9x$

7. $\begin{pmatrix} \dfrac{dx}{dt} \\ \dfrac{dy}{dt} \end{pmatrix} = \begin{pmatrix} 3 & 4 \\ 1 & 0 \end{pmatrix} \begin{pmatrix} x \\ y \end{pmatrix}$

8. $\begin{pmatrix} \dfrac{dx}{dt} \\ \dfrac{dy}{dt} \end{pmatrix} = \begin{pmatrix} 2 & -1 \\ -1 & 1 \end{pmatrix} \begin{pmatrix} x \\ y \end{pmatrix}$

9.
$$\frac{dx}{dt} = 2x + y$$
$$\frac{dy}{dt} = x + y$$

10.
$$\frac{dx}{dt} = -x - 2y$$
$$\frac{dy}{dt} = x - 4y$$

11. Solve the initial-value problem

$$\frac{dx}{dt} = -2x - 2y$$
$$\frac{dy}{dt} = -2x + y,$$

where the initial condition $(x(0), y(0))$ is:

(a)   (1, 0)   (b)   (0, 1)   (c)   (1, −2)

12. Solve the initial-value problem

$$\frac{dx}{dt} = 3x$$
$$\frac{dy}{dt} = x - 2y,$$

where the initial condition $(x(0), y(0))$ is:

(a)   (1, 0)   (b)   (0, 1)   (c)   (2, 2)

13. Solve the initial-value problem

$$\frac{d\mathbf{Y}}{dt} = \begin{pmatrix} -4 & 1 \\ 2 & -3 \end{pmatrix} \mathbf{Y}, \quad \mathbf{Y}(0) = \mathbf{Y}_0,$$

where the initial condition $\mathbf{Y}_0$ is:

(a) $\mathbf{Y}_0 = (1, 0)$ (b) $\mathbf{Y}_0 = (2, 1)$ (c) $\mathbf{Y}_0 = (-1, -2)$

14. Solve the initial-value problem

$$\frac{d\mathbf{Y}}{dt} = \begin{pmatrix} 4 & -2 \\ 1 & 1 \end{pmatrix} \mathbf{Y}, \quad \mathbf{Y}(0) = \mathbf{Y}_0,$$

where the initial condition $\mathbf{Y}_0$ is:

(a) $\mathbf{Y}_0 = (1, 0)$ (b) $\mathbf{Y}_0 = (2, 1)$ (c) $\mathbf{Y}_0 = (-1, -2)$

15. Show that a is the only eigenvalue and that every nonzero vector is an eigenvector for the matrix

$$\mathbf{A} = \begin{pmatrix} a & 0 \\ 0 & a \end{pmatrix}.$$

16. A matrix of the form

$$\mathbf{A} = \begin{pmatrix} a & b \\ 0 & d \end{pmatrix}$$

is called **upper triangular**. Suppose that $b \neq 0$ and $a \neq d$. Find the eigenvalues and eigenvectors of $\mathbf{A}$.

17. A matrix of the form

$$\mathbf{B} = \begin{pmatrix} a & b \\ b & d \end{pmatrix}$$

is called **symmetric**. Show that $\mathbf{B}$ has real eigenvalues and that, if $b \neq 0$, then $\mathbf{B}$ has two distinct eigenvalues.

18. Compute the eigenvalues of a matrix of the form

$$\mathbf{C} = \begin{pmatrix} a & b \\ c & 0 \end{pmatrix}.$$

Compare your results to those of Exercise 16.

19. Consider the second-order equation

$$\frac{d^2y}{dt^2} + p\frac{dy}{dt} + qy = 0,$$

where p and q are positive.

(a) Convert this equation into a first-order, linear system.
(b) Compute the characteristic polynomial of the system.
(c) Find the eigenvalues.
(d) Under what conditions on p and q are the eigenvalues two distinct real numbers?
(e) Verify that the eigenvalues are negative if they are real numbers.

20. For the harmonic oscillator with mass $m = 1$, spring constant $k = 4$, and damping coefficient $b = 5$,

(a) compute the eigenvalues and associated eigenvectors;
(b) for each eigenvalue, pick an associated eigenvector $\mathbf{V}$ and determine the solution $\mathbf{Y}(t)$ with $\mathbf{Y}(0) = \mathbf{V}$;
(c) for each solution derived in part (b), plot its solution curve in the yv-phase plane;
(d) for each solution derived in part (b), plot its $y(t)$- and $v(t)$-graphs; and
(e) for each solution derived in part (b), give a brief description of the behavior of the mass-spring system.

In Exercises 21–24, we return to Exercises 1–4 in Section 2.3. (For convenience, the equations are reproduced below.) For each second-order equation,

(a) convert the equation to a first-order, linear system;

(b) compute the eigenvalues and eigenvectors of the system;

(c) for each eigenvalue, pick an associated eigenvector **V**, and determine the solution **Y**(t) to the system; and

(d) compare the results of your calculations in part (c) with the results that you obtained when you used the guess-and-test method of Section 2.3.

21. $\dfrac{d^2y}{dt^2} + 7\dfrac{dy}{dt} + 10y = 0$ **22.** $\dfrac{d^2y}{dt^2} + 5\dfrac{dy}{dt} + 6y = 0$

23. $\dfrac{d^2y}{dt^2} + 4\dfrac{dy}{dt} + y = 0$ **24.** $\dfrac{d^2y}{dt^2} + 6\dfrac{dy}{dt} + 7y = 0$

25. Verify that the linear system that models the harmonic oscillator with mass $m = 1$, spring constant $k = 4$, and damping coefficient $b = 1$ does not have real eigenvalues. Does this tell you anything about the phase portrait of this system?

3.3 PHASE PORTRAITS FOR LINEAR SYSTEMS WITH REAL EIGENVALUES

In the preceding section we saw that straight-line solutions play a dominant role in finding the general solution of certain linear systems of differential equations. To solve such a system, we first use algebra to compute the eigenvalues and eigenvectors of the coefficient matrix. When we find a real eigenvalue and an associated eigenvector, we can write down the corresponding straight-line solution. Moreover, in the special case where we find two real, distinct eigenvalues, we can write down an explicit formula for the general solution of the system.

The sign of the eigenvalue plays an important role in determining the behavior of the corresponding straight-line solutions. If the eigenvalue is negative, the solution tends to the origin as $t \to \infty$. If the eigenvalue is positive, the solution tends away from the origin as $t \to \infty$. In this section we use the behavior of these straight-line solutions to determine the behavior of all solutions.

Saddles

One common type of linear system features both a positive and negative eigenvalue. For example, consider the linear system

$$\frac{d\mathbf{Y}}{dt} = \mathbf{AY}, \quad \text{where } \mathbf{A} = \begin{pmatrix} -3 & 0 \\ 0 & 2 \end{pmatrix}.$$

This is a particularly simple linear system, since it corresponds to the equations

$$\frac{dx}{dt} = -3x$$

$$\frac{dy}{dt} = 2y.$$

Note that dx/dt depends only on x and dy/dt depends only on y. That is, the system completely decouples. We can solve these two equations independently using methods from Chapter 1. However, in order to understand the geometry more fully, we use the methods of the previous two sections.

As usual, we first compute the eigenvalues of $\mathbf{A}$ by finding the roots of the characteristic polynomial

$$\det(\mathbf{A} - \lambda\mathbf{I}) = \det \begin{pmatrix} -3 - \lambda & 0 \\ 0 & 2 - \lambda \end{pmatrix} = (-3 - \lambda)(2 - \lambda) = 0.$$

Thus the eigenvalues of $\mathbf{A}$ are $\lambda_1 = -3$ and $\lambda_2 = 2$.

Next we compute the eigenvectors. For $\lambda_1 = -3$, we must solve the equation $\mathbf{AV} = -3\mathbf{V}$ for $\mathbf{V}$. If $\mathbf{V}_1 = (x_1, y_1)$, then we have

$$\begin{cases} -3x_1 = -3x_1 \\ 2y_1 = -3y_1. \end{cases}$$

So any nonzero vector $\mathbf{V}$ lying along the line $y = 0$ (the x-axis) in the plane is an eigenvector for $\lambda_1 = -3$. We choose $\mathbf{V}_1 = (1, 0)$. Therefore the solution

$$\mathbf{Y}_1(t) = e^{-3t}\mathbf{V}_1$$

is a straight-line solution whose solution curve is the positive x-axis. The solution tends to the origin as t increases.

In similar fashion we can check that any eigenvector corresponding to $\lambda_2 = 2$ lies along the y-axis. We choose $\mathbf{V}_2 = (0, 1)$ and obtain a second solution

$$\mathbf{Y}_2(t) = e^{2t}\mathbf{V}_2.$$

The general solution is therefore

$$\mathbf{Y}(t) = k_1 e^{-3t}\mathbf{V}_1 + k_2 e^{2t}\mathbf{V}_2 = \begin{pmatrix} k_1 e^{-3t} \\ k_2 e^{2t} \end{pmatrix}.$$

In Figure 3.12 we display the phase portrait for this system. The straight-line solutions lie on the axes, but all other solutions behave differently. In the figure we see that the other solutions seem to tend to infinity asymptotic to the y-axis and to come from infinity asymptotic to the x-axis. To see why, consider a solution $\mathbf{Y}(t)$ that is not a straight-line solution. Then

$$\mathbf{Y}(t) = k_1 e^{-3t}\mathbf{V}_1 + k_2 e^{2t}\mathbf{V}_2,$$

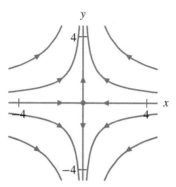

Figure 3.12
Phase portrait for the system

$$\frac{d\mathbf{Y}}{dt} = \mathbf{AY} = \begin{pmatrix} -3 & 0 \\ 0 & 2 \end{pmatrix} \mathbf{Y}.$$

where both k_1 and k_2 are nonzero. When t is large and positive, the term e^{-3t} is very small. Therefore for large positive t, the vector $e^{-3t}\mathbf{V}_1$ in the general solution is negligible, and we have

$$\mathbf{Y}(t) \approx k_2 e^{2t}\mathbf{V}_2 = \begin{pmatrix} 0 \\ k_2 e^{2t} \end{pmatrix}.$$

That is, for large positive values of t, our solution behaves like a straight-line solution on the y-axis.

The opposite is true when we consider large negative values of t. In this case the term e^{2t} is very small, so we have

$$\mathbf{Y}(t) \approx k_1 e^{-3t}\mathbf{V}_1 = \begin{pmatrix} k_1 e^{-3t} \\ 0 \end{pmatrix},$$

which is a straight-line solution along the x-axis.

For example, the particular solution of this system that satisfies $\mathbf{Y}(0) = (1, 1)$ is

$$\mathbf{Y}(t) = \begin{pmatrix} e^{-3t} \\ e^{2t} \end{pmatrix}.$$

The x-coordinate of this solution tends to 0 as $t \to \infty$ and to infinity as $t \to -\infty$. The y-coordinate behaves in the opposite manner (see these $x(t)$- and $y(t)$-graphs in Figure 3.13).

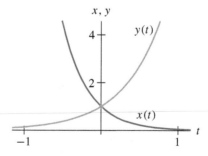

Figure 3.13
The $x(t)$- and $y(t)$-graphs for the solution with initial position $(1, 1)$.

Despite the fact that this example really consists of two one-dimensional differential equations, its phase portrait is entirely new. Along the axes we see the familiar phase lines for one-dimensional equations—a sink along the x-axis and a source on the y-axis. All other solutions tend to infinity as $t \to \pm\infty$. These solutions come from infinity in the direction of the eigenvectors corresponding to the one-dimensional sink, and they tend back to infinity in the direction of the one-dimensional source.

Any linear system for which we have one positive and one negative eigenvalue has similar behavior. An equilibrium point of this form is called a **saddle**. This name is supposed to remind you of a saddle for a horse. The path followed by a drop of water on a horse's saddle resembles the path of a solution of this type of linear system; it approaches the center of the seat in one direction and then veers off toward the ground in another.

Phase portraits for other saddles

The previous example is special in that the eigenvectors lie on the x- and y-axes. In general the eigenvectors for a saddle can lie on any two distinct lines through the origin. This makes the phase portraits and the $x(t)$- and $y(t)$-graphs appear somewhat different in the general case.

For example, consider the system

$$\frac{d\mathbf{Y}}{dt} = \mathbf{BY}, \quad \text{where } \mathbf{B} = \begin{pmatrix} 8 & -11 \\ 6 & -9 \end{pmatrix}.$$

We first compute the eigenvalues of $\mathbf{B}$ by finding the roots of the characteristic polynomial

$$\det(\mathbf{B} - \lambda I) = \det \begin{pmatrix} 8 - \lambda & -11 \\ 6 & -9 - \lambda \end{pmatrix} = (8 - \lambda)(-9 - \lambda) + 66 = \lambda^2 + \lambda - 6 = 0.$$

The roots of this quadratic equation are $\lambda_1 = -3$ and $\lambda_2 = 2$, the eigenvalues of $\mathbf{B}$. These are exactly the same eigenvalues as in the previous example, so the origin is a saddle.

Next we compute the eigenvectors. For $\lambda_1 = -3$, the equations that give the eigenvectors (x_1, y_1) are

$$\begin{cases} 8x_1 - 11y_1 = -3x_1 \\ 6x_1 - 9y_1 = -3y_1. \end{cases}$$

So any nonzero vector that lies along the line $y = x$ in the plane serves as an eigenvector for $\lambda_1 = -3$. We choose $\mathbf{V}_1 = (1, 1)$. Therefore the solution

$$\mathbf{Y}_1(t) = e^{-3t}\mathbf{V}_1$$

is a straight-line solution lying on the line $y = x$. It tends to the origin as t increases.

Similar computations yield eigenvectors corresponding to $\lambda_2 = 2$ lying along the line $6x - 11y = 0$, for example $\mathbf{V}_2 = (11, 6)$. We get a straight-line solution of the form

$$\mathbf{Y}_2(t) = e^{2t}\mathbf{V}_2$$

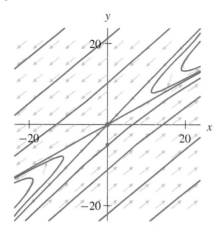

Figure 3.14
The direction field and phase portrait for the system

$$\frac{d\mathbf{Y}}{dt} = \mathbf{BY} = \begin{pmatrix} 8 & -11 \\ 6 & -9 \end{pmatrix} \mathbf{Y}.$$

The eigenvectors lie along the two distinguished lines that run through the first and third quadrants. Although some of the other solution curves look almost straight, they really curve slightly.

that tends away from the origin as t increases. Thus the general solution is

$$\mathbf{Y}(t) = k_1 e^{-3t}\mathbf{V}_1 + k_2 e^{2t}\mathbf{V}_2.$$

As above, we expect that if k_1 and k_2 are nonzero, these solutions come from infinity in the direction of $\mathbf{V}_1$ and tend back to infinity in the direction of $\mathbf{V}_2$. In the phase plane we see these straight-line solutions together with several other solutions (see Figure 3.14). The important point is that once we have the eigenvalues and eigenvectors, we can immediately visualize the entire phase portrait.

$x(t)$- and $y(t)$-graphs

Given an initial condition, we can graph the corresponding $x(t)$- and $y(t)$-graphs by solving the initial-value problem analytically and graphing the result. However, it is useful to realize that a great deal of information about these graphs can be determined directly from the the solution curve in the phase portrait.

For example, consider the initial-value problem

$$\frac{d\mathbf{Y}}{dt} = \mathbf{BY}, \quad \mathbf{Y}_0 = \begin{pmatrix} 0 \\ -5 \end{pmatrix}.$$

After calculating the general solution and doing the appropriate algebra, we see that $x(t) = 11e^{2t} - 11e^{-3t}$ and $y(t) = 6e^{2t} - 11e^{-3t}$, and we can plot their graphs using these formulas.

However, by simply considering the solution curve in the phase plane that corresponds to this initial condition, we can see that both $x(t)$ and $y(t) \to \infty$ at the rate of e^{2t} as $t \to \infty$ because the solution curve is asymptotic to the straight-line solutions with eigenvalue $\lambda_2 = 2$. Moreover, these solutions go to infinity in such a way that $y(t) \approx (6/11)x(t)$ for large t because the eigenvectors corresponding to λ_2 satisfy $y = (6/11)x$.

Similarly, as $t \to -\infty$, both $x(t)$ and $y(t)$ tend to $-\infty$ at a rate of $-e^{-3t}$ as $t \to -\infty$. In addition, $x(t) \approx y(t)$ as $t \to -\infty$ (see Figure 3.15).

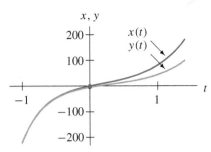

Figure 3.15
The $x(t)$- and $y(t)$-graphs for the solution to

$$\frac{d\mathbf{Y}}{dt} = \begin{pmatrix} 8 & -11 \\ 6 & 9 \end{pmatrix} \mathbf{Y}$$

with the initial condition $(x_0, y_0) = (0, -5)$. Many aspects of these graphs can be determined from the corresponding solution curve in the phase plane.

Sinks

Now consider the system of differential equations

$$\frac{d\mathbf{Y}}{dt} = \mathbf{CY}, \quad \text{where } \mathbf{C} = \begin{pmatrix} -1 & 0 \\ 0 & -4 \end{pmatrix}.$$

The matrix $\mathbf{C}$ has eigenvalues $\lambda_1 = -1$ and $\lambda_2 = -4$. Therefore we expect to have straight-line solutions that tend to the origin as $t \to \infty$.

An eigenvector corresponding to $\lambda_1 = -1$ is $\mathbf{V}_1 = (1, 0)$, and an eigenvector for $\lambda_2 = -4$ is $\mathbf{V}_2 = (0, 1)$. Thus the general solution is

$$\mathbf{Y}(t) = k_1 e^{-t} \mathbf{V}_1 + k_2 e^{-4t} \mathbf{V}_2 = k_1 e^{-t} \begin{pmatrix} 1 \\ 0 \end{pmatrix} + k_2 e^{-4t} \begin{pmatrix} 0 \\ 1 \end{pmatrix} = \begin{pmatrix} k_1 e^{-t} \\ k_2 e^{-4t} \end{pmatrix}.$$

Since each term involves either e^{-t} or e^{-4t}, we know that every solution of this system tends to the origin. In Figure 3.16 we sketch the phase portrait for this system. In this picture we clearly see the straight-line solutions. As predicted, all other solutions tend to the origin. In fact whenever we have a linear system with two negative eigenvalues, all solutions tend to the origin. By analogy with autonomous, first-order equations, we call this type of equilibrium point a **sink**.

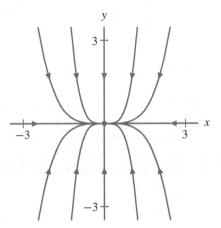

Figure 3.16
The phase portrait for the system

$$\frac{d\mathbf{Y}}{dt} = \mathbf{CY} = \begin{pmatrix} -1 & 0 \\ 0 & -4 \end{pmatrix} \mathbf{Y}.$$

Note that all solution curves tend to the equilibrium point at the origin.

In Figure 3.16 it appears that every solution (with the exception of those on the y-axis) tends to the origin tangent to the x-axis. To see why, consider the general solution

$$\begin{pmatrix} x(t) \\ y(t) \end{pmatrix} = \begin{pmatrix} k_1 e^{-t} \\ k_2 e^{-4t} \end{pmatrix}.$$

If $k_1 \neq 0$, then we can solve for e^{-t} in $x(t) = k_1 e^{-t}$, and we obtain

$$e^{-t} = \frac{x(t)}{k_1}.$$

We then substitute this expression for e^{-t} into the formula for $y(t)$, and we obtain

$$y(t) = k_2 e^{-4t}$$

$$= k_2 (e^{-t})^4$$

$$= k_2 \left(\frac{x(t)}{k_1} \right)^4$$

$$= \frac{k_2}{k_1^4} (x(t))^4.$$

In other words, each solution curve lies along a curve of the form

$$y = K x^4$$

for some constant K if $k_1 \neq 0$. Since these curves are always tangent to the x-axis, we see why all solution curves whose initial conditions are not on the y-axis approach the equilibrium point at the origin along curves that are tangent to the x-axis.

More general sinks

In general, for any linear system with two distinct, negative eigenvalues, we have a similar phase portrait. For example, consider the system of differential equations

$$\frac{d\mathbf{Y}}{dt} = \mathbf{DY}, \quad \text{where } \mathbf{D} = \begin{pmatrix} -2 & -2 \\ -1 & -3 \end{pmatrix}.$$

The matrix $\mathbf{D}$ has eigenvalues $\lambda_1 = -4$ and $\lambda_2 = -1$. For $\lambda_1 = -4$, one eigenvector is $\mathbf{V}_1 = (1, 1)$, and for $\lambda_2 = -1$, one eigenvector is $\mathbf{V}_2 = (-2, 1)$. (Checking this assertion is a good review of eigenvalues and eigenvectors. You should be able to check that these vectors are eigenvectors without recomputing them from scratch.)

Thus this phase portrait has two distinct lines of solutions that tend to the origin, and in fact the general solution is

$$\mathbf{Y}(t) = k_1 e^{-4t} \mathbf{V}_1 + k_2 e^{-t} \mathbf{V}_2$$

$$= k_1 e^{-4t} \begin{pmatrix} 1 \\ 1 \end{pmatrix} + k_2 e^{-t} \begin{pmatrix} -2 \\ 1 \end{pmatrix}$$

$$= \begin{pmatrix} k_1 e^{-4t} - 2k_2 e^{-t} \\ k_1 e^{-4t} + k_2 e^{-t} \end{pmatrix}.$$

Once we know that the eigenvalues for this system are -4 and -1, we know that every term in the general solution has a factor of either e^{-4t} or e^{-t}. Hence every solution tends to the origin as $t \rightarrow \infty$, and the origin is a sink. The long-term behavior of solutions can be determined from the eigenvalues alone (without the eigenvectors or the formula for the general solution).

In Figure 3.17 we sketch the phase portrait for this system. In this picture we clearly see the straight-line solutions. As predicted, all other solutions tend to the origin as well. Again, all solutions with the exception of the straight-line solutions associated with $\lambda_1 = -4$ seem to tend to the origin tangent to the line of eigenvectors for $\lambda_2 = -1$.

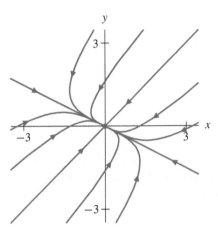

Figure 3.17
Phase portrait for the system

$$\frac{d\mathbf{Y}}{dt} = \mathbf{D}\mathbf{Y} = \begin{pmatrix} -2 & -2 \\ -1 & -3 \end{pmatrix} \mathbf{Y}.$$

All solutions tend to the equilibrium point at the origin, and all solutions with the exception of the straight-line solutions associated to $\lambda_1 = -4$ tend to the origin tangent to the line of eigenvectors for $\lambda_2 = -1$.

Direction of approach to the sink

To understand why solution curves approach the origin in the way that they do, we need to resort to some calculus. We compute the slope of the tangent line to any solution curve and then ask what happens to this slope as $t \rightarrow \infty$. Each solution curve is given by

$$\begin{pmatrix} x(t) \\ y(t) \end{pmatrix} = \begin{pmatrix} k_1 e^{-4t} - 2k_2 e^{-t} \\ k_1 e^{-4t} + k_2 e^{-t} \end{pmatrix}$$

for some choice of constants k_1 and k_2. From calculus we know that the slope of the tangent vector to a curve is given by dy/dx and

$$\frac{dy}{dx} = \frac{dy/dt}{dx/dt}.$$

Since $x(t) = k_1 e^{-4t} - 2k_2 e^{-t}$ and $y(t) = k_1 e^{-4t} + k_2 e^{-t}$, we have

$$\frac{dy/dt}{dx/dt} = \frac{-4k_1 e^{-4t} - k_2 e^{-t}}{-4k_1 e^{-4t} + 2k_2 e^{-t}}.$$

If we take the limit of this expression as $t \to \infty$, we end up with the indeterminate form $\frac{0}{0}$. It is tempting to use L'Hôpital's Rule, but this approach is destined to fail since the derivatives all involve exponential terms as well. The way to compute this limit is to multiply both numerator and denominator by e^t. Then the new expression is

$$\frac{dy/dt}{dx/dt} = \frac{-4k_1 e^{-3t} - k_2}{-4k_1 e^{-3t} + 2k_2}.$$

As $t \to \infty$, both exponential terms in this quotient tend to 0, and we see that the limit is $-k_2/(2k_2) = -1/2$ if $k_2 \neq 0$. That is, these solutions tend to the origin with slopes tending to $-1/2$ or, equivalently, tangent to the line of eigenvectors corresponding to the eigenvalue λ_2.

If $k_2 = 0$, our expression for dy/dx reduces to

$$\frac{dy/dt}{dx/dt} = \frac{-4k_1 e^{-4t}}{-4k_1 e^{-4t}} = 1,$$

which is exactly the slope for the straight-line solutions whose initial conditions lie along the line of eigenvectors associated to $\lambda_1 = -4$.

This discussion of the direction of approach to the equilibrium point may seem technical, but there really is a good qualitative reason that most solutions tend to $(0, 0)$ tangent to the eigenvector corresponding to the eigenvalue -1. Recall that the vector field on the line of eigenvectors corresponding to the eigenvalue λ is simply the scalar product of λ and the position vector. Because $-4 < -1$, the vector field on the line of eigenvectors for the eigenvalue -4 at a given distance from the origin is much longer than those on the line of eigenvectors for the eigenvalue -1. So solutions on the line of eigenvectors for -4 tend to zero much more quickly than those for the eigenvalue -1. In particular, the solution $e^{-4t}\mathbf{V}_1$ tends to $(0, 0)$ more quickly than $e^{-t}\mathbf{V}_2$.

In our general solution

$$\mathbf{Y}(t) = k_1 e^{-4t}\mathbf{V}_1 + k_2 e^{-t}\mathbf{V}_2,$$

if both k_1 and k_2 are nonzero, then the first term tends to the origin more quickly than the second. So when t is sufficiently large, the second term dominates, and we see that most solutions tend to zero along the direction of the eigenvectors for the eigenvalue closer to zero. The only exceptions are the solutions on the line of eigenvectors for the eigenvalue that is more negative. So, as in the previous example, provided that $k_2 \neq 0$, we can write $\mathbf{Y}(t) \approx k_2 e^{-t}\mathbf{V}_2$ as $t \to \infty$.

The case of an arbitrary sink with two eigenvalues $\lambda_1 < \lambda_2 < 0$ is entirely analogous. All solutions tend to the origin, and with the exception of those solutions with initial conditions that are eigenvectors corresponding to λ_1, all solutions tend to $(0, 0)$ tangent to the line of eigenvectors for λ_2.

$x(t)$- and $y(t)$-graphs

Once again we can identify important features of the $x(t)$- and $y(t)$-graphs of solutions directly from the corresponding solution curve in the phase portrait. For example, consider the initial-value problem

$$\frac{d\mathbf{Y}}{dt} = \mathbf{DY}, \quad \mathbf{Y}_0 = \begin{pmatrix} -3 \\ -1 \end{pmatrix}.$$

From the phase portrait, we know that the solution curve tends to the origin tangent to the line $y = (-1/2)x$, and in fact the curve must enter the second quadrant in order to do so. Consequently both $x(t)$ and $y(t)$ tend to zero at the rate of e^{-t}, and $y(t) \approx (-1/2)x(t)$ for large t. Furthermore although the $x(t)$-graph never crosses the t-axis, the $y(t)$-graph does become positive. It attains a (small) maximum value before tending to zero (see Figure 3.18).

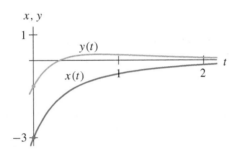

Figure 3.18
The $x(t)$- and $y(t)$-graphs for the solution to $d\mathbf{Y}/dt = \mathbf{DY}$ with the initial condition $(x_0, y_0) = (-3, -1)$. Note that $x(t)$ remains negative for $t \geq 0$ but that $y(t)$ increases and becomes positive before eventually tending to zero. The two functions satisfy $y(t) \approx (-1/2)x(t)$ for large t.

Sources

Consider the system

$$\frac{d\mathbf{Y}}{dt} = \mathbf{EY}, \quad \text{where } \mathbf{E} = \begin{pmatrix} 2 & 2 \\ 1 & 3 \end{pmatrix}.$$

In the previous section we computed that the eigenvalues of this matrix are $\lambda_1 = 4$ and $\lambda_2 = 1$. Also $\mathbf{V}_1 = (1, 1)$ is an eigenvector for the eigenvalue $\lambda_1 = 4$, and $\mathbf{V}_2 = (-2, 1)$ is an eigenvector for the eigenvalue 1. (Remember that you can check these assertions by computing $\mathbf{EV}_1$ and $\mathbf{EV}_2$.) Then $e^{4t}\mathbf{V}_1$ and $e^t\mathbf{V}_2$ are two linearly independent, straight-line solutions, and the general solution is

$$\mathbf{Y}(t) = k_1 e^{4t}\mathbf{V}_1 + k_2 e^t \mathbf{V}_2.$$

Since both eigenvalues of this system are positive, all nonzero solutions move away from the origin as $t \to \infty$.

The phase portrait for this system is shown in Figure 3.19. As in the previous example, we see two distinguished lines that correspond to the straight-line solutions, and all other solutions leave the origin in a direction tangent to the line of eigenvectors

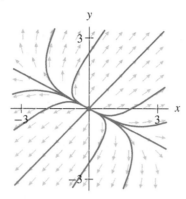

Figure 3.19
Phase portrait for the system

$$\frac{d\mathbf{Y}}{dt} = \mathbf{EY} = \begin{pmatrix} 2 & 2 \\ 1 & 3 \end{pmatrix} \mathbf{Y}.$$

Note that, since $\mathbf{E} = -\mathbf{D}$, we can obtain the phase portrait for this example from the phase portrait for $d\mathbf{Y}/dt = \mathbf{DY}$. The solution curves are identical, but solutions travel away from the origin as $t \to \infty$.

corresponding to the eigenvalue $\lambda_2 = 1$. The reason for this is essentially the same as the reason given for sinks earlier in the section. In fact, the astute reader will note that $\mathbf{E} = -\mathbf{D}$ where $\mathbf{D}$ is the matrix specified in the previous example. Consequently, for the vector field, we have changed merely the direction of the arrows, and not the geometry of the solution curves. For this system instead of considering the behavior as $t \to \infty$, we consider the behavior as $t \to -\infty$. Now the eigenvalue 4 plays the role of the stronger eigenvalue. Solutions involving terms with e^{4t} tend to the origin much more quickly than those involving e^{t} as $t \to -\infty$.

In general, once we know that both eigenvalues of a linear system are positive, we can conclude that all solutions tend away from the origin as t increases. We call the equilibrium point for a linear system with two positive eigenvalues a **source**. All other solutions tend away from the equilibrium point as $t \to \infty$, and all except those on the line of eigenvectors corresponding to λ_1 leave the origin in a direction tangent to the line of eigenvectors corresponding to λ_2.

Stable and Unstable Equilibrium Points

Before considering one more example, we summarize the behavior described earlier in this section.

Three types of equilibrium points
Consider a linear system with two nonzero, real, distinct eigenvalues λ_1 and λ_2.

- If $\lambda_1 < 0 < \lambda_2$, then the origin is a saddle. There are two lines in the phase portrait that correspond to straight-line solutions. Solutions along one line tend toward $(0, 0)$ as t increases, and solutions on the other line tend away from $(0, 0)$. All other solutions come from and go to infinity.
- If $\lambda_1 < \lambda_2 < 0$, then the origin is a sink. All solutions tend to $(0, 0)$ as $t \to \infty$, and most tend to $(0, 0)$ in the direction of the λ_2-eigenvectors.
- If $0 < \lambda_2 < \lambda_1$, then the origin is a source. All solutions except the equilibrium solution go to infinity as $t \to \infty$, and most solution curves leave the origin in the direction of the λ_2-eigenvectors.

A sink is said to be **stable** because nearby initial points yield solutions that tend back toward the equilibrium point as time increases. So if the initial condition is "bumped" a little bit away from the sink, the resulting solution does not stray far away from the initial point. Saddle and source equilibrium points are called **unstable** because there are initial conditions arbitrarily close to the equilibrium point whose solutions move away. Hence a small bump to an initial condition can have dramatic consequences. For a source, every initial condition near the equilibrium point corresponds to a solution that moves away. For a saddle, every initial condition except those that are eigenvectors for the negative eigenvalue (so almost every initial condition) corresponds to a solution that moves away.

If we run time backward, then a source looks like a sink with solutions tending toward it. Similarly, in backward time a sink looks like a source with solutions moving away from it. This is analogous to the situation for phase lines.

The saddle is a new type of equilibrium point—one that cannot occur in one-dimensional systems. Saddles need two dimensions in order to have one direction that is stable (corresponding to the negative eigenvalue) and another that is unstable (corresponding to the positive eigenvalue).

Paul's High Test Coffee versus Bob's Gourmet Tea

Recall the model of the interaction of Paul's and Bob's cafés from Section 3.1. Market research shows that, if a café gets too crowded, then profits tend to decrease—busy students have a limit to how long they can wait in line. Also, if the area near a café is crowded, then all nearby cafés also suffer and their profits decrease. In other words, if Paul's profit become positive, the profits of both his café and of Bob's café tend to decrease, so the parameters a and c should be negative. In the unlikely event that Bob's café even became popular, then his profitability would have the same effect on the profits of both cafés.

For example, let $a = -2$, $b = -3$, $c = -3$, and $d = -2$. We have

$$\frac{d\mathbf{Y}}{dt} = \begin{pmatrix} -2 & -3 \\ -3 & -2 \end{pmatrix} \mathbf{Y}.$$

All the coefficients are negative, so we might be tempted to say that this model predicts that profit for either café is impossible because, whenever one café starts to make money, it makes the rate of change of the profits of both cafés smaller. However, we cannot always trust guesses. We use the tools that we have developed to study this system carefully.

To give an accurate sketch of the phase portrait, we first compute the eigenvalues and eigenvectors. The characteristic polynomial is

$$(-2 - \lambda)(-2 - \lambda) - 9 = \lambda^2 + 4\lambda - 5 = (\lambda - 1)(\lambda + 5),$$

and the eigenvalues are $\lambda_1 = -5$ and $\lambda_2 = 1$. Because one eigenvalue is positive and one is negative, the origin is a saddle (see Figure 3.20). We find an eigenvector for the

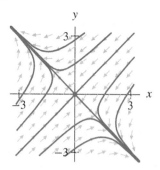

Figure 3.20
Phase portrait for the system

$$\frac{d\mathbf{Y}}{dt} = \begin{pmatrix} -2 & -3 \\ -3 & -2 \end{pmatrix} \mathbf{Y}.$$

The equilibrium point at the origin is a saddle, and most solutions tend to infinity asymptotic to the straight-line solutions whose solution curves lie in the second or fourth quadrants.

eigenvalue $\lambda_1 = -5$ by solving

$$\begin{cases} -2x_1 - 3y_1 = -5x_1 \\ -3x_1 - 2y_1 = -5y_1, \end{cases}$$

and these equations have a line of solutions given by $x_1 = y_1$. So $(1, 1)$ is an eigenvector for the eigenvalue $\lambda_1 = -5$.

For the other eigenvalue $\lambda_2 = 1$, we must solve

$$\begin{cases} -2x_2 - 3y_2 = x_2 \\ -3x_2 - 2y_2 = y_2. \end{cases}$$

These equations have a line of solutions given by $x_2 = -y_2$. So $(-1, 1)$ is an eigenvector for the eigenvalue $\lambda_2 = 1$. We could now use this information to write down the general solution, but it is more useful to use it to sketch the phase portrait. We know that the diagonal $x_1 = y_1$ through the origin contains straight-line solutions and that these solutions tend toward the origin because the eigenvalue $\lambda_1 = -5$ is negative. The other diagonal line through the origin, $x_2 = -y_2$, contains straight-line solutions that move away from $(0, 0)$ as t increases. Every other solution is a linear combination of these two. So the only solutions that tend to $(0, 0)$ are those on the line $x = y$. As $t \to \infty$, all other solutions eventually move away from the origin in either the second or fourth quadrants (see Figure 3.20).

Analysis of the model

This model leads to some startling predictions for Paul's and Bob's profits. Suppose that at $t = 0$ both Paul and Bob are making a profit [$x(0) > 0$ and $y(0) > 0$]. If it happens that Paul and Bob are making *exactly* the same amount, then $x(0) = y(0)$ and the initial point is on the $x = y$ line. The solution with this initial condition tends to the origin as t increases; that is, Paul and Bob both make less and less profit as time increases, both of them tending toward the break-even point $(x, y) = (0, 0)$.

Next consider the case $x(0) > y(0)$ (even by just a tiny amount). Now the initial point is just below the diagonal $x = y$. The corresponding solution at first tends toward $(0, 0)$, but it eventually turns and follows the straight-line solution along the line $x = -y$ into the fourth quadrant. In this case $x(t) \to \infty$ but $y(t) \to -\infty$. In other words, Paul eventually makes a fortune, but Bob loses his shirt. The vector field is very small near $(0, 0)$, so the solution moves slowly when it is near the origin. But eventually it turns the corner and Paul gets rich and Bob loses out (see Figure 3.20).

On the other hand, suppose $y(0)$ is slightly larger than $x(0)$. Then the initial point is just above the line $x = y$. In this case the solution first tends toward $(0, 0)$, but it eventually "turns the corner" and tends toward infinity along the line $x = -y$ in the second quadrant. In this case $x(t) \to -\infty$ (Paul goes broke) and $y(t) \to \infty$ (Bob gets rich; see Figure 3.20).

In this example a tiny change in the initial condition causes a large change in the long-term behavior of the system. We emphasize that the difference in behavior of solutions takes a long time to appear because solutions move very slowly near the equilibrium point. This sensitive dependence on the choice of initial condition is caused by the straight-line solution through the origin. The solutions with $x(0) = y(0) + 0.01$ and $x(0) = y(0) + 0.02$ are both on the same side of the diagonal, so they both behave the same way in the long run. It is only when a small change pushes the initial condition to the other side of the straight-line solution along the diagonal that the big jump in the long-term behavior occurs (see Figure 3.20). For this reason, a straight-line solution of a saddle corresponding to the negative eigenvalue is sometimes called a **separatrix**, because it separates two different types of long-term behavior.

Common Sense versus Computation

The predictions of this model are not at all what we might have expected. The coefficient matrix

$$\begin{pmatrix} -2 & -3 \\ -3 & -2 \end{pmatrix}$$

consists of only negative numbers, so any increase in profits of either café has a negative effect on the rate of change of the profits. Common sense might suggest that neither café will ever show a profit. The behavior of the model is quite different.

One lesson to be learned from this simple-minded example is that, although it is always wise to compare the predictions of a model with common sense, common sense does not replace computation. Models are most valuable when they predict something unexpected.

EXERCISES FOR SECTION 3.3

In Exercises 1–8, we refer to linear systems from the exercises in Section 3.2. Sketch the phase portrait for the system specified.

1. The system in Exercise 1, Section 3.2

2. The system in Exercise 2, Section 3.2

3. The system in Exercise 3, Section 3.2

4. The system in Exercise 6, Section 3.2

5. The system in Exercise 7, Section 3.2

6. The system in Exercise 8, Section 3.2

7. The system in Exercise 9, Section 3.2

8. The system in Exercise 10, Section 3.2

In Exercises 9–12, we refer to initial-value problems from the exercises in Section 3.2. Sketch the solution curves in the phase plane and the $x(t)$- and $y(t)$-graphs for the solutions corresponding to the initial-value problems specified.

9. The initial-value problems in Exercise 11, Section 3.2

10. The initial-value problems in Exercise 12, Section 3.2

11. The initial-value problems in Exercise 13, Section 3.2

12. The initial-value problems in Exercise 14, Section 3.2

In Exercises 13–16, we refer to the second-order equations from the exercises in Section 3.2. Sketch the phase portrait for the second-order equations specified.

13. The second-order equation in Exercise 21, Section 3.2

14. The second-order equation in Exercise 22, Section 3.2

15. The second-order equation in Exercise 23, Section 3.2

16. The second-order equation in Exercise 24, Section 3.2

In Exercises 17–18, we consider the model of Paul's and Bob's cafés from Section 3.1. Suppose Paul and Bob are both operating at the break-even point $(x, y) = (0, 0)$. For the models given below, state what happens if one of the cafés starts to earn or lose just a little bit. That is, will the profits return to 0 for both cafés? If not, does it matter which café starts to earn money first?

17. $\begin{pmatrix} \dfrac{dx}{dt} \\ \dfrac{dy}{dt} \end{pmatrix} = \begin{pmatrix} 2 & 1 \\ 0 & -1 \end{pmatrix} \begin{pmatrix} x \\ y \end{pmatrix}$

18. $\begin{pmatrix} \dfrac{dx}{dt} \\ \dfrac{dy}{dt} \end{pmatrix} = \begin{pmatrix} -2 & -1 \\ -1 & -1 \end{pmatrix} \begin{pmatrix} x \\ y \end{pmatrix}$

19. The slope field for the system

$$\frac{dx}{dt} = -2x + \frac{1}{2}y$$

$$\frac{dy}{dt} = -y$$

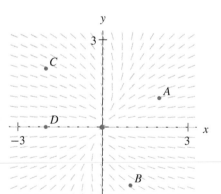

is shown to the right.

(a) Determine the type of the equilibrium point at the origin.

(b) Calculate all straight-line solutions.

(c) Plot the $x(t)$- and $y(t)$-graphs, $(t \geq 0)$, for the initial conditions $A = (2, 1)$, $B = (1, -2)$, $C = (-2, 2)$, and $D = (-2, 0)$.

20. The slope field for the system

$$\frac{dx}{dt} = 2x + 6y$$

$$\frac{dy}{dt} = 2x - 2y$$

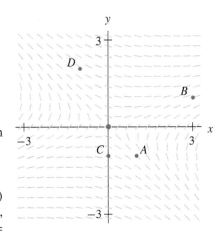

is shown to the right.

(a) Determine the type of the equilibrium point at the origin.

(b) Calculate all straight-line solutions.

(c) Plot the $x(t)$- and $y(t)$-graphs ($t \geq 0$) for the initial conditions $A = (1, -1)$, $B = (3, 1)$, $C = (0, -1)$, and $D = (-1, 2)$.

21. For the harmonic oscillator with mass $m = 1$, spring constant $k = 6$, and damping coefficient $b = 7$,

(a) write the second-order equation and the corresponding system,

(b) compute the characteristic polynomial,

(c) find the eigenvalues, and

(d) discuss the motion of the mass for the initial condition $(y(0), v(0)) = (2, 0)$. (How often does the mass cross the rest position $y = 0$? How quickly does the mass approach the equilibrium?)

22. Consider a harmonic oscillator with mass $m = 1$, spring constant $k = 1$, and damping coefficient $b = 4$. For the initial position $y(0) = 2$, find the initial velocity for which $y(t) > 0$ for all t and $y(t)$ reaches 0.1 most quickly. [*Hint*: It helps to look at the phase plane first.]

In Exercises 23–26, we consider a small pond inhabited by a species of fish. When left alone, the population of these fish settles into an equilibrium population. Suppose a few fish of another species are introduced to the pond. We would like to know if the new species survives and if the population of the native species changes much from its equilibrium population.

To determine the answers to these questions, we create a very simple model of the fish populations. Let $f(t)$ be the population of the native fish, and let f_0 denote the equilibrium population. We are interested in the change of the native fish population from its equilibrium level, so we let $x(t) = f(t) - f_0$; that is, $x(t)$ is the difference of the population of the native fish species from its equilibrium level. Let $y(t)$ denote the population of the introduced species. We note that, because $y(t)$ is an "absolute" population, it does not make sense to have $y(t) < 0$. So if $y(t)$ is ever equal to zero, we say that the introduced species has gone extinct. On the other hand, $x(t)$ can assume both positive and negative values because this variable measures the difference of the native fish population from its equilibrium level.

We are concerned with the behavior of these populations when both variables x and y are small, so the effects of terms involving x^2, y^2, xy, or higher powers are very, very small. Consequently we ignore them in this model (see Section 5.1). Also we know that if $x = y = 0$, then the native fish population is in equilibrium and none of the introduced species are there, so the population does not change; that is, $(x, y) = (0, 0)$ is an equilibrium point. Hence it is reasonable to use a linear system as a model.

For each model:

(a) Discuss briefly what sort of interaction between the species corresponds to the model; that is, whether the introduced fish work to increase or decrease the native fish population, etc.

(b) Decide if the model agrees with the information above about the system. That is, will the population of the native species return to equilibrium if the introduced species is not present?

(c) Sketch the phase plane and describe the solutions of the linear system (using technology and information about eigenvalues and eigenvectors).

(d) State what predictions the model makes about what happens when a small number of the new species is introduced into the lake.

23. $\dfrac{d\mathbf{Y}}{dt} = \begin{pmatrix} -0.2 & -0.1 \\ 0.0 & -0.1 \end{pmatrix} \mathbf{Y}$

24. $\dfrac{d\mathbf{Y}}{dt} = \begin{pmatrix} -0.1 & 0.2 \\ 0.0 & 1.0 \end{pmatrix} \mathbf{Y}$

25. $\dfrac{d\mathbf{Y}}{dt} = \begin{pmatrix} -0.2 & 0.1 \\ 0.0 & -0.1 \end{pmatrix} \mathbf{Y}$

26. $\dfrac{d\mathbf{Y}}{dt} = \begin{pmatrix} 0.1 & 0.0 \\ -0.2 & 0.2 \end{pmatrix} \mathbf{Y}$

27. Consider the linear system

$$\frac{d\mathbf{Y}}{dt} = \begin{pmatrix} -2 & 1 \\ 0 & 2 \end{pmatrix} \mathbf{Y}.$$

(a) Show that $(0, 0)$ is a saddle.

(b) Find the eigenvalues and eigenvectors and sketch the phase plane.

(c) On the phase plane, sketch the solution curves with initial conditions $(1, 0.01)$ and $(1, -0.01)$.

(d) Estimate the time t at which the solutions with initial conditions $(1, 0.01)$ and $(1, -0.01)$ will be 1 unit apart.

3.4 COMPLEX EIGENVALUES

The techniques of the previous sections were based on the geometric observation that, for some linear systems, certain solution curves lie on straight lines in the phase plane. This geometric observation led to the algebraic notions of eigenvalues and eigenvectors. These, in turn, gave us the formulas for the general solution.

Unfortunately these ideas do not work for all linear systems. Geometrically we hit a road block when we encounter linear systems whose direction fields do not show any straight-line solutions (see Figure 3.21). In this case it is the algebra of eigenvalues and eigenvectors that leads to an understanding of the system. Even though the method is different, the goals are the same: Starting with the entries in the coefficient matrix, understand the geometry of the phase portrait, the $x(t)$- and $y(t)$-graphs, and find the general solution.

Complex numbers

In this section and the rest of the book, we use complex numbers extensively. Complex numbers are numbers of the form $x + iy$, where x and y are real numbers and i is the "imaginary" number $\sqrt{-1}$. (There is a brief summary of the properties of complex numbers in Appendix C.)

One word of caution: All mathematicians and almost everyone else denote the imaginary number $\sqrt{-1}$ by the letter i. Electrical engineers use the letter i for the current (because "current" starts with "c"), so they use the letter j for $\sqrt{-1}$.

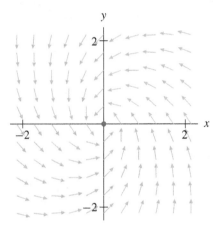

Figure 3.21
The direction field for

$$\frac{d\mathbf{Y}}{dt} = \begin{pmatrix} -2 & -3 \\ 3 & -2 \end{pmatrix} \mathbf{Y}.$$

Apparently there are no straight-line solutions.

A Linear System without Straight-Line Solutions

Consider the system

$$\frac{d\mathbf{Y}}{dt} = \mathbf{AY} = \begin{pmatrix} -2 & -3 \\ 3 & -2 \end{pmatrix} \mathbf{Y}.$$

From the direction field for this system (see Figure 3.21), we see that there are no solution curves that lie on straight lines. Instead, solutions spiral around the origin.

The characteristic polynomial of $\mathbf{A}$ is

$$\det(\mathbf{A} - \lambda \mathbf{I}) = (-2 - \lambda)(-2 - \lambda) + 9,$$

which simplifies to

$$\lambda^2 + 4\lambda + 13.$$

The eigenvalues are the roots of the characteristic polynomial, that is, the solutions of the equation

$$\lambda^2 + 4\lambda + 13 = 0.$$

Hence for this system the eigenvalues are the complex numbers $\lambda_1 = -2 + 3i$ and $\lambda_2 = -2 - 3i$. So how are we going to find solutions, and what information do complex eigenvalues give us?

General Solutions for Systems with Complex Eigenvalues

The most important thing to remember now is: Don't panic. Things are not nearly as complicated as they might seem. We cannot use the geometric ideas of Sections 3.2 and 3.3 to find solution curves that are straight lines because there aren't any straight-line solutions. However, the algebraic techniques we used in those sections work the same way for complex numbers as they do for real numbers. The rules of arithmetic for complex numbers are exactly the same as for real numbers, so all of the computations we did in the previous sections are still valid even if the eigenvalues are complex. Consequently our main observation about solutions of linear systems still holds even if the eigenvalues are complex. That is, given the linear system $d\mathbf{Y}/dt = \mathbf{A}\mathbf{Y}$, if λ is an eigenvalue for $\mathbf{A}$ and $\mathbf{Y}_0 = (x_0, y_0)$ is an eigenvector for the eigenvalue λ, then

$$\mathbf{Y}(t) = e^{\lambda t}\mathbf{Y}_0 = e^{\lambda t}\begin{pmatrix} x_0 \\ y_0 \end{pmatrix} = \begin{pmatrix} e^{\lambda t}x_0 \\ e^{\lambda t}y_0 \end{pmatrix}$$

is a solution. We can easily check this fact by differentiation, as we did before. We have

$$\frac{d\mathbf{Y}}{dt} = \frac{d}{dt}(e^{\lambda t}\mathbf{Y}_0) = \lambda e^{\lambda t}\mathbf{Y}_0 = e^{\lambda t}(\lambda\mathbf{Y}_0) = e^{\lambda t}\mathbf{A}\mathbf{Y}_0 = \mathbf{A}(e^{\lambda t}\mathbf{Y}_0) = \mathbf{A}\mathbf{Y}$$

because $\mathbf{Y}_0$ is an eigenvector with eigenvalue λ, so $\mathbf{Y}(t)$ is a solution. Of course, we need to make sense of the fact that the exponential is now a complex function and the fact that the eigenvector may contain complex entries, but this is no real problem. The important thing to notice here is that this computation is exactly the same if the numbers are real or complex.

Example revisited

For the system

$$\frac{d\mathbf{Y}}{dt} = \mathbf{A}\mathbf{Y}, \quad \text{where } \mathbf{A} = \begin{pmatrix} -2 & -3 \\ 3 & -2 \end{pmatrix},$$

we already know that the eigenvalues are $\lambda_1 = -2 + 3i$ and $\lambda_2 = -2 - 3i$. We now find the eigenvector for $\lambda_1 = -2 + 3i$ just as we would if λ_1 were real, by solving for $\mathbf{Y}_0$ in the system of equations

$$\mathbf{A}\mathbf{Y}_0 = \lambda_1\mathbf{Y}_0.$$

That is, we must find $\mathbf{Y}_0 = (x_0, y_0)$ such that

$$\begin{cases} -2x_0 - 3y_0 = (-2 + 3i)x_0 \\ 3x_0 - 2y_0 = (-2 + 3i)y_0, \end{cases}$$

which can be rewritten as

$$\begin{cases} -3ix_0 - 3y_0 = 0 \\ 3x_0 - 3iy_0 = 0. \end{cases}$$

Just as in the case of real eigenvalues, these equations are redundant. (Multiply both sides of the first equation by i to obtain the second equation and recall that $i^2 = -1$.) Thus the solutions of these equations are all pairs of complex numbers (x_0, y_0) that satisfy $-3ix_0 - 3y_0 = 0$, or $x_0 = iy_0$. If we let $y_0 = 1$, then $x_0 = i$. In other words, the vector $(i, 1)$ is an eigenvector for eigenvalue $\lambda_1 = -2 + 3i$. We can double-check this by computing

$$\mathbf{A}\begin{pmatrix} i \\ 1 \end{pmatrix} = \begin{pmatrix} -2 & -3 \\ 3 & -2 \end{pmatrix}\begin{pmatrix} i \\ 1 \end{pmatrix} = \begin{pmatrix} -2i - 3 \\ 3i - 2 \end{pmatrix} = (-2 + 3i)\begin{pmatrix} i \\ 1 \end{pmatrix}.$$

Thus we know that

$$\mathbf{Y}(t) = e^{(-2+3i)t}\begin{pmatrix} i \\ 1 \end{pmatrix} = \begin{pmatrix} ie^{(-2+3i)t} \\ e^{(-2+3i)t} \end{pmatrix}$$

is a solution of the system.

Obtaining Real-Valued Solutions from Complex Solutions

So this is both good news and bad news. The good news is that we can find solutions to linear systems with complex eigenvalues. The bad news is that these solutions involve complex numbers. If this system is a model of populations, profits, or the position of a mechanical device, then only real numbers make sense. It is hard to imagine "$5i$" predators or a position of "$2 + 3i$" units from the rest position. In other words, the physical meaning of complex numbers is not readily apparent. We have to find a way of producing real solutions from complex solutions.

The key to getting real solutions from complex solutions is **Euler's formula**

$$e^{a+ib} = e^a e^{ib} = e^a(\cos b + i \sin b) = e^a \cos b + ie^a \sin b$$

for all real numbers a and b. Using power series, one can verify that

$$e^{ib} = \cos b + i \sin b,$$

and Euler's formula follows from the laws of exponents (see Appendix C). Euler's formula is how we exponentiate with complex exponents.

Euler's formula is one of the most remarkable identities in all of mathematics. It lets us relate some of the most important functions and constants in a most intriguing way. For example, if $a = 0$ and $b = \pi$, then we obtain

$$e^{\pi i} = e^0 \cos \pi + i e^0 \sin \pi,$$

so

$$e^{\pi i} = -1.$$

That is, when we combine three of the most interesting numbers in mathematics, e, i and π, in the fashion $e^{i\pi}$, we obtain -1.

We use Euler's formula to define a complex-valued exponential function. We have

$$e^{(\alpha+i\beta)t} = e^{\alpha t} e^{i\beta t}$$

$$= e^{\alpha t} (\cos \beta t + i \sin \beta t)$$

$$= e^{\alpha t} \cos \beta t + i e^{\alpha t} \sin \beta t.$$

For the example above, this gives

$$e^{(-2+3i)t} = e^{-2t} \cos 3t + i e^{-2t} \sin 3t.$$

We can now rewrite the solution $\mathbf{Y}(t)$ as

$$\mathbf{Y}(t) = (e^{-2t} \cos 3t + i e^{-2t} \sin 3t) \begin{pmatrix} i \\ 1 \end{pmatrix}$$

$$= \begin{pmatrix} (e^{-2t} \cos 3t + i e^{-2t} \sin 3t)i \\ e^{-2t} \cos 3t + i e^{-2t} \sin 3t \end{pmatrix}$$

$$= \begin{pmatrix} i e^{-2t} \cos 3t - e^{-2t} \sin 3t \\ e^{-2t} \cos 3t + i e^{-2t} \sin 3t \end{pmatrix},$$

which in turn can be broken into

$$\mathbf{Y}(t) = \begin{pmatrix} -e^{-2t} \sin 3t \\ e^{-2t} \cos 3t \end{pmatrix} + i \begin{pmatrix} e^{-2t} \cos 3t \\ e^{-2t} \sin 3t \end{pmatrix}.$$

So far we have only rearranged the solution $\mathbf{Y}(t)$ to isolate the part that involves the number i. We now use the fact that we are dealing with a *linear* system to find the required real solutions.

THEOREM   Suppose $\mathbf{Y}(t)$ is a complex-valued solution to a linear system

$$\frac{d\mathbf{Y}}{dt} = \mathbf{A}\mathbf{Y} = \begin{pmatrix} a & b \\ c & d \end{pmatrix} \mathbf{Y},$$

where the coefficient matrix $\mathbf{A}$ has all real entries (a, b, c, and d are real numbers). Suppose

$$\mathbf{Y}(t) = \mathbf{Y}_{re}(t) + i\,\mathbf{Y}_{im}(t),$$

where $\mathbf{Y}_{re}(t)$ and $\mathbf{Y}_{im}(t)$ are real-valued functions of t. Then $\mathbf{Y}_{re}(t)$ and $\mathbf{Y}_{im}(t)$ are both solutions of the original system $d\mathbf{Y}/dt = \mathbf{A}\mathbf{Y}$.   ∎

It is important to note that there are no i's in the expression $\mathbf{Y}_{im}(t)$. They have been factored out.

To verify this theorem, we use the fact that $\mathbf{Y}(t)$ is a solution. In other words,

$$\frac{d\mathbf{Y}}{dt} = \mathbf{A}\mathbf{Y} \quad \text{for all } t.$$

Now we replace $\mathbf{Y}(t)$ with $\mathbf{Y}_{re}(t) + i\,\mathbf{Y}_{im}(t)$ on both sides of the equation. On the left-hand side, the usual rules of differentiation give

$$\frac{d\mathbf{Y}}{dt} = \frac{d(\mathbf{Y}_{re} + i\mathbf{Y}_{im})}{dt}$$

$$= \frac{d\mathbf{Y}_{re}}{dt} + i\frac{d\mathbf{Y}_{im}}{dt}.$$

On the right-hand side, we use the fact that this is a linear system to obtain that

$$\mathbf{A}\mathbf{Y}(t) = \mathbf{A}(\mathbf{Y}_{re}(t) + i\,\mathbf{Y}_{im}(t))$$

$$= \mathbf{A}\,\mathbf{Y}_{re}(t) + i\mathbf{A}\,\mathbf{Y}_{im}(t).$$

So we have

$$\frac{d\mathbf{Y}_{re}}{dt} + i\frac{d\mathbf{Y}_{im}}{dt} = \mathbf{A}\mathbf{Y}_{re} + i\mathbf{A}\mathbf{Y}_{im}$$

for all t. Two complex numbers are equal only if both their real parts and their imaginary parts are equal. Hence the only way the above equation can hold is if

$$\frac{d\mathbf{Y}_{re}}{dt} = \mathbf{A}\mathbf{Y}_{re} \quad \text{and} \quad \frac{d\mathbf{Y}_{im}}{dt} = \mathbf{A}\mathbf{Y}_{im},$$

and this is exactly what it means to say $\mathbf{Y}_{re}(t)$ and $\mathbf{Y}_{im}(t)$ are solutions of $d\mathbf{Y}/dt = \mathbf{A}\mathbf{Y}$.

Completion of the first example

Recall that, for the system

$$\frac{d\mathbf{Y}}{dt} = \begin{pmatrix} -2 & -3 \\ 3 & -2 \end{pmatrix} \mathbf{Y},$$

we have shown that the complex vector-valued function

$$\mathbf{Y}(t) = \begin{pmatrix} -e^{-2t} \sin 3t \\ e^{-2t} \cos 3t \end{pmatrix} + i \begin{pmatrix} e^{-2t} \cos 3t \\ e^{-2t} \sin 3t \end{pmatrix}$$

is a solution. By taking real and imaginary parts, we know that both the real part

$$\mathbf{Y}_{re}(t) = \begin{pmatrix} -e^{-2t}\sin 3t \\ e^{-2t}\cos 3t \end{pmatrix}$$

and the imaginary part

$$\mathbf{Y}_{im}(t) = \begin{pmatrix} e^{-2t}\cos 3t \\ e^{-2t}\sin 3t \end{pmatrix}$$

are solutions of the original system. They are independent since their initial values $\mathbf{Y}_{re}(0) = (0, 1)$ and $\mathbf{Y}_{im}(0) = (1, 0)$ are independent. So the general solution of this system is

$$\mathbf{Y}(t) = k_1 \begin{pmatrix} -e^{-2t}\sin 3t \\ e^{-2t}\cos 3t \end{pmatrix} + k_2 \begin{pmatrix} e^{-2t}\cos 3t \\ e^{-2t}\sin 3t \end{pmatrix}$$

for constants k_1 and k_2. This can be rewritten in the form

$$\mathbf{Y}(t) = e^{-2t} \begin{pmatrix} -k_1\sin 3t + k_2\cos 3t \\ k_1\cos 3t + k_2\sin 3t \end{pmatrix}.$$

A little gift

Note that in the above example we only needed to compute the eigenvector corresponding to one of the two complex eigenvalues. By breaking the resulting complex-valued solution into its real and imaginary parts, we obtained a pair of independent solutions. So, in this case, using complex arithmetic means we only have to do half as much work. (If a matrix with real coefficients has complex eigenvalues, then the eigenvalues are related in a nice way. There is also a nice relationship among the eigenvectors. See Exercises 17–20.)

Qualitative analysis

The direction field for the system

$$\frac{d\mathbf{Y}}{dt} = \mathbf{AY}, \quad \text{where } \mathbf{A} = \begin{pmatrix} -2 & -3 \\ 3 & -2 \end{pmatrix}$$

indicates that solution curves spiral toward the origin (see Figure 3.22). The corresponding $x(t)$- and $y(t)$-graphs of solutions must alternate between positive and negative values with decreasing amplitude as the solution curve in the phase plane winds around the origin. The pictures of the solution curve and the $x(t)$- and $y(t)$-graphs confirm this, at least to some extent (see Figures 3.22 and 3.23). From these graphs it appears that the solution winds only once around the origin before reaching $(0, 0)$. Actually this solution spirals infinitely often, though these oscillations are difficult to detect. In Figure 3.24 we magnify a small portion of Figure 3.22. Indeed, the solution does continue to spiral.

The formula for the general solution of this system provides us with detailed behavior of this spiral. The oscillation in $x(t)$ and $y(t)$ are caused by the sine and cosine

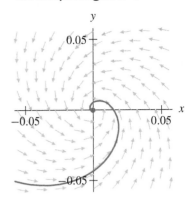

Figure 3.22
A solution curve in the phase plane for

$$\frac{d\mathbf{Y}}{dt} = \begin{pmatrix} -2 & -3 \\ 3 & -2 \end{pmatrix} \mathbf{Y}.$$

Figure 3.23
The $x(t)$- and $y(t)$-graphs of a solution to this differential equation.

Figure 3.24
A magnification of Figure 3.22.

terms. These trigonometric expressions are all of the form $\sin 3t$ and $\cos 3t$, so when t increases through $2\pi/3$, these terms return to their original values. Hence the period of the oscillation around the origin is always $2\pi/3$, no matter how large t is or how close the solution comes to the origin. Meanwhile, solutions must approach the origin because of the exponential term e^{-2t}. This term shows that the amplitude of the oscillations of the $x(t)$- and $y(t)$-graphs decreases at this very fast rate. This also explains why it is difficult to see these oscillations near the origin.

Happily, this kind of description of solutions can be accomplished without resorting to computing the general solution of the system. In fact it can be obtained from the eigenvalues alone.

The Qualitative Behavior of Systems with Complex Eigenvalues

The discussion above can be generalized to any linear system with complex eigenvalues. First, find a complex solution by finding the complex eigenvalues and eigenvectors. Then take the real and imaginary parts of this solution to obtain two independent solutions (see Exercise 19 for the verification that the real and imaginary parts of a solution are independent solutions). Finally, form the general solution in the usual way as a linear combination of the two independent particular solutions. This is sometimes a very tedious process, but it works. If the general solution is what we need, we can find it.

Just as in the case of real eigenvalues, we can tell a tremendous amount about the system from the complex eigenvalues without doing all the detailed computations to obtain the general solution. Suppose $d\mathbf{Y}/dt = \mathbf{AY}$ is a linear system with complex eigenvalues $\lambda_1 = \alpha + i\beta$ and $\lambda_2 = \alpha - i\beta$, $\beta \neq 0$. (Verifying that complex eigenvalues always come in pairs of this form is an interesting exercise—see Exercise 17). Then we know that the complex solutions have the form

$$\mathbf{Y}(t) = e^{(\alpha + i\beta)t}\mathbf{Y}_0,$$

where $\mathbf{Y}_0$ is a (complex) eigenvector of the matrix $\mathbf{A}$. We can rewrite this system as

$$\mathbf{Y}(t) = e^{\alpha t}(\cos\beta t + i\sin\beta t)\mathbf{Y}_0.$$

Because $\mathbf{Y}_0$ is a constant, the real and imaginary parts of the solution $\mathbf{Y}(t)$ are a combination of two types of terms—exponential and trigonometric. The effect of the exponential term on solutions depends on the sign of α. If $\alpha > 0$, then the $e^{\alpha t}$ term increases exponentially as $t \to \infty$, and the solution curve spirals off "toward infinity." If $\alpha < 0$, then the $e^{\alpha t}$ term tends to zero exponentially fast as t increases, so the solutions tend to the origin. If $\alpha = 0$, then the $e^{\alpha t}$ is identically 1 and the solutions oscillate with constant amplitude for all time. That is, the solutions are periodic.

The sine and cosine terms alternate from positive to negative and back again as t increases or decreases, so these terms make $x(t)$ and $y(t)$ oscillate. Hence the solutions in the xy-phase plane spiral around $(0, 0)$. The period of this oscillation is the amount of time it takes to go around once (say from one crossing of the positive x-axis to the next). The period is determined by β. The functions $\sin\beta t$ and $\cos\beta t$ satisfy the equations

$$\sin\beta(t + 2\pi/\beta) = \sin\beta t$$

$$\cos\beta(t + 2\pi/\beta) = \cos\beta t,$$

so increasing t by $2\pi/\beta$ returns $\sin\beta t$ and $\cos\beta t$ to their original values (see Figure 3.25).

We can summarize these observations with the classification on the next page.

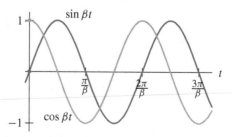

Figure 3.25
Graphs of $\cos\beta t$ and $\sin\beta t$. Note where the graphs cross the t-axis.

Linear systems with complex eigenvalues

Given a linear system with complex eigenvalues $\lambda = \alpha \pm i\beta$, $\beta > 0$, the solution curves spiral around the origin in the phase plane with a period of $2\pi/\beta$. Moreover:

- If $\alpha < 0$, then the solutions spiral toward the origin. In this case the origin is called a **spiral sink**.
- If $\alpha > 0$, then the solutions spiral away from the origin. In this case the origin is called a **spiral source**.
- If $\alpha = 0$, then the solutions are *periodic*. They return exactly to their initial conditions in the phase plane and repeat the same closed curve over and over. In this case the origin is called a **center**.

The question of which way the solutions spiral—clockwise or counterclockwise—can best be answered by looking at the direction field. Knowing just one nonzero vector of the direction field is enough information to determine the direction in which a linear system with complex eigenvalues spirals. For example, if the direction field at $(1, 0)$ points down into the fourth quadrant, then the solutions must spiral in the clockwise direction. If the direction field at $(1, 0)$ points up into the first quadrant, then the solutions must spiral in the counterclockwise direction.

As we have seen, the $\sin \beta t$ and $\cos \beta t$ terms in the solutions make the solutions oscillate with a period of $2\pi/\beta$. This quantity is called the **natural period** of the system. Every solution curve takes this amount of time to cycle once around the origin.

The **natural frequency** is the number of cycles that solutions make in one unit of time. Since each cycle takes one period to complete, the product of the natural frequency and the natural period of the system is 1, and consequently, frequency is the reciprocal of period. In other words, the natural frequency of $\sin \beta t$ and $\cos \beta t$ is $\beta/2\pi$.

Frequencies are often measured in the cycles per second: one cycle per second is one hertz (Hz). For example, Glen and Paul listen to the Red Sox baseball games on an AM radio station that broadcasts at a frequency of 850 kHz (850,000 cycles per second), and the computer processor in Bob's personal computer runs at a speed of 1 GHz (1 billion cycles per second).*

It is sometimes convenient to talk about the frequency of $\sin \beta t$ and $\cos \beta t$ in terms of radians rather than cycles. Since a complete cycle is 2π radians, the natural frequency of $\beta/2\pi$ cycles per unit of time is the same as a frequency of $(\beta/2\pi)(2\pi)$ radians per unit of time. So in terms of radians, the frequency of $\sin \beta t$ and $\cos \beta t$ is simply β. Frequency expressed in this form is often referred to as **angular frequency**.

A word of caution concerning terminology: In some subjects the terms *natural frequency* and *natural period* are reserved for linear systems with eigenvalues having zero real part (that is, only for centers, not spiral sinks or sources—that convention is

---

*See www.howstuffworks.com/radio-spectrum.htm for a discussion of the frequencies that are used by numerous wireless gadgets such as cordless phones, baby monitors, and Global Positioning System receivers.

reasonable because solutions of the system are periodic only for centers). We use the terms *natural period* and *natural frequency* for any linear system with complex eigenvalues.

A Spiral Source

Consider the initial-value problem

$$\frac{d\mathbf{Y}}{dt} = \mathbf{BY}, \quad \mathbf{Y}(0) = \begin{pmatrix} 1 \\ 1 \end{pmatrix}, \quad \text{where } \mathbf{B} = \begin{pmatrix} 0 & 2 \\ -3 & 2 \end{pmatrix}.$$

The eigenvalues are the roots of the characteristic polynomial

$$\det(\mathbf{B} - \lambda\mathbf{I}) = (0 - \lambda)(2 - \lambda) + 6 = \lambda^2 - 2\lambda + 6,$$

so the eigenvalues are $\lambda = 1 \pm i\sqrt{5}$. Since the real parts of the eigenvalues are positive, the origin is a spiral source, and the natural period of the system is $2\pi/\sqrt{5}$. Thus the solution of the initial-value problem oscillates with increasing amplitude and a period of $2\pi/\sqrt{5}$. The direction field (see Figure 3.26) shows that solutions spiral in the clockwise direction around the origin in the xy-phase plane.

To find the formula for the solution of the initial-value problem, we must find an eigenvector (x_0, y_0) for one of the eigenvalues, say $1 + i\sqrt{5}$. In other words, we solve

$$\mathbf{B} \begin{pmatrix} x_0 \\ y_0 \end{pmatrix} = (1 + i\sqrt{5}) \begin{pmatrix} x_0 \\ y_0 \end{pmatrix},$$

which is equivalent to

$$\begin{cases} 2y_0 = (1 + i\sqrt{5})x_0 \\ -3x_0 + 2y_0 = (1 + i\sqrt{5})y_0. \end{cases}$$

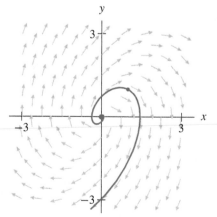

Figure 3.26
Direction field and the solution of the initial-value problem

$$\frac{d\mathbf{Y}}{dt} = \begin{pmatrix} 0 & 2 \\ -3 & 2 \end{pmatrix}\mathbf{Y} \quad \text{and} \quad \mathbf{Y}(0) = \begin{pmatrix} 1 \\ 1 \end{pmatrix}.$$

The solution spirals away from the origin.

Just as in the case of real eigenvalues, these two equations are redundant. (The second equation can be turned into the first by subtracting $2y_0$ from both sides, then multiplying both sides by $-(1 + i\sqrt{5})/3$.) We only need one eigenvector, so we choose any convenient value for x_0 and solve for y_0. If we set $x_0 = 2$, then we must have $y_0 = 1 + i\sqrt{5}$. Hence for the eigenvalue $\lambda = 1 + i\sqrt{5}$, the vector $(2, 1 + i\sqrt{5})$ is an eigenvector.

The corresponding complex solution is

$$\mathbf{Y}_1(t) = e^{(1+i\sqrt{5})t}\begin{pmatrix} 2 \\ 1 + i\sqrt{5} \end{pmatrix}.$$

Rewriting this using Euler's formula we obtain

$$\mathbf{Y}_1(t) = e^t\begin{pmatrix} 2\cos\sqrt{5}\,t \\ \cos\sqrt{5}\,t - \sqrt{5}\sin\sqrt{5}\,t \end{pmatrix} + ie^t\begin{pmatrix} 2\sin\sqrt{5}\,t \\ \sqrt{5}\cos\sqrt{5}\,t + \sin\sqrt{5}\,t \end{pmatrix}.$$

The general solution is

$$\mathbf{Y}(t) = k_1 e^t\begin{pmatrix} 2\cos\sqrt{5}\,t \\ \cos\sqrt{5}\,t - \sqrt{5}\sin\sqrt{5}\,t \end{pmatrix} + k_2 e^t\begin{pmatrix} 2\sin\sqrt{5}\,t \\ \sqrt{5}\cos\sqrt{5}\,t + \sin\sqrt{5}\,t \end{pmatrix},$$

where k_1 and k_2 are arbitrary constants. To solve the initial-value problem, we solve for k_1 and k_2 by setting the general solution at $t = 0$ equal to the initial condition $(1, 1)$, obtaining

$$k_1\begin{pmatrix} 2 \\ 1 \end{pmatrix} + k_2\begin{pmatrix} 0 \\ \sqrt{5} \end{pmatrix} = \begin{pmatrix} 1 \\ 1 \end{pmatrix}.$$

So $k_1 = 1/2$ and $k_2 = 1/(2\sqrt{5})$. The solution of the initial-value problem is

$$\mathbf{Y}(t) = \frac{1}{2}e^t\begin{pmatrix} 2\cos\sqrt{5}\,t \\ \cos\sqrt{5}\,t - \sqrt{5}\sin\sqrt{5}\,t \end{pmatrix} + \frac{1}{2\sqrt{5}}e^t\begin{pmatrix} 2\sin\sqrt{5}\,t \\ \sqrt{5}\cos\sqrt{5}\,t + \sin\sqrt{5}\,t \end{pmatrix}.$$

Centers

Consider an undamped harmonic oscillator with mass $m = 1$, spring constant $k = 2$, and no damping ($b = 0$). The second-order equation is

$$\frac{d^2y}{dt^2} = -2y,$$

and the corresponding linear system is

$$\frac{d\mathbf{Y}}{dt} = \mathbf{CY}, \quad \text{where } \mathbf{C} = \begin{pmatrix} 0 & 1 \\ -2 & 0 \end{pmatrix} \quad \text{and} \quad \mathbf{Y} = \begin{pmatrix} y \\ v \end{pmatrix}.$$

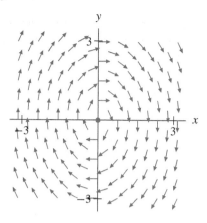

Figure 3.27
Direction field for the undamped harmonic oscillator system

$$\frac{d\mathbf{Y}}{dt} = \mathbf{C}\mathbf{Y}, \quad \text{where } \mathbf{C} = \begin{pmatrix} 0 & 1 \\ -2 & 0 \end{pmatrix}.$$

Although the direction field suggests that the eigenvalues of the system are complex, we cannot determine by looking at the direction field if the origin is a center, a spiral source, or a spiral sink.

The direction field for this system is given in Figure 3.27. We see from this picture that the solution curves encircle the origin. From this we predict that the eigenvalues for this system are complex. It is difficult to determine from the direction field picture if solution curves are periodic or if they spiral very slowly toward or away from the origin. Since these equations model a mechanical system for which we have assumed there is no damping, we might suspect that the solution curves are periodic. We can verify this by computing the eigenvalues for the system. As a bonus, the eigenvalues give us the period of the oscillations.

The eigenvalues for the matrix $\mathbf{C}$ are the roots of its characteristic polynomial

$$\det(\mathbf{C} - \lambda\mathbf{I}) = (0 - \lambda)(0 - \lambda) + 2 = \lambda^2 + 2,$$

which are $\lambda = \pm i\sqrt{2}$. Hence the origin is a center and all solutions are periodic. The imaginary part of the eigenvalue is $\sqrt{2}$, so the natural period of the system is $(2\pi)/\sqrt{2}$. This means that every solution completes one oscillation in $\sqrt{2}\,\pi$ units of time regardless of its initial condition.

In fact, all solution curves for this system lie on ellipses that encircle the origin. To see why, we compute the general solution of the system. Using methods of this section we first find that a complex eigenvector corresponding to the eigenvalue $i\sqrt{2}$ is $(1, i\sqrt{2})$, and we obtain the general solution

$$\mathbf{Y}(t) = k_1 \begin{pmatrix} \cos\sqrt{2}\,t \\ -\sqrt{2}\sin\sqrt{2}\,t \end{pmatrix} + k_2 \begin{pmatrix} \sin\sqrt{2}\,t \\ \sqrt{2}\cos\sqrt{2}\,t \end{pmatrix}.$$

Note that if $k_2 = 0$, we have

$$\begin{pmatrix} x(t) \\ y(t) \end{pmatrix} = \begin{pmatrix} k_1 \cos\sqrt{2}\,t \\ -k_1\sqrt{2}\sin\sqrt{2}\,t \end{pmatrix}.$$

Since we have

$$\frac{(x(t))^2}{k_1^2} + \frac{(y(t))^2}{2k_1^2} = 1,$$

this solution lies on an ellipse.

All solution curves for linear systems for which the origin is a center are ellipses (or circles). However, these ellipses need not have major and minor axes that lie along the x- and y-axes. For example, consider the linear system

$$\frac{d\mathbf{Y}}{dt} = \mathbf{D}\mathbf{Y}, \qquad \text{where } \mathbf{D} = \begin{pmatrix} -3 & 10 \\ -1 & 3 \end{pmatrix}.$$

The eigenvalues of this matrix are roots of $\lambda^2 + 1 = 0$, so $\lambda = \pm i$. The phase portrait consists entirely of ellipses, but they are not "centered" (see Figure 3.28 and Exercise 26).

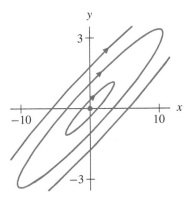

y
3
−10
10
x
−3

Figure 3.28
The phase portrait for the system

$$\frac{d\mathbf{Y}}{dt} = \begin{pmatrix} -3 & 10 \\ -1 & 3 \end{pmatrix} \mathbf{Y}.$$

All of the solution curves are ellipses, but their major and minor axes do not lie on the x- and y-axes.

Paul's and Bob's Cafés Revisited

Recall the linear model

$$\frac{d\mathbf{Y}}{dt} = \mathbf{A}\mathbf{Y}$$

for Paul's and Bob's cafés from Section 3.1. Suppose that the coefficient matrix is

$$\mathbf{A} = \begin{pmatrix} 2 & 1 \\ -4 & -1 \end{pmatrix}.$$

We would like to predict the behavior of solutions to this system with as little computation as possible.

First we compute the eigenvalues from the characteristic polynomial

$$\det(\mathbf{A} - \lambda\mathbf{I}) = (2 - \lambda)(-1 - \lambda) + 4 = \lambda^2 - \lambda + 2 = 0.$$

The roots are $(1 \pm i\sqrt{7})/2$, so we know that solutions spiral around the equilibrium point at $(0, 0)$. Because the real part of the eigenvalues is $1/2$, the origin is a spiral source. This information tells us that every solution (except the equilibrium point $(0, 0)$) spirals away from $(0, 0)$ in bigger and bigger loops as t increases. We can determine the direction (clockwise or counterclockwise) and approximate shape of the solution curves by sketching the phase portrait (see Figure 3.29).

The $x(t)$- and $y(t)$-graphs of solutions oscillate with increasing amplitude. The period of these oscillations is $2\pi/(\sqrt{7}/2) = 4\pi/\sqrt{7} \approx 4.71$, and the amplitude increases like $e^{t/2}$. We sketch the qualitative behavior of the $x(t)$- and $y(t)$-graphs in Figure 3.30.

Either Paul and Bob will stay precisely at the break-even point $(x, y) = (0, 0)$, or the profits and losses of their cafés will go up and down with increasing amplitude (a boom to bust to boom business cycle). Also the equilibrium point at the origin is unstable, so even a tiny profit or loss by either café eventually leads to large oscillations in the profits of both cafés. It would be very difficult to predict this behavior from just looking at the linear system without any computations.

Figure 3.29
Phase portrait for

$$\frac{d\mathbf{Y}}{dt} = \begin{pmatrix} 2 & 1 \\ -4 & -1 \end{pmatrix} \mathbf{Y}.$$

Figure 3.30
$x(t)$- and $y(t)$- graphs of a solution for the system

$$\frac{d\mathbf{Y}}{dt} = \begin{pmatrix} 2 & 1 \\ -4 & -1 \end{pmatrix} \mathbf{Y}.$$

EXERCISES FOR SECTION 3.4

1. Suppose that the 2×2 matrix $\mathbf{A}$ has $\lambda = 1 + 3i$ as an eigenvalue with eigenvector

$$\mathbf{Y}_0 = \begin{pmatrix} 2 + i \\ 1 \end{pmatrix}.$$

Compute the general solution to $d\mathbf{Y}/dt = \mathbf{AY}$.

2. Suppose that the 2×2 matrix $\mathbf{B}$ has $\lambda = -2 + 5i$ as an eigenvalue with eigenvector

$$\mathbf{Y}_0 = \begin{pmatrix} 1 \\ 4 - 3i \end{pmatrix}.$$

Compute the general solution to $d\mathbf{Y}/dt = \mathbf{BY}$.

In Exercises 3–8, each linear system has complex eigenvalues. For each system,

(a) find the eigenvalues;

(b) determine if the origin is a spiral sink, a spiral source, or a center;

(c) determine the natural period and natural frequency of the oscillations,

(d) determine the direction of the oscillations in the phase plane (do the solutions go clockwise or counterclockwise around the origin?); and

(e) using HPGSystemSolver, sketch the xy-phase portrait and the $x(t)$- and $y(t)$-graphs for the solutions with the indicated initial conditions.

3. $\dfrac{d\mathbf{Y}}{dt} = \begin{pmatrix} 0 & 2 \\ -2 & 0 \end{pmatrix} \mathbf{Y}$, with initial condition $\mathbf{Y}_0 = (1, 0)$

4. $\dfrac{d\mathbf{Y}}{dt} = \begin{pmatrix} 2 & 2 \\ -4 & 6 \end{pmatrix} \mathbf{Y}$, with initial condition $\mathbf{Y}_0 = (1, 1)$.

5. $\dfrac{d\mathbf{Y}}{dt} = \begin{pmatrix} -3 & -5 \\ 3 & 1 \end{pmatrix} \mathbf{Y}$, with initial condition $\mathbf{Y}_0 = (4, 0)$

6. $\dfrac{d\mathbf{Y}}{dt} = \begin{pmatrix} 0 & 2 \\ -2 & -1 \end{pmatrix} \mathbf{Y}$, with initial condition $\mathbf{Y}_0 = (-1, 1)$

7. $\dfrac{d\mathbf{Y}}{dt} = \begin{pmatrix} 2 & -6 \\ 2 & 1 \end{pmatrix} \mathbf{Y}$, with initial condition $\mathbf{Y}_0 = (2, 1)$

8. $\dfrac{d\mathbf{Y}}{dt} = \begin{pmatrix} 1 & 4 \\ -3 & 2 \end{pmatrix} \mathbf{Y}$, with initial condition $\mathbf{Y}_0 = (1, -1)$

In Exercises 9–14, the linear systems are the same as in Exercises 3–8. For each system,

(a) find the general solution;

(b) find the particular solution with the given initial value; and

(c) sketch the $x(t)$- and $y(t)$-graphs of the particular solution. (Compare these sketches with the sketches you obtained in the corresponding problem from Exercises 3–8.)

9. $\dfrac{d\mathbf{Y}}{dt} = \begin{pmatrix} 0 & 2 \\ -2 & 0 \end{pmatrix} \mathbf{Y}$, with initial condition $\mathbf{Y}_0 = (1, 0)$

10. $\dfrac{d\mathbf{Y}}{dt} = \begin{pmatrix} 2 & 2 \\ -4 & 6 \end{pmatrix} \mathbf{Y}$, with initial condition $\mathbf{Y}_0 = (1, 1)$.

11. $\dfrac{d\mathbf{Y}}{dt} = \begin{pmatrix} -3 & -5 \\ 3 & 1 \end{pmatrix} \mathbf{Y}$, with initial condition $\mathbf{Y}_0 = (4, 0)$

12. $\dfrac{d\mathbf{Y}}{dt} = \begin{pmatrix} 0 & 2 \\ -2 & -1 \end{pmatrix} \mathbf{Y}$, with initial condition $\mathbf{Y}_0 = (-1, 1)$

13. $\dfrac{d\mathbf{Y}}{dt} = \begin{pmatrix} 2 & -6 \\ 2 & 1 \end{pmatrix} \mathbf{Y}$, with initial condition $\mathbf{Y}_0 = (2, 1)$

14. $\dfrac{d\mathbf{Y}}{dt} = \begin{pmatrix} 1 & 4 \\ -3 & 2 \end{pmatrix} \mathbf{Y}$, with initial condition $\mathbf{Y}_0 = (1, -1)$

15. The following six figures are graphs of functions $x(t)$.

    (a) Which of the graphs can be $x(t)$-graphs for a solution of a linear system with complex eigenvalues?

    (b) For each such graph, give the natural period of the system and classify the equilibrium point at the origin as a spiral sink, a spiral source, or a center.

    (c) For each graph that cannot be an $x(t)$-graph for a solution of a linear system with complex eigenvalues, explain why not.

(i)

(ii)

(iii)

(iv)

(v)

(vi)
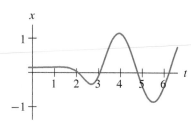

16. Show that a matrix of the form

$$\mathbf{A} = \begin{pmatrix} a & b \\ -b & a \end{pmatrix}$$

with $b \neq 0$ has complex eigenvalues.

17. Suppose that a and b are real numbers and that the polynomial $\lambda^2 + a\lambda + b$ has $\lambda_1 = \alpha + i\beta$ as a root with $\beta \neq 0$. Show that $\lambda_2 = \alpha - i\beta$, the complex conjugate of λ_1, must also be a root. [*Hint*: There are (at least) two ways to attack this problem. Either look at the form of the quadratic formula for the roots, or notice that

$$(\alpha + i\beta)^2 + a(\alpha + i\beta) + b = 0$$

and take the complex conjugate of both sides of this equation.]

18. Suppose that the matrix $\mathbf{A}$ with real entries has complex eigenvalues $\lambda = \alpha + i\beta$ and $\bar{\lambda} = \alpha - i\beta$ with $\beta \neq 0$. Show that the eigenvectors of $\mathbf{A}$ must be complex; that is, show that, if $\mathbf{Y}_0 = (x_0, y_0)$ is an eigenvector for $\mathbf{A}$, then either x_0 or y_0 or both have a nonzero imaginary part.

19. Suppose the matrix $\mathbf{A}$ with real entries has the complex eigenvalue $\lambda = \alpha + i\beta$, $\beta \neq 0$. Let $\mathbf{Y}_0$ be an eigenvector for λ and write $\mathbf{Y}_0 = \mathbf{Y}_1 + i\mathbf{Y}_2$, where $\mathbf{Y}_1 = (x_1, y_1)$ and $\mathbf{Y}_2 = (x_2, y_2)$ have real entries. Show that $\mathbf{Y}_1$ and $\mathbf{Y}_2$ are linearly independent. [*Hint*: Suppose they are not linearly independent. Then $(x_2, y_2) = k(x_1, y_1)$ for some constant k. Then $\mathbf{Y}_0 = (1 + ik)\mathbf{Y}_1$. Then use the fact that $\mathbf{Y}_0$ is an eigenvector of $\mathbf{A}$ and that $\mathbf{AY}_1$ contains no imaginary part.]

20. Suppose the matrix $\mathbf{A}$ with real entries has complex eigenvalues $\lambda = \alpha + i\beta$ and $\bar{\lambda} = \alpha - i\beta$. Suppose also that $\mathbf{Y}_0 = (x_1 + iy_1, x_2 + iy_2)$ is an eigenvector for the eigenvalue λ. Show that $\overline{\mathbf{Y}_0} = (x_1 - iy_1, x_2 - iy_2)$ is an eigenvector for the eigenvalue $\bar{\lambda}$. In other words, the complex conjugate of an eigenvector for λ is an eigenvector for $\bar{\lambda}$.

21. Consider the function $x(t) = e^{-\alpha t} \sin \beta t$, where α and β are positive.

    (a) What is the distance between successive zeros of this function? More precisely, if $t_1 < t_2$ are such that $x(t_1) = x(t_2) = 0$ and $x(t) \neq 0$ for $t_1 < t < t_2$, then what is $t_2 - t_1$?

    (b) What is the distance between the first local maximum and the first local minimum of $x(t)$ for $t > 0$?

    (c) What is the distance between the first two local maxima of $x(t)$ for $t > 0$?

    (d) What is the distance between $t = 0$ and the first local maximum of $x(t)$ for $t > 0$?

22. Show that a function of the form

$$x(t) = k_1 \cos \beta t + k_2 \sin \beta t$$

can be written as

$$x(t) = K \cos(\beta t - \phi),$$

where $K = \sqrt{k_1^2 + k_2^2}$. (Sometimes a solution of a linear system with complex co-efficients is expressed in this form in order to clarify its behavior. The magnitude K gives the *amplitude* of the solution, and the angle ϕ is the *phase* of the solution.) [*Hint*: Pick ϕ such that $K \cos \phi = k_1$ and $K \sin \phi = k_2$.]

23. For the second-order equation

$$\frac{d^2 y}{dt^2} + p\frac{dy}{dt} + qy = 0 :$$

   (a) Write this equation as a first-order linear system.

   (b) What conditions on p and q guarantee that the eigenvalues of the corresponding linear system are complex?

   (c) What relationship between p and q guarantees that the origin is a spiral sink? What relationship guarantees that the origin is a center? What relationship guarantees that the origin is a spiral source?

   (d) If the eigenvalues are complex, what conditions on p and q guarantee that solutions spiral around the origin in a clockwise direction?

24. The slope field for the system

$$\frac{dx}{dt} = -0.9x - 2y$$

$$\frac{dy}{dt} = x + 1.1y$$

is given to the right. Plot the $x(t)$- and $y(t)$-graphs for the initial conditions $A = (1, 1)$ and $B = (-2, 1)$. What do the graphs have in common?

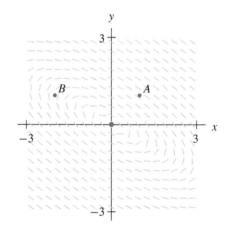

25. (Essay Question) We have seen that linear systems with real eigenvalues can be classified as sinks, sources, or saddles, depending on whether the eigenvalues are greater or less than zero. Linear systems with complex eigenvalues can be classified as spiral sources, spiral sinks, or centers, depending on the sign of the real part of the eigenvalue. Why is there not a type of linear system called a "spiral saddle"?

26. Consider the linear system

$$\frac{d\mathbf{Y}}{dt} = \begin{pmatrix} -3 & 10 \\ -1 & 3 \end{pmatrix} \mathbf{Y}.$$

Show that all solution curves in the phase portrait for this system are elliptical.

3.5 SPECIAL CASES: REPEATED AND ZERO EIGENVALUES

In the previous three sections we discussed the linear systems

$$\frac{d\mathbf{Y}}{dt} = \mathbf{AY}$$

for which the 2×2 matrix $\mathbf{A}$ has either two distinct, nonzero real eigenvalues or a pair of complex-conjugate eigenvalues. In these cases, we were able to use the eigenvalues and eigenvectors to sketch the solutions in the xy-phase plane, to draw the $x(t)$- and $y(t)$-graphs, and to derive an explicit formula for the general solution. We have not yet discussed the case where the characteristic polynomial of $\mathbf{A}$ has only one root (a double root), that is, where $\mathbf{A}$ has only one eigenvalue. In previous sections we also classified the equilibrium point at the origin as a sink, source, saddle, spiral sink, spiral source, or center, depending on the signs of the eigenvalues (or the sign of their real parts). This classification scheme omits the case where zero is an eigenvalue. In this section we modify our methods to handle these remaining cases.

Most quadratic polynomials have two distinct, nonzero roots, so linear systems with only one eigenvalue or with a zero eigenvalue are relatively rare. These systems are sometimes called *degenerate*. Nevertheless, they are still important. These special systems form the "boundaries" between the most common types of linear systems. Whenever we study linear systems that depend on a parameter and the system changes behavior or bifurcates as the parameter changes, these special systems play a crucial role (see Section 3.7).

A System with Repeated Eigenvalues

Consider the linear system

$$\frac{d\mathbf{Y}}{dt} = \mathbf{AY} = \begin{pmatrix} -2 & 1 \\ 0 & -2 \end{pmatrix} \mathbf{Y}.$$

The direction field for this system looks somewhat different from the vector fields we have considered thus far in that there appears to be only one straight line of solutions (note the x-axis in Figure 3.31).

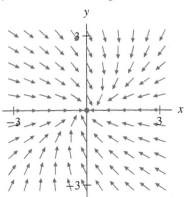

Figure 3.31
Direction field for the system

$$\frac{d\mathbf{Y}}{dt} = \begin{pmatrix} -2 & 1 \\ 0 & -2 \end{pmatrix} \mathbf{Y}.$$

The straight-line solutions are confined to the x-axis. There are no other straight-line solutions.

From an algebraic point of view, this matrix is also unusual. The eigenvalues of this system are the roots of the characteristic polynomial

$$\det(\mathbf{A} - \lambda\mathbf{I}) = (-2 - \lambda)(-2 - \lambda) - 0 = (\lambda + 2)^2,$$

whose only root is $\lambda = -2$. We say that $\lambda = -2$ is a **repeated eigenvalue** of $\mathbf{A}$. We find the associated eigenvectors by solving $\mathbf{A}\mathbf{Y}_0 = -2\mathbf{Y}_0$ for $\mathbf{Y}_0 = (x_0, y_0)$. We have

$$\begin{cases} -2x_0 + y_0 = -2x_0 \\ -2y_0 = -2y_0, \end{cases}$$

which yields $y_0 = 0$. Therefore all eigenvectors corresponding to the eigenvalue $\lambda = -2$ lie on the x-axis, so all straight-line solutions for this system lie on this axis. The vector $(1, 0)$ is an eigenvector associated to $\lambda = -2$, and consequently the function

$$\mathbf{Y}_1(t) = e^{-2t}\begin{pmatrix} 1 \\ 0 \end{pmatrix}$$

is a solution of this system. But this is only one solution and, as we know, we need two independent solutions to obtain the general solution. Bummer.

On the other hand, this is not a complete catastrophe. Our goal is to understand the behavior of the solutions of this system. Writing a formula for the general solution certainly helps, but this is not the only option. We can always study the system using numerical and qualitative techniques.

To obtain a qualitative description of the solutions, we start with the one straight line of solutions. Because the eigenvalue is negative, we know that solutions tend to the origin along this line as t increases. Looking at the direction field (or using Euler's method), we can sketch other solutions (see Figure 3.32). Every solution tends to the origin as t increases, so $(0, 0)$ is a sink. For initial conditions that do not lie on the x-axis, it appears that the corresponding solutions make a turn and arrive at the origin in a direction that is tangent to the x-axis. These solutions look as if they are trying to spiral around the origin, but the line of solutions somehow "gets in the way." In Section 3.7, we see that linear systems with repeated eigenvalues form the "boundary" between linear systems that spiral and those with two independent lines of solutions.

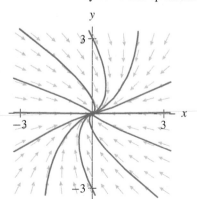

Figure 3.32
Phase portrait for the system

$$\frac{d\mathbf{Y}}{dt} = \begin{pmatrix} -2 & 1 \\ 0 & -2 \end{pmatrix}\mathbf{Y}.$$

Note that all solution curves approach the sink at the origin, although there is only one line of eigenvectors.

The general solution for this example

One way to compute the general solution of this system is consider the component equations

$$\frac{dx}{dt} = -2x + y$$
$$\frac{dy}{dt} = -2y.$$

We spot a lucky break. The variable x does not appear in the equation for dy/dt, and the system partially decouples (see Section 2.4). Considering the equation

$$\frac{dy}{dt} = -2y$$

by itself, we note that the general solution is $y(t) = y_0 e^{-2t}$, where the constant y_0 represents the initial value of $y(t)$. Now that we know $y(t)$, we can substitute it back into the first equation, which becomes

$$\frac{dx}{dt} = -2x + y_0 e^{-2t}.$$

We must be living a good life (or have chosen this example very carefully) because this equation is a first-order, nonhomogeneous, linear equation for $x(t)$. We can solve it using the Extended Linearity Principle and the guessing method discussed in Section 1.8. Its general solution is

$$x(t) = y_0 t e^{-2t} + x_0 e^{-2t},$$

where the constant x_0 is the initial value of $x(t)$. (The reader who does not remember how to compute the formulas for solutions of these equations should take this as a warning to review the analytic techniques developed in Section 1.8.) We can write the general solution in vector form as

$$\mathbf{Y}(t) = \begin{pmatrix} x(t) \\ y(t) \end{pmatrix} = \begin{pmatrix} y_0 t e^{-2t} + x_0 e^{-2t} \\ y_0 e^{-2t} \end{pmatrix}.$$

Now we study this solution closely to see if there is any pattern that can be generalized. If we separate the terms that involve $te^{\lambda t}$ from those that involve only $e^{\lambda t}$, we get

$$\mathbf{Y}(t) = e^{-2t} \begin{pmatrix} x_0 \\ y_0 \end{pmatrix} + t e^{-2t} \begin{pmatrix} y_0 \\ 0 \end{pmatrix}.$$

Notice that the first vector is the initial condition for the solution and that the second vector is an eigenvector unless $y_0 = 0$. If $y_0 = 0$, the second term vanishes, and we obtain a straight-line solution.

At first, the notation involved in the general solution

$$\mathbf{Y}(t) = e^{-2t} \begin{pmatrix} x_0 \\ y_0 \end{pmatrix} + t e^{-2t} \begin{pmatrix} y_0 \\ 0 \end{pmatrix}$$

may seem different than the notation we have used up to this point in this chapter. However, we can think of the arbitrary initial condition (x_0, y_0) as just another pair of parameters. Rather than specify the general solution as a linear combination of two linearly independent solutions, we write the general solution in terms of initial conditions. For the case of repeated eigenvalues, this is the most convenient way to express the general solution.

The Form of the General Solution

This form of the solution provides some hints on how to determine the general solution for a linear system with repeated eigenvalues but only one line of eigenvectors. We examine the situation algebraically. Consider a system of the form

$$\frac{d\mathbf{Y}}{dt} = \mathbf{A}\mathbf{Y},$$

where $\mathbf{A}$ has a repeated eigenvalue λ. Motivated by our example, we try for a solution of the form

$$\mathbf{Y}(t) = e^{\lambda t}\mathbf{V}_0 + te^{\lambda t}\mathbf{V}_1.$$

Note that $\mathbf{Y}(0) = \mathbf{V}_0$, the initial condition of $\mathbf{Y}(t)$.

Given this form for $\mathbf{Y}(t)$, we calculate both sides of the equation $d\mathbf{Y}/dt = \mathbf{A}\mathbf{Y}$. For the left-hand side, we differentiate $\mathbf{Y}(t)$ using the Product Rule. We have

$$\frac{d\mathbf{Y}}{dt} = \lambda e^{\lambda t}\mathbf{V}_0 + e^{\lambda t}\mathbf{V}_1 + t\lambda e^{\lambda t}\mathbf{V}_1$$

$$= e^{\lambda t}(\lambda\mathbf{V}_0 + \mathbf{V}_1) + te^{\lambda t}(\lambda\mathbf{V}_1).$$

For the right-hand side, we multiply $\mathbf{Y}(t)$ by the matrix $\mathbf{A}$ and obtain

$$\mathbf{A}\mathbf{Y} = e^{\lambda t}\mathbf{A}\mathbf{V}_0 + te^{\lambda t}\mathbf{A}\mathbf{V}_1.$$

So the differential equation $d\mathbf{Y}/dt = \mathbf{A}\mathbf{Y}$ becomes

$$e^{\lambda t}(\lambda\mathbf{V}_0 + \mathbf{V}_1) + te^{\lambda t}(\lambda\mathbf{V}_1) = e^{\lambda t}\mathbf{A}\mathbf{V}_0 + te^{\lambda t}\mathbf{A}\mathbf{V}_1.$$

In order for this equality to hold, the $e^{\lambda t}$ term on one side must equal the $e^{\lambda t}$ term on the other side, and the same statement holds for the $te^{\lambda t}$ terms as well (see Exercise 15). Equating the vectors that go with $te^{\lambda t}$ gives

$$\lambda\mathbf{V}_1 = \mathbf{A}\mathbf{V}_1,$$

and equating the vectors that go with $e^{\lambda t}$ yields

$$\lambda\mathbf{V}_0 + \mathbf{V}_1 = \mathbf{A}\mathbf{V}_0.$$

The first equation says that $\mathbf{V}_1$ is an eigenvector unless $\mathbf{V}_1$ is zero, and the second equation says that

$$\mathbf{V}_1 = \mathbf{A}\mathbf{V}_0 - \lambda\mathbf{V}_0 = (\mathbf{A} - \lambda\mathbf{I})\mathbf{V}_0.$$

In other words, the function

$$\mathbf{Y}(t) = e^{\lambda t}\mathbf{V}_0 + te^{\lambda t}\mathbf{V}_1$$

is a solution if and only if $\mathbf{V}_1 = (\mathbf{A} - \lambda\mathbf{I})\mathbf{V}_0$ and $\mathbf{V}_1$ is an eigenvector, unless it is zero.

It seems as if these two conditions must be verified separately, but it turns out that if $\mathbf{A}$ has repeated eigenvalues and $\mathbf{V}_1 = (\mathbf{A} - \lambda\mathbf{I})\mathbf{V}_0$, then $\mathbf{V}_1$ is either the zero vector or an eigenvector (see Exercise 16). If $\mathbf{V}_1$ is zero, then we immediately know that $\mathbf{V}_0$ is an eigenvector. We summarize this calculation in the following theorem.

THEOREM   Suppose $d\mathbf{Y}/dt = \mathbf{AY}$ is a linear system in which the 2×2 matrix $\mathbf{A}$ has a repeated real eigenvalue λ but only one line of eigenvectors. Then the general solution has the form

$$\mathbf{Y}(t) = e^{\lambda t}\mathbf{V}_0 + te^{\lambda t}\mathbf{V}_1,$$

where $\mathbf{V}_0 = (x_0, y_0)$ is an arbitrary initial condition and $\mathbf{V}_1$ is determined from $\mathbf{V}_0$ by

$$\mathbf{V}_1 = (\mathbf{A} - \lambda\mathbf{I})\mathbf{V}_0.$$

If $\mathbf{V}_1$ is zero, then $\mathbf{V}_0$ is an eigenvector and $\mathbf{Y}(t)$ is a straight-line solution. Otherwise, $\mathbf{V}_1$ is an eigenvector.   ■

Warning: Don't make the mistake of thinking that the two separate terms $e^{\lambda t}\mathbf{V}_0$ and $te^{\lambda t}\mathbf{V}_1$ are solutions. Remember that $\mathbf{V}_1$ is determined by $\mathbf{V}_0$. Also, $e^{\lambda t}\mathbf{V}_0$ by itself can only be a solution if $\mathbf{V}_0$ is an eigenvector.

Instant replay

In the example that we did "by hand" above, namely, the system

$$\frac{d\mathbf{Y}}{dt} = \mathbf{AY} = \begin{pmatrix} -2 & 1 \\ 0 & -2 \end{pmatrix}\mathbf{Y},$$

we computed that $\lambda = -2$ is an eigenvalue and that there is only one line of eigenvectors. Following the method outlined in the theorem, we let $\mathbf{V}_0$ be the arbitrary initial condition (x_0, y_0). Then

$$\mathbf{V}_1 = (\mathbf{A} + 2\mathbf{I})\mathbf{V}_0$$

$$= \begin{pmatrix} 0 & 1 \\ 0 & 0 \end{pmatrix}\begin{pmatrix} x_0 \\ y_0 \end{pmatrix}$$

$$= \begin{pmatrix} y_0 \\ 0 \end{pmatrix}.$$

Once again we obtain the general solution

$$\mathbf{Y}(t) = e^{-2t}\begin{pmatrix} x_0 \\ y_0 \end{pmatrix} + te^{-2t}\begin{pmatrix} y_0 \\ 0 \end{pmatrix}.$$

Qualitative Analysis of Systems with Repeated Eigenvalues

Before discussing additional examples, we consider what the general solution tells us about the qualitative behavior of solutions. The form of the general solution is

$$\mathbf{Y}(t) = e^{\lambda t}\mathbf{V}_0 + te^{\lambda t}\mathbf{V}_1,$$

where $\mathbf{V}_1 = (\mathbf{A} - \lambda I)\mathbf{V}_0$ is either an eigenvector or zero. The dependence on t comes in two terms, the $e^{\lambda t}$ term and the $te^{\lambda t}$ term.

If $\lambda < 0$, then both terms tend to zero as t increases, and therefore the equilibrium point at the origin is a sink. To determine the direction at which solutions approach the origin, we factor out the $e^{\lambda t}$ term and obtain

$$\mathbf{Y}(t) = e^{\lambda t}(\mathbf{V}_0 + t\mathbf{V}_1).$$

Since the directions of $\mathbf{Y}(t)$ and $\mathbf{V}_0 + t\mathbf{V}_1$ are the same, we see that the $t\mathbf{V}_1$ term dominates if t is large. Thus the solution tends to the origin in a direction that is tangent to the line of eigenvectors (see Figure 3.33 for typical examples of phase portraits for these systems).

If $\lambda > 0$, then all solutions (except the equilibrium solution) tend to infinity as t increases, so the origin is a source. Again the $te^{\lambda t}$ term dominates for large t if $\mathbf{V}_1$ is not zero.

If we use the formula for the general solution to draw the phase portrait, we see once more that it looks as if the solutions (other than the straight-line solutions) are trying to spiral around the origin. But they cannot spiral since the straight-line solutions get in the way. The usual form of the general solution of systems with one line of eigenvectors is another indication of the fact that these systems lie "between" systems with complex eigenvalues and those with real, yet distinct eigenvalues.

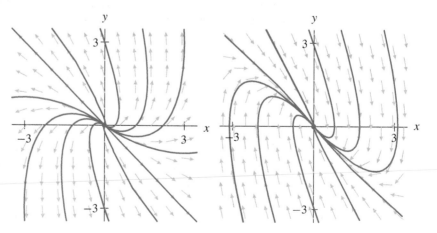

Figure 3.33
Typical phase portraits for systems with repeated eigenvalues.

A Harmonic Oscillator with Repeated Eigenvalues

Consider the harmonic oscillator with mass $m = 1$, spring constant $k = 2$, and damping coefficient $b = 2\sqrt{2}$. The second-order equation that models the motion of the oscillator is

$$\frac{d^2y}{dt^2} + 2\sqrt{2}\frac{dy}{dt} + 2y = 0,$$

and its associated system is

$$\frac{d\mathbf{Y}}{dt} = \mathbf{BY}, \quad \text{where } \mathbf{B} = \begin{pmatrix} 0 & 1 \\ -2 & -2\sqrt{2} \end{pmatrix}, \quad \mathbf{Y} = \begin{pmatrix} y \\ v \end{pmatrix},$$

and $v = dy/dt$. The eigenvalues of this system are the roots of the characteristic polynomial

$$\det(\mathbf{B} - \lambda\mathbf{I}) = (0 - \lambda)(-2\sqrt{2} - \lambda) + 2 = \lambda^2 + 2\sqrt{2}\lambda + 2.$$

Using the quadratic formula, we have

$$\lambda = \frac{-2\sqrt{2} \pm \sqrt{8 - 8}}{2},$$

and consequently $-\sqrt{2}$ is a repeated eigenvalue.

Given the eigenvalue $\lambda = -\sqrt{2}$, we determine the eigenvectors $\mathbf{V} = (y, v)$ by solving $\mathbf{BV} = \lambda\mathbf{V}$, which is

$$\begin{pmatrix} 0 & 1 \\ -2 & -2\sqrt{2} \end{pmatrix}\begin{pmatrix} y \\ v \end{pmatrix} = -\sqrt{2}\begin{pmatrix} y \\ v \end{pmatrix}.$$

So the eigenvectors lie on the line $v = -\sqrt{2}\,y$. For instance, one convenient eigenvector is $\mathbf{V} = (1, -\sqrt{2})$. The phase portrait for this system is shown in Figure 3.34. All solutions approach the origin as t increases, and all solutions are tangent to the line of eigenvectors as they approach the origin.

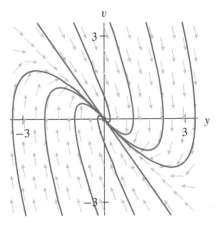

Figure 3.34
Direction field and solution curves for

$$\frac{d\mathbf{Y}}{dt} = \begin{pmatrix} 0 & 1 \\ -2 & -2\sqrt{2} \end{pmatrix}\mathbf{Y}.$$

Note that the solution curves approach the origin tangent to the line

$$v = -\sqrt{2}\,y$$

of eigenvectors.

To find the general solution to this system, we start with an arbitrary initial condition $\mathbf{V}_0 = (y_0, v_0)$. Then

$$\mathbf{V}_1 = (\mathbf{B} + \sqrt{2}\,\mathbf{I})\mathbf{V}_0$$

$$= \begin{pmatrix} \sqrt{2} & 1 \\ -2 & -\sqrt{2} \end{pmatrix} \begin{pmatrix} y_0 \\ v_0 \end{pmatrix}$$

$$= \begin{pmatrix} \sqrt{2}\,y_0 + v_0 \\ -2y_0 - \sqrt{2}\,v_0 \end{pmatrix}.$$

We obtain the general solution

$$\mathbf{Y}(t) = e^{-\sqrt{2}\,t} \begin{pmatrix} y_0 \\ v_0 \end{pmatrix} + t e^{-\sqrt{2}\,t} \begin{pmatrix} \sqrt{2}\,y_0 + v_0 \\ -2y_0 - \sqrt{2}\,v_0 \end{pmatrix}.$$

Paul's and Bob's Cafés One More Time

Recall the model of Paul's and Bob's cafés from Section 3.1. We suppose that the model has the form

$$\frac{d\mathbf{Y}}{dt} = \mathbf{A}\mathbf{Y}, \quad \text{where } \mathbf{A} = \begin{pmatrix} -5 & 1 \\ -1 & -3 \end{pmatrix}.$$

The coefficients imply that if Bob is making money, then Paul's profits will increase. But if Paul is making money, then Bob's profits are hurt. In this model, coffee drinkers don't like tea.

The eigenvalues for this system are the roots of the characteristic polynomial

$$\det(\mathbf{A} - \lambda\mathbf{I}) = \lambda^2 + 8\lambda + 16.$$

There is only one eigenvalue, $\lambda = -4$. Hence the origin is a sink and all solutions tend to $(0, 0)$ as t increases. The profits of both cafés tend toward zero no matter what the initial condition.

To find the eigenvectors, we solve

$$\begin{cases} -5x_1 + y_1 = -4x_1 \\ -x_1 - 3y_1 = -4y_1, \end{cases}$$

which has the line $y_1 = x_1$ as its set of solutions. From this information we conclude that as t increases, every solution tends to $(0, 0)$ tangent to the line $y = x$. As the values of x and y tend to zero, they are almost equal.

By looking at the direction field, we can see which initial conditions lead to solutions that tend to zero along $y = x$ in the first quadrant and which initial conditions

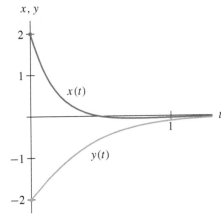

Figure 3.35
Phase portrait for the system

$$\frac{d\mathbf{Y}}{dt} = \begin{pmatrix} -5 & 1 \\ -1 & -3 \end{pmatrix} \mathbf{Y}.$$

Figure 3.36
The $x(t)$- and $y(t)$-graphs for the solution with the initial condition indicated in Figure 3.35.

end up in the third quadrant, along with the corresponding $x(t)$- and $y(t)$-graphs (see Figures 3.35 and 3.36).

Because the solutions tend to the line $y = x$, the two cafés have essentially the same profits or losses over the long term. This conclusion is a little surprising because the coefficients in the model imply that the profits of the two cafés react in very different ways to the profits of the other café.

Effect of a small change in the coefficients

We might wonder what would happen if the coefficients were changed just a little. Suppose Bob decides that he will help Paul (because Paul needs all the help he can get), and he puts up a sign:

Are you tired of the Tea Party?

Have some coffee at Paul's Café.

The more people who come to Bob's café (the larger y is), the more this sign helps Paul. So the parameter b increases from 1 to, say, 1.1. The new system is

$$\frac{d\mathbf{Y}}{dt} = \mathbf{BY} = \begin{pmatrix} -5 & 1.1 \\ -1 & -3 \end{pmatrix} \mathbf{Y}.$$

The eigenvalues of this matrix are complex and have negative real part. Hence the origin is a spiral sink, and all solutions still tend to $(0, 0)$. However, when the solutions are close to $(0, 0)$, they will oscillate instead of tending toward $(0, 0)$ along the line $y = x$. This small change in the system has made a distinct change in the qualitative behavior of the solutions. However, the oscillations are very subtle. In fact, the phase portrait and $x(t)$- and $y(t)$-graphs for $d\mathbf{Y}/dt = \mathbf{BY}$ are indistinguishable from those of $d\mathbf{Y}/dt = \mathbf{AY}$ (compare Figures 3.35 and 3.36 to Figures 3.37 and 3.38).

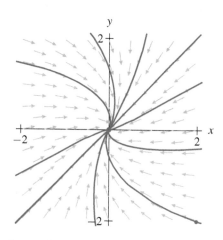

Figure 3.37
Phase portrait for

$$\frac{d\mathbf{Y}}{dt} = \begin{pmatrix} -5 & 1.1 \\ -1 & -3 \end{pmatrix} \mathbf{Y}.$$

Figure 3.38
The $x(t)$- and $y(t)$-graphs for the solution with the initial condition indicated in Figure 3.37.

Systems for Which Every Vector Is an Eigenvector

At this point we can discuss linear systems that have repeated real eigenvalues but only one line of eigenvectors. An example of a system with repeated eigenvalues and more than one line of eigenvectors is given by

$$\frac{d\mathbf{Y}}{dt} = \begin{pmatrix} a & 0 \\ 0 & a \end{pmatrix} \mathbf{Y}.$$

This system has the repeated eigenvalue $\lambda = a$, and *every* nonzero vector is an eigenvector for $\lambda = a$. In this case finding the general solution is equivalent to finding the general solutions of the two equations $dx/dt = ax$ and $dy/dt = ay$, that is, the system completely decouples (see Section 2.4). Because every vector is an eigenvector, every solution curve (except the equilibrium point at the origin) is a ray that approaches or leaves the origin as t increases. If $a > 0$, all the solutions tend to infinity as t increases (a source), whereas if $a < 0$, all solutions tend to zero as t increases (a sink). In the exercises we see that the only systems with one eigenvalue having more than one line of eigenvectors are those whose coefficient matrix is $\lambda\mathbf{I}$, where λ is the eigenvalue (see Exercises 13 and 14). This case is very special.

Systems with Zero as an Eigenvalue

We are close to having a complete understanding of linear systems and their phase portraits. We can classify and sketch the behavior of solutions for the cases of real, complex, and repeated eigenvalues. The only case we have not yet explicitly considered is the case where one or both of the eigenvalues is zero. This case is important because it divides the linear systems with strictly positive eigenvalues (sources) and strictly negative eigenvalues (sinks) from those with one positive and one negative eigenvalue (saddles).

Suppose we have a linear system $d\mathbf{Y}/dt = \mathbf{AY}$, and the matrix $\mathbf{A}$ has eigenvalues $\lambda_1 = 0$ and $\lambda_2 \neq 0$. Suppose $\mathbf{V}_1$ is an eigenvector for λ_1 and $\mathbf{V}_2$ is an eigenvector for λ_2. We have two real, distinct eigenvalues, and all of the algebra we did in Section 3.3 applies, and the general solution is

$$\mathbf{Y}(t) = k_1 e^{\lambda_1 t}\mathbf{V}_1 + k_2 e^{\lambda_2 t}\mathbf{V}_2.$$

But $\lambda_1 = 0$, so $e^{\lambda_1 t} = e^{0t} = 1$ for all t. We obtain

$$\mathbf{Y}(t) = k_1\mathbf{V}_1 + k_2 e^{\lambda_2 t}\mathbf{V}_2.$$

Note that the $k_1\mathbf{V}_1$ term is constant, and therefore all solutions with $k_2 = 0$ are equilibrium solutions. In other words, every point on the line of eigenvectors for the eigenvalue $\lambda_1 = 0$ is an equilibrium point. If $\lambda_2 < 0$, then the second term in the general solution tends to zero as t increases, so the solution

$$\mathbf{Y}(t) = k_1\mathbf{V}_1 + k_2 e^{\lambda_2 t}\mathbf{V}_2,$$

tends to the equilibrium point $k_1\mathbf{V}_1$ along a line parallel to $\mathbf{V}_2$. If $\lambda_2 > 0$, then the solution moves away from the line of equilibrium points as t increases. We have enough information to sketch the phase portraits.

An example with zero as an eigenvalue
Consider the system

$$\frac{d\mathbf{Y}}{dt} = \mathbf{AY}, \quad \text{where } \mathbf{A} = \begin{pmatrix} -3 & 1 \\ 3 & -1 \end{pmatrix}.$$

We compute the eigenvalues from the characteristic polynomial by solving

$$\det(\mathbf{A} - \lambda\mathbf{I}) = (-3 - \lambda)(-1 - \lambda) - 3 = 0,$$

which simplifies to $\lambda^2 + 4\lambda = 0$. The eigenvalues are $\lambda_1 = 0$ and $\lambda_2 = -4$.

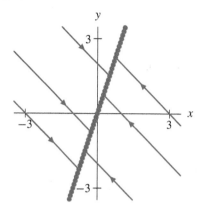

Figure 3.39
Phase portrait for the system

$$\frac{d\mathbf{Y}}{dt} = \begin{pmatrix} -3 & 1 \\ 3 & -1 \end{pmatrix} \mathbf{Y}.$$

Solutions tend toward the line of equilibrium points.

The eigenvectors $\mathbf{V}_1 = (x_1, y_1)$ for $\lambda_1 = 0$ satisfy the equations

$$\begin{cases} -3x_1 + y_1 = 0x_1 \\ 3x_1 - y_1 = 0y_1. \end{cases}$$

They are on the line $y_1 = 3x_1$. For instance, $\mathbf{V}_1 = (1, 3)$ is an eigenvector for $\lambda_1 = 0$. Similarly, the eigenvectors $\mathbf{V}_2 = (x_2, y_2)$ for $\lambda_2 = -4$ satisfy

$$\begin{cases} -3x_2 + y_2 = -4x_2 \\ 3x_2 - y_2 = -4y_2, \end{cases}$$

and these equations can be simplified to $x_2 + y_2 = 0$, so the solutions of these equations are on the line $x_2 = -y_2$. So $\mathbf{V}_2 = (-1, 1)$ is an eigenvector for $\lambda_2 = -4$.

From this information we can draw the phase portrait. There is a line of equilibrium points given by $y = 3x$, and every other solution approaches an equilibrium point on this line by following a line parallel to the line $y = -x$ (see Figure 3.39). The $x(t)$- and $y(t)$-graphs for the solution with initial condition $\mathbf{Y}(0) = (5, 2)$ and $\mathbf{Y}(0) = (1, 0)$ are given in Figure 3.40.

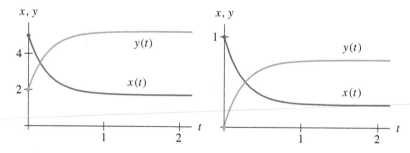

Figure 3.40
Graphs of $x(t)$ and $y(t)$ for solutions of $d\mathbf{Y}/dt = \mathbf{AY}$ with initial positions $(5, 2)$ (left graphs) and $(1, 0)$ (right graphs).

EXERCISES FOR SECTION 3.5

In Exercises 1–4, each of the linear systems has one eigenvalue and one line of eigenvectors. For each system,

(a) find the eigenvalue;

(b) find an eigenvector;

(c) sketch the direction field;

(d) sketch the phase portrait, including the solution curve with initial condition $Y_0 = (1, 0)$; and

(e) sketch the $x(t)$- and $y(t)$-graphs of the solution with initial condition $Y_0 = (1, 0)$.

1. $\dfrac{dY}{dt} = \begin{pmatrix} -3 & 0 \\ 1 & -3 \end{pmatrix} Y$
   2. $\dfrac{dY}{dt} = \begin{pmatrix} 2 & 1 \\ -1 & 4 \end{pmatrix} Y$

3. $\dfrac{dY}{dt} = \begin{pmatrix} -2 & -1 \\ 1 & -4 \end{pmatrix} Y$
   4. $\dfrac{dY}{dt} = \begin{pmatrix} 0 & 1 \\ -1 & -2 \end{pmatrix} Y$

In Exercises 5–8, the linear systems are the same as those in Exercises 1–4. For each system,

(a) find the general solution;

(b) find the particular solution for the initial condition $Y_0 = (1, 0)$; and

(c) sketch the $x(t)$- and $y(t)$-graphs of the solution. (Compare these sketches with the sketches you obtained in the corresponding problem from Exercises 1–4.)

5. $\dfrac{dY}{dt} = \begin{pmatrix} -3 & 0 \\ 1 & -3 \end{pmatrix} Y$
   6. $\dfrac{dY}{dt} = \begin{pmatrix} 2 & 1 \\ -1 & 4 \end{pmatrix} Y$

7. $\dfrac{dY}{dt} = \begin{pmatrix} -2 & -1 \\ 1 & -4 \end{pmatrix} Y$
   8. $\dfrac{dY}{dt} = \begin{pmatrix} 0 & 1 \\ -1 & -2 \end{pmatrix} Y$

9. Given a quadratic $\lambda^2 + \alpha\lambda + \beta$, what condition on α and β guarantees
   (a) that the quadratic has a double root?
   (b) that the quadratic has zero as a root?

10. Evaluate the limit of $te^{\lambda t}$ as $t \to \infty$ if
   (a) $\lambda > 0$
   (b) $\lambda < 0$

   Be sure to justify your answer.

11. Consider the matrix

$$\mathbf{A} = \begin{pmatrix} 0 & 1 \\ -q & -p \end{pmatrix},$$

where p and q are positive. What condition on q and p guarantees:

(a) that A has two real eigenvalues?

(b) that A has complex eigenvalues?

(c) that A has only one eigenvalue and one line of eigenvectors?

12. Let

$$\mathbf{A} = \begin{pmatrix} a & b \\ c & d \end{pmatrix}.$$

Define the trace of $\mathbf{A}$ to be $\text{tr}(\mathbf{A}) = a + d$. Show that $\mathbf{A}$ has only one eigenvalue if and only if $(\text{tr}(\mathbf{A}))^2 - 4\det(\mathbf{A}) = 0$.

13. Suppose

$$\mathbf{A} = \begin{pmatrix} a & b \\ c & d \end{pmatrix}$$

is a matrix with eigenvalue λ such that every nonzero vector is an eigenvector with eigenvalue λ, that is, $\mathbf{AY} = \lambda\mathbf{Y}$ for every vector $\mathbf{Y}$. Show that $a = d = \lambda$ and $b = c = 0$. [*Hint:* Since $\mathbf{AY} = \lambda\mathbf{Y}$ for every $\mathbf{Y}$, try $\mathbf{Y} = (1, 0)$ and $\mathbf{Y} = (0, 1)$.]

14. Suppose λ is an eigenvalue for the matrix

$$\mathbf{A} = \begin{pmatrix} a & b \\ c & d \end{pmatrix},$$

and suppose that there are two linearly independent eigenvectors $\mathbf{Y}_1$ and $\mathbf{Y}_2$ associated to λ. Show that every nonzero vector is an eigenvector with eigenvalue λ. What does this imply about a, b, c, and d?

15. Suppose the two functions

$$\mathbf{Y}_1(t) = e^{\lambda t}\mathbf{V}_0 + te^{\lambda t}\mathbf{V}_1 \quad \text{and} \quad \mathbf{Y}_2(t) = e^{\lambda t}\mathbf{W}_0 + te^{\lambda t}\mathbf{W}_1$$

are equal for all t. Show that $\mathbf{V}_0 = \mathbf{W}_0$ and $\mathbf{V}_1 = \mathbf{W}_1$.

16. Suppose λ_0 is a repeated eigenvalue for the 2×2 matrix $\mathbf{A}$.

(a) Show that $(\mathbf{A} - \lambda_0\mathbf{I})^2 = \mathbf{0}$ (the zero matrix).

(b) Given an arbitrary vector $\mathbf{V}_0$, let $\mathbf{V}_1 = (\mathbf{A} - \lambda_0\mathbf{I})\mathbf{V}_0$. Using the result of part (a), show that $\mathbf{V}_1$ is either an eigenvector of $\mathbf{A}$ or the zero vector.

In Exercises 17–19, each of the given linear systems has zero as an eigenvalue. For each system,

(a) find the eigenvalues;

(b) find the eigenvectors;

(c) sketch the phase portrait;

(d) sketch the $x(t)$- and $y(t)$-graphs of the solution with initial condition $Y_0 = (1, 0)$;

(e) find the general solution; and

(f) find the particular solution for the initial condition $Y_0 = (1, 0)$ and compare it with your sketch from part (d).

17. $\dfrac{d\mathbf{Y}}{dt} = \begin{pmatrix} 0 & 2 \\ 0 & -1 \end{pmatrix} \mathbf{Y}$ 18. $\dfrac{d\mathbf{Y}}{dt} = \begin{pmatrix} 2 & 4 \\ 3 & 6 \end{pmatrix} \mathbf{Y}$ 19. $\dfrac{d\mathbf{Y}}{dt} = \begin{pmatrix} 4 & 2 \\ 2 & 1 \end{pmatrix} \mathbf{Y}$

20. Let $\mathbf{A} = \begin{pmatrix} a & b \\ c & d \end{pmatrix}$.

(a) Show that if one or both of the eigenvalues of $\mathbf{A}$ is zero, then the determinant of $\mathbf{A}$ is zero.

(b) Show that if $\det \mathbf{A} = 0$, then at least one of the eigenvalues of $\mathbf{A}$ is zero.

21. Find the eigenvalues and sketch the phase portraits for the linear systems

(a) $\dfrac{d\mathbf{Y}}{dt} = \begin{pmatrix} 0 & 2 \\ 0 & 0 \end{pmatrix} \mathbf{Y}$ (b) $\dfrac{d\mathbf{Y}}{dt} = \begin{pmatrix} 0 & -2 \\ 0 & 0 \end{pmatrix} \mathbf{Y}$

22. Find the general solution for the linear systems

(a) $\dfrac{d\mathbf{Y}}{dt} = \begin{pmatrix} 0 & 2 \\ 0 & 0 \end{pmatrix} \mathbf{Y}$ (b) $\dfrac{d\mathbf{Y}}{dt} = \begin{pmatrix} 0 & -2 \\ 0 & 0 \end{pmatrix} \mathbf{Y}$

23. Consider the linear system

$$\frac{d\mathbf{Y}}{dt} = \begin{pmatrix} a & 0 \\ 0 & d \end{pmatrix} \mathbf{Y}.$$

(a) Find the eigenvalues.

(b) Find the eigenvectors.

(c) Suppose $a = d < 0$. Sketch the phase portrait and compute the general solution. (What are the eigenvectors in this case?)

(d) Suppose $a = d > 0$. Sketch the phase portrait and compute the general solution.

24. The slope field for the system

$$\frac{dx}{dt} = -3x - y$$

$$\frac{dy}{dt} = 4x + y$$

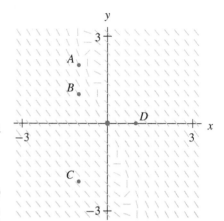

is shown at the right.

(a) Determine the type of the equilibrium point at the origin.

(b) Calculate all straight-line solutions.

(c) Plot the $x(t)$- and $y(t)$-graphs ($t \geq 0$) for the initial conditions $A = (-1, 2)$, $B = (-1, 1)$, $C = (-1, -2)$, and $D = (1, 0)$.

3.6 SECOND-ORDER LINEAR EQUATIONS

Throughout this chapter we have used the harmonic oscillator as an example. We have solved the second-order equation and its associated system of equations in a number of different cases. Now it is time to summarize all that we have learned about this important model.

Second-Order Equations versus First-Order Systems

As we know, the motion of a harmonic oscillator can be modeled by the second-order equation

$$m\frac{d^2y}{dt^2} + b\frac{dy}{dt} + ky = 0,$$

where $m > 0$ is the mass, $k > 0$ is the spring constant, and $b \geq 0$ is the damping coefficient. Since $m \neq 0$, we can also write this equation in the form

$$\frac{d^2y}{dt^2} + p\frac{dy}{dt} + qy = 0,$$

where $p = b/m$ and $q = k/m$ are nonnegative constants, and the corresponding linear system is

$$\frac{d\mathbf{Y}}{dt} = \begin{pmatrix} 0 & 1 \\ -q & -p \end{pmatrix} \mathbf{Y}.$$

As we will see in this section, any method to compute the general solution of the second-order equation also gives the general solution of the associated system, and vice versa. In particular we can use the Linearity Principle to produce new solutions from known ones by adding solutions and by multiplying solutions by constants. Therefore second-order equations of the form

$$a\frac{d^2y}{dt^2} + b\frac{dy}{dt} + cy = 0,$$

where a, b, and c are constants, are said to be **linear**. More precisely, these equations are **homogeneous**, **constant-coefficient**, linear, second-order equations. The constants a, b, and c are the coefficients, and the equation is homogeneous due to the fact that the right-hand side is 0. In Chapter 4 we will study the difference between homogeneous and nonhomogeneous second-order linear equations in detail.

We can find the general solution of the linear system that models the harmonic oscillator by finding the eigenvalues and eigenvectors of the coefficient matrix. The arithmetic is not always pleasant, but the steps are clear. In this section we give a shortcut for finding the general solution of the corresponding second-order equation, and we relate this shortcut to the geometry and qualitative behavior of the solutions of both the second-order equation and the system.

A Free Gift from the Math Department

The shortcut method for finding the general solution of a second-order equation such as

$$\frac{d^2y}{dt^2} + 7\frac{dy}{dt} + 10y = 0,$$

for example, is to guess it. Given what we now know about solutions of the corresponding system, this is not as silly as it sounds. We know that the solutions of the system are often made up of terms of the form $e^{\lambda t}\mathbf{V}$, where λ is an eigenvalue and $\mathbf{V}$ is an eigenvector. Hence if we are trying to guess the solution of the second-order equation, the most natural guess is

$$y(t) = e^{st},$$

where s is a constant to be determined. (From our point of view it makes more sense to use λ as the unknown constant. However, s is commonly used in applications, and for this discussion we follow that custom.) Substituting the guess into the left-hand side of the second-order equation gives

$$\frac{d^2y}{dt^2} + 7\frac{dy}{dt} + 10y = \frac{d^2(e^{st})}{dt^2} + 7\frac{d(e^{st})}{dt} + 10e^{st}$$

$$= s^2e^{st} + 7se^{st} + 10e^{st}$$

$$= (s^2 + 7s + 10)e^{st}.$$

Since e^{st} is never zero, we must have

$$s^2 + 7s + 10 = 0$$

in order for $y(t) = e^{st}$ to be a solution. This quadratic equation has roots $s = -5$ and $s = -2$, so we know that $y_1(t) = e^{-5t}$ and $y_2(t) = e^{-2t}$ are solutions of the differential equation. (If this guess-and-test technique seems familiar, it should. We have already used this procedure to find straight-line solutions to the damped harmonic oscillator in Section 2.3 (see page 185).

Applying the Linearity Principle, we see that any function of the form

$$y(t) = k_1 e^{-5t} + k_2 e^{-2t}$$

is also a solution for any choice of constants k_1 and k_2 (see Exercise 30 for a direct verification of this assertion). To see that this expression is in fact the general solution of the equation, we note that there is a one-to-one correspondence between solutions of

$$\frac{d^2y}{dt^2} + 7\frac{dy}{dt} + 10y = 0$$

and solutions of the associated system

$$\frac{dy}{dt} = v$$

$$\frac{dv}{dt} = -10y - 7v.$$

If we have a solution $\mathbf{Y}(t) = (y(t), v(t))$ to the system, then $y(t)$ is a solution to the second-order equation. If $y(t)$ is a solution to the equation, then we have

$$v(t) = -5k_1 e^{-5t} - 2k_2 e^{-2t},$$

where $v = dy/dt$. If we form the vector function

$$\mathbf{Y}(t) = \begin{pmatrix} y(t) \\ v(t) \end{pmatrix} = \begin{pmatrix} k_1 e^{-5t} + k_2 e^{-2t} \\ -5k_1 e^{-5t} - 2k_2 e^{-2t} \end{pmatrix},$$

we have a solution to the system that can be rewritten in the form

$$\mathbf{Y}(t) = k_1 e^{-5t} \begin{pmatrix} 1 \\ -5 \end{pmatrix} + k_2 e^{-2t} \begin{pmatrix} 1 \\ -2 \end{pmatrix}.$$

Solving the associated system

This form of the solution looks suspiciously familiar. If we write this second-order equation as a system in matrix notation, we obtain

$$\frac{d\mathbf{Y}}{dt} = \begin{pmatrix} 0 & 1 \\ -10 & -7 \end{pmatrix} \mathbf{Y},$$

which has $\lambda^2 + 7\lambda + 10$ as its characteristic polynomial. Note that this quadratic is exactly the same as the one we obtained earlier when we applied our guess-and-test technique to the second-order equation (with s replaced by λ). The eigenvalues for this system are $\lambda_1 = -5$ and $\lambda_2 = -2$. Computing the associated eigenvectors, we find that one eigenvector corresponding to λ_1 is $(1, -5)$ and one eigenvector corresponding to λ_2 is $(1, -2)$. Thus using the eigenvalue/eigenvector methods of this chapter, we obtain

$$\mathbf{Y}(t) = k_1 e^{-5t} \begin{pmatrix} 1 \\ -5 \end{pmatrix} + k_2 e^{-2t} \begin{pmatrix} 1 \\ -2 \end{pmatrix},$$

which is exactly the same general solution we obtained earlier.

If we had chosen different eigenvectors for the system, we would have obtained a slightly different form of the general solution. For example, $(-1, 5)$ is also an eigenvector corresponding to $\lambda_1 = -5$, and $(2, -4)$ is an eigenvector corresponding to $\lambda_1 = -2$. So our general solution may also be written as

$$\mathbf{Y}(t) = k_1 e^{-5t} \begin{pmatrix} -1 \\ 5 \end{pmatrix} + k_2 e^{-2t} \begin{pmatrix} 2 \\ -4 \end{pmatrix}.$$

But these solutions are precisely the same as those already obtained. (Replace k_1 with $-k_1$ and k_2 with $k_2/2$.) There really is no difference between this guessing method and the eigenvalue/eigenvector method.

At this point, you may be wondering when you should use the vector form of the general solution and when you should use the scalar form. In general, if you want to calculate formulas for solutions, the scalar form

$$y(t) = k_1 e^{-5t} + k_2 e^{-2t}$$

is quicker to derive and easier to use. If you want to understand the behavior of solutions qualitatively using the phase plane, then the vector form

$$\mathbf{Y}(t) = k_1 e^{-5t} \begin{pmatrix} 1 \\ -5 \end{pmatrix} + k_2 e^{-2t} \begin{pmatrix} 1 \\ -2 \end{pmatrix}$$

is more appropriate. Of course, once you have the scalar form of the general solution, it is easy to calculate the vector form by differentiating $y(t)$, as was illustrated earlier in this example.

Solving initial-value problems

If we are given an initial-value problem such as

$$\frac{d^2 y}{dt^2} + 7 \frac{dy}{dt} + 10y = 0, \quad y(0) = 2, \ y'(0) = -13,$$

and we just want a formula for the solution, we start with the scalar form of the general solution

$$y(t) = k_1 e^{-5t} + k_2 e^{-2t}.$$

We differentiate $y(t)$, obtaining

$$y'(t) = -5k_1e^{-5t} - 2k_2e^{-2t},$$

and then we evaluate both $y(t)$ and $y'(t)$ at $t = 0$. We get

$$\begin{cases} k_1 + k_2 = 2 \\ -5k_1 - 2k_2 = -13, \end{cases}$$

a system of two equations in the two unknowns k_1 and k_2. Solving for k_1 and k_2, we get $k_1 = 3$ and $k_2 = -1$. So the solution to the initial-value problem is

$$y(t) = 3e^{-5t} - e^{-2t}.$$

Complex Eigenvalues

The method described above works in general for any second-order, linear equation, even those for which the characteristic polynomial has complex roots. For example, consider the second-order equation

$$\frac{d^2y}{dt^2} + 4\frac{dy}{dt} + 13y = 0.$$

As usual, we guess that $y(t) = e^{st}$ is a solution and obtain the characteristic equation

$$s^2 + 4s + 13 = 0.$$

Using the quadratic formula, we obtain the roots

$$s = \frac{-4 \pm \sqrt{16 - 52}}{2} = -2 \pm 3i.$$

Therefore we have a pair of complex solutions of this equation of the form $e^{(-2\pm3i)t}$. As we did with systems with complex eigenvalues (see Section 3.4), let's look more closely at one of these solutions. Consider $y(t) = e^{(-2+3i)t}$. Using Euler's formula, we have

$$y(t) = e^{(-2+3i)t} = e^{-2t}e^{3it} = e^{-2t}(\cos 3t + i \sin 3t) = e^{-2t}\cos 3t + ie^{-2t}\sin 3t.$$

This function is a complex-valued solution to a real differential equation, so just as we argued in the case of systems, the real and imaginary parts of $y(t)$ are themselves solutions of the original equation (see Exercise 31). That is, we have two real solutions given by $y_1(t) = e^{-2t}\cos 3t$ and $y_2(t) = e^{-2t}\sin 3t$. By the Linearity Principle, any linear combination

$$y(t) = k_1e^{-2t}\cos 3t + k_2e^{-2t}\sin 3t$$

of $y_1(t)$ and $y_2(t)$ is also a solution. Note that this calculation illustrates the fact that we can go right from the roots of the characteristic equation, the eigenvalues $-2 \pm 3i$, to the general solution without performing the intermediate calculations each time.

We can also obtain a vector solution to the associated system by differentiating $y(t)$ to obtain $v = dy/dt$. We have

$$\mathbf{Y}(t) = k_1 e^{-2t} \begin{pmatrix} \cos 3t \\ -2\cos 3t - 3\sin 3t \end{pmatrix} + k_2 e^{-2t} \begin{pmatrix} \sin 3t \\ -2\sin 3t + 3\cos 3t \end{pmatrix}.$$

This general solution to the system is exactly what we would have obtained had we used the eigenvalue/eigenvector methods.

The Method of the Lucky Guess

For a linear, second-order equation of the form

$$a\frac{d^2y}{dt^2} + b\frac{dy}{dt} + cy = 0,$$

where a, b, and c are constants, we can compute the characteristic polynomial by guessing that $y(t) = e^{st}$ is a solution. We obtain

$$a\frac{d^2y}{dt^2} + b\frac{dy}{dt} + cy = (as^2 + bs + c)e^{st},$$

and we see that the characteristic polynomial $as^2 + bs + c$ appears as the coefficient of e^{st}. Now that we have made this calculation once, we do not have to repeat it every time. We can just write down the characteristic polynomial immediately from the second-order equation.

In both the eigenvalue/eigenvector method for the system and the lucky guess method for the second-order equation, we must find the roots of the characteristic polynomial in order to compute the general solution. Whatever method we use, once we have the roots (that is, the eigenvalues), we can obtain the general solution. (We have already discussed examples with two distinct real eigenvalues and with complex eigenvalues. Later in this section we will see how to adapt this method in order to treat repeated eigenvalues.)

Finding the general solution via this lucky guess method is very efficient. We obtain the characteristic polynomial immediately from the second-order equation, and we can skip the work involved in finding the eigenvectors of the system. Consequently, we will use this method extensively in Chapter 4 where we will need to solve a number of second-order equations.

Indeed, this method is so efficient that one might be tempted to ask, "Do we really need systems, eigenvalues, eigenvectors, phase planes, and the rest of the ideas of this chapter?" The answer is "no," provided we care only about formulas and not about a qualitative understanding of solutions. It is also important to remember that this method does not generalize well to other linear systems.

A Classification of Harmonic Oscillators

We can now tell the full story about the solutions of the second-order equation

$$m\frac{d^2y}{dt^2} + b\frac{dy}{dt} + ky = 0$$

that models harmonic oscillators (among other things), and in doing so we will have occasion to use both the lucky guess method and the phase plane. Before starting our analysis, it is important to note that the mass m and the spring constant k are always positive but that the damping constant b can be either zero or positive. If $b = 0$, we have no damping and the oscillator is said to be undamped.

The undamped harmonic oscillator

The second-order equation for this case is simply

$$m\frac{d^2y}{dt^2} + ky = 0,$$

and the characteristic polynomial is

$$ms^2 + k = 0.$$

Since m and k are both positive, the eigenvalues are $\pm i\sqrt{k/m}$. This square root comes up so often that it is commonly written as $\omega = \sqrt{k/m}$.

We therefore have complex solutions of the form

$$e^{i\omega t} = \cos\omega t + i\sin\omega t.$$

Both the real and imaginary parts of this expression are solutions of the equation, so the general solution is

$$y(t) = k_1\cos\omega t + k_2\sin\omega t.$$

Each of these functions is a periodic function with period $2\pi/\omega = 2\pi\sqrt{m/k}$ (see Exercise 22 in Section 3.4).

Computing $v = dy/dt$, we obtain the vector form of the solution

$$\mathbf{Y}(t) = k_1\begin{pmatrix} \cos\omega t \\ -\omega\sin\omega t \end{pmatrix} + k_2\begin{pmatrix} \sin\omega t \\ \omega\cos\omega t \end{pmatrix}.$$

Each of these solutions generates an ellipse in the phase plane that begins at the point $(k_1, k_2\omega)$ and travels around the origin in the clockwise direction (see Exercise 20 in Section 2.1). Each solution returns to its initial position after $2\pi/\omega$ units of time. Therefore the quantity $\omega/(2\pi)$ is called the **natural frequency** of the motion (see Section 3.4, page 305). The phase plane and the $y(t)$-graphs illustrate this periodicity (see Figure 3.41).

In terms of the actual undamped mass-spring system, these plots tell us that the mass either remains at rest forever or oscillates around its rest position without ever

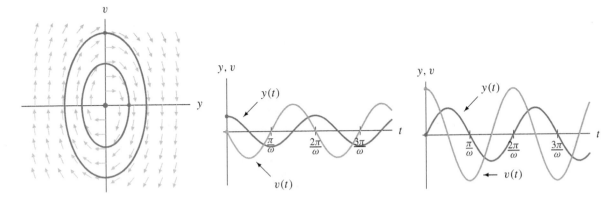

Figure 3.41
Solutions in the phase plane and the $y(t)$- and $v(t)$-graphs corresponding to an undamped harmonic oscillator with natural frequency $\omega/(2\pi)$.

settling down. Without damping, the mass-spring system oscillates forever with the same amplitude and period. This regular behavior is why watches are often made with springs. Of course, physical systems have some damping, which explains why watches need winding every so often.

This type of motion is often called **simple harmonic motion**. One interesting observation about simple harmonic motion is that the period of the motion is determined solely by m and k. Therefore the period is independent of the initial condition (and consequently, the amplitude of the motion.)

Harmonic Oscillators with Damping

If damping is present, the mass-spring system behaves in several different ways, depending on the roots of the characteristic equation. For the harmonic oscillator equation

$$m\frac{d^2y}{dt^2} + b\frac{dy}{dt} + ky = 0,$$

the characteristic equation is

$$ms^2 + bs + k = 0$$

with roots given by the quadratic formula

$$\frac{-b \pm \sqrt{b^2 - 4mk}}{2m}.$$

Thus there are three possibilities for the roots of the characteristic equation.

- If b is small relative to $4km$ (or more precisely, if $b^2 - 4km < 0$), then we have complex roots. The real part of these roots is $-b/(2m)$, which is always negative. In this case the harmonic oscillator is said to be **underdamped**.
- If $b^2 - 4km > 0$, there are two distinct, real roots to this equation. In this case the oscillator is said to be **overdamped**.
- If $b^2 - 4km = 0$, we have repeated roots and the oscillator is said to be **critically damped**.

An underdamped oscillator

If b is relatively small but nonzero, the roots of the characteristic equation are complex with negative real parts. We expect spiraling in the phase plane for this system and corresponding oscillations for the $y(t)$-graphs.

For example, if $m = 1$, $b = 0.2$, and $k = 1.01$, the second-order equation for the motion of the oscillator is

$$\frac{d^2y}{dt^2} + 0.2\frac{dy}{dt} + 1.01y = 0,$$

and the roots of the characteristic polynomial $s^2 + 0.2s + 1.01$ are

$$\frac{-0.2 \pm \sqrt{0.04 - 4.04}}{2} = -0.1 \pm i.$$

Consequently the complex solution is

$$e^{(-0.1\pm i)t} = e^{-0.1t}(\cos t + i \sin t)$$

and the general solution is

$$y(t) = k_1 e^{-0.1t} \cos t + k_2 e^{-0.1t} \sin t.$$

These solutions have a natural period of 2π, but the amplitude of the oscillations decays as time increases (see Figure 3.42). The corresponding motion of the spring is the familiar oscillation about the rest position, but the amplitude of successive oscillations decrease as t increases.

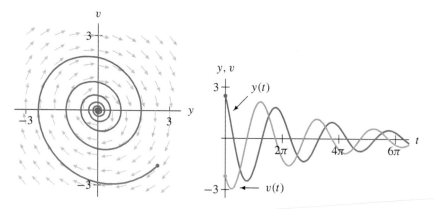

Figure 3.42
Solution in the phase plane and the $y(t)$- and $v(t)$-graphs for the underdamped harmonic oscillator

$$\frac{d^2y}{dt^2} + 0.2\frac{dy}{dt} + 1.01y = 0.$$

An overdamped oscillator

If the damping of the mass-spring system is relatively large, we expect somewhat different behavior for the motion of the mass. For example, if the system is submerged in a vat of peanut butter, we hardly expect the mass to oscillate about its rest position as in the underdamped case.

For example, the characteristic polynomial of the harmonic oscillator modeled by

$$\frac{d^2y}{dt^2} + 3\frac{dy}{dt} + y = 0$$

is $s^2 + 3s + 1 = 0$, and the eigenvalues are

$$s = \frac{-3 \pm \sqrt{5}}{2} \approx -1.5 \pm 1.12.$$

Both of these eigenvalues are real and negative. Hence all solutions of this equation tend to the rest position of the mass as time goes forward. But how do these solutions tend to this position? To answer this, we could write down the general solution of the second-order equation. However, since the answer we seek is a qualitative description of the motion of the oscillator, we can obtain it more directly using qualitative methods.

The system corresponding to the second-order equation above is

$$\frac{d\mathbf{Y}}{dt} = \begin{pmatrix} 0 & 1 \\ -1 & -3 \end{pmatrix} \mathbf{Y}.$$

Suppose $\mathbf{V}_1$ is an eigenvector corresponding to the eigenvalue $(-3 - \sqrt{5})/2$ and $\mathbf{V}_2$ is an eigenvector associated to the eigenvalue $(-3 + \sqrt{5})/2$.

We know that all solutions in the phase plane (except those on the line determined by $\mathbf{V}_1$) tend to the origin tangent to the $\mathbf{V}_2$ direction (see Figure 3.43). In particular, suppose we stretch or compress the spring and release the mass with no initial velocity ($v_0 = 0$). Our solution begins at a point on the y-axis, for example at $(3, 0)$. As t increases, such a solution tends directly to the origin without crossing the y- or v-axes (see Figure 3.43). The position $y(t)$ decreases to zero, and $v(t)$ is always negative

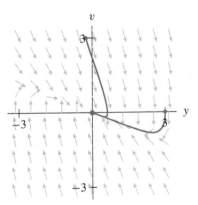

Figure 3.43
The direction field and two solution curves for

$$\frac{d\mathbf{Y}}{dt} = \begin{pmatrix} 0 & 1 \\ -1 & -3 \end{pmatrix} \mathbf{Y}.$$

One solution curve has initial condition $(y_0, v_0) = (3, 0)$, and the other solution curve has initial condition $(y_0, v_0) = (-0.25, 3)$.

(see Figure 3.44). In terms of the mass-spring system, the behavior of this solution means that the mass simply glides to its rest position without oscillating. The damping medium is so thick that the mass does not overshoot the rest position.

However, for other initial conditions, it is possible for the mass to overshoot the rest position. For example, consider the solution to the system with initial condition $(-0.25, 3)$. According to our model, this initial condition corresponds to the situation where the spring is compressed and then released with a nonzero speed in the direction of the rest position. Note that the corresponding solution curve through this point crosses the y-axis and then turns and tends to the origin along the direction of $\mathbf{V}_2$ (see Figure 3.43).

The $y(t)$-graph for this initial condition $(-0.25, 3)$ is displayed in Figure 3.45. This graph shows that $y(t)$ initially increases and passes through $y = 0$ (the t-axis in Figure 3.45). Then $y(t)$ reaches a maximum and slowly decreases to 0 without touching $y = 0$ again.

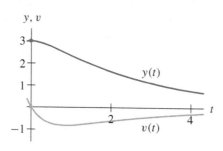

Figure 3.44
The $y(t)$- and $v(t)$-graphs for the solution of the harmonic oscillator system with initial condition $(3, 0)$.

Figure 3.45
The $y(t)$- and $v(t)$-graphs for the solution of the harmonic oscillator system with initial condition $(-0.25, 3)$.

A critically damped oscillator

If the damping coefficient and the spring constant satisfy the equation

$$b^2 - 4km = 0,$$

then the characteristic equation has only one root, $s = -b/(2m)$.

As we know, this condition divides the phase portraits where solutions spiral toward the origin (spiral sinks) from the phase portraits that do not spiral. We call this oscillator "critically" damped because a small change in the damping coefficient changes the nature of the motion of the mass. If we decrease the damping just a tiny amount, the mass oscillates as it approaches its rest position. Increasing the damping puts us in the overdamped case, and there is no possibility of oscillation.

For example, suppose we consider a harmonic oscillator with mass $m = 1$ and spring constant $k = 2$, and we consider different values of the damping coefficient b. Then the second-order equation that models this oscillator is

$$\frac{d^2y}{dt^2} + b\frac{dy}{dt} + 2y = 0.$$

The roots of the characteristic equation $s^2 + bs + 2 = 0$ are

$$\frac{-b \pm \sqrt{b^2 - 8}}{2},$$

and consequently they are complex if $b < 2\sqrt{2}$ and real if $b > 2\sqrt{2}$. Repeated roots occur for $b = 2\sqrt{2}$.

Since we have already discussed the noncritical cases, we concentrate on the case where $b = 2\sqrt{2}$. In this case we know that the system has only one eigenvalue, $s = -\sqrt{2}$, and we know that $y_1(t) = e^{-\sqrt{2}t}$ is one solution of this equation. In order to find the general solution, we need another solution that is not a multiple of $y_1(t)$, and therefore we turn to the method of the lucky guess.

But what should a second guess be? From the characteristic polynomial, we know that the natural guess, $y(t) = e^{st}$, will not be a solution unless $s = -\sqrt{2}$. To determine the desired $y_2(t)$, we can convert this equation to its corresponding linear system. After some calculation, we see that $y_2(t) = te^{-\sqrt{2}t}$ is also a solution (see Exercise 33). Of course, once we have a candidate for $y_2(t)$, we can check that it is a solution by substituting it back into the differential equation. To do so, we calculate

$$\frac{dy_2}{dt} = \left(1 - \sqrt{2}\,t\right)e^{-\sqrt{2}t} \quad \text{and} \quad \frac{d^2y_2}{dt^2} = \left(-2\sqrt{2} + 2t\right)e^{-\sqrt{2}t},$$

and then

$$\frac{d^2y_2}{dt^2} + 2\sqrt{2}\,\frac{dy_2}{dt} + 2y_2 = \left(-2\sqrt{2} + 2t\right)e^{-\sqrt{2}t} + 2\sqrt{2}\left(1 - \sqrt{2}\,t\right)e^{-\sqrt{2}t} + 2te^{-\sqrt{2}t}$$

$$= \left(-2\sqrt{2} + 2t + 2\sqrt{2} - 4t + 2t\right)e^{-\sqrt{2}t}$$

$$= 0.$$

In fact, the general solution of the equation is

$$y(t) = k_1 e^{-\sqrt{2}t} + k_2 t e^{-\sqrt{2}t}.$$

Both $e^{-\sqrt{2}t}$ and $te^{-\sqrt{2}t}$ tend to 0 as t increases (see Exercise 10 in Section 3.5), so solutions tend to the rest position as we expect. Also, these solutions do not involve sines or cosines, so the corresponding solutions do not oscillate about the rest position. This example is also discussed in Section 3.5, where we use eigenvectors to help us plot its phase portrait and, consequently, understand the behavior of solutions (see page 321).

Summary

We now have a complete picture of the behavior of harmonic oscillators modeled by the second-order, linear equation

$$m\frac{d^2y}{dt^2} + b\frac{dy}{dt} + ky = 0.$$

- If $b = 0$, the oscillator is undamped, and the equilibrium point at the origin in the phase plane is a center. All solutions are periodic, and the mass oscillates forever about its rest position. The (natural) period of the oscillations is $2\pi\sqrt{m/k}$.
- If $b > 0$ and $b^2 - 4km < 0$, the oscillator is underdamped. The origin in the phase plane is a spiral sink, and all other solutions spiral toward the origin. The mass oscillates back and forth as it tends to its rest position with period $4m\pi/\sqrt{4km - b^2}$.
- If $b > 0$ and $b^2 - 4km > 0$, the oscillator is overdamped. The origin in the phase plane is a real sink with two distinct eigenvalues. The mass tends to its rest position but does not oscillate.
- If $b > 0$ and $b^2 - 4km = 0$, the oscillator is critically damped. The system has exactly one eigenvalue, which is negative. All solutions tend to the origin tangent to the unique line of eigenvectors. As in the overdamped case, the mass tends to its rest position but does not oscillate.

The four cases just described completely classify the various long-term behaviors of all harmonic oscillators. In the next section we will derive a geometric way to classify these behaviors.

EXERCISES FOR SECTION 3.6

In Exercises 1–6, find the general solution (in scalar form) of the given second-order equation.

1. $\dfrac{d^2y}{dt^2} - 6\dfrac{dy}{dt} - 7y = 0$

2. $\dfrac{d^2y}{dt^2} - \dfrac{dy}{dt} - 12y = 0$

3. $\dfrac{d^2y}{dt^2} + 6\dfrac{dy}{dt} + 9y = 0$

4. $\dfrac{d^2y}{dt^2} - 4\dfrac{dy}{dt} + 4y = 0$

5. $\dfrac{d^2y}{dt^2} + 8\dfrac{dy}{dt} + 25y = 0$

6. $\dfrac{d^2y}{dt^2} - 4\dfrac{dy}{dt} + 29y = 0$

In Exercises 7–12, find the solution of the given initial-value problem.

7. $\dfrac{d^2y}{dt^2} + 2\dfrac{dy}{dt} - 3y = 0$

$y(0) = 6, \ y'(0) = -2$

8. $\dfrac{d^2y}{dt^2} + 4\dfrac{dy}{dt} - 5y = 0$

$y(0) = 11, \ y'(0) = -7$

9. $\dfrac{d^2y}{dt^2} - 4\dfrac{dy}{dt} + 13y = 0$

$y(0) = 1, \ y'(0) = -4$

10. $\dfrac{d^2y}{dt^2} + 4\dfrac{dy}{dt} + 20y = 0$

$y(0) = 2, \ y'(0) = -8$

11. $\dfrac{d^2y}{dt^2} - 8\dfrac{dy}{dt} + 16y = 0$

$y(0) = 3, \ y'(0) = 11$

12. $\dfrac{d^2y}{dt^2} - 4\dfrac{dy}{dt} + 4y = 0$

$y(0) = 1, \ y'(0) = 1$

In Exercises 13–20, consider harmonic oscillators with mass m, spring constant k, and damping coefficient b. For the values specified,

(a) write the second-order differential equation and the corresponding first-order system;

(b) find the eigenvalues and eigenvectors of the linear system;

(c) classify the oscillator (as underdamped, overdamped, critically damped, or undamped) and, when appropriate, give the natural period;

(d) sketch the phase portrait of the associated linear system and include the solution curve for the given initial condition; and

(e) sketch the $y(t)$- and $v(t)$-graphs of the solution with the given initial condition.

13. $m = 1, k = 7, b = 8$, with initial conditions $y(0) = -1, v(0) = 5$

14. $m = 1, k = 8, b = 6$, with initial conditions $y(0) = 1, v(0) = 0$

15. $m = 1, k = 5, b = 4$, with initial conditions $y(0) = 1, v(0) = 0$

16. $m = 1, k = 8, b = 0$, with initial conditions $y(0) = 1, v(0) = 4$

17. $m = 2, k = 1, b = 3$, with initial conditions $y(0) = 0, v(0) = 3$

18. $m = 9, k = 1, b = 6$, with initial conditions $y(0) = 1, v(0) = 1$

19. $m = 2, k = 3, b = 0$, with initial conditions $y(0) = 2, v(0) = -3$

20. $m = 2, k = 3, b = 1$, with initial conditions $y(0) = 0, v(0) = -3$

In Exercises 21–28, consider harmonic oscillators with mass m, spring constant k, and damping coefficient b. (The values of these parameters match up with those in Exercises 13–20). For the values specified,

(a) find the general solution of the second-order equation that models the motion of the oscillator;

(b) find the particular solution for the given initial condition; and

(c) using the equations for the solution of the initial-value problem, sketch the $y(t)$- and $v(t)$-graphs. Compare these graphs to your sketches for the corresponding exercise from Exercises 13–20.

21. $m = 1, k = 7, b = 8$, with initial conditions $y(0) = -1, v(0) = 5$

22. $m = 1, k = 8, b = 6$, with initial conditions $y(0) = 1, v(0) = 0$

23. $m = 1, k = 5, b = 4$, with initial conditions $y(0) = 1, v(0) = 0$

24. $m = 1, k = 8, b = 0$, with initial conditions $y(0) = 1, v(0) = 4$

25. $m = 2, k = 1, b = 3$, with initial conditions $y(0) = 0, v(0) = 3$

26. $m = 9, k = 1, b = 6$, with initial conditions $y(0) = 1, v(0) = 1$

27. $m = 2, k = 3, b = 0$, with initial conditions $y(0) = 2, v(0) = -3$

28. $m = 2, k = 3, b = 1$, with initial conditions $y(0) = 0, v(0) = -3$

29. Construct a table of all the possible harmonic oscillator systems as follows:

    (a) The first column contains the type of oscillator.

    (b) The second column contains the eigenvalue condition that corresponds to this type of system.

    (c) The third column contains the condition on the parameters m, k, and b that is equivalent to the eigenvalue condition.

    (d) The fourth column contains the rate that solutions approach the origin and the natural period of the oscillator (if applicable).

    (e) The fifth column contains sample phase-plane diagrams.

    (f) The sixth column contains typical $y(t)$- and $v(t)$-graphs for solutions.

30. Suppose $y_1(t)$ and $y_2(t)$ are solutions of

$$\frac{d^2y}{dt^2} + p\frac{dy}{dt} + qy = 0.$$

Verify that $y(t) = k_1 y_1(t) + k_2 y_2(t)$ is also a solution for any choice of constants k_1 and k_2.

31. Suppose $y(t)$ is a complex-valued solution of

$$\frac{d^2y}{dt^2} + p\frac{dy}{dt} + qy = 0,$$

where p and q are real numbers. Show that if $y(t) = y_{re}(t) + iy_{im}(t)$, where $y_{re}(t)$ and $y_{im}(t)$ are real valued, then both $y_{re}(t)$ and $y_{im}(t)$ are solutions of the second-order equation.

32. Suppose λ is an eigenvalue for the second-order equation

$$\frac{d^2y}{dt^2} + p\frac{dy}{dt} + qy = 0.$$

Show that $\mathbf{V} = (1, \lambda)$ is an eigenvector for the corresponding first-order system.

33. Suppose the second-order equation

$$\frac{d^2y}{dt^2} + p\frac{dy}{dt} + qy = 0$$

has λ_0 as a repeated eigenvalue.

    (a) Determine the matrix $\mathbf{A}$ for the corresponding linear system $d\mathbf{Y}/dt = \mathbf{AY}$, where $\mathbf{Y} = (y, v)$ and $v = dy/dt$ as usual. Express your answer in terms of λ_0 rather than in terms of p and q.

    (b) Using the method of Section 3.5, find the general solution to the system in part (a).

    (c) Using the result of part (b), determine the general solution of the original second-order equation.

(d) Explain why the general solution obtained in part (c) is the same as

$$y(t) = k_1 e^{\lambda_0 t} + k_2 t e^{\lambda_0 t}.$$

34. Consider a harmonic oscillator with mass $m = 1$ and spring constant $k = 3$, and let the damping coefficient b be a parameter. Then the motion of the oscillator is modeled by the equation

$$\frac{d^2 y}{dt^2} + b \frac{dy}{dt} + 3y = 0.$$

For what value of b does the *typical* solution approach the equilibrium position most rapidly? (The equilibrium position is the point $(y, v) = (0, 0)$ where $v = dy/dt$.)

35. Consider a harmonic oscillator with mass $m = 1$, spring constant $k = 3$, and a (fixed) damping coefficient b. Then the motion of the oscillator is modeled by the equation

$$\frac{d^2 y}{dt^2} + b \frac{dy}{dt} + 3y = 0.$$

What is the quickest rate at which a solution can approach the equilibrium state? (The equilibrium state is the point $(y, v) = (0, 0)$ where $v = dy/dt$. Your answer should depend on the value of b.)

36. An automobile's suspension system consists essentially of large springs with damping. When the car hits a bump, the springs are compressed. It is reasonable to use a harmonic oscillator to model the up-and-down motion, where $y(t)$ measures the amount the springs are stretched or compressed and $v(t)$ is the vertical velocity of the bouncing car.

Suppose that you are working for a company that designs suspension systems for cars. One day your boss comes to you with the results of a market research survey indicating that most people want shock absorbers that "bounce twice" when compressed, then gradually return to their equilibrium position from above. That is, when the car hits a bump, the springs are compressed. Ideally they should expand, compress, and expand, then settle back to the rest position. After the initial bump, the spring would pass through its rest position three times and approach the rest position from the expanded state.

(a) Sketch a graph of the position of the spring after hitting a bump, where $y(t)$ denotes the state of the spring at time t, $y > 0$ corresponds to the spring being stretched, and $y < 0$ corresponds to the spring being compressed.

(b) Explain (politely) why the behavior pictured in the figure is impossible with standard suspension systems that are accurately modeled by the harmonic oscillator system.

(c) What is your suggestion for a choice of a harmonic oscillator system that most closely approximates the desired behavior? Justify your answer with an essay.

37. Suppose material scientists discover a new type of fluid called "magic-finger fluid." Magic-finger fluid has the property that, as an object moves through the fluid, it is accelerated in the direction that it travels ("anti-damping").

   (a) Suppose the force F_{mf} that the magic-finger fluid applies to an object is proportional to the velocity of the object with proportionality constant b_{mf}. Formulate a linear, second-order differential equation for a mass-spring system moving in magic-finger fluid, assuming that the only forces involved are the natural restoring force F_s of the spring (given by Hooke's law) and the "anti-damping" force F_{mf}.

   (b) Convert this mass-spring system to a first-order, linear system.

   (c) Classify the possible behaviors of the linear system you constructed in part (b).

38. Consider a harmonic oscillator with $m = 1$, $k = 2$, and $b = 1$.

   (a) What is the natural period?

   (b) If m is increased slightly, does the natural period increase or decrease? How fast does it increase or decrease?

   (c) If k is increased slightly, does the natural period increase or decrease? How fast does it increase or decrease?

   (d) If b is increased slightly, does the natural period increase or decrease? How fast does it increase or decrease?

39. Suppose we wish to make a clock using a mass and a spring sliding on a table. We arrange for the clock to "tick" whenever the mass crosses $y = 0$. We use a spring with spring constant $k = 2$. If we assume there is no friction or damping ($b = 0$), then what mass m must be attached to the spring so that its natural period is one time unit?

40. As pointed out in the text, an undamped or underdamped harmonic oscillator can be used to make a clock. As in Exercise 39, if we arrange for the clock to tick whenever the mass passes the rest position, then the time between ticks is equal to one-half of the natural period of the oscillator.

   (a) If dirt increases the coefficient of damping slightly for the harmonic oscillator, will the clock run fast or slow?

   (b) Suppose the spring provides slightly less force for a given compression or extension as it ages. Will the clock run fast or slow?

   (c) If grime collects on the harmonic oscillator and slightly increases the mass, will the clock run fast or slow?

   (d) Suppose all of the above occur—the coefficient of damping increases slightly, the spring gets "tired," and the mass increases slightly—will the clock run fast or slow?

3.7 THE TRACE-DETERMINANT PLANE

In the previous sections, we have encountered a number of different types of linear systems of differential equations. At this point, it may seem that there are many different possibilities for these systems, each with its own characteristics. In order to put all of these examples in perspective, it is useful to pause and review the big picture.

One way to summarize everything that we have done so far is to make a table. As we have seen, the behavior of a linear system is governed by the eigenvalues and eigenvectors of the system, so our table should contain the following:

1. The name of the system (spiral sink, saddle, source, ...)

2. The eigenvalue conditions

3. One or two representative phase portraits

For example, we could begin to construct this table as in Table 3.1.

This list is by no means complete. In fact, one exercise at the end of this section is to compile a complete table (see Exercise 1). There are eight other entries.

As is so often the case in mathematics, it is helpful to view information in several different ways. Since we are looking for "the big picture," why not try to summarize the different behaviors for linear systems in a picture rather than a table? One such picture is called the *trace-determinant plane*.

Table 3.1
Partial table of linear systems.

Type	Eigenvalues	Phase Plane	Type	Eigenvalues	Phase Plane
Saddle	$\lambda_1 < 0 < \lambda_2$		Spiral Sink	$\lambda = a \pm ib$ $a < 0, b \neq 0$	
Sink	$\lambda_1 < \lambda_2 < 0$		Spiral Source	$\lambda = a \pm ib$ $a > 0, b \neq 0$	
Source	$0 < \lambda_1 < \lambda_2$		Center	$\lambda = \pm ib$ $b \neq 0$	

Trace and Determinant

Suppose we begin with the linear system $d\mathbf{Y}/dt = \mathbf{AY}$, where $\mathbf{A}$ is the matrix

$$\begin{pmatrix} a & b \\ c & d \end{pmatrix}.$$

The characteristic polynomial for $\mathbf{A}$ is

$$\det(\mathbf{A} - \lambda I) = (a - \lambda)(d - \lambda) - bc = \lambda^2 - (a + d)\lambda + ad - bc.$$

The quantity $a + d$ is called the **trace** of the matrix $\mathbf{A}$ and, as we know, the quantity $ad - bc$ is the determinant of $\mathbf{A}$. So the characteristic polynomial of $\mathbf{A}$ can be written more succinctly as

$$\lambda^2 - T\lambda + D,$$

where $T = a + d$ is the trace of $\mathbf{A}$ and $D = ad - bc$ is the determinant of $\mathbf{A}$. For example, if

$$\mathbf{A} = \begin{pmatrix} 1 & 2 \\ 3 & 4 \end{pmatrix},$$

then the characteristic polynomial is $\lambda^2 - 5\lambda - 2$, since $T = 5$ and $D = 4 - 6 = -2$. (Remember that the coefficient of the λ-term is $-T$. It is a common mistake to put this minus sign in the wrong place or even to forget it entirely.)

Since the characteristic polynomial of $\mathbf{A}$ depends only on T and D, it follows that the eigenvalues of $\mathbf{A}$ also depend only on T and D. If we solve the characteristic polynomial $\lambda^2 - T\lambda + D = 0$, we obtain the eigenvalues

$$\lambda = \frac{T \pm \sqrt{T^2 - 4D}}{2}.$$

From this formula we see immediately that the eigenvalues of $\mathbf{A}$ are complex if $T^2 - 4D < 0$, they are repeated if $T^2 - 4D = 0$, and they are real and distinct if $T^2 - 4D > 0$.

The Trace-Determinant Plane

We can now begin to paint the big picture for linear systems by examining the *trace-determinant* plane. We draw the T-axis horizontally and the D-axis vertically. Then the curve $T^2 - 4D = 0$, or equivalently $D = T^2/4$, is a parabola opening upward in this plane. We call it the *repeated-root* parabola. Above this parabola $T^2 - 4D < 0$, and below it $T^2 - 4D > 0$.

To use this picture, we first compute T and D for a given matrix and then locate the point (T, D) in this plane. Then we can immediately read off whether the eigenvalues are real, repeated, or complex, depending on the location of (T, D) relative to the repeated-root parabola (see Figure 3.46). For example, if

$$\mathbf{A} = \begin{pmatrix} 2 & 3 \\ 1 & 2 \end{pmatrix},$$

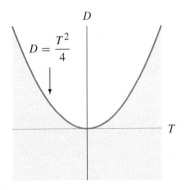

Figure 3.46
The shaded region corresponds to
$T^2 - 4D > 0$.

then $(T, D) = (4, 1)$, and the point $(4,1)$ lies below the curve $T^2 - 4D = 0$ (in this case, $T^2 - 4D = 12 > 0$), so the eigenvalues of A are real and distinct.

Refining the Big Picture

We can actually do much more with the trace-determinant plane. For example, if

$$T^2 - 4D < 0,$$

(the point (T, D) lies above the repeated-root parabola), then we know that the eigenvalues are complex and their real part is $T/2$. We have a spiral sink if $T < 0$, a spiral source if $T > 0$, and a center if $T = 0$. In the trace-determinant plane, the point (T, D) is located above the repeated-root parabola. If (T, D) lies to the left of the D-axis, the corresponding system has a spiral sink. If (T, D) lies to the right of the D-axis, the system has a spiral source. If (T, D) lies on the D-axis, then the system has a center. So our refined picture can be drawn this way (see Figure 3.47).

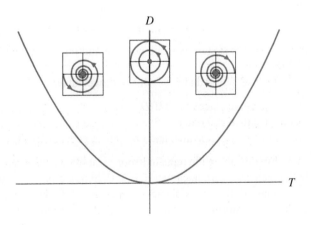

Figure 3.47
Above the repeated-root parabola, we have centers along the D-axis, spiral sources to the right, and spiral sinks to the left.

Real eigenvalues

We can also distinguish different regions in the trace-determinant plane where the linear system has real and distinct eigenvalues. In this case (T, D) lies below the repeated-root parabola. If $T^2 - 4D > 0$, the real eigenvalues are

$$\lambda = \frac{T \pm \sqrt{T^2 - 4D}}{2}.$$

If $T > 0$, the eigenvalue

$$\frac{T + \sqrt{T^2 - 4D}}{2}$$

is the sum of two positive terms and therefore is positive. Thus we only have to determine the sign of the other eigenvalue

$$\frac{T - \sqrt{T^2 - 4D}}{2}$$

to determine the type of the system.

If $D = 0$, then this eigenvalue is 0, so our matrix has one positive and one zero eigenvalue. If $D > 0$, then

$$T^2 - 4D < T^2.$$

Since we are considering the case where $T > 0$, we have

$$\sqrt{T^2 - 4D} < T$$

and

$$\frac{T - \sqrt{T^2 - 4D}}{2} > 0.$$

In this case both eigenvalues are positive, so the origin is a source.

On the other hand, if $T > 0$ but $D < 0$, then

$$T^2 - 4D > T^2,$$

so that

$$\sqrt{T^2 - 4D} > T$$

and

$$\frac{T - \sqrt{T^2 - 4D}}{2} < 0.$$

In this case the system has one positive and one negative eigenvalue, so the origin is a saddle.

In case $T < 0$ and $T^2 - 4D > 0$, we have

- two negative eigenvalues if $D > 0$,
- one negative and one positive eigenvalue if $D < 0$, or
- one negative eigenvalue and one zero eigenvalue if $D = 0$.

Finally, along the repeated-root parabola we have repeated eigenvalues. If $T < 0$, both eigenvalues are negative; if $T > 0$, both are positive; and if $T = 0$, both are zero.

The full picture is displayed in Figure 3.48. Note that this picture gives us some of the same information that we compiled in our table earlier in this section.

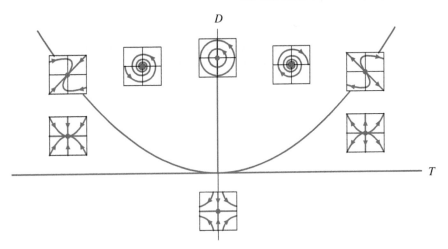

Figure 3.48
The big picture.

The Parameter Plane

The trace-determinant plane is an example of a *parameter plane*. The entries of the matrix **A** are parameters that we can adjust. When these entries change, the trace and determinant of the matrix also change, and our point (T, D) moves around in the parameter plane. As this point enters the various regions in the trace-determinant plane, we should envision the corresponding phase portraits changing accordingly. The trace-determinant plane is very much different from previous pictures we have drawn. It is a picture of a classification scheme of the behavior of all possible solutions to linear systems.

We must emphasize that the trace-determinant plane does not give complete information about the linear system at hand. For example, along the repeated-root parabola we have repeated eigenvalues, but we cannot determine whether we have one or many linearly independent eigenvectors. In order to make that distinction, we must actually calculate the eigenvectors. Similarly, we cannot determine the direction in which solutions wind about the origin if $T^2 - 4D < 0$. For example, both of the matrices

$$\mathbf{A} = \begin{pmatrix} 0 & 1 \\ -1 & 0 \end{pmatrix} \quad \text{and} \quad \mathbf{B} = \begin{pmatrix} 0 & -1 \\ 1 & 0 \end{pmatrix}$$

have trace 0 and determinant 1, but solutions of the system $d\mathbf{Y}/dt = \mathbf{AY}$ wind around the origin in the clockwise direction, whereas solutions of $d\mathbf{Y}/dt = \mathbf{BY}$ travel in the opposite direction.

The Harmonic Oscillator

We can also paint the same picture for the harmonic oscillator. Recall that this second-order equation is given by

$$m\frac{d^2y}{dt^2} + b\frac{dy}{dt} + ky = 0,$$

where $m > 0$ is the mass, $k > 0$ is the spring constant, and $b \geq 0$ is the damping coefficient. As a system we have

$$\frac{d\mathbf{Y}}{dt} = \begin{pmatrix} 0 & 1 \\ -k/m & -b/m \end{pmatrix} \mathbf{Y},$$

so the trace $T = -b/m$ and the determinant $D = k/m$. We plot $T = -b/m$ on the horizontal axis and $D = k/m$ on the vertical axis as before.

Since m and k are positive and b is nonnegative, we are restricted to one-quarter of the picture for general linear systems, namely the second quadrant of the TD-plane. The picture is shown in Figure 3.49.

The repeated-root parabola in this case is $T^2 - 4D = b^2 - 4km = 0$. Above this parabola we have a spiral sink (if $b \neq 0$) or a center (if $b = 0$). Below the repeated-root parabola we have a sink with real distinct eigenvalues. On the parabola, we have repeated negative eigenvalues.

In the language of oscillators introduced in the previous section, if $(-b/m, k/m)$ lies above the repeated-root parabola and $b > 0$, we have an underdamped oscillator, or if $b = 0$, we have an undamped oscillator. If $(-b/m, k/m)$ lies on the repeated-root parabola, the oscillator is critically damped. Below the parabola, the oscillator is overdamped.

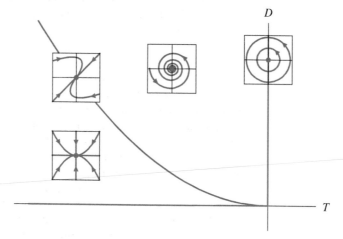

Figure 3.49
The trace-determinant plane for the harmonic oscillator.

Kathleen Alligood (1947–) received her Ph.D. in mathematics at the University of Maryland. She taught at the College of Charleston and at Michigan State University before assuming her current position as Professor of Mathematical Sciences at George Mason University.

Alligood's research centers on the behavior of nonlinear systems but encompasses many of the topics described in this chapter. Nonlinear systems of differential equations may possess sinks, just as linear systems do. However, it need not be the case that all solutions tend to the sink as in the linear case. Often the boundary of the set of solutions that tend to the sink is an extremely complicated mathematical object that contains infinitely many saddle points and their stable curves. Using techniques from topology, fractal geometry, and dynamical systems, Alligood and her coworkers were among the first to analyze the structure of these "fractal basin boundaries."

Navigating the Trace-Determinant Plane

One of the best uses of the trace-determinant plane is in the study of linear systems that depend on parameters. As the parameters change, so do the trace and determinant of the matrix. Consequently, the phase portrait for the system also changes.

Usually, small changes in the parameters do not affect the qualitative behavior of the linear system very much. For example, a spiral sink remains a spiral sink and a saddle remains a saddle. Of course the eigenvalues and eigenvectors change as we vary the parameters, but the basic behavior of solutions remains more or less the same.

The critical loci

There are, however, certain exceptions to this scenario. For example, suppose that a change in parameters forces the point (T, D) to cross the positive D-axis from left to right. The corresponding linear system has changed from a spiral sink to a center and then immediately thereafter to a spiral source. Instead of all solutions tending to the equilibrium point at $(0, 0)$, suddenly we have a center, and then all of the nonequilibrium solutions tend to infinity. That is, the family of linear systems has encountered a bifurcation at the moment the point (T, D) crosses the D-axis.

The trace-determinant plane provides us with a chart of those locations where we can expect significant changes in the phase portrait. There are three such *critical loci*.

The first critical locus is the positive D-axis, as we saw above. A second critical line is the T- axis. If (T, D) crosses this line as our parameters vary, our system moves from a saddle to a sink, a source, or a center (or vice versa). The third critical locus is the repeated-root parabola where spirals turn into real sinks or sources.

There is one point in the trace-determinant plane where many different possibilities arise. If the trace and determinant are both zero, the chart shows that our system can change into any type of system whatsoever. All three of the critical loci meet at this point.

It is helpful to think of these three critical loci as fences. As long as we change parameters so that (T, D) does not pass over one of the fences, the linear system remains "unchanged" in the sense that the qualitative behavior of the solutions does not change. However, passing over a fence changes the behavior dramatically. The system undergoes a bifurcation.

A One-Parameter Family of Linear Systems

Consider the one-parameter family of linear systems $d\mathbf{Y}/dt = \mathbf{AY}$, where

$$\mathbf{A} = \begin{pmatrix} -2 & a \\ -2 & 0 \end{pmatrix},$$

which depends on the parameter a. As a varies, the determinant of this matrix is $2a$, but the trace is always -2. If we vary the parameter a from a large negative number to a large positive number, the corresponding point (T, D) in the trace-determinant plane moves vertically along the straight line $T = -2$ (see Figure 3.50). As a increases, we first travel from the saddle region into the region where we have a real sink. This change occurs when the system admits a zero eigenvalue, which in turn occurs at $a = 0$. As a continues to increase, we next move across the repeated-root parabola, and the system changes from having a sink with real eigenvalues to a spiral sink. This second bifurcation occurs when $T^2 - 4D = 0$, which for this example reduces to $D = 1$. Hence this bifurcation occurs at $a = 1/2$.

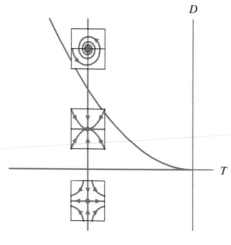

Figure 3.50
Motion in the trace-determinant plane corresponding to the one-parameter family of systems

$$\frac{d\mathbf{Y}}{dt} = \mathbf{AY}, \quad \text{where } \mathbf{A} = \begin{pmatrix} -2 & a \\ -2 & 0 \end{pmatrix}.$$

Bifurcation from sink to spiral sink

Let's investigate how the bifurcation from sink to spiral sink occurs in terms of the phase portraits of the corresponding systems. We need first to compute the eigenvalues and eigenvectors of the system. Of course these quantities depend on a. Since the characteristic polynomial is $\lambda^2 + 2\lambda + 2a = 0$, the eigenvalues are

$$\lambda = \frac{-2 \pm \sqrt{4 - 8a}}{2} = -1 \pm \sqrt{1 - 2a}.$$

As we deduced above, if $a > 1/2$, then $1 - 2a < 0$ and the eigenvalues are complex with negative real part. For $a < 1/2$, the eigenvalues

$$\lambda = -1 \pm \sqrt{1 - 2a}$$

are both real. In particular, if $0 < a < 1/2$, $\sqrt{1 - 2a} < 1$, so both eigenvalues are negative. Hence the origin is a sink with two straight lines of solutions (see Figure 3.51).

If we compute the eigenvectors for the eigenvalue $\lambda = -1 + \sqrt{1 - 2a}$, we find that they lie along the line

$$y = \left(\frac{1 + \sqrt{1 - 2a}}{a}\right) x.$$

Similarly, the eigenvectors corresponding to the eigenvalue $\lambda = -1 - \sqrt{1 - 2a}$ lie along the line

$$y = \left(\frac{1 - \sqrt{1 - 2a}}{a}\right) x.$$

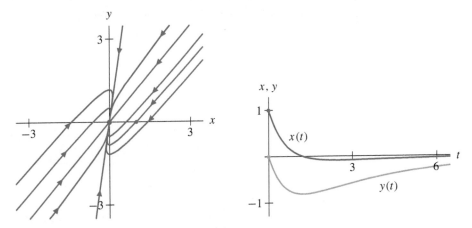

Figure 3.51
Phase portrait and the $x(t)$- and $y(t)$-graphs for the indicated solution for the one-parameter family with $a = 1/4$.

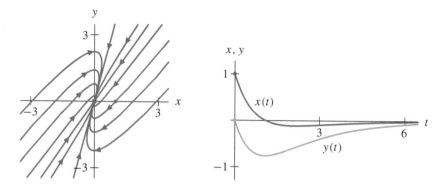

Figure 3.52
Phase portrait and the $x(t)$- and $y(t)$-graphs for the indicated solution for the one-parameter family with $a = 0.4$.

Note that the slopes of both of these lines tend to 2 as a approaches $1/2$. That is, our two straight-line solutions merge to produce a single straight-line solution along the line $y = 2x$ as $a \to 1/2$ (see Figure 3.52).

As a approaches $1/2$, the family of linear systems approaches a linear system with a repeated eigenvalue. At $a = 1/2$, the system is

$$\frac{d\mathbf{Y}}{dt} = \begin{pmatrix} -2 & 1/2 \\ -2 & 0 \end{pmatrix} \mathbf{Y},$$

whose characteristic polynomial is $\lambda^2 + 2\lambda + 1 = 0$. Hence the system has the repeated eigenvalue $\lambda = -1$. This system has a single line of eigenvectors that lie along the line $y = 2x$. The phase portrait and typical $x(t)$-graph are shown in Figure 3.53. Thus we see that the two independent eigenvectors come together to form the single line of eigenvectors as a approaches $1/2$.

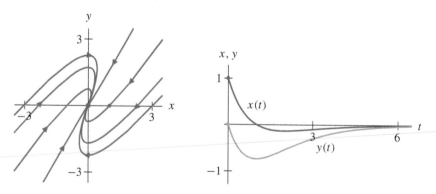

Figure 3.53
Phase portrait and the $x(t)$- and $y(t)$-graphs for the indicated solution for the one-parameter family with $a = 1/2$.

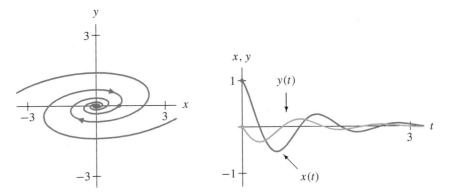

Figure 3.54
Phase portrait and $x(t)$-graph for the indicated solution for the one-parameter family with $a = 10$.

When the parameter crosses the repeated-root parabola, the origin becomes a spiral sink. The real part of the eigenvalue is -1, and the natural period is $2\pi/\sqrt{2a-1}$. For all values of a, solutions spiral toward the origin. If a is very large, solutions approach the origin at the exponential rate of e^{-t} with a very small period. The phase portrait and $x(t)$-graph for $a = 10$ are shown in Figure 3.54.

On the other hand, if a is just slightly larger than $1/2$, solutions still spiral toward the origin. However, the period of the oscillations, which is given by $2\pi/\sqrt{2a-1}$, is very large for a near $1/2$. To observe one oscillation, we must watch a solution for a long time. Since the solutions are tending to the origin at an exponential rate, these oscillations may be very hard to detect (see Figure 3.55, which is almost indistinguishable from Figure 3.53). In applications there may be very little practical difference between a very slowly oscillating solution decaying toward the origin and a solution that does not oscillate.

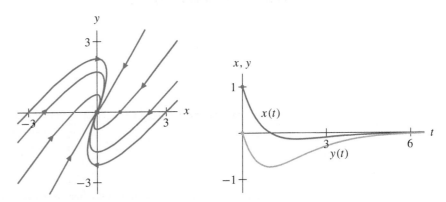

Figure 3.55
Phase portrait and $x(t)$-graph for the indicated solution for the one-parameter family with $a = 0.51$.

EXERCISES FOR SECTION 3.7

1. Construct a table of the possible linear systems as follows:

   (a) The first column contains the type of the system (sink, spiral sink, source, ...), if it has a name.

   (b) The second column contains the condition on the eigenvalues that corresponds to this case.

   (c) The third column contains a small picture of two or more possible phase portraits for this system, and

   (d) The fourth column contains $x(t)$- and $y(t)$-graphs of typical solutions indicated in your phase portraits.

   [*Hint*: The most complete table contains 14 cases. Don't forget the double eigenvalue and zero eigenvalue cases.]

In Exercises 2–7, we consider the one-parameter families of linear systems depending on the parameter a. Each family therefore determines a curve in the trace-determinant plane. For each family,

   (a) sketch the corresponding curve in the trace-determinant plane;

   (b) in a brief essay, discuss different types of behaviors exhibited by the system as a increases along the real line (unless otherwise noted); and

   (c) identify the values of a where the type of the system changes. These are the bifurcation values of a.

2. $\dfrac{d\mathbf{Y}}{dt} = \begin{pmatrix} a & -1 \\ 2 & 0 \end{pmatrix} \mathbf{Y}$

3. $\dfrac{d\mathbf{Y}}{dt} = \begin{pmatrix} a & a^2 + a \\ 1 & a \end{pmatrix} \mathbf{Y}$

4. $\dfrac{d\mathbf{Y}}{dt} = \begin{pmatrix} a & a \\ 1 & 0 \end{pmatrix} \mathbf{Y}$

5. $\dfrac{d\mathbf{Y}}{dt} = \begin{pmatrix} a & \sqrt{1 - a^2} \\ 1 & 0 \end{pmatrix} \mathbf{Y}$

$-1 \le a \le 1$

6. $\dfrac{d\mathbf{Y}}{dt} = \begin{pmatrix} 2 & 0 \\ a & -3 \end{pmatrix} \mathbf{Y}$

7. $\dfrac{d\mathbf{Y}}{dt} = \begin{pmatrix} a & 1 \\ a & a \end{pmatrix} \mathbf{Y}$

8. Consider the two-parameter family of linear systems

$$\frac{d\mathbf{Y}}{dt} = \begin{pmatrix} a & 1 \\ b & 1 \end{pmatrix} \mathbf{Y}.$$

In the ab-plane, identify all regions where this system possesses a saddle, a sink, a spiral sink, and so on. [*Hint*: Draw a picture of the ab-plane and shade each point (a, b) of the plane a different color depending on the type of linear system for that choice (a, b) of parameters.]

9. Consider the two-parameter family of linear systems

$$\frac{d\mathbf{Y}}{dt} = \begin{pmatrix} a & b \\ b & a \end{pmatrix} \mathbf{Y}.$$

In the ab-plane, identify all regions where this system possesses a saddle, a sink, a spiral sink, and so on. [*Hint*: Draw a picture of the ab-plane and shade each point (a, b) of the plane a different color depending on the type of linear system for that choice (a, b) of the parameters.]

10. Consider the two-parameter family of linear systems

$$\frac{d\mathbf{Y}}{dt} = \begin{pmatrix} a & b \\ -b & a \end{pmatrix} \mathbf{Y}.$$

In the ab-plane, identify all regions where this system possesses a saddle, a sink, a spiral sink, and so on. [*Hint*: Draw a picture of the ab-plane and shade each point (a, b) of the plane a different color depending on the type of linear system for that choice (a, b) of parameters.]

In Exercises 11–13, we consider the equation

$$m\frac{d^2y}{dt^2} + b\frac{dy}{dt} + ky = 0$$

that models the motion of a harmonic oscillator with mass m, spring constant k, and damping coefficient b. In each exercise, we fix two values of these three parameters and obtain a one-parameter family of second-order equations. For each one-parameter family,

(a) rewrite the one-parameter family as a one-parameter family of linear systems,

(b) draw the curve in the trace-determinant plane obtained by varying the parameter, and

(c) in a brief essay, discuss the different types of behavior exhibited by this one-parameter family.

11. Consider

$$\frac{d^2y}{dt^2} + b\frac{dy}{dt} + 3y = 0.$$

That is, fix $m = 1$ and $k = 3$, and let $0 \le b < \infty$.

12. Consider

$$\frac{d^2y}{dt^2} + 2\frac{dy}{dt} + ky = 0.$$

That is, fix $m = 1$ and $b = 2$, and let $0 < k < \infty$.

13. Consider

$$m\frac{d^2y}{dt^2} + \frac{dy}{dt} + 2y = 0.$$

That is, fix $b = 1$ and $k = 2$, and let $0 < m < \infty$.

14. Using the DETools program TDPlaneQuiz, describe the path through the trace-determinant plane that was used to produce each animation.

3.8 LINEAR SYSTEMS IN THREE DIMENSIONS

So far, we have studied linear systems with two dependent variables. For these systems, the behavior of solutions and the nature of the phase plane can be determined by computing the eigenvalues and eigenvectors of the 2×2 coefficient matrix. Once we have found two solutions with linearly independent initial conditions, we can give the general solution.

In this section we show that the same is true for linear systems with three dependent variables. The eigenvalues and eigenvectors of the 3×3 coefficient matrix determine the behavior of solutions and the general solution. Three-dimensional linear systems have three eigenvalues, so the list of possible qualitatively distinct phase spaces is longer than for planar systems. Since we must deal with three scalar equations rather than two, the arithmetic can quickly become much more involved. You might want to seek out software or a calculator capable of handling 3×3 matrices.

Linear Independence and the Linearity Principle

The general form of a linear system with three dependent variables is

$$\frac{dx}{dt} = a_{11}x + a_{12}y + a_{13}z$$

$$\frac{dy}{dt} = a_{21}x + a_{22}y + a_{23}z$$

$$\frac{dz}{dt} = a_{31}x + a_{32}y + a_{33}z,$$

where x, y, and z are the dependent variables and the coefficients a_{ij}, $(i, j = 1, 2, 3)$, are constants. We can write this system in matrix form as

$$\frac{d\mathbf{Y}}{dt} = \mathbf{AY},$$

where $\mathbf{A}$ is the coefficient matrix

$$\mathbf{A} = \begin{pmatrix} a_{11} & a_{12} & a_{13} \\ a_{21} & a_{22} & a_{23} \\ a_{31} & a_{32} & a_{33} \end{pmatrix}$$

and $\mathbf{Y}$ is the vector of dependent variables,

$$\mathbf{Y} = \begin{pmatrix} x \\ y \\ z \end{pmatrix}.$$

To specify an initial condition for such a system, we must give three numbers, x_0, y_0, and z_0.

The Linearity Principle holds for linear systems in all dimensions, so if $\mathbf{Y}_1(t)$ and $\mathbf{Y}_2(t)$ are solutions, then $k_1\mathbf{Y}_1(t) + k_2\mathbf{Y}_2(t)$ is also a solution for any constants k_1 and k_2.

Suppose $\mathbf{Y}_1(t)$, $\mathbf{Y}_2(t)$ and $\mathbf{Y}_3(t)$ are three solutions of the linear system

$$\frac{d\mathbf{Y}}{dt} = \mathbf{AY}.$$

If for any point (x_0, y_0, z_0) there exist constants k_1, k_2, and k_3 such that

$$k_1\mathbf{Y}_1(0) + k_2\mathbf{Y}_2(0) + k_3\mathbf{Y}_3(0) = (x_0, y_0, z_0),$$

then the general solution of the system is

$$\mathbf{Y}(t) = k_1\mathbf{Y}_1(t) + k_2\mathbf{Y}_2(t) + k_3\mathbf{Y}_3(t).$$

In order for three solutions $\mathbf{Y}_1(t)$, $\mathbf{Y}_2(t)$, and $\mathbf{Y}_3(t)$ to give the general solution, the three vectors $\mathbf{Y}_1(0)$, $\mathbf{Y}_2(0)$, and $\mathbf{Y}_3(0)$ must point in "different directions"; that is, no one of them can be in the plane through the origin and the other two. In this case the vectors $\mathbf{Y}_1(0)$, $\mathbf{Y}_2(0)$, and $\mathbf{Y}_3(0)$ (and the corresponding solutions) are said to be *linearly independent*. We present an algebraic technique for checking linear independence in the exercises (see Exercises 2 and 3).

An example
Consider the linear system

$$\frac{d\mathbf{Y}}{dt} = \mathbf{AY} = \begin{pmatrix} 0 & 0.1 & 0 \\ 0 & 0 & 0.2 \\ 0.4 & 0 & 0 \end{pmatrix} \begin{pmatrix} x \\ y \\ z \end{pmatrix}.$$

We can check that the functions

$$\mathbf{Y}_1(t) = e^{0.2t} \begin{pmatrix} 1 \\ 2 \\ 2 \end{pmatrix}$$

$$\mathbf{Y}_2(t) = e^{-0.1t} \begin{pmatrix} -\cos\left(\sqrt{0.03}\,t\right) - \sqrt{3}\sin\left(\sqrt{0.03}\,t\right) \\ -2\cos\left(\sqrt{0.03}\,t\right) + 2\sqrt{3}\sin\left(\sqrt{0.03}\,t\right) \\ 4\cos\left(\sqrt{0.03}\,t\right) \end{pmatrix}$$

$$\mathbf{Y}_3(t) = e^{-0.1t} \begin{pmatrix} -\sin\left(\sqrt{0.03}\,t\right) + \sqrt{3}\cos\left(\sqrt{0.03}\,t\right) \\ -2\sin\left(\sqrt{0.03}\,t\right) - 2\sqrt{3}\cos\left(\sqrt{0.03}\,t\right) \\ 4\sin\left(\sqrt{0.03}\,t\right) \end{pmatrix}$$

are solutions by substituting them into the differential equation. For example,

$$\frac{d\mathbf{Y}_1}{dt} = e^{0.2t} \begin{pmatrix} 0.2 \\ 0.4 \\ 0.4 \end{pmatrix}$$

and

$$\mathbf{AY}_1(t) = \begin{pmatrix} 0 & 0.1 & 0 \\ 0 & 0 & 0.2 \\ 0.4 & 0 & 0 \end{pmatrix} e^{0.2t} \begin{pmatrix} 1 \\ 2 \\ 2 \end{pmatrix} = e^{0.2t} \begin{pmatrix} 0.2 \\ 0.4 \\ 0.4 \end{pmatrix},$$

so $\mathbf{Y}_1(t)$ is a solution. The other two functions can be checked similarly (see Exercise 1). We can sketch the solution curves that correspond to these solutions in the three-dimensional phase space (see Figure 3.56).

The initial conditions of these three solutions are $\mathbf{Y}_1(0) = (1, 2, 2)$, $\mathbf{Y}_2(0) = (-1, -2, 4)$, and $\mathbf{Y}_3(0) = (\sqrt{3}, -2\sqrt{3}, 0)$. These vectors are shown in Figure 3.57, where we can see that none of them is in the plane determined by the other two; hence, they are linearly independent.

For example, to find the solution $\mathbf{Y}(t)$ with initial position $\mathbf{Y}(0) = (2, 1, 3)$, we must solve

$$k_1\mathbf{Y}_1(0) + k_2\mathbf{Y}_2(0) + k_3\mathbf{Y}_3(0) = (2, 1, 3),$$

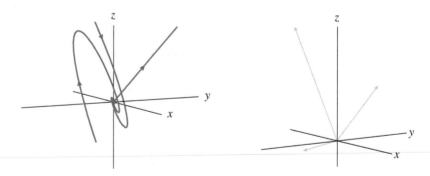

Figure 3.56
The solution curves of $\mathbf{Y}_1(t)$, $\mathbf{Y}_2(t)$, and $\mathbf{Y}_3(t)$.

Figure 3.57
Vectors $\mathbf{Y}_1(0) = (1, 2, 2)$,
$\mathbf{Y}_2(0) = (-1, -2, 4)$, and
$\mathbf{Y}_3(0) = (\sqrt{3}, -2\sqrt{3}, 0)$ in xyz-space.

which is equivalent to

$$
\begin{cases}
k_1 - k_2 + \sqrt{3}k_3 = 2 \\
2k_1 - 2k_2 - 2\sqrt{3}k_3 = 1 \\
2k_1 + 4k_2 = 3.
\end{cases}
$$

We obtain $k_1 = 4/3$, $k_2 = 1/12$ and $k_3 = \sqrt{3}/4$, and the solution is

$$
\mathbf{Y}(t) = \frac{4}{3}\mathbf{Y}_1(t) + \frac{1}{12}\mathbf{Y}_2(t) + \frac{\sqrt{3}}{4}\mathbf{Y}_3(t).
$$

Eigenvalues and Eigenvectors

The method for finding solutions of systems with three dependent variables is the same as that for systems with two variables. We begin by finding eigenvalues and eigenvectors. Suppose we are given a linear system $d\mathbf{Y}/dt = A\mathbf{Y}$, where $\mathbf{A}$ is a 3×3 matrix of coefficients and $\mathbf{Y} = (x, y, z)$. An *eigenvector* for the matrix $\mathbf{A}$ is a nonzero vector $\mathbf{V}$ such that

$$
\mathbf{AV} = \lambda\mathbf{V},
$$

where λ is the *eigenvalue* for $\mathbf{V}$. If $\mathbf{V}$ is an eigenvector for $\mathbf{A}$ with eigenvalue λ, then

$$
\mathbf{Y}(t) = e^{\lambda t}\mathbf{V}
$$

is a solution of the linear system.

The method for finding eigenvalues and eigenvectors for a 3×3 matrix

$$
\mathbf{A} = \begin{pmatrix} a_{11} & a_{12} & a_{13} \\ a_{21} & a_{22} & a_{23} \\ a_{31} & a_{32} & a_{33} \end{pmatrix}
$$

is very similar to that for two-dimensional systems, only requiring more arithmetic. In particular, we need the formula for the determinant of a 3×3 matrix.

DEFINITION   The *determinant* of the matrix $\mathbf{A}$ is

$$
\det \mathbf{A} = a_{11}(a_{22}a_{33} - a_{23}a_{32}) - a_{12}(a_{21}a_{33} - a_{23}a_{31}) + a_{13}(a_{21}a_{32} - a_{22}a_{31}). \quad \blacksquare
$$

Using the 3×3 identity matrix

$$
\mathbf{I} = \begin{pmatrix} 1 & 0 & 0 \\ 0 & 1 & 0 \\ 0 & 0 & 1 \end{pmatrix},
$$

we obtain the *characteristic polynomial* of $\mathbf{A}$ as

$$\det(\mathbf{A} - \lambda\mathbf{I}) = \det \begin{pmatrix} a_{11} - \lambda & a_{12} & a_{13} \\ a_{21} & a_{22} - \lambda & a_{23} \\ a_{31} & a_{32} & a_{33} - \lambda \end{pmatrix}.$$

As in the two-dimensional case, we have:

THEOREM The eigenvalues of a 3×3 matrix $\mathbf{A}$ are the roots of its characteristic polynomial. ∎

To find the eigenvalues of a 3×3 matrix, we must find the roots of a cubic polynomial. This is not as easy as finding the roots of a quadratic. Although there is a "cubic equation" analogous to the quadratic equation for finding the roots of a cubic, it is quite complicated. (It is used by computer algebra packages to give exact values of roots of cubics.) However, in cases where the cubic does not easily factor, we frequently turn to numerical techniques such as Newton's method for finding roots.

To find the corresponding eigenvectors, we must solve a system of three linear equations with three unknowns. Luckily there are many examples of systems that illustrate the possible behaviors in three dimensions and for which the arithmetic is manageable.

A diagonal matrix

The simplest type of 3×3 matrix is a *diagonal* matrix—the only nonzero terms lie on the diagonal. For example, consider the system

$$\frac{d\mathbf{Y}}{dt} = \mathbf{A}\mathbf{Y} = \begin{pmatrix} -3 & 0 & 0 \\ 0 & -1 & 0 \\ 0 & 0 & -2 \end{pmatrix} \begin{pmatrix} x \\ y \\ z \end{pmatrix}.$$

The characteristic polynomial of $\mathbf{A}$ is $(-3 - \lambda)(-1 - \lambda)(-2 - \lambda)$, which is simple because so many of the coefficients of $\mathbf{A}$ are zero. The eigenvalues are the roots of this polynomial, that is, the solutions of

$$(-3 - \lambda)(-1 - \lambda)(-2 - \lambda) = 0.$$

Thus the eigenvalues are $\lambda_1 = -3$, $\lambda_2 = -1$, and $\lambda_3 = -2$.

Finding the corresponding eigenvectors is also not too hard. For $\lambda_1 = -3$, we must solve

$$\mathbf{A}\mathbf{V}_1 = -3\mathbf{V}_1$$

for $\mathbf{V}_1 = (x_1, y_1, z_1)$. The product $\mathbf{A}\mathbf{V}_1$ is

$$\begin{pmatrix} -3 & 0 & 0 \\ 0 & -1 & 0 \\ 0 & 0 & -2 \end{pmatrix} \begin{pmatrix} x_1 \\ y_1 \\ z_1 \end{pmatrix} = \begin{pmatrix} -3x_1 \\ -y_1 \\ -2z_1 \end{pmatrix},$$

and therefore we want to solve

$$\begin{pmatrix} -3x_1 \\ -y_1 \\ -2z_1 \end{pmatrix} = -3 \begin{pmatrix} x_1 \\ y_1 \\ z_1 \end{pmatrix}$$

for x_1, y_1, and z_1. Solutions of this system of three equations with three unknowns are $y_1 = z_1 = 0$ and x_1 may have any (nonzero) value. So, in particular, $(1, 0, 0)$ is an eigenvector for $\lambda_1 = -3$. Similarly, we find that $(0, 1, 0)$ and $(0, 0, 1)$ are eigenvectors for $\lambda_2 = -1$ and $\lambda_3 = -2$, respectively. Note that $(1, 0, 0)$, $(0, 1, 0)$, and $(0, 0, 1)$ are linearly independent.

From these eigenvalues and eigenvectors we can construct solutions of the system

$$\mathbf{Y}_1(t) = e^{-3t} \begin{pmatrix} 1 \\ 0 \\ 0 \end{pmatrix} = \begin{pmatrix} e^{-3t} \\ 0 \\ 0 \end{pmatrix},$$

$$\mathbf{Y}_2(t) = e^{-t} \begin{pmatrix} 0 \\ 1 \\ 0 \end{pmatrix} = \begin{pmatrix} 0 \\ e^{-t} \\ 0 \end{pmatrix},$$

and

$$\mathbf{Y}_3(t) = e^{-2t} \begin{pmatrix} 0 \\ 0 \\ 1 \end{pmatrix} = \begin{pmatrix} 0 \\ 0 \\ e^{-2t} \end{pmatrix}.$$

Because this system is diagonal, we could have gotten this far "by inspection." If we write the system in components

$$\frac{dx}{dt} = -3x$$

$$\frac{dy}{dt} = -y$$

$$\frac{dz}{dt} = -2z,$$

we see that dx/dt depends only on x, dy/dt depends only on y, and dz/dt depends only on z. In other words, the system completely decouples, and each coordinate can be dealt with independently. It is easy to solve these equations.

Now that we have three independent solutions, we can solve any initial-value problem for this system. For example, to find the solution $\mathbf{Y}(t)$ with $\mathbf{Y}(0) = (2, 1, 2)$, we must find constants k_1, k_2, and k_3 such that

$$(2, 1, 2) = k_1 \mathbf{Y}_1(0) + k_2 \mathbf{Y}_2(0) + k_3 \mathbf{Y}_3(0).$$

So $k_1 = 2$, $k_2 = 1$, and $k_3 = 2$, and $\mathbf{Y}(t) = (2e^{-3t}, e^{-t}, 2e^{-2t})$ is the required solution.

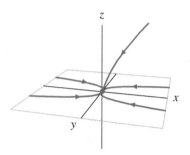

Figure 3.58
Phase space for $d\mathbf{Y}/dt = \mathbf{AY}$ for the diagonal matrix $\mathbf{A}$.

Figure 3.58 is a sketch of the phase space. Note that the coordinate axes are lines of eigenvectors, so they form straight-line solutions. Since all three of the eigenvalues are negative, solutions along all three of the axes tend toward the origin. Because every other solution can be made up as a linear combination of the solutions on the axes, all solutions must tend to the origin and it is natural to call the origin a sink.

Three-dimensional behavior

Before giving a classification of linear systems in three dimensions, we give an example whose qualitative behavior is different from that of any two-dimensional system.

Consider the system

$$\frac{d\mathbf{Y}}{dt} = \mathbf{BY} = \begin{pmatrix} 0.1 & -1 & 0 \\ 1 & 0.1 & 0 \\ 0 & 0 & -0.2 \end{pmatrix} \begin{pmatrix} x \\ y \\ z \end{pmatrix}.$$

The characteristic polynomial of $\mathbf{B}$ is

$$((0.1 - \lambda)(0.1 - \lambda) + 1)(-0.2 - \lambda) = (\lambda^2 - 0.2\lambda + 1.01)(-0.2 - \lambda),$$

so the eigenvalues are $\lambda_1 = -0.2$, $\lambda_2 = 0.1 + i$, and $\lambda_3 = 0.1 - i$. Corresponding to the real negative eigenvalue λ_1, we expect to see a line of solutions that approach the origin in the phase space. By analogy to the two-dimensional case, the complex eigenvalues with positive real part correspond to solutions that spiral away from the origin. This is a "spiral saddle," which is not possible in two dimensions.

We could find the eigenvectors associated with each eigenvalue as above and find the general solution. The eigenvectors for the complex eigenvalues are complex, and to find the real solutions, we would have to take real and imaginary parts, just as in two dimensions. However, we are lucky again, and this system also decouples into

$$\frac{dx}{dt} = 0.1x - y$$

$$\frac{dy}{dt} = x + 0.1y$$

and

$$\frac{dz}{dt} = -0.2z.$$

In the xy-plane, the eigenvalues are $0.1 \pm i$, so the origin is a spiral source. Along the z-axis, all solutions tend toward zero as time increases (see Figure 3.59).

Combining these pictures, we obtain a sketch of the three-dimensional phase space. Note that the z-coordinate of each solution decreases toward zero, while in the xy-plane solutions spiral away from the origin (see Figure 3.60).

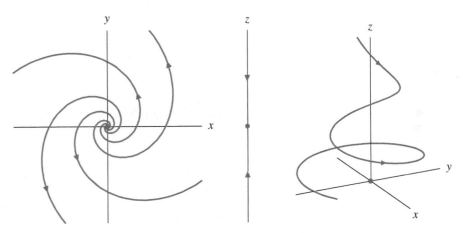

Figure 3.59
Phase plane for xy-system and phase line for z.

Figure 3.60
Phase space for $d\mathbf{Y}/dt = \mathbf{BY}$.

Classification of Three-Dimensional Linear Systems

Although there are more possible types of phase space pictures for three-dimensional linear systems than for two dimensions, the list is still finite. Just as for two dimensions, the nature of the system is determined by the eigenvalues. Real eigenvalues correspond to straight-line solutions that tend toward the origin if the eigenvalue is negative and away from the origin if the eigenvalue is positive. Complex eigenvalues correspond to spiraling. Negative real parts indicate spiraling toward the origin, whereas positive real parts indicate spiraling away from the origin.

Since the characteristic polynomial is a cubic, there are three eigenvalues (which might not all be distinct if there are repeated roots). It is always the case that at least one of the eigenvalues is real. The other two may be real or a complex conjugate pair (see exercises).

The most important types of three-dimensional linear systems can be divided into three categories: sinks, sources, and saddles. Examples of the other cases (which include systems with double eigenvalues and zero eigenvalues) are given in the exercises.

Sinks

We call the equilibrium point at the origin a sink if all solutions tend toward it as time increases. If all three eigenvalues are real and negative, then there are three straight lines of solutions, all of which tend toward the origin. Since every other solution is a linear combination of these solutions, all solutions tend to the origin as time increases (see Figure 3.58).

The other possibility for a sink is to have one real negative eigenvalue and two complex eigenvalues with negative real parts. This means that there is one straight line of solutions tending to the origin and a plane of solutions that spiral toward the origin. All other solutions exhibit both of these behaviors (see Figure 3.61).

Sources

There are two possibilities for sources as well. We can have either three real and positive eigenvalues or one real positive eigenvalue and a complex conjugate pair with positive real parts. An example of such a phase space is given in Figure 3.62. Note that this system looks just like the sink in Figure 3.61 except the directions of the arrows have been reversed, so solutions move away from the origin as time increases.

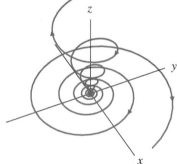

Figure 3.61
Example phase space for spiral sink.

Figure 3.62
Example phase space for spiral source.

Saddles

The equilibrium point at the origin is a saddle if, as time increases to infinity, some solutions tend toward it while other solutions move away from it. This can occur in four different ways. If all the eigenvalues are real, then we could have one positive and two negative or two positive and one negative. In the first case, one positive and two negative, there is one straight line of solutions that tend away from the origin as time increases and a plane of solutions that tend toward the origin as time increases. In the other case, two positive and one negative, there is a plane of solutions that tend away from the origin as time increases and a line of solutions that tend toward the origin as time increases. In both cases, all other solutions will eventually move away from the origin as time increases or decreases (see Figure 3.63).

The other two cases occur if there is only one real eigenvalue and the other two are a complex conjugate pair. If the real eigenvalue is negative and the real parts of the complex eigenvalues are positive, then as time increases there is a straight line of solutions that tend toward the origin and a plane of solutions that tend away from it. All other solutions are a combination of these behaviors, so as time increases they spiral around the straight line of solutions in ever widening loops (see Figure 3.60). The other possibility is that the real eigenvalue is positive and the complex eigenvalues have negative real part. In this case there is a straight line of solutions that tend away from the origin as time increases and a plane of solutions that spiral toward the origin as time increases. Every other solution spirals around the straight line of solutions while moving away from the origin (see Figure 3.64).

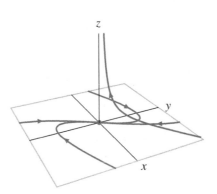

Figure 3.63
Example of a saddle with one positive and two negative eigenvalues.

Figure 3.64
Example of a saddle with one real eigenvalue and a complex conjugate pair of eigenvalues.

An example revisited

We end this section by returning to the example that we used at the start of the section. All of the other examples in this section have been systems that decouple into systems of smaller dimension. Sadly, the general case is not so simple. This example doesn't look too complicated because the coefficient matrix has many zero entries. However, it does not immediately decouple into lower dimensional systems.

Consider the system

$$\frac{d\mathbf{Y}}{dt} = \mathbf{AY} = \begin{pmatrix} 0 & 0.1 & 0 \\ 0 & 0 & 0.2 \\ 0.4 & 0 & 0 \end{pmatrix} \begin{pmatrix} x \\ y \\ z \end{pmatrix}.$$

The characteristic polynomial for $\mathbf{A}$ is $-\lambda^3 + 0.008$, so the eigenvalues are the solutions of

$$-\lambda^3 + 0.008 = 0.$$

That is, the eigenvalues are the cube roots of 0.008. Every number has three cube roots if we consider complex as well as real roots. The cube roots of 0.008 are $\lambda_1 = 0.2$, $\lambda_2 = 0.2e^{2\pi i/3}$, and $\lambda_3 = 0.2e^{-2\pi i/3}$. The last two eigenvalues may be written as $\lambda_2 = -0.1 + i\sqrt{0.03}$ and $\lambda_3 = -0.1 - i\sqrt{0.03}$.

This system is a saddle with one positive real eigenvalue and a complex conjugate pair of eigenvalues with negative real parts. Solutions spiral tightly around the line of eigenvectors associated to the eigenvalue $\lambda_1 = 0.2$. In order to sketch the phase space, we must find the eigenvectors for these eigenvalues.

For $\lambda_1 = 0.2$, the eigenvectors are solutions of

$$\mathbf{A}\mathbf{V}_1 = 0.2\mathbf{V}_1,$$

which is written in coordinates as

$$\begin{cases} 0.1y_1 = 0.2x_1 \\ 0.2z_1 = 0.2y_1 \\ 0.4x_1 = 0.2z_1. \end{cases}$$

In particular $\mathbf{V}_1 = (1/2, 1, 1)$ is one such eigenvector. The vector $\mathbf{V}_1$ can be used to determine an entire line of eigenvectors in space.

To find the plane of solutions that spiral toward the origin, we must find the eigenvectors for $\lambda_2 = -0.1 + i\sqrt{0.03}$. That is, we must solve

$$\mathbf{A}\mathbf{V}_2 = \left(-0.1 + i\sqrt{0.03}\right)\mathbf{V}_2$$

for $\mathbf{V}_2$. In other words,

$$y_2 = \left(-1 + i\sqrt{3}\right)x_2$$
$$2z_2 = \left(-1 + i\sqrt{3}\right)y_2$$
$$4x_2 = \left(-1 + i\sqrt{3}\right)z_2.$$

One eigenvector associated to λ_2 is $\mathbf{V}_2 = (-1+i\sqrt{3}, -2-i2\sqrt{3}, 4)$. The corresponding solution to the system is

$$\mathbf{Y}_2(t) = e^{\left(-0.1+i\sqrt{0.03}\right)t}\left(-1 + i\sqrt{3}, -2 - i2\sqrt{3}, 4\right).$$

We can convert this into two real-valued solutions by taking real and imaginary parts. Since our goal is to find the plane on which solutions spiral, we need only look at the initial point $\mathbf{Y}_2(0) = (-1 + i\sqrt{3}, -2 - i2\sqrt{3}, 4)$. The initial points of the real and imaginary parts are $(-1, -2, 4)$ and $(\sqrt{3}, -2\sqrt{3}, 0)$, respectively. The plane on which solutions spiral toward the origin is the plane made up of all linear combinations of these two vectors. We can use this information to give a fairly accurate sketch of the phase space of this system (see Figure 3.56). We also sketch the graphs of the coordinate functions for one solution (see Figures 3.65 and 3.66). Note that for the example

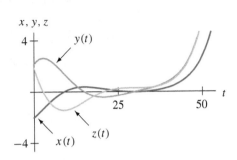

Figure 3.65
Phase space for system
$d\mathbf{Y}/dt = \mathbf{AY}$.

Figure 3.66
Graphs of $x(t)$, $y(t)$ and $z(t)$ for the indicated
solution in Figure 3.65.

solution shown, all three coordinates tend to infinity as t increases because the eigen-vector for the eigenvalue λ_1 has nonzero components for all three variables.

Three linearly independent solutions of this system are given in the first example of this section (see page 361). We can see from this example that linear systems in three dimensions can be quite complicated (even when many of the coefficients are zero). However, the qualitative behavior is still determined by the eigenvalues, so it is possible to classify these systems without completely solving them.

EXERCISES FOR SECTION 3.8

1. Consider the linear system

$$\frac{d\mathbf{Y}}{dt} = \mathbf{AY} = \begin{pmatrix} 0 & 0.1 & 0 \\ 0 & 0 & 0.2 \\ 0.4 & 0 & 0 \end{pmatrix} \begin{pmatrix} x \\ y \\ z \end{pmatrix}.$$

Check that the functions

$$\mathbf{Y}_2(t) = e^{-0.1t} \begin{pmatrix} -\cos\left(\sqrt{0.03}\,t\right) - \sqrt{3}\sin\left(\sqrt{0.03}\,t\right) \\ -2\cos\left(\sqrt{0.03}\,t\right) + 2\sqrt{3}\sin\left(\sqrt{0.03}\,t\right) \\ 4\cos\left(\sqrt{0.03}\,t\right) \end{pmatrix}$$

and

$$\mathbf{Y}_3(t) = e^{-0.1t} \begin{pmatrix} -\sin\left(\sqrt{0.03}\,t\right) + \sqrt{3}\cos\left(\sqrt{0.03}\,t\right) \\ -2\sin\left(\sqrt{0.03}\,t\right) - 2\sqrt{3}\cos\left(\sqrt{0.03}\,t\right) \\ 4\sin\left(\sqrt{0.03}\,t\right) \end{pmatrix}$$

are solutions to the system.

2. If a vector $\mathbf{Y}_3$ lies in the plane determined by the two vectors $\mathbf{Y}_1$ and $\mathbf{Y}_2$, then we can write $\mathbf{Y}_3$ as a linear combination of $\mathbf{Y}_1$ and $\mathbf{Y}_2$. That is,

$$\mathbf{Y}_3 = k_1\mathbf{Y}_1 + k_2\mathbf{Y}_2$$

for some constants k_1 and k_2. But then

$$k_1\mathbf{Y}_1 + k_2\mathbf{Y}_2 - \mathbf{Y}_3 = (0, 0, 0).$$

Show that if

$$k_1\mathbf{Y}_1 + k_2\mathbf{Y}_2 + k_3\mathbf{Y}_3 = (0, 0, 0),$$

with not all of k_1, k_2, and $k_3 = 0$, then the vectors are not linearly independent. [*Hint:* Start by assuming that $k_3 \neq 0$ and show that $\mathbf{Y}_3$ is in the plane determined by $\mathbf{Y}_1$ and $\mathbf{Y}_2$. Then treat the other cases.] Note that this computation leads to the theorem that three vectors $\mathbf{Y}_1$, $\mathbf{Y}_2$, and $\mathbf{Y}_3$ are linearly independent if and only if the only solution of

$$k_1\mathbf{Y}_1 + k_2\mathbf{Y}_2 + k_3\mathbf{Y}_3 = (0, 0, 0)$$

is $k_1 = k_2 = k_3 = 0$.

3. Using the technique of Exercise 2, determine whether or not the following sets of three vectors are linearly independent.

    (a) $(1, 2, 1)$, $(1, 3, 1)$, $(1, 4, 1)$

    (b) $(2, 0, -1)$, $(3, 2, 2)$, $(1, -2, -3)$

    (c) $(1, 2, 0)$, $(0, 1, 2)$, $(2, 0, 1)$

    (d) $(-3, \pi, 1)$, $(0, 1, 0)$, $(-2, -2, -2)$

In Exercises 4–7, consider the linear system $d\mathbf{Y}/dt = \mathbf{A}\mathbf{Y}$ with the coefficient matrix $\mathbf{A}$ specified. Each of these systems decouples into a two-dimensional system and a one-dimensional system. For each exercise,

  (a) compute the eigenvalues,

  (b) determine how the system decouples,

  (c) sketch the two-dimensional phase plane and one-dimensional phase line for the decoupled systems, and

  (d) give a rough sketch of the phase portrait of the system.

4. $\mathbf{A} = \begin{pmatrix} 0 & 1 & 0 \\ -1 & 0 & 0 \\ 0 & 0 & 2 \end{pmatrix}$
        5. $\mathbf{A} = \begin{pmatrix} -2 & 3 & 0 \\ 3 & -2 & 0 \\ 0 & 0 & -1 \end{pmatrix}$

6. $\mathbf{A} = \begin{pmatrix} 1 & 0 & 3 \\ 0 & -1 & 0 \\ -3 & 0 & 1 \end{pmatrix}$
        7. $\mathbf{A} = \begin{pmatrix} 1 & 0 & 0 \\ 0 & 2 & -1 \\ 0 & -1 & 2 \end{pmatrix}$

Exercises 8–9 consider the properties of the cubic polynomial

$$p(\lambda) = \alpha\lambda^3 + \beta\lambda^2 + \gamma\lambda + \delta,$$

where α, β, γ, and δ are real numbers.

8. (a) Show that, if α is positive, then the limit of $p(\lambda)$ as $\lambda \to \infty$ is ∞ and the limit of $p(\lambda)$ as $\lambda \to -\infty$ is $-\infty$.
    (b) Show that, if α is negative, then the limit of $p(\lambda)$ as $\lambda \to \infty$ is $-\infty$ and the limit of $p(\lambda)$ as $\lambda \to -\infty$ is ∞.
    (c) Using the above, show that $p(\lambda)$ must have at least one real root (that is, at least one real number λ_0 such that $p(\lambda_0) = 0$). [*Hint*: Look at the graph of $p(\lambda)$.]

9. Suppose $a + ib$ is a root of $p(\lambda)$ (so $p(a + ib) = 0$). Show that $a - ib$ is also a root. [*Hint*: Remember that a complex number is zero if and only if both its real and imaginary parts are zero. Then compute $p(a + ib)$ and $p(a - ib)$.]

In Exercises 10–13, consider the linear system $d\mathbf{Y}/dt = \mathbf{BY}$ with the coefficient matrix $\mathbf{B}$ specified. These systems do not fit into the classification of the most common types of systems given in the text. However, the equations for dx/dt and dy/dt decouple from dz/dt. For each of these systems,
 (a) compute the eigenvalues,
 (b) sketch the xy-phase plane and the z-phase line, and
 (c) give a rough sketch of the phase portrait of the system.

10. $\mathbf{B} = \begin{pmatrix} -2 & 1 & 0 \\ 0 & -2 & 0 \\ 0 & 0 & -1 \end{pmatrix}$

11. $\mathbf{B} = \begin{pmatrix} -2 & 1 & 0 \\ 0 & -2 & 0 \\ 0 & 0 & 1 \end{pmatrix}$

12. $\mathbf{B} = \begin{pmatrix} -1 & 2 & 0 \\ 2 & -4 & 0 \\ 0 & 0 & -1 \end{pmatrix}$

13. $\mathbf{B} = \begin{pmatrix} -1 & 2 & 0 \\ 2 & -4 & 0 \\ 0 & 0 & 0 \end{pmatrix}$

In Exercises 14–15, consider the linear system $d\mathbf{Y}/dt = \mathbf{CY}$. These systems do not fit into the classification of the most common types of systems given in the text, and they do not decouple into lower-dimensional systems. For each system,
 (a) compute the eigenvalues,
 (b) compute the eigenvectors, and
 (c) sketch (as best you can) the phase portrait of the system. [*Hint*: Use the eigenvalues and eigenvectors and also vectors in the vector field.]

14. $\mathbf{C} = \begin{pmatrix} -2 & 1 & 0 \\ 0 & -2 & 1 \\ 0 & 0 & -2 \end{pmatrix}$

15. $\mathbf{C} = \begin{pmatrix} 0 & 1 & 0 \\ 0 & 0 & 1 \\ 0 & 0 & 0 \end{pmatrix}$

16. For the linear system

$$\frac{d\mathbf{Y}}{dt} = \mathbf{AY} = \begin{pmatrix} 2 & -1 & 0 \\ 0 & -2 & 3 \\ -1 & 3 & -1 \end{pmatrix} \begin{pmatrix} x \\ y \\ z \end{pmatrix} :$$

(a) Show that $\mathbf{V}_1 = (1, 1, 1)$ is an eigenvector of the coefficient matrix by computing $\mathbf{AV}_1$. What is the eigenvalue for this eigenvector?

(b) Find the other two eigenvalues for the matrix $\mathbf{A}$.

(c) Classify the system (source, sink, ...).

(d) Sketch (as best you can) the phase portrait. [*Hint*: Use the other eigenvalues and find the other eigenvectors.]

17. For the linear system

$$\frac{d\mathbf{Y}}{dt} = \mathbf{AY} = \begin{pmatrix} -4 & 3 & 0 \\ 0 & -1 & 1 \\ 5 & -5 & 0 \end{pmatrix} \begin{pmatrix} x \\ y \\ z \end{pmatrix} :$$

(a) Show that $\mathbf{V}_1 = (1, 1, 0)$ is an eigenvector of the coefficient matrix by computing $\mathbf{AV}_1$. What is the eigenvalue for this eigenvector?

(b) Find the other two eigenvalues for the matrix $\mathbf{A}$.

(c) Classify the system (source, sink, ...).

(d) Sketch (as best you can) the phase portrait. [*Hint*: Use the other eigenvalues and find the other eigenvectors.]

18. Consider the linear system

$$\frac{d\mathbf{Y}}{dt} = \mathbf{BY} = \begin{pmatrix} -10 & 10 & 0 \\ 28 & -1 & 0 \\ 0 & 0 & -8/3 \end{pmatrix} \begin{pmatrix} x \\ y \\ z \end{pmatrix} .$$

(This system is related to the Lorenz system studied in Section 2.8, and we will use the results obtained in this exercise when we return to the Lorenz equations in Section 5.5.)

(a) Find the characteristic polynomial and the eigenvalues.

(b) Find the eigenvectors.

(c) Sketch the phase portrait (as best you can).

(d) Comment on how the fact that the system "decouples" helps in the computations and in sketching the phase space.

Many years later, when Glen finally retires from writing math texts, he decides to join his friends and former collaborators Paul and Bob. He opens an ice creme store between Paul's and Bob's cafés. Let $z(t)$ be Glen's profits at time t (with $x(t)$ and $y(t)$ representing Paul's and Bob's profits, respectively). Suppose the three stores affect each other in such a way that

$$\frac{dx}{dt} = -y + z$$
$$\frac{dy}{dt} = -x + z$$
$$\frac{dz}{dt} = z.$$

19. **(a)** If Glen makes a profit, does this help or hurt Paul's and Bob's profits?

    (b) If Paul and Bob are making profits, does this help or hurt Glen's profits?

20. Write this system in matrix form and find the eigenvalues. Use them to classify the system.

21. Suppose that at time $t = 0$, both Paul and Bob are making (equal) small profits, but Glen is just breaking even [$x(0) = y(0)$ are small and positive, but $z(0) = 0$].

    (a) Sketch the solution curve in the xyz-phase space.

    (b) Sketch the $x(t)$-, $y(t)$-, and $z(t)$-graphs of the solution.

    (c) Describe what happens to the profits of each store.

22. Suppose that at time $t = 0$ both Paul and Bob are just breaking even, but Glen is making a small profit [$x(0)$ and $y(0)$ are zero, but $z(0)$ is small and positive].

    (a) Sketch the solution curve in the xyz-phase space.

    (b) Sketch the $x(t)$-, $y(t)$-, and $z(t)$-graphs of the solution.

    (c) Describe what happens to the profits of each store.

REVIEW EXERCISES FOR CHAPTER 3

Short answer exercises: Exercises 1–10 focus on the basic ideas, definitions, and vocabulary of this chapter. Their answers are short (a single sentence or drawing), and you should be able to do them with little or no computation. However, they vary in difficulty, so think carefully before you answer.

1. What are the eigenvalues of the matrix

$$\mathbf{A} = \begin{pmatrix} 1 & 0 \\ 0 & 2 \end{pmatrix}?$$

2. What are the eigenvalues of the matrix

$$\mathbf{B} = \begin{pmatrix} 0 & 1 \\ 2 & 0 \end{pmatrix}?$$

3. Compute the general solution of the system

$$\frac{d\mathbf{Y}}{dt} = \begin{pmatrix} 3 & 0 \\ 0 & -2 \end{pmatrix} \mathbf{Y}$$

and sketch its phase portrait.

4. Which of the following vectors are eigenvectors for the matrix

$$\mathbf{A} = \begin{pmatrix} 1 & 0 \\ 2 & 3 \end{pmatrix}?$$

(i) $\quad \mathbf{Y}_1 = \begin{pmatrix} 0 \\ 0 \end{pmatrix}$

(ii) $\quad \mathbf{Y}_2 = \begin{pmatrix} 2 \\ -2 \end{pmatrix}$

(iii) $\quad \mathbf{Y}_3 = \begin{pmatrix} 1 \\ 1 \end{pmatrix}$

(iv) $\quad \mathbf{Y}_4 = \begin{pmatrix} 0 \\ 1 \end{pmatrix}$

(v) $\quad \mathbf{Y}_5 = \begin{pmatrix} -4 \\ 4 \end{pmatrix}$

(vi) $\quad \mathbf{Y}_6 = \begin{pmatrix} 1 \\ 0 \end{pmatrix}$

5. Consider the harmonic oscillator with mass 1, spring constant 5, and damping coefficient b. Find the values of b for which the system is overdamped, underdamped, critically damped, or undamped.

6. Compute the equilibrium points for the system

$$\frac{d\mathbf{Y}}{dt} = \begin{pmatrix} 0 & 0 \\ 1 & -1 \end{pmatrix} \mathbf{Y}.$$

7. Solve the initial-value problem

$$\frac{d\mathbf{Y}}{dt} = \begin{pmatrix} \pi^2 & 37.4 \\ \sqrt{555} & 8.01234 \end{pmatrix} \mathbf{Y}, \quad \mathbf{Y}(0) = \begin{pmatrix} 0 \\ 0 \end{pmatrix}.$$

8. For each function $y(t)$, determine if it is a solution to the second-order equation

$$\frac{d^2 y}{dt^2} + ky = 0,$$

assuming k satisfies the given condition.

(a) $y(t) = \sin kt, \ k < 0$ (b) $y(t) = 0, \ k > 0$
(c) $y(t) = t^2, \ k = 0$ (d) $y(t) = \sin \sqrt{k}\, t + 2\cos \sqrt{k}\, t, \ k > 0$
(e) $y(t) = e^{kt}, \ k > 0$ (f) $y(t) = e^{\sqrt{-kt}}, \ k < 0$

9. Find a linear system for which the function $\mathbf{Y}(t) = (3\cos 2t, \sin 2t)$ is a solution.

10. Compute the general solution of the system

$$\frac{d\mathbf{Y}}{dt} = \begin{pmatrix} 0 & 1 \\ 0 & 0 \end{pmatrix} \mathbf{Y}$$

and sketch its phase portrait.

True-false: For Exercises 11–18, determine if the statement is true or false. If it is true, explain why. If it is false, provide a counterexample or an explanation.

11. The origin is the only equilibrium point for any linear system.

12. If $\mathbf{Y}_0$ is an eigenvector for a matrix, then so is any nonzero scalar multiple of $\mathbf{Y}_0$.

13. The function $\mathbf{Y}(t) = (\cos 2t, \sin t)$ is not a solution to any linear system.

14. The graph on the right is the graph of the solution of a damped harmonic oscillator.

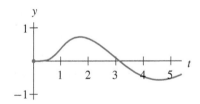

15. The graph on the right is the graph of a typical solution of a damped harmonic oscillator.

16. If k increases, then the time between successive maxima of solutions of the harmonic oscillator $d^2 y/dt^2 + ky = 0$ decreases.

17. Suppose $\mathbf{A}$ is a 2×2 matrix. The linear system $d\mathbf{Y}/dt = \mathbf{AY}$ can have three different straight-line solutions.

18. Suppose $\mathbf{A}$ is a 2×2 matrix. No solution of $d\mathbf{Y}/dt = \mathbf{AY}$ can blow up in finite time.

19. Eight matrices and four phase portraits are given below. For each matrix, form the associated linear system, and determine which system corresponds to each phase portrait. State briefly how you know your choice is correct. You should do this exercise without using technology.

(i) $\begin{pmatrix} 1 & 1 \\ -2 & -1 \end{pmatrix}$ (ii) $\begin{pmatrix} -3 & 1 \\ -1 & 1 \end{pmatrix}$ (iii) $\begin{pmatrix} -3 & 1 \\ -1 & 0 \end{pmatrix}$ (iv) $\begin{pmatrix} -1 & 1 \\ -2 & 1 \end{pmatrix}$

(v) $\begin{pmatrix} 2 & 0 \\ 1 & -1 \end{pmatrix}$ (vi) $\begin{pmatrix} 3 & 1 \\ -1 & 0 \end{pmatrix}$ (vii) $\begin{pmatrix} 0 & 1 \\ -4 & -4 \end{pmatrix}$ (viii) $\begin{pmatrix} -3 & -3 \\ 2 & 1 \end{pmatrix}$

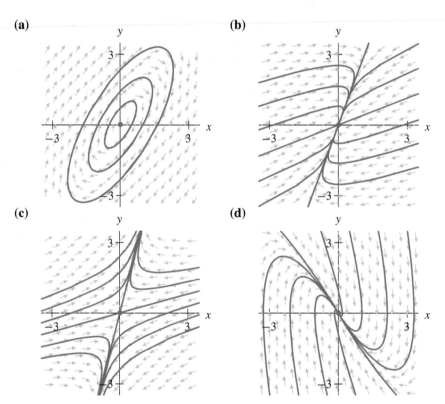

20. Consider the one-parameter family of linear systems

$$\frac{d\mathbf{Y}}{dt} = \begin{pmatrix} 0 & 3a \\ 1 & a \end{pmatrix} \mathbf{Y}.$$

(a) Draw the curve in the trace-determinant plane that is obtained from varying the parameter a.

(b) Determine all bifurcation values of a and briefly discuss the different types of phase portraits that are exhibited in this one-parameter family.

21. Eight matrices and four pairs of $x(t)$- and $y(t)$-graphs of solutions to linear systems are given below. For each matrix, form the associated linear system, and determine which system corresponds to each pair of graphs. State briefly how you know your choice is correct. You should do this exercise without using technology.

(i) $\begin{pmatrix} 3 & 4 \\ 1 & 0 \end{pmatrix}$ (ii) $\begin{pmatrix} 2 & 0 \\ 1 & 5 \end{pmatrix}$

(iii) $\begin{pmatrix} -3 & -2 \\ 0 & -1 \end{pmatrix}$ (iv) $\begin{pmatrix} 0 & -1 \\ 4 & 0 \end{pmatrix}$

(v) $\begin{pmatrix} 1 & 2 \\ -5 & -1 \end{pmatrix}$ (vi) $\begin{pmatrix} 1 & .25 \\ .25 & 1 \end{pmatrix}$

(vii) $\begin{pmatrix} -1.1 & -2 \\ 2 & -1.1 \end{pmatrix}$ (viii) $\begin{pmatrix} -1.1 & -5 \\ 1 & 0.9 \end{pmatrix}$

(a)

(b)

(c)

(d)

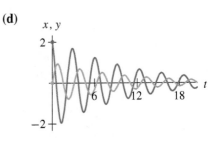

22. **(a)** Give an example of a linear system that has an equilibrium solution $(x(t), y(t))$ such that $x(0) = -1$ and $y(t) = 2x(t)$ for all t.

(b) Give an example of a linear system that has a straight-line solution $(x(t), y(t))$ such that $x(0) = -1$ and $y(t) = 2x(t)$ for all t.

In Exercises 23–26, find the solution (in scalar form) of the given initial-value problem.

23. $\dfrac{d^2y}{dt^2} + 5\dfrac{dy}{dt} + 6y = 0$

$y(0) = 0, \; y'(0) = 2$

24. $\dfrac{d^2y}{dt^2} + 2\dfrac{dy}{dt} + 5y = 0$

$y(0) = 3, \; y'(0) = -1$

25. $\dfrac{d^2y}{dt^2} + 2\dfrac{dy}{dt} + y = 0$

$y(0) = 1, \; y'(0) = 1$

26. $\dfrac{d^2y}{dt^2} + 2y = 0$

$y(0) = 3, \; y'(0) = -\sqrt{2}$

In Exercises 27–32, a linear system and an initial condition are given. For each system,

(a) compute the general solution;

(b) sketch its phase portrait;

(c) solve the initial-value problem; and

(d) sketch the $x(t)$- and $y(t)$-graphs for the solution to the initial-value problem.

27. $\dfrac{d\mathbf{Y}}{dt} = \begin{pmatrix} 1 & 3 \\ 1 & -1 \end{pmatrix} \mathbf{Y}, \quad \mathbf{Y}(0) = \begin{pmatrix} -2 \\ 3 \end{pmatrix}$

28. $\dfrac{d\mathbf{Y}}{dt} = \begin{pmatrix} 4 & 2 \\ 1 & 3 \end{pmatrix} \mathbf{Y}, \quad \mathbf{Y}(0) = \begin{pmatrix} 0 \\ 1 \end{pmatrix}$

29. $\dfrac{d\mathbf{Y}}{dt} = \begin{pmatrix} -2 & 3 \\ -2 & 2 \end{pmatrix} \mathbf{Y}, \quad \mathbf{Y}(0) = \begin{pmatrix} -2 \\ 2 \end{pmatrix}$

30. $\dfrac{d\mathbf{Y}}{dt} = \begin{pmatrix} -3 & 6 \\ -2 & 1 \end{pmatrix} \mathbf{Y}, \quad \mathbf{Y}(0) = \begin{pmatrix} -7 \\ 7 \end{pmatrix}$

31. $\dfrac{d\mathbf{Y}}{dt} = \begin{pmatrix} -3 & 1 \\ -1 & -1 \end{pmatrix} \mathbf{Y}, \quad \mathbf{Y}(0) = \begin{pmatrix} -3 \\ 1 \end{pmatrix}$

32. $\dfrac{d\mathbf{Y}}{dt} = \begin{pmatrix} 0 & 1 \\ -2 & -3 \end{pmatrix} \mathbf{Y}, \quad \mathbf{Y}(0) = \begin{pmatrix} 0 \\ 3 \end{pmatrix}$

LAB 3.1   Bifurcations in Linear Systems

In Chapter 3, we have studied techniques for solving linear systems. Given the coefficient matrix for the system, we can use these techniques to classify the system, describe the qualitative behavior of solutions, and give a formula for the general solution. In this lab we consider a two-parameter family of linear systems. The goal is to better understand how different linear systems are related to each other, or in other words, what bifurcations occur in parameterized families of linear systems.

Consider the linear system

$$\frac{dx}{dt} = ax + by$$
$$\frac{dy}{dt} = -x - y,$$

where a and b are parameters that can take on any real value. In your report, address the following items:

1. For each value of a and b, classify the linear system as source, sink, center, spiral sink, and so forth. Draw a picture of the ab-plane and indicate the values of a and b for which the system is of each type (that is, shade the values of a and b for which the system is a sink red, for which it is a source blue, and so forth). Be sure to describe all of the computations involved in creating this picture.

2. As the values of a and b are changed so that the point (a, b) moves from one region to another, the type of the linear system changes, that is, a bifurcation occurs. Which of these bifurcations is important for the long-term behavior of solutions? Which of these bifurcations corresponds to a dramatic change in the phase plane or the $x(t)$- and $y(t)$-graphs?

Your report: Address the items above in the form of a short essay. Include any computations necessary to produce the picture in Part 1. You may include phase planes and/or graphs of solutions to illustrate your essay, but your answer should be complete and understandable without the pictures.

LAB 3.2   RLC Circuits

We have already seen examples of differential equations that serve as models of simple electrical circuits involving only a resistor, a capacitor, and a voltage source. In this lab we consider slightly more complicated circuits consisting of a resistor, a capacitor, an inductor, and a voltage source (see Figure 3.67). The behavior of the system can be described by specifying the current moving around the circuit and the changes in voltages across each component of the circuit. In this lab we take an axiomatic approach to the relationship between the current and the voltages. Readers interested in more information on the derivation of these laws are referred to texts in electric circuit theory.

Figure 3.67
An RLC circuit.

Following the conventions used by electrical engineers, we let i denote the current moving around the circuit. We let v_T, v_C, and v_L denote the voltages across the voltage source, the capacitor, and the inductor, respectively. Also, we let R denote the resistance, C the capacitance, and L the inductance of the associated components of the circuit (see Figure 3.67). We think of v_T, R, C, and L as parameters set by the person building the circuit. The quantities i, v_C, and v_L depend on time.

We need the following basic relationships between the quantities above. First, Kirchhoff's voltage law states that the sum of the voltage changes around a closed loop must be zero. For our circuit this gives

$$v_T - Ri = v_C + v_L.$$

Next, we need the relationship between current and voltage in the capacitor and the inductor. In a capacitor the current is proportional to the rate of change of the voltage. The proportionality constant is the capacitance C. Hence we have

$$C \frac{dv_C}{dt} = i.$$

In an inductor, the voltage is proportional to the rate of change of the current. The proportionality constant is the inductance L. Hence we have

$$L \frac{di}{dt} = v_L.$$

In this lab we consider the possible behavior of the circuit above for several different input voltages.

In your report, address the following questions:

1. First, set the input voltage to zero, that is, assume $v_T = 0$. Using the three equations above, write a first-order system of differential equations with dependent variables i and v_C. [*Hint*: Use the first equation to eliminate v_L from the third equation. You should have R, C, and L as parameters in your system.]

2. Find the eigenvalues of the resulting system in terms of the parameters R, C, and L. What are the possible phase planes for your system given that R, C, and L are always nonnegative? Sketch the phase plane and the $v_C(t)$- and $i(t)$-graphs for each case.

3. Convert the first-order system of equations from Part 1 into a second-order differential equation involving only v_C (and not i). (This is the form of the equation that you will typically find in electric circuit theory texts.)

4. Repeat Part 1, assuming that v_T is nonzero. The resulting system will have R, C, L, and v_T as parameters.

5. The units used in applications are volts and amps for voltages and currents, ohms for resistors, farads for capacitors, and henrys for inductors. A typical, off-the-shelf circuit might have parameter values $R = 2000$ ohms (or 2 kilo-ohms), $C = 2 \cdot 10^{-7}$ farads (or 0.2 microfarads), and $L = 1.5$ henrys. Assuming zero input, $v_T = 0$, and that the initial values of the current and voltage are $i(0) = 0$ and $v_C(0) = 10$, describe the behavior of the current and voltage for this circuit.

6. Repeat Part 5 using a voltage source of $v_T = 10$ volts.

Your report: Address each of the items above. Show all algebra and justify all steps. In Parts 5 and 6, you may work either analytically or numerically. Give phase portraits and graphs of solutions as appropriate.

LAB 3.3 Measuring Mass in Space

The effects on the human body of prolonged weightlessness during space flights is not completely understood. One important variable that must be monitored is the astronaut's "weight." However, weight refers to the force of gravity on a body. What actually must be measured is body mass. To perform this measurement, a mass-spring system is used, where the mass is the body of the astronaut. The astronaut sits in a special chair attached to springs. The frequency of the oscillation of the astronaut in the chair is measured and from this the mass is computed.*
In your report, address the following items:

1. Suppose the chair has a mass of 20kg. The system is initially calibrated by placing a known mass in the chair and measuring the period of oscillations. Suppose that a 25kg mass placed in the chair results in an oscillation with period of 1.3 seconds per oscillation. We assume that the coefficient of damping of the apparatus is very small (so as a first approximation we assume that there is no damping). What will be the period of oscillations of an astronaut with mass 60kg? What would your frequency of oscillation be?

2. Does it matter whether or not the calibration is done on the earth or in space? (It would be much better if it could be done on the earth since it is expensive to launch 25kg masses into space.)

3. Suppose an error is made during the calibration, and the actual frequency resulting

---

*This information comes from the "Q & A" column of the *New York Times*, Science Section, August 1, 1995, page C6.

when a 25kg mass is placed in the chair is 1.31 seconds instead of 1.3 seconds. How much error then results in the measurement of the mass of astronaut with mass 60kg? With mass 80kg?

4. Suppose a small amount of damping develops in the chair. How seriously does this affect the measurements? How could you determine if damping were present (that is, what measurements would you perform during the calibration phase)?

Your report: Address each of the items above. Show all algebraic computations you perform and justify all assertions. While this lab does not require numerical approximations of solutions (since we can explicitly solve all the equations involved), you may include sketches of solutions and/or computer-generated graphs *if appropriate*.

LAB 3.4 Exploring a Parameter Space

In Lab 3.1, we studied a specific two-parameter family of linear systems. In this lab, we investigate the three-parameter family

$$\frac{dx}{dt} = ax + by$$

$$\frac{dy}{dt} = cx.$$

The goal is to produce a "picture" of abc-space, indicating the regions where this system has various types of behavior (spiral sinks, repeated eigenvalues, saddles, etc.). Making a representation of three-dimensional space is difficult, so be creative.

In your report, you should address the following items:

1. First consider the case where $a = 0$. Compute the eigenvalues for this case and determine the exact bc-values where this system has certain types of phase portraits, for example, spiral sinks, sources, saddles, etc. Draw an accurate picture of the bc-plane, indicating these different regions. Display *all* types that arise. Also indicate where you find special situations such as repeated or zero eigenvalues.

2. Repeat Part 1 for $a = 1$.

3. Describe the behavior of the system for $0 < a < 1$.

4. Repeat Part 1 for $a = -1$.

5. "Draw" a three-dimensional picture, with a as the vertical axis and the bc-plane perpendicular to the vertical axis. Highlight the special cases where your system changes type. Be creative.

Your report: Address the items above, paying particular attention to the quality of your representations of the bc-plane and abc-space.

LAB 3.5 Find Your Own Harmonic Oscillator

In the text, we claim that the harmonic oscillator can be used as a simple model for many different situations. The key ingredients are a restoring force, which pushes the system back toward a rest position, and (perhaps) a damping force.

For this lab, you are to find such a system. It may be mechanical, biological, psychological, political, financial, or whatever. You will have to do some analysis on your system, so you will need to be able to obtain some data on its "motion." You can either record the data yourself or use published data (with proper references).

In your report, address the following items:

1. In a short essay, carefully describe the system you propose to model with a harmonic oscillator equation. State the dependent variable and describe the rest position, the restoring force, and the damping.

2. Collect data on the behavior of your system either by observing your system or from published sources (taking care to give proper references).

3. Use your data to estimate the parameter values appropriate for your system. Classify your system as overdamped, critically damped, or underdamped. Compare solutions of the harmonic oscillator equation with the data you have collected. Describe how well the solutions of the model fit the data and discuss any discrepancies.

Your report: Address the three items above in the form of a short essay. In Part 2, describe how you obtained your data (either citing appropriate references or giving details of your experimental procedure). In Part 3, be sure to address any problems that arose in using the harmonic oscillator as a model. (If your data differs significantly from your solutions, then you may want to adjust the parameter values that you are using or search for another situation that is better modeled by the harmonic oscillator.)

LAB 3.6 A Baby Bottle Harmonic Oscillator

If you performed the previous lab experiment, you undoubtedly saw that harmonic oscillators are everywhere. Here is a harmonic oscillator that one of the authors discovered late one night.

Consider a U-shaped tube or pipe partially filled with fluid. (For example, a common design of baby bottle has the shape shown in Figure 3.68.)

If held upright, the level of fluid will come to rest at an equal height on each side of the bottle. If we tip the bottle quickly to one side, then return it to level, there will be more fluid in one side than in the other. Gravity pulling on this extra fluid will provide a restoring force, pushing the system back toward the rest position where both sides are at an equal level.

Figure 3.68
A U-shaped baby bottle.

In your report, address the following items:

1. Let $y(t)$ denote the height of the fluid in one side of bottle, with $y = 0$ corresponding to the rest position. Assume that the restoring force is proportional to the difference in the mass in the left- and right-hand sides of the bottle. Also assume, as a first approximation, that the damping force is proportional to the velocity. Use this to give a differential equation model of this system based on the harmonic oscillator equation.

2. Buy or borrow such a baby bottle and observe it. Estimate the natural period of the oscillations and the rate of decay of the amplitude of the oscillations. Use the observations to estimate the parameter values for the harmonic oscillator equation model.

3. Compare solutions of the harmonic oscillator equation in Part 2 to your observations. In particular, if you increase the total mass in the bottle but leave the size of the initial displacement unchanged, does the model make the correct prediction of the change of behavior of the fluid in the bottle? For example, does the natural period increase or decrease in this situation?

Your report: Address each of the three questions above in the form of a short essay. In Part 1 carefully describe the relationship between the forces and the differential equation model. In Part 2 describe the experiments you performed to obtain data and the computations necessary for computing the coefficients in the harmonic oscillator equation from your data. In Part 3 discuss any discrepancies between your data and the predictions of the model.

4 FORCING AND RESONANCE

In this chapter we consider some particularly important nonautonomous, second-order equations and systems, along with the mathematical techniques used to study them. These examples are important because they occur in numerous applications and because the mathematical ideas developed to attack them are quite appealing.

We begin with a system called the forced harmonic oscillator. This system models (among other things) a mass attached to a spring sliding on a table. The only forces included in the model are the restoring force of the spring and the damping. Any other force affecting the motion of the mass is referred to as an external force. A harmonic oscillator with external force that depends only on time is a forced harmonic oscillator.

In the first four sections we give techniques for finding solutions of forced harmonic oscillator equations for typical forcing functions, and using these special equations as a guide, we discuss the phenomenon of resonance. In the final section we discuss recent models that explain the collapse of the Tacoma Narrows Bridge.

4.1   FORCED HARMONIC OSCILLATORS

In Sections 2.1 and 2.3 we introduced the harmonic oscillator equation as a model for a physical system of a mass attached to a spring that slides on a table. The mass is subject both to a restoring force provided by the spring and to damping. The restoring force is assumed to be proportional to the displacement, and the damping is assumed to be proportional to the velocity. The proportionality constants are parameters called the spring constant and the coefficient of damping. We obtain a linear, second-order, constant coefficient equation. In Section 3.6 we used the techniques from the study of linear systems to classify the possible behaviors of solutions of the harmonic oscillator equation. Moreover, we described an efficient method for generating the general solution of these equations.

In this chapter we consider the effect of "external forces," that is, forces other than the restoring force and the damping. Such forces may include shaking the table or pushing the mass.

Another device that can be very informally modeled using the harmonic oscillator system is the motion of a crystal glass after being gently tapped with a fork. The position variable y measures the amount the shape of the glass is deformed from its rest position. The restoring force pushes the glass back toward its original shape. After the tap with the fork, the ringing sound is caused by oscillations of the glass around the rest position, so this is an underdamped oscillator. The frequency of the sound is the frequency of the oscillations. As an example of an external force, we imagine an opera singer standing near the glass, singing a fixed note. The sound waves push against the glass and deform it slightly. The size and direction of this external force depend only on time.

In the first four sections of this chapter we consider how external forces affect the motion of a harmonic oscillator. In particular, how can an opera singer break the glass by singing a particular note?

An Equation for the Forced Harmonic Oscillator

Recall that the equations for the harmonic oscillator come from Newton's second law

$$\text{force} = \text{mass} \times \text{acceleration},$$

applied to the motion of a mass attached to a spring, sliding on a table. We let $y(t)$ denote the position of the mass m at time t, with $y = 0$ the rest position. The forces on the mass are the spring force, $-ky$, and the damping, $-b(dy/dt)$. Substituting into Newton's law gives

$$-ky - b\frac{dy}{dt} = m\frac{d^2y}{dt^2},$$

which is often written as

$$\frac{d^2y}{dt^2} + \frac{b}{m}\frac{dy}{dt} + \frac{k}{m}y = 0.$$

The parameters are $m > 0$, $k > 0$, and $b \geq 0$.

To include an external force, we must add another term to the right-hand side of this equation. This forcing term can be any function. We consider external forcing that depends only on time t, so the external force is given by a function $f(t)$. Typical examples of forcing functions include constant functions (for a force pushing at a constant strength against the mass) and $f(t) = e^{-at}$ (for a force whose strength decreases exponentially with time). A particularly important example is $f(t) = \sin \omega t$ or $\cos \omega t$, called sinusoidal forcing with period $2\pi/\omega$ (or frequency $\omega/(2\pi)$). This corresponds to a force that alternately pushes and pulls the mass back and forth in a periodic fashion (like sound waves pushing against a glass).

The new equation is

$$m\frac{d^2y}{dt^2} = -ky - b\frac{dy}{dt} + f(t),$$

which is also written

$$\frac{d^2y}{dt^2} + \frac{b}{m}\frac{dy}{dt} + \frac{k}{m}y = \frac{f(t)}{m}.$$

If we let $p = b/m$, $q = k/m$, and $g(t) = f(t)/m$, then we have

$$\frac{d^2y}{dt^2} + p\frac{dy}{dt} + qy = g(t).$$

This is a second-order, linear, constant-coefficient, **nonhomogeneous**, nonautonomous equation. As in Section 1.8, the adjective *nonhomogeneous* refers to the fact that the right-hand side of the equation is nonzero. We also refer to this equation as a **forced harmonic oscillator**, or a **forced equation**, for short. We often abuse terminology slightly and refer to $g(t)$ as the forcing function. The equation

$$\frac{d^2y}{dt^2} + p\frac{dy}{dt} + qy = 0$$

is said to be the **associated homogeneous** or **unforced** equation.

Our goal for the equation of the forced harmonic oscillator is the same as for the harmonic oscillator. Given the parameters p and q and the forcing function $g(t)$, describe the motion of the mass, produce graphs of the solutions and, if possible, give formulas for solutions. In this section we give examples of a method for finding explicit solutions of the forced harmonic oscillator system for special types of forcing functions.

The method used here is a "lucky guess" method with the impressive name of the *Method of Undetermined Coefficients*. An alternative, somewhat more systematic technique is the method of Laplace transforms given in Chapter 6. Again it must be emphasized that as with nonlinear equations, there is no method that gives the general solution for arbitrary forcing functions. The equations we can solve explicitly are special cases. The goal of this section is to obtain the general solution for a few special cases and to study the behavior of solutions in these cases.

The Extended Linearity Principle

To find the general solution of an equation of the form

$$\frac{d^2y}{dt^2} + p\frac{dy}{dt} + qy = g(t),$$

we take full advantage of the fact that we already know how to find the general solution of the homogeneous equation

$$\frac{d^2y}{dt^2} + p\frac{dy}{dt} + qy = 0.$$

We could convert this second-order equation into a first-order system and find the general solution by computing eigenvalues and eigenvectors. However, in Section 3.6 we pointed out that the eigenvalues and the general solution can be obtained quickly by a guess-and-test method. We use this efficient method throughout this chapter.

The key to computing the general solution of a force harmonic oscillator equation is an extension of the Linearity Principle of Chapter 3.

EXTENDED LINEARITY PRINCIPLE   Consider a nonhomogeneous equation (a forced equation)

$$\frac{d^2y}{dt^2} + p\frac{dy}{dt} + qy = g(t)$$

and its associated homogeneous equation (the unforced equation)

$$\frac{d^2y}{dt^2} + p\frac{dy}{dt} + qy = 0.$$

1. Suppose $y_p(t)$ is a particular solution of the nonhomogeneous equation and $y_h(t)$ is a solution of the associated homogeneous equation. Then $y_h(t) + y_p(t)$ is also a solution of the nonhomogeneous equation.

2. Suppose $y_p(t)$ and $y_q(t)$ are two solutions of the nonhomogeneous equation. Then $y_p(t) - y_q(t)$ is a solution of the associated homogeneous equation.

Therefore, if $k_1 y_1(t) + k_2 y_2(t)$ is the general solution of the homogeneous equation, then

$$k_1 y_1(t) + k_2 y_2(t) + y_p(t)$$

is the general solution of the nonhomogeneous equation.   ■

We can verify this theorem by substituting the given functions into the non-homogeneous equation. To verify the first assertion, we substitute $y_h(t) + y_p(t)$ into the nonhomogeneous equation and get

$$\frac{d^2}{dt^2}(y_h + y_p) + p\frac{d}{dt}(y_h + y_p) + q(y_h + y_p)$$

$$= \left(\frac{d^2 y_h}{dt^2} + \frac{d^2 y_p}{dt^2}\right) + p\left(\frac{dy_h}{dt} + \frac{dy_p}{dt}\right) + q\left(y_h + y_p\right)$$

$$= \left(\frac{d^2 y_h}{dt^2} + p\frac{dy_h}{dt} + qy_h\right) + \left(\frac{d^2 y_p}{dt^2} + p\frac{dy_p}{dt} + qy_p\right)$$

$$= 0 + g(t) = g(t),$$

so $y_h(t) + y_p(t)$ is a solution of the nonhomogeneous equation.

To verify the second assertion, we compute

$$\frac{d^2}{dt^2}(y_p - y_q) + p\frac{d}{dt}(y_p - y_q) + q(y_p - y_q)$$

$$= \left(\frac{d^2 y_p}{dt^2} - \frac{d^2 y_q}{dt^2}\right) + p\left(\frac{dy_p}{dt} - \frac{dy_q}{dt}\right) + q\left(y_p - y_q\right)$$

$$= \left(\frac{d^2 y_p}{dt^2} + p\frac{dy_p}{dt} + qy_p\right) - \left(\frac{d^2 y_q}{dt^2} + p\frac{dy_q}{dt} + qy_q\right)$$

$$= g(t) - g(t) = 0,$$

so $y_p(t) - y_q(t)$ is a solution of the homogeneous equation.

If $k_1 y_1(t) + k_2 y_2(t)$ is the general solution of the homogeneous equation, then the first half of the Extended Linearity Principle says that

$$k_1 y_1(t) + k_2 y_2(t) + y_p(t)$$

is a solution of the nonhomogeneous equation for *any* choice of the constants k_1 and k_2. The second half of the Extended Linearity Principle says that *any* solution $y_q(t)$ of the nonhomogeneous equation can be written as

$$k_1 y_1(t) + k_2 y_2(t) + y_p(t)$$

for some choice of k_1 and k_2. Therefore,

$$k_1 y_1(t) + k_2 y_2(t) + y_p(t)$$

is the general solution of the nonhomogeneous equation.

The Extended Linearity Principle gives us an algorithm for finding the general solution of equations for forced harmonic oscillators:

Step 1 Find the general solution of the associated unforced (homogeneous) second-order equation.

Step 2 Find *one* particular solution of the forced (nonhomogeneous) second-order equation.

Step 3 Add the results of Steps 1 and 2 to obtain the general solution of the forced equation.

Qualitative Implications of the Extended Linearity Principle

The Extended Linearity Principle gives us a powerful tool for understanding the qualitative behavior of the solutions of the *damped* forced harmonic oscillator equations. To see this, we first recall that the coefficients p and q in

$$\frac{d^2y}{dt^2} + p\frac{dy}{dt} + qy = 0$$

satisfy the inequalities $q > 0$ and $p \geq 0$. Whenever damping is present (as it is in any physical mechanical device), the damping coefficient p is positive, and the equilibrium point at the origin of the corresponding system is a sink (see Section 3.6). Consequently every solution of the unforced equation tends to zero.

Adding an external force $g(t)$, the Extended Linearity Principle implies that the equation

$$\frac{d^2y}{dt^2} + p\frac{dy}{dt} + qy = g(t)$$

has the general solution

$$k_1y_1(t) + k_2y_2(t) + y_p(t),$$

where $k_1y_1(t) + k_2y_2(t)$ is the general solution of the unforced equation and $y_p(t)$ is a solution of the forced equation. But we know that $k_1y_1(t) + k_2y_2(t)$ tends to zero. Hence for large t, we must have

$$k_1y_1(t) + k_2y_2(t) + y_p(t) \approx y_p(t).$$

That is, for large t, *all* solutions of the forced harmonic oscillator equation with nonzero damping ($p > 0$) are approximately the same. A schematic of this situation is shown in Figure 4.1. In other words, the initial conditions have no effect on the long-term behavior of the solutions.

Figure 4.1

Typical graphs of solutions of a forced harmonic oscillator equation with nonzero damping ($p > 0$). The solution $y_p(t)$ is shown in dark blue.

Steady-state solution

In the above discussion any solution can play the role of $y_p(t)$. In engineering applications $y_p(t)$ is often called the **forced response**. The term **steady-state response** is used for the behavior of the forced response for large t. (The phrase "steady-state" is somewhat misleading since $y_p(t)$ need not be constant.)

A solution of the unforced harmonic oscillator (with any initial conditions) is called the **natural response**, or *free response*. The discussion above can be restated as: "All solutions of a forced, damped harmonic oscillator approach the steady-state solution because the natural response dies out, leaving only the forced response."

An Example of the Method of Undetermined Coefficients

Consider the equation

$$\frac{d^2y}{dt^2} + 5\frac{dy}{dt} + 6y = e^{-t}.$$

To find the general solution, we follow the procedure suggested by the Extended Linearity Principle (as we did in Section 1.8 for first-order equations). The first step is to find the general solution of the unforced (homogeneous) equation

$$\frac{d^2y}{dt^2} + 5\frac{dy}{dt} + 6y = 0.$$

Using the methods of Section 3.6, the eigenvalues are -2 and -3. This is an overdamped system and the general solution is

$$y_h(t) = k_1 e^{-2t} + k_2 e^{-3t}.$$

The second step is to find one solution $y_p(t)$ of the forced (nonhomogeneous) equation

$$\frac{d^2y}{dt^2} + 5\frac{dy}{dt} + 6y = e^{-t}.$$

The method we use is a slightly sophisticated guess-and-test. Since we have the term e^{-t} on the right-hand side of the equation, a natural first guess is $y_p(t) = e^{-t}$. Substituting this guess into the left-hand side yields

$$\frac{d^2y_p}{dt^2} + 5\frac{dy_p}{dt} + 6y_p = e^{-t} - 5e^{-t} + 6e^{-t} = 2e^{-t},$$

which does not equal the right-hand side of the forced equation.

Our guess e^{-t} is therefore not a solution. However, it is off only by a constant. It is reasonable to hope that a constant multiple of e^{-t} is a solution, but the correct choice of constant is not clear. We avoid the issue for the moment and improve our guess to $y_p(t) = ke^{-t}$, where k is an unknown constant (or an "undetermined coefficient"). By choosing the correct value for k, we hope to make $y_p(t)$ a solution.

Substituting ke^{-t} into the left-hand side of the differential equation, we have

$$\frac{d^2y_p}{dt^2} + 5\frac{dy_p}{dt} + 6y_p = ke^{-t} - 5ke^{-t} + 6ke^{-t} = 2ke^{-t}.$$

In order for $y_p(t)$ to be a solution, we must choose $k = 1/2$. Hence $y_p(t) = e^{-t}/2$ is a solution of the forced equation.

Therefore the general solution of

$$\frac{d^2y}{dt^2} + 5\frac{dy}{dt} + 6y = e^{-t}$$

is the sum of the general solution of the unforced equation plus a particular solution of the forced equation. So the general solution is

$$y(t) = k_1e^{-2t} + k_2e^{-3t} + \frac{e^{-t}}{2}.$$

Several solutions with different initial conditions are shown in Figure 4.2. As predicted, these solutions all approach $y_p(t) = e^{-t}/2$ as t increases.

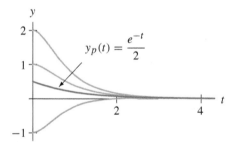

Figure 4.2

Several solutions of

$$\frac{d^2y}{dt^2} + 5\frac{dy}{dt} + 6y = e^{-t}.$$

Note that all of the graphs are asymptotic to $y_p(t) = e^{-t}/2$ as t increases.

The Method of Undetermined Coefficients

Deriving the solution to the unforced equation was discussed in Section 3.6. Thus what is new in the example just completed is the technique used to find one particular solution of the forced equation. We used a lucky-guess method. However, we did not leave everything to luck. The fact that the solution contains a term of the form e^{-t} is not so surprising since this expression forms the right-hand side of the equation. We therefore made an educated guess that the particular solution assumes the form $y_p(t) = ke^{-t}$ and then determined the specific value of k that made $y_p(t)$ a solution.

This "method" for finding particular solutions of forced harmonic oscillator equations is similar in spirit to our guess-and-test method for second-order homogeneous equations. It can be summarized in two steps. First, make a reasonable guess for a solution, but leave some coefficients undetermined. Then substitute the guess into the differential equation and solve for the values of the coefficients that make the guess a solution. This method is called the **Method of Undetermined Coefficients**.

Second Guessing

Whenever we use a guess-and-test method, we run the risk that our first guess (no matter how reasonable) is not a solution. Even leaving coefficients to be determined at the last moment may not be sufficient to obtain a solution. In this situation we just guess again. ("If at first, you don't succeed ….")

Consider the equation

$$\frac{d^2y}{dt^2} + 5\frac{dy}{dt} + 6y = e^{-2t}.$$

This equation differs from the previous example only in the forcing function. The general solution of the unforced equation is still

$$y_h(t) = k_1 e^{-2t} + k_2 e^{-3t}.$$

To find a particular solution of the forced equation, we make the reasonable guess $y_p(t) = ke^{-2t}$, with k as the undetermined coefficient. Substituting into the left-hand side of the forced equation gives

$$\frac{d^2y_p}{dt^2} + 5\frac{dy_p}{dt} + 6y_p = 4ke^{-2t} - 10ke^{-2t} + 6ke^{-2t} = 0.$$

This result is very upsetting. No matter what value we choose for k, substituting $y_p(t)$ into the left-hand side of the forced equation gives zero. Therefore there is no particular solution of the form $y_p(t) = ke^{-2t}$.

Choosing a second guess

The problem with our first guess is that for every k, $y_p(t) = ke^{-2t}$ is a solution of the unforced equation. When we substitute this $y_p(t)$ into the left-hand side of the differential equation, we are guaranteed to get zero. On the other hand, any guess for $y_p(t)$ should contain e^{-2t} as a factor for there to be any hope of obtaining e^{-2t} on the right-hand side.

We have already encountered this difficulty in two similar circumstances—once when we had to second guess for first-order equations (see Section 1.8) and again when we discussed repeated roots of the characteristic equation for homogeneous second-order equations (see Section 3.6). In both cases, we saw that a second, not-quite-so-obvious guess was kte^{-2t}. To see if a guess of this form makes sense in this situation, we first do some numerics. We use a numerical method to find an approximate solution $y(t)$ of the initial-value problem

$$\frac{d^2y}{dt^2} + 5\frac{dy}{dt} + 6y = e^{-2t}, \quad y(0) = 1, \ y'(0) = -2,$$

and we compare the result to e^{-2t}. (Note that we chose the same initial conditions for $y(t)$ as those of e^{-2t}. That is, at $t = 0$, $e^{-2t} = 1$ and $d(e^{-2t})/dt = -2$.) In Figure 4.3, we graph both the solution $y(t)$ and e^{-2t}. Note that the graphs of these two functions are difficult to distinguish. Close examination shows that the graph of the solution $y(t)$ is slightly but consistently above the graph of e^{-2t}. Since we expect that e^{-2t} appears as one factor in $y(t)$, it makes sense to graph the ratio $y(t)/e^{-2t}$ (see Figure 4.4). After an initial period of time, the ratio seems to follow a straight line. That is, for large t, the solution appears to be approximately kte^{-2t}.

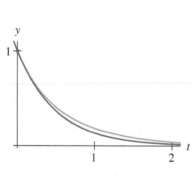

Figure 4.3
Solution of
$d^2y/dt^2 + 5dy/dt + 6y = e^{-2t}$ with
initial conditions $y(0) = 1$, $y'(0) = -2$
graphed with the function e^{-2t}.

Figure 4.4
Ratio y/e^{-2t} of the two graphs in Figure 4.3.

Using this numerical experiment as motivation, we choose $y_p(t) = kte^{-2t}$ as our second guess where k is the undetermined coefficient. Substituting $y_p(t)$ into the left-hand side of the forced equation yields

$$\frac{d^2y_p}{dt^2} + 5\frac{dy_p}{dt} + 6y_p = \frac{d^2(kte^{-2t})}{dt} + 5\frac{d(kte^{-2t})}{dt} + 6kte^{-2t}$$

$$= 4kte^{-2t} - 4ke^{-2t} + 5(-2kte^{-2t} + ke^{-2t}) + 6kte^{-2t}$$

$$= ke^{-2t}$$

Hence if we choose $k = 1$, the function $y_p(t) = te^{-2t}$ is a solution of the forced equation. The general solution of the forced equation is therefore

$$y(t) = k_1e^{-2t} + k_2e^{-3t} + te^{-2t}.$$

Figure 4.5 shows several solutions of the forced equation with different initial conditions. All solutions approach $y_p(t) = te^{-2t}$ (which approaches the t-axis), as predicted.

Figure 4.5
Several solutions of

$$\frac{d^2y}{dt^2} + 5\frac{dy}{dt} + 6y = e^{-2t}.$$

Note that all of the graphs approach $y_P(t) = te^{-2t}$ as t increases.

Rule of thumb for second guessing

The motivation described above is a good example of what is so unsatisfying about lucky-guess methods. While the numerical evidence for multiplying the first guess by t is compelling, it requires a careful eye to find it. In Chapter 6, we give other methods for coming up with this second guess.

It turns out that for linear equations of any order, the two examples we have discussed give a general rule of thumb for how to find particular solutions. The form of the guess depends on the right-hand side of the equation (the forcing function). If the first guess doesn't yield a particular solution (which occurs if the right-hand side is a solution of the homogeneous equation), we try a second guess by multiplying the first guess by t.

An Example from Electric Circuit Theory

We present an example of a nonhomogeneous second-order equation motivated by electric circuit theory (with realistic and rather messy numbers). Consider the resistor-inductor-capacitor (RLC) circuit pictured in Figure 4.6. Suppose that the voltage at the source is $V_S(t) = 2e^{-t/3}$ volts at time t; that is, the strength of the voltage source is decaying at an exponential rate.

Figure 4.6
RLC circuit to be modeled.

Suppose that $R = 4k\Omega = 4000\Omega$, $C = 0.25\mu F = 0.25 \times 10^{-6}F$, and $L = 1.6H$ (where Ω, F, and H stand for ohms, farads, and henrys, standard units in electric circuit theory). The voltage $v(t)$ across the capacitor is a solution of the second-order equation

$$LC\frac{d^2v}{dt^2} + RC\frac{dv}{dt} + v = V_S(t).$$

Using the values given above, we have

$$0.4 \times 10^{-6} \frac{d^2 v}{dt^2} + 10^{-3} \frac{dv}{dt} + v = 2e^{-t/3}.$$

We can put this equation into a more familiar form by multiplying through by 2.5×10^6, obtaining

$$\frac{d^2 v}{dt^2} + 2500 \frac{dv}{dt} + 2.5 \times 10^6 v = 5 \times 10^6 e^{-t/3}.$$

These numbers are imposing but are realistic for the standard units of measure in electric circuits. To find the general solution, we follow the procedure above. First, we find the general solution of the homogeneous equation. Then we find a particular solution of the nonhomogeneous equation.

We begin by computing the general solution of

$$\frac{d^2 v}{dt^2} + 2500 \frac{dv}{dt} + 2.5 \times 10^6 v = 0.$$

The eigenvalues are $(-1.25 \pm i\sqrt{0.9375}) \times 10^3$. Using the approximation 0.968 for $\sqrt{0.9375}$, the general solution is

$$v(t) = k_1 e^{-1250t} \sin(968t) + k_2 e^{-1250t} \cos(968t),$$

where k_1 and k_2 are constants. (Note that the solution decays incredibly quickly to zero, but oscillates very rapidly.)

Now, to find a particular solution of the nonhomogeneous equation

$$\frac{d^2 v}{dt^2} + 2500 \frac{dv}{dt} + 2.5 \times 10^6 v = 5 \times 10^6 e^{-t/3},$$

we guess $v_p(t) = ke^{-t/3}$, where k is the undetermined coefficient. To find the appropriate value of k, we substitute the guess back into the differential equation. The left-hand side is

$$\frac{d^2 v_p}{dt^2} + 2500 \frac{dv_p}{dt} + 2.5 \times 10^6 v_p = \frac{d^2(ke^{-t/3})}{dt^2} + 2500 \frac{d(ke^{-t/3})}{dt} + 2.5 \times 10^6 ke^{-t/3}$$

$$= \frac{k}{9} e^{-t/3} - 2500 \frac{k}{3} e^{-t/3} + 2.5 \times 10^6 ke^{-t/3}$$

$$= \left(\frac{1}{9} - \frac{2500}{3} + 2.5 \times 10^6 \right) ke^{-t/3}.$$

Next, we equate this left-hand side with the right-hand side of the nonhomogeneous equation, which is $5 \times 10^6 e^{-t/3}$. Cancelling the $e^{-t/3}$ term and solving for k, we see that $k \approx 2$. So our particular solution is (approximately) $v_p(t) = 2e^{-t/3}$.

We can now form the general solution

$$v(t) = k_1 e^{-1250t} \sin(968t) + k_2 e^{-1250t} \cos(968t) + 2e^{-t/3},$$

where k_1 and k_2 are constants that we adjust according to the initial conditions. The first two terms of the general solution oscillate very quickly and decrease very quickly to zero. Hence for large t, every solution is close to the particular solution. The graph of the solution that satisfies the initial condition $(v(0), v'(0)) = (1, 0)$ is shown in Figure 4.7. To see the oscillations, we must look very close to $t = 0$ (see Figure 4.8). In the language of circuit theory we say that the natural response (the solution of the homogeneous equation) decays quickly and every solution approaches the forced response, which is $2e^{-t/3}$.

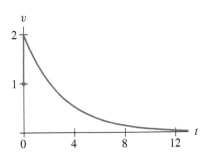

Figure 4.7
Solution of

$$\frac{d^2v}{dt^2} + 2500\frac{dv}{dt} + 2.5 \times 10^6 = 5 \times 10^6 e^{-t/3}$$

with initial condition $v(0) = 1$,
$v'(0) = 0$.

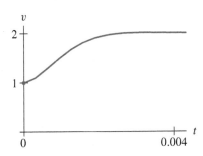

Figure 4.8
The same solution that is graphed in Figure 4.7. Here we show the graph only for values of t that are very close to 0 in order to illustrate the rapid increase from the initial value $v(0) = 1$.

EXERCISES FOR SECTION 4.1

In Exercises 1–8, find the general solution of the given differential equation.

1. $\dfrac{d^2y}{dt^2} - \dfrac{dy}{dt} - 6y = e^{4t}$

2. $\dfrac{d^2y}{dt^2} + 6\dfrac{dy}{dt} + 8y = 2e^{-3t}$

3. $\dfrac{d^2y}{dt^2} - \dfrac{dy}{dt} - 2y = 5e^{3t}$

4. $\dfrac{d^2y}{dt^2} + 4\dfrac{dy}{dt} + 13y = e^{-t}$

5. $\dfrac{d^2y}{dt^2} + 4\dfrac{dy}{dt} + 13y = -3e^{-2t}$

6. $\dfrac{d^2y}{dt^2} + 7\dfrac{dy}{dt} + 10y = e^{-2t}$

7. $\dfrac{d^2y}{dt^2} - 5\dfrac{dy}{dt} + 4y = e^{4t}$

8. $\dfrac{d^2y}{dt^2} + \dfrac{dy}{dt} - 6y = 4e^{-3t}$

In Exercises 9–12, find the solution of the given initial-value problem.

9. $\dfrac{d^2y}{dt^2} + 6\dfrac{dy}{dt} + 8y = e^{-t}$,

$y(0) = y'(0) = 0$

10. $\dfrac{d^2y}{dt^2} + 7\dfrac{dy}{dt} + 12y = 3e^{-t}$,

$y(0) = 2,\ y'(0) = 1$

11. $\dfrac{d^2y}{dt^2} + 4\dfrac{dy}{dt} + 13y = -3e^{-2t}$,

$y(0) = y'(0) = 0$

12. $\dfrac{d^2y}{dt^2} + 7\dfrac{dy}{dt} + 10y = e^{-2t}$,

$y(0) = y'(0) = 0$

In Exercises 13–18,

(a) compute the general solution,

(b) compute the solution with $y(0) = y'(0) = 0$, and

(c) describe the long-term behavior of solutions in a brief paragraph.

13. $\dfrac{d^2y}{dt^2} + 4\dfrac{dy}{dt} + 3y = e^{-t/2}$

14. $\dfrac{d^2y}{dt^2} + 4\dfrac{dy}{dt} + 3y = e^{-2t}$

15. $\dfrac{d^2y}{dt^2} + 4\dfrac{dy}{dt} + 3y = e^{-4t}$

16. $\dfrac{d^2y}{dt^2} + 4\dfrac{dy}{dt} + 20y = e^{-t/2}$

17. $\dfrac{d^2y}{dt^2} + 4\dfrac{dy}{dt} + 20y = e^{-2t}$

18. $\dfrac{d^2y}{dt^2} + 4\dfrac{dy}{dt} + 20y = e^{-4t}$

19. Find the general solution of

$$\frac{d^2y}{dt^2} + 2\frac{dy}{dt} + y = e^{-t}.$$

[*Hint*: Just keep guessing.]

20. One of the most common forcing functions is constant forcing. The equation for a harmonic oscillator with constant forcing is

$$\frac{d^2y}{dt^2} + p\frac{dy}{dt} + qy = c,$$

where p, q, and c are constants. Since we can find the general solution of the unforced equation by our usual methods, we can find the general solution of the forced equation if we can find one particular solution. Find one particular solution of the equation

$$\frac{d^2y}{dt^2} + p\frac{dy}{dt} + qy = c.$$

[*Hint*: Guess a solution that looks like the forcing function.]

In Exercises 21–24, compute
  (a) the general solution and
  (b) the solution with $y(0) = y'(0) = 0$.

21. $\dfrac{d^2y}{dt^2} - 5\dfrac{dy}{dt} + 4y = 5$

22. $\dfrac{d^2y}{dt^2} + 5\dfrac{dy}{dt} + 6y = 2$

23. $\dfrac{d^2y}{dt^2} + 2\dfrac{dy}{dt} + 10y = 10$

24. $\dfrac{d^2y}{dt^2} + 4\dfrac{dy}{dt} + 6y = -8$

The discussion of the qualitative behavior of solutions of equations for damped forced harmonic oscillators relies on the fact that solutions of the unforced equation (natural response) tend to zero as t increases. If damping is not present, then the natural response does not tend to zero and the qualitative behavior of solutions is more complicated (see Section 4.3). However, the Extended Linearity Principle and the Method of Undetermined Coefficients apply to all nonhomogeneous, linear equations.

    In Exercises 25–30,
  (a) compute the general solution,
  (b) compute the solution with $y(0) = y'(0) = 0$, and
  (c) give a rough sketch and describe in a brief paragraph the long-term behavior of the solution in part (b).

25. $\dfrac{d^2y}{dt^2} + 9y = e^{-t}$

26. $\dfrac{d^2y}{dt^2} + 4y = 2e^{-2t}$

27. $\dfrac{d^2y}{dt^2} + 2y = -3$

28. $\dfrac{d^2y}{dt^2} + 4y = e^t$

29. $\dfrac{d^2y}{dt^2} + 9y = 6$

30. $\dfrac{d^2y}{dt^2} + 2y = -e^t$

31. In order to use the Method of Undetermined Coefficients to find a particular solution, we must be able to make a reasonable guess (up to multiplicative constants—the undetermined coefficients) of a particular solution. In this section we discussed exponential forcing. If the forcing is some other type of function, then we must adjust our guess accordingly. For example, for an equation of the form

$$\frac{d^2y}{dt^2} + 4y = -3t^2 + 2t + 3,$$

the forcing function is the quadratic polynomial $g(t) = -3t^2 + 2t + 3$. It is reasonable to guess that a particular solution in this case is also a quadratic polynomial. Hence we guess $y_p(t) = at^2 + bt + c$. The constants a, b, and c are determined by substituting $y_p(t)$ into the equation.

  (a) Find the general solution of the equation specified in this exercise.
  (b) Find the particular solution with the initial condition $y(0) = 2$, $y'(0) = 0$.

In Exercises 32–35,

(a) compute the general solution,

(b) compute the solution with $y(0) = y'(0) = 0$, and

(c) give a rough sketch and describe in a brief paragraph the long-term behavior of the solution in part (b).

32. $\dfrac{d^2y}{dt^2} + 2\dfrac{dy}{dt} = 3t + 2$

33. $\dfrac{d^2y}{dt^2} + 4y = 3t + 2$

34. $\dfrac{d^2y}{dt^2} + 3\dfrac{dy}{dt} + 2y = t^2$

35. $\dfrac{d^2y}{dt^2} + 4y = t - \dfrac{t^2}{20}$

36. We can extend the Method of Undetermined Coefficients in order to solve equations whose forcing functions are sums of several types of functions. More precisely, suppose that $y_1(t)$ is a solution of the equation

$$\frac{d^2y}{dt^2} + p\frac{dy}{dt} + qy = g(t)$$

and that $y_2(t)$ is a solution of the equation

$$\frac{d^2y}{dt^2} + p\frac{dy}{dt} + qy = h(t).$$

Show that $y_1(t) + y_2(t)$ is a solution of the equation

$$\frac{d^2y}{dt^2} + p\frac{dy}{dt} + qy = g(t) + h(t).$$

In Exercises 37–42,

(a) compute the general solution,

(b) compute the solution with $y(0) = y'(0) = 0$, and

(c) describe in a brief paragraph the long-term behavior of the solution in part (b).

37. $\dfrac{d^2y}{dt^2} + 5\dfrac{dy}{dt} + 6y = e^{-t} + 4$

38. $\dfrac{d^2y}{dt^2} + 3\dfrac{dy}{dt} + 2y = e^{-t} - 4$

39. $\dfrac{d^2y}{dt^2} + 6\dfrac{dy}{dt} + 8y = 2t + e^{-t}$

40. $\dfrac{d^2y}{dt^2} + 6\dfrac{dy}{dt} + 8y = 2t + e^{t}$

41. $\dfrac{d^2y}{dt^2} + 4y = t + e^{-t}$

42. $\dfrac{d^2y}{dt^2} + 4y = 6 + t^2 + e^{t}$

4.2   SINUSOIDAL FORCING

In this section we study the forced harmonic oscillator equation

$$\frac{d^2y}{dt^2} + p\frac{dy}{dt} + qy = g(t),$$

where $g(t)$ is a sine or cosine function. This type of external force occurs frequently in applications. Examples include the shaking of a building by an earthquake and the pressure waves of a sound striking a glass. These external forces share two qualitative features. First, they are periodic. That is, they repeat after a definite time T, called the period, so

$$g(t+T) = g(t)$$

for all t. Second, they have zero average. For each period, the push in one direction equals the pull in the other, which can be expressed analytically as

$$\int_0^T g(t)\,dt = 0.$$

The simplest and most familiar functions with this property are $\sin \omega t$ and $\cos \omega t$. The period of these functions is $2\pi/\omega$ and the frequency is $\omega/(2\pi)$.

In this section we apply the method of undetermined coefficients to study damped forced harmonic oscillators with sinusoidal forcing. In Section 4.3 we consider the undamped harmonic oscillator with sinusoidal forcing, and in Section 4.4 we study how the solutions of the damped harmonic oscillator with sinusoidal forcing depend on the forcing frequency and the damping parameter.

Sinusoidal Forcing of a Damped Harmonic Oscillator

To better understand the behavior of a harmonic oscillator with periodic forcing, we begin by finding the general solution of

$$\frac{d^2y}{dt^2} + 2\frac{dy}{dt} + 2y = \sin t$$

following the procedure given in Section 4.1.

The eigenvalues of the unforced equation

$$\frac{d^2y}{dt^2} + 2\frac{dy}{dt} + 2y = 0$$

are $-1 \pm i$. Hence this oscillator is underdamped, and the general solution is

$$y(t) = k_1 e^{-t}\cos t + k_2 e^{-t}\sin t.$$

To find one solution of the forced equation, we use the method of undetermined coefficients. At first glance it seems that $y_p(t) = k\sin t$ would be a reasonable guess for a solution of the forced equation. However, because of the dy/dt term, substituting this guess into the differential equation yields a term containing $\cos t$ on the left-hand side. No choice of constant k will eliminate this term, and there are no cosines on the right-hand side of the equation. We could make a more general guess of

$$y_p(t) = a \cos t + b \sin t$$

with two undetermined coefficients, but it is more efficient to take advantage of our knowledge of complex exponentials.

Complexification

First, consider the equation

$$\frac{d^2y}{dt^2} + 2\frac{dy}{dt} + 2y = e^{it}.$$

A particular solution of this complex differential equation, which we call $y_c(t)$ (c for complex), has a real part and an imaginary part. That is,

$$y_c(t) = y_{re}(t) + iy_{im}(t).$$

Substituting $y_c(t)$ into the differential equation and using Euler's formula gives

$$\frac{d^2(y_{re} + iy_{im})}{dt^2} + 2\frac{d(y_{re} + iy_{im})}{dt} + 2(y_{re} + iy_{im}) = \cos t + i \sin t.$$

Taking real and imaginary parts on both sides of the equation, we have

$$\frac{d^2y_{re}}{dt^2} + 2\frac{dy_{re}}{dt} + 2y_{re} = \cos t \quad \text{and} \quad \frac{d^2y_{im}}{dt^2} + 2\frac{dy_{im}}{dt} + 2y_{im} = \sin t.$$

Hence the imaginary part, $y_{im}(t)$, is a particular solution of the original equation with forcing $\sin t$.

The advantage of using complex exponentials is that the appropriate guess for the particular solution is once again another exponential. For

$$\frac{d^2y}{dt^2} + 2\frac{dy}{dt} + 2y = e^{it},$$

we guess $y_c(t) = ae^{it}$ and solve for the (complex) coefficient a that makes this guess a solution. Substituting into the left-hand side of the differential equation, we obtain

$$\frac{d^2y_c}{dt^2} + 2\frac{dy_c}{dt} + 2y_c = -ae^{it} + 2aie^{it} + 2ae^{it}$$

$$= a(1 + 2i)e^{it}.$$

In order for $y_c(t)$ to be a solution, we must take

$$a = \frac{1}{1 + 2i} = \frac{1 - 2i}{5}$$

(see Appendix C for a review of complex arithmetic). To determine the real and imaginary parts, we write

$$y_c(t) = \frac{1-2i}{5}(\cos t + i \sin t)$$

$$= \left(\frac{1}{5}\cos t + \frac{2}{5}\sin t\right) + i\left(-\frac{2}{5}\cos t + \frac{1}{5}\sin t\right).$$

Taking the imaginary part, we obtain a particular solution of the original equation with $\sin t$ forcing,

$$y_p(t) = -\frac{2}{5}\cos t + \frac{1}{5}\sin t.$$

Hence the general solution of

$$\frac{d^2y}{dt^2} + 2\frac{dy}{dt} + 2y = \sin t$$

is

$$y(t) = k_1 e^{-t}\sin t + k_2 e^{-t}\cos t + \left(-\frac{2}{5}\cos t + \frac{1}{5}\sin t\right).$$

Qualitative analysis

To determine the qualitative behavior of solutions of

$$\frac{d^2y}{dt^2} + 2\frac{dy}{dt} + 2y = \sin t,$$

we note that the unforced equation is an underdamped harmonic oscillator. As pointed out in Section 4.1, the portion of a solution arising from the solution of the unforced oscillator (the natural response) tends to zero, leaving only the particular solution of the forced oscillator (the forced response). From the general solution for this equation, we see that the natural response tends to zero like e^{-t} (see Figure 4.9.) Hence for large t, every solution is close to

$$y_p(t) = -\frac{2}{5}\cos t + \frac{1}{5}\sin t.$$

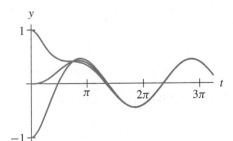

Figure 4.9
Several solutions of the equation

$$\frac{d^2y}{dt^2} + 2\frac{dy}{dt} + 2y = \sin t.$$

The behavior of $y_p(t)$ can be made much more evident by the use of trigonometric identities to combine the sine and cosine into one cosine function (see Exercise 22 on page 313). However, because we are using complex numbers to determine particular solutions, we use an alternative approach involving Euler's formula to combine these two terms.

Consider the solution $y_c(t) = ae^{it}$, where $a = (1 - 2i)/5$, to the complex differential equation. We write a in its polar form as

$$a = |a|e^{i\theta}$$

$$= \frac{1}{\sqrt{5}}e^{i\theta}$$

(see Appendix C for a definition of polar form). In terms of the angle θ, we have

$$y_c(t) = ae^{it}$$

$$= \frac{1}{\sqrt{5}}e^{i\theta}e^{it}.$$

Since we add exponents when we multiply exponentials,

$$y_c(t) = \frac{1}{\sqrt{5}}e^{i(t+\theta)} = \frac{1}{\sqrt{5}}(\cos(t+\theta) + i\sin(t+\theta)).$$

Since our particular solution $y_p(t)$ is the imaginary part of $y_c(t)$, we see that

$$y_p(t) = \frac{1}{\sqrt{5}}\sin(t+\theta).$$

Since it is traditional to use cosine rather than sine in this context and to represent the angle θ in terms of degrees, we recall that

$$\sin\alpha = \cos(\alpha - 90°).$$

Thus

$$y_p(t) = \frac{1}{\sqrt{5}}\cos(t + \theta - 90°),$$

and we need only determine θ to complete the calculation. Since θ is the polar angle associated to the complex number a, we see that

$$\theta = \arctan\left(\frac{-2}{1}\right) \approx -63°.$$

Our particular solution can therefore be expressed as

$$y_p(t) = \frac{1}{\sqrt{5}}\cos(t - 153°).$$

The angle $153°$ is often denoted ϕ.

This form of the equation for $y_p(t)$ is particularly useful. We see without computation that $y_p(t)$ is a periodic function with amplitude $1/\sqrt{5}$ and period 2π. We also see that the graph of $y_p(t)$ can be obtained from the graph of cosine by shifting it to the right by $153°$ and "stretching" it by a factor of $1/\sqrt{5}$. The angle ϕ is called the **phase angle**.

It is useful to note that the complex number a, which is the undetermined coefficient in this computation, determines both the amplitude and the phase angle of the forced response. The amplitude is just the magnitude $|a|$ of a, and the phase angle can be calculated immediately from the polar angle of a. In Section 4.4, we study how the amplitude and phase shift depend on the parameters for a periodically forced harmonic oscillator.

Physical interpretation

Think of the harmonic oscillator as modeling a mass attached to a spring sliding on a table with damping and the external force arising from a periodic tilting of the table. The result above can be summarized by saying that the long-term behavior of the mass is to slide back and forth with the same frequency as the tilting of the table. Writing the particular solution of the forced equation in the form $A \cos(t + \phi)$, we can read off the amplitude A and the phase angle ϕ.

The behavior of the solutions in the above example is typical of the long-term behavior of solutions of equations for damped harmonic oscillators with sine or cosine forcing. All solutions approach a periodic solution (the forced response), which oscillates with the same period as the forcing. It is vital to remember that while the forcing function and the forced response look similar, they are not the same. The amplitude and phase of the forced response depend in a complicated way on the parameters of the harmonic oscillator and on the forcing frequency. We consider this dependence in detail in Section 4.4.

The Phase Portrait of a Forced Harmonic Oscillator

So far, all of our pictures of solutions in this chapter have been of the solution as a function of time. We have used our knowledge of linear systems and linear, constant-coefficient, second-order equations to help find the general solution of forced harmonic oscillator equations, but we have not converted the forced equation into a first-order system.

The reason that we have not done this conversion is that a forced harmonic oscillator

$$\frac{d^2y}{dt^2} + p\frac{dy}{dt} + qy = g(t)$$

and its corresponding first-order system

$$\frac{dy}{dt} = v$$

$$\frac{dv}{dt} = -qy - pv + g(t)$$

are nonautonomous. The tangent vector of a solution curve in the phase plane depends on the position (y, v) *and* on the time t, so the vector field changes with time. Moreover, because the vector field changes with time, two solutions with the same (y, v) values at different times can follow different paths. Thus solution curves can cross in the yv-plane without violating the Uniqueness Theorem.

Converting the second-order nonautonomous equation to a first-order system is still useful. There are no elementary numerical methods that apply to second-order equations, but Euler's method (as well as others, see Chapter 7) is essentially the same for autonomous and nonautonomous first-order systems. Also, the solution curves in the yv-plane can help in understanding the behavior of solutions.

To demonstrate this and to gain more practice with the method of undetermined coefficients, we consider the equation

$$\frac{d^2y}{dt^2} + 2\frac{dy}{dt} + 10y = 4\cos 2t.$$

To find the general solution, we first note that the eigenvalues of the unforced equation

$$\frac{d^2y}{dt^2} + 2\frac{dy}{dt} + 10y = 0$$

are $-1 \pm 3i$. Hence the general solution of the unforced equation is

$$y(t) = k_1 e^{-t} \cos 3t + k_2 e^{-t} \sin 3t.$$

To find a particular solution of the forced equation, we complexify and obtain

$$\frac{d^2y}{dt^2} + 2\frac{dy}{dt} + 10y = 4e^{2it}.$$

The usual guess is $y_c(t) = ae^{2it}$. Substituting this guess into the left-hand side of the differential equation yields

$$\frac{d^2y_c}{dt^2} + 2\frac{dy_c}{dt} + 10y_c = -4ae^{2it} + 4aie^{2it} + 10ae^{2it}$$
$$= a(6 + 4i)e^{2it}$$

In order for ae^{2it} to be a solution, we must have

$$a(6 + 4i) = 4.$$

Therefore

$$a = \frac{4}{6 + 4i} = \frac{4(6 - 4i)}{52} = \frac{6 - 4i}{13},$$

so

$$y_c(t) = \frac{6-4i}{13}e^{2it}$$

$$= \frac{6-4i}{13}(\cos 2t + i\sin 2t)$$

$$= \left(\frac{6}{13}\cos 2t + \frac{4}{13}\sin 2t\right) + i\left(-\frac{4}{13}\cos 2t + \frac{6}{13}\sin 2t\right).$$

A particular solution $y_p(t)$ of the original equation with forcing $4\cos 2t$ is the real part of $y_c(t)$, which is

$$y_p(t) = \frac{6}{13}\cos 2t + \frac{4}{13}\sin 2t.$$

Hence the general solution of the forced equation

$$\frac{d^2y}{dt^2} + 2\frac{dy}{dt} + 10y = 4\cos 2t$$

is

$$y(t) = k_1e^{-t}\cos 3t + k_2e^{-t}\sin 3t + \frac{6}{13}\cos 2t + \frac{4}{13}\sin 2t,$$

which can be rewritten as

$$y(t) = k_1e^{-t}\cos 3t + k_2e^{-t}\sin 3t + \frac{2\sqrt{13}}{13}\cos(2t + \phi),$$

where the phase shift $\phi \approx -34°$.

Qualitative analysis: Graphs of solutions

We know that for any choice of initial conditions, the term

$$k_1e^{-t}\cos 3t + k_2e^{-t}\sin 3t$$

in the general solution quickly tends to zero. Hence every solution is approximately

$$\frac{2\sqrt{13}}{13}\cos(2t - 34°)$$

for large t. This function is periodic in t with period π and amplitude $2\sqrt{13}/13 \approx 0.55$. If we rewrite the forced equation as a first-order system, we obtain

$$\frac{dy}{dt} = v$$

$$\frac{dv}{dt} = -10y - 2v + 4\cos 2t.$$

If $y(t)$ is a solution of the second-order equation, then $(y(t), v(t))$ is a solution of the system. We can graph solutions in both the ty- and yv-planes.

For example, to find the solution that satisfies $y(0) = y'(0) = 0$, we must solve

$$\begin{cases} k_1 + \dfrac{6}{13} = 0 \\ -k_1 + 3k_2 + \dfrac{8}{13} = 0. \end{cases}$$

Solving these equations, we find $k_1 = -6/13$ and $k_2 = -14/39$, and the solution with these initial conditions is

$$y(t) = -\frac{6}{13}e^{-t}\cos 3t - \frac{14}{39}e^{-t}\sin 3t + \frac{6}{13}\cos 2t + \frac{4}{13}\sin 2t.$$

We can graph this solution in the ty-plane and the corresponding solution of the system in the yv-plane (see Figure 4.10). Both these pictures show the solution oscillating around $y = 0$. After a short initial time period, the solution is essentially periodic with period π.

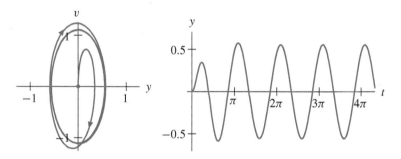

Figure 4.10
Graph of the solution of $d^2y/dt^2 + 2dy/dt + 10y = 4\cos 2t$ with initial conditions $y(0) = y'(0) = 0$ in the yv- and ty-planes.

As a second example, consider the solution with $y(0) = 2$ and $y'(0) = 0$. As above, we can solve for k_1 and k_2, obtaining $k_1 = 20/13$ and $k_2 = 4/13$. The graphs of this solution in the ty- and yv-plane are shown in Figure 4.11. Again, after a short time period, both representations of the solution settle into periodic behavior, oscillating around $y = 0$. In this case, we can clearly see the solution in the yv-plane crossing over itself. As noted above, this is not a violation of the Uniqueness Theorem because the system is nonautonomous.

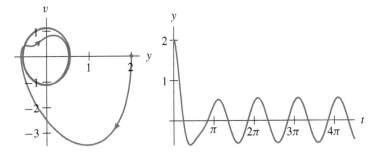

Figure 4.11
Graph of the solution of $d^2y/dt^2 + 2dy/dt + 10y = 4\cos 2t$ with initial
conditions $y(0) = 2$ and $y'(0) = 0$ in the yv- and ty-planes.

Interpretation of the phase portraits

Looking at the phase portrait, we can make the following qualitative observations. The
restoring force and the damping are proportional to y and $v = dy/dt$, respectively.
For y and v large, the external force is much smaller than the restoring and damping
forces of the unforced oscillator. Hence, far from the origin, solutions in the yv-plane
look very similar to solutions of the corresponding unforced oscillator; that is, solutions
tend toward the origin (see Figure 4.12). However, when y and v are close to zero, the
external force is as large or larger than the restoring and damping forces. In this part of
the yv-plane, the external force overcomes the damping and pushes the solution away
from the origin (see Figure 4.13). Solutions tend toward the region of the yv-plane
where the external forces and the damping "balance."

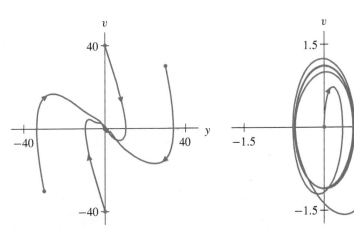

Figure 4.12
Four solutions of
$d^2y/dt^2 + 2dy/dt + 10y = 4\cos 2t$ with
initial conditions far from the origin.

Figure 4.13
Two solutions of
$d^2y/dt^2 + 2dy/dt + 10y = 4\cos 2t$ with
initial conditions close to the origin.

EXERCISES FOR SECTION 4.2

In Exercises 1–10, find the general solution of the given equation.

1. $\dfrac{d^2y}{dt^2} + 3\dfrac{dy}{dt} + 2y = \cos t$

2. $\dfrac{d^2y}{dt^2} + 3\dfrac{dy}{dt} + 2y = 5\cos t$

3. $\dfrac{d^2y}{dt^2} + 3\dfrac{dy}{dt} + 2y = \sin t$

4. $\dfrac{d^2y}{dt^2} + 3\dfrac{dy}{dt} + 2y = 2\sin t$

5. $\dfrac{d^2y}{dt^2} + 6\dfrac{dy}{dt} + 8y = \cos t$

6. $\dfrac{d^2y}{dt^2} + 6\dfrac{dy}{dt} + 8y = -4\cos 3t$

7. $\dfrac{d^2y}{dt^2} + 4\dfrac{dy}{dt} + 13y = 3\cos 2t$

8. $\dfrac{d^2y}{dt^2} + 4\dfrac{dy}{dt} + 20y = -\cos 5t$

9. $\dfrac{d^2y}{dt^2} + 4\dfrac{dy}{dt} + 20y = -3\sin 2t$

10. $\dfrac{d^2y}{dt^2} + 2\dfrac{dy}{dt} + y = \cos 3t$

In Exercises 11–14, find the solution of the given initial-value problem.

11. $\dfrac{d^2y}{dt^2} + 6\dfrac{dy}{dt} + 8y = \cos t$
$y(0) = y'(0) = 0$

12. $\dfrac{d^2y}{dt^2} + 6\dfrac{dy}{dt} + 8y = 2\cos 3t$
$y(0) = y'(0) = 0$

13. $\dfrac{d^2y}{dt^2} + 4\dfrac{dy}{dt} + 20y = -3\sin 2t$
$y(0) = y'(0) = 0$

14. $\dfrac{d^2y}{dt^2} + 2\dfrac{dy}{dt} + y = 2\cos 2t$
$y(0) = y'(0) = 0$

15. To find a particular solution of a forced equation with sine or cosine forcing, we solved the corresponding complexified equation. This method is particularly efficient, but there are other approaches. Find a particular solution via the Method of Undetermined Coefficients for the equation

$$\frac{d^2y}{dt^2} + 3\frac{dy}{dt} + y = \cos 3t$$

   (a) by using the guess $y_p(t) = a\cos 3t + b\sin 3t$ (with a and b as undetermined coefficients) and

   (b) by using the guess $y_p(t) = A\cos(3t + \phi)$ (with A and ϕ as undetermined coefficients).

16.   **(a)** Show that if $y_p(t)$ is a solution of

$$\frac{d^2y}{dt^2} + p\frac{dy}{dt} + qy = g(t),$$

then $ky_p(t)$ is a solution of

$$\frac{d^2y}{dt^2} + p\frac{dy}{dt} + qy = kg(t)$$

for any constant k.

(b) Find the general solution of

$$\frac{d^2y}{dt^2} + 6\frac{dy}{dt} + 8y = 5\cos t.$$

[*Hint*: See Exercise 5.]

17. Six second-order equations and four $y(t)$-graphs are given below. The equations are all of the form

$$\frac{d^2y}{dt^2} + p\frac{dy}{dt} + qy = \cos \omega t$$

for various values of the parameters p, q, and ω. For each $y(t)$-graph, determine the second-order equation for which $y(t)$ is a solution, and state briefly how you know your choice is correct. You should do this exercise without using technology.

(i) $p = 5, q = 3, \omega = 1$ (ii) $p = 1, q = 3, \omega = 1$ (iii) $p = 5, q = 2, \omega = 2$
(iv) $p = 1, q = 3, \omega = 2$ (v) $p = 5, q = 1, \omega = 3$ (vi) $p = 1, q = 1, \omega = 3$

(a)

(b)

(c)

(d)

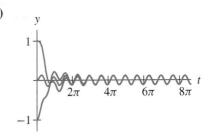

18.   **(a)** Find the general solution of

$$\frac{d^2y}{dt^2} + 4\frac{dy}{dt} + 20y = 3 + 2\cos 2t.$$

[*Hint*: See Exercise 36 in Section 4.1.]

(b) Discuss the long-term behavior of solutions of this equation.

19. **(a)** Find the general solution of

$$\frac{d^2 y}{dt^2} + 4\frac{dy}{dt} + 20y = e^{-t}\cos t.$$

[*Hint*: Consider the forcing function to be the real part of a complex exponential.]

(b) Discuss the long-term behavior of solutions of this equation.

20. **(a)** Find the general solution of

$$\frac{d^2 y}{dt^2} + 4\frac{dy}{dt} + 20y = e^{-2t}\sin 4t.$$

[*Hint*: Keep guessing.]

(b) Discuss the long-term behavior of solutions of this equation.

In Exercises 21 and 22, we explore the different types of notation used to express solutions of harmonic oscillator equations with trigonometric forcing functions. Sadly, the differences in terminology sometimes make it quite difficult for mathematicians and engineers to communicate, even though we are talking about the same things.

21. Given a function $g(t) = a\cos \omega t + b\sin \omega t$, where a and b are constants, show that $g(t)$ is the real part of the complex function

$$ke^{i\phi}e^{i\omega t}.$$

for some k and ϕ. [*Hint*: Use Euler's formula.]

Remark: The complex expression $ke^{i\phi}$ is called a **phasor**. If we know that $g(t)$ has the form $k\cos(\omega t + \phi)$, then we need know only the constants k and ϕ—the amplitude and the phase—to know the function g. Hence we can use the phasor $ke^{i\phi}$ as a notation for the function $g(t) = ke^{i\phi}e^{i\omega t}$.

22. Show that if $k_1 e^{i\phi_1}$ is the phasor representing $g_1(t)$ and $k_2 e^{i\phi_2}$ is the phasor representing $g_2(t)$ (both with frequency $\omega/(2\pi)$), then the phasor representing the sum $g_1(t) + g_2(t)$ is $k_1 e^{i\phi_1} + k_2 e^{i\phi_2}$. [*Hint*: Write the functions $g_1(t)$ and $g_2(t)$ as the real parts of the product of the phasors and $e^{i\omega t}$, add them, and simplify.]

Remark: Phasor terminology is widely used in electric circuit theory, in which an algebra of phasors is developed suited to dealing with currents and voltages in circuits with periodic voltage sources.

23. In Exercise 22 in Section 3.4, we showed that a function of the form

$$y(t) = k_1 \cos \beta t + k_2 \sin \beta t$$

can be written as $y(t) = K\cos(\beta t - \phi)$, where

$$K = \sqrt{k_1^2 + k_2^2}.$$

Verify this identity by writing the complex number $k_1 + ik_2$ in its polar form $Ke^{i\phi}$ and then calculating the real part of the product $(k_1 - ik_2)e^{i\beta t}$ in two different ways. [*Hint*: What's the polar form of $k_1 - ik_2$?]

4.3 UNDAMPED FORCING AND RESONANCE

In Section 4.2 we considered the effect of periodic forcing on damped harmonic oscil-lators. In this section we turn our attention to periodically forced undamped oscillators. Despite the fact that physical systems always have some damping, there is a good rea-son to pay attention to the undamped case. If the physical system we are modeling has sufficiently small damping, then the undamped equation is a good approximation, at least for a certain time interval. As a consequence, the behavior of solutions of the un-damped equation gives us considerable qualitative insight into the behavior of solutions if the damping is small.

The Extended Linearity Principle from Section 4.1 applies for any forced har-monic oscillator equation, damped or undamped. The general solution is the sum of the general solution of the unforced equation (natural response) and a particular solution of the forced equation (forced response). For the forced damped harmonic oscillator, we know that the damping term causes the natural response to tend to zero. Thus ev-ery solution approaches the forced response as time increases, regardless of the initial conditions. This is not the case for undamped equations. The detailed behavior of solu-tions depends on the initial conditions for all time. It is, however, still possible to give a qualitative description of solutions that applies for all initial conditions.

In this section we restrict attention to the case of sinusoidal forcing of undamped equations. Of particular interest is the way that the qualitative description of solutions depends on the frequency of the forcing term. A very dramatic change in the qualitative behavior of the solutions occurs as the frequency of the forcing function approaches the natural frequency of the equation. This phenomenon is called *resonance*.

An Undamped Harmonic Oscillator with Sinusoidal Forcing

We begin with a careful study of a particular undamped harmonic oscillator equation with sinusoidal forcing given by

$$\frac{d^2y}{dt^2} + 2y = \cos \omega t.$$

This equation contains a parameter ω that controls the frequency of the external forcing. The frequency of the forcing is $\omega/(2\pi)$. We first use the method of Section 4.1 to find the general solution of this equation.

The eigenvalues of the unforced equation

$$\frac{d^2y}{dt^2} + 2y = 0$$

are $\pm i\sqrt{2}$, so the general solution is

$$y_h(t) = k_1 \cos \sqrt{2}\,t + k_2 \sin \sqrt{2}\,t.$$

Since there is no damping, solutions oscillate for all time with the natural frequency $\sqrt{2}/(2\pi)$. The amplitude of the oscillation is determined by the values of k_1 and k_2, which in turn are determined by the initial conditions.

To find one solution of the forced equation

$$\frac{d^2y}{dt^2} + 2y = \cos \omega t,$$

we use the Method of Undetermined Coefficients. We could use a combination of trigonometric functions as our guess of the particular solution, but as we will see, there are advantages to complexifying the equation as we did in Section 4.2. We consider

$$\frac{d^2y}{dt^2} + 2y = e^{i\omega t}.$$

The real part of a solution of this equation is a solution of the equation with forcing $\cos \omega t$. As usual, we guess $y_c(t) = ae^{i\omega t}$ and solve for a. Substituting $y_c(t)$ into the left-hand side of the equation, we find

$$\frac{d^2y_c}{dt^2} + 2y_c = -a\omega^2 e^{i\omega t} + 2ae^{i\omega t}$$

$$= a(2 - \omega^2)e^{i\omega t}.$$

In order for $y_c(t)$ to be a solution, we must have $a = 1/(2 - \omega^2)$, so

$$y_c(t) = \frac{1}{2 - \omega^2} e^{i\omega t} = \frac{1}{2 - \omega^2} (\cos \omega t + i \sin \omega t).$$

Hence a solution $y_p(t)$ of the equation

$$\frac{d^2y}{dt^2} + 2y = \cos \omega t$$

is the real part of $y_c(t)$, so

$$y_p(t) = \frac{1}{2 - \omega^2} \cos \omega t.$$

The general solution of the forced equation is therefore

$$y(t) = k_1 \cos \sqrt{2}\, t + k_2 \sin \sqrt{2}\, t + \frac{1}{2 - \omega^2} \cos \omega t.$$

Note that the solution depends on ω, which we think of as a parameter. We must omit the case $\omega = \sqrt{2}$ since this value makes the denominator $2 - \omega^2$ vanish. We deal with this case later in this section.

Qualitative Analysis

With the general solution in hand, we can consider the qualitative behavior of solutions of

$$\frac{d^2y}{dt^2} + 2y = \cos \omega t.$$

First note that, as predicted, the unforced response does not tend to zero. Also, for all t, the solution is a sum of sine and cosine terms of frequency $\sqrt{2}/(2\pi)$ and $\omega/(2\pi)$. For some values of t, both the forced and unforced responses are positive or both are negative, so they reinforce each other. For other values of t, the forced and unforced responses have opposite sign and cancel each other. We see these effects in the graphs of typical solutions (see Figures 4.14 and 4.15).

Figure 4.14
Solution of $d^2y/dt^2 + 2y = \cos \omega t$
for $\omega = 0.5$, $y(0) = y'(0) = 0$.

Figure 4.15
Solution of $d^2y/dt^2 + 2y = \cos \omega t$
for $\omega = 1.2$, $y(0) = y'(0) = 0$.

Beating

There is additional structure to the solutions that is more evident if $\omega = 1.2$ than if $\omega = 0.5$ (compare Figures 4.14 and 4.15). If $\omega = 1.2$, the amplitudes of the oscillations increase and decrease in a very regular pattern. This phenomenon is called **beating**. Beating occurs when the natural response and forced response have approximately the same frequency. This is the case if $\omega = 1.2$, since the natural frequency, $\sqrt{2}/(2\pi) \approx 1.4/(2\pi)$, and the forcing frequency, $1.2/(2\pi)$, are close.

If these two frequencies are close, the two responses tend to reinforce or cancel each other over long time intervals. We can see this by looking at the natural and forced responses separately for a particular choice of initial conditions (see Figure 4.16). For certain time intervals these graphs oscillate in unison, and so their sum has large amplitude. Over other intervals the functions have opposite signs, and their sum is smaller. Incidentally, you can hear the phenomenon of beating when listening to a piano or a guitar that is slightly out of tune.

Figure 4.16
In the plot on the left, we graph the natural and forced responses for the initial-value problem $d^2y/dt^2 + 2y = \cos \omega t$ with $\omega = 1.2$, $y(0) = y'(0) = 0$. On the right, we graph the sum of these two functions. This sum is the desired solution.

We can determine the period and frequency of the beats by using complex exponentials. We have already determined that the general solution of the differential equation is

$$y(t) = k_1 \cos \sqrt{2}\, t + k_2 \sin \sqrt{2}\, t + a \cos \omega t,$$

where $a = 1/(2 - \omega^2)$. Now we consider the particular solution $y(t)$ that satisfies the initial conditions $y(0) = 0$ and $y'(0) = 0$. These initial values imply that $k_1 = -a$ and $k_2 = 0$. Hence the solution to the initial-value problem is

$$y(t) = a \left(\cos \omega t - \cos \sqrt{2}\, t \right).$$

To simplify the notation involved in our computation with exponentials, we let

$$\alpha = \frac{\omega + \sqrt{2}}{2} \quad \text{and} \quad \beta = \frac{\omega - \sqrt{2}}{2}.$$

That is, α is the average of ω and $\sqrt{2}$, and β is one-half of the difference between ω and $\sqrt{2}$. Using Euler's formula, we can think of $y(t)$ as the real part of the complex function

$$y_c(t) = a \left(e^{i\omega t} - e^{i\sqrt{2}\, t} \right).$$

Note that $\omega t = (\alpha + \beta)t$ and $\sqrt{2}\, t = (\alpha - \beta)t$. Therefore,

$$y_c(t) = a \left(e^{i(\alpha+\beta)t} - e^{i(\alpha-\beta)t} \right),$$

and using the usual rules for manipulating exponents, we have

$$y_c(t) = a e^{i\alpha t} (e^{i\beta t} - e^{-i\beta t}).$$

Since $e^{i\beta t} = \cos \beta t + i \sin \beta t$ and $e^{-i\beta t} = \cos \beta t - i \sin \beta t$, we get

$$e^{i\beta t} - e^{-i\beta t} = 2i \sin \beta t,$$

and therefore

$$y_c(t) = a e^{i\alpha t} (2i \sin \beta t).$$

Recalling that the desired solution $y(t)$ is the real part of $y_c(t)$, we calculate

$$y_c(t) = a(\cos \alpha t + i \sin \alpha t)(2i \sin \beta t)$$
$$= a(-2 \sin \alpha t\ \sin \beta t + 2i \cos \alpha t\ \sin \beta t),$$

and obtain

$$y(t) = -2a\, [\sin \alpha t]\, [\sin \beta t]$$

$$= -2a \left[\sin \left(\frac{\omega + \sqrt{2}}{2} \right) t \right] \left[\sin \left(\frac{\omega - \sqrt{2}}{2} \right) t \right].$$

If $\omega \approx \sqrt{2}$, then the frequency of $\sin((\omega - \sqrt{2})t/2)$ is quite small. This is the frequency of the beats. Since the frequency is small and the period is the reciprocal of the frequency, the period of the function $\sin((\omega - \sqrt{2})t/2)$ is large. For example, if $\omega = 1.2$, the frequency of the beats is

$$\frac{\sqrt{2} - 1.2}{4\pi} \approx 0.017,$$

and the period is $4\pi/(\sqrt{2} - 1.2) \approx 58$. This slow oscillation is represented by the dashed outline in Figure 4.15.

On the other hand, the frequency of the term $\sin((\omega + \sqrt{2})t/2)$ is $(\sqrt{2} + \omega)/(4\pi)$, which is the average of the natural and forcing frequencies. If $\omega = 1.2$, this frequency is $(\sqrt{2} + 1.2)/(4\pi) \approx 0.21$, which is relatively large. The period is $4\pi/(\sqrt{2} + 1.2) \approx 4.76$. This period and frequency of the fast oscillations is visible in Figure 4.15.

Dependence of amplitude on forcing frequency

There is one more important feature of the general solution of

$$\frac{d^2y}{dt^2} + 2y = \cos \omega t$$

that we can obtain from the formula

$$y(t) = k_1 \cos \sqrt{2}\, t + k_2 \sin \sqrt{2}\, t + \frac{1}{2 - \omega^2} \cos \omega t.$$

The precise behavior of a solution depends on the values of k_1 and k_2, and these values are determined by the initial conditions. However, every solution contains the term

$$\frac{1}{2 - \omega^2} \cos \omega t.$$

The amplitude of the oscillations of this function is given by $1/|2 - \omega^2|$ (see Figure 4.17).

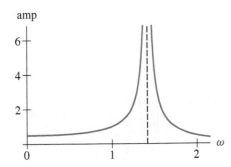

Figure 4.17
Graph of $1/|2 - \omega^2|$, the amplitude of the oscillations of solutions of

$$\frac{d^2y}{dt^2} + 2y = \cos \omega t$$

for large t.

From Figure 4.17, we can make some predictions about the behavior of unforced solutions. If ω is very large, then the coefficient $a = 1/(2 - \omega^2)$ is small. In this case solutions of the forced equation

$$\frac{d^2y}{dt^2} + 2y = \cos \omega t$$

are very close to solutions of the unforced equation. A large value of ω means the external force has very short period and large frequency. The above discussion predicts that such an external force with large frequency has only a small effect on solutions. The harmonic oscillator does not have enough time to respond to a push in one direction before the sign of the external force changes. A typical example is shown in Figure 4.18.

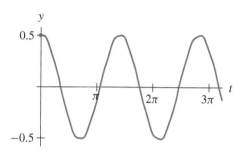

Figure 4.18
Solution of

$$\frac{d^2y}{dt^2} + 2y = \cos 10t$$

for initial conditions $y(0) = 0.5$, $y'(0) = 0$.

On the other hand, if ω is close to $\sqrt{2}$, $1/|2-\omega^2|$ is very large since the denominator is small. In this case the external force has a dramatic effect on solutions. Since the forcing frequency and the natural frequency are nearly equal, the external force pushes and pulls the harmonic oscillator with almost the same frequency as the unforced oscillator. This is called *near resonant* forcing. In this case the period of a beat is very long, and the natural response and forced response reinforce each other for long stretches of time. Even solutions with relatively small initial conditions eventually oscillate with very large amplitude (see Figure 4.19).

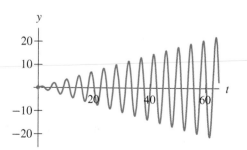

Figure 4.19
Solution of

$$\frac{d^2y}{dt^2} + 2y = \cos 1.4t$$

with $y(0) = y'(0) = 0$.

Resonance

In our study of solutions of

$$\frac{d^2y}{dt^2} + 2y = \cos \omega t,$$

we have avoided the case $\omega = \sqrt{2}$. That is, we have avoided the case where the frequency of the external force is exactly the same as the natural frequency. Forcing whose frequency is the same as the natural frequency of the oscillator is called **resonant forcing**, and the oscillator is said to be in **resonance**. Avoiding this case was necessary for algebraic reasons, since we encountered $2 - \omega^2$ in the denominator in the general solution. From the earlier discussion for the case $\omega \approx \sqrt{2}$, we suspect that the solutions have qualitatively interesting behavior in the case of resonant forcing.

Qualitative behavior

Setting $\omega = \sqrt{2}$, the differential equation becomes

$$\frac{d^2y}{dt^2} + 2y = \cos \sqrt{2}\, t.$$

We can predict the behavior of solutions of this equation in two ways. First, we can study solutions of the forced equation

$$\frac{d^2y}{dt^2} + 2y = \cos \omega t$$

for ω very close to $\sqrt{2}$. As pointed out above, the frequency of the beats decreases and the amplitude of the forced response increases as ω approaches $\sqrt{2}$. A sequence of solutions with ω approaching $\sqrt{2}$ is shown in Figure 4.20.

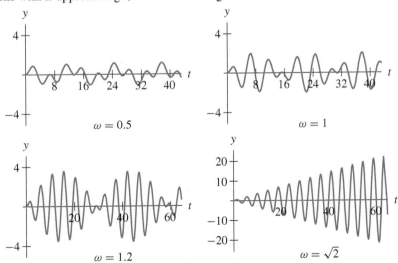

Figure 4.20
Solutions of $d^2y/dt^2 + 2y = \cos \omega t$ with $y(0) = y'(0) = 0$ for $\omega = 0.5$, $\omega = 1$, $\omega = 1.2$, and $\omega = \sqrt{2}$.

Also in Figure 4.20, we see a numerical solution of

$$\frac{d^2y}{dt^2} + 2y = \cos\sqrt{2}\,t.$$

From these pictures we can make the reasonable prediction that the forced response of the undamped harmonic oscillator with resonant forcing oscillates with an amplitude that increases for all time.

The general solution in the resonant case

Next we compute the general solution for

$$\frac{d^2y}{dt^2} + 2y = \cos\sqrt{2}\,t.$$

The general solution of the unforced equation is still

$$k_1\cos\sqrt{2}\,t + k_2\sin\sqrt{2}\,t.$$

To find a particular solution of the forced equation, we proceed as before and consider

$$\frac{d^2y}{dt^2} + 2y = e^{i\sqrt{2}t}.$$

The natural first guess of $y_c(t) = ae^{i\sqrt{2}t}$ does not yield a solution for any value of a because it is a solution of the unforced equation. In Figure 4.20, we see that the solutions of the forced equation still oscillate, but the amplitude of the oscillation grows linearly with t. This motivates our second guess $y_c(t) = ate^{i\sqrt{2}t}$; that is, we multiply our first guess by t. (This guess also agrees with the rule-of-thumb for second-guessing in Section 4.1.) Using

$$\frac{dy_c}{dt} = a\left(e^{i\sqrt{2}t} + i\sqrt{2}\,te^{i\sqrt{2}t}\right)$$

and

$$\frac{d^2y_c}{dt^2} = 2a\left(i\sqrt{2}e^{i\sqrt{2}t} - te^{i\sqrt{2}t}\right),$$

we substitute $y_c(t)$ into the left-hand side of the differential equation and obtain

$$\frac{d^2y_c}{dt^2} + 2y_c = 2a\left(i\sqrt{2}e^{i\sqrt{2}t} - te^{i\sqrt{2}t}\right) + 2ate^{i\sqrt{2}t}$$

$$= 2ai\sqrt{2}e^{i\sqrt{2}t}.$$

Since this left-hand side must equal $e^{i\sqrt{2}t}$ if $y_c(t)$ is a solution, we have

$$a = \frac{1}{2i\sqrt{2}},$$

which can also be written as $a = -i/(2\sqrt{2})$ because $1/i = -i$. Thus

$$y_c(t) = -\frac{i}{2\sqrt{2}} t\, e^{i\sqrt{2}t}$$

$$= -\frac{i}{2\sqrt{2}} t\, \left(\cos\sqrt{2}t + i\sin\sqrt{2}t\right).$$

Since the particular solution $y_p(t)$ we desire is the real part of $y_c(t)$, we have

$$y_p(t) = \frac{1}{2\sqrt{2}} t\sin\sqrt{2}t,$$

and the general solution of

$$\frac{d^2y}{dt^2} + 2y = \cos\sqrt{2}t$$

is

$$k_1\cos\sqrt{2}t + k_2\sin\sqrt{2}t + \frac{1}{2\sqrt{2}} t\sin\sqrt{2}t.$$

From the general solution we can see that our prediction of the qualitative behavior of solutions is correct. Every solution oscillates with the natural frequency (which in this case is the same as the forcing frequency). For small times t, the solution resembles a solution of the unforced oscillator because the particular solution is close to zero for small t. For large t, the term

$$\frac{1}{2\sqrt{2}} t\,\sin\sqrt{2}t$$

dominates the other terms, so the amplitude of the oscillation grows linearly (see Figure 4.21).

Figure 4.21
Solution of

$$\frac{d^2y}{dt^2} + 2y = \cos\sqrt{2}t$$

with $y(0) = y'(0) = 0$. The dashed lines are $y = \pm t/(2\sqrt{2})$.

The Millennium Bridge

The Millennium Bridge is the first new bridge to span the Thames River in central London in over 100 years. It opened to the public in June 2000 but was quickly closed. The 320-meter footbridge experienced *horizontal* oscillations of surprisingly high amplitude when many people were crossing.*

---

*See *New Scientist*, Vol. 167, July 8, 2000, p. 14.

Subsequent studies by the designers of the bridge revealed two important facts. First, the bridge has a large amplitude response to horizontal forcing at a frequency of one cycle per second (one Hz). Since people typically walk at a rate of two steps per second, the time between successive steps with the left foot is about one second. Of course, the right foot also pushes with the same frequency, so we should wonder about resonant forcing. Perhaps the slight forces that are caused by walking could provide a periodic forcing at precisely the resonant frequency.

Even if it had occurred to someone to consider this forcing term on the dynamics of the bridge, it probably would have been quickly dismissed as unimportant. The forces caused by a single pedestrian are tiny compared to the other horizontal forces on the bridge. While a large number of people marching together could provide a significant forcing term, tourists don't generally march in time.

A video of pedestrians crossing the bridge led to another surprising observation. They were, in fact, marching in time. In order to keep their balance on the bridge which was oscillating slightly left and right, they tended to time their steps with the oscillations. The result was a significant amount of forcing at the resonant frequency, which caused a larger forced response.

While the Millennium Bridge is a light and flexible structure, it is not entirely without damping. In the next section, we consider the interplay between the damping coefficient and the frequency of the forcing. We will see that the amplitude of the forced response does not tend to infinity if damping is present. However, it can be large if damping is small.

The analysis above is based on a purely linear model for the dynamics of the bridge. We must also remind ourselves that the linear model is only accurate in a certain regime. Outside this regime, when the loads are too great or the initial conditions too large, we must work with more accurate nonlinear models. We will review this lesson when we consider the Tacoma Narrows Bridge (see Section 4.5).

Moral

The key feature of resonance is that the amplitude of the forced response is very large *if the frequency of the forcing is very close to the natural frequency*. Resonance can be dangerous, as in the case of the Millennium Bridge. On the other hand, being able to selectively amplify a particular forcing frequency can be extremely useful. Amplifying a particular frequency of a radio signal is perhaps the most common example of the usefulness of resonance. In either case, it is clear we must be aware of the subtle dependence of the long-term behavior of a forced system on the specific values of the parameters.

EXERCISES FOR SECTION 4.3

In Exercises 1–8, compute the general solution of the given equation.

1. $\dfrac{d^2y}{dt^2} + 9y = \cos t$

2. $\dfrac{d^2y}{dt^2} + 9y = 5\sin 2t$

3. $\dfrac{d^2y}{dt^2} + 4y = -\cos\dfrac{t}{2}$

4. $\dfrac{d^2y}{dt^2} + 4y = 3\cos 2t$

5. $\dfrac{d^2y}{dt^2} + 9y = 2\cos 3t$

6. $\dfrac{d^2y}{dt^2} + 3y = 2\cos 9t$

7. $\dfrac{d^2y}{dt^2} + 3y = \cos 3t$

8. $\dfrac{d^2y}{dt^2} + 5y = 5\sin 5t$

In Exercises 9–14, compute the solution of the given initial-value problem.

9. $\dfrac{d^2y}{dt^2} + 9y = \cos t$
$y(0) = y'(0) = 0$

10. $\dfrac{d^2y}{dt^2} + 4y = 3\cos 2t$
$y(0) = y'(0) = 0$

11. $\dfrac{d^2y}{dt^2} + 5y = 3\cos 2t$
$y(0) = y'(0) = 0$

12. $\dfrac{d^2y}{dt^2} + 9y = \sin 3t$
$y(0) = 1,\ y'(0) = -1$

13. $\dfrac{d^2y}{dt^2} + 9y = 2\cos 3t$
$y(0) = 2,\ y'(0) = -9$

14. $\dfrac{d^2y}{dt^2} + 4y = \sin 3t$
$y(0) = 2,\ y'(0) = 0$

In Exercises 15–18, for the equation specified,
 (a) determine the frequency of the beats,
 (b) determine the frequency of the rapid oscillations, and
 (c) use the information from parts (a) and (b) to give a rough sketch of a typical solution. [*Hint*: You should be able to do this sketch with no further calculation.]

15. $\dfrac{d^2y}{dt^2} + 4y = \cos\dfrac{9t}{4}$

16. $\dfrac{d^2y}{dt^2} + 11y = 2\cos 3t$

17. $\dfrac{d^2y}{dt^2} + 5y = 3\cos 2t$

18. $\dfrac{d^2y}{dt^2} + 6y = \cos 2t$

19. Consider the equation

$$\dfrac{d^2y}{dt^2} + 12y = 3\cos 4t + 2\sin t.$$

 (a) Compute the general solution. [*Hint*: See Exercise 36 in Section 4.1.]
 (b) Compute the solution with $y(0) = y'(0) = 0$.
 (c) Sketch the graph of the solution of part (b).
 (d) In a short essay, describe the behavior of the solution of part (b). How could you have predicted this behavior in advance?

20. Commercials for a maker of recording tape show a crystal glass being broken by the sound of a recording of an opera singer's voice. Is this a good test of the quality of the tape? Why or why not?

21. Six second-order equations and four $y(t)$-graphs are given below. For each $y(t)$-graph, determine the second-order equation for which $y(t)$ is a solution, and state briefly how you know your choice is correct. You should do this exercise without using technology.

(i) $\dfrac{d^2y}{dt^2} + 16y = 10$ (ii) $\dfrac{d^2y}{dt^2} + 16y = -10$

(iii) $\dfrac{d^2y}{dt^2} + 16y = 5\cos 3t$ (iv) $\dfrac{d^2y}{dt^2} + 14y = 2\cos 4t$

(v) $\dfrac{d^2y}{dt^2} + 16y = \dfrac{1}{2}\cos 4t$ (vi) $\dfrac{d^2y}{dt^2} + 2\dfrac{dy}{dt} + 16y = \cos 4t$

(a)

(b)

(c)

(d)

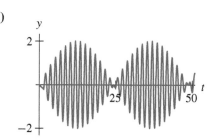

22. Legend has it that the marching band of a major university is forbidden from playing the song "Hey, Jude." The reason is that whenever the band played this song, the fans would stomp their feet in time to the music and after the second chorus, the stadium structure would begin to rock quite noticeably. The administration insisted that there was no danger that the stadium would collapse, but banned the song as a precaution. No other song has this effect on the stadium (although, presumably, the fans will stomp to other songs). Give a possible explanation of the stadium structure's violent reaction to the stomping accompanying the song "Hey, Jude" and to no other song.

23. Consider a unit mass sliding on a frictionless table attached to a spring, with spring constant $k = 16$. Suppose the mass is lightly tapped by a hammer every T seconds.

Mass tapped periodically with a hammer.

Suppose that the first tap occurs at time $t = 0$ and before that time the mass is at rest. Describe what happens to the motion of the mass for the following choices of the tapping period T:

(a) $T = 1$ **(b)** $T = 3/2$ **(c)** $T = 2$ **(d)** $T = 5/2$ **(e)** $T = 3$

24. Suppose the suspension system of the average car can be fairly well modeled by an underdamped harmonic oscillator with a natural period of 2 seconds. How far apart should speed bumps be placed so that a car traveling at 10 miles per hour over several bumps will bounce more and more violently with each bump?

4.4 AMPLITUDE AND PHASE OF THE STEADY STATE

In this section we return to the equation

$$\frac{d^2y}{dt^2} + p\frac{dy}{dt} + qy = \cos \omega t$$

for the periodically forced damped harmonic oscillator. Our goal is to establish a quantitative relationship between the behavior of the solutions and the parameters—particularly ω, which determines the forcing frequency, and p, which determines the damping. As we saw in Section 4.2, the long-term behavior of every solution is the same. It is determined by a particular solution $y_p(t)$ of the forced equation, which can be written in the form

$$y_p(t) = A\cos(\omega t + \phi).$$

Hence for large t, all solutions oscillate with the same frequency as the forcing. This is the *steady-state* behavior of solutions. The amplitude of the oscillation is A and the *phase angle* is ϕ. In this section we find expressions for A and ϕ as functions of ω, p, and q.

For small damping (p near zero), we expect the solutions to behave approximately the same as in the undamped case, at least for a bounded amount of time. The amplitude A of the forced response is large if the forcing frequency is close to the natural frequency of the unforced equation.

Behavior of the Steady State

To find a particular solution of the forced equation

$$\frac{d^2y}{dt^2} + p\frac{dy}{dt} + qy = \cos\omega t,$$

we follow the techniques introduced in Section 4.2 and use the Method of Undetermined Coefficients applied to the complexified equation

$$\frac{d^2y}{dt^2} + p\frac{dy}{dt} + qy = e^{i\omega t}.$$

Next, we guess $y_c(t) = ae^{i\omega t}$ and determine the value of a that makes $y_c(t)$ a solution. Substituting $y_c(t)$ into the left-hand side of the differential equation, we have

$$\frac{d^2y_c}{dt^2} + p\frac{dy_c}{dt} + qy_c = -a\omega^2 e^{i\omega t} + p(ia\omega e^{i\omega t}) + q(ae^{i\omega t})$$

$$= a\left((q-\omega^2) + ip\omega\right)e^{i\omega t}.$$

In order for $y_c(t)$ to be a solution, we must take

$$a = \frac{1}{b}$$

where b is the complex number $(q-\omega^2) + ip\omega$.

As we saw in Section 4.2, we can determine the amplitude A and phase angle ϕ of $y_p(t)$ from the polar form

$$a = |a|e^{i\phi} = Ae^{i\phi}$$

of the complex number a. (See the appendix on complex numbers for a review of the polar representation of a complex number.) In this case, since $a = 1/b$, we have

$$|a| = \frac{1}{|b|}$$

and the polar angle ϕ of a is the negative of the polar angle of b. Therefore the complex number

$$a = \frac{1}{b} = \frac{1}{(q-\omega^2) + i(p\omega)}$$

encapsulates the amplitude and phase angle of the forced response.

This is neat. Given any damped harmonic oscillator with forcing $\cos\omega t$, we can obtain the long-term behavior of the solution for any ω directly from the values of the parameters p, q, and ω.

Amplitude and Phase

Given $y_p(t) = A\cos(\omega t + \phi)$, it is convenient to have a formulas for A and ϕ. Since $A = |a| = 1/|b|$ and

$$|b| = \sqrt{(q - \omega^2)^2 + p^2\omega^2},$$

we have

$$A = \frac{1}{\sqrt{(q - \omega^2)^2 + p^2\omega^2}}.$$

Moreover we are assuming that both p and ω are positive, and thus the complex number b has a positive imaginary part. We can therefore consider its polar angle to be between 0 and π. Furthermore, since the polar angle of $a = 1/b$ is the negative of the polar angle of b, we consider ϕ to be the angle between $-\pi$ and 0 such that

$$\tan\phi = \frac{-p\omega}{q - \omega^2}.$$

Amplitude

To recap, for large t, we know that all solutions of

$$\frac{d^2y}{dt^2} + p\frac{dy}{dt} + qy = \cos\omega t$$

oscillate with the same amplitude A and with the same frequency, $\omega/(2\pi)$, as the forcing function. The formula for the amplitude of these oscillations is

$$A = \frac{1}{\sqrt{(q - \omega^2)^2 + p^2\omega^2}}.$$

In Figure 4.22 we show the graph of A as a function of ω for $q = 2$ and several values of p. The figures for other values of q are similar. If p is small, the damping is small. It is not surprising to see that the amplitude A is large for $\omega \approx \sqrt{2}$, since if $p = 0$, this value of ω corresponds to resonant forcing. As p increases, two things happen: the maximum value of A decreases, and the value of ω where A attains its maximum moves to the left. (It is a good review of max-min problems to compute the maximum value of A and where it is attained. See Exercise 6.)

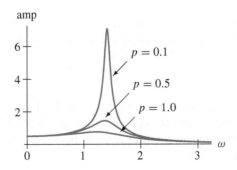

Figure 4.22
Graphs of the amplitude

$$A = \frac{1}{\sqrt{(q - \omega^2)^2 + p^2\omega^2}}$$

with $q = 2$ and $p = 0.1$, $p = 0.5$, and $p = 1.0$.

We can visualize the amplitude A in several ways. For example, if we fix q, then we can graph A as a function of the two variables p and ω (see Figure 4.23). Again, the most noticeable feature of this graph is the explosion of A near resonance for p near zero and ω near $\sqrt{q}$.

Another view of the amplitude is obtained by fixing the angular frequency ω of the forcing and graphing A as a function of p and q. Such a graph is shown in Figure 4.24. This graph shows that, by choosing p and q appropriately, we can obtain a large amplitude response. Ideally, we would like to see the graph of A as a function of all three parameters ω, p, and q, but this graph would live in four dimensions.

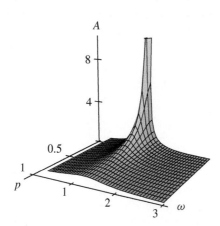

Figure 4.23
Graph of amplitude A as a function of p and ω for $q = 2$.

Figure 4.24
Graph of amplitude A as a function of p and q for $\omega = 1$.

Phase angle

A particular solution of the equation

$$\frac{d^2y}{dt^2} + p\frac{dy}{dt} + qy = \cos \omega t$$

can be written in the form

$$y_p(t) = A \cos(\omega t + \phi),$$

and we can think of the long-term behavior of solutions of this equation as translations and scalings of the external forcing $\cos \omega t$. The scaling is accomplished by multiplication by the amplitude A. The translation corresponds to the addition of the phase angle ϕ. The graph of

$$A \cos(\omega t + \phi) = A \cos\left(\omega\left(t + \frac{\phi}{\omega}\right)\right)$$

is obtained by sliding the graph of $A \cos \omega t$ by ϕ/ω to the right if $\phi/\omega < 0$ and to the left if $\phi/\omega > 0$.

By convention, we choose the phase angle ϕ between $-\pi$ and 0 so that it satisfies the condition

$$\tan \phi = \frac{-p\omega}{q - \omega^2}.$$

Ideally, we would like to see the graph of ϕ as a function of all three parameters ω, p, and q, but this graph also requires four dimensions. In Figure 4.25 we fix $q = 2$ and sketch the graph of ϕ as a function of ω for several values of p. Similar graphs for other values of q are qualitatively the same. Note that when p is small, the phase angle ϕ stays close to zero for ω small. The graph of ϕ has a large slope when it passes through $-\pi/2 = -90°$ for $\omega = \sqrt{q}$. For ω large, ϕ is asymptotic to $-\pi$. In other words, for an external force with small frequency, solutions are almost in phase with the forcing. For large forcing frequency, solutions are almost half a period out of phase with the forcing.

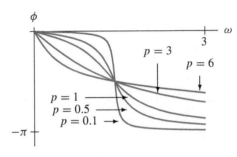

Figure 4.25
Phase angle ϕ for $q = 2$ with $p = 0.1$, $p = 0.5$, $p = 1.0$, $p = 3.0$, and $p = 6.0$. Note that the phase angle ϕ stays close to zero for small ω if p is small. The graph of ϕ has a large slope when it passes through $-\pi/2 = -90°$ for $\omega = \sqrt{q}$. For ω large, ϕ is asymptotic to $-\pi$.

More Parameters

There are two more parameters that we may wish to consider—the amplitude and phase of the forcing function. That is, we consider

$$\frac{d^2y}{dt^2} + p\frac{dy}{dt} + qy = F\cos(\omega t + \theta),$$

where F and θ are constants. Luckily these two new parameters add little complication to the discussion.

First, suppose $y_p(t)$ is a solution of the forced equation

$$\frac{d^2y}{dt^2} + p\frac{dy}{dt} + qy = g(t)$$

for some function $g(t)$. If F is a constant, then it follows from the linearity of the differential equation that the product $Fy_p(t)$ is a solution of

$$\frac{d^2y}{dt^2} + p\frac{dy}{dt} + qy = Fg(t)$$

(see Exercise 16 in Section 4.2). Multiplying the external force by a constant multiplies the forced response by the same constant.

Second, suppose again that $y_p(t)$ is a solution of the forced equation

$$\frac{d^2y}{dt^2} + p\frac{dy}{dt} + qy = g(t)$$

for some function $g(t)$. If θ is a constant, then $y_p(t + \theta)$ is a solution of the differential equation

$$\frac{d^2y}{dt^2} + p\frac{dy}{dt} + qy = g(t + \theta).$$

This assertion can be verified by defining a new independent variable $\tau = t + \theta$ and checking that $y_p(\tau)$ is a solution of

$$\frac{d^2y}{d\tau^2} + p\frac{dy_p}{d\tau} + qy_p = g(\tau)$$

(that is, we just shift the t-axis—see Exercise 3).

An Application for Washing Machines

Adjusting the parameters for a forced harmonic oscillator, particularly the damping parameter, can have a dramatic effect on the amplitude of the forced response. Recent advances in material science have made it possible to build dashpots that provide a damping coefficient that can be adjusted in "real time." The dashpot is filled with a fluid that changes its characteristics when subjected to a magnetic field. The "active suspension systems" built using these fluids are beginning to appear in everyday devices, and one of the first applications of this technology is to the problem of rattling washing machines (see also Lab 2.5 in Chapter 2).*

The drum of a washing machine must spin over a wide range of speeds in order to wash the clothes and to remove excess water during the spin cycle. To isolate the rest of the washing machine from the high-frequency periodic forcing this spinning causes, the drum of a front-loading washer is suspended inside the machine casing by springs (see Figure 4.26). During the spin cycle, the drum imparts a periodic forcing to the springs. This forcing can have large amplitude if the load is "unbalanced" in the drum, and it occurs over a range of frequencies.

As with any mass-spring system, the forced response becomes most dramatic at the resonant frequency of the springs. The range of frequencies for the spinning drum is quite large, so the periodic forcing caused by the rotation of the drum passes through the resonant frequency of the springs several times for each load of wash. In order to keep the washing machine from shaking too violently at the resonant frequency, dashpots are also attached to the drum. As we noted earlier, high damping greatly reduces

*M. J. Chrzan, J. D. Carlson, "MR Fluid Sponge Devices and Their Use in Vibration Control of Washing Machines," Proceedings of the 8th Annual Symposium on Smart Structures and Materials, Society of Photo-Optical Instrumentation Engineers. Also, go to www.lord.com/mr and click on Applications.

Figure 4.26
Schematic of a front loading washing machine with
drum springs and dampers holding the drum.

the amplitude of resonant oscillations. So very stiff dashpots with a large damping coefficient solve the problem of oscillations at resonance. However, very stiff dashpots
effectively connect the drum of the washer rigidly to the casing. When the drum spins
very rapidly, dashpots with high damping coefficients allow vibrations to be transmitted from the drum to the rest of the washer. This rattling can be as damaging as the
shaking at the resonant frequency. Traditionally, a compromise is chosen between low
damping,which allows resonant oscillations to occur and high damping, which trans-
mits vibration caused by the high-frequency spinning. A better solution is to install a
dashpot that provides a damping coefficient that can be adjusted according to the frequency of the forcing.

To get a clearer understanding of the nature of this problem, think more carefully
about the physical system. Unlike our previous harmonic oscillators, neither end of the
spring is fixed in this system. As a very first model, we consider a mass, of size $m = 1$
for convenience, with horizontal position $y(t)$ and a handle with horizontal position
$z(t)$. The mass and the handle are connected by a spring and a dashpot. The spring
has spring constant k and rest length y_0. The dashpot provides a damping coefficient b
(see Figure 4.27). We think of the handle as the drum and the mass as the casing of the

Figure 4.27
Mass attached to a handle by a spring and dashpot.
The mass and handle are free to move horizontally.

washer. Moving the handle back and forth corresponds to the periodic forcing created by the rotation of the drum.

Suppose the horizontal motion of the handle is prescribed, that is, the function $z(t)$ is given. The force necessary to make the position of the handle equal $z(t)$ at time t is provided by the one who holds the handle. The forces acting on the mass are the restoring force from the spring and the damping from the dashpot. The spring force is proportional to the difference between the rest length of the spring y_0 and its actual length $y(t) - z(t)$. The damping force of the dashpot is proportional to the rate at which the mass and handle are moving together or apart. In other words, the damping is proportional to the difference $dy/dt - dz/dt$. The equation governing the motion of the mass is

$$\frac{d^2 y}{dt^2} = -k(y - z - y_0) - b\left(\frac{dy}{dt} - \frac{dz}{dt}\right).$$

The motion of the handle represents the oscillations of the drum of the washer, and we assume that it is simple periodic motion of the form $z(t) = A\cos\omega t$. Consequently, the equation for $y(t)$ is

$$\frac{d^2 y}{dt^2} = -ky - b\frac{dy}{dt} + kA\cos\omega t - b\omega A\sin\omega t + ky_0.$$

The constant term ky_0 on the right provides a displacement of $y(t)$ by a constant. Since we are only concerned with the amplitude of oscillations, we can ignore this term (see Exercise 8 for a careful justification of this assertion). Our model equation is

$$\frac{d^2 y}{dt^2} + b\frac{dy}{dt} + ky = kA\cos\omega t - b\omega A\sin\omega t.$$

Fixing k and b, we can compute the forced response, obtaining

$$y(t) = A\frac{k^2 - k\omega^2 + b^2\omega^2}{(k - \omega^2)^2 + b^2\omega^2}\cos\omega t + A\frac{b\omega^3}{(k - \omega^2)^2 + b^2\omega^2}\sin\omega t$$

(see Exercise 9). Since our goal is to get an idea of the behavior of solutions of this system, not to test a particular brand of machine, we fix $k = 1$ for convenience. Hence, our particular solution is

$$y(t) = A\frac{1 - \omega^2 + b^2\omega^2}{(1 - \omega^2)^2 + b^2\omega^2}\cos\omega t + A\frac{b\omega^3}{(1 - \omega^2)^2 + b^2\omega^2}\sin\omega t.$$

From this solution we can see what happens as ω varies for various choices of b. If b is small, $y(t)$ has very large amplitude for ω near the resonant angular frequency $\omega \approx 1$. The response will have much larger amplitude than the amplitude A of the forcing term. On the other hand, if b is very large, then the amplitude of $y(t)$ will be close to the amplitude of the forcing until ω is very large (see Figure 4.28).

Now, suppose we install dashpots with adjustable damping. Clearly, we want the damping coefficient to be large for ω near the resonant frequency and small for ω larger

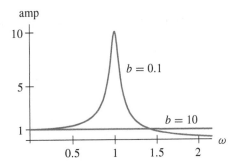

Figure 4.28
Amplitude of the forced response for $A = 1$
and two values of b, $b = 0.1$ and $b = 10.0$.

than the resonant frequency. That is, we assume that b is a function of ω given by

$$b(\omega) = \begin{cases} 10, & \text{if } \omega < 2, \\ 0.1, & \text{if } \omega \geq 2, \end{cases}$$

where 0.1 and 10 are chosen as convenient "small" and "large" numbers. With this choice of $b(\omega)$, the amplitude of the forced response $y(t)$ as a function of ω is given in Figure 4.29. At least in theory, the method works beautifully. In experiments with a real washing machine, Chrzan and Carlson were able to obtain good control of the motion of the washer near resonance and a slight reduction in shaking for higher frequencies using this approach.

Of course, we need not choose $b(\omega)$ to be this particular function. We may get better behavior if we choose $b(\omega)$ so that it continuously varies from large to small as ω increases. Finding a good choice for $b(\omega)$ is an example of a problem in control theory.

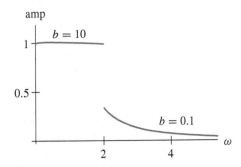

Figure 4.29
Amplitude of the forced response with $b(\omega)$
large for small ω and $b(\omega)$ small for large ω.

A Sobering Reminder

We have amassed a tremendous amount of information concerning the solutions of the equation

$$\frac{d^2y}{dt^2} + p\frac{dy}{dt} + qy = \cos \omega t,$$

especially if $p \geq 0$ and $q > 0$. It is important to remember that our calculations rely on the special form of this equation. Only the fact that the equation is linear allows us to use the Extended Linearity Principle to decompose the general solution into the general solution of the unforced equation plus a particular solution of the forced equation. Only the fact that the forcing function is a cosine function allows us to find a particular solution and manipulate it as we have done in the last three sections.

Nevertheless, the specific nature of the equations considered in this chapter does not diminish the importance of the calculations that we have done in the last four sections. We have learned a tremendous amount about particular examples. This is important for two reasons. First, as we have said many times, this particular differential equation is an excellent model for many different physical systems. Second, understanding this equation gives us a starting place for studying other types of equations—not all periodic forcing functions are sines or cosines and not all differential equations are linear.

EXERCISES FOR SECTION 4.4

1. The glass harmonica is a musical instrument invented by Ben Franklin. It consists of a nested set of crystal glass bowls with a rod running down the middle. The rod is supported at the ends and the bowls are spun rapidly using a foot pedal or a motor. The musician rubs her slightly damp finger around the edge of the glass and the resulting vibrations of the glass make a very pure tone. The harder the musician pushes against the glass, the louder the note (but the frequency stays the same). Explain how the glass harmonica works.

2. Suppose an opera singer can break a glass by singing a particular note.

    (a) Will the singer have to sing a higher or a lower note to break an identical glass that is half full of water?

    (b) Suppose both notes are within the singer's range. Will it be harder or easier to break the glass when it is half full of water?

3. Given that $y_p(t)$ is a solution of

$$\frac{d^2y}{dt^2} + p\frac{dy}{dt} + qy = g(t),$$

show that $y_p(t + \theta)$ is a solution of

$$\frac{d^2y}{dt^2} + p\frac{dy}{dt} + qy = g(t + \theta).$$

4. Consider the harmonic oscillator equation with two forcing terms

$$\frac{d^2y}{dt^2} + p\frac{dy}{dt} + qy = \cos\omega_1 t + \cos\omega_2 t.$$

Suppose $p > 0$ is fixed.

(a) How should q be chosen so that the term in the forced response with angular frequency ω_1 has the largest amplitude?

(b) How large is the amplitude of the term in the forced response with angular frequency ω_2?

5. Consider the differential equation studied in Exercise 4 and fix q so that the amplitude of the ω_1-term is greatest (see part (a) of Exercise 4).

(a) What is the ratio of the amplitude of the ω_1-term to the amplitude of the ω_2-term?

(b) How does this ratio depend on p? (For example, what happens as $p \to 0$ and $p \to \infty$, etc.?)

6. For large t, every solution of

$$\frac{d^2y}{dt^2} + p\frac{dy}{dt} + qy = \cos\omega t$$

oscillates with angular frequency ω and amplitude A given by

$$A(\omega, p, q) = \frac{1}{\sqrt{(q - \omega^2)^2 + p^2\omega^2}}.$$

That is, the amplitude A is a function of the parameters ω, p, and q.

(a) Compute $\partial A/\partial\omega$.

(b) For fixed p and q, let $M(p, q)$ denote the maximum value of $A(\omega, p, q)$ as a function of ω. Compute an expression for $M(p, q)$. [*Hint*: This is a max-min problem from calculus.]

7. For the function $M(p, q)$ of Exercise 6,

(a) set $q = 1$ and plot $M(p, q)$ as a function of p, and

(b) explain why $M(p, q)$ is proportional to $1/p$ as $p \to 0$.

8. Consider the forced harmonic oscillator equation

$$\frac{d^2y}{dt^2} + b\frac{dy}{dt} + ky = g(t) + y_0,$$

where the forcing is made up of two parts, constant forcing y_0 and forcing $g(t)$ that changes over time.

(a) Let $w(t) = y(t) - y_0/k$. Rewrite the forced harmonic oscillator equation in terms of the new variable w.

(b) In what ways are the solutions of the two equations the same?

9. In this section we derived the equation

$$\frac{d^2y}{dt^2} + b\frac{dy}{dt} + ky = kA\cos\omega t - b\omega A\sin\omega t,$$

where k and b are constants. Use the Method of Undetermined Coefficients to find a particular solution of this equation. [*Hint*: Note that the parameters k and b appear on both sides of the equation. Since they are constants, this causes no more complication but does require careful bookkeeping.]

10. The solution you found in Exercise 9 depends on the parameters k and b. Describe the qualitative behavior of each solution in the following situations in terms of the motion of the "mass-spring with handle" model in this section (assume $\omega \approx 1$):

(a) k large, b and ω small but positive.

(b) b large, k and ω small but positive.

11. The particular solution you found in Exercise 9 also depends on the parameter ω. Describe the qualitative behavior of the solution in terms of the motion of the "mass-spring with handle" when k and b are bounded (for example, less than 1) and ω is very large.

12. In this section we computed a particular solution of the equation

$$\frac{d^2y}{dt^2} + p\frac{dy}{dt} + qy = \cos\omega t$$

of the form

$$y_p(t) = A\cos(\omega t + \phi)$$

where the phase angle ϕ satisfies the equation

$$\tan\phi = \frac{-p\omega}{q - \omega^2}$$

and $-180° < \phi < 0$. The angle ϕ is a function $\phi(\omega, p, q)$ of the parameters ω, p, and q.

(a) Compute $\partial\phi/\partial\omega$.

(b) Compute $\partial^2\phi/\partial\omega^2$.

(c) For $q = 2$ and several values of p near zero, find the value of ω where ϕ changes most rapidly.

[*Hint*: This exercise is an excellent opportunity to test the power of symbolic differentiation packages.]

4.5 THE TACOMA NARROWS BRIDGE

On July 1, 1940, the $6 million Tacoma Narrows Bridge opened for traffic. On November 7, 1940, during a windstorm, the bridge broke apart and collapsed. During its short stand, the structure, a suspension bridge more than a mile long, became known as "Galloping Gertie" because the roadbed oscillated dramatically in the wind. The collapse of the bridge proved to be a scandal in more ways than one, including the fact that because the insurance premiums had been embezzled, the bridge was uninsured.*

The roadbed of a suspension bridge hangs from vertical cables that are attached to cables strung between towers (see Figure 4.30 for a schematic picture). If we think of the vertical cables as long springs, then it is tempting to model the oscillations of the roadbed with a harmonic oscillator equation. We can think of the wind as somehow providing periodic forcing. It is very tempting to say, "Aha, the collapse must be due to resonance."

It turns out that things are not quite so simple. We know that to cause dramatic effects, the forcing frequency of a forced harmonic oscillator must be very close to its natural frequency. The wind seldom behaves in such a nice way for very long, and it would be very bad luck indeed if the oscillations caused by the wind happened to have a frequency almost exactly the same as the natural frequency of the bridge.

Recent research on the dynamics of suspension bridges (by two mathematicians, A. C. Lazer and P. J. McKenna♣) indicates that the linear harmonic oscillator does not make an accurate model of the movement of a suspension bridge. The vertical cables do act like springs when they are stretched. That is, when the roadbed is below its rest position, the cables pull up. However, when the roadbed is significantly above its rest position, the cables are slack, so they do not push down. Hence the roadbed feels less

Figure 4.30
Schematic of a suspension bridge.

---

*The story of the bridge and its collapse can be found in Martin Braun, *Differential Equations and Their Applications*, Springer-Verlag, 1993, p. 173, and Matthys Levy and Mario Salvadori, *Why Buildings Fall Down: How Structures Fail*, W. W. Norton Co., 1992, p. 109.

♣See "Large-amplitude Periodic Oscillations in Suspension Bridges: Some New Connections with Nonlinear Analysis" by A. C. Lazer and P. J. McKenna, in *SIAM Review*, Vol. 32, No. 4, 1990, pp. 537–578.

force trying to pull it back into the rest position when it is pushed up than when it is pulled down (see Figure 4.31).

In this section, we study a model for a system with these properties. The system of equations we study was developed by Lazer and McKenna from more complicated models of oscillations of suspension bridges. This system gives considerable insight into the possible behaviors of suspension bridges and even hints at how they can be made safer.

Figure 4.31
Close-up of the vertical cable when the roadbed is above and below the rest position.

Derivation of the Equations

The model we consider for the motion of the bridge uses only one variable to describe the position of the bridge. We assume that the bridge oscillates up and down as in Figure 4.31. We let $y(t)$ (measured in feet or meters) denote the vertical position of the center of the bridge, with $y = 0$ corresponding to the position where the cables are taut but not stretched. We let $y < 0$ correspond to the position in which the cables are stretched and $y > 0$ correspond to the position in which the cables are slack (see Figure 4.32). Of course, using one variable to study the motion of the bridge ignores many possible motions, and we comment on other models at the end of this section.

To develop a model for $y(t)$, we consider the forces that act on the center of the bridge. Gravity provides a constant force in the negative direction of y. We also assume that the cables provide a force that pulls the bridge up when $y < 0$ and that is proportional to y. On the other hand, when $y > 0$, the cable provides no force. When $y \neq 0$, there is also a restoring force that pulls y back toward $y = 0$ due to the stretching of the roadbed. Finally, there will also be some damping, which is assumed to be proportional to dy/dt. We choose units so that the mass of the bridge is 1.

Based on these assumptions, the equation developed by Lazer and McKenna to model a suspension bridge on a calm day (no wind) is

$$\frac{d^2y}{dt^2} + \alpha \frac{dy}{dt} + \beta y + c(y) = -g.$$

Figure 4.32
Positions of the bridge corresponding to $y > 0$ and $y < 0$.

The first term is the vertical acceleration. The second term, $\alpha(dy/dt)$, arises from the damping. Since suspension bridges are relatively flexible structures, we assume that α is small. The term βy accounts for the force provided by the material of the bridge pulling the bridge back toward $y = 0$. The function $c(y)$ accounts for the pull of the cable when $y < 0$ (and the lack thereof when $y \geq 0$), and therefore it is given by

$$c(y) = \begin{cases} \gamma y, & \text{if } y < 0; \\ 0, & \text{if } y \geq 0. \end{cases}$$

The constant g represents the force due to gravity.

We can convert this to a system in the usual way, obtaining

$$\frac{dy}{dt} = v$$

$$\frac{dv}{dt} = -\beta y - c(y) - \alpha v - g.$$

This is an autonomous system. Simplifying the right-hand side of this system by combining the $-\beta y$ term with the terms in $c(y)$, we obtain

$$\frac{dy}{dt} = v$$

$$\frac{dv}{dt} = -h(y) - \alpha v - g,$$

where $h(y)$ is the piecewise-defined function

$$h(y) = \begin{cases} ay, & \text{if } y < 0; \\ by, & \text{if } y \geq 0, \end{cases}$$

and $a = \beta + \gamma$ and $b = \beta$.

To study this example numerically, we choose particular values of the parameters. (These values are not motivated by any particular bridge.) Following Lazer and McKenna, we take $a = 17$, $b = 13$, $\alpha = 0.01$, and $g = 10$. So the system we study is

$$\frac{dy}{dt} = v$$

$$\frac{dv}{dt} = -h(y) - 0.01v - 10,$$

where

$$h(y) = \begin{cases} 17y, & \text{if } y < 0; \\ \\ 13y, & \text{if } y \geq 0. \end{cases}$$

We can easily compute that this system has only one equilibrium point, which is given by $(y, v) = (-10/17, 0)$. The y-coordinate of this equilibrium point is negative because gravity forces the bridge to sag a little, stretching the cables. Numerical results indicate that solutions spiral toward the equilibrium point very slowly. This is what we expect because there is a small amount of damping present. The direction field and a typical solution curve are shown in Figure 4.33.

The behavior of the solutions indicate that the bridge oscillates around the equilibrium position. The amplitude of the oscillation dies out slowly due to the small amount of damping. Because the forces controlling the motion change abruptly along $y = 0$, solutions also change direction when they cross from the left half-plane to the right half-plane.

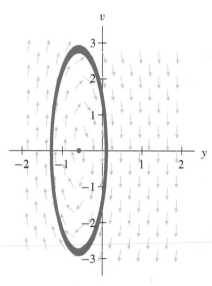

Figure 4.33
Direction field and typical solution for the system

$$\frac{dy}{dt} = v$$

$$\frac{dv}{dt} = -h(y) - 0.01v - 10.$$

There is a spiral sink at $(y, v) = (-10/17, 0)$. Solution curves spiral toward the equilibrium point very slowly.

The effect of wind

To add the effect of wind into the model, we add an extra term to the right-hand side of the equation. The effect of the wind is very difficult to quantify. Not only are there gusts of more or less random duration and strength, but also the way in which the wind interacts with the bridge can be very complicated. Even if we assume that the wind has constant speed and direction, the effect on the bridge need not be constant. As air moves past the bridge, swirls or vortices (like those at the end of an oar in water) form above and below the roadbed. When they become large enough, these vortices "break off," causing the bridge to rebound. Hence even a constant wind can give a periodic push to the bridge.

Despite these complications, we assume, for simplicity, that the wind provides a forcing term of the form $\lambda \sin \mu t$. Since it is unlikely that turbulent winds will give a forcing term with constant amplitude λ or constant frequency $\mu/(2\pi)$, we will look for behavior of solutions that persist for a range of λ- and μ-values.

The system with forcing is given by

$$\frac{dy}{dt} = v$$

$$\frac{dv}{dt} = -h(y) - 0.01v - 10 + \lambda \sin \mu t.$$

This is a fairly simple model for the complicated behavior for a bridge moving in the wind, but we see below that even this simple model has solutions that behave in a surprising way.

Behavior of Solutions

We have seen that a linear system with damping and sinusoidal forcing has one periodic solution to which every other orbit tends as time increases, the steady-state solution. In other words, no matter what the initial conditions, the long-term behavior of the system will be the same. The amplitude and frequency of this periodic solution are determined by the amplitude and frequency of the forcing term (see Section 4.2).

The behavior of the system

$$\frac{dy}{dt} = v$$

$$\frac{dv}{dt} = -h(y) - 0.01v - 10 + \lambda \sin \mu t$$

is quite different. We now describe the results of a numerical study of these solutions of this system carried out by Glover, Lazer, and McKenna.[*]

If, for example, we choose $\mu = 4$ and λ very small ($\lambda < 0.05$), then every solution tends toward a periodic solution with small magnitude near $y = -10/17$ (see Figure 4.34). For this periodic solution, $y(t)$ is negative for all t. Since this solution

Figure 4.34
Solution of the system with small forcing.

[*]See "Existence and Stability of Large-scale Nonlinear Oscillations in Suspension Bridges" by J. Glover, A. C. Lazer, and P. J. McKenna, *ZAMP*, Vol. 40, 1989, pp. 171–200.

Figure 4.35
Schematic of the bridge oscillating in light winds.

never crosses $y = 0$, it behaves just like the solution of the forced linear system

$$\frac{dy}{dt} = v$$

$$\frac{dv}{dt} = -17y - 0.01v - 10 + \lambda \sin 4t.$$

In terms of the behavior of the bridge, this means that in light winds, we expect the bridge to oscillate with small amplitude. Gravity keeps the bridge sagging downward and the cables are always stretched somewhat (see Figure 4.35). In this range, modeling the cables as linear springs is reasonable.

As λ increases, a new phenomenon is observed. Initial conditions near $(y, v) = (-10/17, 0)$ still yield solutions that oscillate with small amplitude (see Figure 4.36). However, if $y(0) = -10/17$ but $v(0)$ is large, solutions can behave differently. There is another periodic solution that oscillates around $y = -10/17$, but with (relatively) large amplitude (see Figure 4.37).

Figure 4.36
Solution of the forced system with larger forcing than in Figure 4.34 and initial conditions near the equilibrium.

Figure 4.37
Solution of the forced system with the same large forcing as in Figure 4.36 but with initial conditions farther from the equilibrium.

This has dramatic implications for the behavior of the bridge. If the initial displacement is small, then we expect to see small oscillations as before. However, if a gust of wind gives the bridge a kick large enough to cause it to rise above $y = 0$, then the cables will go slack and the linear model will no longer be accurate (see Figure 4.38). In this situation the bridge can start oscillating with much larger amplitude, and these oscillations *do not die out*. So, in a moderate wind ($\lambda > 0.06$), a single strong gust could suddenly cause the bridge to begin oscillating with much larger amplitude, perhaps with devastating consequences.

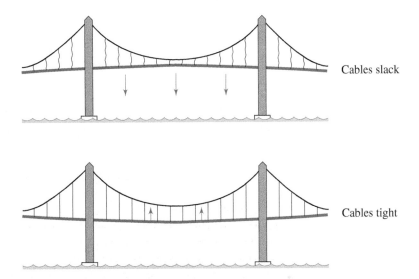

Cables slack

Cables tight

Figure 4.38
Schematic of large-amplitude oscillations of the bridge.

Varying the parameters

As mentioned above, because the effects of the wind are not particularly regular, we should investigate the behavior of solutions as λ and μ are varied. It turns out that the large-amplitude periodic solution persists for a fairly large range of λ and μ. This means that even in winds with uneven velocity and direction, the sudden jump in behavior to a persistent oscillation with large amplitude is possible.

Does This Explain the Tacoma Narrows Bridge Disaster?

As with any simple model of a complicated system, a note of caution is in order. To construct this model, we have made a number of simplifying assumptions. These include, but are not limited to, assuming that the bridge oscillates in one piece. The bridge can oscillate in two or more sections (see Figure 4.39). To include this in our model, we would have to include a new independent variable for the position along the bridge. The resulting model is a partial differential equation.

Figure 4.39
More complicated forms of oscillation of a suspension bridge.

Another factor we have ignored is that the roadbed of the bridge has width as well as length. The final collapse of the Tacoma Narrows Bridge was preceded by violent twisting motions of the roadbed, alternately stretching and loosening the cables on either side of the road. Analysis of a model including the width gives considerable insight into the final moments before the bridge's collapse (see the paper by Lazer and McKenna cited on page 439).

This being said, the simple model discussed above still helps a great deal in understanding the behavior of the bridge. If this simple system can feature the surprising appearance of large-amplitude periodic solutions, then it is not at all unreasonable to expect that more complicated and more accurate models will also exhibit this behavior. So this model does what it is supposed to do: It tells us what to look for when studying the behavior of a flexible suspension bridge.

Joseph McKenna (1948 –   ), born in Dublin, Ireland, received his Ph.D. in mathematics at the University of Michigan. For most of his professional life he has been involved in research in differential equations, applying them to diverse problems in soil physics, fluid dynamics, optics, biology, and most recently on flexing in bridges and ships.

His work with A. C. Lazer, described in this section, has received considerable attention in both the mathematics and engineering communities. Their research directly contradicts the long-standing view that resonance phenomena caused the collapse of the Tacoma Narrows Bridge. They have also suggested several alternative types of differential equations that govern the motion of such suspension bridges, including the equation described in this section as well as a more complicated model involving nonlinear equations somewhat like those we will discuss in the next chapter.

McKenna is currently Professor of Mathematics at the University of Connecticut.

EXERCISES FOR SECTION 4.5

For Exercises 1–3, recall that our simple model of a suspension bridge is

$$\frac{d^2y}{dt^2} + \alpha \frac{dy}{dt} + \beta y + c(y) = -g,$$

where α is the coefficient of damping, β is a parameter corresponding to the stiffness of the roadbed, the function $c(y)$ accounts for the pull of the cables, and g is the gravitational constant. For each of the following modifications of bridge design,

(a) discuss which parameters are changed, and

(b) discuss how you expect a change in the parameter values to affect the solutions. (For example, does the modification make the system look more or less like a linear system?)

1. The "stiffness" of the roadbed is increased, for example, by reinforcing the concrete or adding extra material that makes it harder for the roadbed to bend.

2. The coefficient of damping is increased.

3. The strength of the cables is increased.

4. The figure below is a schematic for an alternate suspension bridge design called the Lazer-McKenna light flexible long span suspension bridge. Why does this design avoid the problems of the standard suspension bridge design? Discuss this in a paragraph and give model equations similar to those in the text for this design.

A schematic of the Lazer-McKenna light flexible long span suspension bridge.

In Exercises 5–8, we consider another application of the ideas in this section. Lazer and McKenna observe that the equations they study may also be used to model the up-and-down motion of an object floating in water, which can rise completely out of the water. This has serious implications for the behavior of a ship in heavy seas.

Suppose we have a cube made of a light substance floating in water. Gravity always pulls the cube downward. The cube floats at an equilibrium level at which the mass of the water displaced equals the mass of the cube. If the cube is higher or lower than the equilibrium level, then there is a restoring force proportional to the size of the displacement. We assume that the bottom and top of the cube stay parallel to the surface of the water at all times and that the system has a small amount of damping.

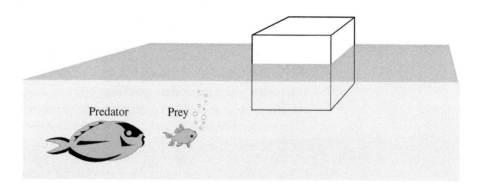

Cube floating in water.

5. Write a differential equation model for the up-and-down motion of the cube, assuming that it always stays in contact with the water and is never completely submerged.

6. Write a differential equation for the up-and-down motion of the cube, assuming that it always stays in contact with the water, but that it can be completely submerged.

7. Write a differential equation model for the up-and-down motion of the cube, assuming that it will never be completely submerged but can rise completely out of the water by some distance.

8. (a) Adjust each of the models in Exercises 5–7 to include the effect of waves on the motion of the cube (assuming the top and bottom remain parallel to the average water level).
    (b) Discuss the implications of the behavior of solutions of this system considered in the text for the motion of the cube.

REVIEW EXERCISES FOR CHAPTER 4

Short answer exercises: Exercises 1–10 focus on the basic ideas, definitions, and vocabulary of this chapter. Their answers are short (a single sentence or drawing), and you should be able to do them with little or no computation. However, they vary in difficulty, so think carefully before you answer.

1. Find one solution of the forced harmonic oscillator

$$m\frac{d^2y}{dt^2} + b\frac{dy}{dt} + ky = 1.$$

2. For which values of ω is the forced harmonic oscillator $d^2y/dt^2 + 4y = \cos \omega t$ in resonance?

3. What is the frequency of the steady-state solution of the equation

$$\frac{d^2y}{dt^2} + 3\frac{dy}{dt} + y = 4\cos 2t \, ?$$

4. Find all equilibrium solutions of the equation $d^2y/dt^2 + 4y = \sin t$.

5. Is there an analogue of a steady-state response for a first-order equation of the form

$$\frac{dy}{dt} + \lambda y = \cos \omega t$$

if $\lambda > 0$? If so, what is it? If not, why not?

6. Is there an analogue of resonance in a first-order equation of the form

$$\frac{dy}{dt} + \lambda y = \cos \omega t$$

if $\lambda > 0$? If so, what is it? If not, why not?

7. Consider the equation

$$\frac{d^2y}{dt^2} + 2\frac{dy}{dt} + 4y = 2\cos 3t.$$

Suppose that $y_1(t)$ is the solution with initial condition $(y_1(0), y_1'(0)) = (1, 0)$ and that $y_2(t)$ is the solution with initial condition $(y_2(0), y_2'(0)) = (100, 0)$. Give a rough estimate of the smallest value of T such that

$$|y_1(t) - y_2(t)| \leq 1 \text{ for all } t \geq T.$$

8. If you walk down a flight of stairs at one particular speed holding a full cup of coffee, the coffee sloshes (painfully) onto your hand. However, at slower or faster speeds, the coffee stays in the cup. Why?

9. Suppose you have a machine whose vibrations can be accurately modeled with a harmonic oscillator with almost no damping but you do not know the mass or spring constant. If you can observe the vibrations of the machine under external forcing given by $\cos \omega t$ for various values of ω, what can you say about the mass and spring constant?

10. Suppose a friend solves the initial-value problem

$$\frac{d^2y}{dt^2} + 5\frac{dy}{dt} + 4y = \sin 2t - \cos 2t, \quad y(0) = 1, \ y'(0) = 1,$$

and the solution has large amplitude oscillations when t is large. How do you imme-diately know that your friend has made a mistake?

True-false: For Exercises 11–14, determine if the statement is true or false. If it is true, explain why. If it is false, provide a counterexample or an explanation.

11. There is exactly one solution $y(t)$ of

$$\frac{d^2y}{dt^2} + \frac{dy}{dt} + 6y = \cos t$$

such that $-1 < y(t) < 1$ for all t.

12. If $y_1(t)$ is a solution of

$$\frac{d^2y}{dt^2} + \frac{dy}{dt} + 4y = \cos 3t,$$

then $2y_1(t)$ is a solution of

$$\frac{d^2y}{dt^2} + \frac{dy}{dt} + 4y = 2\cos 3t.$$

13. The amplitude of oscillations of the forced response to

$$\frac{d^2y}{dt^2} + ky = \cos \sqrt{k}\, t$$

doubles every k units of time t.

14. When you drive down a bumpy road, the vibrations you experience are probably due to resonance.

In Exercises 15–22, find the general solution of the given differential equation.

15. $\dfrac{d^2y}{dt^2} + 6\dfrac{dy}{dt} + 8y = e^{-t}$

16. $\dfrac{d^2y}{dt^2} + 7\dfrac{dy}{dt} + 12y = 3e^{-2t}$

17. $\dfrac{d^2y}{dt^2} - 2\dfrac{dy}{dt} - 3y = e^{3t}$

18. $\dfrac{d^2y}{dt^2} + \dfrac{dy}{dt} - 2y = 5e^{-2t}$

19. $\dfrac{d^2y}{dt^2} + 6\dfrac{dy}{dt} + 8y = 5$

20. $\dfrac{d^2y}{dt^2} - \dfrac{dy}{dt} - 6y = 6t + 3e^{4t}$

21. $\dfrac{d^2y}{dt^2} - 4\dfrac{dy}{dt} + 13y = 5\cos 4t$

22. $\dfrac{d^2y}{dt^2} + 3y = 2t + \cos 4t$

23. Eight second-order equations and four $y(t)$-graphs are given below. For each $y(t)$-graph, determine the second-order equation for which $y(t)$ is a solution, and state briefly how you know your choice is correct. You should do this exercise without using technology.

(i) $\dfrac{d^2y}{dt^2} + 16y = 0$

(ii) $\dfrac{d^2y}{dt^2} + 5\dfrac{dy}{dt} + y = 5\cos 2t$

(iii) $\dfrac{d^2y}{dt^2} + 5\dfrac{dy}{dt} + y = 5\cos 4t$

(iv) $2\dfrac{d^2y}{dt^2} - \dfrac{dy}{dt} + 10y = 0$

(v) $2\dfrac{d^2y}{dt^2} + \dfrac{dy}{dt} + 10y = 0$

(vi) $\dfrac{d^2y}{dt^2} + 3y = \cos 11t$

(vii) $\dfrac{d^2y}{dt^2} + 9y = 0$

(viii) $\dfrac{d^2y}{dt^2} + 11y = \cos 3t$

(a)

(b)

(c)

(d)

In Exercises 24–27,

(a) find the general solution of the given differential equation, and

(b) compute the amplitude and phase angle of the steady-state solution.

24. $\dfrac{d^2y}{dt^2} + 6\dfrac{dy}{dt} + 13y = 2\cos 3t$

25. $\dfrac{d^2y}{dt^2} + 2\dfrac{dy}{dt} + 3y = \cos 2t$

26. $\dfrac{d^2y}{dt^2} + 4\dfrac{dy}{dt} + 4y = 2\cos 3t$

27. $\dfrac{d^2y}{dt^2} + 4\dfrac{dy}{dt} + 3y = 5\sin 2t$

LAB 4.1 Two Magnets and a Spring Revisited

In this lab we again study the motion of a mass that can slide on the x-axis (see Lab 2.1). The mass is attached to a spring that has its other end attached to the point $(0, 2)$ on the y-axis. In addition, the mass is made of iron and is attracted to two magnets of equal strength—one located at the point $(-1, -a)$ and the other at $(1, -a)$ (see Figure 4.40).

We assume that the spring obeys Hooke's Law, and the magnets attract the mass with a force proportional to the inverse of the square of the distance of the mass to the magnet (the inverse square law). In Lab 2.1, we observed a subtle dependence of the solutions on the position of the magnets.

In this lab, we consider the effect of an external forcing term on this system. We can think of the external force as a wind that gusts, alternately blowing the mass to the left and the right. More precisely, we study the solutions of the nonautonomous second-order equation

$$\frac{d^2x}{dt^2} = -0.3x - \frac{x-1}{((x-1)^2 + a^2)^{3/2}} - \frac{x+1}{((x+1)^2 + a^2)^{3/2}} + b\cos t,$$

where b is the amplitude of the forcing. This equation is very complicated, so you are expected to carry out your analysis numerically. In order to observe the effects of the forcing, you will have to follow the solutions over intervals of time that are at least as long as several periods of the forcing function.

Address the following items in your report:

1. Recall and summarize the behavior of the unforced system ($b = 0$), the one you studied in Lab 2.1, Briefly describe both the phase portraits and the motion of the mass along the x-axis for the two positions of the magnets $a = 0.5$ and $a = 2$.

Figure 4.40
Schematic of a mass sliding on the x-axis
attached to a spring and attracted by two magnets.

2. Study solutions to the forced system for magnets that are far from the x-axis assuming the amplitude of the forcing is small, for example, $a = 2$ and $b = 0.1$. How do solutions differ from those in Part 1? Express your conclusions in terms of the solution curves in the phase plane and in terms of the motion of the mass along the x-axis.

3. Study solutions of the forced system for magnets that are close to the x-axis assuming the amplitude of the forcing is small, for example, $a = 0.5$ and $b = 0.1$. How do solutions differ from the unforced case? Express your conclusions in terms of the solution curves in the phase plane and in terms of the motion of the mass along the x-axis. Pay particular attention to solutions whose initial conditions are near the origin in the phase plane.

In your report, pay particular attention to the physical interpretation of the solutions in terms of the possible motions of the mass as it slides along the x-axis. Include graphs and phase portraits to illustrate your discussion, but pictures alone are not sufficient.

LAB 4.2 A Periodically Forced RLC Circuit

In this lab we continue the study of simple RLC circuits that we began in a lab in Chapter 3 (see page 381). The circuit is shown in Figure 4.41. The parameters are the resistance R, the capacitance C, and the inductance L. The dependent variables we use are v_C, the voltage across the capacitor, and i, the current. (We follow the engineering convention of representing current as i. This variable does *not* represent the square root of -1.) In this lab we consider circuits in which the voltage source $v_T = v_T(t)$ is a time-dependent forcing term.

Figure 4.41
An RLC circuit.

From the lab in Chapter 3, we know that the voltage v_C and current i satisfy the system of differential equations

$$\frac{dv_C}{dt} = \frac{i}{C}$$

$$\frac{di}{dt} = -\frac{v_C}{L} - \frac{R}{L}i - \frac{v_T(t)}{L},$$

which is more commonly written as the second-order equation

$$LC\frac{d^2v_C}{dt^2} + RC\frac{dv_C}{dt} + v_C = v_T(t).$$

In this lab, we consider the possible behavior of solutions to this equation (or the corresponding system) when $v_T(t) = a \sin \omega t$.

Consider the following questions:

1. Assuming R, C, and L are all nonnegative, what types of long-term behavior are possible for solutions of the equation

$$LC\frac{d^2v_C}{dt^2} + RC\frac{dv_C}{dt} + v_C = a \sin \omega t?$$

Describe how the behavior of solutions depends on the parameters a and ω.

2. In a typical circuit R is on the order of 1000, C is on the order of 10^{-6}, and L is on the order of 1. Does this information help in limiting the possible behaviors of solutions?

3. Describe the solutions for various values of a and ω if $R = 2000$, $C = 2 \times 10^{-7}$, and $L = 1.5$.

Your report: Address each of the items above, justifying all statements and showing all details. Give graphs of solutions as appropriate.

LAB 4.3 The Tacoma Narrows Bridge

In Section 4.5, we discussed a model of the Tacoma Narrows Bridge. This model is based on the observation that the cables of the bridge can be reasonably modeled as springs when stretched but they do not exert a restoring force when contracted. This one-sided Hooke's Law may remind you of Lab 2.4 where we considered a mass attached to a spring and a rubber band. In this lab, we extend the study of the system in Lab 2.4 by adding periodic forcing. As noted in Section 4.5, this model displays very interesting (and scary) behavior, particularly when the solutions are interpreted in terms of the behavior of a suspension bridge. Since this nonlinear model displays nontrivial behavior, it is reasonable to suspect that more complicated models could display the same complicated solutions.

In this lab we compare the usual forced harmonic oscillator

$$\frac{d^2y}{dt^2} + b\frac{dy}{dt} + k_1y = 10 + 0.1 \sin \omega t$$

to the forced mass-spring system with a rubber band added. The equation for the forced system is

$$\frac{d^2y}{dt^2} + b\frac{dy}{dt} + k_1y + k_2h(y) = 10 + 0.1 \sin \omega t,$$

where $h(y)$ is defined piecewise as

$$h(y) = \begin{cases} y, & \text{if } y \geq 0, \\ 0, & \text{if } y < 0. \end{cases}$$

Recall from Lab 2.4 that the term $k_1 y$ is due to the force from the spring, $b(dy/dt)$ represents damping, the 10 on the right-hand side is (a rough approximation to) the constant force of gravity, and $k_2 h(y)$ is the force of a rubber band, which is a restoring force proportional to the displacement when stretched but no force when contracted (see Lab 2.4 and Figure 4.42). The new term, $\sin \omega t$, represents periodic external forcing.

Figure 4.42
A mass-spring system and a mass-spring system with a rubber band.

We first review the behavior of the standard forced harmonic oscillator (no rubber band present), particularly as the parameters b and ω vary. Then we compare these solutions to the solutions of the forced mass-spring system with rubber band for some specific parameter values. The numerics of this lab are delicate and require patience.

In your report, you should address the following items:

1. (Undamped, forced harmonic oscillator) First consider solutions to

$$\frac{d^2 y}{dt^2} + k_1 y = 10 + 0.1 \sin \omega t$$

with a fixed value of the parameter k_1 where $12 \leq k_1 \leq 13$. You may use numerical methods or the Method of the Lucky Guess to estimate the amplitude of the solutions for a particular choice of initial condition and various values of ω where $0 < \omega \leq 5$.

2. (Damped, forced oscillator) Using the same value of k_1 as in Part 1, study the solutions of

$$\frac{d^2 y}{dt^2} + b\frac{dy}{dt} + k_1 y = 10 + 0.1 \sin \omega t.$$

You may use analytic or numerical methods to study the relationship between the amplitude of the solutions for $0 \leq \omega \leq 5$ for different values of b. Graph the amplitude as a function of ω for different b values. What happens to these graphs as $b \to 0$?

3. (Small-amplitude periodic solutions for the forced mass-spring system with a rubber band) Using the same value of k_1 as in Part 2, fix $b = 0.01$ and $\omega = 4$. Also fix a value of k_2 where $4.5 \leq k_2 \leq 5$, and calculate a periodic solution with small amplitude for the equation

$$\frac{d^2y}{dt^2} + b\frac{dy}{dt} + k_1 y + k_2 h(y) = 10 + 0.1 \sin \omega t.$$

Use analytic methods to calculate the solution and numerical methods to verify your result. Provide its $y(t)$-graph and its solution curve in the phase plane. Be sure to specify its initial condition.

4. (Large-amplitude periodic solutions for the forced mass-spring system with a rubber band) Surprisingly, the equation in Part 3 has a second "large amplitude" periodic solution for parameter values in the range we are considering. Using the same values of k_1, b, ω, and k_2 as in Part 3, search for initial conditions that lead to this large amplitude solution. This is delicate work. You will need to try many different initial conditions and allow the numerical method to work for a long time interval (on the order of $0 \leq t \leq 1000$). Then check to see if the solution has approached a periodic solution. To help you in your search, we include Figure 4.43. It is a picture of a square in the first quadrant of the yv-plane. Solutions with initial conditions in the blue bands approach the small-amplitude periodic orbit over the long term. Solutions with initial conditions in the black bands do not approach the small amplitude solution for $0 \leq t \leq 1000$. This figure was produced for $k_1 = 13$, $k_2 = 4$, $b = 0.01$, and $\omega = 4$, but it should give you some idea of how to start your search with your values of k_1 and k_2.

Your report: The text of your report should address the items above. For each item, describe the solutions of the given equations and compare them to the solutions of the equations in previous items. Be sure to relate your conclusions about solutions to the vertical motion of the system.

Figure 4.43
Solutions with initial conditions in the blue bands approach the small-amplitude periodic orbit. Solutions with initial conditions in the black bands do not approach the small-amplitude solution for $0 \leq t \leq 1000$.

5 NONLINEAR SYSTEMS

In this chapter we study nonlinear autonomous systems. In Chapter 3 we saw that, using a combination of analytic and geometric techniques, we can understand linear systems completely. Here we combine these linear techniques with some additional qualitative methods to tackle nonlinear systems. While these techniques do not allow us to determine the phase portraits of all nonlinear systems, we see that we are able to handle some important nonlinear systems.

We first show how a nonlinear system can be approximated near an equilibrium point by a linear system. This process, known as "linearization," is one of the most frequently used techniques in applications. By studying the linear approximation, we can surmise the behavior of solutions of the nonlinear system, at least near the equilibrium point.

Next we give a qualitative method for extracting more information from direction fields. By looking at where one component of the direction field is zero (so the direction field is either vertical or horizontal), we obtain curves called "nullclines," which subdivide the phase space. When combined with linearization of equilibrium points, the nullclines can, in some cases, yield a complete description of the possible long-term behaviors of solutions.

In the remainder of the chapter, we study special types of models and the nonlinear systems associated with them. These special nonlinear systems are important both because they arise in applications and because the techniques involved in their analysis are delightful.

5.1   EQUILIBRIUM POINT ANALYSIS

From our work in Chapter 3, we are able to understand the solutions of linear systems both qualitatively and analytically. Unfortunately, nonlinear systems are in general much less amenable to the analytic and algebraic techniques that we have developed, but we can use the mathematics of linear systems to understand the behavior of solutions of nonlinear systems near their equilibrium points.

The Van der Pol Equation

To illustrate how to analyze the behavior of solutions near an equilibrium point, we begin with a simple but important nonlinear system—the Van der Pol system. Recall that the Van der Pol system is

$$\frac{dx}{dt} = y$$

$$\frac{dy}{dt} = -x + (1 - x^2)y.$$

We studied this system numerically in Section 2.5 (see page 198), and its direction field and phase portrait are shown in Figure 5.1.

The only equilibrium point of this system is the origin, so let's examine how solutions near the origin behave. The direction field shows that the solutions circle around the origin, and if we plot numerical approximations of solutions near the origin, we see a picture that is reminiscent of a spiral source (see Figure 5.2).

We can understand why solutions spiral away from the origin by approximating the Van der Pol system with another system that is much easier for us to analyze. Although the system is nonlinear, there is only one nonlinear term, the x^2y term in the

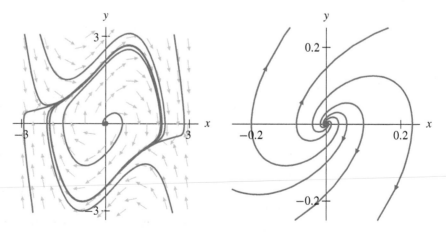

Figure 5.1
Direction field and phase portrait for
the Van der Pol system.

Figure 5.2
Phase portrait for the Van der Pol
system near the origin.

equation for dy/dt. If x and y are small, then this term is much smaller than any of the other terms in the equation. For example, if both x and y are 0.1, then the x^2y term is 0.001, which is significantly smaller than either x or y. If both x and y are 0.01, then $x^2y = 10^{-6}$, which is again much smaller than either x or y. Perhaps we can approximate the Van der Pol system by one in which we simply neglect the x^2y term, at least if both x and y are close to 0. If we drop this term from the system, then we are left with

$$\frac{dx}{dt} = y$$
$$\frac{dy}{dt} = -x + y,$$

which is a linear system. Consequently, the techniques of Chapter 3 apply. The eigenvalues of the linear system are $(1 \pm \sqrt{3}\,i)/2$, and since they are complex with a real part that is positive, we know that the solutions of the linear system spiral away from the origin.

The linear system and the Van der Pol system have vector fields that are very close to each other near the equilibrium point at the origin. Since solutions of the linear system spiral away from the origin, it is not surprising that solutions of the Van der Pol system that start near the origin also spiral away.

The technique we applied above is called **linearization**. Near the equilibrium point, we approximate the nonlinear system by an appropriate linear system. For initial conditions near the equilibrium point, the solutions of the nonlinear system and the linear approximation remain close at least for some interval of time.

A Competing Species Model

Let x and y denote the populations of two species that compete for resources. An increase in either species has an adverse effect on the growth rate of the other species. An example of a model of such a system is

$$\frac{dx}{dt} = 2x\left(1 - \frac{x}{2}\right) - xy$$
$$\frac{dy}{dt} = 3y\left(1 - \frac{y}{3}\right) - 2xy.$$

Although the terms involved in these equations are based on reasonable assumptions, we choose coefficients to simplify our discussion rather than to model any particular species.

Note that, for a given value of x, if y increases then the $-xy$ term causes dx/dt to decrease. Similarly, for a given value of y, if x increases then the $-2xy$ term causes dy/dt to decrease. An increase in the population of either species causes a decrease in the rate of growth of the other species.

Qualitative analysis

We begin our analysis of this system by noting that if $y = 0$, we have $dy/dt = 0$. In other words, if the y's are extinct, they stay extinct. If $y = 0$,

$$\frac{dx}{dt} = 2x\left(1 - \frac{x}{2}\right),$$

which is a logistic population model with a carrying capacity of 2. The phase line of this equation agrees with the x-axis of the phase plane. In particular, $(0, 0)$ and $(2, 0)$ are equilibrium points of the system. Similarly, if $x = 0$, we have $dx/dt = 0$, so the phase line of

$$\frac{dy}{dt} = 3y\left(1 - \frac{y}{3}\right)$$

agrees with the y-axis of the phase plane of the system, and $(0, 3)$ is another equilibrium point.

By the Uniqueness Theorem, solutions with initial conditions in the first quadrant must remain in the first quadrant for all time. That is, the axes coincide with solution curves, and the Uniqueness Theorem guarantees that solutions cannot cross. Because this model refers to populations and negative populations do not make much sense, we limit our attention to solutions that are contained in the first quadrant only.

We find the equilibrium points by solving for x and y in the system of equations

$$\begin{cases} 2x\left(1 - \dfrac{x}{2}\right) - xy = 0 \\ 3y\left(1 - \dfrac{y}{3}\right) - 2xy = 0, \end{cases}$$

which can be rewritten in the form

$$\begin{cases} x(2 - x - y) = 0 \\ y(3 - y - 2x) = 0. \end{cases}$$

The first equation is satisfied if $x = 0$ or if $2 - x - y = 0$, and the second equation is satisfied if either $y = 0$ or if $3 - y - 2x = 0$.

Suppose first that $x = 0$. Then the equation $y = 0$ yields an equilibrium point at the origin, and the equation $3 - y - 2x = 0$ yields an equilibrium point at $(0, 3)$.

Now suppose that $2 - x - y = 0$. Then the equation $y = 0$ yields an equilibrium point at $(2, 0)$, and the equation $3 - y - 2x = 0$ yields an equilibrium point at $(1, 1)$. (Solve the equations $2 - x - y = 0$ and $3 - y - 2x = 0$ simultaneously.) Hence the equilibrium points are $(0, 0)$, $(0, 3)$, $(2, 0)$, and $(1, 1)$.

Linearization of this system about $(1, 1)$

The equilibrium point $(1, 1)$ is of particular interest. Its existence indicates that it is possible for these two species to coexist in equilibrium. If we numerically compute the phase portrait for this system (see Figure 5.3), solutions seem to have only three different types of long-term behaviors. Some solutions tend to $(2, 0)$, some tend to $(0, 3)$, and others tend to $(1, 1)$.

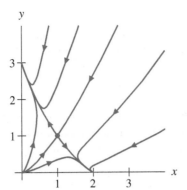

Figure 5.3
Phase portrait for the system

$$\frac{dx}{dt} = 2x\left(1 - \frac{x}{2}\right) - xy$$

$$\frac{dy}{dt} = 3y\left(1 - \frac{y}{3}\right) - 2xy.$$

Two important questions remain. First, what solutions tend to the equilibrium point $(1, 1)$? In particular, is the set of these solutions large enough that we could hope to find an example of such a solution in nature? Second, what separates the initial conditions that yield solutions for which x tends to zero from those solutions for which y tends to zero? To answer these questions, we study the system near the equilibrium point $(1, 1)$ using linearization.

Linear systems always have an equilibrium point at the origin. Hence the first step in comparing the nonlinear system near the equilibrium point $(1, 1)$ to a linear system is to move the equilibrium point to the origin via a change of variables. Once the equilibrium point is at the origin, we can use the same ideas as we did in the Van der Pol example to identify the linear approximation.

To move the equilibrium point to the origin, we introduce two new variables, u and v, by the formulas $u = x - 1$ and $v = y - 1$. Note that u and v are both close to 0 when (x, y) is close to $(1, 1)$. To obtain the system in the new variables, we first compute

$$\frac{du}{dt} = \frac{d(x-1)}{dt} = \frac{dx}{dt}$$

$$\frac{dv}{dt} = \frac{d(y-1)}{dt} = \frac{dy}{dt}.$$

The right-hand sides of the system in the new variables are given by

$$\begin{aligned}\frac{dx}{dt} &= 2x\left(1 - \frac{x}{2}\right) - xy \\ &= 2(u+1)\left(1 - \frac{u+1}{2}\right) - (u+1)(v+1) \\ &= -u - v - u^2 - uv,\end{aligned}$$

and

$$\begin{aligned}\frac{dy}{dt} &= 3y\left(1 - \frac{y}{3}\right) - 2xy \\ &= 3(v+1)\left(1 - \frac{v+1}{3}\right) - 2(u+1)(v+1) \\ &= -2u - v - 2uv - v^2.\end{aligned}$$

In terms of the new variables, we have

$$\frac{du}{dt} = -u - v - u^2 - uv$$

$$\frac{dv}{dt} = -2u - v - 2uv - v^2.$$

As we expect, the origin is an equilibrium point for this system. The expression for du/dt involves the linear terms $-u$ and $-v$ and the nonlinear terms $-u^2$ and $-uv$. For dv/dt the linear terms are $-2u$ and $-v$, and the nonlinear terms are $-2uv$ and $-v^2$. Near the origin the nonlinear terms are much smaller than the linear terms, and we therefore approximate the nonlinear system near $(u, v) = (0, 0)$ with the linear system

$$\frac{du}{dt} = -u - v$$

$$\frac{dv}{dt} = -2u - v.$$

The eigenvalues of this system are $-1 \pm \sqrt{2}$. One of these numbers is positive and the other is negative, so $(u, v) = (0, 0)$ is a saddle point for the linear system (see Figure 5.4). Since the linear and nonlinear systems are approximately the same, we expect the phase portrait for the nonlinear system near the equilibrium point $(x, y) = (1, 1)$ to look like the uv-phase portrait of the linear system.

We conclude that there are only two curves of solutions in the xy-plane that tend toward the equilibrium point $(1, 1)$ as t increases. Hence the solutions that tend to $(1, 1)$ form a very small set. Even if initial conditions are chosen exactly on this curve, arbitrarily small perturbations can push the initial conditions to one side or the other. Since our model is only a very simplified version of the dynamics of the populations, leaving out innumerable sources of such small perturbations, we do not expect to see solutions leading to the $(1, 1)$ equilibrium point in nature.

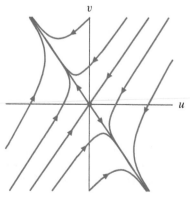

Figure 5.4
Phase portrait for

$$\frac{du}{dt} = -u - v$$

$$\frac{dv}{dt} = -2u - v,$$

the linear approximation of the competitive system near $(x, y) = (1, 1)$, which is the same point as $(u, v) = (0, 0)$.

On the other hand, the curve of solutions that tend to $(1, 1)$ do play an important role in the study of this system. Looking at the phase plane, we note that this curve divides the first quadrant into two regions. In one region one species survives, and in the other region the other species survives (see Figure 5.3).

A Nonpolynomial Example

The two examples above have vector fields that are polynomials. When the vector field is a polynomial and the equilibrium point under consideration is the origin, it is very easy to identify which terms are linear and which are nonlinear. However, not all vector fields are polynomials.

Consider the system

$$\frac{dx}{dt} = y$$
$$\frac{dy}{dt} = -y - \sin x.$$

This system is a model for the motion of a damped pendulum, and we will study it extensively in Sections 5.3 and 5.4. The equilibria of this system occur at the points $(x, y) = (0, 0), (\pm\pi, 0), (\pm 2\pi, 0), \ldots$. Suppose we want to study the solutions that are close to the equilibrium point at $(0, 0)$. Since dy/dt includes a $\sin x$ term, it is not immediately clear what the linear terms of this system are. However, from calculus we know that the power series expansion of $\sin x$ is

$$\sin x = x - \frac{x^3}{3!} + \frac{x^5}{5!} - \cdots.$$

Therefore we can write

$$\frac{dx}{dt} = y$$
$$\frac{dy}{dt} = -y - \left(x - \frac{x^3}{3!} + \frac{x^5}{5!} \cdots\right).$$

Near the origin we drop the nonlinear terms, and we are left with the linear system

$$\frac{dx}{dt} = y$$
$$\frac{dy}{dt} = -y - x.$$

The eigenvalues of this system are $(-1 \pm \sqrt{3}\,i)/2$. Since these eigenvalues are complex with negative real parts, we expect the corresponding equilibrium point for the nonlinear system to be a spiral sink (see Figures 5.5 and 5.6).

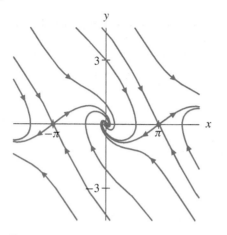

Figure 5.5
Phase portrait for the system

$$\frac{dx}{dt} = y$$
$$\frac{dy}{dt} = -y - \sin x.$$

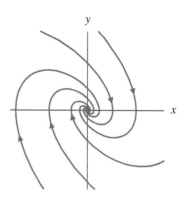

Figure 5.6
Phase portrait for the system

$$\frac{dx}{dt} = y$$
$$\frac{dy}{dt} = -y - x.$$

Linearization

The next step is to make the process of linearization more orderly. In the preceding examples, we used a change of variables to move the equilibrium point to the origin. Then we identified the linear and nonlinear terms using calculus (if necessary).

Consider the general form of a nonlinear system

$$\frac{dx}{dt} = f(x, y)$$
$$\frac{dy}{dt} = g(x, y).$$

Suppose that (x_0, y_0) is an equilibrium point for this system. We wish to understand what happens to solutions near (x_0, y_0)—that is, to linearize the system near (x_0, y_0). We introduce new variables

$$u = x - x_0$$
$$v = y - y_0$$

that move the equilibrium point to the origin. If x and y are close to the equilibrium point (x_0, y_0), then both u and v are close to 0.

Since $x = u + x_0$ and $y = v + y_0$ and the numbers x_0 and y_0 are constants, the system written in terms of u and v is

$$\frac{du}{dt} = \frac{d(x - x_0)}{dt} = \frac{dx}{dt} = f(x, y) = f(x_0 + u, y_0 + v)$$

$$\frac{dv}{dt} = \frac{d(y - y_0)}{dt} = \frac{dy}{dt} = g(x, y) = g(x_0 + u, y_0 + v).$$

Therefore we have

$$\frac{du}{dt} = f(x_0 + u, y_0 + v)$$

$$\frac{dv}{dt} = g(x_0 + u, y_0 + v).$$

If $u = v = 0$, the right-hand side of this system vanishes, so we have moved the equilibrium point to the origin in the uv-plane.

Now we would like to be able to eliminate the "higher-order" or nonlinear terms in the expressions for du/dt and dv/dt. Since these expressions may include exponentials, logarithms, and trigonometric functions, it is not always clear what the linear terms are. In this case it is necessary to study f and g more closely.

The basic idea of differential calculus is that it is possible to study a function by studying the "best linear approximation" of that function. For functions of two variables, the best linear approximation at a particular point is given by the tangent plane. Hence we have

$$f(x_0 + u, y_0 + v) \approx f(x_0, y_0) + \left[\frac{\partial f}{\partial x}(x_0, y_0) \right] u + \left[\frac{\partial f}{\partial y}(x_0, y_0) \right] v,$$

where the right-hand side of this equation is the equation for the tangent plane to the graph of f at (x_0, y_0). (This expression is also the first-degree Taylor polynomial approximation of f.)

Thus we can rewrite the system for du/dt and dv/dt as

$$\frac{du}{dt} = f(x_0, y_0) + \left[\frac{\partial f}{\partial x}(x_0, y_0) \right] u + \left[\frac{\partial f}{\partial y}(x_0, y_0) \right] v + \dots$$

and

$$\frac{dv}{dt} = g(x_0, y_0) + \left[\frac{\partial g}{\partial x}(x_0, y_0) \right] u + \left[\frac{\partial g}{\partial y}(x_0, y_0) \right] v + \dots ,$$

where "..." stands for the terms that make up the difference between the tangent plane and the function. These are precisely the terms we wish to ignore when forming the linear approximation of the system. Since $f(x_0, y_0) = 0$ and $g(x_0, y_0) = 0$, we can use matrix notation to write the system more succinctly as

$$\begin{pmatrix} \dfrac{du}{dt} \\ \dfrac{dv}{dt} \end{pmatrix} = \begin{pmatrix} \dfrac{\partial f}{\partial x}(x_0, y_0) & \dfrac{\partial f}{\partial y}(x_0, y_0) \\ \dfrac{\partial g}{\partial x}(x_0, y_0) & \dfrac{\partial g}{\partial y}(x_0, y_0) \end{pmatrix} \begin{pmatrix} u \\ v \end{pmatrix} + \dots .$$

The 2×2 matrix of partial derivatives in this expression is called the **Jacobian matrix** of the system at (x_0, y_0).

Hence the **linearized system** at the equilibrium point (x_0, y_0) is

$$\begin{pmatrix} \dfrac{du}{dt} \\ \dfrac{dv}{dt} \end{pmatrix} = \begin{pmatrix} \dfrac{\partial f}{\partial x}(x_0, y_0) & \dfrac{\partial f}{\partial y}(x_0, y_0) \\ \dfrac{\partial g}{\partial x}(x_0, y_0) & \dfrac{\partial g}{\partial y}(x_0, y_0) \end{pmatrix} \begin{pmatrix} u \\ v \end{pmatrix} .$$

We use this "linearized" system to study the behavior of solutions of the nonlinear system near the equilibrium point (x_0, y_0). Note that we need only know the partial derivatives of the components of the vector field at the equilibrium point to create the linearized system. We do not need to compute the change of variables moving the equilibrium point to the origin.

As always, the derivative of a nonlinear function provides only a local approximation. Hence the solutions of the linearized system are close to solutions of the nonlinear system only near the equilibrium point. How close to the equilibrium point we must be for the linear approximation to be any good depends on the size of the nonlinear terms.

More Examples of Linearization

Consider the nonlinear system

$$\frac{dx}{dt} = -2x + 2x^2$$
$$\frac{dy}{dt} = -3x + y + 3x^2.$$

There are two equilibrium points for this system—$(0, 0)$ and $(1, 0)$. To understand solutions that start near these points, we first compute the Jacobian matrix

$$\begin{pmatrix} \dfrac{\partial f}{\partial x}(x, y) & \dfrac{\partial f}{\partial y}(x, y) \\ \dfrac{\partial g}{\partial x}(x, y) & \dfrac{\partial g}{\partial y}(x, y) \end{pmatrix} = \begin{pmatrix} -2 + 4x & 0 \\ -3 + 6x & 1 \end{pmatrix}$$

since $f(x, y) = -2x + 2x^2$ and $g(x, y) = -3x + y + 3x^2$. At the two equilibrium points, we have

$$\begin{pmatrix} \dfrac{\partial f}{\partial x}(0, 0) & \dfrac{\partial f}{\partial y}(0, 0) \\ \dfrac{\partial g}{\partial x}(0, 0) & \dfrac{\partial g}{\partial y}(0, 0) \end{pmatrix} = \begin{pmatrix} -2 & 0 \\ -3 & 1 \end{pmatrix}$$

and

$$\begin{pmatrix} \dfrac{\partial f}{\partial x}(1,0) & \dfrac{\partial f}{\partial y}(1,0) \\[2ex] \dfrac{\partial g}{\partial x}(1,0) & \dfrac{\partial g}{\partial y}(1,0) \end{pmatrix} = \begin{pmatrix} 2 & 0 \\ 3 & 1 \end{pmatrix}.$$

Near $(0, 0)$ the phase portrait for the nonlinear system should resemble that of the linearized system

$$\frac{d\mathbf{Y}}{dt} = \begin{pmatrix} -2 & 0 \\ -3 & 1 \end{pmatrix} \mathbf{Y}.$$

The eigenvalues of this linear system are -2 and 1, so the origin is a saddle. We can compute that $(0, 1)$ is an eigenvector for the eigenvalue 1 and that $(1, 1)$ is an eigenvector for the eigenvalue -2. Using this information we can sketch the phase portrait for the linear system (see Figure 5.7).

At the other equilibrium point, $(1, 0)$, the linearized system is

$$\frac{d\mathbf{Y}}{dt} = \begin{pmatrix} 2 & 0 \\ 3 & 1 \end{pmatrix} \mathbf{Y}.$$

Here the eigenvalues are 2 and 1, so the origin is a source for this system. Using the fact that $(0, 1)$ is an eigenvector for the eigenvalue 1 and $(1, 3)$ is an eigenvector for the eigenvalue 2, we can sketch the phase portrait (see Figure 5.8).

Near the two equilibrium points, the phase portrait for the nonlinear system resembles that of the linearized systems. Solution curves (numerically approximated by the computer) are shown in Figure 5.9. If we magnify the phase plane near $(0, 0)$ and $(1, 0)$, we see that the solution curves do indeed look like those of the corresponding linearized systems (see Figure 5.9).

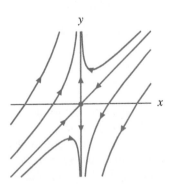

Figure 5.7
Phase portrait for the system

$$\frac{d\mathbf{Y}}{dt} = \begin{pmatrix} -2 & 0 \\ -3 & 1 \end{pmatrix} \mathbf{Y}.$$

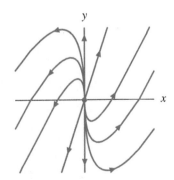

Figure 5.8
Phase portrait for the system

$$\frac{d\mathbf{Y}}{dt} = \begin{pmatrix} 2 & 0 \\ 3 & 1 \end{pmatrix} \mathbf{Y}.$$

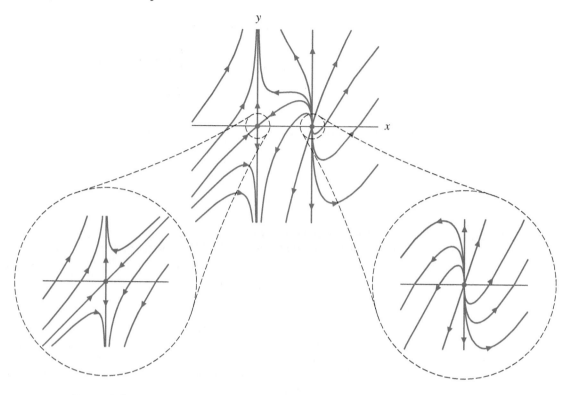

Figure 5.9
Solution curves and magnifications near the equilibrium points for the system

$$\frac{dx}{dt} = -2x + 2x^2$$

$$\frac{dy}{dt} = -3x + y + 3x^2.$$

Classification of Equilibrium Points

The fundamental idea underlying the technique of linearization is to use a linear system to approximate the behavior of solutions of a nonlinear system near an equilibrium point. The solutions of the nonlinear system near the equilibrium point are close to solutions of the approximating linear system, at least for a short time interval. For most systems, the information gained by studying the linearization is enough to determine the long-term behavior of solutions of the nonlinear system near the equilibrium point.

For example, consider the system

$$\frac{dx}{dt} = f(x, y)$$

$$\frac{dy}{dt} = g(x, y).$$

Suppose (x_0, y_0) is an equilibrium point and let

$$\mathbf{J} = \begin{pmatrix} \dfrac{\partial f}{\partial x}(x_0, y_0) & \dfrac{\partial f}{\partial y}(x_0, y_0) \\ \dfrac{\partial g}{\partial x}(x_0, y_0) & \dfrac{\partial g}{\partial y}(x_0, y_0) \end{pmatrix}$$

be the Jacobian matrix at (x_0, y_0). The linearized system is

$$\begin{pmatrix} \dfrac{du}{dt} \\ \dfrac{dv}{dt} \end{pmatrix} = \mathbf{J} \begin{pmatrix} u \\ v \end{pmatrix}.$$

If all of the eigenvalues of the Jacobian matrix are negative real numbers or complex numbers with negative real parts, then $(u, v) = (0, 0)$ is a sink for the linear system and all solutions approach $(u, v) = (0, 0)$ as $t \to \infty$. For the nonlinear system, solutions that start near the equilibrium point $(x, y) = (x_0, y_0)$ approach it as $t \to \infty$. Hence we say that (x_0, y_0) is a **sink**. If the eigenvalues are complex, then (x_0, y_0) is said to be a **spiral sink**.

Similarly, if the Jacobian matrix has only positive eigenvalues or complex eigenvalues with positive real parts, then solutions with initial conditions near the equilibrium point (x_0, y_0) move away from (x_0, y_0) as t increases. The equilibrium point (x_0, y_0) of the nonlinear system is said to be a **source**. If the eigenvalues are complex, then (x_0, y_0) is called a **spiral source**.

If the Jacobian matrix has one positive and one negative eigenvalue, then the equilibrium point (x_0, y_0) is called a **saddle** for the nonlinear system. As for a linear system with a saddle equilibrium point at the origin, there are exactly two curves of solutions that approach the equilibrium point as t increases and exactly two curves of solutions that approach the equilibrium point as t decreases (see Figures 5.7 and 5.9). For the nonlinear system, these curves of solutions need not be straight lines. All other solutions with initial position near (x_0, y_0) move away as t increases and as t decreases.

A Reminder
It is important to remember that this classification of equilibrium points for nonlinear systems says nothing about the behavior of solutions of the nonlinear system with initial positions far from (x_0, y_0).

Separatrices

The four special solution curves that tend toward a saddle equilibrium point as $t \to \infty$ or as $t \to -\infty$ are called **separatrices**. (One of these curves by itself is called a *separatrix*.) Separatrices are of special importance because they separate solutions with different behaviors. The two separatrices on which solutions tend toward the saddle as $t \to \infty$ are called **stable separatrices**, while those on which solutions tend toward the saddle as $t \to -\infty$ are called **unstable separatrices**.

In the system

$$\frac{dx}{dt} = -2x + 2x^2$$

$$\frac{dy}{dt} = -3x + y + 3x^2$$

studied above, the origin is a saddle (see Figures 5.7 and 5.9). The stable separatrix that tends toward the origin separates the strip of the phase plane bounded by the lines $x = 0$ and $x = 1$ into two parts. Initial conditions in this strip that are above the separatrix yield solutions where $y(t) \to \infty$ as t increases, while initial conditions in this strip that are below the separatrix yield solutions where $y(t) \to -\infty$ as t increases (see Figure 5.10).

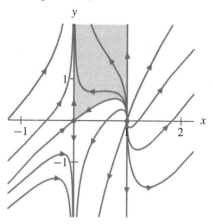

Figure 5.10
Separatrices of $(0, 0)$ for the system

$$\frac{dx}{dt} = -2x + 2x^2$$

$$\frac{dy}{dt} = -3x + y + 3x^2,$$

and regions of the strip between $x = 0$ and $x = 1$ with different long-term behaviors.

When Linearization Fails

Unfortunately, for some equilibria of some systems, the information given by the linearized system is not enough to determine the complete behavior of solutions of the nonlinear system near the equilibrium point.

For example, consider the system

$$\frac{dx}{dt} = y - (x^2 + y^2)x$$

$$\frac{dy}{dt} = -x - (x^2 + y^2)y.$$

The origin is an equilibrium point for this system, and its linearized system is

$$\begin{pmatrix} \dfrac{du}{dt} \\ \dfrac{dv}{dt} \end{pmatrix} = \begin{pmatrix} 0 & 1 \\ -1 & 0 \end{pmatrix} \begin{pmatrix} u \\ v \end{pmatrix}.$$

The eigenvalues of this linear system are $\pm i$, and hence it is a center. All nonzero solutions of the linearized system are periodic. In fact, each solution curve is a circle centered at the origin.

However, there are no periodic solutions for the nonlinear system. To see why, consider the vector field as a sum of two vector fields, the linear vector field $\mathbf{V}_1(x, y) = (y, -x)$ and the nonlinear vector field $\mathbf{V}_2(x, y) = (-(x^2 + y^2)x, -(x^2 + y^2)y)$. The linear vector field $\mathbf{V}_1$ corresponds to the linearized system. It is always tangent to circles centered at the origin. On the other hand, the vector field $\mathbf{V}_2$ always points directly toward $(0, 0)$ since it is a scalar multiple of the field $(-x, -y)$. (The scalar is the positive number $x^2 + y^2$.) The sum of $\mathbf{V}_1(x, y)$ and $\mathbf{V}_2(x, y)$ is a vector field that always has a negative radial component. Thus, solutions to the nonlinear system spiral toward $(0, 0)$ (see Figure 5.11).

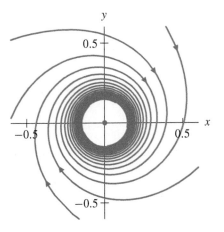

Figure 5.11
The solution curves for the system

$$\frac{dx}{dt} = y - (x^2 + y^2)x$$

$$\frac{dy}{dt} = -x - (x^2 + y^2)y$$

spiral slowly toward the equilibrium point at the origin even though the linearization at the origin is a center.

Note that if we merely change the signs of the higher-order terms in the above system, then the resulting system

$$\frac{dx}{dt} = y + (x^2 + y^2)x$$

$$\frac{dy}{dt} = -x + (x^2 + y^2)y$$

has the same linearization near $(0, 0)$, but now solutions spiral away from the origin.

In this example the solutions of the nonlinear system near the origin and the solutions of the linearized system are still approximately the same, at least for a short amount of time. However, since the linearized system is a center, any small perturbation can change the long-term behavior of the solutions. Even the very small perturbation caused by the inclusion of the nonlinear terms can turn the center into a spiral sink or a spiral source.

Fortunately there are only two situations in which the long-term behavior of solutions near an equilibrium point of the nonlinear system and its linearization can differ. One is when the linearized system is a center. The other is when the linearized system has zero as an eigenvalue (see Exercises 17–19). In every other case, the long-term behavior of solutions of the nonlinear system near an equilibrium point is the same as the solutions of its linearization.

EXERCISES FOR SECTION 5.1

1. Consider the three systems

(i)
$$\frac{dx}{dt} = 2x + y$$
$$\frac{dy}{dt} = -y + x^2$$

(ii)
$$\frac{dx}{dt} = 2x + y$$
$$\frac{dy}{dt} = y + x^2$$

(iii)
$$\frac{dx}{dt} = 2x + y$$
$$\frac{dy}{dt} = -y - x^2.$$

All three have an equilibrium point at $(0, 0)$. Which two systems have phase portraits with the same "local picture" near $(0, 0)$? Justify your answer. [*Hint*: Very little computation is required for this exercise, but be sure to give a complete justification.]

2. Consider the following three systems:

(i)
$$\frac{dx}{dt} = 3 \sin x + y$$
$$\frac{dy}{dt} = 4x + \cos y - 1$$

(ii)
$$\frac{dx}{dt} = -3 \sin x + y$$
$$\frac{dy}{dt} = 4x + \cos y - 1$$

(iii)
$$\frac{dx}{dt} = -3 \sin x + y$$
$$\frac{dy}{dt} = 4x + 3 \cos y - 3.$$

All three have an equilibrium point at $(0, 0)$. Which two systems have phase portraits with the same "local picture" near $(0, 0)$? Justify your answer. [*Hint*: Very little computation is required for this exercise, but be sure to give a complete justification.]

3. Consider the system

$$\frac{dx}{dt} = -2x + y$$
$$\frac{dy}{dt} = -y + x^2.$$

(a) Find the linearized system for the equilibrium point $(0, 0)$.
(b) Classify $(0, 0)$ (as either a source, sink, center, ...).
(c) Sketch the phase portrait for the linearized system near $(0, 0)$.
(d) Repeat parts (a)–(c) for the equilibrium point at $(2, 4)$.

4. Consider the system

$$\frac{dx}{dt} = -x$$
$$\frac{dy}{dt} = -4x^3 + y.$$

(a) Show that the origin is the only equilibrium point.
(b) Find the linearized system at the origin.
(c) Classify the linearized system and sketch its phase portrait.

5. Consider the system in Exercise 4.

    (a) Find the general solution of the equation $dx/dt = -x$. [*Hint*: This is as easy as it looks.]

    (b) Using the solution to part (a) in place of x, find the general solution of the equation

$$\frac{dy}{dt} = -4x^3 + y.$$

    [*Hint*: Note that this system is partially decoupled, so this part of the exercise is really asking you to follow the techniques described in Sections 1.8 and 2.4.]

    (c) Use the results from parts (a) and (b) to form the general solution of the system.

    (d) Find the solution curves of the system that tend toward the origin as $t \to \infty$.

    (e) Find the solution curves of the system that tend toward the origin as $t \to -\infty$.

    (f) Sketch the solution curves in the phase plane corresponding to these solutions. These are the separatrices.

    (g) Compare the sketch of the linearized system that you obtained in Exercise 4 with a sketch of the separatrix solutions for the equilibrium point at the origin for this system. In what ways are the two pictures the same? How do they differ?

6. For the competing species population model

$$\frac{dx}{dt} = 2x\left(1 - \frac{x}{2}\right) - xy$$

$$\frac{dy}{dt} = 3y\left(1 - \frac{y}{3}\right) - 2xy$$

studied in this section, we showed that the equilibrium point $(1, 1)$ is a saddle.

    (a) Find the linearized system near each of the other equilibrium points.

    (b) Classify each equilibrium point (as either a source, a sink, a saddle, ...).

    (c) Sketch the phase portrait of each linearized system.

    (d) Give a brief description of the phase portrait near each equilibrium point of the nonlinear system.

In Exercises 7–16, we restrict attention to the first quadrant ($x, y \geq 0$). For each system,

  (a) find and classify all equilibria,

  (b) sketch the phase portrait of the system near each equilibrium point, and

  (c) use `HPGSystemSolver` to compare the actual phase portrait to the phase portraits of the linearizations. (You may need to set $\Delta t = 0.0001$ for some of these systems.)

7.
$$\frac{dx}{dt} = x(-x - 3y + 150)$$
$$\frac{dy}{dt} = y(-2x - y + 100)$$

8.
$$\frac{dx}{dt} = x(10 - x - y)$$
$$\frac{dy}{dt} = y(30 - 2x - y)$$

9.
$$\frac{dx}{dt} = x(100 - x - 2y)$$
$$\frac{dy}{dt} = y(150 - x - 6y)$$

10.
$$\frac{dx}{dt} = x(-x - y + 100)$$
$$\frac{dy}{dt} = y(-x^2 - y^2 + 2500)$$

11.
$$\frac{dx}{dt} = x(-x - y + 40)$$
$$\frac{dy}{dt} = y(-x^2 - y^2 + 2500)$$

12.
$$\frac{dx}{dt} = x(-4x - y + 160)$$
$$\frac{dy}{dt} = y(-x^2 - y^2 + 2500)$$

13.
$$\frac{dx}{dt} = x(-8x - 6y + 480)$$
$$\frac{dy}{dt} = y(-x^2 - y^2 + 2500)$$

14.
$$\frac{dx}{dt} = x(2 - x - y)$$
$$\frac{dy}{dt} = y(y - x^2)$$

15.
$$\frac{dx}{dt} = x(2 - x - y)$$
$$\frac{dy}{dt} = y(y - x)$$

16.
$$\frac{dx}{dt} = x(x - 1)$$
$$\frac{dy}{dt} = y(x^2 - y)$$

17. Consider the system

$$\frac{dx}{dt} = -x^3$$
$$\frac{dy}{dt} = -y + y^2.$$

It has equilibrium points at $(0, 0)$ and $(0, 1)$.

  (a) Find the linearized system at $(0, 0)$.
  (b) Find the eigenvalues and eigenvectors and sketch the phase portrait of the linearized system at $(0, 0)$.
  (c) Find the linearized system at $(0, 1)$.
  (d) Find the eigenvalues and eigenvectors and sketch the phase portrait of the linearized system at $(0, 1)$.
  (e) Sketch the phase portrait of the nonlinear system. [*Hint*: The system decouples, so first draw a phase line for each of the individual equations.]
  (f) Why do the phase portraits for the linearized systems and the phase portrait for the nonlinear system near the equilibrium points look so different?

18. If a nonlinear system depends on a parameter, then the equilibrium points can change as the parameter varies. In other words, as the parameter changes, a bifurcation can occur. Consider the one-parameter system family of systems

$$\frac{dx}{dt} = x^2 - a$$
$$\frac{dy}{dt} = -y(x^2 + 1),$$

where a is the parameter.

(a) Show that the system has no equilibrium points if $a < 0$.

(b) Show that the system has two equilibrium points if $a > 0$.

(c) Show that the system has exactly one equilibrium point if $a = 0$.

(d) Find the linearization of the equilibrium point for $a = 0$ and compute the eigenvalues of this linear system.

Remark: The system changes from having no equilibrium points to having two equilibrium points as the parameter a is increased through $a = 0$. We say that the system has a bifurcation at $a = 0$, and that $a = 0$ is a bifurcation value of the parameter.

19. Continuing the study of the nonlinear system given in Exercise 18,

(a) use the direction field to sketch the phase portrait for the system if $a = -1$,

(b) use the direction field and the linearization at the equilibrium point to sketch the phase portrait for $a = 0$, and

(c) use the direction field and the linearization at the equilibrium points to sketch the phase portrait for $a = 1$.

Remark: The transition from a system with no equilibrium points to a system with one saddle and one sink via a system with one equilibrium point with zero as an eigenvalue is typical of bifurcations that create equilibria.

In Exercises 20–25, each system depends on the parameter a. In each exercise,

(a) find all equilibrium points,

(b) determine all values of a at which a bifurcation occurs, and

(c) in a short paragraph complete with pictures, describe the phase portrait at, before, and after each bifurcation value.

20. $$\frac{dx}{dt} = y - x^2$$
$$\frac{dy}{dt} = y - a$$

21. $$\frac{dx}{dt} = y - x^2$$
$$\frac{dy}{dt} = a$$

22. $$\frac{dx}{dt} = y - x^2$$
$$\frac{dy}{dt} = y - x - a$$

23. $$\frac{dx}{dt} = y - ax^3$$
$$\frac{dy}{dt} = y - x$$

24. $$\frac{dx}{dt} = y - x^2 + a$$
$$\frac{dy}{dt} = y + x^2 - a$$

25. $$\frac{dx}{dt} = y - x^2 + a$$
$$\frac{dy}{dt} = y + x^2$$

26. The system

$$\frac{dx}{dt} = x(-x - y + 70)$$

$$\frac{dy}{dt} = y(-2x - y + a)$$

is a model for a pair of competing species for which dy/dt depends on the parameter a. Find the two bifurcation values of a. Describe the fate of the x and y populations before and after each bifurcation.

27. Suppose two species X and Y are to be introduced to an island. It is known that the two species compete, but the precise nature of their interaction is unknown. We assume that the populations $x(t)$ and $y(t)$ of X and Y, respectively, are modeled by a system

$$\frac{dx}{dt} = f(x, y)$$

$$\frac{dy}{dt} = g(x, y).$$

   (a) Suppose $f(0, 0) = g(0, 0) = 0$; that is, $(0, 0)$ is an equilibrium point. What does this say about the ability of X and Y to migrate to the island?

   (b) Suppose that a small population consisting solely of one species reproduces rapidly. What can you conclude about the values of $\partial f / \partial x$ and $\partial g / \partial y$ at $(0, 0)$?

   (c) Since X and Y compete for resources, the presence of either of the species will decrease the rate of growth of the population of the other. What does this say about $\partial f / \partial y$ and $\partial g / \partial x$ at $(0, 0)$?

   (d) Using the assumptions from parts (a)–(c), what type(s) of equilibrium point could $(0, 0)$ possibly be? [There may be more than one possibility. If so, specify all of them.]

   (e) For each of the possibilities listed in part (d), sketch a possible phase portrait near $(0, 0)$.

Be sure to justify all answers.

28. For the two species X and Y of Exercise 27, suppose that both X and Y reproduce very slowly. Also suppose that competition between these two species is very intense.

   (a) What can you conclude about $\partial f / \partial x$ and $\partial g / \partial y$ at $(0, 0)$?

   (b) What can you conclude about $\partial f / \partial y$ and $\partial g / \partial x$ at $(0, 0)$?

   (c) What type(s) of equilibrium point can $(0, 0)$ possibly be? [There may be more than one possibility. If so, specify all of them.]

   (d) For each of the possibilities listed in part (c), sketch a possible phase portrait near $(0, 0)$.

Remember to justify all answers.

29. For the species X and Y in Exercises 27 and 28, suppose that X reproduces very quickly if it is on the island with no Y's present and that species Y reproduces slowly if there are no X's present. Also suppose that the growth rate of species X is decreased a relatively large amount by the presence of Y but that species Y is indifferent to X's population.

(a) What can you say about $\partial f/\partial x$ and $\partial g/\partial y$ at $(0, 0)$?

(b) What can you say about $\partial f/\partial y$ and $\partial g/\partial x$ at $(0, 0)$?

(c) What are the possible type(s) for the equilibrium point at $(0, 0)$? [There may be more than one possibility. If so, specify all of them.]

(d) For each of the types listed above, sketch the phase portrait near $(0, 0)$.

Be sure to justify all answers.

30. Suppose two similar countries Y and Z are engaged in an arms race. Let $y(t)$ and $z(t)$ denote the size of the stockpiles of arms of Y and Z, respectively. We model this situation with the system of differential equations

$$\frac{dy}{dt} = h(y, z)$$
$$\frac{dz}{dt} = k(y, z).$$

Suppose that all we know about the functions h and k are the two assumptions:

- If country Z's stockpile of arms is not changing, then any increase in size of Y's stockpile of arms results in a decrease in the rate of arms building in country Y. The same is true for country Z.
- If either country increases its stockpile, the other responds by increasing its rate of arms production.

(a) What do the assumptions imply about $\partial h/\partial y$ and $\partial k/\partial z$?

(b) What do the assumptions imply about $\partial h/\partial z$ and $\partial k/\partial y$?

(c) What types of equilibrium points are possible for this system? Justify your answer. [*Hint*: Suppose you have an equilibrium point. What do your results in parts (a) and (b) imply about the Jacobian matrix at that equilibrium point?]

5.2 QUALITATIVE ANALYSIS

The process of linearization discussed in Section 5.1 gives us a powerful technique for understanding the behavior of solutions of a nonlinear system near an equilibrium point. Unfortunately it provides "local" information only—information that can be used only near equilibrium points. (Making predictions based on linearizations far away from equilibria can have drastic consequences—see Section 4.5.)

So far our only general techniques for the study of the behavior of nonlinear systems away from equilibrium points are numerical. Indeed, a careful numerical study of a system can give considerable insight into the behavior of its solutions. Unfortunately it is difficult to know if enough initial conditions have been tested to observe all of the possible behaviors of solutions. In this section we develop qualitative techniques that can be used in combination with linearization and numerics.

Competing Species

Recall the system

$$\frac{dx}{dt} = 2x\left(1 - \frac{x}{2}\right) - xy$$

$$\frac{dy}{dt} = 3y\left(1 - \frac{y}{3}\right) - 2xy,$$

where x and y are populations of two species that compete for resources (see Section 5.1).

In Section 5.1 we determined that the equilibrium points are $(0, 0)$, $(0, 3)$, $(2, 0)$, and $(1, 1)$. By linearizing, we found that the point $(1, 1)$ is a saddle. There is one curve of solutions that approach $(1, 1)$ as $t \to \infty$, and this curve separates the phase plane into two regions. The use of numerical methods suggests that solutions that do not tend to $(1, 1)$ tend either to $(0, 3)$ or $(2, 0)$ as t increases (see Figure 5.12). To verify this observation and to better understand the behavior of solutions, we look more closely at the direction field.

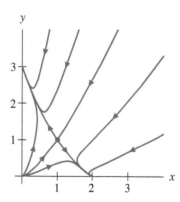

Figure 5.12
Phase portrait for the competing species system

$$\frac{dx}{dt} = 2x\left(1 - \frac{x}{2}\right) - xy$$

$$\frac{dy}{dt} = 3y\left(1 - \frac{y}{3}\right) - 2xy.$$

This computer-generated phase portrait suggests that solutions that do not tend to $(1, 1)$ tend either to $(0, 3)$ or to $(2, 0)$.

Nullclines

With a bit more qualitative analysis, we can give a much more complete description of the behavior of solutions. One tool for this analysis is the **nullcline**.

DEFINITION   For the system

$$\frac{dx}{dt} = f(x, y)$$

$$\frac{dy}{dt} = g(x, y),$$

the *x-nullcline* is the set of points (x, y) where $f(x, y)$ is zero—that is, the level curve where $f(x, y)$ is zero. The *y-nullcline* is the set of points where $g(x, y)$ is zero.

Along the x-nullcline, the x-component of the vector field is zero, and consequently the vector field is vertical. It points either straight up or straight down. Similarly, on the y-nullcline, the y-component of the vector field is zero, so the vector field is horizontal. It points either left or right. Because both $f(x, y)$ and $g(x, y)$ must be zero at an equilibrium point, the intersections of the nullclines are the equilibrium points.

To show how nullclines can be used to help in the qualitative analysis of systems, we consider our competing species example

$$\frac{dx}{dt} = 2x\left(1 - \frac{x}{2}\right) - xy$$
$$\frac{dy}{dt} = 3y\left(1 - \frac{y}{3}\right) - 2xy.$$

Recall that we are interested only in what happens in the first quadrant $(x, y \geq 0)$, since this is a population model. The x-nullcline is the set of points (x, y) that satisfy

$$\frac{dx}{dt} = 2x\left(1 - \frac{x}{2}\right) - xy = 0.$$

Since this equation is equivalent to $x(2 - x - y) = 0$, the x-nullcline consists of two lines, $x = 0$ and $y = -x + 2$. On these lines the x-component of the vector field is zero. Thus the vector field is vertical along these lines. In the remainder of the phase plane the x-component of the vector field is either positive or negative. If $dx/dt > 0$, then solutions move toward the right. If $dx/dt < 0$, then solutions move toward the left. In Figure 5.13 we mark part of the x-nullcline with vertical lines as a reminder that the vector field is vertical along the nullcline. We can label the regions off the x-nullcline as either "right" or "left" depending on whether dx/dt is positive or negative.

Similarly, the y-nullcline is the set of points where $dy/dt = 0$. This is the set of points that satisfy

$$3y\left(1 - \frac{y}{3}\right) - 2xy = y(3 - y - 2x) = 0.$$

This set also consists of two lines, $y = 0$ and $y = -2x + 3$. On these lines the y-component of the vector field is zero, so the vector field is horizontal. On the rest

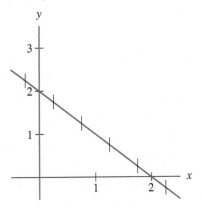

Figure 5.13
The x-nullclines for the system

$$\frac{dx}{dt} = 2x\left(1 - \frac{x}{2}\right) - xy$$
$$\frac{dy}{dt} = 3y\left(1 - \frac{y}{3}\right) - 2xy.$$

of the phase plane, either $dy/dt > 0$ and solutions move upward, or $dy/dt < 0$ and solutions move downward. In Figure 5.14 we mark part of the y-nullcline with horizontal segments as a reminder that the vector field is horizontal along the nullcline. We label the regions off the y-nullcline as either "up" or "down" depending on the sign of dy/dt.

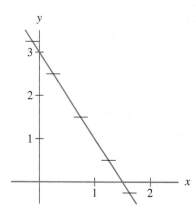

Figure 5.14
The y-nullclines for the system

$$\frac{dx}{dt} = 2x\left(1 - \frac{x}{2}\right) - xy$$

$$\frac{dy}{dt} = 3y\left(1 - \frac{y}{3}\right) - 2xy.$$

Analysis using nullclines

We combine the x- and y-nullclines in Figure 5.15. The equilibrium points occur at the intersections of the x- and y-nullclines. The nullclines divide the first quadrant into four regions labeled A, B, C, and D. We can use this picture to give a detailed analysis of the behavior of solutions of this system.

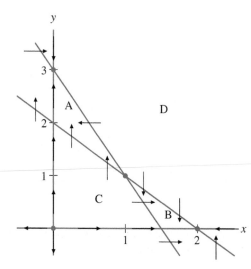

Figure 5.15
The x- and y-nullclines for

$$\frac{dx}{dt} = 2x\left(1 - \frac{x}{2}\right) - xy$$

$$\frac{dy}{dt} = 3y\left(1 - \frac{y}{3}\right) - 2xy.$$

The nullclines divide the first quadrant into four regions marked A, B, C, and D.

First, consider the triangular region A (see Figure 5.15). The segment $0 < y < 3$ on the y-axis is a solution curve, and the vector field on the other two sides of A point into A. Hence a solution that begins in region A at time zero will remain in A for all positive time. There is no way the solution can leave. Region A lies in the "left" ($dx/dt < 0$) and the "up" ($dy/dt > 0$) regions in the phase plane, so we can label it "left-up." That is, as t increases, solutions in region A must move toward the upper-left corner of the region—toward the equilibrium point $(0, 3)$.

Similarly, solutions cannot leave region B, and they move "right-down" as t increases. Hence all solutions in region B tend to the equilibrium point $(2, 0)$ as t increases. In Figure 5.16, we display two solutions of the competing species model together with their $x(t)$- and $y(t)$-graphs.

Solutions in region C move "right-up." There are three possibilities for what happens to these solutions. They may leave region C and enter A, they may leave C and enter B, or they may approach the equilibrium point $(1, 1)$. Since the point $(1, 1)$ is

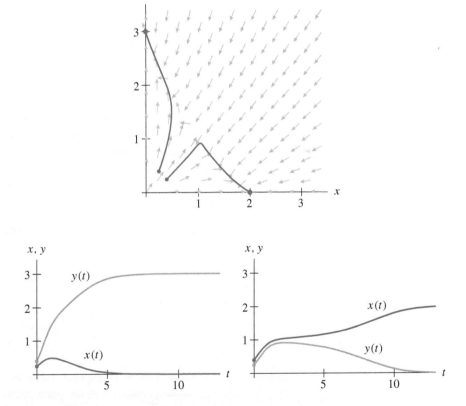

Figure 5.16
Two solutions in the phase plane for the competing species system above with their corresponding $x(t)$- and $y(t)$-graphs.

a saddle, the solutions that enter A are divided from those that enter B by the stable separatrix of $(1, 1)$.

Similarly, solutions in region D move left-down, tending toward regions A, B, or the equilibrium point $(1, 1)$. Again, the stable separatrix of $(1, 1)$ separates the solutions that enter A (and tend to $(0, 3)$) from those that enter B (and tend to $(2, 0)$).

This analysis gives us a fairly complete qualitative picture of the behavior of solutions of the competing species model. We know that most solutions tend to an equilibrium population with one species extinct and the other at its carrying capacity (see Figure 5.16). The stable separatrix of the saddle $(1, 1)$ divides the two long-term behaviors.

Important observations

In this example both nullclines consist of straight lines. However, in general a nullcline can be any type of curve, and we will soon study examples with such nullclines.

Also note that there are two very different kinds of nullclines in this example. For the nullclines along the x- and y-axes, the vector field was (coincidentally) tangent to these lines. Consequently, solutions that start on these lines stay on them forever. We can use techniques from Section 1.6 to analyze these solutions completely, since these nullclines really are phase lines for a one-dimensional equation.

The vector field is not tangent to the other two nullclines in this example. It points across these lines, and as a result these nullclines give us information only about the direction of solutions as they cross the nullcline.

It is particularly important to note the difference between a straight-line solution (as discussed in Chapter 3) and a nullcline. A straight-line solution is a solution curve that corresponds to a real eigenvector for a linear system. A nullcline is a curve along which the vector field is either purely horizontal or purely vertical. It is possible for a nullcline and a straight-line solution to coincide for some linear systems, but in general nullclines and straight-line solutions are different.

Nullclines That Are Not Lines

As a somewhat more complicated example of how nullclines are used, consider

$$\frac{dx}{dt} = 2x \left(1 - \frac{x}{2} \right) - xy$$

$$\frac{dy}{dt} = y \left(\frac{9}{4} - y^2 \right) - x^2 y.$$

We can still interpret this as a competing species model because the growth rate of each species decreases when the population of the other increases. In this model, the additional complications in the equation for dy/dt are for the sake of illustration of our techniques. As is usual with population models, we consider only solutions in the first quadrant.

The x-nullcline satisfies the equation

$$x(2 - x - y) = 0,$$

which consists of the two lines $x = 0$ and $y = -x + 2$. The y-nullcline is the set of points (x, y) that satisfy

$$y\left(-x^2 - y^2 + \frac{9}{4}\right) = 0.$$

These points lie either on the line $y = 0$ or on the circle $x^2 + y^2 = 9/4 = (3/2)^2$. The intersection points of the x- and y-nullclines give the equilibrium points $(0, 0)$, $(0, 3/2)$, $(1 + \sqrt{2}/4, 1 - \sqrt{2}/4) \approx (1.35, 0.65)$, $(1 - \sqrt{2}/4, 1 + \sqrt{2}/4) \approx (0.65, 1.35)$, and $(2, 0)$ (see Figure 5.17). (The point $(0, -3/2)$ is also an equilibrium point, but since it lies outside the first quadrant, it is not important to our analysis.)

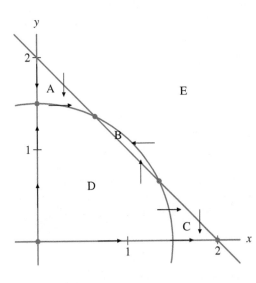

Figure 5.17
Nullclines for the system

$$\frac{dx}{dt} = 2x\left(1 - \frac{x}{2}\right) - xy$$

$$\frac{dy}{dt} = y\left(\frac{9}{4} - y^2\right) - x^2 y.$$

The nullclines separate the first quadrant into five regions.

As in the previous example, the x- and y-axes consist of solution curves. If $y = 0$, then $dy/dt = 0$ and we have

$$\frac{dx}{dt} = 2x\left(1 - \frac{x}{2}\right),$$

which is the same logistic equation as in the previous example. If $x = 0$, we have $dx/dt = 0$ and

$$\frac{dy}{dt} = y\left(-y^2 + \frac{9}{4}\right).$$

The phase line for $dy/dt = y(-y^2 + 9/4)$ has equilibrium points at $y = 0$, $y = 3/2$ (and $y = -3/2$ but we are not considering $y < 0$). On the phase line, $y = 0$ is a source and $y = 3/2$ is a sink (see Figure 5.18).

The x- and y-nullclines divide the first quadrant into 5 regions labeled A, B, C, D, and E in Figure 5.17. A solution that enters in regions A, B, or C remains in that region as t increases since the vector field on the boundaries of these regions never points out of these regions. By labeling the regions with the direction of the vector field (such as "right-up" in region D), we see that solutions in A and B tend toward the equilibrium point $(0.65, 1.35)$, whereas those in region C tend toward $(2, 0)$. In Figure 5.18 we sketch the phase plane and $x(t)$- and $y(t)$-graphs for two of these solutions. Solutions in regions D and E either enter one of the regions A, B, or C or else they tend to one of the equilibrium points $(0.65, 1.35)$ or $(1.35, 0.65)$. Again we must have separatrix solutions that divide the solutions that tend to $(0.65, 1.35)$ from those that tend to $(2, 0)$. By linearizing at $(1.35, 0.65)$, we can confirm that it is a saddle.

Using qualitative analysis involving nullclines, we can conclude that all solutions in this model tend toward equilibrium points as t increases, as in the previous model. The choice of which equilibrium point the solution tends toward depends on the location of the initial condition. Unlike the previous example, there is a large set of solutions that tend to the equilibrium point $(0.65, 1.35)$, where the populations of both

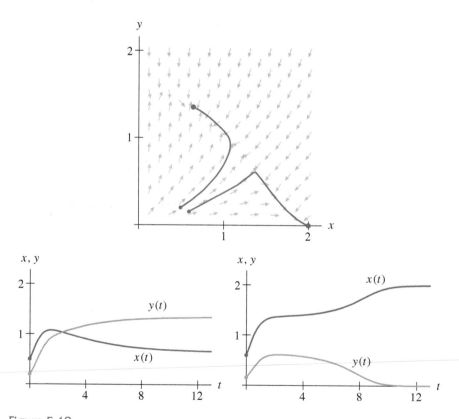

Figure 5.18
Phase plane for modified competing species system with $x(t)$- and $y(t)$-graphs for indicated solutions.

species are positive (mutual coexistence). The only solutions that approach the equilibrium $(0, 3/2)$ are those on the y-axis. On the other hand, there is a large set of initial conditions that give solutions approaching the equilibrium point $(2, 0)$, where the y-species is extinct and the x-species is at its carrying capacity.

Using All Our Tools

Consider the system

$$\frac{dx}{dt} = x + y - x^3$$
$$\frac{dy}{dt} = -0.5x.$$

Systems of this form arise in the study of nerve cells. Roughly speaking, the variable $x(t)$ represents voltage across the boundary of a nerve cell at time t, and $y(t)$ represents the permeability of the cell wall at time t. A rapid change in x corresponds to the nerve cell "firing."

Information from the nullclines

The x-nullcline for this system is the curve $y = -x + x^3$. Above this curve the x-component of the vector field is positive, and below it the x-component is negative. Hence, above the x-nullcline solutions move to the right, and below the x-nullcline solutions move to the left. The y-nullcline is the line $x = 0$, the y-axis. To the right of the y-nullcline we have $dy/dt < 0$, and to the left of the y-nullcline we have $dy/dt > 0$. Hence, in the right half of the phase plane solutions move down, and in the left half plane solutions move up (see Figure 5.19). The x- and y-nullclines divide the phase plane into four regions. Using the above qualitative analysis, we can conclude that all solutions must circulate clockwise around the origin—the only equilibrium point of the system (see Figure 5.20).

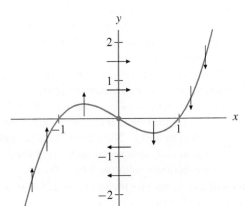

Figure 5.19
The x- and y-nullclines for the system

$$\frac{dx}{dt} = x + y - x^3$$
$$\frac{dy}{dt} = -0.5x.$$

The nullclines for this system separate the plane into four regions.

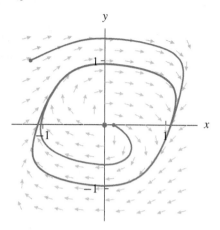

Figure 5.20
Solutions of the system

$$\frac{dx}{dt} = x + y - x^3$$

$$\frac{dy}{dt} = -0.5x.$$

Note that solutions with initial conditions close to the origin spiral outward, while those with initial conditions far from the origin spiral inward.

Information from linearization
The linearized system at the origin is

$$\frac{dx}{dt} = x + y$$
$$\frac{dy}{dt} = -0.5x,$$

which has eigenvalues $(1 \pm i)/2$. Hence the origin is a spiral source.

This analysis is applicable only near the origin. Since the term we dropped to obtain the linearization is $-x^3$, the linear approximation is no longer valid once this term has significant size. Once x^3 is comparable in size to x, it is not safe to use the linearization to study the nonlinear system.

Information from numerical approximations of solutions
To get a more detailed idea of the behavior of solutions of this system, we use numerical methods to compute some approximate solutions. From the linearization at the origin, we know that initial conditions near the origin yield solutions that spiral outward. If we take an initial condition far from the origin, the numerics show that the resulting solution spirals inward.

From the Uniqueness Theorem, we know that solution curves never cross. Hence the solutions that start near the origin must eventually stop spiraling outward; otherwise they would cross the inward spiraling solutions. Between the outward- and inward-spiraling solutions, there must be at least one solution that spirals neither outward nor inward. This solution is periodic. Numerical evidence indicates that there is only one periodic solution and that all other solutions (except the equilibrium solution at the origin) spiral toward this periodic solution. The $x(t)$- and $y(t)$-graphs of two solutions are shown in Figure 5.21. From this picture we see that both of these solutions converge toward the same type of periodic behavior.

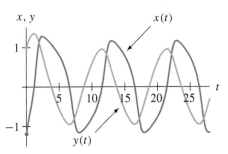

Figure 5.21
The $x(t)$- and $y(t)$-graphs for two solutions of the system

$$\frac{dx}{dt} = x + y - x^3$$

$$\frac{dy}{dt} = -0.5x.$$

The initial condition for the first solution is close to the origin, whereas the initial condition for the second solution lies outside the solution curve associated to the periodic solution. Note that both graphs indicate that these two solutions behave the same over the long term.

Mathematical Toolbox

In this section we have analyzed three first-order systems of differential equations, employing analytical, qualitative, and numerical techniques. As we have seen, the analysis of systems of differential equations is much more difficult than the analysis of differential equations with only one dependent variable. We need many different types of techniques to handle systems, and we must be willing to use whichever technique is appropriate. As with any craft, the ability to choose the appropriate tool for the problem at hand is a crucial skill.

EXERCISES FOR SECTION 5.2

In Exercises 1–3, sketch the x- and y-nullclines of the system specified. Then find all equilibrium points. Using the direction of the vector field between the nullclines, describe the possible fate of the solution curves corresponding to the initial conditions (a), (b), and (c).

1.
$$\frac{dx}{dt} = 2 - x - y$$
$$\frac{dy}{dt} = y - x^2$$

   (a) $x_0 = 2$, $y_0 = 1$

   (b) $x_0 = 0$, $y_0 = -1$

   (c) $x_0 = 0$, $y_0 = 0$

2.
$$\frac{dx}{dt} = 2 - x - y$$
$$\frac{dy}{dt} = y - |x|$$

   (a) $x_0 = -1$, $y_0 = 1$

   (b) $x_0 = 2$, $y_0 = 1$

   (c) $x_0 = 2$, $y_0 = 2$

3.
$$\frac{dx}{dt} = x(x - 1)$$
$$\frac{dy}{dt} = x^2 - y$$

   (a) $x_0 = -1$, $y_0 = 0$

   (b) $x_0 = 0.8$, $y_0 = 0$

   (c) $x_0 = 1$, $y_0 = 3$

4. The **Volterra-Lotka system** for two competitive species is

$$\frac{dx}{dt} = x(-Ax - By + C)$$

$$\frac{dy}{dt} = y(-Dx - Ey + F)$$

where $x, y \geq 0$ and where the parameters A–F are all positive.

(a) Can there ever be more than one equilibrium point with both $x > 0$ and $y > 0$ (that is, where the species "coexist in equilibrium")? If so, give an example of values of A–F where the species can coexist. If not, why not?

(b) What conditions on the parameters A–F guarantee that there is at least one equilibrium point with $x > 0$ and $y > 0$?

In Exercises 5–14, we restrict attention to the first quadrant $(x, y \geq 0)$. For each system,

(a) sketch the nullclines,

(b) sketch the phase portrait, and

(c) write a brief paragraph describing the possible behaviors of solutions. [*Hint*: You may wish to use information obtained in Exercises 7–16 of Section 5.1.]

5.
$$\frac{dx}{dt} = x(-x - 3y + 150)$$
$$\frac{dy}{dt} = y(-2x - y + 100)$$

6.
$$\frac{dx}{dt} = x(10 - x - y)$$
$$\frac{dy}{dt} = y(30 - 2x - y)$$

7.
$$\frac{dx}{dt} = x(100 - x - 2y)$$
$$\frac{dy}{dt} = y(150 - x - 6y)$$

8.
$$\frac{dx}{dt} = x(-x - y + 100)$$
$$\frac{dy}{dt} = y(-x^2 - y^2 + 2500)$$

9.
$$\frac{dx}{dt} = x(-x - y + 40)$$
$$\frac{dy}{dt} = y(-x^2 - y^2 + 2500)$$

10.
$$\frac{dx}{dt} = x(-4x - y + 160)$$
$$\frac{dy}{dt} = y(-x^2 - y^2 + 2500)$$

11.
$$\frac{dx}{dt} = x(-8x - 6y + 480)$$
$$\frac{dy}{dt} = y(-x^2 - y^2 + 2500)$$

12.
$$\frac{dx}{dt} = x(2 - x - y)$$
$$\frac{dy}{dt} = y(y - x^2)$$

13.
$$\frac{dx}{dt} = x(2 - x - y)$$
$$\frac{dy}{dt} = y(y - x)$$

14.
$$\frac{dx}{dt} = x(x - 1)$$
$$\frac{dy}{dt} = y(x^2 - y)$$

15. Some species live in a "cooperative" manner—each species helping the other to survive and prosper (for example, flowers and honeybees).

    (a) How would you alter the Volterra-Lotka system described in Exercise 4 to obtain a model of cooperative species?

    (b) What are the nullclines for your cooperative system? Are there equilibrium points? Are there conditions on the parameters that guarantee that there are equilibrium points with both x and y positive?

Exercises 16–20 refer to the models of chemical reactions created in Exercises 25–30 of Section 2.1. Here $a(t)$ is the amount of substance A in a solution, and $b(t)$ is the amount of substance B in a solution at time t. We only need to consider nonnegative $a(t)$ and $b(t)$.

    For each system,

  (a) sketch the nullclines and indicate the direction of the vector field along the nullcline,

  (b) label the regions in the ab-phase plane created by the nullclines and determine which general direction the vector field points in each region (that is, increasing or decreasing a, increasing or decreasing b), and

  (c) identify the regions that solutions cannot leave and determine the fate of solutions in these regions as time increases.

16.
$$\frac{da}{dt} = -\frac{ab}{2}$$
$$\frac{db}{dt} = -\frac{ab}{2}$$

17.
$$\frac{da}{dt} = 2 - \frac{ab}{2}$$
$$\frac{db}{dt} = \frac{3}{2} - \frac{ab}{2}$$

18.
$$\frac{da}{dt} = 2 - \frac{ab}{2} - \frac{a^2}{3}$$
$$\frac{db}{dt} = \frac{3}{2} - \frac{ab}{2}$$

19.
$$\frac{da}{dt} = 2 - \frac{ab}{2} + \frac{b^2}{6}$$
$$\frac{db}{dt} = \frac{3}{2} - \frac{ab}{2} - \frac{b^2}{3}$$

20.
$$\frac{da}{dt} = 2 - \frac{ab}{2} - \frac{ab^2}{3}$$
$$\frac{db}{dt} = \frac{3}{2} - \frac{ab}{2} - \frac{2ab^2}{3}$$

Exercises 21–23 refer to the system

$$\frac{dx}{dt} = y$$
$$\frac{dy}{dt} = x - x^2.$$

21. Sketch the nullclines and find the direction of the vector field along the nullclines.

22. Show that there is at least one solution in each of the second and fourth quadrants that tends to the origin as $t \to \infty$. Similarly, show that there is at least one solution in each of the first and third quadrants that tends to the origin as $t \to -\infty$.

23. Find the linearized system near the equilibrium points $(0, 0)$ and $(1, 0)$. Use information obtained from the linearized systems and Exercises 21 and 22 to describe the phase portrait. What is it that you still do not know about the phase portrait?

5.3 HAMILTONIAN SYSTEMS

As we have emphasized many times, nonlinear systems of differential equations are almost impossible to solve explicitly. We have also seen that solution curves of systems may behave in many different ways and that there are no qualitative techniques that are guaranteed to work in all cases. Fortunately there are certain special types of nonlinear systems that arise often in practice and for which there are special techniques that enable us to gain some understanding of the phase portrait. In this and the next section we will discuss two of these special types of nonlinear systems. But first, we pause for a story.

How This Book Came to Be

Paul and Glen have been writing a differential equations textbook for the past ten years. They want their book to be filled with brilliant and witty new ideas about differential equations, but they are finding that new ideas are hard to come by.

More troubling to them is the following observation that they have both made over the years. Whenever Paul comes to the office in the morning with a witty new idea, they both work feverishly on his idea, and at first their creative juices flow. They work diligently, but eventually the idea doesn't pan out. So their enthusiasm wanes and with it, their creativity. On the other hand, whenever Glen arrives in the morning with a new idea, no matter how trivial, something different happens. They again both work feverishly. Sometimes Paul's enthusiasm wanes, but Glen is always excited. There is good give and take; they never give up. At least something comes of the idea and the book progresses. Now this bothered Paul and Glen. Why should their creative energy depend so critically on who has the first idea? They decided to model their plight with a system of differential equations.

They let $x(t)$ denote Glen's and $y(t)$ Paul's level of enthusiasm at time t. Now, $x(t)$ and $y(t)$ are difficult to measure because of a lack of standardized units of enthusiasm. However, it is clear that $y(t) > 0$ means that Paul is enthusiastic, whereas $y(t) < 0$ means that Paul is glum. When $y(t) = 0$, Paul just sits there, neither happy nor sad, and the situation is similar for Glen.

Both Glen and Paul have observed that Glen's enthusiasm changes at a rate directly proportional to Paul's level of enthusiasm. When Paul is excited, Glen gets more enthusiastic, but when Paul is glum, Glen loses his verve. A simple equation catching this behavior is

$$\frac{dx}{dt} = y.$$

Paul is a bit more difficult to categorize. When Glen is mildly enthusiastic, Paul's enthusiasm goes up. But when Glen gets wildly excited, Paul starts to lose enthusiasm. Apparently, when the torrent of ideas spilling out of Glen becomes too great, Paul gets a headache and tunes Glen out. On the other hand, when Glen is down, Paul is really glum. For the rate of change of Paul's enthusiasm we use

$$\frac{dy}{dt} = x - x^2.$$

The graph of $dy/dt = x - x^2$ as a function of x captures exactly Paul's enthusiasm level (see Figure 5.22). We can see that $dy/dt > 0$ if $0 < x < 1$, but $dy/dt < 0$ otherwise. The system of differential equations they settle on is

$$\frac{dx}{dt} = y$$

$$\frac{dy}{dt} = x - x^2.$$

Figure 5.22
The graph of $dy/dt = x - x^2$
as a function of x.

The direction field for this system is shown in Figure 5.23. There are two equilibrium points, one at the origin and one at $x = 1$, $y = 0$. The technique of linearization from Section 5.1 can be used to study the solutions near the equilibrium points. At the origin the Jacobian matrix is

$$\begin{pmatrix} 0 & 1 \\ 1 & 0 \end{pmatrix},$$

which has eigenvalues ± 1. Hence the origin is a saddle.

The equilibrium point at $(x, y) = (1, 0)$ has Jacobian matrix

$$\begin{pmatrix} 0 & 1 \\ -1 & 0 \end{pmatrix},$$

which has eigenvalues $\pm i$. The linearized system is a center. As we saw in Section 5.1, this is one of the cases when the long-term behavior of solutions of the nonlinear system near the equilibrium point is not completely determined by the linearization.

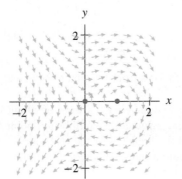

Figure 5.23
The direction field for

$$\frac{dx}{dt} = y$$
$$\frac{dy}{dt} = x - x^2.$$

The system has two equilibrium points, and by linearizing we know that the equilibrium point at the origin is a saddle.

The behavior of the nonlinear system near the equilibrium point $(1, 0)$ could be that of a spiral sink, a spiral source, or a center.

Numerical approximations of solutions give the phase portrait that is shown in Figure 5.24, which implies that solutions behave in a very regular way. The solutions with initial conditions near $(1, 0)$ seem to form closed loops corresponding to periodic solutions. Also, the unstable and stable separatrices emerging from the saddle at the origin into the first and fourth quadrants seem to form a single loop.

The qualitative techniques we have studied so far do not allow us to predict that this system will have such regular solution curves. Also, because the numerical approximations of solutions are only approximations, we should be cautious. Distinguishing solution curves that form closed loops from those that spiral very slowly can be very difficult. We would like to have techniques that can be used to verify the special behavior of this system.

Conserved quantities

Paul and Glen were so mystified by the behavior of their system that they decided to show it to their friend Bob. Bob exclaimed that he had seen this system before and that it had a conserved quantity. Noting the confused looks he got from Paul and Glen, he explained:

DEFINITION   A real-valued function $H(x, y)$ of the two variables x and y is a **conserved quantity** for a system of differential equations if it is constant along all solution curves of the system. That is, if $(x(t), y(t))$ is a solution of the system, then $H(x(t), y(t))$ is constant. In other words,

$$\frac{d}{dt} H(x(t), y(t)) = 0. \quad \blacksquare$$

Bob remembered that

$$H(x, y) = \frac{1}{2}y^2 - \frac{1}{2}x^2 + \frac{1}{3}x^3$$

is a conserved quantity for the system in question. To check this, suppose $(x(t), y(t))$ is a solution of the system. Then we compute

$$\frac{d}{dt} H(x(t), y(t)) = \left(\frac{\partial H}{\partial x}\right)\left(\frac{dx}{dt}\right) + \left(\frac{\partial H}{\partial y}\right)\left(\frac{dy}{dt}\right)$$
$$= (-x + x^2)(y) + (y)(x - x^2)$$
$$= 0,$$

where the first equality follows from the Chain Rule, and the second equality uses the fact that $(x(t), y(t))$ is a solution to replace dx/dt with y and dy/dt with $x - x^2$.

Consequently, the solution curves always lie along the level curves of H, which are shown in Figure 5.25. Around the equilibrium point $(1, 0)$, the level curves of H form closed circles, and the branches of the level curve emanating from the origin toward the right connect into a loop. This agrees with our analysis of the phase portrait for the system (see Figure 5.24).

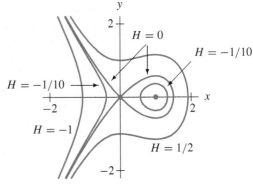

Figure 5.24
The phase portrait of the system $dx/dt = y$ and $dy/dt = x - x^2$.

Figure 5.25
The level curves of H. Compare the solution curves shown in Figure 5.24 to these level curves.

The fact that the right-hand branches of the stable and unstable separatrices of the origin form a single loop is very special. This type of solution curve is a **saddle connection**. Inside the saddle connection, all solution curves are periodic, whereas outside the saddle connection, all solutions have the property that both $x(t) \to -\infty$ and $y(t) \to -\infty$ as $t \to \infty$. This phase portrait explains, to Paul and Glen's great relief, why their daily productivity depends so crucially on who has the first idea. If $y(0) > 0$ but $x(0) = 0$, the solution curve $(x(t), y(t)) \to (-\infty, -\infty)$ as $t \to \infty$. But if $x(0)$ is positive and not too large ($0 < x(0) < 3/2$) and $y(0) = 0$, then both $x(t)$ and $y(t)$ are periodic, with $x(t) > 0$ for all t. We sketch the $x(t)$- and $y(t)$-graphs for two such initial conditions near $(0, 0)$ in Figure 5.26.

This explanation was a source of great satisfaction to both Paul and Glen. So much was their happiness that they invited Bob to become a coauthor of their book. Bob reluctantly agreed, but only after Glen and Paul promised to listen to *The Marriage of Figaro* in its entirety.

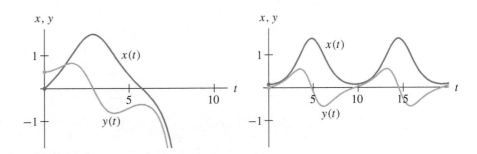

Figure 5.26
The $x(t)$- and $y(t)$-graphs on the left correspond to an initial condition for which $x(0) = 0$ and $y(0)$ is positive. The $x(t)$- and $y(t)$-graphs on the right correspond to an initial condition for which $0 < x(0) < 1$ and $y(0) = 0$

Hamiltonian Systems

What makes the preceding analysis work is the fact that the system of differential equations modeling Glen's and Paul's enthusiasm levels is a special kind of system called a Hamiltonian system (named for William Rowan Hamilton [1805–1865], an Irish mathematician).

DEFINITION   A system of differential equations is called a **Hamiltonian system** if there exists a real-valued function $H(x, y)$ such that

$$\frac{dx}{dt} = \frac{\partial H}{\partial y}$$

$$\frac{dy}{dt} = -\frac{\partial H}{\partial x}$$

for all x and y. The function H is called the *Hamiltonian function* for the system.   ∎

Note that H is always a conserved quantity for such a system. We can verify this by letting $(x(t), y(t))$ be any solution of the system. Then

$$\frac{d}{dt} H(x(t), y(t)) = \left(\frac{\partial H}{\partial x}\right)\left(\frac{dx}{dt}\right) + \left(\frac{\partial H}{\partial y}\right)\left(\frac{dy}{dt}\right)$$

$$= \left(\frac{\partial H}{\partial x}\right)\left(\frac{\partial H}{\partial y}\right) + \left(\frac{\partial H}{\partial y}\right)\left(-\frac{\partial H}{\partial x}\right)$$

$$= 0.$$

The first equality is the Chain Rule, and the second equality uses the fact that the system is Hamiltonian and that $(x(t), y(t))$ is a solution to replace dx/dt with $\partial H/\partial y$ and dy/dt with $-\partial H/\partial x$.

So, as above, solution curves of the system lie along the level curves of H. Sketching the phase portrait for a Hamiltonian system is the same as sketching the level sets of the Hamiltonian function.

Examples of Hamiltonian Systems: The Harmonic Oscillator

Recall that the undamped harmonic oscillator system is

$$\frac{dy}{dt} = v$$

$$\frac{dv}{dt} = -qy,$$

where q is a positive constant. If we let

$$H(y, v) = \frac{1}{2}v^2 + \frac{q}{2}y^2,$$

then

$$\frac{dy}{dt} = v = \frac{\partial H}{\partial v}$$

and

$$\frac{dv}{dt} = -qy = -\frac{\partial H}{\partial y}.$$

Hence the undamped harmonic oscillator system is a Hamiltonian system. The level sets of the function H are ellipses in the yv-plane which correspond to the phase portrait of the undamped harmonic oscillator (see Section 3.4). This Hamiltonian function is sometimes called the **energy function** for the oscillator.

The Nonlinear Pendulum

Consider a pendulum made of a light rigid rod of length l with a ball at one end of mass m. The ball is called the *bob*, and the rigid rod is called the *arm* of the pendulum. We assume that all the mass of the pendulum is in the bob, neglecting the mass of the rod. The other end of the rigid rod is attached to the wall in such a way that it can turn through an entire circle in a plane perpendicular to the ground. The position of the bob at time t is given by an angle $\theta(t)$, which we choose to measure in the counterclockwise direction with $\theta = 0$ corresponding to the downward vertical axis (see Figure 5.27).

We assume that there are only two forces acting on the pendulum: gravity and friction. The constant gravitational force equal to mg is in the downward direction, where g is the acceleration of gravity near earth ($g \approx 9.8$ m/s^2). Only the component of this force tangent to the circle of motion affects the motion of the pendulum. This component is $-mg \sin \theta$ (see Figure 5.28). There is also a force due to friction, which we assume to be proportional to the velocity of the bob.

The position of the bob at time t is given by the point $(l \sin \theta(t), -l \cos \theta(t))$ on the circle of radius l (remember that $\theta = 0$ corresponds to straight down). The speed of the bob is the length of the velocity vector, which is $l d\theta/dt$. The component of the acceleration that points along the direction of the motion of the bob has length $l d^2\theta/dt^2$. We take the force due to friction to be proportional to the velocity, so this force is $-b(l d\theta/dt)$, where $b > 0$ is a parameter that corresponds to the *coefficient of damping*.

Figure 5.27
A pendulum with rod length l and angle θ.

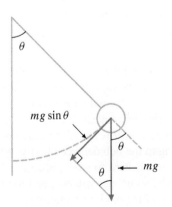

Figure 5.28
Decomposition of the force of gravity into components along the pendulum arm and tangent to the circle of motion of the pendulum bob.

Using Newton's second law, $F = ma$, we obtain the equation of motion

$$-bl\frac{d\theta}{dt} - mg\sin\theta = ml\frac{d^2\theta}{dt^2},$$

which is often written as

$$\frac{d^2\theta}{dt^2} + \frac{b}{m}\frac{d\theta}{dt} + \frac{g}{l}\sin\theta = 0.$$

We can rewrite this equation as a first-order system in the usual manner by letting the variable v represent the angular velocity $d\theta/dt$. The corresponding system is

$$\frac{d\theta}{dt} = v$$

$$\frac{dv}{dt} = -\frac{b}{m}v - \frac{g}{l}\sin\theta.$$

Remember that θ is an angular variable, so that $\theta = 0, 2\pi, 4\pi, \ldots$ all represent the same position—the pendulum at its lowest point. The values $\theta = \pi, 3\pi, 5\pi, \ldots$ occur when the pendulum is at the highest point of its circular motion.

The ideal pendulum

In practice there is always a force due to friction acting on a pendulum. For the moment, however, suppose there is no friction in this model. This is an "ideal" case that does not occur in the real world. However, it is not an unreasonable model for a very well-built and well-lubricated pendulum.

When no friction is present, the coefficient b vanishes. For convenience we suppose that the pendulum arm has unit length; that is, we suppose $l = 1$. (We consider the effect of adjusting the length of the arm in Exercises 4–7.) The equations of motion of the ideal pendulum with unit length arm are

$$\frac{d\theta}{dt} = v$$

$$\frac{dv}{dt} = -g\sin\theta.$$

The first step is to find the equilibrium points. We must have $v = 0$ from the first equation and $\sin\theta = 0$ from the second. Consequently, the equilibruim points are the points $(\theta, v) = (n\pi, 0)$, where n is any integer. If θ is an even multiple of π and $v = 0$, the bob hangs motionless in a downward position, an obvious equilibrium position for the pendulum. If θ is an odd multiple of π and $v = 0$, there is also a rest position corresponding to the bob balanced perfectly motionless in an upright position. This kind of equilibrium is hard to see in practice: Just when you manage to balance the pendulum perfectly, someone twitches, and the resulting current of air moves the pendulum out of equilibrium and into motion in one direction or the other.

Shortly we will give a method for determining if a given system is a Hamiltonian system and, if so, how to compute the Hamiltonian. However, for the moment we use the rabbit-out-of-the-hat method and claim that the differential equations for the ideal pendulum form a Hamiltonian system, with the Hamiltonian function given by

$$H(\theta, v) = \frac{1}{2}v^2 - g\cos\theta.$$

We check by computing the partial derivatives of H, that is,

$$\frac{\partial H}{\partial \theta} = g\sin\theta = -\frac{dv}{dt} \quad \text{and} \quad \frac{\partial H}{\partial v} = v = \frac{d\theta}{dt}.$$

We can describe the graph of the function H by noting that, for each fixed θ, the $v^2/2$ term implies that the graph of H is a parabola in the v-direction. The critical points occur where $v = 0$ and $g\sin\theta = 0$. In other words, $(\theta, v) = (0, 0)$, $(\pm\pi, 0)$, $(\pm 2\pi, 0), \ldots$. The critical points where θ is an even multiple of π are local minima, and those where θ is an odd multiple of π are saddle points of the graph.

The level curves of H are plotted in Figure 5.29. We know that the vector field is always tangent to these curves. The θ component of the vector field equals v, so the vector field points to the right when $v > 0$ and to the left when $v < 0$. Using this fact, we can assign directions along these curves and thereby determine the phase portrait. This is shown in Figure 5.30.

There are three different types of solution curves present in the phase portrait. These three types are shown schematically in Figure 5.31 and labeled A, B, and C. Around the equilibrium points at $(\theta, v) = (0, 0)$, $(\pm 2\pi, 0)$, $(\pm 4\pi, 0), \ldots$, we find periodic solutions traversed in the clockwise direction. These are the type A solutions in Figure 5.31. Since the value of θ along these solution curves never reaches $\pi, 3\pi, \ldots$, the pendulum never passes through the upright position. Consequently, the pendulum simply oscillates back and forth periodically, with the maximum and minimum values of θ determined by where the solution curve crosses the θ-axis. This is the usual swinging motion we associate with a pendulum. The graph of $\theta(t)$ for such a solution is shown in Figure 5.32.

On the other hand, a solution curve such as B in Figure 5.31 corresponds to the pendulum rotating forever in a counterclockwise direction. Note that $v \neq 0$ along such

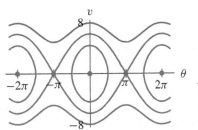

Figure 5.29
Level curves for the ideal pendulum.

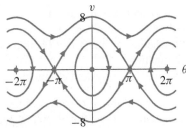

Figure 5.30
Phase portrait for the ideal pendulum.

Figure 5.31
Special solution curves.

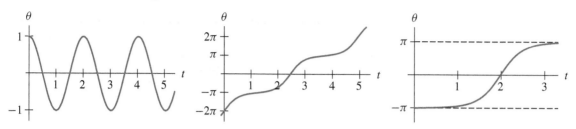

Figure 5.32
Graphs of $\theta(t)$ for the three different types of solutions represented in Figure 5.31.

a solution curve, so the pendulum never reaches a point where its velocity is 0. The graph of $\theta(t)$ for such a solution is increasing for all t if $v > 0$ and decreasing for all t if $v < 0$ (see Figure 5.32).

The intermediate types of solutions (see C in Figure 5.31) are separatrices of saddle equilibrium points forming saddle connections. These solutions tend toward and come from the upright equilibrium position as time tends to $\pm\infty$. The $\theta(t)$-graph of such a solution is shown in Figure 5.32. In order to be on such a solution, we must choose the initial angle and velocity perfectly. With slightly too high an initial velocity, the pendulum swings past the vertical position; with slightly too low an initial velocity, the pendulum falls back. The separatrices of the saddle equilibrium points separate the "oscillating" solutions, which oscillate back and forth, from the "rotating" solutions, which swing around and around.

The value of the Hamiltonian function for the ideal pendulum at a particular point (y, v) is called the **energy**. The physical principle of conservation of energy applies to the ideal pendulum in the same way that it does to the undamped harmonic oscillator. This is one way we could predict that the undamped harmonic oscillator and the ideal pendulum are Hamiltonian systems. A more mathematical approach is given next.

Finding Hamiltonian Systems

Hamiltonian systems are special kinds of systems of differential equations in several senses. As we have seen, we can "solve" a Hamiltonian system in the plane in a qualitative sense once we know the Hamiltonian function. All we need to do is plot the level curves of H and sketch in the directions to determine the phase portrait. Unfortunately, Hamiltonian systems are fairly rare. Given a system, we would like to be able to determine whether or not it is a Hamiltonian system and, if it is, to determine the Hamiltonian function.

Suppose we have a system of equations

$$\frac{dx}{dt} = f(x, y)$$
$$\frac{dy}{dt} = g(x, y)$$

and wish to check whether it is Hamiltonian. We ask whether there exists a function $H(x, y)$ such that for all (x, y),

$$f(x, y) = \frac{\partial H}{\partial y}$$

and

$$g(x, y) = -\frac{\partial H}{\partial x}$$

If such a function H exists (and has continuous second partial derivatives), then

$$\frac{\partial^2 H}{\partial x \partial y} = \frac{\partial^2 H}{\partial y \partial x}.$$

Therefore if the system is Hamiltonian, then

$$\frac{\partial f}{\partial x} = \frac{\partial^2 H}{\partial x \partial y} = \frac{\partial^2 H}{\partial y \partial x} = -\frac{\partial g}{\partial y}.$$

That is, to check whether a system may be Hamiltonian, we compute $\partial f/\partial x$ and $\partial g/\partial y$ and check if $\partial f/\partial x = -\partial g/\partial y$. If this equation does not hold for all (x, y), then the system is not Hamiltonian.

For example, consider the nonlinear system

$$\frac{dx}{dt} = f(x, y) = x + y^2$$

$$\frac{dy}{dt} = g(x, y) = y^2 - x.$$

We compute

$$\frac{\partial f}{\partial x} = 1 \neq -2y = -\frac{\partial g}{\partial y}.$$

Therefore this system is not a Hamiltonian system.

Constructing Hamiltonian functions
For the system

$$\frac{dx}{dt} = f(x, y)$$

$$\frac{dy}{dt} = g(x, y),$$

if

$$\frac{\partial f}{\partial x} = -\frac{\partial g}{\partial y},$$

then the system is Hamiltonian. We can verify this by actually constructing the Hamiltonian function as follows: If we are to have

$$f(x, y) = \frac{\partial H}{\partial y},$$

then integrating both sides of this equation with respect to y, we must have

$$H(x, y) = \int f(x, y)\, dy$$

up to a "constant of integration" that may depend on x. That is, we can write

$$H(x, y) = \int f(x, y)\, dy + \phi(x),$$

where ϕ is some function of x alone that is to be determined. To find ϕ, we simply differentiate $H(x, y)$ with respect to x and equate the result with $-g(x, y)$. We find

$$\frac{\partial H}{\partial x} = \frac{\partial}{\partial x} \int f(x, y)\, dy + \phi'(x) = -g(x, y).$$

That is,

$$\phi'(x) = -g(x, y) - \frac{\partial}{\partial x} \int f(x, y)\, dy.$$

Integration of the right-hand side with respect to x then determines ϕ and consequently the Hamiltonian H.

Two examples

Recall the system of differential equations governing Glen and Paul's book-writing dilemma:

$$\frac{dx}{dt} = f(x, y) = y$$

$$\frac{dy}{dt} = g(x, y) = x - x^2.$$

Since

$$\frac{\partial f}{\partial x} = 0 = -\frac{\partial g}{\partial y},$$

we know this system is Hamiltonian. To find H we first integrate f with respect to y, finding

$$H(x, y) = \int y\, dy = \frac{1}{2}y^2 + \phi(x).$$

Next we must have

$$x - x^2 = -\frac{\partial H}{\partial x}$$

$$= -\frac{\partial}{\partial x}\left(\frac{1}{2}y^2 + \phi(x)\right)$$

$$= -\phi'(x).$$

Zhihong Xia (1962 –) received his bachelor's degree in astronomy from Nanjing University, China, and his doctorate in mathematics from Northwestern University. In his doctoral thesis, he combined his mathematical skills and astronomical background to settle a 100-year-old question in celestial mechanics. Using a variety of analytic and qualitative techniques involving Hamiltonian systems, he showed that certain solutions of the nonlinear system of differential equations known as the n-body problem could escape to infinity in finite time.

In 1993, Xia received the prestigious Blumenthal Award for the research in his doctoral thesis. He has taught at Harvard University and Georgia Institute of Technology. He is currently professor of mathematics at Northwestern University.

Integrating $\phi'(x) = -x + x^2$ with respect to the variable x, we find

$$\phi(x) = -\frac{1}{2}x^2 + \frac{1}{3}x^3,$$

so

$$H(x, y) = \frac{1}{2}y^2 - \frac{1}{2}x^2 + \frac{1}{3}x^3.$$

This is exactly the function we found earlier. Of course, we could add an arbitrary constant to H, but this would not change the shape of the level curves of H.

As another example, consider the system

$$\frac{dx}{dt} = f(x, y) = -x \sin y + 2y$$

$$\frac{dy}{dt} = g(x, y) = -\cos y.$$

This system is Hamiltonian since

$$\frac{\partial f}{\partial x} = -\sin y = -\frac{\partial g}{\partial y}$$

for all (x, y). So we first integrate f with respect to y, finding

$$H(x, y) = x \cos y + y^2 + \phi(x).$$

Then we must have

$$-\cos y = -\frac{\partial H}{\partial x}$$

$$= -\frac{\partial}{\partial x}\left(x \cos y + y^2 + \phi(x) \right)$$

$$= -\cos y - \phi'(x).$$

Therefore we can choose ϕ to be any constant, say 0, and our Hamiltonian function is

$$H(x, y) = x \cos y + y^2.$$

We study this system more closely in Exercise 3.

Equilibrium Points of Hamiltonian Systems

Hamiltonian systems have a number of special properties not shared by general systems of differential equations. For example, suppose (x_0, y_0) is our equilibrium point for the Hamiltonian system

$$\frac{dx}{dt} = \frac{\partial H}{\partial y}$$
$$\frac{dy}{dt} = -\frac{\partial H}{\partial x}.$$

The Jacobian matrix at this equilibrium point is given by

$$\begin{pmatrix} \dfrac{\partial^2 H}{\partial x \partial y} & \dfrac{\partial^2 H}{\partial y^2} \\ -\dfrac{\partial^2 H}{\partial x^2} & -\dfrac{\partial^2 H}{\partial y \partial x} \end{pmatrix},$$

where each of these partial derivatives is evaluated at (x_0, y_0). Since

$$\frac{\partial^2 H}{\partial x \partial y} = \frac{\partial^2 H}{\partial y \partial x},$$

the Jacobian matrix assumes the form

$$\begin{pmatrix} \alpha & \beta \\ \gamma & -\alpha \end{pmatrix},$$

where $\alpha = \partial^2 H / \partial x \partial y$, $\beta = \partial^2 H / \partial y^2$, and $\gamma = -\partial^2 H / \partial x^2$.

The characteristic polynomial of this matrix is

$$(\alpha - \lambda)(-\alpha - \lambda) - \beta\gamma = \lambda^2 - \alpha^2 - \beta\gamma,$$

and consequently the eigenvalues are $\lambda = \pm\sqrt{\alpha^2 + \beta\gamma}$. Thus we see that there are only three possibilities for the eigenvalues:

1. If $\alpha^2 + \beta\gamma > 0$, both eigenvalues are real and have opposite signs.

2. If $\alpha^2 + \beta\gamma < 0$, both eigenvalues are imaginary with real part equal to zero.

3. If $\alpha^2 + \beta\gamma = 0$, then 0 is the only eigenvalue.

In Case 1 we know that the equilibrium point must be a saddle because its linearized system is a saddle. Unfortunately, the linearization technique is not conclusive in Cases 2 and 3. As we saw in Section 5.1, purely imaginary eigenvalues can correspond to spiral sinks and sources as well as to centers.

However, since the solution curves of a Hamiltonian system must lie on the level sets of the Hamiltonian function, having nonzero purely imaginary eigenvalues (Case 2) does imply that the equilibrium point is a nonlinear center if the system is Hamiltonian.

Case 3 is harder to describe because there are many possibilities. Nevertheless, from the geometry of the level sets of the Hamiltonian function, we do know that a Hamiltonian system never has sinks or sources of any type. Consequently we can sometimes see that a system is not Hamiltonian from its phase portrait. For example, consider the two phase portraits shown in Figure 5.33. The phase portrait on the left does not come from a Hamiltonian system because it has an equilibrium point that is a sink. The phase portrait on the right might come from a Hamiltonian system, but the only way to determine if it is Hamiltonian is to study the formulas for the system, as we have done earlier in this section.

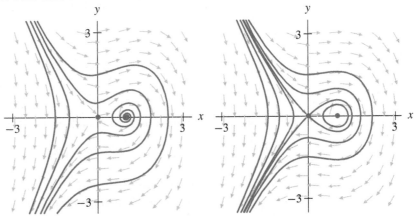

Figure 5.33
The phase portrait on the left cannot be a Hamiltonian system, whereas the phase portrait on the right might be Hamiltonian.

EXERCISES FOR SECTION 5.3

1. For the system

$$\frac{dx}{dt} = y$$

$$\frac{dy}{dt} = x^3 - x,$$

(a) show that the system is Hamiltonian with Hamiltonian function

$$H(x, y) = \frac{y^2}{2} + \frac{x^2}{2} - \frac{x^4}{4},$$

(b) sketch the level sets of H, and

(c) sketch the phase portrait for the system. Include a description of all equilibrium points and any saddle connections.

2. For the system

$$\frac{dx}{dt} = x\cos(xy)$$

$$\frac{dy}{dt} = -y\cos(xy),$$

(a) show that the system is a Hamiltonian system with Hamiltonian function

$$H(x, y) = \sin(xy),$$

(b) sketch the level sets of H, and

(c) sketch the phase portrait of the system. Include a description of all equilibrium points and any saddle connections.

3. For the system

$$\frac{dx}{dt} = -x\sin y + 2y$$

$$\frac{dy}{dt} = -\cos y,$$

(a) check that the system is a Hamiltonian system with Hamiltonian function

$$H(x, y) = x\cos y + y^2,$$

(b) sketch the level sets of H, and

(c) sketch the phase portrait of the system. Include a description of all equilibrium points and any saddle connections.

[*Hint*: In part (b), take advantage of whatever technology you have available for drawing level sets. In part (c), use `HPGSystemSolver`.]

In Exercises 4–8, we continue the study of the ideal pendulum system with bob mass m and arm length l given by

$$\frac{d\theta}{dt} = v$$

$$\frac{dv}{dt} = -\frac{g}{l}\sin\theta.$$

4.   **(a)** What is the linearization of the ideal pendulum system above at the equilibrium point $(0, 0)$?

   (b) Using $g = 9.8$ m/s^2, how should l and m be chosen so that small swings of the pendulum have period 1 second?

5. For the linearization of the ideal pendulum above at $(0, 0)$, the period of the oscillation is independent of the amplitude. Does the same statement hold for the ideal pendulum itself? Is the period of oscillation the same no matter how high the ideal pendulum swings? If not, will the period be shorter or longer for high swings?

6. An ideal pendulum clock—a clock containing an ideal pendulum that "ticks" once for each swing of the pendulum arm—keeps perfect time when the pendulum makes very high swings. Will the clock run slow or fast if the amplitude of the swings is very small?

7.   **(a)** If the arm length of the ideal pendulum is doubled from l to $2l$, what is the effect on the period of small amplitude swinging solutions?

   (b) What is the rate of change of the period of small amplitude swings as l is varied?

8. Will an ideal pendulum clock that keeps perfect time on earth run slow or fast on the moon?

9. Which of the following phase portraits could possibly be the phase portrait of a Hamiltonian system? Explain your answer.

(a)

(b)

(c)

(d)

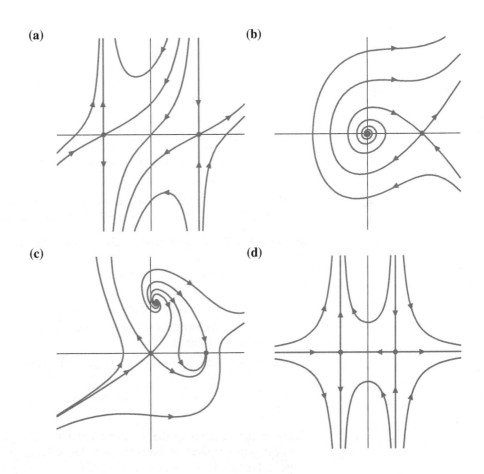

In Exercises 10–13, is the given system Hamiltonian? If so, find a Hamiltonian function. If not, why not?

10.
$$\frac{dx}{dt} = \sin x \cos y$$
$$\frac{dy}{dt} = 2x - \cos x \sin y$$

11.
$$\frac{dx}{dt} = x - 3y^2$$
$$\frac{dy}{dt} = -y$$

12.
$$\frac{dx}{dt} = 1$$
$$\frac{dy}{dt} = y$$

13.
$$\frac{dx}{dt} = x \cos y$$
$$\frac{dy}{dt} = -y \cos x$$

14. Consider a system of the form

$$\frac{dx}{dt} = F(y)$$
$$\frac{dy}{dt} = G(x).$$

That is, dx/dt depends only on y, and dy/dt depends only on x. Show that this system is Hamiltonian. What is the Hamiltonian function?

15. Consider the nonlinear system

$$\frac{dx}{dt} = 1 - y^2$$
$$\frac{dy}{dt} = x(1 + y^2).$$

(a) Is this a Hamiltonian system?

(b) Use HPGSystemSolver to plot the phase portrait of this system. Do you think that there is conserved quantity for this system? Can you find one?

(c) Suppose we multiply the vector field by the function $f(x, y) = 1/(1 + y^2)$. The new system is

$$\frac{dx}{dt} = \frac{1 - y^2}{1 + y^2}$$
$$\frac{dy}{dt} = x.$$

Is this system Hamiltonian?

(d) Can you now find a conserved quantity for the original system?

(e) Why does this method work?

16. Consider the nonlinear system

$$\frac{dx}{dt} = -yx^2$$
$$\frac{dy}{dt} = x + 1.$$

(a) Is this a Hamiltonian system?

(b) Can you find a conserved quantity? [*Hint*: Use the technique of Exercise 15 and multiply this system by a function that yields a new system in the form of Exercise 14.]

17. Consider the system of differential equations

$$\frac{dx}{dt} = 1 - y^2$$
$$\frac{dy}{dt} = x(2 - y).$$

Find a conserved quantity for this system.

In Exercises 18–20, we study bifurcations of parameterized families of Hamiltonian systems. Because the equilibrium points of Hamiltonian systems assume a special form, their bifurcations do also.

18. Consider the system

$$\frac{dx}{dt} = y$$
$$\frac{dy}{dt} = x^2 - a,$$

where a is a parameter.

(a) Verify that this system is a Hamiltonian system with the Hamiltonian function

$$H(x, y) = \frac{y^2}{2} - \frac{x^3}{3} + ax.$$

(b) Show that the system has equilibrium points at $(x, y) = (\pm\sqrt{a}, 0)$ if $a \geq 0$ and no equilibrium points if $a < 0$. Thus, $a = 0$ is a bifurcation value of the parameter.

(c) Linearize the system at each of the equilibrium points and determine the behavior of solutions near the equilibrium points.

(d) Sketch the level curves of H (and hence the phase portrait of the system) for $a = -1$, $a = -0.5$, $a = 0$, $a = 0.5$, and $a = 1$. [*Hint*: Think about the effect of changing a on the graph of H. It is worthwhile to employ whatever graphing technology you have available.]

(e) In a brief paragraph, describe the bifurcation that takes place at $a = 0$.

19. Consider the system of differential equations

$$\frac{dx}{dt} = x^2 + 2xy$$
$$\frac{dy}{dt} = a - 2xy - y^2,$$

which depend on the parameter a. Describe in detail the bifurcation that occurs at $a = 0$.

20. Suppose Glen and Paul decide to modify the model of their enthusiasm levels because Paul notes that his enthusiasm level also goes down at a constant rate over time. Therefore the new equations are

$$\frac{dx}{dt} = y$$
$$\frac{dy}{dt} = x - x^2 - a.$$

(a) Describe in detail any major bifurcations that occur in this family.

(b) Suppose $a = 1$. What happens to Glen and Paul's book no matter how enthusiastic both authors start out?

5.4 DISSIPATIVE SYSTEMS

The Hamiltonian systems discussed in the previous section are obviously idealized systems. In the real world, pendulums do not swing forever in a periodic motion. Rather, they eventually wind down; energy is dissipated. In this section we will discuss in detail these types of dissipative systems. We begin by modifying the ideal pendulum by considering a small damping term.

The Nonlinear Pendulum with Friction

Recall that the second-order equation governing the motion of the pendulum is

$$\frac{d^2\theta}{dt^2} + \frac{b}{m}\frac{d\theta}{dt} + \frac{g}{l}\sin\theta = 0,$$

where b is the coefficient of damping, m is the mass of the pendulum bob, g is the acceleration of gravity ($g \approx 9.8$ m/s^2), and l is the length of the pendulum arm (see page 496). All of the parameters are positive. The term

$$\frac{b}{m}\frac{d\theta}{dt}$$

results from damping due, for example, to air resistance or friction at the pivot point of the pendulum arm. We assumed that $b = 0$ in our idealized system. In this section, we assume that $b \neq 0$.

Introducing the velocity $v = d\theta/dt$, this second-order differential equation can be written as a nonlinear system in the usual manner,

$$\frac{d\theta}{dt} = v$$

$$\frac{dv}{dt} = -\frac{b}{m}v - \frac{g}{l}\sin\theta.$$

This system is no longer Hamiltonian, since our test (see page 499) for the existence of a Hamiltonian function fails because

$$\frac{\partial}{\partial\theta}(v) = 0 \quad \text{but} \quad -\frac{\partial}{\partial v}\left(-\frac{b}{m}v - \frac{g}{l}\sin\theta\right) = \frac{b}{m} \neq 0.$$

The equilibrium points and nullclines

We begin the study of the damped pendulum system

$$\frac{d\theta}{dt} = v$$

$$\frac{dv}{dt} = -\frac{b}{m}v - \frac{g}{l}\sin\theta$$

by finding the equilibrium points. As for the ideal pendulum, the equilibrium points occur at $(\theta, v) = (0, 0), (\pm\pi, 0), (\pm2\pi, 0), \ldots$.

The Jacobian matrix of the vector field at (θ, v) is

$$\begin{pmatrix} 0 & 1 \\ -\frac{g}{l}\cos\theta & -\frac{b}{m} \end{pmatrix}.$$

If $(\theta, v) = (0, 0), (\pm2\pi, 0), (\pm4\pi, 0), \ldots$ (the equilibrium points where the pendulum is hanging downward), this matrix is

$$\begin{pmatrix} 0 & 1 \\ -\frac{g}{l} & -\frac{b}{m} \end{pmatrix}.$$

The eigenvalues of this matrix are roots of the characteristic equation

$$-\lambda\left(-\frac{b}{m} - \lambda\right) + \frac{g}{l} = 0,$$

which simplifies to

$$\lambda^2 + \frac{b}{m}\lambda + \frac{g}{l} = 0.$$

Hence the eigenvalues are

$$-\frac{b}{2m} \pm \sqrt{\left(\frac{b}{2m}\right)^2 - \frac{g}{l}}.$$

There are three different cases, depending on the sign of $(b/(2m))^2 - g/l$. If this quantity is negative, then the eigenvalues of the Jacobian matrix are complex. The real part is $-b/(2m)$, which is negative, so in this case the equilibrium is a spiral sink. If $(b/(2m))^2 - g/l > 0$, we find two real distinct eigenvalues. Since

$$0 < \left(\frac{b}{2m}\right)^2 - \frac{g}{l} < \left(\frac{b}{2m}\right)^2,$$

we have that

$$\frac{b}{2m} > \sqrt{\left(\frac{b}{2m}\right)^2 - \frac{g}{l}},$$

and both eigenvalues are negative. Finally, if $(b/(2m))^2 - g/l = 0$, we have a repeated eigenvalue $-b/(2m)$, which is again negative.

Thus in each case we find that the equilibrium points at $(\theta, v) = (0, 0)$, $(\pm 2\pi, 0)$, $(\pm 4\pi, 0)$, ... are sinks. If we focus attention on the case of a pendulum with only a small amount of friction, then b is very small. For b sufficiently small, we have that $(b/(2m))^2 - g/l$ is negative and the points $(\theta, v) = (0, 0)$, $(\pm 2\pi, 0)$, $(\pm 4\pi, 0)$, ... are spiral sinks. We deal only with this case here (see the exercises for the other cases).

At the other equilibrium points $(\theta, v) = (\pm \pi, 0)$, $(\pm 3\pi, 0)$, ..., which correspond to the pendulum standing upright, the Jacobian matrix is

$$\begin{pmatrix} 0 & 1 \\ \frac{g}{l} & -\frac{b}{m} \end{pmatrix},$$

and this is the coefficient matrix for the linearized system at these points. The eigenvalues are

$$-\frac{b}{2m} \pm \sqrt{\left(\frac{b}{2m}\right)^2 + \frac{g}{l}}.$$

These eigenvalues are always real because b, m, l, and g are all positive. The eigenvalue

$$-\frac{b}{2m} + \sqrt{\left(\frac{b}{2m}\right)^2 + \frac{g}{l}}$$

is positive, whereas the other eigenvalue,

$$-\frac{b}{2m} - \sqrt{\left(\frac{b}{2m}\right)^2 + \frac{g}{l}},$$

is negative. Hence the equilibrium points $(\theta, v) = (\pm \pi, 0)$, $(\pm 3\pi, 0)$, ... are saddles.

To find the θ- and v-nullclines, note that $d\theta/dt = 0$ if $v = 0$, so the vector field points up or down along the θ-axis. The v-nullcline is where $dv/dt = 0$, which occurs on the curve

$$v = -\frac{mg}{bl}\sin\theta;$$

so the vector field points in a horizontal direction along this curve. We can determine the direction of the vector field in a region between two nullclines by checking the direction at a single point (see Figure 5.34).

Coupled with our analysis at the equilibrium points, we can now describe the behavior of solution curves near the equilibrium points as shown in Figure 5.35. Note that the solution curves circulate about the equilibrium points at $(\theta, v) = (0, 0)$, $(\pm 2\pi, 0)$, $(\pm 4\pi, 0)$, ... in a clockwise direction. We can verify that the stable and unstable separatrices at $(\theta, v) = (\pm\pi, 0)$, $(\pm 3\pi, 0)$, ... point in the directions indicated in Figure 5.35 by finding the linearization at these points or by using the direction field (see Figure 5.34).

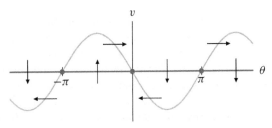

Figure 5.34
Nullclines for the system

$$\frac{d\theta}{dt} = v$$
$$\frac{dv}{dt} = -\frac{b}{m}v - \frac{g}{l}\sin\theta.$$

Figure 5.35
Phase portrait near the equilibrium points for the system

$$\frac{d\theta}{dt} = v$$
$$\frac{dv}{dt} = -\frac{b}{m}v - \frac{g}{l}\sin\theta.$$

The Effects of Dissipation

At this point Figure 5.35 contains all the information about the phase portrait of the pendulum with small damping available by the qualitative techniques that we have at our disposal. We do not have the benefit of a conserved quantity as we did in the frictionless case, so we cannot determine the entire phase portrait. However, using the Hamiltonian function

$$H(\theta, v) = \frac{1}{2}v^2 - \frac{g}{l}\cos\theta$$

that we developed in the last section for the ideal pendulum, we can complete the picture. If $(\theta(t), v(t))$ is a solution of the ideal pendulum system (with $b = 0$), then

$$\frac{d}{dt} H(\theta(t), v(t)) = 0.$$

If $b \neq 0$, this is no longer true. However, if $(\theta(t), v(t))$ is a solution of the pendulum system with $b \neq 0$, we have

$$\frac{dH}{dt} = \frac{\partial H}{\partial \theta} \frac{d\theta}{dt} + \frac{\partial H}{\partial v} \frac{dv}{dt}$$

$$= \left(\frac{g}{l} \sin\theta\right) v + v \left(-\frac{b}{m} v - \frac{g}{l} \sin\theta\right)$$

$$= -\frac{b}{m} v^2$$

$$\leq 0.$$

That is, the function H is no longer constant along solution curves of the system, but rather $H(\theta(t), v(t))$ is decreasing whenever $v(t) \neq 0$. Hence solution curves in the θv-phase plane cross level sets of H moving from larger to smaller H values.

This provides us with a strategy for determining the phase portrait. First, we draw the level curves of H as we did before (see Figure 5.36). If the value of b/m is small and the velocity v is small, the value of H decreases slowly along solutions. We can sketch the phase portrait by sketching curves that flow in the same general direction as the solution curves of the ideal pendulum but that move from level curves of higher H value toward lower H value (see Figure 5.37).

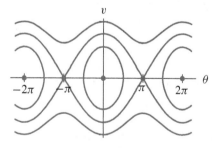

Figure 5.36
Level curves of

$$H = \frac{1}{2} v^2 - \frac{g}{l} \cos\theta.$$

Figure 5.37
Phase portrait for the system

$$\frac{d\theta}{dt} = v$$

$$\frac{dv}{dt} = \frac{b}{m} v - \frac{g}{l} \sin\theta.$$

The fact that H is decreasing along solution curves lets us determine the fate of stable and unstable separatrices emanating from saddle points: The unstable separatrices must fall into adjacent sinks, whereas the stable separatrices come from "infinity" (see Figure 5.37). Also, the equilibrium point analysis we performed earlier agrees with what we have just done. The whole picture begins to fit together in a neat package, with the local analysis near equilibrium points fully complementing the global structure provided by all the level curves of H. Finally, if we graph $\theta(t)$ for a typical solution curve, we see that the pendulum eventually tends to oscillate about the downward-pointing rest position, as it should (see Figure 5.38).

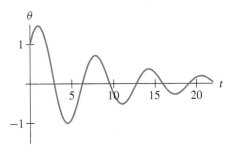

Figure 5.38
A $\theta(t)$-graph for a typical solution to the system

$$\frac{d\theta}{dt} = v$$

$$\frac{dv}{dt} = \frac{b}{m}v - \frac{g}{l}\sin\theta.$$

Lyapunov Functions

The function H above plays a very different role for the damped pendulum than it does for the ideal pendulum. Solutions of the damped pendulum system move across level sets of H from higher to lower values. The function H is called a **Lyapunov function**, after the Russian mathematician Aleksandr Mikhailovich Lyapunov (1857–1918). This idea can be generalized as:

DEFINITION   A function $L(x, y)$ is called a **Lyapunov function** for a system of differential equations if, for every solution $(x(t), y(t))$ that is not an equilibrium solution of the system,

$$\frac{d}{dt}L(x(t), y(t)) \leq 0$$

for all t with strict inequality except for a discrete set of t's.   ■

So the value of a Lyapunov function never increases and usually decreases along a nonequilibrium solution. A Lyapunov function can be a great help in drawing the phase portrait for a system. Solution curves must cross level sets of the Lyapunov function from higher to lower values.

The Damped Harmonic Oscillator

Consider the damped harmonic oscillator system

$$\frac{dy}{dt} = v$$

$$\frac{dv}{dt} = -qy - pv,$$

where q and p are positive constants. If $p = 0$, we saw in Section 5.3 (page 494) that the system is Hamiltonian, with Hamiltonian function

$$H(y, v) = \frac{1}{2}v^2 + \frac{q}{2}y^2.$$

If $p > 0$ and $(y(t), v(t))$ is a solution of the system, then we can compute the rate of change of H along the solution by

$$\frac{d}{dt}H(y(t), v(t)) = \frac{\partial H}{\partial y}\frac{dy}{dt} + \frac{\partial H}{\partial v}\frac{dv}{dt}$$

$$= qy \cdot v + v(-qy - pv)$$

$$= -pv^2$$

$$\leq 0.$$

Hence $H(y(t), v(t))$ decreases at a nonzero rate (except when $v = 0$). This implies that H is a Lyapunov function for the damped harmonic oscillator system. The level sets of H are ellipses in the yv-plane, so solutions of the damped harmonic oscillator system (with $p > 0$) cross these level sets from larger to smaller values of H. Since H has one global minimum at the origin and no other critical points, all solutions must tend toward the origin as time increases (see Figures 5.39 and 5.40). This agrees with our analysis of the damped harmonic oscillator from the point of view of linear systems.

As we mentioned before (see page 495), the value of H at a point (y, v) is called the energy of the harmonic oscillator in position y with velocity v. Saying that H decreases along solutions of the damped harmonic oscillator system is a precise way of saying that the damping dissipates energy. The same holds for the pendulum with friction.

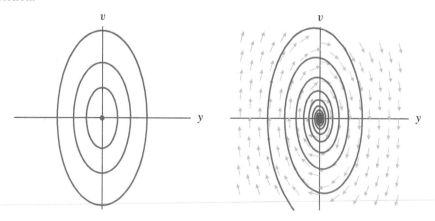

Figure 5.39
Level sets for $H(y, v) = \frac{1}{2}(v^2 + qy^2)$. With the exception of the point at the origin, each level set is an ellipse.

Figure 5.40
Phase portrait for the damped harmonic oscillator with p small.

The Tuned-Mass Damper

So far, all of our examples of Hamiltonian systems have been two-dimensional. We are able to draw accurate pictures of level sets of the Hamiltonian energy function and understand the relationship between the energy and the solutions of the associated system. When we study higher-dimensional systems, we can also use this relationship, even though the pictures are too difficult to draw.

In Section 2.5, we studied the dynamics of a swaying building. For small oscillations, the back-and-forth motion of the top of the building can be modeled by a harmonic oscillator where the building itself provides the restoring force. However, when the amplitude of the oscillations becomes large, gravity can play a role. Once past some critical angle, the building is not able to return to its upright position.

For obvious reasons, the architects who design tall buildings want to minimize any oscillations of the top of the building. On the other hand, using less and lighter materials reduces the cost of construction. In recent years, engineers have developed a new method of limiting the sway in tall, light, flexible buildings. The idea is to put a large mass at the top of the building. This mass is attached to the building by springs so that it can slide back and forth. The mass is also attached to a damper that absorbs energy and converts it to heat. When the building starts to sway, the mass at the top also starts to sway, and the damping of the motion of the mass removes energy from the entire system. The sizes of the mass and the springs are carefully chosen (or tuned) to minimize the swaying of the building. Consequently this system is called a "tuned-mass damper." To understand the behavior of this system and to see how the mass and spring strengths are chosen, we consider a simple model.

We think of the building as an oscillator. The horizontal sway of the building is modeled by the horizontal sliding of a mass-spring system where one end of the spring is attached to a fixed wall. The tuned-mass damper is modeled by another mass, attached only to the first mass via a spring and a dashpot (see Figure 5.41).

Let x_1 and x_2 represent the horizontal positions of the springs measured from the wall, m_1 and m_2 be the masses, and k_1 and k_2 be the spring constants. We first model the system without the dashpot, so there is no damping.

Let L_1 and L_2 denote the rest lengths of the two springs. Then the first spring exerts a force of $-k_1(x_1 - L_1)$. Since the second spring provides a force that pushes the two masses in opposite directions, its force is $\pm k_2(x_2 - x_1 - L_2)$. Using $F = ma$, we have the two equations

$$m_1 \frac{d^2 x_1}{dt^2} = -k_1(x_1 - L_1) + k_2(x_2 - x_1 - L_2)$$

$$m_2 \frac{d^2 x_2}{dt^2} = -k_2(x_2 - x_1 - L_2).$$

Figure 5.41
A mass-spring system with a tuned-mass damper.

Note that the only force acting on the second mass is the force due to the second spring, since this is the only spring to which it is attached.

Since both equations are second order, the corresponding first-order system is four-dimensional. To get a complete picture of a solution, we would need a curve in four-dimensional space or four graphs—one for each variable. Luckily, this system is Hamiltonian, and the existence of the Hamiltonian gives us a way to begin understanding the behavior of solutions.

Let

$$p_1 = m_1 \frac{dx_1}{dt} \quad \text{and} \quad p_2 = m_2 \frac{dx_2}{dt}$$

be the momenta of the two masses (that is, the product of the mass and the velocity). Then we can write

$$\frac{dx_1}{dt} = \frac{p_1}{m_1}$$

$$\frac{dx_2}{dt} = \frac{p_2}{m_2}$$

$$\frac{dp_1}{dt} = -k_1(x_1 - L_1) + k_2(x_2 - x_1 - L_2)$$

$$\frac{dp_2}{dt} = -k_2(x_2 - x_1 - L_2).$$

We can see that this system is Hamiltonian with Hamiltonian function

$$H(x_1, x_2, p_1, p_2) = \frac{p_1^2}{2m_1} + \frac{p_2^2}{2m_2} + \frac{k_1}{2}(x_1 - L_1)^2 + \frac{k_2}{2}(x_2 - x_1 - L_2)^2.$$

(This function comes from the kinetic and potential energy of both masses and springs.) We can check directly that

$$\frac{dx_1}{dt} = \frac{\partial H}{\partial p_1}$$

$$\frac{dx_2}{dt} = \frac{\partial H}{\partial p_2}$$

$$\frac{dp_1}{dt} = -\frac{\partial H}{\partial x_1}$$

$$\frac{dp_2}{dt} = -\frac{\partial H}{\partial x_2}.$$

Also, if $(x_1(t), x_2(t), p_1(t), p_2(t))$ is a solution, we can check that the value of Hamiltonian evaluated along the solution, $H(x_1(t), x_2(t), p_1(t), p_2(t))$, is constant as t varies (see Exercise 23).

Even with this information, it is not easy to see how solutions behave. However, if any of the variables $x_1, x_2, p_1,$ or p_2 is large, then H is large. Hence, once we know the initial condition $(x_1(0), x_2(0), p_1(0), p_2(0))$ and, consequently, its corresponding energy $H(x_1(0), x_2(0), p_1(0), p_2(0))$, then we know that all four of the variables remain bounded for all time for this solution. While we may not know which spring is oscillating the most (or "has most of the energy") at any particular moment, we do know that the maximum stretching of either spring is bounded without an external source of energy.

Now we add the effect of the dashpot to the system. Since the dashpot is attached between the two masses, it affects the system only when the distance between the two masses is changing. If we let b denote the damping coefficient, then the force provided by the damper is

$$\pm b \left(\frac{dx_2}{dt} - \frac{dx_1}{dt} \right),$$

which in terms of the momenta is

$$\pm b \left(\frac{p_2}{m_2} - \frac{p_1}{m_1} \right).$$

Again, we have the plus/minus sign, since the damping force affects the two masses in opposite directions. The new system of differential equations is

$$m_1 \frac{d^2 x_1}{dt^2} = -k_1(x_1 - L_1) + k_2(x_2 - x_1 - L_2) + b \left(\frac{dx_2}{dt} - \frac{dx_1}{dt} \right)$$

$$m_2 \frac{d^2 x_2}{dt^2} = -k_2(x_2 - x_1 - L_2) - b \left(\frac{dx_2}{dt} - \frac{dx_1}{dt} \right),$$

which is equivalent to the first-order system

$$\frac{dx_1}{dt} = \frac{\partial H}{\partial p_1}$$

$$\frac{dx_2}{dt} = \frac{\partial H}{\partial p_2}$$

$$\frac{dp_1}{dt} = -\frac{\partial H}{\partial x_1} + b \left(\frac{p_2}{m_2} - \frac{p_1}{m_1} \right)$$

$$\frac{dp_2}{dt} = -\frac{\partial H}{\partial x_2} - b \left(\frac{p_2}{m_2} - \frac{p_1}{m_1} \right).$$

This system is no longer Hamiltonian. Indeed, the whole point of adding the damper is to remove energy from the system so that H is no longer constant along solutions. If $(x_1(t), x_2(t), p_1(t), p_2(t))$ is a solution of the new system, then we compute

$$\frac{dH(x_1(t), x_2(t), p_1(t), p_2(t))}{dt} = -b \left(\frac{p_2}{m_2} - \frac{p_1}{m_1} \right)^2 = -b \left(\frac{dx_2}{dt} - \frac{dx_1}{dt} \right)^2.$$

Hence $dH/dt \leq 0$ along solutions, and $dH/dt < 0$ whenever the distance between the two masses is changing. In other words, the energy decreases whenever the second mass is moving relative to the first.

This is exactly what we want. Suppose the wind starts our building swaying back and forth. The mass, usually kept on one of the top floors, starts to move relative to the building and energy is removed from this motion by the dampers.

While we now understand how the system works, there are still many choices that need to be made when the system is designed. The parameters derived from the building, m_1 and k_1, are usually set by the architects and engineers. The parameters that need to be chosen for the damper are m_2, b, and k_2. We want the damping coefficient b to be large so that when the distance between m_1 and m_2 changes, there is a rapid loss of energy. Of course, this energy must go somewhere, so b must not be so large that the energy drained from the swaying motion, which is usually converted to heat, cannot be dissipated effectively. We must choose m_2 so that the second mass actually does oscillate with respect to the first. A small mass attached to the first mass by a strong damper will act as if it is rigidly attached and no change in distance between the two masses will occur. On the other hand, m_2 should not be too large because it actually sits in the building (usually near the top floor). Finally, we must choose k_2 both to maximize the rate at which the second mass begins to oscillate with respect to the first mass and to maximize the amplitude of the oscillations of the second mass.

In order to do a numerical simulation that demonstrates the behavior of solutions, we choose convenient numbers for the parameters. We let $m_1 = 1$ and $k_1 = 1$. We choose m_2 and b as large as possible under the constraints noted above (that is, the second mass must sit at the top of the building and we must be able to dissipate the heat generated by the damping). For convenience, let $m_2 = 0.05$ and $b = 0.1$. The last parameter to be chosen is k_2. We wish to choose it so that the maximum amount of energy is dissipated when the building sways, that is, when m_1 oscillates. If we think of the motion of the first mass as a forcing term on the mass-spring system connecting m_1 and m_2, then a reasonable choice of k_2 emerges. We choose k_2 so that the second spring is in resonance with the first. Then the oscillations of the first spring force the second at its resonant frequency, causing relatively large amplitude oscillations and hence rapid loss of energy due to damping.

To test this choice, we consider the solution of the system with m_1, k_1, m_2, and b as above for the initial condition $x_1(0) = 10$, $p_1(0) = 0$, $x_2(0) = 0$, $p_2(0) = 0$, that is, if a sudden gust of wind or a shock from an earthquake displaces the top of the building. Figure 5.42 shows the time necessary for the amplitude of the oscillations of $x_1(t)$ to reach and stay below 2.5 versus the choice of k_2. Choosing the best possible value of k_2 is called "tuning" the damper.

Perhaps the worst scenario for a tall flexible building is an earthquake that provides a forcing term with frequency equal to the natural frequency of the building. In Exercise 24, we consider how well a tuned-mass damper can handle this sort of forcing.

We have not shown the details of the solution or the precise motion of the masses. Careful study of the solutions can give insight into how it "feels" to be on top of a building with a tuned-mass damper. However, even without this detailed information, we have obtained a great deal of information about the way a tuned-mass damper works.

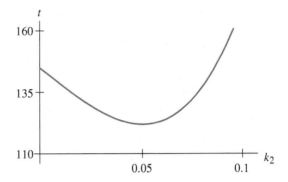

Figure 5.42
The time necessary for the amplitude of the oscillations of $x_1(t)$ to reach and stay below 2.5 as a function of k_2.

The mathematical lesson here is that even though we cannot effectively draw the solutions of a four-dimensional system, we can still use the techniques we have developed to understand certain aspects of the behavior of solutions. As systems become more complicated and have more independent variables, qualitative tools such as studying the Hamiltonian take on greater significance.

Gradient Systems

Not all systems have Lyapunov functions. Finding a Lyapunov function for a system, even if we somehow know that the system has a Lyapunov function, can be extremely challenging. However, for some modeling problems the physical system being modeled provides a motivation for a "lucky guess" of a Lyapunov function. This is the case for the damped pendulum and the damped harmonic oscillator. The Hamiltonian function of the undamped system becomes a Lyapunov function if damping is added.

In some cases the Lyapunov function is an integral part of the construction of the system. Such systems are called **gradient systems**. We begin with an example.

Navigation by smell
Many animal species use smell to navigate through their environments. A lobster, for example, can use its antennae to detect very small concentrations of chemicals in the water around it (that is, it "smells" the water). The lobster uses this ability to determine which direction a smell is coming from and hence find food in the murky deep. Although it is not clear how lobsters accomplish this, one possible method is by determining local variations in the concentration and moving in the direction in which the concentration increases fastest.*

---

*The ability of lobsters to do this "computation" in turbulent fluids is only beginning to be duplicated by technology. See Lipkin, R., "Tracking Undersea Scent," *Science News*, *147*, 5 (1994): 78, for a discussion of a "robo-lobster."

To describe how we can design a mechanical lobster that navigates by smell, we begin by assuming our lobster can only move on a flat two-dimensional plane. Let $S(x, y)$ equal the concentration at (x, y) of the chemicals making up the smell of a dead fish. We know from vector calculus that at (x, y) the direction in which S increases the fastest is given by the gradient vector

$$\nabla S(x, y) = \left(\frac{\partial S}{\partial x}, \frac{\partial S}{\partial y} \right).$$

This defines a vector field on the xy-plane. We assume that our model lobster always moves in the direction that increases the smell the fastest. Hence the velocity of our lobster's motion points in the same direction as the gradient of S. If we let $(x(t), y(t))$ denote the location of the lobster at time t, then the velocity vector is given by

$$\left(\frac{dx}{dt}, \frac{dy}{dt} \right) = \nabla S(x(t), y(t)).$$

Written in terms of scalar equations, we have

$$\frac{dx}{dt} = \frac{\partial S}{\partial x}$$
$$\frac{dy}{dt} = \frac{\partial S}{\partial y}.$$

We use this system of equations to determine our lobster's motion. If the initial position is (x_0, y_0) at time $t = 0$, then the solution curve in the phase plane satisfying this initial condition is our prediction of a lobster's path.

Two dead fish

Suppose $S(x, y)$ is defined for $-2 \le x \le 2$ and $-2 \le y \le 2$ and is given by the formula

$$S(x, y) = \frac{x^2}{2} - \frac{x^4}{4} - \frac{y^2}{2} + 8.$$

The graph of S is shown in Figure 5.43. We restrict attention to the region of the xy-plane with x and y between 2 and -2 because this function S takes on negative values when x or y is large, and negative concentrations are not physically meaningful. More realistic, but more algebraically complicated choices for S are considered in the exercises.

We can see (or compute using the techniques of vector calculus) that S has a local maximum at each of the points $(-1, 0)$ and $(1, 0)$ and a saddle point at $(0, 0)$. A function of this sort might arise if there were a smelly dead fish at each of the points $(-1, 0)$ and $(1, 0)$.

The gradient of S gives rise to a vector field and, as above, this specifies a system of differential equations

$$\frac{dx}{dt} = \frac{\partial S}{\partial x} = x - x^3$$
$$\frac{dy}{dt} = \frac{\partial S}{\partial y} = -y.$$

Figure 5.43
Graph of the function
$S(x, y) = x^2/2 - x^4/4 - y^2/2 + 8$.

By construction, the velocity vector of a solution of this system is equal to the gradient of S and hence points in the direction for which S increases the fastest. If $(x(t), y(t))$ is a solution, we compute the rate of change of S along the solution by

$$\frac{d}{dt} S(x(t), y(t)) = \frac{\partial S}{\partial x} \frac{dx}{dt} + \frac{\partial S}{\partial y} \frac{dy}{dt}$$

$$= (x - x^3)(x - x^3) + (-y)(-y)$$

$$= (x - x^3)^2 + y^2$$

$$\geq 0.$$

Hence S increases at every point where $\nabla S \neq 0$. These are precisely the points where the vector field of the system is zero—that is, the equilibrium points of the system.

To sketch the phase portrait of the system

$$\frac{dx}{dt} = \frac{\partial S}{\partial x} = x - x^3$$
$$\frac{dy}{dt} = \frac{\partial S}{\partial y} = -y,$$

we begin by sketching the level sets of S (see Figure 5.44). The direction field points in the direction of the gradient of S, which is perpendicular to the level sets. To sketch the solution curves, we draw curves always perpendicular to the level sets of S moving in the direction of increasing S (see Figure 5.45).

We can verify by linearization that the equilibrium points $(\pm 1, 0)$ are sinks, while the point $(0, 0)$ is a saddle (see the exercises). Lobsters that start with $x(0) < 0$ tend toward the equilibrium point $(-1, 0)$, whereas those with $x(0) > 0$ tend toward $(1, 0)$.

If a lobster starts precisely on the y-axis ($x(0) = 0$), then it will tend toward the saddle equilibrium point at the origin, unable to decide between the two fish.

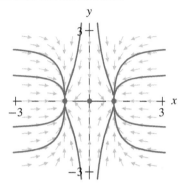

Figure 5.44
Level sets of

$$S(x, y) = \frac{x^2}{2} - \frac{x^4}{4} - \frac{y^2}{2} + 8.$$

Figure 5.45
Phase portrait for the system

$$\frac{dx}{dt} = \frac{\partial S}{\partial x} = x - x^3$$

$$\frac{dy}{dt} = \frac{\partial S}{\partial y} = -y.$$

General Form of Gradient Systems

The vector field for the system in the preceding example is the gradient of the function S; hence, it is not unreasonable to call it a gradient system.

DEFINITION   A system of differential equations is called a **gradient system** if there is a function G such that

$$\frac{dx}{dt} = \frac{\partial G}{\partial x}$$

$$\frac{dy}{dt} = \frac{\partial G}{\partial y}$$

for all (x, y).   ■

As in the example, if $(x(t), y(t))$ is a solution for a gradient system with vector field given by the gradient of $G(x, y)$, then

$$\frac{d}{dt}G(x(t), y(t)) = \frac{\partial G}{\partial x}\frac{dx}{dt} + \frac{\partial G}{\partial y}\frac{dy}{dt}$$

$$= \left(\frac{\partial G}{\partial x}\right)^2 + \left(\frac{\partial G}{\partial y}\right)^2$$

$$\geq 0.$$

That is, the value of G increases along every orbit except at critical points of G. The function $-G$ decreases along orbits and hence is a Lyapunov function for this system.

Properties of Gradient Systems

Just as in the case of Hamiltonian systems, gradient systems enjoy several special properties that make their phase portraits relatively simple. For one thing, gradient systems cannot possess periodic solutions. Since the gradient function must increase along all nonequilibrium solutions, a solution curve can never return to its starting place.

There are also restrictions on the types of equilibrium points that occur in a gradient system. Suppose

$$\frac{dx}{dt} = \frac{\partial G}{\partial x}$$
$$\frac{dy}{dt} = \frac{\partial G}{\partial y}$$

is a gradient system. Then the Jacobian matrix associated with this vector field is

$$\begin{pmatrix} \dfrac{\partial^2 G}{\partial x^2} & \dfrac{\partial^2 G}{\partial y \partial x} \\[2mm] \dfrac{\partial^2 G}{\partial x \partial y} & \dfrac{\partial^2 G}{\partial y^2} \end{pmatrix}.$$

Because the mixed partial derivatives of G are equal, this matrix assumes the form

$$\begin{pmatrix} \alpha & \beta \\ \beta & \gamma \end{pmatrix},$$

where $\alpha = \partial^2 G / \partial x^2$, $\beta = \partial^2 G / \partial y \partial x$, and $\gamma = \partial^2 G / \partial y^2$. The eigenvalues are roots of the equation

$$\lambda^2 - (\alpha + \gamma)\lambda + \alpha\gamma - \beta^2 = 0,$$

which are

$$\frac{\alpha + \gamma \pm \sqrt{(\alpha + \gamma)^2 - 4\alpha\gamma + 4\beta^2}}{2}.$$

Simplifying, we see that the eigenvalues equal

$$\frac{1}{2}(\alpha + \gamma) \pm \frac{1}{2}\sqrt{(\alpha - \gamma)^2 + 4\beta^2}.$$

Since the term $(\alpha - \gamma)^2 + 4\beta^2 \geq 0$, it follows that these eigenvalues never have an imaginary part; that is, the eigenvalues are always real. Gradient systems do not have spiral sinks, spiral sources, or centers.

It is important to note that not all systems of differential equations that possess Lyapunov functions are gradient systems. For example, the linear system

$$\frac{dx}{dt} = -x + y$$
$$\frac{dy}{dt} = -x - y$$

has a Lyapunov function given by $L(x, y) = x^2 + y^2$. To verify this assertion, we compute the derivative of L along a solution $(x(t), y(t))$ by

$$\frac{d}{dt}L(x(t), y(t)) = 2x\frac{dx}{dt} + 2y\frac{dy}{dt}$$

$$= 2x(-x + y) + 2y(-x - y)$$

$$= -2(x^2 + y^2)$$

$$\leq 0.$$

Since $dL/dt = 0$ only at the origin, this function decreases along all nonequilibrium solution curves. But L cannot be a gradient system, since the eigenvalues at $(0,0)$ are complex $(-1 \pm i)$; so the origin is a spiral sink, which cannot occur in gradient systems.

EXERCISES FOR SECTION 5.4

1. Consider the system

$$\frac{dx}{dt} = -x^3$$

$$\frac{dy}{dt} = -y^3.$$

(a) Verify that $L(x, y) = (x^2 + y^2)/2$ is a Lyapunov function for the system.

(b) Sketch the level sets of L.

(c) What can you conclude about the phase portrait of the system from the information given in parts (a) and (b) above? (Sketch the phase portrait and write a short essay describing what you know about the phase portrait and how you know it.)

2. Consider the system

$$\frac{dx}{dt} = y$$

$$\frac{dy}{dt} = -x - \frac{y}{4} + x^2.$$

(a) Verify that the function

$$L(x, y) = \frac{y^2}{2} + \frac{x^2}{2} - \frac{x^3}{3}$$

is a Lyapunov function for the system.

(b) Sketch the level sets of L.

(c) What can you conclude about the phase portrait of the system from the information in parts (a) and (b)? (Sketch the phase portrait and write a short essay describing what you know about the phase portrait and how you know it.)

3. Consider the system

$$\frac{dx}{dt} = y$$

$$\frac{dy}{dt} = -4x - 0.1y.$$

(a) Verify that all solutions tend toward the origin as t increases, and sketch the phase portrait. [*Hint*: The system is linear.]

(b) Verify that $L(x, y) = x^2 + y^2$ is *not* a Lyapunov function for the system.

(c) Verify that $K(x, y) = 2x^2 + y^2/2$ is a Lyapunov function for the system.

In Exercises 4–11, we consider the damped pendulum system

$$\frac{d\theta}{dt} = v$$

$$\frac{dv}{dt} = -\frac{g}{l}\sin\theta - \frac{b}{m}v,$$

where b is the damping coefficient, m is the mass of the bob, l is the length of the arm, and g is the acceleration of gravity ($g \approx 9.8$ m/s^2).

4. What relationship must hold between the parameters b, m, and l for the period of a small swing back and forth of the damped pendulum to be one second?

5. Suppose we have a pendulum clock that uses a slightly damped pendulum to keep time (that is, b is positive but $b \approx 0$). The clock "ticks" each time the pendulum arm crosses $\theta = 0$. If the mass of the pendulum bob is increased, does the clock run fast or slow?

6. Suppose we have a pendulum clock that uses only a slightly damped pendulum to keep time. The clock "ticks" each time the pendulum arm crosses $\theta = 0$.

(a) As the clock "winds down" (so the amplitude of the swings decreases), does the clock run slower or faster?

(b) If the initial push of the pendulum is large so that the pendulum swings very close to the vertical, will the clock run too fast or too slow?

7. For fixed values of b and l, for what values of the mass m will the pendulum be usable as a clock?

8. Suppose we take $l = 9.8$ m (so $g/l = 1$), $m = 1$, and b large, say $b = 4$. For the damped pendulum system above and with this choice of parameter values, do the following.

(a) Find the eigenvalues and eigenvectors of the linearized system at the equilibrium point $(0, 0)$.

(b) Find the eigenvalues and eigenvectors of the linearized system at the equilibrium point $(\pi, 0)$.

(c) Sketch the phase portrait near the equilibrium points.

(d) Sketch the entire phase portrait. [*Hint*: Begin by sketching the level sets of H as in the text.]

9. Suppose we have a pendulum clock that uses a slightly damped oscillator to keep time. Suppose that the clock "ticks" each time the pendulum arm crosses $\theta = 0$, but the arm must reach a height of $\theta = \pm 0.1$ to record the swing (that is, if the entire swing takes place with $-0.1 < \theta < 0.1$, then the clock doesn't tick). Suppose one tick is one second. In terms of the parameters b, m, and l, give a rough estimate of how long the clock can keep accurate time. Comment on why pendulum clocks must be wound.

10. **(a)** For the slightly damped pendulum ($b > 0$ but b close to zero), find the set of all initial conditions $(\theta(0), v(0))$ that execute exactly two complete revolutions for $t > 0$ (that is, pass the vertical position exactly twice) before settling into a back-and-forth swinging motion. Sketch the phase portrait for the slightly damped pendulum and shade these initial conditions.

   (b) Repeat part (a) for solutions that execute exactly five complete revolutions for $t > 0$ before settling into back-and-forth swinging motion.

11. Suppose that rather than adding damping to the ideal pendulum, we add a small amount of "antidamping"; that is, we take b slightly negative in the damped pendulum system. Physically this would mean that whenever the velocity is nonzero, the pendulum is accelerated in the direction of motion.

   (a) Linearize and classify the equilibrium points in this situation.

   (b) Sketch the phase portrait for this system.

   (c) Describe in a brief paragraph the behavior of a solution with initial condition near $\theta = v = 0$.

12. Let $G(x, y) = x^3 - 3xy^2$.

   (a) What is the gradient system with vector field given by the gradient of G?

   (b) Sketch the graph of G and the level sets of G.

   (c) Sketch the phase portrait of the gradient system in part (a).

13. Let $G(x, y) = x^2 - y^2$.

   (a) What is the gradient system with vector field given by the gradient of G?

   (b) Classify the equilibrium point at the origin. [*Hint*: This system is linear.]

   (c) Sketch the graph of G and the level sets of G.

   (d) Sketch the phase portrait of the gradient system in part (a).

   Remark: This is why saddle equilibrium points are called *saddles*.

14. Let $G(x, y) = x^2 + y^2$.

   (a) What is the gradient system with vector field given by the gradient of G?

   (b) Classify the equilibrium point at the origin. [*Hint*: The system is linear.]

   (c) Sketch the graph of G and the level sets of G.

   (d) Sketch the phase portrait of the gradient system in part (a).

15. For the "two dead fish" example given by the system

$$\frac{dx}{dt} = x - x^3$$
$$\frac{dy}{dt} = -y,$$

(a) find the linearized system for the equilibrium point at the origin and verify that the origin is a saddle,

(b) find the linearized system for the equilibrium point at $(1, 0)$ and verify that this point is a sink,

(c) from the eigenvalues and eigenvectors of the system in part (b), determine from which direction the model lobster will approach the equilibrium point $(1, 0)$, and

(d) check that the linearized system at the equilibrium point $(-1, 0)$ is the same as that at $(1, 0)$.

16. The system for the "two dead fish" example

$$\frac{dx}{dt} = x - x^3$$
$$\frac{dy}{dt} = -y$$

has the special property that the equations "decouple"; that is, the equation for dx/dt depends only on x and the equation for dy/dt depends only on y.

(a) Sketch the phase lines for the dx/dt and dy/dt equations.

(b) Using these phase lines, sketch the phase portrait of the system.

17. Suppose the smell of a bunch of dead fish in the region $-2 \le x \le 2, -2 \le y \le 2$ is given by the function

$$S(x, y) = x^2 + y^2 - \frac{x^4 + y^4}{4} - 3x^2y^2 + 100.$$

(a) What is the gradient system whose vector field is the gradient of S?

(b) Using `HPGSystemSolver`, sketch the phase portrait for this system.

(c) How many dead fish are there, and where are they?

(d) Using the results from part (b), sketch the level sets of S.

(e) Why is the model not realistic for large values of x or y?

18. A reasonable model for the smell at (x, y) of a dead fish located at (x_1, y_1) is given by

$$S_1(x, y) = \frac{1}{(x - x_1)^2 + (y - y_1)^2 + 1}.$$

That is, S_1 is given by 1 over the distance to the dead fish squared plus 1.

(a) Form the function S giving the total smell from three dead fish located at $(1, 0)$, $(-1, 0)$, and $(0, 2)$.

(b) Sketch the level sets of S.

(c) Sketch the phase portrait of the gradient system

$$\frac{dx}{dt} = \frac{\partial S}{\partial x}$$
$$\frac{dy}{dt} = \frac{\partial S}{\partial y}.$$

(d) Write out explicitly the formulas for the right-hand sides of the equations in part (c).

(e) Why did we use distance squared plus 1 instead of just distance squared in the definition of S_1? Why did we use distance squared plus 1 instead of just distance plus 1 in the definition of S_1?

19. Suppose

$$\frac{dx}{dt} = f(x, y)$$
$$\frac{dy}{dt} = g(x, y)$$

is a gradient system. That is, there exists a function $G(x, y)$ such that $f = \partial G/\partial x$ and $g = \partial G/\partial y$.

(a) Verify that if f and g have continuous partial derivatives, then $\partial f/\partial y = \partial g/\partial x$ for all (x, y).

(b) Use part (a) to show that the system

$$\frac{dx}{dt} = x^2 + 3xy$$
$$\frac{dy}{dt} = 2x + y^3$$

is *not* a gradient system.

20. The following phase portrait cannot occur for a gradient system. Why not?

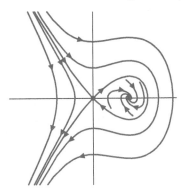

21. Let

$$H(y, v) = \frac{1}{2}v^2 + V(y)$$

for some function V and consider the associated Hamiltonian system

$$\frac{dy}{dt} = \frac{\partial H}{\partial v} = v$$

$$\frac{dv}{dt} = -\frac{\partial H}{\partial y} = -\frac{dV}{dy}.$$

Let k be a positive constant. Give a physical interpretation of the relationship between the Hamiltonian system and the system

$$\frac{dy}{dt} = v$$

$$\frac{dv}{dt} = -\frac{dV}{dy} - kv.$$

Show that H is a Lyapunov function for this system.

22. Consider the Hamiltonian system

$$\frac{dx}{dt} = \frac{\partial H}{\partial y}$$

$$\frac{dy}{dt} = -\frac{\partial H}{\partial x}$$

and the gradient system

$$\frac{dx}{dt} = \frac{\partial H}{\partial x}$$

$$\frac{dy}{dt} = \frac{\partial H}{\partial y},$$

where H is the same function for both systems. How are the two phase portraits related?

23. Check that

$$H(x_1, x_2, p_1, p_2) = \frac{p_1^2}{2m_1} + \frac{p_2^2}{2m_2} + \frac{k_1}{2}(x_1 - L_1)^2 + \frac{k_2}{2}(x_2 - x_1 - L_2)^2.$$

is constant along solutions of

$$\frac{dx_1}{dt} = \frac{p_1}{m_1}$$

$$\frac{dx_2}{dt} = \frac{p_2}{m_2}$$

$$\frac{dp_1}{dt} = -k_1(x_1 - L_1) + k_2(x_2 - x_1 - L_2)$$

$$\frac{dp_2}{dt} = -k_2(x_2 - x_1 - L_2).$$

24. Suppose a building with a tuned-mass damper is in an earthquake. Adapting our model from the section, we have

$$\frac{dx_1}{dt} = \frac{p_1}{m_1}$$

$$\frac{dx_2}{dt} = \frac{p_2}{m_2}$$

$$\frac{dp_1}{dt} = -k_1(x_1 - L_1) + k_2(x_2 - x_1 - L_2) + b\left(\frac{p_2}{m_2} - \frac{p_1}{m_1}\right) + A\cos\omega t$$

$$\frac{dp_2}{dt} = -k_2(x_2 - x_1 - L_2) - b\left(\frac{p_2}{m_2} - \frac{p_1}{m_1}\right),$$

where $A\cos\omega t$ is the added forcing term. How does the energy H behave with respect to time for a solution of this system for various values of k_2? First, compute the energy as a function of time, then graph this function. Use the parameter values of the example in the section ($m_1 = 1$, $k_1 = 1$, $m_2 = 0.05$, $b = 0.1$) and assume $A = 1$.

5.5 NONLINEAR SYSTEMS IN THREE DIMENSIONS

We saw in Section 2.8 that solutions of differential equations with three dependent variables are curves in three-dimensional space. These curves can loop around each other in very complicated ways. In Section 3.8 we studied the behavior of linear systems with three dependent variables. The behavior of linear systems can be determined by the eigenvalues and eigenvectors. However, the list of possible behaviors is much longer than for planar systems.

In this section, we consider two examples of nonlinear systems in three dimensions. The first is a population model with three species forming a food chain. For the second example, we return to the Lorenz equations studied in Section 2.8. We can use the technique of linearization of equilibrium points and numerical approximation (Euler's method) to obtain a little bit of information about these systems. As we have emphasized in the previous sections, there are very few all-purpose tools for differential equations in three and higher dimensions. In fact, these systems are an area of active research in mathematics.

A Food Chain Model

We have studied the behavior of populations living in isolation and of pairs of species interacting in predator-prey, cooperative, and competitive systems. These systems only scratch the surface of the types of complicated interactions found in nature. One possibility is the formation of a food chain of three or more species. A single species can be both predator and prey. An example that has been studied recently involves the balsam fir tree, the moose, and the wolf. The fir trees are eaten by the moose, and the moose (particularly young and infirm individuals) are eaten by wolves. A natural question is whether or not changes in the wolf population affect the tree population. Recent studies of the tree/moose/wolf population in Isle Royale National Park* indicate that changes in the wolf population may affect the trees.

We form a model of a system with three species, each one eaten by the next. For convenience, we call these species trees, moose, and wolves in analogy to the example above. Let

$$x(t) = \text{population of trees at time } t,$$
$$y(t) = \text{population of moose at time } t, \text{ and}$$
$$z(t) = \text{population of wolves at time } t.$$

We assume that each of the populations in isolation can be modeled with a logistic equation and that the effect of interaction between species is proportional to the product of the populations. The behavior of solutions will depend on the parameters chosen for the growth rate, carrying capacity, and effect of interaction. In order to get an idea of how solutions of systems of this form behave, we begin by taking all of the parameters to be 1. (This certainly is not the case for trees, moose, and wolves, but it makes the arithmetic that follows much more pleasant.) Our model is

$$\frac{dx}{dt} = x(1-x) - xy$$
$$\frac{dy}{dt} = y(1-y) + xy - yz$$
$$\frac{dz}{dt} = z(1-z) + yz.$$

---

*See McLaren, B. E. and Peterson, R. O., "Wolves, Moose and Tree Rings on Isle Royale," *Science*, Vol. 266, 1994, p. 1555.

Note that the growth of the tree population is decreased by the presence of moose (the $-xy$ term), the growth of the moose population is increased by the trees (the $+xy$ term) but decreased by the wolves (the $-yz$ term), and the growth of the wolf population is increased by the moose (the $+yz$ term).

We can find the equilibrium points of this system by setting the right-hand sides of the equations to zero and solving for x, y, and z. The equilibrium points are $(0, 0, 0)$, $(1, 0, 0)$, $(0, 1, 0)$, $(0, 0, 1)$, $(1, 0, 1)$, and $(2/3, 1/3, 4/3)$. Of these, only the point $(2/3, 1/3, 4/3)$ has all three coordinates nonzero, so the three species can coexist in equilibrium at these populations. The Jacobian matrix for this system at (x, y, z) is

$$\begin{pmatrix} 1 - 2x - y & -x & 0 \\ y & 1 - 2y + x - z & -y \\ 0 & z & 1 - 2z + y \end{pmatrix}.$$

So at the equilibrium point $(2/3, 1/3, 4/3)$, the Jacobian is

$$\begin{pmatrix} -2/3 & -2/3 & 0 \\ 1/3 & -1/3 & -1/3 \\ 0 & 4/3 & -4/3 \end{pmatrix}.$$

The characteristic polynomial of the matrix is

$$-\left(\lambda^3 + \frac{7}{3}\lambda^2 + \frac{20}{9}\lambda + \frac{8}{9} \right),$$

and the eigenvalues are

$$\lambda_1 = -1 \quad \text{and} \quad \lambda_2, \lambda_3 = -\frac{2}{3} \pm \frac{2}{3}i.$$

Since λ_1 is negative and λ_2 and λ_3 have negative real part, this equilibrium point is a sink, and all solutions with initial conditions sufficiently close to this point will tend toward it as t increases. In fact numerical simulations show that every solution with initial conditions in the first octant (where x, y, and z are all positive) tend to this equilibrium point. Hence we expect that over the long run the populations will settle at these values and the species will coexist in equilibrium.

This model does not include many factors in the environment that can have a dramatic effect on the populations. For example, in 1981 a disease decreased the wolf population dramatically. More recently, the particularly harsh winter of 1996–97 caused a crash in the moose population.* Constructing a model that includes such factors is at least as difficult as long-range prediction of the weather. However, we see below that we can study the effect of such events by adjusting parameters in the simple model.

---

*See "Winter Devastates Island's Moose," by Les Line in the *New York Times* Science Section, April 1, 1997.

Effect of a disease within the top predator population

Suppose that a disease affects the population of wolves, the predator at the top of the food chain. Suppose this disease kills a certain small fraction of the population each year. We can adjust our model to include this element by adding the term $-\gamma z$ to the dz/dt equation, where γ is a parameter indicating the fraction of the population killed by the disease per unit time. The new model is

$$\frac{dx}{dt} = x(1-x) - xy$$

$$\frac{dy}{dt} = y(1-y) + xy - yz$$

$$\frac{dz}{dt} = z(1-z) - \gamma z + yz.$$

We can again compute the equilibrium points for this system, which will now depend on the value of γ. The only equilibrium point with all three coordinates nonzero is $((2-\gamma)/3, (1+\gamma)/3, (4-2\gamma)/3))$. Assuming that γ is small, this point will still be a sink. We can ask what effect adjusting γ has on the populations. To do this we compute the derivative with respect to γ of each of the coordinates of the equilibrium point. For $z = (4-2\gamma)/3$, the derivative with respect to γ is

$$\frac{d}{d\gamma}\left(\frac{4-2\gamma}{3}\right) = -\frac{2}{3},$$

which is negative. This is as we expect, since an increase in the effect of the disease should decrease the number of wolves. For $y = (1+\gamma)/3$, the derivative with respect to γ is

$$\frac{d}{d\gamma}\left(\frac{1+\gamma}{3}\right) = \frac{1}{3},$$

which is positive. So an increase in γ is good for the moose population, again as we would expect. Finally, for $x = (2-\gamma)/3$, the derivative with respect to γ is

$$\frac{d}{d\gamma}\left(\frac{2-\gamma}{3}\right) = -\frac{1}{3},$$

which is negative. An increase in γ decreases the equilibrium population of trees.

There are two important remarks to be made. First, although it is easy to make simple models of systems representing long food chains, it is not so easy to find such systems in nature. Most predators have a variety of prey species to choose from. When one prey species is in short supply, the predator changes its diet.

Second, care must be taken when modifying a system to account for changing situations. Although the discussion above might seem obvious from common sense, if we had chosen to modify the equations to take into account the wolf disease by reducing the growth-rate parameter of the wolf population, then we would obtain different results (see Exercise 7).

Lorenz Equations

The Lorenz equations are

$$\frac{dx}{dt} = \sigma(y - x)$$

$$\frac{dy}{dt} = \rho x - y - xz$$

$$\frac{dz}{dt} = -\beta z + xy,$$

where σ, ρ, and β are parameters. In Section 2.8 we fixed the parameter values to $\sigma = 10$, $\beta = 8/3$, and $\rho = 28$ to obtain the system studied by Lorenz:

$$\frac{dx}{dt} = 10(y - x)$$

$$\frac{dy}{dt} = 28x - y - xz$$

$$\frac{dz}{dt} = -\frac{8}{3}z + xy.$$

Numerical approximations that we studied in that section showed that solutions looped around the phase space in a very complicated way (see Figure 5.46). With the aid of the technique of linearization near the equilibrium points, we can gain a little more insight into the behavior of these orbits. (However, the full story will have to wait for Chapter 8.)

The equilibrium points are $(0, 0, 0)$, $(6\sqrt{2}, 6\sqrt{2}, 27)$, and $(-6\sqrt{2}, -6\sqrt{2}, 27)$. We will linearize around each of these points to determine the local phase-space picture.

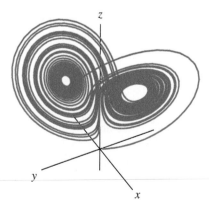

Figure 5.46
One solution curve of the Lorenz system in xyz-phase space.

The Jacobian matrix at (x, y, z) is

$$\begin{pmatrix} -10 & 10 & 0 \\ 28 - z & -1 & -x \\ y & x & -8/3 \end{pmatrix}.$$

At the origin the Jacobian is

$$\begin{pmatrix} -10 & 10 & 0 \\ 28 & -1 & 0 \\ 0 & 0 & -8/3 \end{pmatrix},$$

so the linear system that approximates the Lorenz system near the origin is

$$\frac{dx}{dt} = -10x + 10y$$

$$\frac{dy}{dt} = 28x - y$$

$$\frac{dz}{dt} = -\frac{8}{3}z.$$

This system decouples, since the equations for dx/dt and dy/dt do not depend on z, and the equation for dz/dt does not depend on x or y. We have already studied this system in Section 3.8 and found that the origin is a sink in the z direction and a saddle in the xy-plane. In the full three-dimensional picture, the origin is a saddle with a plane of initial conditions tending toward the origin as t increases and a line of initial conditions tending toward the origin as t decreases. The picture of the phase space is reproduced in Figure 5.47.

Next we consider the equilibrium point $(6\sqrt{2}, 6\sqrt{2}, 27)$. The Jacobian at this point is

$$\begin{pmatrix} -10 & 10 & 0 \\ 1 & -1 & -6\sqrt{2} \\ 6\sqrt{2} & 6\sqrt{2} & -8/3 \end{pmatrix}.$$

The eigenvalues are $\lambda_1 \approx -13.8$, $\lambda_2 \approx 0.094 + 10.2i$, and $\lambda_3 \approx 0.094 - 10.2i$, so this point is a spiral saddle that has a line of solutions that tend toward the equilibrium point as t increases and a plane of solutions that spiral toward the equilibrium point as t decreases. Solutions approach the equilibrium point quickly along the direction of the eigenvector of the negative eigenvalue, then spiral slowly away along the plane corresponding to the complex eigenvalues. By computing the eigenvectors, we can obtain the orientation of the straight line of solutions and the plane of spiraling solutions. This yields a good representation of the phase space near the equilibrium point, which we sketch in Figure 5.47.

We can compute the linearization for the equilibrium point $(-6\sqrt{2}, -6\sqrt{2}, 27)$ in the same manner, and we find that it is also a spiral saddle. These computations give us the "local" picture of the phase space near the equilibrium points (see Figure 5.47). Placing this next to a picture of a solution of the full nonlinear Lorenz system, we start to see some order in the way the solution behaves (see Figure 5.48). The solution approaches one of the equilibrium points ($\pm 6\sqrt{2}, \pm 6\sqrt{2}, 27$) along the straight line corresponding to the negative eigenvalue. When it gets close to the equilibrium point, it begins to spiral away. When the spiral is large enough, the solution becomes involved with the saddle at the origin and either returns to repeat its previous pattern or goes to the equilibrium point "on the other side."

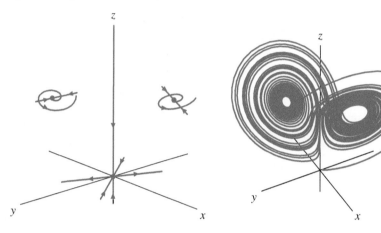

Figure 5.47
Local phase-space pictures around the three equilibrium points.

Figure 5.48
A solution curve of the nonlinear Lorenz system.

Reality Check

The study of the local behavior near the equilibrium points has given us a bit more insight into the complicated behavior of solutions of the Lorenz system. However, the local pictures do not tell the whole story. This is what makes the study of this and other three-dimensional systems so hard. Calculus and linearization techniques tell us a great deal about small parts of the phase space, but the behavior of solutions also depends on what is going on "globally." The development of the tools needed to study these systems as well as understand individual models is an active area of mathematical research.

EXERCISES FOR SECTION 5.5

Exercises 1–4 refer to the three-species food chain model

$$\frac{dx}{dt} = x(1 - x) - xy$$

$$\frac{dy}{dt} = y(1 - y) + xy - yz$$

$$\frac{dz}{dt} = z(1 - z) + yz$$

from the text. In the section we studied the linearization for the equilibrium point where all three species coexisted. The other equilibrium points are listed below. For each point,

(a) find the Jacobian at the point,

(b) find the eigenvalues and eigenvectors,

(c) classify the equilibrium point,

(d) sketch the phase space of the linearized system, and

(e) discuss in a few sentences what happens to a solution with initial value near the equilibrium point. There will be several cases, depending on which populations are nonzero. You need not consider negative populations, but consider all combinations of zero and positive populations that are near the equilibrium point.

1. $(1, 0, 0)$ **2.** $(0, 1, 0)$ **3.** $(0, 0, 1)$ **4.** $(1, 0, 1)$

5. Discuss why there is an equilibrium point at $(1, 0, 1)$ with trees and wolves coexisting at their carrying capacities, but there are not equilibrium points at $(0, 1, 1)$ and $(1, 1, 0)$.

(a) First give the "mathematical" reason by studying the differential equations.

(b) Relate this to the model to see if your conclusions are "physically reasonable."

6. Suppose that a moose disease enters the region of study and that βy moose per time unit are killed by the disease (where β is small).

(a) Modify the model to include the moose disease.

(b) What are the equilibrium points for the new model (recall β is small)?

(c) How does an increase in β affect the equilibrium populations of trees, moose, and wolves where all three species coexist?

7. In the section we modified the original tree-moose-wolf model to include the effect of a wolf disease by subtracting a term ζy from the wolf population. It is tempting to claim that the effect of a disease among the wolves changes the growth-rate parameter of the wolf population. That is, to change the dz/dt equation to

$$\frac{dz}{dt} = \zeta z(1 - z) + yz,$$

where ζ is a constant near 1. The equations for dx/dt and dy/dt remain unchanged.

(a) Find the equilibrium points of this new system [*Hint*: The equilibrium points will depend on the parameter ζ. Remember, ζ is close to 1.]

(b) What effect does a decrease in ζ have on the equilibrium population where all three species coexist?

(c) Is the change in the equilibrium point what you expected? Why or why not?

8. Besides bad weather during the winter of 1996-97, the moose population on Isle Royale was also attacked by winter wood ticks.* We can think of the ticks as another predator on moose.

   (a) Suppose the ticks eat only moose blood. Incorporate the effect of the ticks into the model. [*Hint*: Introduce a new dependent variable.]

   (b) Discuss the equilibria of your new model.

5.6   PERIODIC FORCING OF NONLINEAR SYSTEMS AND CHAOS

Thus far in this chapter we have discussed only autonomous nonlinear systems. On the other hand, in Chapter 4 we encountered many important second-order equations that were nonautonomous. In particular, periodically forced harmonic oscillator equations depended explicitly on time and exhibited quite interesting behavior. For example, in Section 4.2, we found that periodic forcing affects the amplitude of the oscillating solutions and could even cause the system to "blow up" in the case of resonant forcing. Also, in Section 4.5 we studied a particular nonlinear system with periodic forcing (the Tacoma Narrows bridge) and saw that it had surprising behavior as well. In this section we continue the study of periodically forced nonlinear systems.

As in the Lorenz system, these systems are examples of systems that often exhibit *chaotic* behavior. In this context, the word chaotic means that solutions exhibit infinitely many different qualitative behaviors. Moreover, these different solutions are packed closely together, so that any change in the initial conditions has a radical effect on the long-term behavior of the solution. For these reasons, it is impossible to find analytic solutions of a chaotic system since there are so many different types of solutions. Also, a complete qualitative analysis of solutions is impossible for the same reason. Instead, we offer a few glimpses of the startling behavior of these systems so that we can recognize "chaos" when we see it in other systems. We note that these systems are the subject of active research in mathematics. If our presentation seems incomplete, it is because the complete story has yet to be discovered.

A Periodically Forced Duffing's Equation

We begin with an example of a nonlinear system that we "completely understand," that is, we can draw the phase portrait and describe the behavior of every solution. We then add a periodic forcing term to this system and develop techniques for gaining insight for the resulting nonautonomous system.

The nonlinear system we consider is a Duffing's equation of the form

$$\frac{dy}{dt} = v$$

$$\frac{dv}{dt} = y - y^3.$$

---

*See "Winter Devastates Island's Moose," by Les Line, *New York Times* Science Section, April 1, 1997.

This is a Hamiltonian system with energy $H(y, v) = v^2/2 - y^2/2 + y^4/4$. Hence we can draw the phase portrait for this system by drawing the level curves of H (see Figure 5.49). The system has two center equilibrium points at $(\pm 1, 0)$ and a saddle equilibrium point at $(0, 0)$. The solutions that tend to the saddle as $t \to -\infty$ also tend to it as $t \to +\infty$. All other solutions are periodic, either looping around one of the centers or looping around all three equilibrium points.

We modify this system by adding a sinusoidal forcing term to obtain the system

$$\frac{dy}{dt} = v$$

$$\frac{dv}{dt} = y - y^3 + \epsilon \sin t.$$

Note that the amplitude of the forcing is given by ϵ and the period is 2π (see the exercises for the interpretation of this system as a second-order equation).

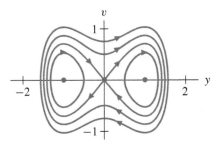

Figure 5.49
Phase portrait for the system

$$\frac{dy}{dt} = v$$

$$\frac{dv}{dt} = y - y^3.$$

The Return Map

As we pointed out in Section 4.1, the usual tools for dealing with systems—the vector field, direction field, and phase portrait—are not relevant to nonautonomous systems because the vector field changes with time. We really need a three-dimensional picture with y-, v-, and t-axes. We would then think of solutions as moving along curves in the three-dimensional yvt-space. An attempt to render such a drawing is given in Figure 5.50. This figure contains only three solutions and yet it is fairly difficult to

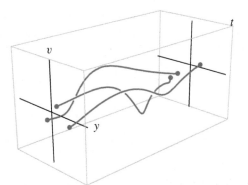

Figure 5.50
Three solutions of the system

$$\frac{dy}{dt} = v$$

$$\frac{dv}{dt} = y - y^3 + \epsilon \sin t$$

in the yvt-space.

visualize. Also, to understand the long-term behavior of solutions, we would have to extend the t-axis quite far. We need to find a better way to depict solutions.

To represent the solutions of this system, we use an idea that is due to the person who first saw and understood the ramifications of chaotic systems, the mathematician Henri Poincaré. The goal is to replace the three-dimensional picture in Figure 5.50 by a two-dimensional picture. Fix an initial point $y(0) = y_0$, $v(0) = v_0$. In Figure 5.51 this point is represented by a point in the plane $t = 0$. If we draw the plane $t = T$ for some later time T in the yvt-space, we can ask where the solution that starts at (y_0, v_0) at time $t = 0$ meets this plane. By following the solution curve until time $t = T$, we can determine where this meeting point is (see Figure 5.51).

Now we take advantage of the fact that the forcing term is periodic. Take $T = 2\pi$. The solution starting at $t = 0$ intersects the plane $t = 2\pi$ at some point $(y_1, v_1) = (y(2\pi), v(2\pi))$. One way to follow this solution further is to extend the t-axis. However, because the forcing is periodic, the differential equations are periodic with period 2π. Hence the curve that starts at (y_1, v_1) when $t = 2\pi$ is a translation by 2π time units of the curve that starts at (y_1, v_1) on the plane $t = 0$. So we can follow the solution for another 2π time units without adding to the t-axis. After 2π more time units, the solution hits the plane $t = 2\pi$ again, this time at a point (y_2, v_2), which is equal to $(y(4\pi), v(4\pi))$. We can repeat the process by moving the point (y_2, v_2) to the plane $t = 0$, and then following the solution for another 2π time units (see Figure 5.52).

This picture is still very complicated when t is large since there are many strands of the solution running from $t = 0$ to $t = 2\pi$. However, the behavior of the solution can be recovered from a simpler picture. Suppose we turn this picture so that the t-axis goes directly into the paper. Then we see only the points where the solution pierces the plane $t = 0$, that is, the points $(y_0, v_0) = (y(0), v(0))$, $(y_1, v_1) = (y(2\pi), v(2\pi))$, $(y_2, v_2) = (y(4\pi), v(4\pi))$, ... This sequence of points is enough to give a good idea of the long-term behavior of the solution.

We thus replace the picture of the solution curves in the three-dimensional phase space with a function that takes a given initial value (y_0, v_0) in the plane $t = 0$ to its "first return" point (y_1, v_1), also in the plane $t = 0$. This function is called the **return**

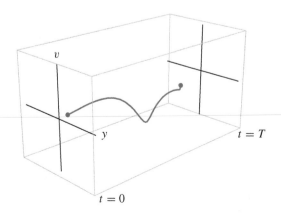

Figure 5.51
Initial point on the plane $t = 0$ and the intersection of the solution on the plane $t = T$.

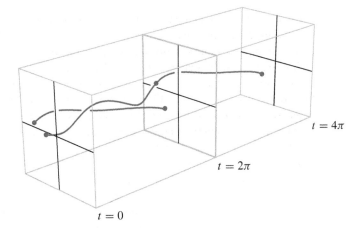

Figure 5.52
The phase space
from $0 \leq t \leq 4\pi$.
Note that the section
from 2π to 4π is just
a translation of the
section from 0 to 2π.

map or **Poincaré return map**; we think of points in the plane $t = 0$ as being taken or "mapped" to the next point of intersection with this plane by the return map. By repeatedly applying this function, we see the points where the solution pierces the plane in succession.

Return map for the unforced system

To illustrate this procedure, we first consider a simple case. Let $\epsilon = 0$, so that we have the unforced system

$$\frac{dy}{dt} = v$$

$$\frac{dv}{dt} = y - y^3.$$

This is an autonomous system, but we use the ideas above to give a picture of the return map in this special case. If we start with an initial condition near the equilibrium point at $(1, 0)$, then the solution in yvt-space spirals around this equilibrium point (see Figure 5.53). The return map for this solution picks out points on this solution curve

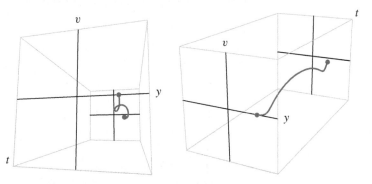

Figure 5.53
In the yvt-space, two views of a solution with initial point near $(1, 0)$ for the unforced system.

at 2π time intervals and thus yields a sequence of points on a loop around the $(1, 0)$. If we take a different initial condition farther from the $(1, 0)$, the solution spirals at a different rate around the $(1, 0)$, and the resulting return map yields a sequence of points that progress more slowly around the $(1, 0)$. The equilibrium points of the system give solutions that always return to the same point on the plane. Such points are called **fixed points** for the Poincaré return map. Finally, the solutions that tend from the saddle equilibrium point back to itself give a sequence of points that progress from the origin and back to the origin via the Poincaré return map (see Figure 5.54).

Figure 5.54
The yvt-space picture of a solution that tends from the saddle at $(0, 0)$ as $t \to -\infty$ back to $(0, 0)$ as $t \to +\infty$.

In Figure 5.55, we depict the results of applying the return map to several different initial conditions in the unforced system. In this case the Poincaré return map picture looks very much like the phase portrait. Initial conditions near the $(\pm 1, 0)$ give solutions that stay on closed loops around $(\pm 1, 0)$, so in the Poincaré map picture these solutions give sequences of points on loops around the $(\pm 1, 0)$. The distribution of points around these loops is different at different radii because the rate at which the solutions spiral around the origin decreases as the radius increases. The orbits connecting the saddle fixed point to itself give sequences running from the origin back to the origin.

Just as with the phase space, it is very instructive to watch these pictures as they are computed. The order in which the sequence of points appears for the return map gives much more information than the static picture. However, we can still use the completed Poincaré map picture to draw conclusions about solutions.

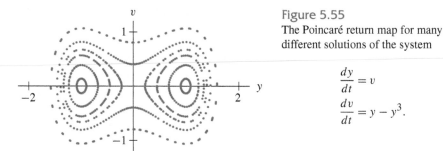

Figure 5.55
The Poincaré return map for many different solutions of the system

$$\frac{dy}{dt} = v$$

$$\frac{dv}{dt} = y - y^3.$$

Return Map for the Forced Nonlinear System

We now return to the forced nonlinear system

$$\frac{dy}{dt} = v$$

$$\frac{dv}{dt} = y - y^3 + \epsilon \sin t.$$

To compute solutions numerically we need to choose a value of ϵ. For the sake of illustration we take $\epsilon = 0.06$; other values of ϵ are considered in the exercises. Our goal is to use the Poincaré return map pictures to interpret the behavior of solutions of this system.

We fix an initial condition near the center $(1, 0)$, compute the solution in the yvt-space and then the Poincaré return map picture. The resulting image is given in Figure 5.56. The $y(t)$-graph of this solution is given in Figure 5.57. We can use this picture to predict the behavior of the solution: At least for times that are multiples of 2π, the position of the solution stays relatively near $(1, 0)$.

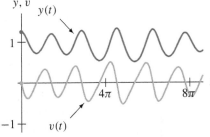

Figure 5.56
Poincaré return map for a solution with initial point near $(1, 0)$ for the forced Duffing system with $\epsilon = 0.06$.

Figure 5.57
The $y(t)$- and $v(t)$-graphs of the solutions in Figure 5.56.

The solutions that start near the saddle at the origin are much more interesting. The Poincaré return map for such a solution is given in Figure 5.58. Instead of a simple curve connecting the origin to itself all on one side of the phase plane, we see a cloud of points on both sides of the origin.

Figure 5.58 allows us to predict that a solution that starts near the origin behaves in a wild way. The y-coordinate takes on both positive and negative values. Moreover, because the points appear to follow no particular pattern, we can predict that the $y(t)$-graph of this solution oscillates without any particular pattern. This indeed is what we observe in Figure 5.59.

Figure 5.58
Poincaré return map for a solution with initial point near the origin for the forced nonlinear system with $\epsilon = 0.06$.

Figure 5.59
The $y(t)$- and $v(t)$-graphs of the solutions in Figure 5.58.

We can get some idea of why this solution behaves the way it does by looking again at the phase portrait of the unforced system (see Figure 5.60). A solution that starts near the origin, say just to the right of $(0, 0)$ near the separatrix leaving the origin, moves away from the origin for a while. Then it loops back and returns close to $(0, 0)$. The forcing term is very small, so it has little effect on the solution when it is far from $(0, 0)$, since the vector field has relatively large magnitude in this region. So we expect that the solution will return to this vicinity near the incoming separatrix. When the solution is close to the origin, the vector field of the unforced system is small, so the forcing becomes more important. When the solution is close to $(0, 0)$, if the forcing term is pushing the solution up (that is, $0.06 \sin t > 0$), then the solution moves toward the region $v > 0$ and hence follows the outgoing separatrix back into the right half-plane. If, on the other hand, the forcing term is negative when the solution is close to the origin, then it is possible that there is enough "push" to move the solution below the incoming separatrix. In this situation the solution proceeds into the left half-plane. Once in the left half-plane, the solution makes a loop around the left equilibrium point.

So with each loop, the solution of the forced system must "make a choice" when it returns to a neighborhood of the origin. The solution can go into either the right or left half-plane. Which direction the solution goes depends on the position of the solution relative to the origin and the sign of the forcing term when it arrives near the origin. So the choice depends on timing.

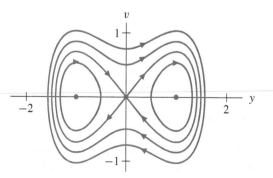

Figure 5.60
Phase portrait for the unforced system

$$\frac{dy}{dt} = v$$

$$\frac{dv}{dt} = y - y^3.$$

As a consequence, we cannot predict what happens when the initial condition is changed slightly. The solution moves slowly when it is near $(0, 0)$, so a slight change of initial condition that pushes the solution closer to the origin makes a significant difference in the amount of time the solution spends near $(0, 0)$. This in turn affects the time at which the solution returns to a neighborhood of the origin, and hence it can affect which side of the yv-plane it next enters. Hence, a slight difference in the initial condition can make a radical difference in the long-term behavior of the solution. This is demonstrated in Figure 5.61, which shows the $y(t)$-graphs of two solutions with initial conditions that are very close together. The solutions stay close for a while, but at some later time they are far enough apart so that they make different decisions about which way to turn when they are near the origin. After this time the solutions are radically different.

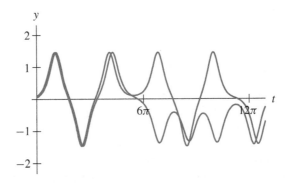

Figure 5.61
The $y(t)$-graphs of two solutions whose initial points are very close together.

Reality check

This sort of behavior is rather unnerving. We know that the system is "deterministic." The behavior of solutions is completely determined by the right-hand side of the system of differential equations. However, when we look at solutions, they behave in a way that does not seem to have any particular pattern. We have even anthropomorphized the solutions, saying things like "the solution decides which way to go." Solutions don't think. They don't have to think because their behavior is determined by the right-hand side of the differential equation. What is going on is that a slight change in the initial conditions can cause a huge change in the long-term behavior of the system.

While this sort of behavior seems unusual compared with the differential equations we have studied in previous sections, it is not at all unusual in nature. Physical systems like the flow of water in a turbulent stream, the weather patterns on the earth, and even the flipping of a coin, all behave this way. These systems are deterministic in that they follow strict laws of physics. This does not mean that they are predictable. A small change in initial conditions can make a radical difference in their behavior.

It is even dangerous to trust numerical simulations of these systems. We know that every numerical method gives only approximations of solutions. There are small errors in each step of the simulation. For a system like the one above, a small error in the numerical approximation gives us an approximate solution that is slightly different from the intended solution. But nearby solutions can have radically different long-term behavior. So a numerical simulation can give results that are very far from the desired solution.

This is one reason that long-range (beyond five days) weather prediction is usually not very accurate. Incomplete knowledge of weather systems and errors in numerical simulations yield predictions that can be far from correct.

The Periodically Forced Pendulum

As a second example of a periodically forced nonlinear system, we return to the system modeling the motion of a pendulum. We can think of a pendulum sitting on a table that is being shaken periodically. It turns out that much of the behavior observed above also occurs for this system.

The equations
The equations of the periodically forced pendulum with mass 1 and arm length 1 are

$$\frac{d\theta}{dt} = v$$

$$\frac{dv}{dt} = -g \sin\theta + \epsilon \sin t,$$

where g is the gravitational constant. The forcing term $\epsilon \sin t$ models an external force that periodically pushes the pendulum clockwise and counterclockwise with amplitude ϵ and period 2π. For convenience we assume that units of time and distance have been chosen so that $g = 1$ and our system is

$$\frac{d\theta}{dt} = v$$

$$\frac{dv}{dt} = -\sin\theta + \epsilon \sin t.$$

The return map
We construct the return map for the forced pendulum system exactly as above. The period of the forcing term is again 2π, so we follow solutions in θvt-space starting on the plane $t = 0$ and marking where they cross the plane $t = 2\pi$.

For the examples below, we fix $\epsilon = 0.01$; other values of ϵ are considered in the exercises. In Figure 5.62 we show the Poincaré return map for a solution with initial condition near $(0, 0)$. The resulting thick loop corresponds to a solution that oscillates with varying amplitude. The $\theta(t)$-graph of the same solution in Figure 5.63 shows this oscillation.

Initial conditions near $(0, 0)$ for the forced pendulum system correspond to starting the pendulum at a small angle and with small velocity. In this situation the unforced

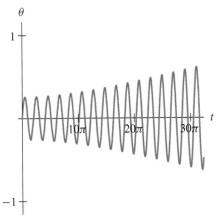

Figure 5.62
Poincaré return map for the periodically forced pendulum system with $\epsilon = 0.01$ for a solution with initial condition near $(0, 0)$.

Figure 5.63
Graph in the $t\theta$-plane of the solution in Figure 5.62. (The amplitude of the oscillation remains bounded for all t.)

(undamped) pendulum oscillates forever with small constant amplitude. The addition of the forcing term means that, just as for the forced harmonic oscillator, the forcing sometimes pushes with the direction of motion, making the pendulum swing higher, and sometimes pushes against the direction of motion, making the pendulum swing less. Unlike the harmonic oscillator, the period of the pendulum swing depends on the amplitude. Hence forcing adds and subtracts energy from the system in a way that, over the long term, is quite complicated. (This is not so evident from the picture because the forcing is small.)

Solutions near the saddle equilibrium points

Figure 5.64 shows the Poincaré return map for a solution of the periodically forced pendulum equation ($\epsilon = 0.01$) with initial condition near $(-\pi, 0)$. These points form a "cloud" with no particular structure. Also, the θ-coordinate becomes quite large. This means that the pendulum arm has completely rotated many times in one direction.

Just as in the forced Duffing system, whenever the solution approaches a saddle equilibrium point it must "decide" which way to turn. If it stays in the upper half-plane, then the θ-coordinate of the solution increases by a multiple of 2π before returning to the vicinity of another saddle. If it chooses to go into the lower half-plane, then the θ-coordinate decreases by 2π before another choice is made. If we graph the θ-coordinate of the solution in the $t\theta$-plane, we see that it moves in a very wild way. In particular, it is possible that the θ-coordinate of the solution can become very large positive (or negative) if the pendulum repeatedly "decides" to rotate the same direction (see Exercises 5–8).

The decision of which way the solution turns when it is close to a saddle depends on the sign of the forcing term. At that time the behavior of the solution depends very

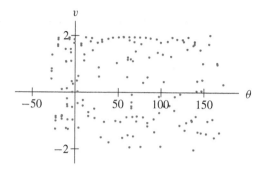

Figure 5.64
Poincaré map for the periodically forced pendulum
system with $\epsilon = 0.01$ for a solution with initial
point near $(-\pi, 0)$.

delicately on the initial condition. If we start two solutions near $(-\pi, 0)$ with almost the same initial condition, then eventually they split apart and become radically different (see Figure 5.65).

From the pictures above, we can deduce the existence of some interesting behavior in the forced pendulum system. A solution of the forced pendulum system with initial condition near $(-\pi, 0)$ corresponds to an initial placement of the pendulum that is almost vertical but with very little velocity. Of course the pendulum swings down. During the swing the forcing term has very little effect on the motion of the pendulum.

When the pendulum swings up to the almost vertical position, it slows down again and the effect of the forcing term is more pronounced. When the pendulum is near the top of the swing, the forcing term either "pushes it over the top" so that it makes another turn in the same direction, or "pulls it back" so that it swings back the way it came. Which way the pendulum goes depends on the sign of the forcing term, which in turn depends on the time the pendulum arrives at the top of its swing. Since

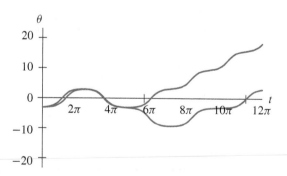

Figure 5.65
The $\theta(t)$-graphs for two solutions of the periodically
forced pendulum equation with almost equal initial
conditions.

the pendulum moves very slowly near the top of the swing, a small change in initial conditions can make a big change in the timing and hence cause a change in direction of the pendulum.

We emphasize that the sort of physical argument given above is not meant to replace the Poincaré return map pictures and the analysis of the solutions. Although the physical argument makes sense, it does not say whether a forcing term with $\epsilon = 0.01$ is large enough to cause this sort of behavior.

Moral

The moral of this section is that even systems like the pendulum that we "understand" can become very complicated when additional terms like periodic forcing are added. The periodically forced pendulum system does not appear at first glance to be all that complicated, but we see from the Poincaré return map that its solutions behave quite unpredictably. A small change in the initial position frequently has a radical effect on the behavior of the solution.

If this sort of behavior, which we now call **chaos**, can be observed in a system as simple as the periodically forced pendulum, it should not be surprising that it can also be found in nature. This is not a new discovery. Henri Poincaré first considered the possibility of the existence of "chaos" in nature in the 1880s. He was studying the motion of a small planet (an asteroid) under the gravitational influence of a star (the sun) and a large planet (Jupiter). Poincaré developed the return map to investigate the behavior of this system. What is remarkable is that Poincaré did not have the advantage of looking at numerical simulations of solutions as we do today. Nevertheless, he could see that the return map would behave in a very complicated fashion, and he wisely said he would not attempt to draw it.

EXERCISES FOR SECTION 5.6

In Exercises 1–4, Poincaré return map pictures are given for four different orbits with four different values of ϵ and initial conditions for the periodically forced Duffing equation

$$\frac{dy}{dt} = v$$

$$\frac{dv}{dt} = y - y^3 + \epsilon \sin t.$$

described in the text. Also, four $y(t)$-graphs for solutions of this system are given.

(a) Match the Poincaré return map picture with the $y(t)$-graph.

(b) Describe in a brief essay how you made the match and describe the qualitative behavior of the solution.

(i)

(ii)

(iii)

(iv)

1. For $\epsilon = 0.1$, $y(0) = 1.1$, $v(0) = 0$.

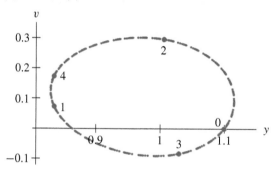

Poincaré return map with 500 iterates. The first four returns are indicated.

2. For $\epsilon = 0.4$, $y(0) = 1.1$, $v(0) = 0$.

Poincaré return map with 500 iterates. The first four returns are indicated.

3. For $\epsilon = 0.1$, $y(0) = 1.6$, $v(0) = 0$.

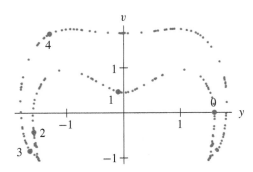

Poincaré return map with 200 iterates. The first
four returns are indicated.

4. For $\epsilon = 0.5$, $y(0) = 1.6$, $v(0) = 0$.

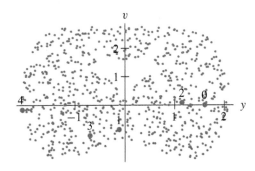

Poincaré return map with 800 iterates. The first
four returns are indicated.

In Exercises 5–8, the Poincaré return map pictures are given for four different orbits
with four different values of ϵ and initial conditions for the periodically forced pendulum system

$$\frac{d\theta}{dt} = v$$

$$\frac{dv}{dt} = -\sin\theta + \epsilon \sin t$$

described in the text. Also, four $\theta(t)$-graphs for solutions of this system are given.
   (a) Match the Poincaré return map picture with the $\theta(t)$-graph.
   (b) Describe in a brief essay how you made the match and describe the qualitative be-
      havior of the solution.
   (c) Describe the behavior of the pendulum arm when it follows the indicated solution.

(i)

(ii)

(iii)

(iv)

5. For $\epsilon = 0.1$, $\theta(0) = .2$, $v(0) = 0$.

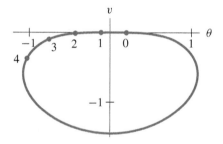

Poincaré return map with 1,000 iterates. The
first four returns are indicated.

6. For $\epsilon = 0.5$, $\theta(0) = .2$, $v(0) = 0$.

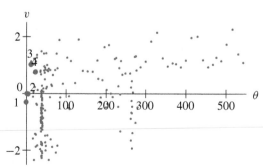

Poincaré return map with 250 iterates. The first four
returns are indicated.

7. For $\epsilon = 0.1$, $\theta(0) = -1.06$, $v(0) = 0$.

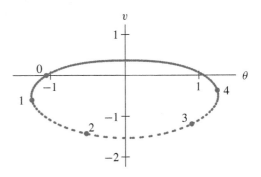

Poincaré return map with 400 iterates. The first four returns are indicated.

8. For $\epsilon = 0.5$, $\theta(0) = -1.06$, $v(0) = 0$.

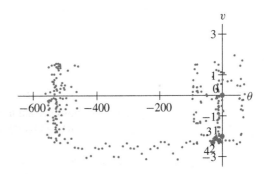

Poincaré return map with 250 iterates. The first four returns are indicated.

9. The Poincaré return map for a solution of the periodically forced harmonic oscillator

$$\frac{dy}{dt} = v$$

$$\frac{dv}{dt} = -3y + 0.2 \sin t$$

is given on the next page.

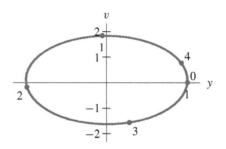

Poincaré return map for the forced harmonic oscillator system with initial conditions $y(0) = 1$, $v(0) = 0$.

(a) Find the solution with the initial condition $y(0) = 1$, $v(0) = 0$.

(b) Discuss why the Poincaré return map for this solution looks the way it does.

(c) What would the Poincaré return map for the solution with $y(0) = 4$, $v(0) = 0$ look like?

(d) Discuss why the Poincaré return map for the forced harmonic oscillator system is different from the Poincaré map of the forced pendulum.

10. Describe and sketch the Poincaré return map for the solution of the forced harmonic oscillator system

$$\frac{dy}{dt} = v$$

$$\frac{dv}{dt} = -4y + 0.1 \sin 2t.$$

[*Hint:* You should be able to do this problem qualitatively, with only a small amount of computation. What is the natural frequency of the system?]

REVIEW EXERCISES FOR CHAPTER 5

Short answer exercises: Exercises 1–10 focus on the basic ideas, definitions, and vocabulary of this chapter. Their answers are short (a single sentence or drawing), and you should be able to do them with little or no computation. However, they vary in difficulty, so think carefully before you answer.

1. For the nonlinear system $dx/dt = x + xy$ and $dy/dt = y^2 - 2y$, what is the linearized system for the equilibrium point at the origin?

2. For the nonlinear system in Exercise 1, determine the type (sink, source, ...) of the equilibrium point at the origin.

3. For the nonlinear system $dx/dt = x^2 + \sin 3x$ and $dy/dt = 2y - \sin xy$, what is the linearized system for the equilibrium point at the origin?

4. For the nonlinear system in Exercise 3, determine the type (sink, source, ...) of the equilibrium point at the origin.

5. Sketch the nullclines for the system $dx/dt = x - y$ and $dy/dt = x^2 + y^2 - 2$.

6. Is the system $dx/dt = 2xy + y^2$ and $dy/dt = x^2 + y^2$ a Hamiltonian system?

7. Is the system in Exercise 6 a gradient system?

8. Describe three different possible long-term behaviors for a solution $\mathbf{Y}(t)$ of a two-dimensional system $d\mathbf{Y}/dt = \mathbf{F}(\mathbf{Y})$.

9. Suppose that a two-dimensional system $d\mathbf{Y}/dt = \mathbf{F}(\mathbf{Y})$ has exactly one equilibrium point and that point is a source. Describe two different possible long-term behaviors for a solution with an initial condition that is near the equilibrium point.

10. Suppose that the origin is an equilibrium point of a two-dimensional system and that zero is not an eigenvalue for the linearized system at the origin. Suppose also that $dy/dt > 0$ on the positive x-axis and $dx/dt > 0$ on the positive y-axis. What type of equilibrium point can the origin be?

True-false: For Exercises 11–14, determine if the statement is true or false. If it is true, explain why. If it is false, provide a counterexample or an explanation.

11. The equilibrium points of a first-order system occur at the intersection of the x- and y-nullclines.

12. The x- and y-nullclines of a system are never identical.

13. If (x_0, y_0) is an equilibrium point of the system $dx/dt = f(x, y)$ and $dy/dt = g(x, y)$, then the eigenvalues of the linearized system at (x_0, y_0) are the partial derivatives $\partial f/\partial x$ and $\partial g/\partial y$ evaluated at (x_0, y_0).

14. For a system of the form $dx/dt = f(x)$ and $dy/dt = g(x, y)$ with $f'(x) > 0$ for all x, every equilibrium point is a source.

For the systems given in Exercises 15–18,
  (a) find and classify all equilibria,
  (b) sketch the nullclines, and
  (c) sketch the phase portrait.

15. $\dfrac{dx}{dt} = x - 3y^2$

$\dfrac{dy}{dt} = x - 3y - 6$

16. $\dfrac{dx}{dt} = 10 - x^2 - y^2$

$\dfrac{dy}{dt} = 3x - y$

17. $\dfrac{dx}{dt} = 4x - x^2 - xy$

$\dfrac{dy}{dt} = 6y - 2y^2 - xy$

18. $\dfrac{dx}{dt} = xy$

$\dfrac{dy}{dt} = x + y - 1$

Exercises 19–24 concern the two-parameter family of systems

$$\frac{dx}{dt} = x^2 - a$$
$$\frac{dy}{dt} = y^2 - b,$$

where a and b are parameters.

19. Find and classify all equilibrium points assuming that $a > 0$ and $b > 0$.

20. Sketch the nullclines and the phase portrait assuming that $a > 0$ and $b > 0$.

21. Repeat Exercises 19 and 20 assuming that $a = 0$ and $b > 0$. In addition, describe the bifurcation that occurs as $a \to 0$ from above.

22. Repeat Exercises 19 and 20 assuming that $a > 0$ and $b = 0$. In addition, describe the bifurcation that occurs as $b \to 0$ from above.

23. Repeat Exercises 19 and 20 assuming that $a = 0$ and $b = 0$.

24. Draw a picture in the ab-plane that represents the different types of phase portraits that occur for this system for all values of a and b.

25. Consider the system

$$\frac{dx}{dt} = y^2 - x^2 - 1$$
$$\frac{dy}{dt} = 2xy.$$

  (a) Find and classify all equilibrium points.
  (b) Sketch the nullclines and phase portrait.
  (c) Is this system a Hamiltonian system? If so, calculate the Hamiltonian function.
  (d) Is this system a gradient system? If so, calculate a function whose gradient yields the system.

26. Consider the differential equation

$$\frac{d^2x}{dt^2} + 2\frac{dx}{dt} - 3x + x^3 = 0.$$

(a) Convert this second-order equation into a first-order system and calculate its equilibrium points.

(b) Classify these equilibrium points using linearization.

27. In Exercise 26 of Section 3.4, we showed that the solution curves of the linear system

$$\frac{d\mathbf{Y}}{dt} = \begin{pmatrix} -3 & 10 \\ -1 & 3 \end{pmatrix} \mathbf{Y}$$

lie on ellipses of the form $x^2 - 6xy + 10y^2 = k$, where k is a constant determined by the initial condition. Reproduce that result by calculating a Hamiltonian function for the system.

28. Consider the linear system

$$\frac{d\mathbf{Y}}{dt} = \begin{pmatrix} a & b \\ c & d \end{pmatrix} \mathbf{Y}.$$

(a) For which values of a, b, c, and d is the system Hamiltonian? Calculate the Hamiltonian function for these values.

(b) For which values of a, b, c, and d is the system a gradient system? For these values, calculate a function whose gradient yields the system.

(c) Can the linear system be Hamiltonian and a gradient system for identical values of a, b, c, and d?

(d) Are there values of a, b, c, and d for which the linear system is neither Hamiltonian nor a gradient system?

29. In Exercise 37 of the Review Exercises of Chapter 2, we studied a simple model of a glider. Setting the drag parameter $D = 1$, we have the equations

$$\frac{d\theta}{dt} = \frac{s^2 - \cos\theta}{s}$$

$$\frac{ds}{dt} = -\sin\theta - s^2,$$

where θ represents the angle of the nose of the glider with the horizon and $s > 0$ represents its speed.

(a) Calculate the equilibrium points for this system, and

(b) classify them using linearization.

LAB 5.1 Hard and Soft Springs

In this lab, we continue our study of second-order equations by considering "nonlinear springs." In Sections 2.1 and 2.3, we developed the model of a spring based on Hooke's law. Hooke's law asserts that the restoring force of a spring is proportional to its displacement, and this assumption leads to the second-order equation

$$m\frac{d^2y}{dt^2} + ky = 0.$$

Since the resulting differential equation is linear, we say that the spring is linear. In this case the restoring force is $-ky$. In addition, we assume that the friction or damping force is proportional to the velocity. The resulting second-order equation is

$$m\frac{d^2y}{dt^2} + b\frac{dy}{dt} + ky = 0.$$

Hooke's law is an idealized model that works well for small oscillations. In fact the restoring force of a spring is roughly linear if the displacement of the spring from its equilibrium position is small, but it is generally more accurate to model the restoring force by a cubic of the form $-ky + ay^3$, where a is small relative to k. If a is negative, the spring is said to be hard, and if a is positive, the spring is soft. In this lab we consider the behavior of hard and soft springs for particular values of the parameters. (Your instructor will tell you which parameter value(s) from Table 5.1 to use.)

In your report, you should analyze the phase planes and $y(t)$- and $v(t)$-graphs to describe the long-term behavior of the solutions to the equations:

1. (Hard spring with no damping) The first equation that you should study is the hard spring with no damping; that is, $b = 0$ and $a = a_1$. Examine solutions using both their graphs and the phase plane. Consider the periods of the periodic solutions that have the initial condition $v(0) = 0$. Sketch the graph of the period as a function of the initial condition $y(0)$. Is there a minimum period? A maximum period? If so, how do you interpret these extrema?

2. (Hard spring with damping) Now use the given value of b and $a = a_1$ to introduce damping into the discussion. What happens to the long-term behavior of solutions in this case? Determine the value of the damping parameter that separates the underdamped case from the overdamped case.

3. (Soft spring with no damping) Consider the soft spring that corresponds to the positive value a_2 of a. Over what range of y-values is this model reasonable? Consider the periods of the periodic solutions that have the initial condition $v(0) = 0$. Sketch the graph of the period as a function of the initial condition $y(0)$. Is there a minimum period? A maximum period? Use the phase portrait to help justify your answer.

4. (Soft spring with damping) Using the given values of b and $a = a_2$, what happens to the long-term behavior of solutions in this case? Determine the value of the damping parameter that separates the underdamped case from the overdamped case.

5. From a physical point of view, what's the difference between a hard spring and a soft spring?

Your report: Address each of the five items in the form of a short essay. You may illustrate your essay with phase portraits and graphs of solutions. However, your essay should be complete and understandable without the pictures. Make sure you relate the behavior of the solutions to the motion of the associated mass and spring systems.

Table 5.1
Choices for the parameter values. Assume the mass $m = 1$ unless you are told otherwise by your instructor.

Choice	k	b	a_1	a_2
1	0.1	0.15	−0.005	0.005
2	0.2	0.20	−0.008	0.008
3	0.3	0.20	−0.009	0.009
4	0.2	0.20	−0.005	0.005
5	0.1	0.10	−0.005	0.005
6	0.3	0.20	−0.007	0.007
7	0.3	0.15	−0.007	0.007
8	0.1	0.15	−0.004	0.004
9	0.2	0.15	−0.005	0.005
10	0.3	0.20	−0.008	0.008

LAB 5.2 Higher Order Approximations of the Pendulum

In previous chapters, we studied the behavior of second-order, homogeneous linear equations (like the harmonic oscillator) by reducing them to first-order linear systems. This "reduction" technique can be applied to nonlinear equations as well, and in this lab we study the ideal pendulum and approximations to the pendulum using this technique.

In the text we modeled the ideal pendulum by the second-order, nonlinear equation

$$\frac{d^2\theta}{dt^2} + \frac{g}{l}\sin\theta = 0,$$

where θ is the angle from the vertical, g is the gravitational constant ($g = 32$ ft/s^2), and l is the length of the rod of the pendulum, that is, the radius of the circle on which the mass travels. In this lab we compare the results of numerical simulation of this model with the results obtained from two approximations to this model. The first approximation is a linear approximation given by

$$\frac{d^2\theta}{dt^2} + \frac{g}{l}\theta = 0.$$

The second approximation is a cubic approximation

$$\frac{d^2\theta}{dt^2} + \frac{g}{l}\left(\theta - \frac{\theta^3}{6}\right) = 0.$$

Recall from calculus that the expression $\theta - \theta^3/6$ represents the first two terms in the power series expansion of $\sin\theta$ about $\theta = 0$. We are especially interested in how close the solutions of the approximations of the ideal pendulum equation are to the original ideal pendulum equation. In particular, we are interested in how closely the periods of the periodic orbits of the approximations of the pendulum equation relate to the periods of the periodic orbits of original equation. Your instructor will tell you what value of the parameter l (the length of the pendulum arm) you should use. Your report should include:

1. A phase portrait analysis for all three equations. Compare and contrast these phase portraits from the point of view of how well the linear and cubic equations approximate the ideal pendulum.

2. In order to study how the periods of the periodic orbits are related, consider the one-parameter family of initial conditions parameterized by θ_0, where $\theta(0) = \theta_0$ and $\theta'(0) = 0$ (no initial velocity). In other words, you should study the various solutions that begin at a given angle with zero velocity. For what intervals of initial conditions do the periods of the periodic orbits of

$$\frac{d^2\theta}{dt^2} + \frac{g}{l}\theta = 0$$

and

$$\frac{d^2\theta}{dt^2} + \frac{g}{l}\left(\theta - \frac{\theta^3}{6}\right) = 0$$

closely approximate the periods of the periodic orbits of the ideal pendulum? (The computation of the periods in the linear approximation can be done exactly using the techniques of Chapter 3. Analytic techniques exist for computing the periods of the periodic orbits of the other two equations, but in this lab you should work numerically.) You should plot graphs of the period as a function of θ_0 using a relatively small table (5, 10, or 15 entries) of periods obtained using direct numerical simulation of the model.

3. Another family of initial conditions is $\theta(0) = 0$ and $\theta'(0) = v_0$. In this family, the initial velocity is the parameter. Initially the pendulum points straight down with a given velocity v_0. What changes from your results in Part 2 above?

4. Suppose you are a clockmaker who makes clocks based on the motion of a pendulum. For each of the three equations, what would you do to double the period of the oscillation?

Your report: Address each of the items above in the form of a short essay. Be as systematic as possible when collecting data, and present this data in a concise and clear format. You may illustrate your essay with phase portraits and graphs of solutions or of the data that you collect. However, your essay should be complete and understandable without the pictures.

LAB 5.3   A Family of Predator-Prey Equations

In this laboratory exercise, you will study a one-parameter family of nonlinear, first-order systems consisting of predator-prey equations. The family is

$$\frac{dx}{dt} = 9x - \alpha x^2 - 3xy$$
$$\frac{dy}{dt} = -2y + xy,$$

where $\alpha \geq 0$ is a parameter. In other words, for different values of α we have different systems. The variable x is the population (in some scaled units) of prey, and y is the population of predators. For a given value of α, we want to understand what happens to these populations as $t \to \infty$.

You should investigate the phase portraits of these equations for various values of α in the interval $0 \leq \alpha \leq 5$. To get started, you might want to try $\alpha = 0, 1, 2, 3, 4$, and 5. Think about what the phase portrait means in terms of the evolution of the x and y populations. Where are the equilibrium points? What does linearization tell you about their types? What happens to a typical solution curve? Also, consider the behavior of the special solutions where either $x = 0$ or $y = 0$ (solution curves lying on the x- or y-axes).

Determine the bifurcation values of α—that is, the values of α where nearby α's lead to "different" behaviors in the phase portrait. For example, $\alpha = 0$ is a bifurcation value because for $\alpha = 0$, the long-term behavior of the populations is dramatically different than the long-term behavior of the populations if α is slightly positive. The technique of linearization suggests bifurcation values.

Your report: After you have determined all of the bifurcation values for α in the interval $0 \leq \alpha \leq 5$, study enough specific values of α to be able to discuss all of the various population evolution scenarios for these systems. In your report, you should describe these scenarios using the phase portraits and $x(t)$- and $y(t)$-graphs. Your report should include:

1. A brief discussion of the significance of the various terms in the system. For example, what does the $9x$ represent? What does the $3xy$ term represent?

2. A discussion of all bifurcations including the bifurcation at $\alpha = 0$. For example, a bifurcation occurs between $\alpha = 3$ and $\alpha = 5$. What does this bifurcation mean for the predator population?

Address the questions above in the form of a short essay, and support your assertions with selected illustrations. (Please remember that although one good illustration may be worth 1000 words, 1000 illustrations are usually worth nothing.)

LAB 5.4 The Glider

Consider a glider flying in the xy-plane (see Figure 5.66). Let $s(t)$ be the speed of the glider along its path at time t and $\theta(t)$ be the angle of the velocity vector of the path with the x-direction at time t. Note that, since the body of the glider points in the direction of motion, θ is also the angle between the glider and the x-direction.

Figure 5.66
The angle θ for the motion of a glider.

The forces involved are gravity, lift provided by the wings (a force perpendicular to the velocity vector), and drag (a force parallel but in the opposite direction to the velocity vector). Using $F = ma$, we can obtain equations for d^2x/dt^2 and d^2y/dt^2 and then derive the system

$$\frac{d\theta}{dt} = \frac{s^2 - \cos\theta}{s}$$

$$\frac{ds}{dt} = -\sin\theta - Ds^2.$$

(The derivation involves several changes of coordinates, including a change of time scale, and is an excellent exercise for your friends who are studying classical mechanics.) Note that this model assumes that both the lift and drag are proportional to the square of the velocity.*

The most remarkable thing about this system is that there is only one parameter, D. This parameter is the drag force caused by air resistance divided by the lift, the "drag-to-lift ratio." The designer of a glider tries to maximize lift while minimizing drag, so the parameter D can be viewed as a measure of the quality of the design (a small value for D is preferred). In this lab we consider the solutions of this model and their relationship to the flight of the glider.

Your report should address the following items:

1. Study the solutions of the system above with $D = 0$ (that is, for the perfect glider with zero drag). Are there equilibrium points? What is the physical interpretation

---

*This model appears in the book *Theory of Oscillators* by A. A. Andronov, A. A. Vitt, and S. E. Khaikin, Dover Publishing, 1987. Other excellent sources are *Nonlinear Dynamics and Chaos*, by S. Strogatz, Perseus Press, 1994, and *Interactive Differential Equations Workbook* by B. West, S. Strogatz, J. M. Dill, and J. Cantwell, Addison Wesley Interactive, 1997.

(in terms of the path of the glider) of the equilibria? How do the initial conditions relate to the launch of the glider and how does the flight path change with different initial conditions? Show that the quantity $C(\theta, s) = s^3 - 3s \cos \theta$ is a conserved quantity for this system. What does the function C tell you about the solution curves of the system? Are there periodic solutions?

2. Repeat your analysis for values of D between 0 and 4 (that is, for increasing drag-to-lift ratio). How do the phase portraits change as D changes? How do the possible glider paths change as D increases?

3. Apply the theory of linearization to classify all equilibria for values of D in the interval $0 \leq D \leq 4$. Determine the bifurcation values of D.

4. Reconstructing the motion of the glider from the equations: Given a value of D and an initial condition (θ_0, s_0), the motion of the glider is determined from the equations. Show how one can start with values of D and (θ_0, s_0) and obtain the path of the glider. Be precise. (One good way to do this part of the project is to write the code that you would need to draw the path of the glider in some convenient programming language.)

5. Why is it more natural to think of the "phase cylinder" for this system rather than the phase plane? What changes if you analyze the system using a phase cylinder in place of a phase plane?

6. Construct a paper glider and relate test flights to your answers in Parts 1, 2 and 3. (Note: Gliders made from higher-quality paper demonstrate the dynamics much better. You may wish to consult the paper airplane literature for design suggestions. For example, J. M. Collins, *The Gliding Flight*, Ten Speed Press, Berkeley, 1989.)

In your report, pay particular attention to the relationship between the geometry of solutions in the phase plane and the motion of the glider.

6 LAPLACE TRANSFORMS

In this chapter we study an analytic technique for finding formulas for solutions of certain differential equations using an operation called the Laplace transform. This technique is particularly effective on linear, constant-coefficient equations and is quite different from our previous methods. The Laplace transform lets us replace the operations of integration and differentiation with algebraic computations. As a technique for solving initial-value problems, the Laplace transform is sometimes more and sometimes less efficient than our previous techniques.

The importance of the Laplace transform is that it is useful in several different types of applications. For example, the Laplace transform lets us deal efficiently with linear, constant-coefficient differential equations that have discontinuous forcing functions. These discontinuities include simple jumps that model the action of a switch. Using Laplace transforms, we can also make a meaningful mathematical model of the impulse force provided by a hammer blow or an explosion. Laplace transforms also find solutions for a forced harmonic oscillator using solutions generated by a completely different forcing term. This technique is useful when working with a system that is modeled by an unknown linear, constant-coefficient differential equation for which we provide the forcing function (see Section 6.5). Finally, Laplace transforms give us a way of extending our qualitative analysis of homogeneous linear equations using eigenvalues to nonhomogeneous linear equations (see Section 6.6).

6.1   LAPLACE TRANSFORMS

Integral Transforms

In this chapter we study a tool, the **Laplace transform**, for solving differential equations. The Laplace transform is one of many different types of *integral transforms*. In general, integral transforms address the question: How much is a given function $y(t)$ "like" a particular standard function?

For example, if $y(t)$ represents a radio signal, we might want to compare it to the function $\sin \omega t$, a sine wave with frequency $\omega/(2\pi)$. By adjusting the parameter ω, we could test how well $y(t)$ fits sine waves with different frequencies. Ideally, for each value of ω, we would like a number that indicates how much $y(t)$ is like $\sin \omega t$. One way to accomplish this comparison is to compute the integral

$$\int_{-N}^{N} y(t) \sin \omega t \, dt$$

for large N. If $y(t)$ is oscillating with frequency $\omega/(2\pi)$ and is positive when $\sin \omega t$ is positive, then this integral is very large. If $y(t)$ has some other frequency, then the signs of $y(t)$ and $\sin \omega t$ differ at some times t, so there is cancellation in the integral and its value is smaller.

To use this idea in differential equations, it is natural to compare $y(t)$ to the function that comes up most often, the exponential function. In fact we could use the complex exponential and compute

$$\int_{-\infty}^{\infty} y(t) \, e^{-zt} \, dt,$$

where $z = s + i\omega$ is a complex parameter. The larger the value of this integral for a particular z, the more $y(t)$ is "like" e^{zt}. In particular, if $y(t) = e^{zt}$, then the integrand is the constant function 1, and the integral is infinite.

In practice it turns out to be easier to write z in terms of its real and imaginary parts as $z = s + i\omega$ and to compute the two transforms

$$\int_{-\infty}^{\infty} y(t) \, e^{-st} \, dt \quad \text{and} \quad \int_{-\infty}^{\infty} y(t) \, e^{-i\omega t} \, dt$$

separately. The integral

$$\int_{-\infty}^{\infty} y(t) \, e^{-i\omega t} \, dt$$

is called the *Fourier transform* of the function $y(t)$, and its value at a particular value of ω is a measure of the extent to which $y(t)$ is oscillating with frequency $\omega/(2\pi)$. The imaginary part of this quantity is the comparison of $y(t)$ to $\sin \omega t$ discussed above, and the real part is the comparison of $y(t)$ to $\cos \omega t$. Although the Fourier transform has many important applications in differential equations, it is not discussed in this chapter. Rather, we focus on the integral $\int y(t) \, e^{-st} \, dt$.

Laplace Transforms

The Laplace transform of a (given) function $y(t)$ uses integration to compare $y(t)$ to the exponential functions e^{st}.

DEFINITION   The **Laplace transform** function Y of the function y is defined by

$$Y(s) = \int_0^\infty y(t)\, e^{-st}\, dt$$

for all numbers s for which this improper integral converges.   ∎

For example, if $y(t) = e^{2t}$, then its Laplace transform $Y(s)$ is determined by evaluating the improper integral

$$Y(s) = \int_0^\infty e^{2t}\, e^{-st}\, dt$$

$$= \int_0^\infty e^{(2-s)t}\, dt$$

$$= \lim_{b\to\infty} \int_0^b e^{(2-s)t}\, dt$$

$$= \lim_{b\to\infty} \left[\frac{1}{2-s} e^{(2-s)t} \Big|_0^b \right]$$

$$= \frac{1}{2-s} \lim_{b\to\infty} \left[e^{(2-s)b} - e^0 \right].$$

Since

$$\frac{1}{2-s} \lim_{b\to\infty} e^{(2-s)b} = \begin{cases} \infty, & \text{if } s < 2; \\ 0, & \text{if } s > 2, \end{cases}$$

we see that the improper integral for $Y(s)$ does not converge if $s \leq 2$ and that

$$Y(s) = \frac{1}{s-2} \quad \text{if } s > 2.$$

In other words, the Laplace transform function $Y(s)$ for the function $y(t) = e^{2t}$ is

$$Y(s) = \begin{cases} \dfrac{1}{s-2}, & \text{if } s > 2; \\ \text{undefined}, & \text{if } s \leq 2. \end{cases}$$

This computation involves a number of technical details with improper integrals that are important, but don't let them cause you to lose touch with the underlying idea involved in the definition. The Laplace transform $Y(s)$ associated to $y(t)$ is a function that measures how close $y(t)$ is to the exponential functions e^{st} for all values of s. In this example we found that the Laplace transform of e^{2t} is a function $Y(s)$ that is very small if s is much larger than 2, but as s approaches 2, the values of $Y(s)$ increase until they become unbounded at $s = 2$.

Surprisingly, we can use this idea to solve differential equations. As we will soon see, we can use the Laplace transform to convert a differential equation into an algebraic equation, which is often easier to solve. This conversion is similar to the translation of an English sentence into a sentence in Chinese. Both sentences have the same meaning, but the words and grammar are very different. The function $y(t)$ represents a phenomenon in terms of the independent variable t "in the time domain," and the function $Y(s)$ represents the same phenomenon in the s-domain.

More formally, we say that the Laplace transform defines a mathematical transformation that takes a function $y(t)$ into its transformed function $Y(s)$. From a strict mathematical point of view, this transform is just one (particularly nice) function that converts a function $y(t)$ into a new function $Y(s)$, and we use the script letter $\mathcal{L}$ to represent this function. In other words, we say that $Y = \mathcal{L}[y]$. In particular, our computation of the Laplace transform of e^{2t} is often written as

$$\mathcal{L}[e^{2t}] = \frac{1}{s-2} \quad \text{for } s > 2.$$

This way of writing the Laplace transform is somewhat sloppy in that it assumes that the independent variable of the transformed function is the variable s, but this lack of precision hardly ever causes confusion.

In order to be certain that the Laplace transform exists (that is, the improper integral converges) for at least some values of s, we must restrict attention to continuous or piecewise continuous functions $y(t)$ for which there are positive constants K and M (that depend on y) such that $|y(t)| < e^{Mt}$ for $t \geq K$. Such functions are said to have no more than "exponential growth." As a practical matter, most functions encountered in applications have this property.

Also note that the Laplace transform is defined as an integral over the interval $0 \leq t < \infty$ (rather than over the entire real line $-\infty < t < \infty$). If the Laplace transform were to be evaluated over the entire real line, then the restrictions on the function $y(t)$ in order to ensure convergence of the integral would be much more stringent. In most applications we are interested primarily in the future ($t \geq 0$) anyway.

Computation of Laplace Transforms of Exponential Functions

For a function $y(t)$ and a given number s, the value of the Laplace transform at s measures the degree to which the function $y(t)$ resembles the function e^{st} on the interval $t \geq 0$. To check this assertion, we compute the Laplace transform of an arbitrary exponential function. If $y(t) = e^{at}$, then

$$\mathcal{L}[e^{at}] = \int_0^\infty e^{at} e^{-st} \, dt$$

$$= \int_0^\infty e^{(a-s)t} \, dt.$$

Recall that this improper integral is actually a limit of integrals as the upper limit of integration goes to infinity. Hence

$$\mathcal{L}[e^{at}] = \lim_{b \to \infty} \int_0^b e^{(a-s)t}\, dt$$

$$= \lim_{b \to \infty} \left[\frac{1}{a-s} e^{(a-s)t} \Big|_0^b \right]$$

$$= \frac{1}{a-s} \lim_{b \to \infty} \left[e^{(a-s)b} - e^0 \right]$$

$$= \frac{1}{s-a} \quad \text{if } s > a.$$

If $s \leq a$, then the improper integral diverges. Strictly speaking, the Laplace transform of e^{at} is the rational function $1/(s-a)$ restricted to the interval $s > a$. Note that our earlier calculation of $\mathcal{L}[e^{2t}]$ was simply a special case of this calculation.

There is another special case of this computation that is worth mentioning at this point. Note that if $a = 0$, then $e^{at} = 1$ for all t. Consequently we have also computed the Laplace transform of the function that is constantly 1. Since $a = 0$, we have

$$\mathcal{L}[1] = \frac{1}{s} \quad \text{for } s > 0.$$

We often need the results of these computations as well as the transforms of other frequently encountered functions. Therefore, we provide **a table of Laplace transforms as well as a list of important properties of this transformation on page 626**.

Properties of the Laplace Transform

There are many transforms that convert one function into another, but the Laplace transform has one very special property that is the basis for its success in solving differential equations.

LAPLACE TRANSFORM OF DERIVATIVES   Given a function $y(t)$ with Laplace transform $\mathcal{L}[y]$, the Laplace transform of dy/dt is

$$\mathcal{L}\left[\frac{dy}{dt} \right] = s\mathcal{L}[y] - y(0). \qquad \blacksquare$$

To verify this theorem we use the definition of $\mathcal{L}[dy/dt]$ and compute

$$\mathcal{L}\left[\frac{dy}{dt} \right] = \int_0^\infty \frac{dy}{dt} e^{-st}\, dt.$$

Using integration by parts with the choices $u = e^{-st}$ and $dv = (dy/dt)\, dt$, we have $du = -se^{-st}\, dt$ and $v = y(t)$, and therefore

$$\mathcal{L}\left[\frac{dy}{dt} \right] = \lim_{b \to \infty} \left[y(t)\, e^{-st} \Big|_0^b \right] + \int_0^\infty y(t)\, s e^{-st}\, dt.$$

Our earlier assumption that $y(t)$ has at most exponential growth is important here. Since $|y(t)| < e^{Mt}$ for some constant M,

$$\lim_{t \to \infty} y(t) e^{-st} = 0$$

for sufficiently large s. Hence,

$$\mathcal{L}\left[\frac{dy}{dt}\right] = -y(0) + \int_0^\infty y(t) \, s e^{-st} \, dt$$

$$= -y(0) + s\mathcal{L}[y].$$

This formula is the fundamental property of the Laplace transform that lets us essentially replace the calculus operation of differentiation in the t-domain with the algebraic operation of multiplication by s in the s-domain. Of course, this description is not quite right since we also have to remember to subtract $y(0)$, but in any case we think of the Laplace transform as an operation that turns a problem in calculus into a problem in algebra.

Turning a problem in differential equations into a problem in algebra would not be very useful if the Laplace transform did not have reasonable algebraic properties. However, due to its definition in terms of an integral, the Laplace transform has nice linearity properties.

LINEARITY OF THE LAPLACE TRANSFORM   Given functions f and g and a constant c,

$$\mathcal{L}[f + g] = \mathcal{L}[f] + \mathcal{L}[g]$$

$$\mathcal{L}[cf] = c\mathcal{L}[f].$$

In other words, the transform "operator" $\mathcal{L}$ is a linear operator.   ∎

To verify these properties, we use the linearity properties of integration. That is,

$$\mathcal{L}[f + g] = \int_0^\infty (f(t) + g(t)) \, e^{-st} \, dt$$

$$= \int_0^\infty f(t) \, e^{-st} \, dt + \int_0^\infty g(t) \, e^{-st} \, dt$$

$$= \mathcal{L}[f] + \mathcal{L}[g],$$

and

$$\mathcal{L}[cf] = \int_0^\infty cf(t) \, e^{-st} \, dt = c \int_0^\infty f(t) \, e^{-st} \, dt = c\mathcal{L}[f].$$

With the derivative formula and the linearity of $\mathcal{L}$ established, we can now use the Laplace transform to solve differential equations.

Solving Differential Equations Using the Laplace Transform

Consider the initial-value problem

$$\frac{dy}{dt} = y - 4e^{-t}, \quad y(0) = 1.$$

We first study this problem using qualitative techniques so that we know what to expect for the analytic form of the solution.

Qualitative analysis

The slope field for

$$\frac{dy}{dt} = y - 4e^{-t}$$

is given in Figure 6.1. It indicates that the solution with $y(0) = 1$ decreases for t near zero. Once the solution crosses below $y = 0$, both terms on the right-hand side of the differential equation are negative, so the solution continues to decrease. For large t, the equation is close to the equation

$$\frac{dy}{dt} = y,$$

and hence we expect the solution $y(t) \to -\infty$ as $t \to \infty$.

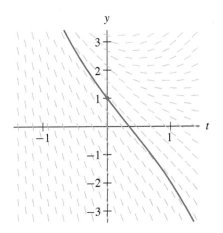

Figure 6.1
Slope field for $dy/dt = y - 4e^{-t}$ and the graph of the solution with $y(0) = 1$.

Solution using Laplace transforms

This equation is linear, so we could use the techniques of Sections 1.8 or 1.9. However, the Laplace transform provides an alternate method of solution.

Starting with the initial-value problem

$$\frac{dy}{dt} = y - 4e^{-t}, \quad y(0) = 1,$$

the first step is to take the Laplace transform of both sides

$$\mathcal{L}\left[\frac{dy}{dt}\right] = \mathcal{L}[y - 4e^{-t}].$$

Then using the previous formula for the Laplace transform of a derivative to simplify the left-hand side and the linearity of Laplace transform on the right-hand side, we obtain

$$s\mathcal{L}[y] - y(0) = \mathcal{L}[y] - 4\mathcal{L}[e^{-t}],$$

and we substitute the initial condition $y(0) = 1$ to get

$$s\mathcal{L}[y] - 1 = \mathcal{L}[y] - 4\mathcal{L}[e^{-t}].$$

Earlier we computed that $\mathcal{L}[e^{at}] = 1/(s - a)$, so we can apply this formula with $a = -1$ to obtain

$$s\mathcal{L}[y] - 1 = \mathcal{L}[y] - \frac{4}{s + 1}.$$

The unknown of the original differential equation is the function y, so we solve this equation for the Laplace transform of y, obtaining

$$\mathcal{L}[y] = \frac{1}{s - 1} - \frac{4}{(s - 1)(s + 1)}.$$

These calculations yield the Laplace transform of the solution of the initial-value problem. Note that we used only arithmetic to obtain $\mathcal{L}[y]$. (The calculus is hidden in the computation of the Laplace transforms of exponentials.) In a certain sense we have now solved the initial-value problem.

Unfortunately we are not looking for the Laplace transform $\mathcal{L}[y]$ of the solution; we really want the actual solution $y(t)$. Somehow we must "undo" or take an "inverse" Laplace transform. That is, we must figure out what function $y(t)$ has the function

$$\frac{1}{s - 1} - \frac{4}{(s - 1)(s + 1)}$$

as its Laplace transform.

Inverse Laplace Transform

To use Laplace transforms to solve a differential equation, we first compute the Laplace transform of both sides of the equation. Next we solve for the Laplace transform of the solution. Finally, to find the solution, we find a function with the given Laplace transform. The last step is called taking the **inverse Laplace transform**. The notation for this inverse transform is $\mathcal{L}^{-1}$, that is,

$$\mathcal{L}^{-1}[F] = f \quad \text{if and only if} \quad \mathcal{L}[f] = F.$$

There is a uniqueness property for inverse Laplace transforms which states that, if f is a continuous function with $\mathcal{L}[f] = F$, then f is the only continuous function whose Laplace transform is F. This uniqueness property allows us to say "the" inverse Laplace transform of F instead of "an" inverse Laplace transform of F.

Because the Laplace transform is a linear operator, that is,

$$\mathcal{L}[f + g] = \mathcal{L}[f] + \mathcal{L}[g] \quad \text{and} \quad \mathcal{L}[cf] = c\mathcal{L}[f]$$

for any functions $f(t)$ and $g(t)$ and any constant c, it follows that the inverse Laplace transform is also a linear operator,

$$\mathcal{L}^{-1}[f + g] = \mathcal{L}^{-1}[f] + \mathcal{L}^{-1}[g] \quad \text{and} \quad \mathcal{L}^{-1}[cf] = c\mathcal{L}^{-1}[f].$$

Linearity is important because it allows us to compute the inverse Laplace transform of a complicated sum by computing the inverse Laplace transform of each summand.

Inverse Laplace transforms are generally computed in much the same way as anti-derivatives in calculus are computed. We work with a small list of "known" transforms, and to compute the inverse Laplace transform of a complicated function $F(s)$, we first break it into a sum of functions whose inverse transforms we already know.

Examples of inverse Laplace transforms

When we applied the Laplace transform to the initial-value problem earlier in this section, we arrived at the expression

$$\mathcal{L}[y] = \frac{1}{s - 1} - \frac{4}{(s - 1)(s + 1)}$$

for the Laplace transform of the solution y of the given initial-value problem. Hence, we find y by computing the inverse Laplace transform

$$y = \mathcal{L}^{-1}\left[\frac{1}{s - 1} - \frac{4}{(s - 1)(s + 1)}\right].$$

By linearity we have

$$y = \mathcal{L}^{-1}\left[\frac{1}{s - 1}\right] - \mathcal{L}^{-1}\left[\frac{4}{(s - 1)(s + 1)}\right],$$

and from the formula for $\mathcal{L}[e^{at}]$ we know that

$$\mathcal{L}^{-1}\left[\frac{1}{s - 1}\right] = e^t.$$

To compute the inverse Laplace transform

$$\mathcal{L}^{-1}\left[\frac{4}{(s - 1)(s + 1)}\right]$$

we must do a little algebra. The idea is to rewrite the term

$$\frac{4}{(s - 1)(s + 1)}$$

as a combination of functions that are known Laplace transforms. In this case (and quite often when using Laplace transforms) we use the technique of partial fractions. That is, we write

$$\frac{4}{(s-1)(s+1)} = \frac{A}{s-1} + \frac{B}{s+1}$$

and solve for the constants A and B, obtaining

$$\frac{4}{(s-1)(s+1)} = \frac{2}{s-1} - \frac{2}{s+1}.$$

Each of the terms on the right-hand side is recognizable as the Laplace transform of an exponential function. Hence

$$\mathcal{L}^{-1}\left[\frac{4}{(s-1)(s+1)}\right] = \mathcal{L}^{-1}\left[\frac{2}{s-1}\right] - \mathcal{L}^{-1}\left[\frac{2}{s+1}\right]$$

$$= 2\mathcal{L}^{-1}\left[\frac{1}{s-1}\right] - 2\mathcal{L}^{-1}\left[\frac{1}{s+1}\right]$$

$$= 2e^t - 2e^{-t}.$$

Completion of the initial-value problem

We showed above that the Laplace transform of the solution y of the initial-value problem

$$\frac{dy}{dt} = y - 4e^{-t}, \quad y(0) = 1$$

is

$$\mathcal{L}[y] = \frac{1}{s-1} - \frac{4}{(s-1)(s+1)}.$$

Hence

$$y = \mathcal{L}^{-1}\left[\frac{1}{s-1} - \frac{4}{(s-1)(s+1)}\right]$$

$$= \mathcal{L}^{-1}\left[\frac{1}{s-1}\right] - \mathcal{L}^{-1}\left[\frac{4}{(s-1)(s+1)}\right]$$

$$= e^t - \left(2e^t - 2e^{-t}\right)$$

$$= -e^t + 2e^{-t}.$$

is the solution. We see that $y(t) \to -\infty$ as $t \to \infty$, as we predicted earlier.

An RC Circuit Example

Consider the RC circuit in Figure 6.2. We let $v_c(t)$ denote the voltage across the capacitor, R the resistance, C the capacitance, and $V(t)$ the voltage supplied by the voltage source. From electric circuit theory we know that the voltage $v_c(t)$ satisfies the differential equation

$$RC\frac{dv_c}{dt} + v_c = V(t).$$

Figure 6.2
RC circuit.

Suppose the voltage source $V(t)$ has constant value 3 and the initial voltage across the capacitor is $v_c(0) = 4$. With the quantities $R = 2$ and $C = 1$ as in Figure 6.2 (rather unrealistic quantities if we use the usual units of ohms for R, farads for C, and volts for V), the initial-value problem is

$$2\frac{dv_c}{dt} + v_c = 3, \quad v_c(0) = 4.$$

Qualitative analysis
Rewriting this initial-value problem in the standard form

$$\frac{dv_c}{dt} = -\frac{v_c}{2} + \frac{3}{2}, \quad v_c(0) = 4,$$

we see that this equation is autonomous with a sink at $v_c = 3$ and no other equilibrium points. Hence, the solution of the initial-value problem tends to $v_c = 3$ as $t \to \infty$, which is confirmed by the slope field and phase line (see Figure 6.3).

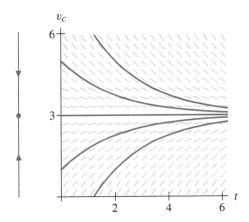

Figure 6.3
Phase line, slope field, and graphs of various solutions of $dv_c/dt = (-v_c + 3)/2$.

Solution using Laplace transforms
The differential equation

$$\frac{dv_c}{dt} = -\frac{v_c}{2} + \frac{3}{2}$$

is both separable and linear, but in this section we choose to use the method of Laplace transforms to find a formula for the solution. First, taking the Laplace transform of both sides of the equation gives

$$\mathcal{L}\left[\frac{dv_c}{dt}\right] = -\frac{1}{2}\mathcal{L}[v_c] + \mathcal{L}\left[\frac{3}{2}\right].$$

Using the formula for $\mathcal{L}[dv_c/dt]$, simplifying, and substituting the initial value $v_c(0) = 4$ yields

$$s\mathcal{L}[v_c] - 4 = -\frac{1}{2}\mathcal{L}[v_c] + \frac{3}{2}\mathcal{L}[1].$$

We have seen earlier that $\mathcal{L}[1] = 1/s$. Therefore we have

$$s\mathcal{L}[v_c] - 4 = -\frac{1}{2}\mathcal{L}[v_c] + \frac{3}{2s}.$$

Solving for $\mathcal{L}[v_c]$ yields

$$\left(s + \frac{1}{2}\right)\mathcal{L}[v_c] = 4 + \frac{3}{2s},$$

which implies that

$$\mathcal{L}[v_c] = \frac{4}{s + 1/2} + \frac{3}{2s(s + 1/2)}$$

$$= \frac{4}{s + 1/2} + \frac{3}{2}\frac{1}{s(s + 1/2)}.$$

To compute the inverse Laplace transform, we use the partial fractions decomposition

$$\frac{1}{s(s + 1/2)} = \frac{2}{s} - \frac{2}{s + 1/2}.$$

We obtain

$$\mathcal{L}[v_c] = \frac{4}{s + 1/2} + \frac{3}{2}\left(\frac{2}{s} - \frac{2}{s + 1/2}\right)$$

$$= \frac{1}{s + 1/2} + \frac{3}{s}.$$

Hence,

$$v_c(t) = \mathcal{L}^{-1}\left[\frac{1}{s + 1/2}\right] + \mathcal{L}^{-1}\left[\frac{3}{s}\right]$$

$$= e^{-t/2} + 3.$$

Note that this solution is consistent with the phase line and slope field for this equation.

EXERCISES FOR SECTION 6.1

In Exercises 1–4, compute the Laplace transform of the given function from the definition.

1. $f(t) = 3$ (a constant function)

2. $g(t) = t$

3. $h(t) = -5t^2$

4. $k(t) = t^5$

5. Verify that

$$\mathcal{L}[t^n] = \frac{n!}{s^{n+1}} \quad (s > 0).$$

[*Hint*: A rigorous derivation of this formula requires mathematical induction.]

6. Using

$$\mathcal{L}[t^n] = \frac{n!}{s^{n+1}} \quad (s > 0),$$

give a formula for the Laplace transform of an arbitrary nth degree polynomial

$$p(t) = a_0 + a_1 t + a_2 t^2 + \cdots + a_n t^n,$$

where the a_i's are constants.

In Exercises 7–14, find the inverse Laplace transform of the given function.

7. $\dfrac{1}{s - 3}$

8. $\dfrac{5}{3s}$

9. $\dfrac{2}{3s + 5}$

10. $\dfrac{14}{(3s + 2)(s - 4)}$

11. $\dfrac{4}{s(s + 3)}$

12. $\dfrac{5}{(s - 1)(s - 2)}$

13. $\dfrac{2s + 1}{(s - 1)(s - 2)}$

14. $\dfrac{2s^2 + 3s - 2}{s(s + 1)(s - 2)}$

In Exercises 15–24,
  (a) compute the Laplace transform of both sides of the equation;
  (b) substitute the initial conditions and solve for the Laplace transform of the solution;
  (c) find a function whose Laplace transform is the same as the solution; and
  (d) check that you have found the solution of the initial-value problem.

15. $\dfrac{dy}{dt} = -y + e^{-2t}, \quad y(0) = 2$

16. $\dfrac{dy}{dt} + 5y = e^{-t}, \quad y(0) = 2$

17. $\dfrac{dy}{dt} + 7y = 1, \quad y(0) = 3$

18. $\dfrac{dy}{dt} + 4y = 6, \quad y(0) = 0$

19. $\dfrac{dy}{dt} + 9y = 2, \quad y(0) = -2$

20. $\dfrac{dy}{dt} = -y + 2, \quad y(0) = 4$

21. $\dfrac{dy}{dt} = -y + e^{-2t}, \quad y(0) = 1$

22. $\dfrac{dy}{dt} = 2y + t, \quad y(0) = 0$

23. $\dfrac{dy}{dt} = -y + t^2, \quad y(0) = 1$

24. $\dfrac{dy}{dt} + 4y = 2 + 3t, \quad y(0) = 1$

25. Find the general solution of the equation

$$\frac{dy}{dt} = 2y + 2e^{-3t}.$$

(This equation is linear, but please use the method of Laplace transforms.)

26. Suppose $g(t) = \int f(t)\,dt$; that is, $g(t)$ is the antiderivative of $f(t)$. Express the Laplace transform of $g(t)$ in terms of the Laplace transform of $f(t)$.

27. All of the examples in this section and all the differential equations in this exercise set are linear. Trying to use the Laplace transform on even the simplest nonlinear equation leads quickly to headaches. Try using the Laplace transform to find the solution of the initial-value problem

$$\frac{dy}{dt} = y^2, \quad y(0) = 1.$$

Give a short explanation outlining where you got stuck and why.

6.2 DISCONTINUOUS FUNCTIONS

In Section 6.1 we saw that using Laplace transforms to find the solutions of a linear differential equation involves entirely different ideas than our previous methods. The operations of differentiation and integration are replaced with algebra. However, Laplace transforms are not a panacea. They apply only to linear equations and even though they replace calculus with algebra, the algebra can be very complicated.

Given these limitations, it is important to ask why we need another method for solving linear equations. The remainder of this chapter is devoted to applications of the Laplace transform that let us study new types of equations and that give us new insights into familiar equations.

In applications discontinuous functions arise naturally. For example, the sudden introduction of a new species or disease affecting a population or the turning on or off of a light switch are discontinuous phenomena. Differential equations containing discontinuous functions are difficult to handle analytically using our previous methods, but Laplace transforms can sometimes tame these discontinuities, as the following examples show.

Laplace transform of a Heaviside function
For $a \geq 0$, let $u_a(t)$ denote the function

$$u_a(t) = \begin{cases} 0, & \text{if } t < a; \\ 1, & \text{if } t \geq a. \end{cases}$$

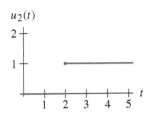

$u_2(t)$

Figure 6.4
Graph of the Heaviside
function $u_2(t)$.

Thus $u_a(t)$ has a discontinuity at $t = a$ where it jumps from 0 to 1. For example, the function $u_2(t)$ has a discontinuity at $t = 2$, where it jumps from 0 to 1 (see Figure 6.4). A function of this form is called a step function, or **Heaviside function** (named for the engineer Oliver Heaviside). It is useful when modeling discontinuous processes such as turning on a light switch.

The Laplace transform of $u_a(t)$ is

$$\mathcal{L}[u_a] = \int_0^\infty u_a(t)\, e^{-st}dt.$$

To compute this integral, we use the definition of $u_a(t)$ and break the computation into two pieces,

$$\mathcal{L}[u_a] = \int_0^a u_a(t)\, e^{-st}dt + \int_a^\infty u_a(t)\, e^{-st}dt.$$

The first integral is zero because $u_a(t)$ is zero for $t < a$. We can simplify the second integral because $u_a(t) = 1$ for $t \geq a$, hence

$$\mathcal{L}[u_a] = \int_a^\infty e^{-st}dt$$

$$= \lim_{b \to \infty} \int_a^b e^{-st}dt$$

$$= \lim_{b \to \infty} \left. \frac{e^{-st}}{-s} \right|_a^b$$

$$= \lim_{b \to \infty} \frac{e^{-sb}}{-s} - \frac{e^{-as}}{-s}$$

$$= 0 + \frac{e^{-as}}{s}.$$

We have established the formula

$$\mathcal{L}[u_a] = \frac{e^{-as}}{s}.$$

Even though the original function u_a has a discontinuity at $t = a$, the Laplace transform is continuous for all $s > 0$. This *smoothing property* is a very useful aspect of Laplace transforms.

A Differential Equation with a Discontinuity

Consider the initial-value problem

$$\frac{dy}{dt} = -y + u_3(t), \quad y(0) = 2.$$

We can rewrite this differential equation in the form

$$\frac{dy}{dt} = \begin{cases} -y, & \text{if } t < 3; \\ -y + 1, & \text{if } t \geq 3. \end{cases}$$

Qualitative analysis

Since this equation really consists of a pair of autonomous first-order equations, we can use the qualitative methods of Chapter 1 to analyze the behavior of solutions. The slope field is shown in Figure 6.5. For $t < 3$ the equation $dy/dt = -y$ has a sink at $y = 0$, and all solutions approach this equilibrium point. For $t \geq 3$ the equation $dy/dt = -y + 1$ has a sink at $y = 1$, which attracts all other solutions. Hence the solution of our initial-value problem initially decreases toward $y = 0$. Then, at $t = 3$, the solution begins to approach $y = 1$. If $y(3) < 1$, then the solution switches from decreasing to increasing at $t = 3$. If $y(3) > 1$, the solution continues to decrease toward $y = 1$ for all $t > 3$.

In order to decide if the solution of the initial-value problem increases or decreases toward 1 as $t \to \infty$, we must compute the value of the solution at time $t = 3$. The initial-value problem

$$\frac{dy}{dt} = -y, \quad y(0) = 2$$

has solution $y(t) = 2e^{-t}$. Hence $y(3) = 2e^{-3}$ and we see that $y(3) < 1$. Consequently the solution increases for $t > 3$ toward $y = 1$.

Note that to obtain even a qualitative description of the solution we had to find the analytic form of the solution up to time $t = 3$, when the term $u_3(t)$ "turns on."

Solution using Laplace transforms

As in Section 6.1, to solve the initial-value problem

$$\frac{dy}{dt} = -y + u_3(t), \quad y(0) = 2$$

using Laplace transforms, we first take the Laplace transform of both sides of the

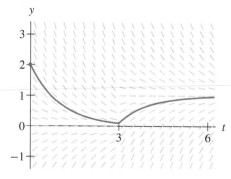

Figure 6.5
Slope field for

$$\frac{dy}{dt} = -y + u_3(t)$$

and graph of solution with $y(0) = 2$. Note that this solution is decreasing for $0 \leq t \leq 3$, increasing for $t \geq 3$, and $y(t) \to 1$ as $t \to \infty$.

differential equation (and use the linearity of the Laplace transform) to obtain

$$\mathcal{L}\left[\frac{dy}{dt}\right] = -\mathcal{L}[y] + \mathcal{L}[u_3].$$

Using the rule for the Laplace transform of a derivative and our computation of $\mathcal{L}[u_3] = e^{-3s}/s$, we have

$$s\mathcal{L}[y] - y(0) = -\mathcal{L}[y] + \frac{e^{-3s}}{s}.$$

Substituting $y(0) = 2$ and solving for $\mathcal{L}[y]$ gives

$$\mathcal{L}[y] = \frac{2}{s+1} + \frac{e^{-3s}}{s(s+1)}.$$

Hence the solution is

$$y = \mathcal{L}^{-1}\left[\frac{2}{s+1}\right] + \mathcal{L}^{-1}\left[\frac{e^{-3s}}{s(s+1)}\right].$$

Now

$$\mathcal{L}^{-1}\left[\frac{2}{s+1}\right] = 2e^{-t},$$

but to compute

$$\mathcal{L}^{-1}\left[\frac{e^{-3s}}{s(s+1)}\right]$$

we need another rule for computing Laplace transforms.

Shifting the Origin on the t-Axis

Given a function $f(t)$, suppose that we wish to consider a function $g(t)$, which is the same as the function $f(t)$, but shifted so that time $t = 0$ for f corresponds to some later time, say $t = a$, for g. For time $t < a$ we let $g(t) = 0$. (So g is the same as f except that g is "turned on" at time $t = a$.) An efficient way to write $g(t)$ is

$$g(t) = u_a(t)f(t-a).$$

Note that

$$g(a) = u_a(a)f(a-a) = f(0)$$

and, if $b > 0$,

$$g(a+b) = u_a(a+b)f(b) = f(b)$$

as desired. For example, if $f(t) = e^{-t}$ and $a = 3$, then $g(t) = u_3(t)e^{-(t-3)}$ (see Figures 6.6 and 6.7).

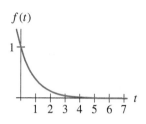

Figure 6.6
Graph of $f(t) = e^{-t}$.

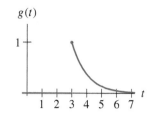

Figure 6.7
Graph of $g(t) = u_3(t)e^{-(t-3)}$.

To compute the Laplace transform of $g(t)$, we return to the definition

$$\mathcal{L}[g] = \int_0^\infty g(t)\, e^{-st} dt.$$

Using the fact that $g(t) = 0$ for $t < a$ and that $g(t) = f(t-a)$ for $t \geq a$, we obtain

$$\mathcal{L}[g] = \int_a^\infty f(t-a)\, e^{-st} dt.$$

The u-substitution $u = t - a$ in the integral gives

$$\mathcal{L}[g] = \int_0^\infty f(u)\, e^{-s(u+a)} du$$

$$= e^{-sa} \int_0^\infty f(u)\, e^{-su} du$$

$$= e^{-sa} \mathcal{L}[f] = e^{-sa} F(s).$$

Hence we can express the Laplace transform of $g(t) = u_a(t) f(t-a)$ in terms of the Laplace transform of $f(t)$ by the following rule:

SHIFTING THE ORIGIN ON THE t-AXIS   If $\mathcal{L}[f] = F(s)$, then

$$\mathcal{L}[u_a(t) f(t-a)] = e^{-as} F(s). \quad \blacksquare$$

(This rule, along with the other rules we develop for Laplace transforms, can be found in the table on page 626.)

For the example just given, if $g(t) = u_3(t)e^{-(t-3)}$, then the Laplace transform of $g(t)$ is

$$\mathcal{L}[g] = e^{-3s} \mathcal{L}[e^{-t}] = \frac{e^{-3s}}{s+1}.$$

Completion of the initial-value problem

For the initial-value problem

$$\frac{dy}{dt} = -y + u_3(t), \quad y(0) = 2,$$

we showed that

$$y = \mathcal{L}^{-1}\left[\frac{2}{s+1}\right] + \mathcal{L}^{-1}\left[\frac{e^{-3s}}{s(s+1)}\right].$$

Using partial fractions, we write

$$\frac{1}{s(s+1)} = \frac{1}{s} - \frac{1}{s+1},$$

so

$$\mathcal{L}^{-1}\left[\frac{e^{-3s}}{s(s+1)}\right] = \mathcal{L}^{-1}\left[\frac{e^{-3s}}{s}\right] - \mathcal{L}^{-1}\left[\frac{e^{-3s}}{s+1}\right]$$

$$= u_3(t) - u_3(t)e^{-(t-3)}.$$

Hence

$$y(t) = 2e^{-t} + u_3(t)\left(1 - e^{-(t-3)}\right).$$

Note that the second term is nonzero only for $t > 3$. That is, this term "turns on" when the $u_3(t)$-term in the differential equation "turns on" (see Figure 6.5).

An RC Circuit with an Exponentially Decaying Voltage Source

Consider the RC circuit in Figure 6.8 with voltage source

$$V(t) = 2u_4(t)e^{-(t-4)}.$$

The voltage source turns on with a voltage 2 at time $t = 4$ and then decays exponentially as t increases (see Figure 6.9).

The differential equation for the voltage v_c across the capacitor is

$$RC\frac{dv_c}{dt} + v_c = V(t).$$

Suppose the initial voltage across the capacitor is $v_c(0) = 5$. Taking the (unrealistic) values $R = 1$ and $C = 1/3$ and using $V(t)$ given above, we have the initial-value problem

$$\frac{1}{3}\frac{dv_c}{dt} + v_c = 2u_4(t)e^{-(t-4)}, \quad v_c(0) = 5.$$

Figure 6.8
RC circuit.

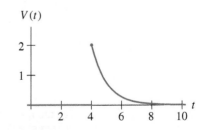

Figure 6.9
Voltage source $V(t) = 2u_4(t)e^{-(t-4)}$.

Multiplying both sides by 3 and moving the v_c term to the right-hand side, we obtain

$$\frac{dv_c}{dt} = -3v_c + 6u_4(t)e^{-(t-4)}, \quad v_c(0) = 5.$$

Qualitative analysis

Again we can rewrite this differential equation as

$$\frac{dv_c}{dt} = \begin{cases} -3v_c, & \text{if } t < 4; \\ -3v_c + 6e^{-(t-4)}, & \text{if } t \geq 4. \end{cases}$$

The solution with initial condition $v_c(0) = 5$ for $0 \leq t < 4$ is $v_c(t) = 5e^{-3t}$. This solution decreases toward $v_c = 0$ quickly as t increases. For $t \geq 4$ the positive term $6e^{-(t-4)}$ causes solutions near $v_c = 0$ to increase. However, because $6e^{-(t-4)}$ tends to zero as t increases, solutions eventually tend to zero. Hence we expect the solution of the initial-value problem to decrease quickly toward $v_c = 0$ for $0 \leq t < 4$, increase for t slightly larger than 4, and then decrease again toward zero as t increases. This behavior is confirmed by the slope field and the graph of the solution in Figure 6.10.

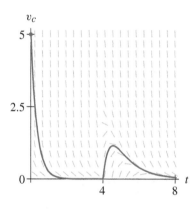

Figure 6.10
Slope field for

$$\frac{dv_c}{dt} = -3v_c + 6u_4(t)e^{-(t-4)}$$

and graph of solution with $v_c(0) = 5$. Note that the solution of the initial-value problem decreases quickly toward $v_c = 0$ for $0 \leq t < 4$, increases for t slightly larger than 4, and then decreases again toward zero as $t \to \infty$.

Solution using Laplace transforms

Starting with the initial-value problem

$$\frac{dv_c}{dt} = -3v_c + 6u_4(t)e^{-(t-4)}, \quad v_c(0) = 5,$$

we take Laplace transforms of both sides of the differential equation to obtain

$$\mathcal{L}\left[\frac{dv_c}{dt}\right] = -3\mathcal{L}[v_c] + 6\mathcal{L}[u_4(t)e^{-(t-4)}],$$

which simplifies to

$$s\mathcal{L}[v_c] - v_c(0) = -3\mathcal{L}[v_c] + \frac{6e^{-4s}}{s+1}.$$

Substituting $v_c(0) = 5$ and solving for $\mathcal{L}[v_c]$ gives

$$\mathcal{L}[v_c] = \frac{5}{s+3} + \frac{6e^{-4s}}{(s+3)(s+1)}.$$

Thus

$$v_c = \mathcal{L}^{-1}\left[\frac{5}{s+3}\right] + \mathcal{L}^{-1}\left[\frac{6e^{-4s}}{(s+3)(s+1)}\right].$$

Using the partial fractions decomposition

$$\frac{6}{(s+3)(s+1)} = \frac{-3}{s+3} + \frac{3}{s+1},$$

we have

$$v_c = \mathcal{L}^{-1}\left[\frac{5}{s+3}\right] - 3\mathcal{L}^{-1}\left[\frac{e^{-4s}}{s+3}\right] + 3\mathcal{L}^{-1}\left[\frac{e^{-4s}}{s+1}\right].$$

Hence the solution of the initial-value problem is

$$v_c(t) = 5e^{-3t} - 3u_4(t)e^{-3(t-4)} + 3u_4(t)e^{-(t-4)}.$$

For $0 \le t < 4$ the solution is $v_c(t) = 5e^{-3t}$, which decreases quickly toward zero. For $t \ge 4$ the solution is

$$v_c(t) = 5e^{-3t} - 3e^{-3(t-4)} + 3e^{-(t-4)}.$$

Each term in this solution tends to 0 as t increases, as predicted (see Figure 6.10).

EXERCISES FOR SECTION 6.2

1. For $a \ge 0$, let $g_a(t)$ denote the function

$$g_a(t) = \begin{cases} 1, & \text{if } t < a; \\ 0, & \text{if } t \ge a. \end{cases}$$

(a) Give an expression for $g_a(t)$ using the Heaviside function $u_a(t)$.
(b) Compute $\mathcal{L}[g_a]$.

2. (a) Suppose $a \ge 0$. Compute the Laplace transform of the function

$$r_a(t) = \begin{cases} 0, & \text{if } t < a; \\ k(t - a) & \text{if } t \ge a. \end{cases}$$

[*Hint:* Your solution should contain a and k as parameters.]
(b) Sketch the graph of $r_a(t)$ and comment on why it is called a *ramp function*.

3. Suppose $a \geq 0$. Compute the Laplace transform of the function

$$g_a(t) = \begin{cases} t/a, & \text{if } t < a; \\ 1, & \text{if } t \geq a \end{cases}$$

[*Hint*: Your solution should contain a as a parameter.]

In Exercises 4–7, find the inverse Laplace transform of the given function.

4. $\dfrac{e^{-2s}}{s-3}$

5. $\dfrac{e^{-3s}}{(s-1)(s-2)}$

6. $\dfrac{4e^{-2s}}{s(s+3)}$

7. $\dfrac{14e^{-s}}{(3s+2)(s-4)}$

In Exercises 8–13, solve the given initial-value problem.

8. $\dfrac{dy}{dt} = u_2(t), \quad y(0) = 3$

9. $\dfrac{dy}{dt} + 9y = u_5(t), \quad y(0) = -2$

10. $\dfrac{dy}{dt} + 7y = u_2(t), \quad y(0) = 3$

11. $\dfrac{dy}{dt} = -y + u_2(t)e^{-2(t-2)}, \quad y(0) = 1$

12. $\dfrac{dy}{dt} = -y + 2u_3(t), \quad y(0) = 4$

13. $\dfrac{dy}{dt} = -y + u_1(t)(t-1), \quad y(0) = 2$

14. Solve the initial-value problem

$$\frac{dy}{dt} = -2y + u_2(t)e^{-t}, \quad y(0) = 3.$$

[*Warning*: This is harder than it looks.]

15. Suppose $a \geq 0$. Find the general solution of

$$\frac{dy}{dt} = -y + u_a(t).$$

[*Hint*: Your solution should contain a as a parameter.]

16. Suppose $f(t)$ is a periodic function with period T; that is,

$$f(t+T) = f(t) \quad \text{for all } t.$$

Show that

$$\mathcal{L}[f] = \frac{1}{1-e^{-Ts}} \int_0^T f(t)\,e^{-st}\,dt.$$

17. Compute the Laplace transform of the square wave

$$w(t) = \begin{cases} 1, & \text{if } 2n \le t < 2n+1 \\ & \text{for some integer } n; \\ -1, & \text{if } 2n+1 \le t < 2n+2 \\ & \text{for some integer } n. \end{cases}$$

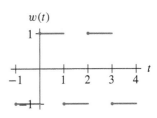

A square wave with period 2.

18. Compute the Laplace transform of the saw-tooth function

$$z(t) = t - [t],$$

where $[t]$ denotes the greatest integer less than or equal to t. [*Hint*: Use the formula developed in Exercise 16.]

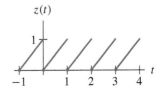

A sawtooth wave with period 1.

19. Consider the initial-value problem

$$\frac{dy}{dt} = -y + w(t), \quad y(0) = 0,$$

where $w(t)$ is the square wave in Exercise 17.

(a) Compute the Laplace transform of the solution.

(b) Describe the qualitative behavior of this solution. [*Hint*: Knowing the Laplace transform of the solution may not help with this part, but see Section 6.6.]

20. Consider the initial-value problem

$$\frac{dy}{dt} = -y + z(t), \quad y(0) = 0,$$

where $z(t)$ is the sawtooth wave in Exercise 18.

(a) Compute the Laplace transform of the solution.

(b) Describe the qualitative behavior of this solution. [*Hint*: Knowing the Laplace transform of the solution may not help with this part, but see Section 6.6.]

6.3 SECOND-ORDER EQUATIONS

In this section we extend the method of Laplace transforms to second-order, constant-coefficient, forced linear equations, that is, equations of the form

$$\frac{d^2y}{dt^2} + p\frac{dy}{dt} + qy = f(t),$$

where p and q are constants. We have dealt with equations of this form for particular forcing functions $f(t)$ in Chapter 4. Laplace transforms let us study a larger class of forcing functions, including some with discontinuities.

To begin, we recall that the functions $\sin \omega t$, $\cos \omega t$, $e^{at} \sin \omega t$, and $e^{at} \cos \omega t$ appear very often in the study of second-order equations. Hence our first task is to compute the Laplace transform of these functions.

Laplace Transform of Sine and Cosine

To compute the Laplace transform of $\sin \omega t$, we could use the definition and evaluate the integral

$$\mathcal{L}[\sin \omega t] = \int_0^\infty \sin \omega t \, e^{-st} \, dt.$$

Another approach is to use complex exponentials. These alternatives lead to good exercises in integration by parts and partial fractions with complex numbers, respectively (see Exercises 1–4 and 19–21).

The method we use to compute $\mathcal{L}[\sin \omega t]$ is to take advantage of the fact that we already know that $y(t) = \sin \omega t$ is the solution of the initial-value problem

$$\frac{d^2 y}{dt^2} + \omega^2 y = 0, \quad y(0) = 0, \quad y'(0) = \omega.$$

Hence we can compute $\mathcal{L}[\sin \omega t]$ by computing the Laplace transform of the solution of this initial-value problem.

First note that two applications of the rule for the Laplace transform of a derivative yields

$$\mathcal{L}\left[\frac{d^2 y}{dt^2}\right] = s\mathcal{L}\left[\frac{dy}{dt}\right] - y'(0)$$

$$= s(s\mathcal{L}[y] - y(0)) - y'(0)$$

$$= s^2 \mathcal{L}[y] - sy(0) - y'(0).$$

To solve the initial-value problem

$$\frac{d^2 y}{dt^2} + \omega^2 y = 0, \quad y(0) = 0, \quad y'(0) = \omega$$

using Laplace transforms, we first take the Laplace transform of both sides of the differential equation, which yields

$$\mathcal{L}\left[\frac{d^2 y}{dt^2}\right] + \omega^2 \mathcal{L}[y] = \mathcal{L}[0].$$

Simplifying and using the fact that $\mathcal{L}[0] = 0$, we have

$$s^2 \mathcal{L}[y] - sy(0) - y'(0) + \omega^2 \mathcal{L}[y] = 0.$$

Substituting the initial conditions $y(0) = 0$ and $y'(0) = \omega$ gives

$$s^2 \mathcal{L}[y] - \omega + \omega^2 \mathcal{L}[y] = 0$$
$$\left(s^2 + \omega^2\right) \mathcal{L}[y] = \omega,$$

and therefore

$$\mathcal{L}[y] = \frac{\omega}{s^2 + \omega^2}.$$

But we already know that the solution of the initial-value problem is $y(t) = \sin \omega t$, so we have

$$\mathcal{L}[\sin \omega t] = \frac{\omega}{s^2 + \omega^2}.$$

To compute $\mathcal{L}[\cos \omega t]$ we could repeat this computation using the initial-value problem

$$\frac{d^2 y}{dt^2} + \omega^2 y = 0, \quad y(0) = 1, \quad y'(0) = 0$$

whose solution is $y(t) = \cos \omega t$ (see Exercise 5). Alternately, we can use the fact that

$$\cos \omega t = \frac{1}{\omega} \frac{d(\sin \omega t)}{dt}.$$

Taking the Laplace transform of both sides of this equation yields

$$\mathcal{L}[\cos \omega t] = \frac{1}{\omega} \mathcal{L}\left[\frac{d(\sin \omega t)}{dt}\right]$$

$$= \frac{1}{\omega}(s \mathcal{L}[\sin \omega t] - \sin 0)$$

$$= \frac{s}{\omega} \mathcal{L}[\sin \omega t]$$

$$= \frac{s}{\omega} \frac{\omega}{s^2 + \omega^2},$$

so

$$\mathcal{L}[\cos \omega t] = \frac{s}{s^2 + \omega^2}.$$

(These formulas are included in the table of Laplace transforms on page 626.)

Shifting the Origin on the s-Axis

Next we consider the Laplace transforms of the functions $e^{at} \sin \omega t$ and $e^{at} \cos \omega t$. In general, the Laplace transform of a product is quite complicated. However, if one of the factors is an exponential, that factor combines nicely with the factor e^{-st} in the definition of the transform.

Suppose we are given a function $f(t)$ and we know that its Laplace transform is $F(s)$. To compute the Laplace transform of $e^{at} f(t)$, we recall the definition and write

$$\mathcal{L}[e^{at} f(t)] = \int_0^\infty e^{at} f(t) e^{-st} dt$$

$$= \int_0^\infty f(t) e^{-(s-a)t} dt$$

$$= F(s - a).$$

We obtain our last important rule for the Laplace transform:

SHIFTING THE ORIGIN ON THE s-AXIS   If $\mathcal{L}[f] = F(s)$, then

$$\mathcal{L}[e^{at} f(t)] = F(s - a).$$

In other words, multiplying $f(t)$ by e^{at} corresponds to replacing the argument s of its Laplace transform by $s - a$. ∎

(This rule, along with the other important rules for Laplace transforms, can be found in the table on page 626.)

Two examples that use this rule
We know that

$$\mathcal{L}[\cos 2t] = \frac{s}{s^2 + 4},$$

so to compute the Laplace transform of $e^{-3t} \cos 2t$, we replace s by $s + 3$ on the right-hand side twice. We obtain

$$\mathcal{L}[e^{-3t} \cos 2t] = \frac{s + 3}{(s + 3)^2 + 4} = \frac{s + 3}{s^2 + 6s + 13}.$$

We use this rule to compute inverse Laplace transforms as well. For example, to compute

$$\mathcal{L}^{-1}\left[\frac{1}{s^2 + 2s + 5} \right]$$

we first note that the roots of $s^2 + 2s + 5$ are $-1 \pm 2i$. If the denominator had real roots, we could factor it and use partial fractions to compute the inverse transform. Since the roots in this case are complex, we write

$$s^2 + 2s + 5 = (s + 1)^2 + 4.$$

(Recall that this algebraic computation is called "completing the square." For a review of this technique, see Exercises 11–14.) Next, we note that

$$\frac{1}{(s + 1)^2 + 4} = F(s + 1),$$

where

$$F(s) = \frac{1}{s^2 + 4}.$$

We know that

$$\mathcal{L}^{-1}\left[\frac{1}{s^2+4}\right] = \frac{1}{2}\mathcal{L}^{-1}\left[\frac{2}{s^2+4}\right] = \frac{1}{2}\sin 2t.$$

So, using the rule, we have

$$\mathcal{L}^{-1}\left[\frac{1}{(s+1)^2+4}\right] = \frac{1}{2}e^{-t}\sin 2t.$$

A Forced Harmonic Oscillator

In Chapter 4 the forced harmonic oscillator equation was introduced as a model of a mass attached to a spring sliding back and forth on a table with an external force caused by, for example, tilting the table. As a first example of the Laplace transform method for such an equation, consider the initial-value problem

$$\frac{d^2y}{dt^2} + 4y = 3\cos t, \quad y(0) = y'(0) = 0.$$

This equation models a unit mass sliding on a frictionless table with spring constant 4 and a periodic external forcing function of $3\cos t$.

Qualitative analysis
The characteristic polynomial for the unforced harmonic oscillator

$$\frac{d^2y}{dt^2} + 4y = 0$$

is $s^2+4 = 0$. Hence the eigenvalues are $s = \pm 2i$, and the natural period is $2\pi/2 = \pi$. Since the forcing has period 2π, the forced equation

$$\frac{d^2y}{dt^2} + 4y = 3\cos t$$

is not a resonant system, and every solution remains bounded for all time. In fact solutions are combinations of trigonometric functions with period π and 2π.

Solution using Laplace transforms
To use the Laplace transform on the initial-value problem

$$\frac{d^2y}{dt^2} + 4y = 3\cos t, \quad y(0) = y'(0) = 0,$$

we take the Laplace transform of both sides of the differential equation, obtaining

$$\mathcal{L}\left[\frac{d^2y}{dt^2}\right] + 4\mathcal{L}[y] = 3\mathcal{L}[\cos t].$$

Simplifying, we have

$$s^2\mathcal{L}[y] - sy(0) - y'(0) + 4\mathcal{L}[y] = \frac{3s}{s^2+1}.$$

Substituting the initial conditions yields

$$(s^2 + 4)\mathcal{L}[y] = \frac{3s}{s^2 + 1},$$

which gives

$$\mathcal{L}[y] = \frac{3s}{(s^2 + 4)(s^2 + 1)}.$$

Therefore, the desired solution is

$$y(t) = \mathcal{L}^{-1}\left[\frac{3s}{(s^2 + 4)(s^2 + 1)}\right].$$

Using partial fractions, we have

$$y(t) = \mathcal{L}^{-1}\left[\frac{-s}{s^2 + 4}\right] + \mathcal{L}^{-1}\left[\frac{s}{s^2 + 1}\right],$$

which gives

$$y(t) = -\cos 2t + \cos t.$$

This solution is bounded and behaves as predicted (see Figure 6.11).

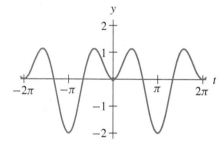

Figure 6.11
Graph of the solution of the initial-value problem

$$\frac{d^2 y}{dt^2} + 4y = 3\cos t$$

with $y(0) = y'(0) = 0$.

An Oscillator with Discontinuous Forcing

Next we consider the initial-value problem

$$\frac{d^2 y}{dt^2} + 2\frac{dy}{dt} + 5y = h(t), \quad y(0) = y'(0) = 0,$$

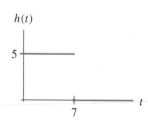

Figure 6.12
Graph of the function $h(t)$.

where $h(t)$ is given by

$$h(t) = \begin{cases} 5, & \text{if } t < 7; \\ 0, & \text{if } t \geq 7; \end{cases}$$

(see Figure 6.12). Note that $h(t) = 5(1 - u_7(t))$. This differential equation is a model for a unit mass attached to a spring with spring constant 5 sliding on a table with damping coefficient 2 (see Chapter 4).

We can interpret the initial conditions and forcing function as follows: Suppose at time $t = 0$, the mass is held at rest at $y = 0$. When $t < 7$, the table is tilted so gravity provides a force of 5 units stretching the spring. At time $t = 7$, the table is suddenly returned to the level position (see Figure 6.13).

Figure 6.13
Schematic of a mass-spring system on a tilted table ($t < 7$) and a level table ($t > 7$).

Qualitative analysis

The characteristic polynomial of the equation

$$\frac{d^2y}{dt^2} + 2\frac{dy}{dt} + 5y = 0$$

is $s^2 + 2s + 5 = 0$, and the eigenvalues are $s = -1 \pm 2i$. Hence this oscillator is underdamped. When $0 < t < 7$, the mass slides down the table and oscillates about an equilibrium position for which the spring is slightly stretched. For $t > 7$, when the table is level, the mass slides back toward the original rest position and oscillates with exponentially decreasing amplitude.

Solution using Laplace transforms

We begin as before by taking the Laplace transform of both sides of the differential equation. Using linearity, we obtain

$$\mathcal{L}\left[\frac{d^2y}{dt^2}\right] + 2\mathcal{L}\left[\frac{dy}{dt}\right] + 5\mathcal{L}[y] = \mathcal{L}[h(t)].$$

Hence

$$\left(s^2\mathcal{L}[y] - sy(0) - y'(0)\right) + 2\left(s\mathcal{L}[y] - y(0)\right) + 5\mathcal{L}[y] = \mathcal{L}[h(t)].$$

Collecting terms, substituting the initial conditions $y(0) = 0$ and $y'(0) = 0$, and evaluating

$$\mathcal{L}[h(t)] = 5\mathcal{L}[1 - u_7(t)] = 5\left(\frac{1}{s} - \frac{e^{-7s}}{s}\right),$$

we have

$$(s^2 + 2s + 5)\mathcal{L}[y] = 5\left(\frac{1}{s} - \frac{e^{-7s}}{s}\right).$$

Solving for $\mathcal{L}[y]$, we obtain

$$\mathcal{L}[y] = \frac{5}{s(s^2 + 2s + 5)} - \frac{5e^{-7s}}{s(s^2 + 2s + 5)}.$$

In other words,

$$y = \mathcal{L}^{-1}\left[\frac{5}{s(s^2 + 2s + 5)} - \frac{5e^{-7s}}{s(s^2 + 2s + 5)}\right].$$

To compute the right-hand side, we first use the partial fractions decomposition

$$\frac{5}{s(s^2 + 2s + 5)} = \frac{1}{s} - \frac{s + 2}{s^2 + 2s + 5}.$$

We know that $\mathcal{L}^{-1}[1/s] = 1$, so we only need to work with the second term. The quadratic denominator in this term has complex roots, and we first complete the square $s^2 + 2s + 5 = (s + 1)^2 + 4$. In order to compute the inverse Laplace transform, we rewrite

$$\frac{s + 2}{s^2 + 2s + 5} = \frac{s + 2}{(s + 1)^2 + 4}$$

$$= \frac{s + 1}{(s + 1)^2 + 4} + \frac{1}{(s + 1)^2 + 4}$$

$$= \frac{s + 1}{(s + 1)^2 + 4} + \frac{1}{2}\frac{2}{(s + 1)^2 + 4}.$$

Thus,

$$\mathcal{L}^{-1}\left[\frac{s + 2}{s^2 + 2s + 5}\right] = e^{-t}\cos 2t + \frac{1}{2}e^{-t}\sin 2t = e^{-t}\left(\cos 2t + \frac{1}{2}\sin 2t\right),$$

and

$$\mathcal{L}^{-1}\left[\frac{5}{s(s^2 + 2s + 5)}\right] = \mathcal{L}^{-1}\left[\frac{1}{s} - \frac{s + 2}{s^2 + 2s + 5}\right]$$

$$= 1 - e^{-t}\left(\cos 2t + \frac{1}{2}\sin 2t\right).$$

The inverse Laplace transform of

$$\frac{5e^{-7s}}{s(s^2 + 2s + 5)}$$

involves the Heaviside function $u_7(t)$ and the inverse Laplace transform that we just calculated. Thus

$$\mathcal{L}^{-1}\left[\frac{5e^{-7s}}{s(s^2 + 2s + 5)}\right] = u_7(t)\left(1 - e^{-(t-7)}\left(\cos 2(t - 7) + \frac{1}{2}\sin 2(t - 7)\right)\right).$$

The solution to our original initial-value problem is

$$y(t) = 1 - e^{-t}\left(\cos 2t + \tfrac{1}{2}\sin 2t\right)$$

$$-u_7(t)\left(1 - e^{-(t-7)}\left(\cos 2(t-7) + \tfrac{1}{2}\sin 2(t-7)\right)\right).$$

Let $v = dy/dt$. When $0 \le t < 7$, the solution oscillates toward the equilibrium position $(y, v) = (1, 0)$, an equilibrium point if the table is tilted. For $t > 7$, the terms that do not involve the exponential factors cancel, and the remaining terms tend to oscillate with decreasing amplitude to the equilibrium point $(y, v) = (0, 0)$ as predicted (see Figure 6.14).

Note that the graph of $y(t)$ is differentiable at $t = 7$, but the graph of $v(t)$ has a corner at $t = 7$. Since the forcing "turns off" at $t = 7$, the second derivative of y has a jump discontinuity, and the first derivative has a corner.

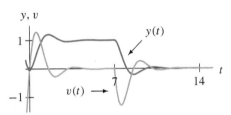

Figure 6.14
The $y(t)$- and $v(t)$-graphs of the solution of the initial-value problem

$$\frac{d^2y}{dt^2} + 2\frac{dy}{dt} + 5y = h(t),$$

with $y(0) = 0$ and $y'(0) = 0$.

Sinusoidal Forcing

We can imagine many different types of forcing that can suddenly appear or disappear. For example, suppose that we again have a mass attached to a spring on a table. For $t \le 5$ we suppose the table is horizontal, so there is no external force on the mass. At $t = 5$, an earthquake begins to shake the table, rocking it back and forth, providing a periodic forcing term.

As an example, let the mass $m = 1$ and the spring constant $k = 4$. We assume the oscillator is undamped, so $b = 0$. For the forcing term we take the function $3u_5(t)\sin(t-5)$, which has amplitude 3 and period 2π and begins at time $t = 5$. The differential equation is

$$\frac{d^2y}{dt^2} + 4y = 3u_5(t)\sin(t-5).$$

We take initial conditions $y(0) = 1$ and $y'(0) = 0$.

Qualitative analysis

For $t < 5$, this equation represents an unforced, undamped harmonic oscillator with natural period π, so the solution with initial conditions $y(0) = 1$ and $y'(0) = 0$ oscillates with constant amplitude. At time $t = 5$, the forcing turns on. The period of the forcing is 2π, so this system is not a resonant one and all solutions are bounded. In fact, all solutions are combinations of trigonometric functions with periods π and 2π.

Solution using Laplace transforms

To apply the Laplace transform method to the initial-value problem

$$\frac{d^2y}{dt^2} + 4y = 3u_5(t)\sin(t-5), \quad y(0) = 1, \quad y'(0) = 0,$$

we transform both sides of the differential equation, obtaining

$$\mathcal{L}\left[\frac{d^2y}{dt^2}\right] + 4\mathcal{L}[y] = \mathcal{L}[3u_5(t)\sin(t-5)].$$

Thus

$$s^2\mathcal{L}[y] - sy(0) - y'(0) + 4\mathcal{L}[y] = \frac{3e^{-5s}}{s^2+1}.$$

Substituting the initial conditions and simplifying, we obtain

$$(s^2+4)\mathcal{L}[y] - s = \frac{3e^{-5s}}{s^2+1}.$$

So

$$y = \mathcal{L}^{-1}\left[\frac{s}{s^2+4}\right] + \mathcal{L}^{-1}\left[\frac{3e^{-5s}}{(s^2+1)(s^2+4)}\right].$$

Using partial fractions, we can write the second term on the right as

$$\frac{3e^{-5s}}{(s^2+1)(s^2+4)} = \frac{e^{-5s}}{s^2+1} - \frac{e^{-5s}}{s^2+4},$$

which gives

$$y = \mathcal{L}^{-1}\left[\frac{s}{s^2+4}\right] + \mathcal{L}^{-1}\left[\frac{e^{-5s}}{s^2+1}\right] - \mathcal{L}^{-1}\left[\frac{e^{-5s}}{s^2+4}\right].$$

Using

$$\mathcal{L}^{-1}\left[\frac{s}{s^2+4}\right] = \cos 2t, \quad \mathcal{L}^{-1}\left[\frac{1}{s^2+1}\right] = \sin t,$$

and

$$\mathcal{L}^{-1}\left[\frac{2}{s^2+4}\right] = \sin 2t,$$

we obtain

$$y(t) = \cos 2t + u_5(t)(\sin(t-5) - \tfrac{1}{2}\sin(2(t-5))).$$

The $y(t)$- and $v(t)$-graphs for this solution are shown in Figure 6.15. Note that the solution behaves as predicted. It agrees with $\cos 2t$ until $t = 5$, at which time the forcing is turned on. There is no discontinuity in the second derivative because $3u_5(t)\sin(t-5)$ vanishes at $t = 5$. The third derivative d^3y/dt^3 is discontinuous at $t = 5$, but this discontinuity is very difficult to see in the graphs.

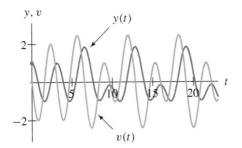

Figure 6.15
The $y(t)$- and $v(t)$-graphs for the solution of the initial-value problem

$$\frac{d^2y}{dt^2} + 4y = 3u_5(t)\sin(t-5),$$

where $y(0) = 1$ and $y'(0) = 0$.

Resonant Forcing

Finally, we consider the case of an undamped harmonic oscillator with resonant forcing. In Chapter 4 we saw that the behavior of resonant oscillators is much different from that of nonresonant cases. Solutions oscillate with an amplitude that grows linearly with time, and hence they are unbounded (see Figure 6.16). Finding formulas for solutions for a resonant oscillator by the methods of Chapter 4 involve "second-guessing" and can be tedious. With the Laplace transform method, finding the solution of a resonant equation is almost as easy as the nonresonant case once we add one additional Laplace transform to our list of known transforms.

Figure 6.16
Graph of the solution of the undamped oscillator with resonant forcing

$$\frac{d^2y}{dt^2} + 9y = \cos 3t$$

with $y(0) = y'(0) = 0$.

As an example, we consider

$$\frac{d^2y}{dt^2} + 9y = \cos 3t, \quad y(0) = y'(0) = 0.$$

Both the natural period and the forcing period of this undamped oscillator are $2\pi/3$, so this is a resonant oscillator.

We begin, as always, by taking the Laplace transform of both sides of the differential equation, which yields

$$\mathcal{L}\left[\frac{d^2y}{dt^2}\right] + 9\mathcal{L}[y] = \mathcal{L}[\cos 3t].$$

With the given initial conditions, we have

$$s^2\mathcal{L}[y] + 9\mathcal{L}[y] = \frac{s}{s^2+9},$$

which reduces to

$$\mathcal{L}[y] = \frac{s}{(s^2 + 9)^2}.$$

Hence

$$y(t) = \mathcal{L}^{-1}\left[\frac{s}{(s^2 + 9)^2}\right].$$

This is the first time that we have encountered a Laplace transform of the form

$$\frac{s}{(s^2 + w^2)^2},$$

and the method of partial fractions does not help because this fraction is already in its partial fractions form. In this case we use the results of Exercise 35, and we note that

$$\mathcal{L}[t \sin \omega t] = \frac{2\omega s}{(s^2 + \omega^2)^2}.$$

In our case we write

$$y(t) = \mathcal{L}^{-1}\left[\frac{s}{(s^2 + 9)^2}\right] = \frac{1}{6}\mathcal{L}^{-1}\left[\frac{6s}{(s^2 + 9)^2}\right],$$

so $y(t) = \frac{1}{6} t \sin 3t$ (see Figure 6.16). The only difference between this computation and the nonresonant examples above is that we used a different entry from the table when finding the inverse Laplace transform. We must remember that, even though the computation is similar, the long-term behavior of the solution of a resonant equation is vastly different from a nonresonant equation.

Analysis of the Laplace Transform Method

If it were a technique only for finding closed-form solutions for constant-coefficient, linear equations, the Laplace transform would quickly go the way of the dinosaur and the slide rule. Laplace transforms do replace calculus computations with algebraic computations, but computer algebra systems replace calculus computations with typing.

The real power of Laplace transforms comes in their unique approach to differential equations. We have already seen that Laplace transforms help us solve differential equations involving discontinuities. In the next section Laplace transforms are used in an even more fundamental way in order to model a forcing term that corresponds to a sudden push like a tap of a hammer or a jolt from an explosion. Laplace transforms are essential to the construction of a mathematically meaningful model of such a forcing term.

In Section 6.5 we use the ability of Laplace transforms to convert calculus into algebra to obtain a method of comparing the solutions arising from a harmonic oscillator with different forcing functions. In Section 6.6 we use Laplace transforms to obtain a generalization of the notion of eigenvalue of a constant-coefficient linear equation that applies to forced (nonautonomous) equations. It is these theoretical uses of the Laplace transform that make this technique worth learning.

EXERCISES FOR SECTION 6.3

In Exercises 1–4, compute the Laplace transform of the given function by computing the integral. (These exercises are mainly a review of integration by parts.)

1. $\sin \omega t$

2. $\cos \omega t$

3. $e^{at} \sin \omega t$

4. $e^{at} \cos \omega t$

5. Use the fact that $y(t) = \cos \omega t$ is the solution of the initial-value problem

$$\frac{d^2 y}{dt^2} + \omega^2 y = 0, \quad y(0) = 1, \quad y'(0) = 0$$

to verify the formula for $\mathcal{L}[\cos \omega t]$.

In Exercises 6–10, compute the Laplace transform of the given function. [*Hint*: One way to do these exercises is to use integration by parts. Alternately, in Exercise 6, differentiate both sides of the formula for $\mathcal{L}[\cos \omega t]$ with respect to ω. The others can be handled similarly.]

6. $f(t) = t \sin \omega t$ **7.** $f(t) = t \cos \omega t$ **8.** $f(t) = te^{at}$

9. $f(t) = t^2 e^{at}$ **10.** $f(t) = t^n e^{at}$

In Exercises 11–14, write the given quadratic in the form $(s+\alpha)^2 + \beta^2$ (that is, complete the square).

11. $s^2 + 2s + 10$

12. $s^2 - 4s + 5$

13. $s^2 + s + 1$

14. $s^2 + 6s + 10$

In Exercises 15–18, compute the inverse Laplace transform of the given functions using the results of Exercises 11–14.

15. $\dfrac{1}{s^2 + 2s + 10}$

16. $\dfrac{s}{s^2 - 4s + 5}$

17. $\dfrac{2s + 3}{s^2 + s + 1}$

18. $\dfrac{s + 1}{s^2 + 6s + 10}$

19. Compute (from the definition) the Laplace transform

$$\mathcal{L}\left[e^{(a+ib)t}\right],$$

where a and b are real constants.

20. Suppose $y(t)$ is complex-valued and write $y(t) = y_{\text{re}}(t) + iy_{\text{im}}(t)$ where $y_{\text{re}}(t)$ and $y_{\text{im}}(t)$ are the real and imaginary parts of $y(t)$ respectively. Show that $\mathcal{L}[y_{\text{re}}]$ is the real part of $\mathcal{L}[y]$ and $\mathcal{L}[y_{\text{im}}]$ is the imaginary part of $\mathcal{L}[y]$.

21. Use Exercises 19 and 20 to compute $\mathcal{L}\left[e^{at}\cos\omega t\right]$ and $\mathcal{L}\left[e^{at}\sin\omega t\right]$.

22. Compute

$$\mathcal{L}^{-1}\left[\frac{1}{s^2+2s+5}\right]$$

as follows:

(a) Verify that $s^2+2s+5=(s+1+2i)(s+1-2i)$.

(b) Use partial fractions to write

$$\frac{1}{s^2+2s+5}=\frac{A}{s+1+2i}+\frac{B}{s+1-2i},$$

where A and B are (complex) constants.

(c) Use Exercise 19 to take the inverse Laplace transform of each term computed in part (b).

(d) Simplify the sum of the inverse Laplace transforms obtained in part (c) to obtain the desired inverse Laplace transform.

In Exercises 23–26, use partial fractions and complex exponentials to compute the inverse Laplace transform of the given functions. (Compare your solutions with your solutions to Exercises 15–18.)

23. $\dfrac{1}{s^2+2s+10}$

24. $\dfrac{s}{s^2-4s+5}$

25. $\dfrac{2s+3}{s^2+s+1}$

26. $\dfrac{s+1}{s^2+6s+10}$

In Exercises 27–34,

(a) compute the Laplace transform of both sides of the differential equation,

(b) substitute in the initial conditions and simplify to obtain the Laplace transform of the solution, and

(c) find the solution by taking the inverse Laplace transform.

27. $\dfrac{d^2y}{dt^2}+4y=8,\quad y(0)=11,\quad y'(0)=5$

28. $\dfrac{d^2y}{dt^2}-y=e^{2t},\quad y(0)=1,\quad y'(0)=-1$

29. $\dfrac{d^2y}{dt^2}-4\dfrac{dy}{dt}+5y=2e^t,\quad y(0)=3,\quad y'(0)=1$

30. $\dfrac{d^2y}{dt^2}+6\dfrac{dy}{dt}+13y=13u_4(t),\quad y(0)=3,\quad y'(0)=1$

31. $\dfrac{d^2y}{dt^2}+4y=\cos 2t,\quad y(0)=-2,\quad y'(0)=0$

32. $\dfrac{d^2 y}{dt^2} + 3y = u_4(t)\cos(5(t-4)), \quad y(0) = 0, \quad y'(0) = -2$

33. $\dfrac{d^2 y}{dt^2} + 4\dfrac{dy}{dt} + 9y = 20u_2(t)\sin(t-2), \quad y(0) = 1, \quad y'(0) = 2$

34. $\dfrac{d^2 y}{dt^2} + 3y = w(t), \quad y(0) = 2, \quad y'(0) = 0, \quad$ where

$$w(t) = \begin{cases} t, & \text{if } 0 \le t < 1; \\ \\ 1, & \text{if } t \ge 1. \end{cases}$$

35. (a) Show that, if $\mathcal{L}[f] = F(s)$, then $\mathcal{L}[tf(t)] = -dF/ds$.
   (b) Using part (a), derive the formula for $\mathcal{L}[t\sin\omega t]$ that is used in this section.

6.4 DELTA FUNCTIONS AND IMPULSE FORCING

We have seen that Laplace transforms make it possible for us to work with equations that have discontinuous forcing terms. In this section we consider another type of discontinuous forcing function known as an impulse function.

Impulse Forcing

Impulse forcing is the term used to describe a very quick push or pull on a system, such as a blow with a hammer or the effect of an explosion. For example, suppose we have an unforced oscillator that satisfies the equation

$$\frac{d^2 y}{dt^2} + 2\frac{dy}{dt} + 26y = 0.$$

This equation models a unit mass attached to a spring with spring constant 26 sliding on a table with damping coefficient 2 (see page 184).

Now suppose we strike the mass with a hammer once at time $t = 4$ (see Figure 6.17). We can write the forced equation as

$$\frac{d^2 y}{dt^2} + 2\frac{dy}{dt} + 26y = g(t),$$

where

$$g(t) = \begin{cases} \text{very big}, & \text{when } t = 4; \\ \\ 0, & \text{when } t \ne 4. \end{cases}$$

To make this precise enough to have any hope of obtaining a solution, we need some sort of formula for $g(t)$. As a first attempt to derive such a formula, we might assume that the hammer delivers a large constant force over a short time interval. More precisely, we assume that the hammer is in contact with the mass during the time interval from $4 - \Delta t$ to $4 + \Delta t$ for some small time $\Delta t > 0$, and that the force on the mass

Figure 6.17
Harmonic oscillator struck by hammer.

during this interval is the constant k. Then the forcing function becomes

$$g_{\Delta t}(t) = \begin{cases} k, & \text{if } 4 - \Delta t \le t \le 4 + \Delta t; \\ 0, & \text{otherwise.} \end{cases}$$

Here we think of both Δt and k as parameters. The parameters k and Δt are closely related. The external force is only nonzero for a time interval of length $2\Delta t$. Hence the smaller we take Δt, the larger we should take k in order to deliver the same total push.

We choose $k = 1/(2\Delta t)$ so that k becomes large as $\Delta t \to 0$. Therefore

$$g_{\Delta t}(t) = \begin{cases} \dfrac{1}{2\Delta t}, & \text{if } 4 - \Delta t \le t \le 4 + \Delta t; \\ 0, & \text{otherwise.} \end{cases}$$

With this choice of k, the area under the graph of $g_{\Delta t}(t)$ is the same as the area of a rectangle with base $2\Delta t$ and height $1/(2\Delta t)$. That is, the area is 1 no matter what we choose for Δt (see Figure 6.18).

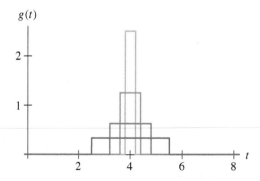

Figure 6.18
Graph of the function $g_{\Delta t}$ for several values of Δt.

Since we are modeling a hammer strike that is delivered very quickly, we would like to make the time interval Δt as small as possible. So we consider what happens as $\Delta t \to 0$. Informally, as $\Delta t \to 0$, we are compressing the same amount of force into a shorter and shorter time interval.

An "instantaneous" force would be represented by the limit as $\Delta t \to 0$. But

$$\lim_{\Delta t \to 0} g_{\Delta t}(t)$$

is a confusing object. For any $t \neq 4$, the limit of $g_{\Delta t}(t)$ as $\Delta t \to 0$ is 0 because for small Δt, the time t is outside the interval on which $g_{\Delta t}(t)$ is positive. On the other hand,

$$\lim_{\Delta t \to 0} g_{\Delta t}(4)$$

does not exist because $g_{\Delta t}(4) \to \infty$ as $\Delta t \to 0$. The "function"

$$\delta_4(t) = \lim_{\Delta t \to 0} g_{\Delta t}(t)$$

is called the **Dirac delta function**. This "function" is zero for all values of t except $t = 4$, where it is infinite.

To a mathematician, this "definition" is worse than useless. According to this definition there is no way to tell the difference between $\delta_4(t)$ and $2\delta_4(t)$ because both of these "functions" are zero for $t \neq 4$ and infinite for $t = 4$. On the other hand, the definite integral of $\delta_4(t)$ seems to make sense. Indeed, the total area under the graph of $g_{\Delta t}$ is always 1, no matter how small we take Δt. So it looks as though we may use Laplace transforms to rescue the Dirac delta function, turning it from nonsense into a useful mathematical object.

The Laplace Transform of a Dirac Delta Function

To make sense of this mess, we turn to the Laplace transform. For any $\Delta t > 0$ we can compute

$$\mathcal{L}[g_{\Delta t}] = \int_0^\infty g_{\Delta t}(t)\, e^{-st} dt$$

$$= \int_{4-\Delta t}^{4+\Delta t} \frac{1}{2\Delta t}\, e^{-st} dt$$

$$= \left(\frac{1}{2\Delta t}\right)\left(\frac{e^{-st}}{-s}\right)\Bigg|_{t=4-\Delta t}^{t=4+\Delta t}$$

$$= \left(\frac{1}{2\Delta t}\right)\left(-\frac{e^{-s(4+\Delta t)} - e^{-s(4-\Delta t)}}{s}\right)$$

$$= \left(\frac{e^{-4s}}{s}\right)\left(\frac{e^{s\Delta t} - e^{-s\Delta t}}{2\Delta t}\right).$$

Taking the limit of this quantity as $\Delta t \to 0$, we have

$$\lim_{\Delta t \to 0} \mathcal{L}[g_{\Delta t}] = \lim_{\Delta t \to 0} \left(\frac{e^{-4s}}{s} \right) \left(\frac{e^{s\Delta t} - e^{-s\Delta t}}{2\Delta t} \right)$$

$$= \left(\frac{e^{-4s}}{s} \right) \lim_{\Delta t \to 0} \left(\frac{e^{s\Delta t} - e^{-s\Delta t}}{2\Delta t} \right)$$

$$= e^{-4s}$$

(see Exercise 1).

The limit of the Laplace transform of $g_{\Delta t}$ as $\Delta t \to 0$ is a perfectly normal function. As long as we take Laplace transform *before* taking the limit, everything stays nice. Hence we define the Laplace transform of $\delta_4(t)$ as

$$\mathcal{L}[\delta_4] = e^{-4s}.$$

We could just as easily put the impulse force at time $t = a$, obtaining a Dirac delta function $\delta_a(t)$ that "is zero for $t \neq a$ and ∞ for $t = a$." We have the following definition:

DEFINITION   The Laplace transform of $\delta_a(t)$ is $\mathcal{L}[\delta_a] = e^{-as}$.   ∎

We use $\delta_a(t)$ to model the force delivered by a tap of a hammer at time $t = a$. As we observed above, as a function, $2\delta_a(t)$ is impossible to distinguish from $\delta_a(t)$. However, taking the Laplace transforms, we use linearity to see that

$$\mathcal{L}[2\delta_a] = 2\mathcal{L}[\delta_a] = 2e^{-as}.$$

Using this, we can effectively model a hammer tap of twice the original strength. In fact we can treat the Laplace transform of the $\delta_a(t)$ in the same way we treat any other Laplace transform.

Impulse Forcing of a Damped Harmonic Oscillator

We now return to the harmonic oscillator equation with mass 1, spring constant 26, and coefficient of damping 2, which is struck by a hammer at $t = 4$. For this example we choose initial conditions $y(0) = 1$ and $y'(0) = 0$. That is, the spring is stretched one unit from its rest position and then released with velocity 0. Using the Dirac delta function, we can write the initial-value problem

$$\frac{d^2y}{dt^2} + 2\frac{dy}{dt} + 26y = \delta_4(t), \quad y(0) = 1, \quad y'(0) = 0.$$

Qualitative analysis

For $0 \leq t < 4$ we have an unforced, underdamped harmonic oscillator, since the characteristic equation has roots $-1 \pm 5i$. The graph of the solution with initial condition

$y(0) = 1$ and $y'(0) = 0$ oscillates with decreasing amplitude (see Figure 6.19). At $t = 4$ the impulse force is, speaking informally, infinite. Because the force is the product of mass and acceleration, d^2y/dt^2 is infinite at $t = 4$. Hence the velocity $v = dy/dt$ has a discontinuity at $t = 4$ and the solution $y(t)$ abruptly changes in amplitude. After $t = 4$ we again have an underdamped oscillator.

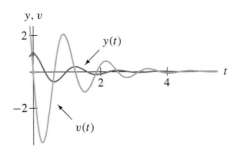

y, v

Figure 6.19
The $y(t)$- and $v(t)$-graphs for the solution of unforced oscillator

$$\frac{d^2y}{dt^2} + 2\frac{dy}{dt} + 26y = 0$$

with initial conditions $y(0) = 1$ and $y'(0) = 0$.

Unfortunately this is only a rough qualitative description of the behavior of the solution using mathematically suspicious phrases like "the acceleration is infinite." To see what is really going on, we use the Laplace transform.

Solution using Laplace transform
Taking the Laplace transform of both sides of the equation, we obtain

$$\mathcal{L}\left[\frac{d^2y}{dt^2}\right] + 2\mathcal{L}\left[\frac{dy}{dt}\right] + 26\mathcal{L}[y] = \mathcal{L}[\delta_4].$$

Using the formulas for the Laplace transform of a derivative and second derivative and solving for $\mathcal{L}[y]$, we find

$$s^2\mathcal{L}[y] - sy(0) - y'(0) + 2s\mathcal{L}[y] - 2y(0) + 26\mathcal{L}[y] = \mathcal{L}[\delta_4].$$

Substituting the initial conditions and evaluating $\mathcal{L}[\delta_4]$, we have

$$s^2\mathcal{L}[y] - s + 2s\mathcal{L}[y] - 2 + 26\mathcal{L}[y] = e^{-4s}.$$

Solving for $\mathcal{L}[y]$ yields

$$\mathcal{L}[y] = \frac{s+2}{s^2 + 2s + 26} + \frac{e^{-4s}}{s^2 + 2s + 26}.$$

Hence

$$y(t) = \mathcal{L}^{-1}\left[\frac{s+2}{s^2 + 2s + 26}\right] + \mathcal{L}^{-1}\left[\frac{e^{-4s}}{s^2 + 2s + 26}\right]$$

$$= \mathcal{L}^{-1}\left[\frac{s+2}{(s+1)^2 + 25}\right] + \frac{1}{5}\mathcal{L}^{-1}\left[\frac{5e^{-4s}}{(s+1)^2 + 25}\right],$$

and the solution is $y(t) = e^{-t}\cos 5t + \frac{1}{5}e^{-t}\sin 5t + \frac{1}{5}u_4(t)e^{-(t-4)}\sin(5(t-4))$.

The solution has the predicted form (see Figure 6.20). Note that the $v(t)$-graph has a jump discontinuity at $t = 4$, just as the impulse force is applied. Thus the slope of the tangent line to the graph of $y(t)$ changes abruptly at $t = 4$. Note that there is a "corner" in the graph of $y(t)$ at $t = 4$ (see Figure 6.21).

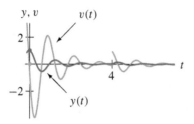

Figure 6.20
The $y(t)$- and $v(t)$-graphs for $d^2y/dt^2 + 2dy/dt + 26y = \delta_4(t)$ with $y(0) = 1$ and $y'(0) = 0$.

Figure 6.21
Magnification of the corner and jump discontinuity of the $y(t)$- and $v(t)$-graphs at $t = 4$.

An Undamped Oscillator with Impulse Forcing

In applications it is common to suppose that the impulse force is applied at $t = 0$. For example, consider the undamped oscillator

$$\frac{d^2y}{dt^2} + 4y = \delta_0(t).$$

This is a model of a unit mass attached to a spring with spring constant 4, sliding on a frictionless table. We assume that, for all $t < 0$, the mass is at the equilibrium point, at rest at $y = 0$. When $t = 0$, the mass is given a sharp push in the positive y-direction. So at $t = 0$ we have $y(0) = 0$, but $y'(0)$ is undefined since the acceleration at this time is infinite. To indicate that prior to $t = 0$ the mass was at rest, we write the initial condition $y'(0) = 0^-$. Thus we consider the initial-value problem

$$\frac{d^2y}{dt^2} + 4y = \delta_0(t), \quad y(0) = 0, \quad y'(0) = 0^-.$$

Qualitative analysis
Since there is no external force for $t > 0$, once the mass is set in motion at time $t = 0$ it oscillates forever with constant period and amplitude. The period of oscillation is the natural period (which is π). To determine the amplitude of the oscillations, we compute the solution.

Solution using Laplace transforms
Taking the Laplace transform of both sides of the differential equation gives

$$\mathcal{L}\left[\frac{d^2y}{dt^2}\right] + 4\mathcal{L}[y] = \mathcal{L}[\delta_0].$$

Simplifying and using the fact that $\mathcal{L}[\delta_0] = e^{-0s} = 1$, we find

$$s^2\mathcal{L}[y] - sy(0) - y'(0) + 4\mathcal{L}[y] = 1.$$

Substituting the initial conditions $y(0) = y'(0) = 0$ (just before the push of the impulse force) and solving for $\mathcal{L}[y]$ yields

$$\mathcal{L}[y] = \frac{1}{s^2 + 4},$$

so

$$y(t) = \mathcal{L}^{-1}\left[\frac{1}{s^2 + 4}\right].$$

Hence the solution is

$$y(t) = \tfrac{1}{2}\sin 2t.$$

That is, the solution oscillates with amplitude $1/2$ for $t > 0$ (see Figure 6.22).

We should be careful to note that the formula that we just obtained is the solution for $t > 0$. The solution for all t is

$$y(t) = \begin{cases} \frac{1}{2}\sin 2t, & \text{if } t > 0; \\ 0, & \text{if } t \leq 0. \end{cases}$$

Note that the derivative of this solution does not exist at $t = 0$ (see Figure 6.22). For $t < 0$ the velocity is identically 0, but for t slightly positive the velocity is close to 1. Thus $v(t)$ has a jump discontinuity at $t = 0$. This observation agrees with our earlier qualitative analysis since the acceleration is infinite at $t = 0$.

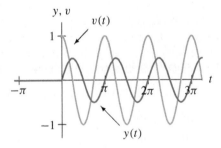

Figure 6.22
The $y(t)$- and $v(t)$-graphs of the solution of the initial-value problem

$$\frac{d^2y}{dt^2} + 4y = \delta_0(t),$$

with $y(0) = y'(0) = 0^-$.

Periodic Impulse Forcing

The qualitative approach can be useful, even when complicated forcing terms with discontinuities are involved. As an example, consider an undamped oscillator that we tap with a hammer every a time units, that is, at times $t = a, t = 2a, t = 3a, \ldots$. This leads to an initial-value problem of the form

$$\frac{d^2y}{dt^2} + y = \sum_{n=1}^{\infty} \delta_{na}(t), \quad y(0) = 0, \quad y'(0) = 0.$$

The long-term effect of this striking will depend on what the mass is doing when the strike arrives. If the mass is moving away from the hammer, then the hammer strike will speed up the mass; but if the mass is moving toward the hammer, the blow will slow it down or, perhaps, reverse its direction.

This situation is similar to that of sinusoidal forcing—if the period of the forcing is the same as the natural period of the oscillator, then the impact of the hammer will always catch the mass in the same state. Hence a resonance phenomenon will occur and the amplitudes of the oscillations will increase without bound. If the forcing period and natural period are different, then the strikes of the hammer will sometimes slow the mass and sometimes speed it up.

EXERCISES FOR SECTION 6.4

1. Compute the limit

$$\lim_{\Delta t \to 0} \left(\frac{e^{s\Delta t} - e^{-s\Delta t}}{2\Delta t} \right).$$

In Exercises 2–5, solve the given initial-value problem.

2. $\dfrac{d^2 y}{dt^2} + 3y = 5\delta_2(t), \quad y(0) = 0, \quad y'(0) = 0$

3. $\dfrac{d^2 y}{dt^2} + 2\dfrac{dy}{dt} + 5y = \delta_3(t), \quad y(0) = 1, \quad y'(0) = 1$

4. $\dfrac{d^2 y}{dt^2} + 2\dfrac{dy}{dt} + 2y = -2\delta_2(t), \quad y(0) = 2, \quad y'(0) = 0$

5. $\dfrac{d^2 y}{dt^2} + 2\dfrac{dy}{dt} + 3y = \delta_1(t) - 3\delta_4(t), \quad y(0) = 0, \quad y'(0) = 0$

6. **(a)** Discuss the qualitative behavior of the solution of the initial-value problem

$$\frac{d^2 y}{dt^2} + 2\frac{dy}{dt} + 3y = \delta_4(t), \quad y(0) = 1, \quad y'(0) = 0.$$

(b) Compute the solution of this initial-value problem.

(c) Graph this solution. Write a paragraph comparing your description of the qualitative behavior with the graph.

7. There is a relationship between the Heaviside function $u_a(t)$ and the Dirac delta function $\delta_a(t)$ that can be observed from their Laplace transforms.

(a) Show that, for $a > 0$, $\mathcal{L}[\delta_a] = s\mathcal{L}[u_a] - u_a(0)$.

(b) What relationship does this suggest between $u_a(t)$ and $\delta_a(t)$?

(c) Is there any way we could understand this relationship in terms of everyday calculus (without Laplace transforms)?

8. Calculate the Laplace transform of

$$g(t) = \sum_{n=1}^{\infty} \delta_{na}(t),$$

where $a > 0$. [*Hint*: See Exercise 16 in Section 6.2.]

9. **(a)** Find the Laplace transform of the solution of the initial-value problem

$$\frac{d^2 y}{dt^2} + 2y = \sum_{n=1}^{\infty} \delta_n(t), \quad y(0) = 0, \quad y'(0) = 0.$$

(b) Find the solution of the initial-value problem in part (a) (that is, take the inverse Laplace transform of your answer in part (a)).

(c) What is the long-term qualitative behavior of this solution?

10. **(a)** Find the Laplace transform of the solution of the initial-value problem

$$\frac{d^2 y}{dt^2} + y = \sum_{n=1}^{\infty} \delta_{2n\pi}(t), \quad y(0) = 0, \quad y'(0) = 0.$$

(b) Find the solution of the initial-value problem in part (a) (that is, take the inverse Laplace transform of your answer to part (a)).

(c) What is the long-term qualitative behavior of this solution?

6.5 CONVOLUTIONS

When using Laplace transforms to find the solution of an initial-value problem, the hardest step is usually the last—computing the inverse Laplace transform. The difficulty arises when we must break a complicated product into the sum of two or more simpler terms using partial fractions. While the arithmetic involved is elementary, it can be very complicated. The probability of making a careless error is high (at least it is for these authors). It would be nice if there was a product rule for inverse Laplace transforms, that is, some way to compute the inverse Laplace transform of a product from the inverse Laplace transform of each of its factors. In this section we derive such a rule.

Unfortunately, using the product rule is almost always more complicated than doing the partial fractions decomposition. Nevertheless, the product rule is important because it has other unexpected applications. In particular, it lets us compute solutions of a forced harmonic oscillator equation using the solutions of the same harmonic oscillator with a different forcing function.

The Product Rule for Inverse Laplace Transforms

We want to compute the inverse Laplace transform of a product, that is

$$\mathcal{L}^{-1}[F(s)G(s)]$$

assuming we know $\mathcal{L}^{-1}[F] = f(t)$ and $\mathcal{L}^{-1}[G] = g(t)$.

From the definition of the Laplace transform, we have

$$F(s) = \int_0^\infty f(\tau) e^{-s\tau} \, d\tau \quad \text{and} \quad G(s) = \int_0^\infty g(u) e^{-su} \, du.$$

(Note that we are using τ and u as the variables of integration. This is just notation, but it will be useful in the computation that follows.) The product of $F(s)$ and $G(s)$ is

$$F(s)G(s) = \left(\int_0^\infty f(\tau) e^{-s\tau} \, d\tau \right) \left(\int_0^\infty g(u) e^{-su} \, du \right).$$

The first integral does not depend on u, so it can be moved inside the second integral. We get

$$F(s)G(s) = \int_0^\infty \left(\int_0^\infty f(\tau) e^{-s\tau} \, d\tau \right) g(u) e^{-su} \, du.$$

Therefore, the product $F(s)G(s)$ is the double integral

$$F(s)G(s) = \int_0^\infty \int_0^\infty f(\tau)g(u) e^{-s(\tau+u)} \, d\tau \, du.$$

Next, we change variables, replacing the variable τ with the new variable $t = \tau + u$ for each fixed u. Note that $dt = d\tau$, that $0 < \tau < \infty$ implies $u < t < \infty$, and that $\tau = t - u$. We obtain

$$F(s)G(s) = \int_0^\infty \int_u^\infty f(t - u)g(u) e^{-st} \, dt \, du$$

(see Figure 6.23 for the region of integration). Reversing the order of integration in the double integral gives

$$F(s)G(s) = \int_0^\infty \int_0^t f(t - u)g(u) e^{-st} \, du \, dt.$$

If we isolate the terms that contain u, we get

$$F(s)G(s) = \int_0^\infty \left(\int_0^t f(t - u)g(u) \, du \right) e^{-st} \, dt.$$

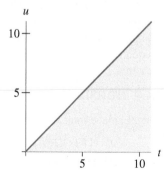

Figure 6.23
Region of integration for the double integral

$$F(s)G(s) = \int_0^\infty \int_u^\infty f(t - u)g(u) e^{-st} \, dt \, du$$

$$= \int_0^\infty \int_0^t f(t - u)g(u) e^{-st} \, du \, dt.$$

Writing $F(s)G(s)$ in this way gives a surprising result. If we define a function $h(t)$ by

$$h(t) = \int_0^t f(t-u)g(u) \, du,$$

then

$$F(s)G(s) = \int_0^\infty h(t) \, e^{-st} \, dt.$$

In other words,

$$F(s)G(s) = \mathcal{L}[h].$$

We have expressed the product of $F(s)$ and $G(s)$ as a Laplace transform. Taking the inverse Laplace transform of both sides of this equation gives

$$\mathcal{L}^{-1}[F(s)G(s)] = h(t) = \int_0^t f(t-u)g(u) \, du.$$

This computation motivates the definition of convolution.

DEFINITION   The **convolution** $f * g$ of two functions $f(t)$ and $g(t)$ is the function defined by

$$(f * g)(t) = \int_0^t f(t-u)g(u) \, du. \quad \blacksquare$$

With this notation we can rewrite the results of our computation as

$$\mathcal{L}^{-1}[F(s)G(s)] = f * g,$$

which is equivalent to

$$\mathcal{L}^{-1}[F(s)G(s)] = \mathcal{L}^{-1}[F] * \mathcal{L}^{-1}[G].$$

In words, the rule is "The inverse Laplace transform of a product is the convolution of the inverse Laplace transforms." Stated in terms of the Laplace transform rather than the inverse transform, we have

$$\mathcal{L}[f * g] = \mathcal{L}[f]\mathcal{L}[g].$$

At first glance, it appears from the formula that $f * g$ is not the same as $g * f$ since the integrals look different. However, we know that

$$\mathcal{L}^{-1}[F(s)G(s)] = \mathcal{L}^{-1}[G(s)F(s)],$$

so this cannot be the case (see Exercise 6).

A disappointing example

As a first attempt to use this rule, we consider

$$\mathcal{L}^{-1}\left[\frac{3s}{(s^2+1)(s^2+4)}\right] = \mathcal{L}^{-1}\left[\left(\frac{3}{s^2+1}\right)\left(\frac{s}{s^2+4}\right)\right].$$

Now

$$\mathcal{L}^{-1}\left[\frac{3}{s^2+1}\right] = 3\mathcal{L}^{-1}\left[\frac{1}{s^2+1}\right] = 3\sin t$$

and

$$\mathcal{L}^{-1}\left[\frac{s}{s^2+4}\right] = \cos 2t.$$

Hence, applying the product rule, we have

$$\mathcal{L}^{-1}\left[\left(\frac{3}{s^2+1}\right)\left(\frac{s}{s^2+4}\right)\right] = (3\sin t) * (\cos 2t)$$

$$= \int_0^t 3\sin(t-u)\cos 2u \, du.$$

This integral is computable but rather nasty (see Exercise 5).

On the other hand, computing this inverse Laplace transform via partial fractions gives

$$\mathcal{L}^{-1}\left[\frac{3s}{(s^2+1)(s^2+4)}\right] = \mathcal{L}^{-1}\left[\frac{s}{s^2+1}\right] - \mathcal{L}^{-1}\left[\frac{s}{s^2+4}\right],$$

so

$$\mathcal{L}^{-1}\left[\frac{3s}{(s^2+1)(s^2+4)}\right] = \cos t - \cos 2t.$$

This example is typical. While doing the partial fractions decomposition is tedious, it is less tedious than computing the convolution. However, for what it's worth, we do get the rather surprising formula

$$\cos t - \cos 2t = 3\int_0^t \sin(t-u)\cos 2t \, du,$$

since both sides have the same Laplace transform.

Delta Forcing and Convolution

It appears that the product rule above is not very useful for computing inverse Laplace transforms. However, convolution does have applications that are interesting and important. For example, consider the second-order equation

$$\frac{d^2y}{dt^2} + p\frac{dy}{dt} + qy = f(t),$$

where p and q are constants and $f(t)$ is a forcing function.

Before we deal with the general forcing term $f(t)$, we consider the special case where $f(t) = \delta_0(t)$, the Dirac delta function at $t = 0$. We consider the initial-value problem

$$\frac{d^2 y}{dt^2} + p\frac{dy}{dt} + qy = \delta_0(t), \quad y(0) = 0, \quad y'(0) = 0^-,$$

where the superscript "$-$" indicates that $y'(t) = 0$ for $t < 0$ (see page 606). To keep the notation straight later on, we let $\zeta(t)$ denote the solution of this initial-value problem. (In electric circuit theory, $\zeta(t)$ is called the *impulse response*.)

To obtain an expression for $\zeta(t)$ we take the Laplace transform of both sides of the differential equation and get

$$\mathcal{L}\left[\frac{d^2\zeta}{dt^2}\right] + p\mathcal{L}\left[\frac{d\zeta}{dt}\right] + q\mathcal{L}[\zeta] = \mathcal{L}[\delta_0].$$

Using the rules for the Laplace transform of a derivative and the fact that $\mathcal{L}[\delta_0] = 1$, we obtain

$$s^2 \mathcal{L}[\zeta] - s\zeta(0) - \zeta'(0) + ps\mathcal{L}[\zeta] - p\zeta(0) + q\mathcal{L}[\zeta] = 1.$$

Substituting the initial conditions and solving for $\mathcal{L}[\zeta]$ yields

$$\mathcal{L}[\zeta] = \frac{1}{s^2 + ps + q}.$$

Now let's return to the case of general forcing

$$\frac{d^2 y}{dt^2} + p\frac{dy}{dt} + qy = f(t), \quad y(0) = y'(0) = 0.$$

Using the Laplace transform exactly as above, we compute that the Laplace transform of the solution $y(t)$ is

$$\mathcal{L}[y] = \frac{\mathcal{L}[f]}{s^2 + ps + q}.$$

Now we can take advantage of the product rule for inverse Laplace transforms by writing

$$\mathcal{L}[y] = \frac{1}{s^2 + ps + q}\mathcal{L}[f] = \mathcal{L}[\zeta]\mathcal{L}[f].$$

So the solution of the initial-value problem with $y(0) = y'(0) = 0$ and forcing function $f(t)$ is

$$y(t) = \mathcal{L}^{-1}[\mathcal{L}[\zeta]\mathcal{L}[f]] = (\zeta * f)(t).$$

We can compute the solution $y(t)$ knowing only the forcing function $f(t)$ and the solution $\zeta(t)$ for the differential equation with the same left-hand side but with the forcing function $\delta_0(t)$. If we know $\zeta(t)$ and $f(t)$ we do not even need to know the coefficients p and q. That is, *we do not need to know the differential equation!*

A computation using convolutions

Consider the initial-value problem

$$\frac{d^2y}{dt^2} + 4y = \delta_0(t), \quad y(0) = y'(0) = 0.$$

Letting $\zeta(t)$ denote the solution, we can compute that

$$\mathcal{L}[\zeta] = \frac{1}{s^2+4}$$

as usual, and therefore, $\zeta(t) = \frac{1}{2}\sin 2t$.

We can now use the observation just made to see that the solution $y(t)$ of

$$\frac{d^2y}{dt^2} + 4y = f(t), \quad y(0) = y'(0) = 0$$

with new forcing function $f(t)$ has as its Laplace transform

$$\mathcal{L}[y] = \mathcal{L}[\zeta]\mathcal{L}[f] = \frac{1}{s^2+4}\mathcal{L}[f].$$

So

$$y(t) = \left(\tfrac{1}{2}\sin 2t\right) * f = \int_0^t \tfrac{1}{2}\sin(2(t-u))f(u)\,du.$$

For example, if $f(t) = e^{-t}$, then the solution of the initial-value problem

$$\frac{d^2y}{dt^2} + 4y = e^{-t}, \quad y(0) = y'(0) = 0$$

is

$$y(t) = \left(\tfrac{1}{2}\sin 2t\right) * e^{-t} = \int_0^t \tfrac{1}{2}\sin(2(t-u))e^{-u}\,du.$$

This is an unpleasant integral to do by hand, but a computer algebra system gives

$$y(t) = \tfrac{1}{5}e^{-t} - \tfrac{1}{5}\cos 2t + \tfrac{1}{10}\sin 2t.$$

(We can also derive the solution to this initial-value problem directly using the methods of Section 4.1.)

Again, we have replaced the usual method for computing solutions with a convolution integral. As we have already noted, this may not decrease the difficulty of the computation. However, this technique differs fundamentally from our previous methods. We are able to compute the solution of

$$\frac{d^2y}{dt^2} + 4y = f(t), \quad y(0) = y'(0) = 0$$

using only the solution of

$$\frac{d^2y}{dt^2} + 4y = \delta_0(t), \quad y(0) = y'(0) = 0.$$

That is, we can compute new solutions (with new forcing functions) from old solutions without referring back to the differential equation.

Computing Solutions from Experimental Data

Suppose we have a "black box," perhaps an RLC circuit or a mechanical device, which we know can be modeled by an equation of the form

$$\frac{d^2y}{dt^2} + p\frac{dy}{dt} + qy = 0,$$

but we do not know the coefficients p and q. Suppose the box has an input where we can apply any forcing function $f(t)$ we choose. Suppose also that the box has an output where we can record the solution of the initial-value problem

$$\frac{d^2y}{dt^2} + p\frac{dy}{dt} + qy = f(t), \quad y(0) = y'(0) = 0$$

(see Figure 6.24).

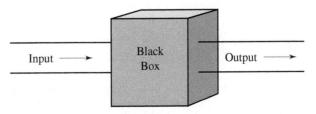

Input →

Black Box

Output →

Figure 6.24
Black box with input and output.

The previous discussion gives us a method for predicting the output $y(t)$ for a given input $f(t)$. First, we do an experiment. We input the delta function $\delta_0(t)$ and carefully record the output $\zeta(t)$ that satisfies

$$\frac{d^2\zeta}{dt^2} + p\frac{d\zeta}{dt} + q\zeta = \delta_0(t), \quad \zeta(0) = 0, \quad \zeta'(0) = 0^-.$$

Then we numerically compute the Laplace transform $\mathcal{L}[\zeta]$.

For the input forcing function $f(t)$, let $y(t)$ denote the output. That is, $y(t)$ satisfies the initial-value problem

$$\frac{d^2y}{dt^2} + p\frac{dy}{dt} + qy = f(t), \quad y(0) = y'(0) = 0.$$

Thus we obtain $\mathcal{L}[y] = \mathcal{L}[\zeta]\mathcal{L}[f]$, and therefore

$$y(t) = (\zeta * f)(t) = \int_0^t \zeta(t-u)f(u)\,du.$$

Since we know both ζ and f, at least numerically, we can numerically compute approximations for $\mathcal{L}[y]$ and y. That is, we can predict the output for a given input from data from a completely different input. This corresponds to solving an initial-value problem from another solution obtained experimentally. We can find a solution of an initial-value problem without ever knowing the differential equation.

Sobering Reminder

From the point of view of differential equations, the above discussion can be a bit distressing. We have found a way to compute solutions from data, completely avoiding the differential equation (we do not even need to know what the differential equation is). However, you should remember that this technique requires a system (black box) that can be modeled by a special type of differential equation—a linear, constant-coefficient equation. Also, all of the discussion above requires that the initial conditions be zero (the rest position with zero input). This corresponds to a device that quickly returns to its rest position after the forcing is removed. Finally, it was only by understanding the differential equation and the Laplace transform that we could derive this method.

While these methods let us compute formulas or numerical approximations of solutions, it gives us little insight into the qualitative behavior of solutions. Nevertheless, Laplace transforms can be used in the qualitative study of linear, constant-coefficient equations. We turn to this topic in the next section.

EXERCISES FOR SECTION 6.5

In Exercises 1–4, compute the convolution $f * g$ for the given functions f and g.

1. $f(t) = 1$ and $g(t) = e^{-t}$

2. $f(t) = e^{-at}$ and $g(t) = e^{-bt}$

3. $f(t) = \cos t$ and $g(t) = u_2(t)$

4. $f(t) = u_2(t)$ and $g(t) = u_3(t)$

5. Compute $(3 \sin t) * (\cos 2t)$ by computing the integral. [*Hint*: Use trigonometric identities first.]

6. Show that convolution is a commutative operation. In other words, show that for any two functions f and g, $f * g = g * f$.

7. Suppose the solution $\zeta(t)$ of the initial-value problem

$$\frac{d^2 y}{dt^2} + p\frac{dy}{dt} + qy = \delta_0(t), \quad y(0) = y'(0) = 0$$

has a Laplace transform $\mathcal{L}[\zeta]$ whose value at $s = 0$ is $1/5$ and whose value at $s = 2$ is $1/17$. Find p and q.

8. Verify that the solution $\eta(t)$ of the initial-value problem

$$\frac{dy}{dt} + ay = f(t), \quad y(0) = 0$$

has Laplace transform $\mathcal{L}[\eta] = \mathcal{L}[\zeta]\mathcal{L}[f]$, where $\zeta(t)$ is the solution of the initial-value problem

$$\frac{dy}{dt} + ay = \delta_0(t), \quad y(0) = 0^-.$$

9. Let $\zeta(t)$ be the solution of the initial-value problem

$$\frac{d^2y}{dt^2} + p\frac{dy}{dt} + qy = \delta_0(t), \quad y(0) = y'(0) = 0^-.$$

Let a and b be constants and let $f(t)$ be an arbitrary function.

(a) Find an expression for the Laplace transform of the solution of the initial-value problem

$$\frac{d^2y}{dt^2} + p\frac{dy}{dt} + qy = 0, \quad y(0) = a, \quad y'(0) = 0$$

in terms of a, p, q, and $\mathcal{L}[\zeta]$.

(b) Find an expression for the Laplace transform of the solution of the initial-value problem

$$\frac{d^2y}{dt^2} + p\frac{dy}{dt} + qy = 0, \quad y(0) = 0, \quad y'(0) = b$$

in terms of b, p, q, and $\mathcal{L}[\zeta]$.

(c) Find an expression for the Laplace transform of the solution of the initial-value problem

$$\frac{d^2y}{dt^2} + p\frac{dy}{dt} + qy = f(t), \quad y(0) = a, \quad y'(0) = b$$

in terms of a, b, p, q, $\mathcal{L}[f]$, and $\mathcal{L}[\zeta]$.

10. Suppose we know $\eta(t)$, the solution of the initial-value problem

$$\frac{d^2y}{dt^2} + p\frac{dy}{dt} + qy = u_0(t), \quad y(0) = y'(0) = 0.$$

Find a formula for the solution $y(t)$ of the initial-value problem

$$\frac{d^2y}{dt^2} + p\frac{dy}{dt} + qy = f(t), \quad y(0) = y'(0) = 0$$

in terms of η and f.

11. Suppose $y_1(t)$ is the solution of the initial-value problem

$$\frac{d^2y}{dt^2} + p\frac{dy}{dt} + qy = f_1(t), \quad y(0) = y'(0) = 0.$$

(a) Compute an expression for $\mathcal{L}[y_1]$.

(b) Suppose $y_2(t)$ is the solution of the initial-value problem

$$\frac{d^2y}{dt^2} + p\frac{dy}{dt} + qy = f_2(t), \quad y(0) = y'(0) = 0.$$

for a different forcing function $f_2(t)$. Show that

$$\frac{\mathcal{L}[f_2]}{\mathcal{L}[y_2]} = \frac{\mathcal{L}[f_1]}{\mathcal{L}[y_1]}.$$

(c) Show that

$$\mathcal{L}[y_2] = \mathcal{L}[f_2]\frac{\mathcal{L}[y_1]}{\mathcal{L}[f_1]}.$$

(This implies that we could use the the solution with any forcing function and zero initial conditions to compute solutions for other forcing functions.)

6.6   THE QUALITATIVE THEORY OF LAPLACE TRANSFORMS

As we have stressed many times, finding the solution of a differential equation means considerably more than finding a formula. Understanding the qualitative behavior of the solution is frequently much more important than the formula. So far we have used the Laplace transform mainly to find formulas for solutions. In this section we use the Laplace transform to study the qualitative behavior of solutions.

We have used Laplace transforms on constant-coefficient, linear equations. In Chapter 3 we saw that we could solve autonomous, linear equations using eigenvalues and eigenvectors. More important, using the eigenvalues, we could give a qualitative description of solutions without nearly as much arithmetic as was involved in finding a formula for the general solution.

The concept of eigenvalues, as we have presented it, does not naturally extend to nonhomogeneous equations such as the forced harmonic oscillator. On the other hand, the Laplace transform works equally well with homogeneous and nonhomogeneous equations. There is an extensive theory that uses the Laplace transform to extend the idea of eigenvalues to nonhomogeneous equations in order to study the qualitative nature of solutions. We study a small part of this qualitative theory in this section.

Homogeneous Second-Order Equations

We begin our investigation of the use of Laplace transforms in the qualitative theory of differential equations by studying a familiar friend, the second-order homogeneous equation

$$\frac{d^2y}{dt^2} + p\frac{dy}{dt} + qy = 0.$$

We have several methods available that solve this equation and that are easier to use than Laplace transforms, but our goal now is to learn more about Laplace transforms.

Taking the Laplace transform of both sides of this equation, we obtain

$$\mathcal{L}\left[\frac{d^2y}{dt^2}\right] + p\mathcal{L}\left[\frac{dy}{dt}\right] + q\mathcal{L}[y] = 0.$$

Using the rules concerning Laplace transforms of derivatives, this equation becomes

$$\left(s^2\mathcal{L}[y] - sy(0) - y'(0)\right) + p\left(s\mathcal{L}[y] - y(0)\right) + q\mathcal{L}[y] = 0,$$

which can be simplified to

$$(s^2 + ps + q)\mathcal{L}[y] - (s + p)y(0) - y'(0) = 0.$$

Thus the Laplace transform of the solution with initial conditions $y(0)$ and $y'(0)$ is

$$\mathcal{L}[y] = \frac{(s+p)y(0)}{s^2 + ps + q} + \frac{y'(0)}{s^2 + ps + q}.$$

The important point to notice here is that the denominator of the Laplace transform of the solution is the quadratic polynomial

$$s^2 + ps + q.$$

We have seen this quadratic in two other contexts already. It is the characteristic polynomial obtained from the differential equation by guessing a solution of the form $y(t) = e^{st}$. It is also the characteristic polynomial of the linear system

$$\frac{dy}{dt} = v$$

$$\frac{dv}{dt} = -qy - pv$$

that corresponds to this second-order equation. This is not a coincidence.

If the roots of the characteristic polynomial (which are the eigenvalues of the system) are λ and μ, then we can write

$$s^2 + ps + q = (s - \lambda)(s - \mu).$$

We first treat the case in which λ and μ are real. The solution y is given by

$$y = \mathcal{L}^{-1}\left[\frac{(s+p)y(0)}{(s-\lambda)(s-\mu)}\right] + \mathcal{L}^{-1}\left[\frac{y'(0)}{(s-\lambda)(s-\mu)}\right].$$

Using partial fractions, we can break this into a sum of fractions with denominators $(s - \lambda)$ and $(s - \mu)$. Every term of the solution will have a factor involving either $e^{\lambda t}$ or $e^{\mu t}$. Qualitatively we know that if λ and μ are both positive, then the origin is a source; if they are both negative, then the origin is a sink; and if one is positive and the other is negative, then the origin is a saddle.

If $\lambda = \alpha + i\beta$ and $\mu = \alpha - i\beta$, then

$$(s - \lambda)(s - \mu) = (s - (\alpha + i\beta))(s - (\alpha - i\beta))$$

$$= ((s - \alpha) - i\beta)((s - \alpha) + i\beta)$$

$$= (s - \alpha)^2 + \beta^2.$$

In this case we can write y as the inverse Laplace transform of functions with the quadratic $(s - \alpha)^2 + \beta^2$ in the denominator and with either a constant or a constant multiple of s in the numerator. The inverse Laplace transforms of these terms are multiples of either $e^{\alpha t} \sin \beta t$ or $e^{\alpha t} \cos \beta t$. Again this is exactly what we expect. In this case the solutions oscillate and the long-range behavior is determined by the sign of α. The amplitude increases if α is positive and decreases if α is negative. The frequency of oscillation is determined by β and is given by $\beta/(2\pi)$.

We can summarize this discussion as follows: For homogeneous, second-order, constant-coefficient equations, the qualitative behavior of solutions is determined by the values of s for which the denominator of the Laplace transform of the solution is zero. (These values of s are precisely the same as the eigenvalues of the corresponding linear system.) This conclusion motivates some terminology.

DEFINITION   Suppose $F(s)$ is a rational function, that is,

$$F(s) = \frac{G(s)}{H(s)},$$

where $G(s)$ and $H(s)$ are polynomials with no common factors. The **poles** of F are the values of s for which $H(s) = 0$. If

$$F(s) = \frac{G_1(s)}{H_1(s)} + \frac{G_1(s)}{H_1(s)} + \cdots + \frac{G_n(s)}{H_n(s)}$$

is a sum of rational functions, then the poles of F are found by first rewriting F as a single fraction, canceling any common factors in the numerator and denominator, and then finding the poles of the resulting rational function.   ∎

We can summarize this computation by saying that, for a homogeneous, second-order, constant-coefficient, linear equation, the poles of the Laplace transform of the general solution are the same as the eigenvalues of the corresponding linear system.

Nonhomogeneous Second-Order Equations

When we consider a nonhomogeneous linear differential equation, the qualitative techniques that we used for homogeneous linear equations and systems no longer apply. A nonautonomous system is very different from an autonomous system because the vector field for the corresponding system changes with time.

However, when using the Laplace transform, there is not much difference between a homogeneous and a nonhomogeneous equation. The arithmetic for nonhomogeneous equations is slightly more complicated, but the basic method is the same. Hence we are led to consider the poles of the Laplace transform of solutions of nonhomogeneous equations in hopes of obtaining the same sort of qualitative information we can get from the eigenvalues of a homogeneous equation.

For example, consider the initial-value problem

$$\frac{d^2y}{dt^2} + 2\frac{dy}{dt} + 2y = e^{-t/10}, \quad y(0) = 4, \quad y'(0) = 1.$$

Taking the Laplace transform of both sides of the equation and solving for $\mathcal{L}[y]$, we obtain

$$\mathcal{L}[y] = \frac{(s+2)y(0)}{s^2 + 2s + 2} + \frac{y'(0)}{s^2 + 2s + 2} + \frac{1}{(s^2 + 2s + 2)(s + \frac{1}{10})},$$

and substituting the initial conditions gives

$$\mathcal{L}[y] = \frac{4(s+2)}{s^2 + 2s + 2} + \frac{1}{s^2 + 2s + 2} + \frac{1}{(s^2 + 2s + 2)(s + \frac{1}{10})}.$$

To find the poles, we note that the least common denominator of the sum for $\mathcal{L}[y]$ is
$$(s^2 + 2s + 2)(s + \tfrac{1}{10}).$$

The roots of $s^2 + 2s + 2$ are $-1 \pm i$, and so the poles are $s = -1 \pm i$ and $s = -\tfrac{1}{10}$. Using partial fractions, we can rewrite the Laplace transform of the solution as a sum of terms with denominators
$$s + \tfrac{1}{10} \quad \text{and} \quad s^2 + 2s + 2 = (s + 1)^2 + 1$$

From the table of Laplace transforms, we see that the only terms that can appear in the solution are $e^{-t/10}$, $e^{-t} \sin t$, and $e^{-t} \cos t$.

Just as for the homogeneous case, the poles of the Laplace transform, $s = -\tfrac{1}{10}$ and $s = -1 \pm i$, tell us the qualitative behavior of the solution. The solution decreases toward zero because all of the poles are negative or have negative real parts. The solution is not monotonic but oscillates with a natural period of 2π because of the complex poles. Finally, the rate at which the solution approaches zero is determined by the exponential term with the exponent whose real part is closest to zero. Hence, this solution approaches zero at the same rate as $e^{-t/10}$.

This qualitative information agrees with the graph of the solution given in Figure 6.25, for the oscillations contributed by the $\sin t$ and $\cos t$ terms are hardly visible. What is remarkable is that the description we obtain using the poles of the Laplace transform is very similar to the description of solutions of linear, homogeneous equations that we derive using the eigenvalues. We informally think of the poles of the Laplace transform of solutions as an extension of the idea of eigenvalues to nonhomogeneous equations.

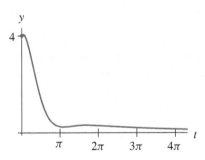

Figure 6.25
Graph of the solution to the initial-value problem
$$\frac{d^2 y}{dt^2} + 2\frac{dy}{dt} + 2y = e^{-t/10}$$
with initial conditions $y(0) = 4$ and $y'(0) = 1$.

Classification Using Poles

Collecting these ideas, we can state some rules concerning the relation between the poles of the Laplace transform of a solution and the qualitative behavior of that solution.

If all of the poles are either negative or have negative real parts, then the solution tends toward the origin as t increases. The rate of approach is exponential with exponent equal to the real part of the pole closest to zero. If one or more of the poles is positive or has positive real part, then the solution is unbounded. It tends to infinity exponentially if the pole with largest real part is real. If the pole with the largest real part is complex, then the solution oscillates and the amplitude of the oscillation grows exponentially. We emphasize that so far we have talked only about solutions that tend to infinity or to zero at an exponential rate. It is traditional in this area to call differential equations for which all solutions tend to zero as t increases *stable*. Differential equations for which one or more solutions tend to infinity as t increases are called *unstable*.

The part of the complex plane to the left of the imaginary axis is called the left half-plane. The left half-plane contains the negative real numbers and complex numbers that have negative real part. The positive reals and the complex numbers with positive real part make up the right half-plane. The condition for stability is efficiently summarized by:

All the poles in the left half-plane implies stability.

and

One or more poles in the right half-plane implies instability.

We summarize this information qualitatively in Figure 6.26. In the special case in

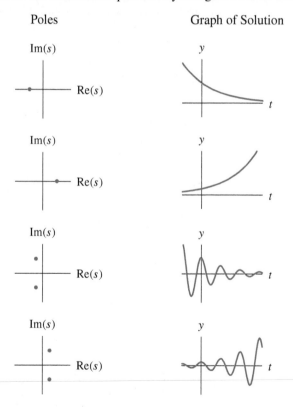

Figure 6.26
Schematic representation of the information contained in the poles of the Laplace transform. Each picture on the left gives the location of the poles in the complex s-plane while the picture on the right is the $y(t)$-graph of the corresponding solution.

which all of the poles lie on the imaginary axis, the situation is more complicated. The solution may oscillate or we may encounter resonance. We will not deal further with this special but important case (however, see Exercises 5–8).

Another Example with a Moral

Consider the differential equation

$$\frac{d^2y}{dt^2} + \frac{dy}{dt} + 3y = u_2(t)e^{-(t-2)/10}\sin(t-2).$$

This is an underdamped harmonic oscillator with a sinusoidal forcing term that is turned on at time $t = 2$ and that decreases exponentially as t increases. Taking the Laplace transform, we obtain

$$s^2\mathcal{L}[y] - sy(0) - y'(0) + s\mathcal{L}[y] - y(0) + 3\mathcal{L}[y] = \mathcal{L}[u_2(t)e^{-(t-2)/10}\sin(t-2)].$$

Solving for $\mathcal{L}[y]$ gives

$$\mathcal{L}[y] = \frac{(s+1)y(0)}{s^2+s+3} + \frac{y'(0)}{s^2+s+3} + \frac{e^{-2s}}{((s+\frac{1}{10})^2+1)(s^2+s+3)}.$$

The poles are $(-1 \pm i\sqrt{11})/2$ and $-\frac{1}{10} \pm i$. The poles $(-1 \pm i\sqrt{11})/2$ are the eigenvalues of the unforced equation and represent the natural response of the system. The poles $-\frac{1}{10} \pm i$ represent the forced response of the system. From this we see that the natural response decays exponentially to zero (like $e^{-t/2}$), and the solution approaches a steady-state oscillation that has period 2π and amplitude decreasing more slowly (like $e^{-t/10}$—see Figure 6.27).

However, we must be careful. If we change the equation to

$$\frac{d^2y}{dt^2} + \frac{dy}{dt} + 3y = (1 - u_2(t))e^{-(t-2)/10}\sin(t-2),$$

then we can compute that the poles of the Laplace transform are exactly the same as before. However, the second equation has a forcing term that turns off at time $t = 2$, so the long-term behavior of solutions is the same as for the unforced, underdamped harmonic oscillator; that is, it tends to zero relatively quickly (like $e^{-t/2}$—see Figure 6.28).

The moral is that one should always expect the unexpected. Relying blindly on a technique, without examining the equation or the underlying question (or physical system, if there is one), invites disaster.

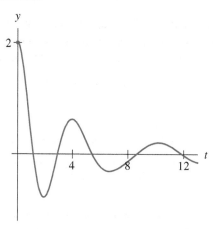

Figure 6.27
Solution of $d^2y/dt^2 + dy/dt + 3y = u_2(t)e^{-(t-2)/10}\sin(t-2)$ with initial conditions $y(0) = 2$ and $y'(0) = 0$.

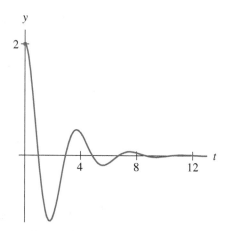

Figure 6.28
Solution of $d^2y/dt^2 + dy/dt + 3y = (1 - u_2(t))e^{-(t-2)/10}\sin(t-2)$ with initial conditions $y(0) = 2$ and $y'(0) = 0$.

EXERCISES FOR SECTION 6.6

In Exercises 1–4,
   (a) compute the Laplace transform of the solution,
   (b) find the poles of the Laplace transform of the solution, and
   (c) discuss the behavior of the solution.

1. $\dfrac{d^2y}{dt^2} + 2\dfrac{dy}{dt} + 2y = e^{-2t}\sin 4t, \quad y(0) = 2, \quad y'(0) = -2$

2. $\dfrac{d^2y}{dt^2} + \dfrac{dy}{dt} + 5y = u_2(t)\sin(4(t-2)), \quad y(0) = -2, \quad y'(0) = 0$

3. $\dfrac{d^2y}{dt^2} + \dfrac{dy}{dt} + 8y = (1 - u_4(t))\cos(t-4), \quad y(0) = 0, \quad y'(0) = 0$

   [*Hint*: Recall that $\cos(t-4) = (\cos t)(\cos 4) + (\sin t)(\sin 4)$.]

4. $\dfrac{d^2y}{dt^2} + \dfrac{dy}{dt} + 3y = (1 - u_2(t))e^{-(t-2)/10}\sin(t-2), \quad y(0) = 1, \quad y'(0) = 2$

   [*Hint*: Recall that $\sin(t-2) = (\cos 2)(\sin t) - (\sin 2)(\cos t)$.]

5. **(a)** Compute the Laplace transform of the solution of the initial-value problem

$$\frac{d^2y}{dt^2} + 16y = 0, \quad y(0) = 1, \quad y'(0) = 1.$$

(b) Compute the poles of the Laplace transform of the solution.

(c) Use this to formulate a conjecture on what having poles on the imaginary axis for the Laplace transform implies about the qualitative behavior of the solution.

6. (a) Compute the Laplace transform of the solution of the initial-value problem

$$\frac{d^2y}{dt^2} + 4y = \sin 2t, \quad y(0) = 0, \quad y'(0) = 0.$$

(b) Compute the poles of the Laplace transform of the solution.

(c) Use this to formulate a conjecture concerning what the occurrence of "double poles" on the imaginary axis implies about the qualitative behavior of solutions.

7. (a) Compute the Laplace transform for the solution of the initial-value problem

$$\frac{d^2y}{dt^2} + 2\frac{dy}{dt} + y = 0, \quad y(0) = 1, \quad y'(0) = 2.$$

(b) What are the poles of the Laplace transform?

(c) Use this to formulate a conjecture concerning what the occurrence of a "double pole" on the real axis in the Laplace transform implies about the qualitative behavior of solutions to homogeneous equations.

8. (a) Compute the Laplace transform of the solution of

$$\frac{d^2y}{dt^2} + 16y = t, \quad y(0) = 1, \quad y'(0) = 1.$$

(b) Compute the poles of the Laplace transform of the solution.

(c) Use this to formulate a conjecture on what having a "double pole" at zero for the Laplace transform implies about the qualitative behavior of the solution.

9. Let $w(t)$ be the square wave described in Exercise 17 of Section 6.2. Consider the initial-value problem

$$\frac{d^2y}{dt^2} + 20\frac{dy}{dt} + 200y = w(t), \quad y(0) = 1, \quad y'(0) = 0.$$

(a) Compute the Laplace transform of the solution of this initial-value problem.

(b) What are the poles of the solution?

(c) Describe the long-term behavior of the solution.

10. Let $z(t)$ be the sawtooth wave described in Exercise 18 of Section 6.2. Consider the initial-value problem

$$\frac{d^2y}{dt^2} + 20\frac{dy}{dt} + 200y = z(t) \quad y(0) = 1, \quad y'(0) = 0.$$

(a) Compute the Laplace transform of the solution of this initial-value problem.

(b) What are the poles of the solution?

(c) Describe the long-term behavior of the solution.

Table 6.1
Frequently Encountered Laplace Transforms.

$y(t) = \mathcal{L}^{-1}[Y]$	$Y(s) = \mathcal{L}[y]$	$y(t) = \mathcal{L}^{-1}[Y]$	$Y(s) = \mathcal{L}[y]$
$y(t) = e^{at}$	$Y(s) = \dfrac{1}{s-a} \quad (s > a)$	$y(t) = t^n$	$Y(s) = \dfrac{n!}{s^{n+1}} \quad (s > 0)$
$y(t) = \sin \omega t$	$Y(s) = \dfrac{\omega}{s^2 + \omega^2}$	$y(t) = \cos \omega t$	$Y(s) = \dfrac{s}{s^2 + \omega^2}$
$y(t) = e^{at} \sin \omega t$	$Y(s) = \dfrac{\omega}{(s-a)^2 + \omega^2}$	$y(t) = e^{at} \cos \omega t$	$Y(s) = \dfrac{s-a}{(s-a)^2 + \omega^2}$
$y(t) = t \sin \omega t$	$Y(s) = \dfrac{2\omega s}{(s^2 + \omega^2)^2}$	$y(t) = t \cos \omega t$	$Y(s) = \dfrac{s^2 - \omega^2}{(s^2 + \omega^2)^2}$
$y(t) = u_a(t)$	$Y(s) = \dfrac{e^{-as}}{s} \quad (s > 0)$	$y(t) = \delta_a(t)$	$Y(s) = e^{-as}$

Table 6.2
Rules for Laplace Transforms:
Given functions $y(t)$ and $w(t)$ with $\mathcal{L}[y] = Y(s)$ and $\mathcal{L}[w] = W(s)$ and constants α and a.

Rule for Laplace Transform	Rule for Inverse Laplace Transform
$\mathcal{L}\left[\dfrac{dy}{dt}\right] = s\mathcal{L}[y] - y(0) = sY(s) - y(0)$	
$\mathcal{L}[y + w] = \mathcal{L}[y] + \mathcal{L}[w] = Y(s) + W(s)$	$\mathcal{L}^{-1}[Y + W] = \mathcal{L}^{-1}[Y] + \mathcal{L}^{-1}[W] = y(t) + w(t)$
$\mathcal{L}[\alpha y] = \alpha \mathcal{L}[y] = \alpha Y(s)$	$\mathcal{L}^{-1}[\alpha Y] = \alpha \mathcal{L}^{-1}[Y] = \alpha y(t)$
$\mathcal{L}[u_a(t)y(t - a)] = e^{-as}\mathcal{L}[y] = e^{-as}Y(s)$	$\mathcal{L}^{-1}[e^{-as}Y] = u_a(t)y(t - a)$
$\mathcal{L}[e^{at}y(t)] = Y(s - a)$	$\mathcal{L}^{-1}[Y(s - a)] = e^{at}\mathcal{L}^{-1}[Y] = e^{at}y(t)$

REVIEW EXERCISES FOR CHAPTER 6

Short answer exercises: Exercises 1–10 focus on the basic ideas, definitions, and vocabulary of this chapter. Their answers are short (a single sentence or drawing), and you should be able to do them with little or no computation. However, they vary in difficulty, so think carefully before you answer.

1. Calculate $\mathcal{L}^{-1}\left[\dfrac{4}{s^2 - 1}\right]$.

2. Find a formula for $\mathcal{L}\left[\dfrac{d^3 y}{dt^3}\right]$ in terms of $\mathcal{L}[y]$.

3. What is $\mathcal{L}[y]$ for the solution $y(t)$ to the initial-value problem

$$\frac{d^2 y}{dt^2} + 5y = 0, \quad y(0) = 1, \quad y'(0) = 0?$$

4. Calculate $\displaystyle\int_0^\infty (1 - u_4(t))t\, dt$.

5. Calculate $\mathcal{L}[y]$ for the function

$$y(t) = \begin{cases} 1, & \text{if } t < 6; \\ 0, & \text{if } t \geq 6. \end{cases}$$

6. Write the function

$$y(t) = \begin{cases} 1, & \text{if } t < 1; \\ \sin t, & \text{if } 1 \leq t < \pi; \\ 2, & \text{if } t \geq \pi \end{cases}$$

in terms of the Heaviside function.

7. What is the value of the improper integral $\displaystyle\int_0^\infty \delta_2(t)e^{-2t}\, dt$?

8. Solve the initial-value problem $dy/dt = \delta_1(t), \quad y(0) = 0$.

9. Solve the initial-value problem

$$\frac{dy}{dt} = \delta_1(t) + \delta_2(t) + \delta_3(t) + \ldots, \quad y(0) = 0,$$

and sketch the graph of $y(t)$.

10. Solve the initial-value problem

$$\frac{dy}{dt} = \delta_1(t) - \delta_2(t) + \delta_3(t) - \delta_4(t) + \ldots, \quad y(0) = 0,$$

and sketch the graph of $y(t)$.

11. Six functions $y(t)$ and twelve functions $Y(s)$ are given below. For each function $y(t)$, match it with its Laplace transform $Y(s)$. You should do this exercise without consulting the table of Laplace transforms.

(a) $y(t) = \sin 2t$

(b) $y(t) = e^{2t} \cos 2t$

(c) $y(t) = e^{2t} - e^{-2t}$

(d) $y(t) = \sin(t - \pi)$

(e) $y(t) = \cos 2t$

(f) $y(t) = u_2(t)e^{3(t-2)}$

(i) $Y(s) = \dfrac{4}{s^2 - 4}$

(ii) $Y(s) = \dfrac{4}{s^2 + 4}$

(iii) $Y(s) = \dfrac{2}{s^2 + 4}$

(iv) $Y(s) = \dfrac{2s}{s^2 + 4}$

(v) $Y(s) = \dfrac{s}{s^2 - 4}$

(vi) $Y(s) = \dfrac{-1}{s^2 + 1}$

(vii) $Y(s) = \dfrac{s}{s^2 + 4}$

(viii) $Y(s) = \dfrac{e^{-2s}}{s + 3}$

(ix) $Y(s) = \dfrac{e^{-2s}}{s - 3}$

(x) $Y(s) = \dfrac{s}{s^2 - 4s + 8}$

(xi) $Y(s) = \dfrac{s - 2}{s^2 - 4s + 8}$

(xii) $Y(s) = \dfrac{e^{-\pi s}}{s^2 + 1}$

True-false: For Exercises 12–17, determine if the statement is true or false. If it is true, explain why. If it is false, provide a counterexample or an explanation.

12. The Laplace transform turns differentiation into multiplication.

13. The Laplace transform of the function $\delta_1(t)$ is a differentiable function.

14. If $y_1(t)$ is the solution of the initial-value problem

$$\frac{d^2 y}{dt^2} + 2\frac{dy}{dt} + 6y = \delta_3(t), \quad y(0) = 0, \quad y'(0) = 0,$$

then $2y_1(t)$ is the solution of the initial-value problem

$$\frac{d^2 y}{dt^2} + 2\frac{dy}{dt} + 6y = \delta_6(t), \quad y(0) = 0, \quad y'(0) = 0.$$

15. The solution of the initial-value problem

$$\frac{d^2 y}{dt^2} + 4y = \delta_\pi(t) + \delta_{2\pi}(t) + \delta_{4\pi}(t) + \ldots, \quad y(0) = 0, \quad y'(0) = 0$$

oscillates with increasing and unbounded amplitude.

16. The function

$$y(t) = \mathcal{L}^{-1}\left[\frac{7s}{(s^2 + 1)(s^2 + 3)(s^2 + 5)}\right]$$

satisfies $y(t) \to \infty$ as $t \to \infty$.

17. The function

$$y(t) = \mathcal{L}^{-1} \left[\frac{3s + 5}{s^4 + 6s^2 + 9} \right]$$

satisfies $-15 \le y(t) \le 15$ for all t.

In Exercises 18–23, find the inverse Laplace transform of the given function.

18. $\dfrac{3}{s^2 + 5s + 6}$

19. $\dfrac{s + 16}{s^2 + 2s - 8}$

20. $\dfrac{2s + 3}{s^2 - 2s + 4}$

21. $\dfrac{(5s - 12)e^{-3s}}{s^2 - 5s + 6}$

22. $\dfrac{5s^2 - 27s + 49}{(s - 2)(s^2 - 6s + 13)}$

23. $\dfrac{1}{s^2 - 4s + 4}$

Given a forced, second-order, linear equation such as we studied in Chapter 4, we now have two ways to solve it. We can use the Extended Linearity Principle along with a guessing technique or we can use the Laplace transform. For the initial-value problems given in Exercises 24 and 25,

  (a) find the solution using the method of Chapter 4,

  (b) find the solution using the Laplace transform, and

  (c) explain which method you prefer.

24. $\dfrac{d^2 y}{dt^2} - 2\dfrac{dy}{dt} = 4, \quad y(0) = -1, \quad y'(0) = 2$

25. $\dfrac{d^2 y}{dt^2} - 4\dfrac{dy}{dt} + 4y = e^{2t}, \quad y(0) = 1, \quad y'(0) = 5$

In Exercises 26–30, solve the given initial-value problem.

26. $\dfrac{dy}{dt} - 3y = h(t), \quad y(0) = 0, \quad \text{where} \quad h(t) = \begin{cases} 6, & \text{if } t < 3; \\ 0, & \text{if } t \ge 3. \end{cases}$

27. $\dfrac{dy}{dt} - 4y = 50u_2(t) \sin(3(t - 2)), \quad y(0) = 5$

28. $\dfrac{d^2 y}{dt^2} + 4\dfrac{dy}{dt} + 7y = h(t), \quad y(0) = 3, \quad y'(0) = 0, \quad \text{where} \quad h(t) = \begin{cases} 1, & \text{if } t < 3; \\ 0, & \text{if } t \ge 3. \end{cases}$

29. $\dfrac{d^2 y}{dt^2} + 2\dfrac{dy}{dt} + 3y = \delta_4(t), \quad y(0) = 2, \quad y'(0) = 1$

30. $\dfrac{d^2 y}{dt^2} + 2\dfrac{dy}{dt} + 5y = \delta_3(t) + u_6(t), \quad y(0) = 1, \quad y'(0) = 2$

LAB 6.1 Poles

In Section 6.6 we discussed the relationship between the poles of the Laplace transform of a solution and the qualitative behavior of the solution. The general rule of thumb is that the real part of the poles gives the exponential growth and decay rate of the solution, whereas the imaginary part of the poles gives the period of oscillation. This makes it easy to pick out exponential growth of solutions (poles with positive real part) or to determine the exponential rate of decay (pole with negative real part closest to zero).

While exponential growth and decay are the most dramatic aspects of solutions, there can also be linear growth. This occurs, for example, with resonant forcing of undamped harmonic oscillator equations. This behavior can be identified in the poles of the solution. The goal of this lab is to make those identifications.

Your report should address the following items:

1. Find the poles of the solutions of each of the following initial-value problems:

(a) $\dfrac{dy}{dt} + y = e^{-t}, \quad y(0) = 0$

(b) $\dfrac{dy}{dt} + y = t, \quad y(0) = 0$

(c) $\dfrac{dy}{dt} + y = t^2, \quad y(0) = 0$

(d) $\dfrac{d^2y}{dt^2} + 9y = \sin 3t, \quad y(0) = y'(0) = 0$

(e) $\dfrac{d^2y}{dt^2} + 9y = t \quad y(0) = y'(0) = 0$

(f) $\dfrac{d^2y}{dt^2} + 9y = t^2 \quad y(0) = y'(0) = 0$

(g) $\dfrac{d^2y}{dt^2} + 3\dfrac{dy}{dt} + 2y = e^{-t} \quad y(0) = y'(0) = 0$

(h) $\dfrac{d^2y}{dt^2} + 2\dfrac{dy}{dt} + y = e^{-t} \quad y(0) = y'(0) = 0$

2. For each initial-value problem in Part 1, find the solution (either numerically or analytically). Determine the rate of growth or decay of the solution. [*Hint*: Determining the exact rate of growth can be subtle if you are working numerically. For equations (a), (g), and (h), you should compare the solution to e^{-t}.]

3. Based on your results in Parts 1 and 2, form a conjecture regarding the relationship between multiple poles and growth or decay rates of solutions.

4. Compute a formula for

$$\mathcal{L}^{-1}\left[\frac{d}{ds}\mathcal{L}[f]\right]$$

in terms of $f(t)$. Write a short essay relating this computation to your work in Parts 1, 2, and 3.

Your report: In Part 3 state as clear and concise a conjecture as possible that is consistent with the results of Parts 1 and 2. Then make sure your conjecture is consistent with your formula in Part 4.

LAB 6.2 Convolutions

In Section 6.5 we saw how it is possible to compute the solution of a nonhomogeneous, constant-coefficient, linear differential equation

$$\frac{d^2y}{dt^2} + p\frac{dy}{dt} + qy = f(t), \quad y(0) = y'(0) = 0$$

for the forcing function $f(t)$ from the solution of

$$\frac{d^2y}{dt^2} + p\frac{dy}{dt} + qy = \delta_0(t), \quad y(0) = y'(0) = 0^-.$$

In this process we did not need to know the constants p and q. This technique applies to linear, constant-coefficient equations of any order.

In this lab we ask you to carry out these computations. Table 6.3 contains solutions of initial-value problem

$$\frac{d^2y}{dt^2} + qy = \delta_0(t), \quad y(0) = 0^-,$$

$$\frac{d^2y}{dt^2} + p\frac{dy}{dt} + qy = \delta_0(t), \quad y(0) = y'(0) = 0^-,$$

or

$$\frac{d^3y}{dt^3} + r\frac{d^2y}{dt^2} + p\frac{dy}{dt} + qy = \delta_0(t), \quad y(0) = y'(0) = y''(0) = 0^-$$

for various choices of constants r, p, and q. From this data, you can compute solutions for different forcing functions by computing convolutions. You will need to develop a method for computing approximate values of convolutions from data, (i.e., use numerical integration techniques for the data above). The arithmetic involved in computing the convolutions numerically can be oppressive, so it is worthwhile to take advantage of whatever technology you have available.

In your report, address the following questions:

1. For each of the given solutions, compute the approximate solution if the delta function on the right-hand side is replaced with $f(t) = \cos 4t$. Initial conditions remain zero.

2. For each of the given solutions, compute the approximate solution if the delta function on the right-hand side is replaced with $g(t) = 3u_1(t) + u_2(t) - 4u_3(t)$. Initial conditions remain zero.

3. For each of the given solutions, compute the approximate solution if the delta function on the right-hand side is replaced with $h(t) = e^{-t}$. Initial conditions remain zero.

4. For each of the given solutions, determine the differential equation. That is, determine the order of the equation and approximate values of the coefficients.

Your report: Summarize your technique for computing the convolutions and present the solutions both as graphs and tables. Since these are numerical calculations, you should also discuss the accuracy of your answers.

Table 6.3
Solutions for initial conditions zero and forcing $\delta_0(t)$.

Time	$y_1(t)$	$y_2(t)$	$y_3(t)$	$y_4(t)$	Time	$y_1(t)$	$y_2(t)$	$y_3(t)$	$y_4(t)$
0.1	0.0854	0.6065	0.0042	0.0905	2.1	−0.0195	0.0000	0.0074	0.2572
0.2	0.1425	0.3679	0.0136	0.1637	2.2	−0.0169	0.0000	0.0070	0.2438
0.3	0.1779	0.2231	0.0243	0.2222	2.3	−0.0139	0.0000	0.0064	0.2306
0.4	0.1927	0.1353	0.0336	0.2681	2.4	−0.0109	0.0000	0.0058	0.2177
0.5	0.1921	0.0821	0.0398	0.3033	2.5	−0.0080	0.0000	0.0052	0.2052
0.6	0.1801	0.0498	0.0424	0.3293	2.6	−0.0054	0.0000	0.0046	0.1931
0.7	0.1604	0.0302	0.0417	0.3476	2.7	−0.0031	0.0000	0.0041	0.1814
0.8	0.1361	0.0183	0.0384	0.3595	2.8	−0.0012	0.0000	0.0036	0.1703
0.9	0.1100	0.0111	0.0337	0.3659	2.9	0.0002	0.0000	0.0032	0.1596
1.0	0.0840	0.0067	0.0283	0.3679	3.0	0.0013	0.0000	0.0028	0.1494
1.1	0.0597	0.0041	0.0231	0.3662	3.1	0.0020	0.0000	0.0025	0.1396
1.2	0.0381	0.0025	0.0186	0.3614	3.2	0.0024	0.0000	0.0023	0.1304
1.3	0.0198	0.0015	0.0149	0.3543	3.3	0.0026	0.0000	0.0021	0.1217
1.4	0.0050	0.0009	0.0123	0.3452	3.4	0.0025	0.0000	0.0019	0.1135
1.5	−0.0062	0.0006	0.0105	0.3347	3.5	0.0023	0.0000	0.0018	0.1057
1.6	−0.0141	0.0003	0.0094	0.3230	3.6	0.0021	0.0000	0.0016	0.0984
1.7	−0.0191	0.0002	0.0088	0.3106	3.7	0.0017	0.0000	0.0015	0.0915
1.8	−0.0217	0.0001	0.0084	0.2975	3.8	0.0014	0.0000	0.0013	0.0850
1.9	−0.0223	0.0001	0.0081	0.2842	3.9	0.0010	0.0000	0.0012	0.0789
2.0	−0.0214	0.0000	0.0078	0.2707	4.0	0.0007	0.0000	0.0011	0.0733

7 NUMERICAL METHODS

Errors in numerical approximations of solutions to initial-value problems are inevitable. Usually these errors can be controlled by reducing the step size, but reducing the step size increases the arithmetic. Consequently, there is a trade-off between accuracy and computation time.

We begin this chapter with an analysis of the errors involved with Euler's method to obtain an understanding of the relationship between error and step size. We see what is meant by the statement that Euler's method is a first-order approximation scheme.

In the second and third sections, we give methods that achieve a given accuracy with fewer steps and less arithmetic. The methods we consider are improved Euler and Runge-Kutta. These two methods generally produce much more accurate results for a given step size.

Computers and calculators can use only a finite amount of memory to store each number in a computation. However, most numbers have infinite decimal expansions. Hence computers and calculators make small errors in each calculation due to the representation of numbers by the machines involved. In the final section we consider the effect of these errors on our approximations.

7.1   NUMERICAL ERROR IN EULER'S METHOD

Of the three approaches to the study of differential equations—the analytic, the qualitative, and the numeric—the most commonly used technique in science and engineering is numerical approximation. This comes as no surprise since there are many efficient software packages available that automate a great deal of the process of obtaining numerical solutions. Even when relatively simple analytic or qualitative techniques exist, there is a temptation to "just let the computer do it."

There is nothing wrong with this mode of operation, as long as two important points are kept in mind. First, the computer gives numbers and graphs, not interpretation and insight. Understanding the qualitative behavior of an approximate solution and its implications for the physical system under consideration is still the job of the human.

Second, numerical methods give *approximate* solutions. There is always a certain degree of error, and if the time interval over which the solution is appropriated is lengthened, the error will usually increase. While the error may remain insignificant for some differential equations, we have seen "chaotic" examples where small errors amplify very quickly (see Sections 2.8, 5.6, 8.4, and 8.5).

So far we have avoided any discussion of the errors that arise with numerical procedures such as Euler's method. However, if we are going to rely on computers, we should be aware of these errors, and we should know how to determine whether our appropriate solutions have errors that are within acceptable ranges.

When we use a computer to approximate the solution of a differential equation, there are two very different ways in which we introduce error into the computation. One is the error that is inherent in the approximation scheme that we use, and the other is the error that results from the use of finite arithmetic during the computations. In this chapter we primarily discuss the errors that arise from the approximation procedure. However, in Section 7.4 we discuss some of the interesting and practical issues that stem from the arithmetic of the computer.

Step Size and Euler's Method

Recall from Section 1.4 that Euler's method is an iterative scheme that approximates the solution to an initial-value problem of the form

$$\frac{dy}{dt} = f(t, y), \quad y(t_0) = y_0.$$

The approximate solution is given by the iteration

$$y_{k+1} = y_k + f(t_k, y_k)\,\Delta t,$$

where Δt is the step size and y_0 is the initial value $y(t_0)$. There is a simple relationship between the step size Δt, the length of the interval over which we would like to approximate the solution, and the number of steps involved in the computations. In particular, if we want to approximate the solution over the interval from t_0 to t_n (using n steps), we divide the total length of the interval into n equal-length subintervals. In other words,

$$\Delta t = \frac{t_n - t_0}{n}.$$

The rule of thumb discussed in Section 1.4 is that the smaller the step size, the more accurate the approximate solution. However, the smaller the step size, the larger the number of steps, so the longer it takes the computer to approximate the solution. It is useful to be able to predict roughly how large a step size to take in order to achieve a particular accuracy.

We begin by studying how the accuracy of Euler's method changes with step size for a particular example. When we introduced Euler's method for first-order equations in Section 1.4, we considered the initial-value problem

$$\frac{dy}{dt} = -2ty^2, \quad y(0) = 1.$$

This equation is separable and the solution is

$$y(t) = \frac{1}{1 + t^2}.$$

Since we have a formula for the solution, we can compare the results of Euler's method for different step sizes to the exact values of the solution.

In Section 1.4 we applied Euler's method to this initial-value problem over the interval $0 \le t \le 2$ using $\Delta t = 0.1$ (20 steps). For this differential equation the Euler's method iteration scheme is

$$y_{k+1} = y_k - 2t_k y_k^2 \Delta t,$$

and we obtained the approximation $y_{20} = 0.1933\ldots$, which is a good approximation of the exact value $y(2) = 1/(1 + 2^2) = 0.2$. Increasing the number of steps to 2000 ($\Delta t = 0.001$) yields the more accurate approximation $y_{2000} = 0.199937\ldots$, which is an even better approximation of $y(2) = 0.2$. In Section 1.4 we simply observed that increasing the number of steps in the overall computation seems to improve the accuracy. Now we would like to be more precise about exactly how the accuracy is improved as the number of steps is increased.

First, we take a closer look at the result of the computation that used 2000 steps over the interval $0 \le t \le 2$. Denote the error in the computation by e_{2000}. We have

$$e_{2000} = |0.2 - y_{2000}| = |0.2 - 0.199937| = 0.000063.$$

Now suppose we double the number of steps. After starting at the same initial condition and taking 4000 steps with $\Delta t = 0.0005$, we obtain the approximation $y_{4000} = 0.199969$. In this case we have the error

$$e_{4000} = |0.2 - y_{4000}| = |0.2 - 0.199969| = 0.000031.$$

In this example we see that the ratio of the errors is

$$\frac{e_{4000}}{e_{2000}} = \frac{0.000031}{0.000063} \approx 0.492.$$

When we doubled the number of steps in this example, we halved the error.

This straightforward computation leads us to wonder how the accuracy of the approximation depends on the step size and the number of steps. Since Euler's method is so easy to implement, we can play around with various values to see what happens. Using the computer we first compute the value of y_n for various choices of the number n of steps. Then we calculate the error e_n for these values. In Figure 7.1 we plot the results for $n = 100, 200, 300, \ldots, 6000$. This graph is a plot of the accuracy for Euler's method applied to the initial-value problem

$$\frac{dy}{dt} = -2ty^2, \quad y(0) = 1,$$

as a function of step size. If we ignore the scales on the axes for a moment, we note that this graph looks especially familiar. In fact if we let n denote the total number of steps, this graph looks remarkably like the graph of $1/n$. Indeed, in Figure 7.2 we plot a graph of a function of the form K/n. (Actually $K = 0.126731$, but it is not important how we came up with this number.) Note the agreement between the data points and the graph of this function. The remainder of this section is devoted to an explanation of why these results are typical.

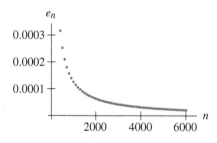

Figure 7.1
The error in Euler's method as a function of the number n of steps used in the approximation. Euler's method is applied to $dy/dt = -2ty^2$, $y(0) = 1$, over the interval $0 \le t \le 2$.

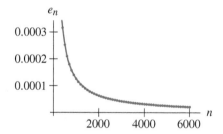

Figure 7.2
The error in Euler's method from Figure 7.1 fit to the function $0.126731/n$. Note that it is almost impossible to distinguish the graph of $0.126731/n$ from the data points e_n.

The Error in the First Step of the Method

Given an initial-value problem of the form

$$\frac{dy}{dt} = f(t, y), \quad y(t_0) = y_0,$$

and a step size Δt, we know that the value of y_1 in the first Euler step is determined by assuming that the graph of the solution is a line segment that starts at (t_0, y_0) and has slope $f(t_0, y_0)$. First, we need to determine how to estimate the error made in this initial step, and we do so using Taylor's Theorem with remainder.

Taylor's Theorem says that

$$y(t_1) = y(t_0) + y'(t_0)\,\Delta t + y''(\xi_1)\frac{(\Delta t)^2}{2}$$

for some number ξ_1 between t_0 and t_1. Since we know that $y(t)$ is a solution to the initial-value problem, we have $y'(t_0) = f(t_0, y_0)$ and $y(t_0) = y_0$. In other words,

$$y(t_1) = y_0 + f(t_0, y_0)\,\Delta t + y''(\xi_1)\frac{(\Delta t)^2}{2}.$$

Euler's method corresponds to dropping the $(\Delta t)^2$ term, that is, taking the linear approximation of the solution from t_0 to t_1. Thus the difference $|y(t_1) - y_1|$ between the first Euler approximation y_1 and the actual value $y(t_1)$ of the solution is

$$\left| y''(\xi_1)\frac{(\Delta t)^2}{2} \right|.$$

It is convenient to have some notation for this term, so we denote it τ_1 to represent the **truncation** error. It is called the truncation error because it results from a "truncation" of the Taylor series for $y(t_1)$. Euler's method is based on the Taylor approximation of $y(t_1)$, but it throws out all nonlinear terms (terms of degree 2 or higher).

For the first step of Euler's method, the truncation error is the error, but that won't be true for subsequent steps. So we introduce notation for the **total error** after k steps. By definition, the total error after k steps is the difference between the kth Euler approximation and the actual value $y(t_k)$ at time t_k, and we denote it by e_k. More precisely,

$$e_k = |y(t_k) - y_k|.$$

Note that for the first step ($k = 1$), the truncation error and the total error are identical. That is,

$$e_1 = \tau_1.$$

Usually the truncation error and the total error agree only at the first step.

At this point it is natural to wonder whether the formula

$$\tau_1 = \left| y''(\xi_1)\frac{(\Delta t)^2}{2} \right|$$

is something that we can really compute if we don't already know the solution $y(t)$. The key point to keep in mind is the fact that $y(t)$ is a solution to the differential equation. Since we know that $dy/dt = f(t, y)$, we can compute the second derivative of $y(t)$

using the multivariable Chain Rule. Since

$$\frac{dy}{dt} = f(t, y(t)),$$

we have

$$\frac{d^2y}{dt^2} = \frac{\partial f}{\partial t}\frac{dt}{dt} + \frac{\partial f}{\partial y}\frac{dy}{dt}$$

$$= \frac{\partial f}{\partial t} + \frac{\partial f}{\partial y}f(t, y).$$

Therefore we can bound the error e_1 using an upper bound M_1 for the quantity

$$\frac{\partial f}{\partial t} + \frac{\partial f}{\partial y}f(t, y)$$

along the graph of $y(t)$ between $y(t_0)$ and $y(t_1)$. Since we are given $f(t, y)$, we can actually compute the quantities in this expression.

Suppose M_1 is a constant that is large enough so that

$$\left|\frac{\partial f}{\partial t} + \frac{\partial f}{\partial y}f(t, y)\right| < M_1$$

along the graph of $y(t)$ (or at least the part of the graph we are considering). If we are able to find such a bound M_1, then the error in the first step is bounded according to the inequality

$$e_1 = \tau_1 \leq M_1 \frac{(\Delta t)^2}{2}$$

(see Figure 7.3).

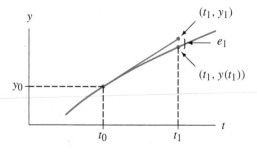

Figure 7.3
The error e_1 in the first step of Euler's method.

The Error in the Second Step

Unfortunately, once we have taken the first step, we are no longer on the graph of $y(t)$, and consequently our error analysis is more complicated. Basically there are two sources of error in the second step. If we knew the value of $y(t_1)$ exactly, then the error e_2 in the second step would again be the truncation error τ_2. In other words, from Taylor's Theorem again we have

$$y(t_2) = y(t_1) + y'(t_1)\,\Delta t + y''(\xi_2)\frac{(\Delta t)^2}{2}$$

for some number ξ_2 between t_1 and t_2, and using the differential equation to replace $y'(t_1)$ with $f(t_1, y(t_1))$, we would have

$$y(t_2) = y(t_1) + f(t_1, y(t_1))\,\Delta t + y''(\xi_2)\frac{(\Delta t)^2}{2}.$$

The truncation error at this step is therefore

$$\tau_2 = \left| y''(\xi_2)\frac{(\Delta t)^2}{2} \right|.$$

Comparing the formula

$$y(t_2) = y(t_1) + f(t_1, y(t_1))\,\Delta t + y''(\xi_2)\frac{(\Delta t)^2}{2}$$

from Taylor's Theorem with the formula

$$y_2 = y_1 + f(t_1, y_1)\,\Delta t$$

for the second approximation y_2 from Euler's method, we see that, not only are we missing the quadratic term, but Euler's method uses the slope $f(t_1, y_1)$ at (t_1, y_1), rather than the slope $f(t_1, y(t_1))$ at $(t_1, y(t_1))$ (see Figure 7.4). Therefore, when we estimate the error e_2, we have

$$e_2 = |y(t_2) - y_2|$$

$$\leq |y(t_1) - y_1| + |f(t_1, y(t_1)) - f(t_1, y_1)|\,\Delta t + |y''(\xi_2)|\frac{(\Delta t)^2}{2}$$

$$\leq e_1 + |f(t_1, y(t_1)) - f(t_1, y_1)|\,\Delta t + M_1\frac{(\Delta t)^2}{2}$$

if we can find a bound M_1 on the second derivative as we did when we estimated τ_1. In other words, the total error after the second step is bounded by the total error after the first step, the truncation error τ_2 associated to the second step, and a third term, $|f(t_1, y(t_1)) - f(t_1, y_1)|\,\Delta t$, which accounts for the fact that we are using the slope $f(t_1, y_1)$ at the point (t_1, y_1) rather than the slope $f(t_1, y(t_1))$ at the point $(t_1, y(t_1))$ on

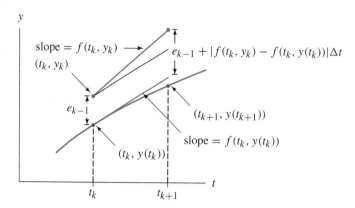

Figure 7.4
The relationship between the error e_{k-1} and the error e_k.

the actual solution curve. Figure 7.4 illustrates the contribution that each of these three terms makes to the overall error.

To obtain an estimate for the error e_2 in a relatively convenient form, we need to bound the difference between $f(t_1, y(t_1))$ and $f(t_1, y_1)$. Since these two values of f come from points along the line $t = t_1$, we have the estimate

$$|f(t_1, y(t_1)) - f(t_1, y_1)| \leq M_2 \cdot e_1,$$

where M_2 is the maximum of the partial derivative $\partial f / \partial y$ along the line segment between $(t_1, y(t_1))$ and (t_1, y_1). We therefore obtain the estimate

$$e_2 \leq e_1 + M_2 e_1 \, \Delta t + M_1 \frac{(\Delta t)^2}{2} = (1 + M_2 \, \Delta t) e_1 + M_1 \frac{(\Delta t)^2}{2}.$$

The analysis of the error e_k at the kth step is essentially the same as the analysis that we just completed for the second step. If M_1 is a bound for the second derivative along the graph of the solution and if M_2 is a bound for the partial derivative $\partial f / \partial y$, then the error e_k is bounded by

$$e_k \leq (1 + M_2 \, \Delta t) e_{k-1} + M_1 \frac{(\Delta t)^2}{2}.$$

Note that this formula is recursive in that it expresses e_k in terms of e_{k-1} (it is really a discrete dynamical system). The term

$$M_1 \frac{(\Delta t)^2}{2}$$

accounts for the truncation error τ_k at the kth step, and the term

$$(1 + M_2 \, \Delta t) e_{k-1}$$

represents the magnification of the error at the $k - 1$st step by a factor due to the fact that we are computing the right-hand side of the differential equation at the Euler approximation (t_k, y_k) rather than at the actual point $(t_k, y(t_k))$ on the solution.

The Error After n Steps

If we are going to approximate the solution $y(t)$ using n steps, then we would definitely like a compact formula for the error e_n in terms of quantities that are easy to compute directly from the differential equation, that is, directly from the right-hand side $f(t, y)$. The recursive formula that we just derived does not really fit these criteria. However, in Exercise 11 we show that, if we find maximums for the expressions

$$M_1 = \left| \frac{\partial f}{\partial t} + \frac{\partial f}{\partial y} f(t, y) \right| \quad \text{and} \quad M_2 = \left| \frac{\partial f}{\partial y} \right|$$

that come up in the estimates, then we can use this recursive formula to bound the total error in Euler's method. Over the interval $t_0 \le t \le t_n$, we can derive a bound of the form

$$e_n \le C \cdot \Delta t,$$

where C is a constant that is determined by the values of M_1, M_2, and the total length of the t-interval over which the solution is approximated. Note that the value of C *does not depend* on the number of steps used in Euler's method. Moreover, since $\Delta t = (t_n - t_0)/n$, we can rewrite this estimate as

$$e_n \le \frac{K}{n},$$

where K is the product of the constant C and the length of the total interval $t_0 \le t \le t_n$. From our point of view, K is just another constant that does not depend on the number of steps used in Euler's method.

What's most important is that we have theoretical bounds that can be expressed either as

$$e_n \le C \cdot \Delta t \quad \text{or as} \quad e_n \le \frac{K}{n}.$$

Roughly speaking, if we halve Δt, then we halve the error in Euler's method. Expressed in terms of the number of steps, this estimate says that if we double the number of steps, then we double the accuracy of the approximation.

The Initial-Value Problem $dy/dt = -2ty^2$, $y(0) = 1$

At the beginning of this section, we compared the results of Euler's method with the analytic solution to the initial-value problem

$$\frac{dy}{dt} = -2ty^2, \quad y(0) = 1.$$

We saw that the error was roughly proportional to the step size Δt. This is consistent with the theoretical estimates that we just discussed. However, using the theoretical estimates mentioned above (and developed in more detail in Exercise 11) to choose the appropriate step size to guarantee a desired error is usually overkill.

To see why, let's consider Euler's method applied to the initial-value problem

$$\frac{dy}{dt} = -2ty^2, \quad y(0) = 1,$$

more carefully. We have been estimating the solution over the interval $0 \leq t \leq 2$. Without even trying to derive an analytic solution to the problem, we know that the solution is decreasing (why?) and that it is never zero, because $y(t) = 0$ for all t is an equilibrium solution. Therefore we know that the graph of the actual solution stays in the rectangle

$$\{(t, y) \mid 0 \leq t \leq 2, \ 0 \leq y \leq 1\}$$

in the ty-plane. Using ideas similar to those discussed above, we can determine that the approximate values $y_1, y_2, y_3, \ldots, y_n$ produced by Euler's method must also lie in the same rectangle. In theory we can now estimate the expressions M_1 and M_2 involving the relevant partial derivatives over this rectangle and come up with a bound on the total error in the approximation. In fact, we do exactly this in Exercise 12. In the end we obtain the bound

$$e_n \leq (16, 661, 456) \, \Delta t.$$

In other words, we need to take roughly 34 million steps to guarantee an accuracy of 1 over this interval of length 2. Looking at the data from our experiments at the beginning of the section, we see that this is an awful estimate. With ten steps we get an error that is less than 0.015.

Why is the theoretical estimate so bad? The problem comes from trying to estimate the total error e_n using upper bounds on M_1 and M_2 that hold throughout the rectangle even though it turns out that the actual solution and the Euler approximates remain close, and their graphs avoid the places where the values of M_1 and M_2 are large. In general, the problem is that we don't have a very good idea of where the graph of the solution is going to be until we make an accurate approximation.

In practice the theoretical bounds for the total error are replaced by a running total of estimates of the error that may be introduced at each step. Although this analysis is not 100% trustworthy, it tends to give a better indication of the accuracy of our approximation.

To illustrate the point, we return to the initial-value problem

$$\frac{dy}{dt} = -2ty^2, \quad y(0) = 1,$$

and we estimate the solution over the interval $0 \leq t \leq 2$ with 100 steps. Since we have the analytic solution in this case, we can compute the error $e_k = |y(t_k) - y_k|$ at each step. The results of this calculation are shown in Figure 7.5. In addition, we keep a running estimate of the total error by appealing to the recursive formula

$$e_k \leq (1 + M_2 \Delta t) e_{k-1} + M_1 \frac{(\Delta t)^2}{2}$$

derived above. However, at each step we estimate

$$M_1 = \left| \frac{\partial f}{\partial t} + \frac{\partial f}{\partial y} f(t, y) \right| \quad \text{and} \quad M_2 = \left| \frac{\partial f}{\partial y} \right|$$

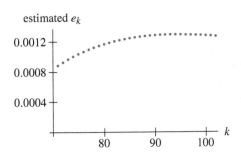

Figure 7.5
The cumulative error e_k after the kth step in the Euler approximation for the initial-value problem

$$\frac{dy}{dt} = -2ty^2, \quad y(0) = 1,$$

over the interval $0 \le t \le 2$ with $n = 100$ steps.

based on the current value of y_k. That is, after we compute y_k from (t_{k-1}, y_{k-1}), we then calculate M_1 at the point $(t_{k-1} + \Delta t/2, (y_{k-1} + y_k)/2)$ and M_2 at the point $(t_{k-1}, (y_{k-1} + y_k)/2)$. (These points are "halfway" points among the set of all possibilities.) Then given the resulting values for M_1 and M_2, we use the approximation

$$e_k \approx (1 + M_2 \Delta t) e_{k-1} + M_1 \frac{(\Delta t)^2}{2}.$$

This gives us a running estimate for the total error e_k, and the estimate is based on the same principles that justify our theoretical results. A plot of these estimates is shown in Figure 7.6. Note that this graph is almost identical to the graph of the actual error as shown in Figure 7.5.

In general this type of local estimation of the errors involved in the calculation is much easier to achieve. In fact the calculation can be left to the computer, and the estimates of the errors e_k can be one of the results of the computation.

Figure 7.6
Estimates of the errors e_k as shown in Figure 7.5 based on the evaluation of relevant partial derivatives.

The Order of Euler's Method

The most important lesson of this section is that the size of the error in Euler's method for an initial-value problem over a fixed interval of time is proportional to the step size, or in other words, it is inversely proportional to the number of steps that we use over a fixed interval. We say that Euler's method is a **first-order** method because the error is, at worst, proportional to the first power of the step size—note the 1 in the exponent of the bound

$$e_n \le C \cdot (\Delta t)^1.$$

In the next two sections we introduce methods that have higher orders. Higher-order methods are more typical of professionally written numerical solvers.

EXERCISES FOR SECTION 7.1

1. Reproduce Figure 7.1 by performing the following calculations:

   (a) Solve the initial-value problem $dy/dt = -2ty^2$, $y(0) = 1$.

   (b) Given the interval $0 \leq t \leq 2$ and $n = 20$, calculate the Euler approximation y_{20} to $y(2)$. (Double-check your calculations by comparing your results to those given at the beginning of the section.)

   (c) Calculate the total error e_{20} involved in your approximation.

   (d) Repeat the calculations in parts (b) and (c) using $n = 1000, 2000, \ldots, 6000$.

   (e) If your computer is up to the task, compute e_n for additional intermediate values of n. (Figure 7.1 was done with n ranging from 100 to 6000 by steps of 100.)

   (f) Some computer math systems can determine the best fit for this data to a function of the form K/n. If your computer can do this, determine the best value of K for this initial-value problem. Plot the data and the graph of K/n on the same axes.

In Exercises 2–5, repeat the steps in Exercise 1 to arrive at plots like Figures 7.1 and 7.2 for the initial-value problem and interval specified.

2. $\dfrac{dy}{dt} = 1 - y$, $\quad y(0) = 0$, $\quad 0 \leq t \leq 1$

3. $\dfrac{dy}{dt} = ty$, $\quad y(0) = 1$, $\quad 0 \leq t \leq \sqrt{2}$

4. $\dfrac{dy}{dt} = -y^2$, $\quad y(0) = 1/2$, $\quad 0 \leq t \leq 2$

5. $\dfrac{dy}{dt} = 1 - t + 3y$, $\quad y(0) = 1$, $\quad 0 \leq t \leq 1$

6. In this section we state that the error e_{2000} is 0.000063 for the Euler approximation of the initial-value problem $dy/dt = -2ty^2$, $y(0) = 1$ over the interval $0 \leq t \leq 2$. Estimate how many steps n are needed to obtain an approximation y_n that has an error e_n no larger than 0.000001.

7. Reproduce Figure 7.6 by performing the following calculations:

   (a) Calculate the expressions

   $$M_1 = \frac{\partial f}{\partial t} + \frac{\partial f}{\partial y} f(t, y) \quad \text{and} \quad M_2 = \frac{\partial f}{\partial y}$$

   for $f(t, y) = -2ty^2$.

   (b) Given the initial-value problem

   $$\frac{dy}{dt} = -2ty^2, \quad y(0) = 1,$$

the interval $0 \le t \le 2$, and $n = 100$ steps, calculate the point (t_1, y_1) that corresponds to the first step of Euler's method. Evaluate the quantity M_1 at the point $(t_0 + \Delta t/2, (y_0 + y_1)/2)$, and use this value to estimate the error e_1. How does your estimate compare with the true error e_1?

(c) Calculate the point (t_2, y_2) that corresponds to the second step of Euler's method. Evaluate M_1 at the point $(t_1 + \Delta t/2, (y_1 + y_2)/2)$ and M_2 at the point $(t_1, (y_1 + y_2)/2)$. Use these values to estimate the error e_2. How does your estimate compare with the true error e_2?

(d) Repeat part (c) for the remaining 98 steps.

(e) Make two plots—one that plots the true errors e_k versus k and the other that plots the estimated error versus k.

In Exercises 8 and 9, repeat parts (b)–(d) of Exercise 7 to estimate the errors e_k for the initial-value problem and interval specified. Why don't we ask for a comparison of this estimate of e_k with the true value of e_k?

8. $\dfrac{dy}{dt} = t - y^3, \quad y(0) = 1, \quad 0 \le t \le 1$

9. $\dfrac{dy}{dt} = \sin ty, \quad y(0) = 3, \quad 0 \le t \le 3$

10. Using what you learned in calculus, show that $1 + \alpha < e^\alpha$ if $\alpha > 0$.

11. In this exercise we derive the theoretical bounds mentioned in the section. Let R be a rectangle $\{(t, y) \mid a \le t \le b, c \le y \le d\}$ in the ty-plane and suppose that $f(t, y)$ is continuously differentiable on R. Given an initial-value problem

$$\frac{dy}{dt} = f(t, y), \quad y(t_0) = y_0,$$

and an interval $t_0 \le t \le t_n$ such that the point (t_0, y_0) is in R and $t_n \le b$, then we can bound the error e_n involved in the Euler approximation assuming the Euler approximate values $y_1, y_2, \ldots, y_n$ all lie between c and d. To do so, let

$$M_1 = \max \left| \frac{\partial f}{\partial t} + \frac{\partial f}{\partial y} f \right| \text{ on } R$$

and let

$$M_2 = \max \left| \frac{\partial f}{\partial y} \right| \text{ on } R.$$

(a) Show that

$$e_1 \le M_1 \frac{(\Delta t)^2}{2}.$$

(b) Show that

$$e_2 \le e_1 + M_2 e_1 \Delta t + M_1 \frac{(\Delta t)^2}{2}$$

and explain the significance of each of these three terms.

(c) Explain why

$$e_{k+1} \leq (1 + M_2 \Delta t) e_k + M_1 \frac{(\Delta t)^2}{2}.$$

(d) Let $K_1 = 1 + M_2 \Delta t$ and $K_2 = M_1 (\Delta t)^2 / 2$. Show that

$$e_3 \leq (K_1^2 + K_1 + 1) K_2.$$

(e) Show that

$$e_n \leq \left(K_1^{n-1} + K_1^{n-2} + \cdots + K_1 + 1 \right) K_2.$$

(f) Explain why

$$K_1^{n-1} + K_1^{n-2} + \cdots + K_1 + 1 = \frac{K_1^n - 1}{K_1 - 1},$$

and thus

$$e_n \leq \left(\frac{K_1^n - 1}{K_1 - 1} \right) K_2.$$

(g) Using the definitions of K_1 and K_2, show that

$$e_n \leq \frac{M_1}{2M_2} \left((1 + M_2 \Delta t)^n - 1 \right) \Delta t.$$

(h) Use the result of Exercise 10 to conclude

$$e_n \leq \frac{M_1}{2M_2} \left(e^{(M_2 \Delta t) n} - 1 \right) \Delta t.$$

(i) Finally, conclude that

$$e_n \leq \frac{M_1}{2M_2} \left(e^{M_2 (t_n - t_0)} - 1 \right) \Delta t.$$

(j) Explain why this justifies the inequality $e_n \leq C \cdot \Delta t$ given in this section.

(k) In what way is this bound for the total error different from the "estimates" shown in Figure 7.6 and computed in Exercises 7–9?

12. Given the initial-value problem

$$\frac{dy}{dt} = -2ty^2, \quad y(0) = 1,$$

and the interval $0 \leq t \leq 2$, carry out the theoretical bound on e_n given by the results of Exercise 11 as follows:

(a) Let R be the rectangle $\{(t, y) \mid 0 \leq t \leq 2, 0 \leq y \leq 1\}$. Determine the maximum values for M_1 and M_2 on R.

(b) Using the results of Exercise 11, derive the constant C for which we are certain that the inequality $e_n \leq C \cdot \Delta t$ holds.

(c) Using the value of C determined in part (b), find K such that $e_n \leq K / n$.

(d) Explain why these two estimates are so conservative compared to what we know from computations such as those given at the beginning of the section.

7.2   IMPROVING EULER'S METHOD

Euler's method is a convenient numerical algorithm in many ways. It is easy to understand and to implement. However, for numerical work where a high degree of accuracy is essential, Euler's method is not the algorithm of choice. There are algorithms that usually are more accurate and that require fewer calculations to attain that accuracy. In this section and the subsequent one, we present two of these algorithms in order to illustrate how we can derive and implement more accurate algorithms.

Higher-Order Algorithms

As we mentioned in the last section, we can interpret Euler's method in terms of Taylor approximation. Given a solution to the initial-value problem

$$\frac{dy}{dt} = f(t, y), \quad y(t_0) = y_0,$$

then its Taylor series

$$y(t_1) = y_0 + y'(t_0)\Delta t + \frac{y''(t_0)}{2}(\Delta t)^2 + \ldots$$

can be rewritten as

$$y(t_1) = y_0 + f(t_0, y_0)\Delta t + \frac{y''(t_0)}{2}(\Delta t)^2 + \ldots.$$

Euler's method can be interpreted as the approximation that results from truncating this series at the linear term. That is,

$$y(t_1) \approx y_1 = y_0 + f(t_0, y_0)\Delta t.$$

One way to improve the accuracy of Euler's method is to use more of the Taylor series. Rather than truncating at the linear term, we include the quadratic term and obtain a more accurate approximation. In other words, we can approximate $y(t_1)$ by a new y_1 where

$$y(t_1) \approx y_1 = y_0 + f(t_0, y_0)\Delta t + \frac{y''(t_0)}{2}(\Delta t)^2.$$

To do this we need to know $y''(t_0)$, the second derivative of the solution. The only information we have about the solution y is that it satisfies the differential equation $dy/dt = f(t, y)$. We can differentiate both sides to obtain an expression for $y''(t_0)$, that is

$$\frac{d^2y}{dt^2} = \frac{\partial f}{\partial t}\frac{dt}{dt} + \frac{\partial f}{\partial y}\frac{dy}{dt}$$

$$= \frac{\partial f}{\partial t} + \frac{\partial f}{\partial y}\, f(t, y).$$

In principle there is no problem performing this calculation and using the results to implement a numerical algorithm that is more accurate than Euler's method. In practice this is seldom done because the calculation requires knowledge of the partial derivatives of f, and if we want to program a "black box" in a traditional computer language such as Fortran or C, we would have to provide these partial derivatives as well as the original function f from the differential equation. These days there exist computer languages that can calculate these derivatives, so this restriction is no longer a problem. Nevertheless, some very clever algorithms have been developed that provide approximation schemes with equivalent accuracy without using the partial derivatives of f, and these algorithms are commonly used. They will be the ones that we focus on in this chapter.

Numerical Approximation and Numerical Integration

To understand how these methods were developed, it is useful to think about Euler's method in terms of the numerical integration techniques that are used to approximate the definite integral. By the Fundamental Theorem of Calculus, we know that the solution of the initial-value problem

$$\frac{dy}{dt} = f(t, y), \quad y(t_0) = y_0$$

satisfies the equation

$$y(t_1) = y(t_0) + \int_{t_0}^{t_1} f(\tau, y(\tau))\, d\tau$$

because $dy/dt = f(t, y)$. In other words,

$$y(t_1) - y(t_0) = \int_{t_0}^{t_1} f(\tau, y(\tau))\, d\tau.$$

This is the "integral equation" equivalent of the original differential equation. In Euler's method we approximate this difference by the step

$$y_1 - y_0 = f(t_0, y_0)\, \Delta t.$$

Geometrically this approximation is really the approximation of the integral—the "area" under the graph of $f(t, y(t))$—by the area of a rectangle with height $f(t_0, y_0)$ and width Δt (see Figure 7.7). We use the value of the derivative $y'(t_0) = f(t_0, y_0)$ at the left-hand endpoint of the interval $t_0 \leq t \leq t_1$ to approximate the value of $y(t_1)$. Thus Euler's method is analogous to approximating a Riemann integral by left-hand Riemann sums.

From your younger days when you took calculus, you may recall that the left-hand Riemann sums converge to the value of the integral as the number of subdivisions increases but that there are better ways to approximate this integral. A simple variation on left-hand approximation is trapezoidal approximation. In this case we try to approximate the area under the graph of $f(t, y(t))$ by trapezoids. The width of each trapezoid

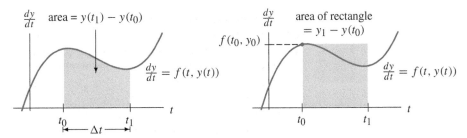

Figure 7.7
The first step of Euler's method interpreted as a Riemann approximation to an area.

is Δt, and the heights are given by the values $f(t_0, y(t_0))$, $f(t_1, y(t_1))$, $f(t_2, y(t_2))$, ... (see Figure 7.8). The area of the kth trapezoid is

$$\left(\frac{f(t_{k-1}, y(t_{k-1})) + f(t_k, y(t_k))}{2}\right) \Delta t.$$

It follows that

$$y(t_k) \approx y(t_{k-1}) + \left(\frac{f(t_{k-1}, y(t_{k-1})) + f(t_k, y(t_k))}{2}\right) \Delta t.$$

This formula for $y(t_k)$ suggests the approximation scheme

$$y_k = y_{k-1} + \left(\frac{f(t_{k-1}, y_{k-1}) + f(t_k, y_k)}{2}\right) \Delta t.$$

There is only one thing wrong with this scheme: The number y_k appears on both sides of the equation. In other words, you need to know y_k to compute y_k. (But if you already knew it, you wouldn't need the formula.) We have to get rid of the y_k on the right-hand side of the approximation scheme.

One way to turn this equation into a useful approximation scheme is to replace the y_k on the right-hand side with some other reasonable value. At this point the only way we know to approximate y_k is to use Euler's method, so that is what we do. In other words, we replace y_k on the right-hand side by the Euler approximation to y_k.

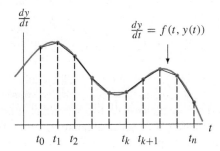

Figure 7.8
Approximating the area under the graph of

$$\frac{dy}{dt} = f(t, y(t))$$

using trapezoidal approximation.

This yields the approximation scheme

$$y_k = y_{k-1} + \left(\frac{f(t_{k-1}, y_{k-1}) + f(t_k, y_{k-1} + f(t_{k-1}, y_{k-1})\Delta t)}{2} \right) \Delta t.$$

This complicated formula can be more clearly thought of as a sequence of steps.

Improved Euler's Method

Given the initial condition $y(t_0) = y_0$ and the step size Δt, compute the point (t_{k+1}, y_{k+1}) from (t_k, y_k) as follows:

1. Use the differential equation to compute the slope $m_k = f(t_k, y_k)$.

2. Calculate the value $\tilde{y}_{k+1}$ that results from one application of Euler's method. That is,

$$\tilde{y}_{k+1} = y_k + f(t_k, y_k)\Delta t.$$

3. Calculate $t_{k+1} = t_k + \Delta t$ and use the differential equation to compute the slope $n_k = f(t_{k+1}, \tilde{y}_{k+1})$ at the point $(t_{k+1}, \tilde{y}_{k+1})$.

4. Compute y_{k+1} by

$$y_{k+1} = y_k + \left(\frac{m_k + n_k}{2} \right) \Delta t.$$

Improved Euler's Method for $f(t, y) = -2ty^2$, $y(0) = 1$

In Section 1.4 we applied Euler's method to the initial-value problem

$$\frac{dy}{dt} = -2ty^2, \quad y(0) = 1,$$

and in the last section we used this example to illustrate the typical errors involved in Euler's method. Now let's compare our previous results with the results from improved Euler's method.

We begin with $\Delta t = 0.1$. In this example the function $f(t, y) = -2ty^2$, and its value at the initial condition is $f(0, 1) = 0$. Therefore the initial slope is $m_0 = 0$. One step of Euler's method yields the point $(t_1, \tilde{y}_1) = (0.1, 1.0)$. Improved Euler's method uses the value of $f(t, y)$ at this new point to help determine the value of y_1. In other words, we compute $n_0 = f(0.1, 1.0) = -0.2$, and average this slope with m_0 to obtain the slope that we use to compute y_1. In this case the average is

$$\frac{m_0 + n_0}{2} = \frac{0.0 - 0.2}{2} = -0.1.$$

Hence we use this slope to calculate y_1. We obtain

$$y_1 = y_0 + (-0.1)\Delta t$$

$$= 1.0 + (-0.1)(0.1) = 0.99.$$

Before computing another step by hand, we compare the result of this calculation to the result that Euler's method provides. At the initial condition the differential equation yields the value 0 for the slope. Therefore Euler's method assumes that the slope over the subinterval $0 \le t \le 0.1$ vanishes, and therefore $y_1 = y_0 = 1.0$. On the other hand, improved Euler's method considers the slope at the point $(0.1, 1.0)$ as well as the slope at $(0.0, 1.0)$. When these two slopes are averaged, we obtain a (small) negative value for the slope to use over the subinterval $0 \le t \le 0.1$. Using this average slope makes sense because we can see from the differential equation that $f(t, y) = -2ty^2$ is negative for all nonzero t throughout the subinterval. Of course, there is no reason to believe that this average of the two slopes is going to be a better choice than the slope from Euler's method, but it is a good idea to try this average.

The process is basically the same for the next step. The first step yielded the point $(t_1, y_1) = (0.1, 0.99)$. Consequently, we use the differential equation to produce the slope that Euler's methods uses. We have

$$m_1 = f(t_1, y_1) = f(0.1, 0.99) = -0.19602.$$

So, starting at $(0.1, 0.99)$, we take one step of Euler's method via the computation

$$\tilde{y}_2 = y_1 + m_1 \, \Delta t$$

$$= 0.99 + (-0.19602)(0.1)$$

$$= 0.970398.$$

Thus for our other slope n_1, we compute $n_1 = f(t_2, \tilde{y}_2) = -0.37669$. We average these two slopes and obtain

$$\frac{m_1 + n_1}{2} = \frac{(-0.19602) + (-0.37669)}{2} = -0.286344.$$

Using this average we can now compute the second step

$$y_2 = y_1 + (-0.286344) \, \Delta t$$

$$= 0.99 + (-0.286344)(0.1)$$

$$= 0.961366.$$

That's enough by hand. Turning to the computer, we obtain the results shown in Table 7.1.

Table 7.1
Improved Euler's method for $dy/dt = -2ty^2$ with $\Delta t = 0.1$.

k	t_k	y_k
0	0	1
1	0.1	0.99
2	0.2	0.961366
3	0.3	0.917246
$\vdots$	$\vdots$	$\vdots$
19	1.9	0.217670
20	2.0	0.200695

Recall that we know the exact solution $y(t) = 1/(1 + t^2)$ for this initial-value problem, and its value at $t = 2$ is $y(2) = 0.2$. Thus the error in this computation is

$$|0.2 - 0.200695| = 0.000695.$$

When we used Euler's method with the same step size, we obtained the approximation $y(2) \approx 0.193342$ (see page 635). The error in that approximation was

$$|0.2 - 0.193342| = 0.00658.$$

Note that improved Euler's method is roughly ten times more accurate than Euler's method in this example.

Cost

For a given step size, using improved Euler's method instead of Euler's method yields a good improvement in accuracy. The down side is that each step of improved Euler's method involves two steps of Euler's method. The first step is used to get the value of $\tilde{y}_{k+1}$, and the second computes y_{k+1}. Each step of improved Euler's method takes roughly twice the amount of arithmetic as one step of Euler's method.

In the example above, we gained ten times the accuracy by using the improved Euler's method over the Euler's method at a cost of doing twice the arithmetic. This is a good return (particularly if a computer is doing the arithmetic).

The Order of Improved Euler

In the previous section we saw that Euler's method has order 1. In other words, the error is roughly proportional to the reciprocal of the number of subdivisions employed in the approximation. In order to compare the two methods, we need to know how the accuracy of improved Euler's method behaves as a function of the step size. As in the previous section, we consider the initial-value problem

$$\frac{dy}{dt} = -2ty^2, \quad y(0) = 1,$$

which has the exact solution $y(t) = 1/(1 + t^2)$. Using the computer we approximate the value $y(2) = 0.2$ twice. First we compute the approximation with 1000 steps. The error e_{1000} in this approximation is

$$e_{1000} = 2.59 \times 10^{-7}.$$

Then we repeat the computation using twice as many steps. In this case we get a more accurate answer with an error of

$$e_{2000} = 6.48 \times 10^{-8}.$$

These errors are definitely small, but to get a sense of how this method depends on the step size, we compute the ratio

$$\frac{e_{1000}}{e_{2000}} \approx 4.003.$$

We see that, if we double the number of steps (that is, halve the step size), we decrease the error by a factor of 4 ($= 2^2$). Consequently we say that improved Euler's method is a numerical method with order 2.

Using estimates like the ones that we discussed in the last section, it can be shown that, given appropriate bounds on f and its derivatives, the error for improved Euler's method behaves like a quadratic function in its step size. That is, the error in a given approximation using improved Euler's method is proportional to $(\Delta t)^2$. Since Δt is simply the length of the t-interval in question divided by the number of steps n, we can also express the error as

$$e_n = \frac{K}{n^2}.$$

for some constant K. In Figure 7.9, we graph the error as a function of the number of steps for Euler's method and improved Euler's method. This plot illustrates the advantage of a second-order method over a first-order method.

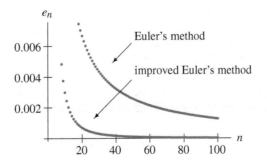

e_n

Euler's method

improved Euler's method

0.006

0.004

0.002

20   40   60   80   100

n

Figure 7.9
The errors related to Euler's method and improved Euler's method for the initial-value problem

$$\frac{dy}{dt} = -2ty^2, \quad y(0) = 1,$$

over the interval $0 \le t \le 2$.

EXERCISES FOR SECTION 7.2

In Exercises 1–8, use improved Euler's method with the given number n of steps to approximate the solution to the initial-value problem specified. Your answer should include a table of approximate values of the dependent variable. It should also include a sketch of the graph of the approximate solution. These initial-value problems also appear in the exercise set of Section 1.4. Compare the graphs that you get from improved Euler's method with those that you get from Euler's method. If your computer system has a built-in routine for the numerical solution of differential equations, compare these graphs with those that come from the built-in numerical solver.

1. $\dfrac{dy}{dt} = 2y + 1,\quad y(0) = 3,\quad 0 \le t \le 2,\quad n = 4$

2. $\dfrac{dy}{dt} = t - y^2,\quad y(0) = 1,\quad 0 \le t \le 1,\quad n = 4$

3. $\dfrac{dy}{dt} = y^2 - 4t,\quad y(0) = 0.5,\quad 0 \le t \le 2,\quad n = 8$

4. $\dfrac{dy}{dt} = \sin y,\quad y(0) = 1,\quad 0 \le t \le 3,\quad n = 6$

5. $\dfrac{dw}{dt} = (3 - w)(w + 1),\quad w(0) = 0,\quad 0 \le t \le 5,\quad n = 10$

6. $\dfrac{dy}{dt} = e^{2/y},\quad y(0) = 2,\quad 0 \le t \le 2,\quad n = 4$

7. $\dfrac{dy}{dt} = e^{2/y},\quad y(1) = 2,\quad 1 \le t \le 3,\quad n = 4$

8. $\dfrac{dy}{dt} = y^2 - y^3,\quad y(0) = 0.2,\quad 0 \le t \le 10,\quad n = 100$

In Exercises 9–12, use improved Euler's method to approximate the solution to the given initial-value problem over the interval specified. For each initial-value problem,

(a) determine an analytic solution (see Exercises 2–5 in Section 7.1); and

(b) using improved Euler's method with $n = 4$, determine an approximate value y_4 and the total error e_4.

(c) How many steps of improved Euler's method would you expect to use to obtain an approximation for which $e_n \le 0.0001$?

9. $\dfrac{dy}{dt} = 1 - y,\quad y(0) = 0,\quad 0 \le t \le 1$

10. $\dfrac{dy}{dt} = ty,\quad y(0) = 1,\quad 0 \le t \le \sqrt{2}$

11. $\dfrac{dy}{dt} = -y^2,\quad y(0) = 1/2,\quad 0 \le t \le 2$

12. $\dfrac{dy}{dt} = 1 - t + 3y, \quad y(0) = 1, \quad 0 \le t \le 1$

13. In this section we showed that the total error e_{20} in the approximation of the initial-value problem $dy/dt = -2ty^2$, $y(0) = 1$, over the interval $0 \le t \le 2$ is

$$e_{20} = 0.000695.$$

(a) How many steps should you use to get an approximate value y_n that has an associated error e_n no greater than 0.0001?

(b) Calculate that approximate value y_n.

(c) Calculate the total error in that case.

In Exercises 14–17, produce a plot like Figure 7.9 for the given initial-value problem over the interval specified. See the directions to Exercise 1 in Section 7.1 for more details. However, restrict the values of n to numbers in the interval $10 \le n \le 100$.

14. $\dfrac{dy}{dt} = 1 - y, \quad y(0) = 0, \quad 0 \le t \le 1$

15. $\dfrac{dy}{dt} = ty, \quad y(0) = 1, \quad 0 \le t \le \sqrt{2}$

16. $\dfrac{dy}{dt} = -y^2, \quad y(0) = 1/2, \quad 0 \le t \le 2$

17. $\dfrac{dy}{dt} = 1 - t + 3y, \quad y(0) = 1, \quad 0 \le t \le 1$

7.3 THE RUNGE-KUTTA METHOD

Improved Euler's method shows us that if we intelligently modify Euler's method and do a little bit more work in each step, we can get a significantly more accurate approximation to our solution. This observation leads us to wonder if there is a "standard" method for the numerical approximation of solutions of differential equations. We want an algorithm that minimizes the number of computations that have to be performed while it maximizes the accuracy.

In fact for most practical purposes, there is such an algorithm. It is called the (fourth-order) Runge-Kutta method. It is named after the two German mathematicians who developed the method approximately 100 years ago. There are more sophisticated algorithms that are required for specialized problems, but the Runge-Kutta method is often the numerical method used for the numerical approximation of solutions.

The Runge-Kutta Method

In improved Euler's method applied to the differential equation $dy/dt = f(t, y)$, we used an average of two slopes to compute each value. In other words, to compute y_k

from y_{k-1}, we used the average of values of the right-hand side $f(t, y)$ at $t = t_{k-1}$ and $t = t_k$. By analogy with numerical integration, improved Euler's method is similar to the trapezoidal rule.

For numerical integration, there are algorithms that are usually more efficient than the trapezoidal rule. Simpson's rule approximates the area under the graph using parabolic interpolation and generally gives a better estimate for the integral. In fact Simpson's rule can be interpreted as a weighted average of values where twice as much weight is given to the values of the function at the midpoints of the subintervals than is given to the endpoints of the subintervals. The Runge-Kutta method is similar to Simpson's rule in the way that it considers a weighted average.

We begin with a description of the algorithm and then illustrate its implementation and accuracy using the example from previous sections. Refer to Figure 7.10 as we describe how the various slopes used in the method are derived.

To calculate the value y_{k+1} from y_k, we use four slopes given by the function $f(t, y)$ that defines the differential equation. These slopes are denoted m_k, n_k, q_k, and p_k. We also need several intermediate variables. Thus the bookkeeping in the Runge-Kutta method is more formidable than improved Euler's method, so these variables wear many different hats.

The four slopes are determined successively as follows:

1. The first slope m_k is calculated just as in Euler's method; that is, $m_k = f(t_k, y_k)$.

2. In the improved Euler's method we used m_k to produce a second slope corresponding to a point with $t = t_{k+1} = t_k + \Delta t$. The Runge-Kutta method does almost the same thing except that it goes only halfway along the t-axis to $\tilde{t} = t_k + \Delta t/2$. That is, we use m_k to produce a point $(\tilde{t}, \tilde{y}_k)$ where

$$\tilde{y}_k = y_k + m_k \frac{\Delta t}{2}$$

(see Figure 7.10(a)). Once we have determined this point, we use the function $f(t, y)$ to determine the second slope n_k by

$$n_k = f(\tilde{t}, \tilde{y}_k).$$

3. Now we repeat the previous step of the algorithm where we use the slope n_k in place of the slope m_k. In other words, we go from (t_k, y_k) to the line $t = \tilde{t}$ along a line of slope n_k. Thus we obtain a new number $\hat{y}_k$ where

$$\hat{y}_k = y_k + n_k \frac{\Delta t}{2}$$

(see Figure 7.10(b)). Given this point on the line $t = \tilde{t}$, we compute the third slope q_k by

$$q_k = f(\tilde{t}, \hat{y}_k).$$

4. Finally, we obtain our fourth slope by using q_k to produce a point on the line $t = t_{k+1}$. We get

$$\overline{y}_k = y_k + q_k \Delta t$$

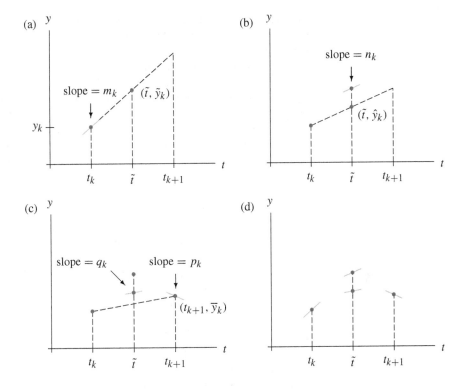

Figure 7.10
Part (d) illustrates the four slopes that are used in the Runge-Kutta method. Parts (a)–(c) illustrate how the points at which the slopes are evaluated are determined.

(see Figure 7.10(c)). Once we have this fourth point we compute the slope at that point by

$$p_k = f(t_{k+1}, \overline{y}_k).$$

5. Now that we have collected the four slopes (see Figure 7.10(d)), we take a weighted average and compute the next step using this average slope. We count each of the slopes that come from the points with $t = \tilde{t}$ twice as heavily as we count the other two slopes. In other words, our weighted average is

$$\frac{m_k + 2n_k + 2q_k + p_k}{6}.$$

The actual step that we take is therefore

$$y_{k+1} = y_k + \left(\frac{m_k + 2n_k + 2q_k + p_k}{6} \right) \Delta t.$$

This seems complicated, but it works well. Basically the slopes are chosen so that we are approximating the solution up to terms of order 4. As we shall see, this gives us an algorithm that is very efficient.

The Initial-Value Problem $dy/dt = -2ty^2$, $y(0) = 1$

Let's illustrate a couple of steps of the algorithm using our old friend

$$\frac{dy}{dt} = -2ty^2, \quad y(0) = 1.$$

In the previous section, we approximated the solution to this initial-value problem using improved Euler's method with $\Delta t = 0.1$. Now we use the same Δt with the Runge-Kutta method.

We begin with the initial slope

$$m_0 = f(t_0, y_0) = f(0, 1) = 0.$$

This slope determines

$$\tilde{y}_0 = y_0 + m_0 \frac{\Delta t}{2} = 1 + 0 \left(\frac{0.1}{2} \right) = 1.$$

Given $\tilde{y}_0$ and $\tilde{t} = t_0 + \Delta t/2 = 0 + 0.05 = 0.05$, we obtain the second slope we need by

$$n_0 = f(\tilde{t}, \tilde{y}_0) = f(0.05, 1.0) = -0.1.$$

In turn this slope determines the y-value $\hat{y}_0$ via

$$\hat{y}_0 = y_0 + n_0 \frac{\Delta t}{2}$$

$$= 1.0 + (-0.1)(0.05)$$

$$= 0.995.$$

Applying $f(t, y)$ to the point $(\tilde{t}, \hat{y}_0)$ determines a third slope

$$q_0 = f(\tilde{t}, \hat{y}_0) = -0.0990025,$$

and using this slope, we obtain the fourth point with y-value $\overline{y}_0$ by

$$\overline{y}_0 = y_0 + q_0 \Delta t = 1.0 + (-0.0990025)(0.1) = 0.9901.$$

Given this fourth point, we compute the slope at that point by

$$p_0 = f(t_{k+1}, \overline{y}_0) = -0.19606.$$

Now that we have these four slopes, we are able to compute the weighted average that we use to compute the slope that determines the first step. We have

$$\frac{m_0 + 2n_0 + 2q_0 + p_0}{6} = \left(\frac{0 + 2(-0.1) + 2(-0.0990025) + -0.19606}{6} \right)$$

$$= -0.0990108.$$

This weighted average yields the first step

$$y_1 = y_0 - 0.0990108 \, \Delta t = 1.0 + (-0.0990108)(0.1) = 0.990099.$$

(If you try to reproduce these calculations on your calculator and you get a slightly different answer, don't be concerned. Your calculator may carry a different number of significant digits in the calculation, and this difference may affect your computation slightly.)

Doing these calculations by hand is definitely not the way to go now that we have easy access to programmable calculators, spreadsheets, and general systems for doing mathematics by computers such as Maple, *Mathematica*, and MATLAB. Later in this section we provide codes that can be used in various computing environments to implement Runge-Kutta. They are not too complicated, and using them definitely beats doing these computations by hand.

Computing 20 steps via the Runge-Kutta method yields the results shown in Table 7.2. Note that the approximation of $y(2) = 0.2$ is very good, especially considering the relatively large step size that we used. Compare these results to those found in Table 7.1. The error obtained in improved Euler's method was 0.000658, whereas the error here is 0.000001. Switching to the Runge-Kutta method in this example improves the accuracy by a factor of 658. (Actually the improvement is more like a factor of 1000, but that is not clear from Table 7.2 because we show the results only to six significant digits.)

Table 7.2
Runge-Kutta method for $dy/dt = -2ty^2$ with $\Delta t = 0.1$.

k	t_k	y_k
0	0	1
1	0.1	0.990099
2	0.2	0.961538
3	0.3	0.917431
$\vdots$	$\vdots$	$\vdots$
19	1.9	0.216920
20	2.0	0.200001

The Order of the Runge-Kutta Method

Euler's method has order 1 and improved Euler's method has order 2. Now let's see if we are right about Runge-Kutta's order. To find the order of the Runge Kutta method, we recall that the solution of the initial-value problem

$$\frac{dy}{dt} = -2ty^2, \quad y(0) = 1$$

is $y(t) = 1/(1 + t^2)$. If we approximate this solution over the interval $0 \le t \le 2$ using the Runge-Kutta method with ten steps, we find that the error in this approximation is

$$e_{10} = 1.095 \times 10^{-5},$$

and if we approximate the solution using 20 steps, we get an error of

$$e_{20} = 6.541 \times 10^{-7}.$$

The ratio

$$\frac{e_{10}}{e_{20}} = 16.75$$

indicates that if we double the number of steps, the error improves by roughly a factor of 16 (= 2^4). This is what we expect from a numerical procedure that has order 4.

Programming Runge-Kutta

Although the Runge-Kutta method seems somewhat complicated at first, it really isn't especially difficult to program. Here we give three implementations—one for a graphing calculator, one for *Mathematica*, and one in the C programming language. Some of the exercises at the end of this section require that you implement the method using whatever technology is available to you. These three implementations should give you some guidance as to how to proceed.

The TI graphing calculator

This code assumes that the initial t-value is stored in A, the end value is stored in B, the number of steps is stored in N, the initial y-value is stored in Y, and $Y_1 = f(X,Y)$. The code uses X to represent t and Y to represent y in our notation.

```
:(B - A)/N -> H
:A -> X
:0 -> I
:Lbl 1
:Disp Y
:Pause
:Y -> Z
:Y1 -> C
:X + H/2 -> X
:Z + C*H/2 -> Y
:Y1 -> D
:Z + D*H/2 -> Y
:Y1 -> E
:X + H/2 -> X
:Z + E*H -> Y
:Y1 -> F
:Z + (C + 2*D + 2*E + F)*H/6 -> Y
:IS>(I,N)
:Goto 1
```

Mathematica code

The *Mathematica* code is more compact, but using it effectively requires some knowledge of commands such as ListPlot and NestList. We begin by defining a function that takes one step of the Runge-Kutta method.

```
RungeKuttaStep[f_, vars_, {tk_, yk_}, deltat_] :=
 Module[{k1, k2, k3, k4, tmiddle, tend},
    k1 = f /. {vars[[1]] -> tk, vars[[2]] -> yk};
    tmiddle = tk + deltat/2;
    k2 = f /. {vars[[1]] -> tmiddle,
                 vars[[2]] -> yk + k1 deltat/2};
    k3 = f /. {vars[[1]] -> tmiddle,
                 vars[[2]] -> yk + k2 deltat/2};
    tend = tk + deltat;
    k4 = f /. {vars[[1]] -> tend,
                 vars[[2]] -> yk + k3 deltat};
    {tend, yk + (k1 + 2 k2 + 2 k3 + k4) deltat/6}
    ]
```

For example, the command

```
RungeKuttaStep[-2 t y^2, {t, y}, {0,1}, 0.1]
```

returns

```
{0.1, 0.990099}
```

We then approximate the solution to the initial-value problem

$$\frac{dy}{dt} = -2ty^2, \quad y(0) = 1$$

over the interval $0 \le t \le 2$ using 20 steps by the command

```
RKresults =
    NestList[RungeKuttaStep[-2 t y^2, {t, y}, #, 0.1]&,
              {0, 1}, 20]
```

This command stores the resulting points in the variable RKresults, and hence we are able to plot the approximate solution using ListPlot. For example,

```
ListPlot[RKresults,
      PlotJoined -> True,
      AxesOrigin -> {0,0}]
```

C code for Runge-Kutta

The following is a C program that applies the Runge-Kutta method to the initial-value problem

$$\frac{dy}{dt} = -2ty^2, \quad y(0) = 1.$$

It does the approximation over the interval $0 \le t \le 2$ with 20 steps. You should change the values of T0, Tn, NUMSTEPS, and Y0 at the top and the definition of $f(t, y)$ in the return statement to apply the code to a different initial-value problem.

```c
/*

                    Runge-Kutta approximation of
                          dy/dt = f(t,y)
*/


/* need to set these values and the function f(t,y) below
   before compilation */

#define T0 0.0
#define Tn 2.0
#define NUMSTEPS 20
#define Y0 1.0

#include <stdio.h>
#include <math.h>

/*
   this subroutine contains the formula for the
   first-order differential equation,
   that is, dy/dt = f(t,y)
*/

double f(t,y)
     double t,y;
{
  return(-2.0 * t * y * y);
}



main()
{
  double tk = T0, tmid, yk = Y0,
       k1, k2, k3, k4,
          deltat = (double)(Tn - T0)/NUMSTEPS,
          f();
  int stepnum = 0;
```

```
printf("%f %f\n", tk, yk);
while (++stepnum <= NUMSTEPS){
 k1 = f(tk, yk);
 tmid = tk + deltat/2.0;
 k2 = f(tmid, yk + k1 * deltat/2.0);
 k3 = f(tmid, yk + k2 * deltat/2.0);
 tk = tk + deltat;
 k4 = f(tk, yk + k3 * deltat);
 yk = yk + (k1 + 2.0 * k2 + 2.0 * k3 + k4) * deltat/6.0;
 printf("%f %f\n", tk, yk);
 }
}
```

The Runge-Kutta Method for Systems

The generalization of the Runge-Kutta method to first-order systems is straightforward if we take advantage of our vector notation for systems. Moreover, once we implement Runge-Kutta for systems, we also have an efficient method for the numerical study of higher-order differential equations, since higher-order equations can be rewritten as first-order systems. In fact this is the only way to study second- and higher-order differential equations numerically.

Consider the first-order system

$$\frac{d\mathbf{Y}}{dt} = \mathbf{F}(t, \mathbf{Y}), \quad \mathbf{Y}(t_0) = \mathbf{Y}_0.$$

To apply the Runge-Kutta method to this system, we need to compute four vectors $\mathbf{K}_1$, $\mathbf{K}_2$, $\mathbf{K}_3$, and $\mathbf{K}_4$ at each step of the algorithm. These vectors are analogous to the four slopes that we computed for first-order equations, and they are defined in essentially the same way. Each of these vectors also depends on the index k, but we will not indicate this dependence to keep the formulas more readable. That is, suppose we have the point $(t_k, \mathbf{Y}_k)$. Then the vectors are

$$\mathbf{K}_1 = \mathbf{F}(t_k, \mathbf{Y}_k)$$

$$\mathbf{K}_2 = \mathbf{F}(t_k + \Delta t/2, \mathbf{Y}_k + (\Delta t/2)\mathbf{K}_1)$$

$$\mathbf{K}_3 = \mathbf{F}(t_k + \Delta t/2, \mathbf{Y}_k + (\Delta t/2)\mathbf{K}_2)$$

$$\mathbf{K}_4 = \mathbf{F}(t_k + \Delta t, \mathbf{Y}_k + \Delta t \, \mathbf{K}_3).$$

Once we have these vectors, then we can form a weighted average just as in the case of first-order equations. We get the vector

$$\mathbf{K} = \left(\frac{\mathbf{K}_1 + 2\mathbf{K}_2 + 2\mathbf{K}_3 + \mathbf{K}_4}{6} \right),$$

and we use this vector $\mathbf{K}$ to make the step from $(t_k, \mathbf{Y}_k)$ to

$$(t_{k+1}, \mathbf{Y}_{k+1}) = (t_k + \Delta t, \mathbf{Y}_k + \Delta t \, \mathbf{K}).$$

If we are plotting solution curves in the phase plane, then we simply plot the points $\mathbf{Y}_k$.

Figures 7.11 and 7.12 compare the results of Euler's method with the Runge-Kutta method on the predator-prey system we discussed in Section 2.1. Note that although Euler's method is not accurate enough to catch the closed orbits in the phase plane, the Runge-Kutta method does a nice job of capturing the periodic nature of this system.

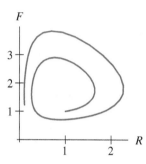

Figure 7.11
Euler's method for the solution with initial condition $\mathbf{Y}(0) = (1, 1)$ with $\Delta t = 0.1$ and 80 steps for the predator-prey system.

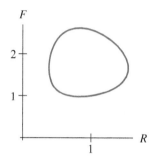

Figure 7.12
Runge-Kutta applied to the solution with initial condition $\mathbf{Y}(0) = (1, 1)$ with $\Delta t = 0.1$ and 80 steps for the same system as Figure 7.11.

EXERCISES FOR SECTION 7.3

In Exercises 1–5, use the Runge-Kutta method with the given number n of steps to approximate the solution to the initial-value problem specified. Your answer should include a table of approximate values of the dependent variable. It should also include a sketch of the graph of the approximate solution. These initial-value problems also appear in the exercise set of Section 7.2. Compare the graphs that you get from the Runge-Kutta method to those that come from Euler's method and improved Euler's method. If your computer has a built-in routine for the numerical solution of differential equations, compare these graphs with those that come from the built-in numerical solver.

1. $\dfrac{dy}{dt} = 2y + 1, \quad y(0) = 3, \quad 0 \le t \le 2, \quad n = 4$

2. $\dfrac{dy}{dt} = t - y^2, \quad y(0) = 1, \quad 0 \le t \le 1, \quad n = 4$

3. $\dfrac{dy}{dt} = (3 - y)(y + 1), \quad y(0) = 0, \quad 0 \le t \le 5, \quad n = 10$

4. $\dfrac{dy}{dt} = e^{2/y}$, $y(0) = 2$, $0 \le t \le 2$, $n = 4$

5. $\dfrac{dy}{dt} = e^{2/y}$, $y(1) = 2$, $1 \le t \le 3$, $n = 4$

6. Consider the initial-value problem

$$\frac{dy}{dt} = -2ty^2, \quad y(0) = 1$$

over the interval $0 \le t \le 2$.

    (a) Calculate the Runge-Kutta approximation to the solution with $n = 4$ steps.

    (b) Calculate the total error e_4 associated with this approximation.

    (c) How many steps are necessary to approximate the solution with an error of less than 0.0001?

7. Consider the predator-prey system

$$\frac{dR}{dt} = 2R - 1.2RF$$

$$\frac{dF}{dt} = -F + 1.2RF$$

and the initial condition $(R_0, F_0) = (1.0, 1.0)$.

    (a) Calculate the Runge-Kutta approximation to this system for five steps with $\Delta t = 1.0$. Plot the results in the phase plane.

    (b) With the aid of a calculator or a computer, repeat the calculation in part (a) with $\Delta t = 0.1$ and $n = 80$ steps. Plot the result in the phase plane.

    (c) Using the results of part (b), plot the $R(t)$- and $F(t)$-graphs for this solution.

8. Use Runge-Kutta to approximate the solution to the second-order equation

$$\frac{d^2y}{dt^2} + (y^2 - 3)\frac{dy}{dt} + 5y = 0$$

that satisfies the initial condition $(y(0), y'(0)) = (1, 0)$. Use $\Delta t = 0.1$ and $n = 100$. Plot your results both in the yv-phase plane (where $v = y'$) and as the graph of $y(t)$.

7.4 THE EFFECTS OF FINITE ARITHMETIC

In theory the errors involved in the use of numerical approximation procedures decrease as the step sizes decrease, and our preference for the Runge-Kutta method stems from the fact that this method is a fourth-order method that isn't too complicated to implement. It is tempting to assume that the computer or calculator is doing exactly what it was told to do. Unfortunately this is not the case. Floating-point arithmetic has only a finite amount of accuracy, and this limitation introduces another source of error in each step of the numerical method.

In this section we briefly take a look at a couple of informative examples so that you are aware of some of the pitfalls that are associated with computer arithmetic. As always, we start with our favorite example so that we know the exact solution, and then we discuss what should be done when we do not know the solution. Of course, the latter case is the situation in which we typically employ these algorithms.

The Initial-Value Problem $dy/dt = -2ty^2$, $y(0) = 1$ Yet Again

In the previous section we found that the Runge-Kutta method gives us a very accurate approximation to the solution $y(t) = 1/(1 + t^2)$ over the interval $0 \leq t \leq 2$ using only the step size $\Delta t = 0.1$. This is a fairly large step size. In theory we can make the error as small as we like by decreasing the step size. To see if this works in practice, we try an experiment. (Such experiments are an excellent use of home computers, a much better use of electricity than the drawing of flying toasters.)

We start by subdividing the interval $0 \leq t \leq 2$ into two subintervals (so there are two steps and $\Delta t = 1.0$), and we calculate the error e_2. The Runge-Kutta method with $\Delta t = 1.0$ yields the approximation $y_2 = 0.182588$ of the exact value $y(2) = 0.2$, and therefore the error is

$$e_2 = |0.2 - 0.182588| = 0.0174118.$$

It's remarkable that Runge-Kutta gives such an accurate approximation with such a large Δt.

The theory says that, if we double the number of steps in the Runge-Kutta method, the error should decrease by a factor of 16. Therefore we double the number of steps (and halve the step size) and recompute the approximate solution. Repeating this procedure, doubling the number of steps each time, we obtain the data in Table 7.3.

Table 7.3
The error in the value of $y(2)$ associated with the Runge-Kutta method for $dy/dt = -2ty^2$, $y(0) = 1$ for 2^n subdivisions of the interval $0 \leq t \leq 2$.

n	2^n	e_{2^n}	n	2^n	e_{2^n}
1	2	0.01741182739635213639	10	1024	0.00000000000008953949
2	4	0.00040567218499903968	11	2048	0.00000000000000552336
3	8	0.00002714430679354174	12	4096	0.00000000000000080491
4	16	0.00000161801719431032	13	8192	0.00000000000000024980
5	32	0.00000009769105269175	14	16384	0.00000000000000094369
6	64	0.00000000598843688526	15	32768	0.00000000000000072164
7	128	0.00000000037049122104	16	65536	0.00000000000000077716
8	256	0.00000000002303576774	17	131072	0.00000000000000144329
9	512	0.00000000000143582368	18	262144	0.00000000000000355271

Table 7.4
The error associated with the Runge-Kutta method for $dy/dt = -2ty^2$ for 2^n subdivisions.

n	2^n	e_{2^n}	n	2^n	e_{2^n}
12	4096	0.0000000000000080491	20	1048576	0.0000000000000224820
13	8192	0.0000000000000024980	21	2097152	0.0000000000000621725
14	16384	0.0000000000000094369	22	4194304	0.0000000000001271205
15	32768	0.0000000000000072164	23	8388608	0.0000000000001187939
16	65536	0.0000000000000077716	24	16777216	0.0000000000001057487
17	131072	0.0000000000000144329	25	33554432	0.0000000000002037259
18	262144	0.0000000000000355271	26	67108864	0.0000000000000949241
19	524288	0.0000000000000130451			

Note that the error improves as expected for the first 10 doublings. Note that, for $n = 13$, we have an approximation that agrees with the exact value to 15 decimal places. Then, at $n = 14$, the error starts to increase. Calculating more steps does not seem to help (see Table 7.4).

The Effects of Finite Arithmetic

At a certain point the efficiency of the numerical method is defeated by the effects of round-off errors in the arithmetic of the computer. Unless we are lucky for some reason, we cannot expect to get an answer that is any more accurate than the arithmetic allows. This is like the concept of significant digits in the sciences.

There are various ways to go about modeling the accuracy of the computer arithmetic, and like our models they lead to somewhat different predictions about the effects of finite arithmetic. For our purposes we'll adopt one of the standard models that says the total error, denoted by $e(\Delta t)$, as a function of step size Δt for the Runge-Kutta method is roughly

$$e(\Delta t) = c_1(\Delta t)^4 + \frac{c_2}{\Delta t},$$

where c_1 and c_2 are constants determined by the differential equation, its initial condition, and interval of which the solution is being approximated. The first term, $c_1(\Delta t)^4$, is the term that says that Runge-Kutta is a fourth-order method. The second term, $c_2/(\Delta t)$, is the term that measures the round-off errors for small values of Δt. Presumably, the constant c_2 is very small, and thus Δt must be very small before the effects of this term are noticeable.

We can observe the effects of round-off if we repeat the calculations done above with single-precision arithmetic (see Table 7.5). Note that, in this case, we are better off with 128 subdivisions than we are with millions of subdivisions. In fact, by the time we get to 16 million subdivisions, our approximation is about as accurate as it was with 2 subdivisions.

Table 7.5

The error associated to the Runge-Kutta method implemented in single precision for $dy/dt = -2ty^2$ for 2^n subdivisions.

n	2^n	e_{2^n}	n	2^n	e_{2^n}
1	2	0.01741183	13	8192	0.00000031
2	4	0.00040567	14	16384	0.00000002
3	8	0.00002714	15	32768	0.00000001
4	16	0.00000161	16	65536	0.00000035
5	32	0.00000012	17	131072	0.00000324
6	64	0.00000002	18	262144	0.00000173
7	128	0.00000001	19	524288	0.00002292
8	256	0.00000007	20	1048576	0.00000132
9	512	0.00000001	21	2097152	0.00062698
10	1024	0.00000005	22	4194304	0.00052327
11	2048	0.00000012	23	8388608	0.00506006
12	4096	0.00000004	24	16777216	0.01590731

What to Do in Practice

Table 7.6

Runge-Kutta applied to the initial-value problem $dy/dt = y^2 - t$, $y(1) = 1$.

n	$y_n \approx y(3)$
2	−1.39687991
4	−1.47455096
8	−1.49300539
16	−1.49384940
32	−1.49389482
64	−1.49389744
128	−1.49389744
256	−1.49389791
512	−1.49389756

In practice, we want to use these numerical procedures when we do not know analytic representations for the solutions of a differential equation. Since it is usually impossible to carry out meaningful estimates from the theory, how do we decide what is a reasonable strategy to follow? If we accept the assertion that the error is roughly determined by an expression of the form

$$e(\Delta t) = c_1 (\Delta t)^4 + \frac{c_2}{\Delta t},$$

then we want to use a value of Δt that minimizes the first term as much as possible before the second term starts to have an effect. However, since we don't know the actual value of the solution, how can we predict where this minimum is going to occur?

We compute multiple approximations where we double the number of steps, and we look for our approximations to "converge" to a certain value. For example, if we apply Runge-Kutta to the initial-value problem

$$\frac{dy}{dt} = y^2 - t, \quad y(1) = 1,$$

over the interval $1 \leq t \leq 3$ with various numbers of subdivisions, we get the results given in Table 7.6.

Note that we have the approximation $y(3) \approx -1.49389744$ with both 64 and 128 subdivisions. Given the repetition, we compute one more approximation with 256 subdivisions, and we are fairly confident that the actual value of the solution to five decimal places is -1.49389. It's also a relatively safe bet that the next digit in the decimal expansion is 7, but if we need a more precise answer, we repeat the calculations using more precision.

EXERCISES FOR SECTION 7.4

In Exercises 1–3, use Runge-Kutta to produce a table similar to Table 7.6 to approximate the value of the solution at the right-hand endpoint of the interval specified. What number of steps corresponds to the optimal approximation?

1. $\dfrac{dy}{dt} = t - y^3$, $\quad y(0) = 2$, $\quad 0 \le t \le 1$

2. $\dfrac{dy}{dt} = \sin ty$, $\quad y(0) = 3$, $\quad 0 \le t \le 3$

3. $\dfrac{dy}{dt} = (\cos y)(t^2 - y)$, $\quad y(0) = 0$, $\quad 0 \le t \le 2$

REVIEW EXERCISES FOR CHAPTER 7

True-false: For Exercises 1–5, determine if the statement is true or false. If it is true, explain why. If it is false, provide a counterexample or an explanation.

1. In a first-order method, halving the step size typically halves the error.

2. In a second-order method, doubling the number of steps usually doubles the accuracy.

3. In a fourth-order method, doubling the number of steps usually quadruples the accuracy.

4. Improved Euler's method is a second-order method.

5. Runge-Kutta is a third-order method.

6. Consider the initial-value problem

$$\frac{dy}{dt} = t - y^3, \quad y(0) = 1$$

over the interval $0 \le t \le 1$.

(a) Calculate the Euler approximation to the solution using $n = 4$ steps and draw its graph.

(b) Calculate the improved Euler approximation to the solution using $n = 4$ steps and draw its graph.

(c) Calculate the Runge-Kutta approximation to the solution using $n = 4$ steps and draw its graph.

(d) Using a calculator or computer, repeat parts (a)–(c) using $n = 100$ steps.

7. Consider the competing species system

$$\frac{dx}{dt} = 2x - x^2 - xy$$
$$\frac{dy}{dt} = 3y - y^2 - 2xy$$

and the initial condition $(x_0, y_0) = (3, 3)$.

(a) Using a calculator or computer, calculate the Euler approximation to the solution over the interval $0 \le t \le 5$ using $\Delta t = 0.1$. Plot the result in the xy-phase plane and plot the corresponding $x(t)$- and $y(t)$-graphs.

(b) Repeat part (a) using improved Euler's method.

(c) Repeat part (a) using Runge-Kutta.

LAB 7.1 Errors of Numerical Approximations

In this chapter, we studied three different methods for finding numerical approximations of solutions of differential equations: Euler, improved Euler, and Runge-Kutta. The accuracy of these methods depends on the step size Δt. In this lab we obtain additional insight into the relationship between the error in numerical approximations and the step size by studying some special equations and solutions.

In your report, address the following items:

1. Use Euler's method to approximate the solution of the initial-value problem

$$\frac{dy}{dt} = y^2 - 4ty + 4t^2 - 4y + 8t - 3, \quad y(0) = -1.$$

How does the approximation change when the step size is changed? Interpret your results. [*Hint*: Consider the slope field for this equation (see page 733).]

2. Use Euler's method, improved Euler's method, and Runge-Kutta to approximate the solution of the initial-value problem

$$\frac{dy}{dt} = 2\sqrt{y}, \quad y(0) = 1.$$

Find the formula for the solution and evaluate the error as a function of the step size for each of the methods. Are any of the methods better or worse than you would have expected? [*Hint*: The geometry of the slope field will not be as helpful here, but the special form of the formula of the solution is helpful.]

3. Repeat Part 2 for the initial-value problem

$$\frac{dy}{dt} = y + t^3 - 6, \quad y(0) = 0.$$

Your report: In your report, you should address each of the items above. You should *not* include any tables of numerical solutions. You may include a limited number of tables and/or graphs of how the error in your approximate solutions depends on the step size. The goal of this lab is to interpret why the error depends on the step size the way it does.

LAB 7.2 Lost in Space

The motion of planets and satellites in space is governed by the Newton's laws of motion and gravitational attraction. Newton's law of motion states that the total force is the product of the mass and the acceleration, and Newton's law of gravitation asserts that the force due to gravity between two bodies is inversely proportional to the distance between the bodies. This force is an attracting force, pulling each body directly toward the other. As in the case of the harmonic oscillator in Chapter 2 and the pendulum in Chapter 5, these laws give rise to second-order differential equations for the positions of the bodies as a function of time.

It is frequently said that the differential equations governing the motion of the moon and the planets were the first differential equations. The ability to compute the position of heavenly bodies (like the moon) was of great financial importance during the time of Newton. Knowing the precise location of the moon and planets at any given time was essential in navigation of the open sea. These equations are still of great theoretical importance in the development of techniques for the study of differential equations. They are also of great practical importance for space navigation. The type of numerical approximation techniques we have discussed are used for computing orbits of satellites to the moon and planets (and back—see the movie *Apollo 13*).

In this lab, suppose you are on a mission in deep space, and after a misunderstanding, your roommate leaves you adrift in your space suit with no fuel. Not being completely without a heart, your roommate also leaves behind an escape capsule. However, you and the escape capsule are 100 meters apart and you are both stationary. "Not to worry," you think, "two objects attract each other via the force of gravity, so the escape capsule and I will come together eventually." Suppose your mass (in the space suit) is 150 kg and the mass of the escape capsule is 10^6 kg. How long will it take you to reach the capsule?

First we need the differential equations. Since both you and the capsule start at rest and the only force is gravity, which acts along the line connecting you with the capsule, both you and the capsule will remain on that line. Consequently we only need one number to denote the relative positions of you and the capsule. Let $x_1(t)$ be the position of the capsule at time t and $x_2(t)$ denote your position at time t. We assume that coordinates have been chosen so that $x_2 > x_1$. We are most interested in the distance between you and the capsule, so let $x(t) = x_2(t) - x_1(t)$. By Newton's laws we know that the product of your mass and your acceleration is equal to the force of gravity between you and the capsule and that this force is proportional to the product of your masses divided by the square of the distance between you. That is,

$$150 \frac{d^2 x_2}{dt^2} = -G \frac{150 \times 10^6}{|x_2 - x_1|^2},$$

where the minus sign indicates that the force is pulling in the direction of decreasing x_2 and G is the "gravitation constant." This is the same as

$$\frac{d^2 x_2}{dt^2} = -G \frac{10^6}{|x_2 - x_1|^2}.$$

Similarly, the acceleration of the capsule is given by

$$\frac{d^2 x_1}{dt^2} = G \frac{150}{|x_2 - x_1|^2}.$$

Subtracting these equations and using $x = x_2 - x_1$, we obtain the differential equation

$$\frac{d^2 x}{dt^2} = -G \frac{10^6 + 150}{x^2}$$

for $x(t)$. This is a nonlinear, second-order differential equation. The gravitation constant G has been carefully measured by experiment and is given by

$$G \approx 6.6732 \times 10^{-11} \frac{\text{m}^3}{(\text{kg})(\text{sec})^2}.$$

In your report, address the following items:

1. Convert the second-order differential equation

$$\frac{d^2x}{dt^2} = -G\frac{10^6 + 150}{x^2}$$

into a first-order system.

2. Using numerical techniques, approximate the time necessary for you to reach the escape capsule, given that you started 100 meters away and both you and the capsule have initial velocity 0. At what velocity are you and the capsule coming together when you are 1 meter apart? (You may use whatever numerical method you like.)

3. Can you verify that your computation of the time until you reach the escape capsule is accurate to within 1 second.

Your report: In your report you should address these three items. Be sure to explain your methods, particularly for Part 3. You may use graphs and/or charts to illustrate your methods, but do not submit long lists of numbers.

8 DISCRETE DYNAMICAL SYSTEMS

In this chapter we begin the study of a completely different type of model for processes that evolve in time, namely discrete dynamical systems, or difference equations. Unlike differential equations, these models are well suited to situations in which changes occur at specific times, rather than continuously. For example, we often measure population growth at specific times, such as the year end of a reproductive cycle.

The study of discrete dynamical systems most often involves the process of iteration. To *iterate* means to repeat a procedure numerous times. In discrete dynamics the process that is repeated is the application of a mathematical function. We have encountered several types of iterative processes already. For example, Euler's method for solving a differential equation involves iteration.

As was the case with differential equations, we see that finding exact solutions of discrete systems is rarely possible. As usual we rely on a combination of numerical, analytic, and graphical techniques to understand these systems. Even with these sophisticated techniques, we are unable to describe even one-dimensional, nonlinear iterations completely. The reason is that many discrete nonlinear systems behave in a complex and unpredictable manner. This phenomenon, called chaos by mathematicians, is discussed at length in Sections 8.4 and 8.5.

8.1   THE DISCRETE LOGISTIC EQUATION

The population models we have studied so far all have the property that the rate of change of the population is continuous. This is a reasonable assumption for species that reproduce quickly with respect to the time scale we are considering. However, for some species, all births occur in the spring and most deaths occur during the winter; hence the assumption of continuous population change is not valid. Instead, we should think of time as discrete. If we measure the population once a summer, then we will have an accurate estimate of the population for the entire year.

We begin our study of discrete time systems by recasting several of the population models we discussed in Chapter 1 as discrete dynamical systems.

Exponential Growth Model

The **exponential growth** model is a simple but unrealistic model of population growth. We discuss it first because it illustrates the main ideas behind discrete dynamics. In this model we assume that the population of a certain species in the next generation (or other time period) is directly proportional to the population in the current generation. We let P_n denote the population of the species at the end of the nth time period. The assumption says that, for each time n, the population at the end of the $n + 1$st time step, P_{n+1}, is proportional to the population at the end of the previous time step, P_n. That is,

$$P_{n+1} = k P_n,$$

where k is the proportionality constant that determines the growth rate. This equation is an example of a **discrete dynamical system**, or a **difference equation**. Unlike the differential equation for unlimited population growth,

$$\frac{dP}{dt} = kP,$$

which tells us the rate of change of the population, this difference equation tells us directly the population P_{n+1} in the succeeding generation, provided we know P_n. Thus if we know P_0, the initial population, we can determine the population in each succeeding generation by simply computing the expression $k P_n$ at each stage. There are no integrals to evaluate, no solution curves to approximate. All we have to do is repeatedly calculate the right-hand side of the equation, using the output of the previous calculation as the input for the next. This repetition is what we call **iteration**. Thus if we know the initial population P_0, we then compute

$$P_1 = k P_0$$
$$P_2 = k P_1 = k^2 P_0$$
$$P_3 = k P_2 = k^3 P_0$$
$$\vdots$$
$$P_n = k P_{n-1} = k^n P_0.$$

We see that the population at generation n is determined by the formula $P_n = k^n P_0$. Therefore, if $P_0 > 0$, we conclude the following: If $k > 1$, the population explodes, since

$$\lim_{n \to \infty} k^n = \infty.$$

On the other hand, if $k < 1$, the population dies out, since

$$\lim_{n \to \infty} k^n = 0.$$

Finally, if $k = 1$, the population never changes and $P_n = P_0$.

This model has the same drawbacks as the differential equation version. Although it may work well for small populations in large environments, if the population grows at all, then the model predicts unlimited growth. Thus we modify this model to account for a limited environment.

The Logistic Difference Equation

To make the exponential growth model somewhat more realistic, we add some assumptions that account for overcrowding, just as we did with the logistic differential equation in Chapter 1. The assumptions we make are:

- The population at the end of the next generation is proportional to the population at the end of the current generation when the population is very small.
- If the population is too large, then all resources will be used and the entire population will die out in the next generation and extinction will result.

This last assumption is slightly different from the one we made for the logistic differential equation. Here we assume that there is a maximum population level M that, when reached, results in extinction of the population in the next generation. Thus M is called the **annihilation parameter**. If the population ever reaches M, the species is doomed.

One model that reflects these assumptions is

$$P_{n+1} = k P_n \left(1 - \frac{P_n}{M} \right).$$

As before, P_n denotes the population at the end of generation n and P_0 is the initial population. Note that if P_n is small, the term $(1 - P_n/M)$ is approximately 1. So the difference equation becomes $P_{n+1} \approx k P_n$, which is the exponential growth model. On the other hand, if $P_n \geq M$, then $P_{n+1} \leq 0$; that is, the population is nonpositive. We interpret this to mean that the species is extinct.

Rather than deal with the large numbers that often arise in population models, we will assume that P_n represents the percentage or fraction of this maximum population alive at generation n. That is, we assume that $M = 1$ and that P_n lies between 0 and 1, with $P_n = 0$ (or P_n negative) representing extinction and $P_n = 1$ representing the maximum population level. Thus the model becomes

$$P_{n+1} = k P_n (1 - P_n),$$

which we call the **discrete logistic equation**, or **logistic difference equation**. As before, k is a parameter that depends on the specific species under investigation.

As always, we must insert the caveat that this is an extremely naive model for population growth. We have neglected all kinds of other factors that affect P_n, including the effects of predators, cyclical diseases, and the variable nature of the food supply. Nevertheless, this model does provide many more scenarios for changes in population than does the logistic differential equation.

Some predictions of the model

As an example of the various possibilities we encounter in the discrete logistic model, we sample the output of this model for a few k-values. Suppose we begin with a population that is exactly half the maximum population allowed, that is, $P_0 = 0.5$. Then depending on k, we find very different results when we compute successive values of P_n. Table 8.1 lists the populations (using only 4 significant digits) when $k = 0.5, 1.5, 2, 3.2, 3.5$, and 3.9. Note that these different k-values yield very different behaviors for the population. When $k = 0.5$, the population tends gradually toward extinction.

Table 8.1
Successive populations of the logistic model with $P_0 = 0.5$ and $k = 0.5, 1.5, 2, 3.2, 3.5$, and 3.9.

n	$k = 0.5$	$k = 1.5$	$k = 2$	$k = 3.2$	$k = 3.5$	$k = 3.9$
1	0.1250	0.3750	0.5000	0.8000	0.8750	0.9750
2	0.0546	0.3515	0.5000	0.5120	0.3828	0.0950
3	0.0258	0.3419	0.5000	0.7995	0.8269	0.3355
4	0.0125	0.3375	0.5000	0.5128	0.5008	0.8694
5	0.0062	0.3354	0.5000	0.7995	0.8749	0.4426
6	0.0030	0.3343	0.5000	0.5130	0.3828	0.9621
7	0.0015	0.3338	0.5000	0.7995	0.8269	0.1419
8	7.7×10^{-4}	0.3335	0.5000	0.5130	0.5008	0.4750
9	3.8×10^{-4}	0.3334	0.5000	0.7995	0.8749	0.9725
10	1.9×10^{-4}	0.3333	0.5000	0.5130	0.3828	0.1040
11	9.6×10^{-5}	0.3333	0.5000	0.7995	0.8269	0.3634
12	4.8×10^{-5}	0.3333	0.5000	0.5130	0.5008	0.9022
13	2.4×10^{-5}	0.3333	0.5000	0.7995	0.8749	0.3438
14	1.2×10^{-5}	0.3333	0.5000	0.5130	0.3828	0.8799
15	6.0×10^{-6}	0.3333	0.5000	0.7995	0.8269	0.4120
16	3.0×10^{-6}	0.3333	0.5000	0.5130	0.5008	0.9448
17	1.5×10^{-6}	0.3333	0.5000	0.7995	0.8749	0.2033
18	7.5×10^{-7}	0.3333	0.5000	0.5130	0.3828	0.6316
19	3.7×10^{-7}	0.3333	0.5000	0.7995	0.8269	0.9073
20	1.9×10^{-7}	0.3333	0.5000	0.5130	0.5008	0.3278

When $k = 1.5$, the population seems to level out and approach an equilibrium state. When $k = 2$, the population never changes and remains fixed at 0.5. When $k = 3.2$, we see a different result: The population eventually oscillates back and forth between two different values. The population is high one year, approximately 0.7995, low the next, approximately 0.513, and then repeats cyclically. When $k = 3.5$, we see a similar cyclic behavior, but now the populations eventually repeat every four years instead of two. Finally, when $k = 3.9$, there is no apparent pattern to the successive populations.

Iteration

For the discrete logistic model, finding successive populations is the same as iterating a quadratic function of the form

$$L_k(x) = kx(1-x).$$

This function depends on the parameter k and is often called the **logistic function**. To iterate this function, we begin with an initial population P_0 and then compute in succession

$$P_1 = L_k(P_0)$$
$$P_2 = L_k(P_1)$$
$$P_3 = L_k(P_2)$$
$$\vdots$$
$$P_n = L_k(P_{n-1})$$

and so forth. The list of numbers P_0, P_1, P_2, ... that results from this iteration is called the **orbit** of P_0 under the function L_k. The initial value P_0 is sometimes called the **seed**, or initial condition, for the orbit.

In discrete dynamics the basic goal is to predict the fate of orbits for a given function. That is, the main question is: What happens to the numbers that constitute the orbit as n tends to infinity?

Sometimes predicting the fate of orbits is easy. For example, if $F(x) = x^2$, then we can easily determine what happens to all orbits. For example, the seeds $x_0 = 0$ and $x_0 = 1$ are **fixed points**, since $F(0) = 0$ and $F(1) = 1$. That is, the orbit of 0 is the constant sequence $0, 0, 0, 0, \ldots$ as is the orbit of 1: $1, 1, 1, \ldots$. The orbit of -1 under $F(x) = x^2$ is slightly different: This orbit is **eventually fixed** since $F(-1) = 1$, which is a fixed point. The orbit of -1 is $-1, 1, 1, 1, \ldots$.

For any other seed x_0, there are only two possibilities for the orbit of x_0: Either the orbit tends to the fixed point at 0, or the orbit tends to infinity. For example, if $x_0 = 1/2$, the orbit is

$$x_0 = 1/2$$
$$x_1 = 1/4$$
$$x_2 = 1/16$$

$$x_3 = 1/256$$

$$\vdots$$

$$x_n = 1/2^{2^n},$$

which tends to 0 as n tends to infinity. The fate of the orbit is the same for any other seed x_0 with $|x_0| < 1$.

If $|x_0| > 1$, successive applications of the squaring function $F(x) = x^2$ yield larger and larger results. For example, if $x_0 = 2$, we have

$$x_0 = 2$$

$$x_1 = 4$$

$$x_2 = 16$$

$$x_3 = 256$$

$$\vdots$$

$$x_n = 2^{2^n},$$

and we see that the orbit of x_0 tends to infinity.

Cycles

In a typical discrete dynamical system, there are often many different types of orbits. For example, if $G(x) = x^2 - 1$, then the orbit of 0 lies on a **cycle of period** 2, or a **periodic orbit of period** 2, since

$$x_0 = 0$$

$$x_1 = -1$$

$$x_2 = 0$$

$$x_3 = -1$$

$$\vdots$$

This orbit is the repeating sequence $0, -1, 0, -1, \ldots$.

If we choose the seed $x_0 = \sqrt{2}$, then this orbit is **eventually periodic** since we have

$$x_0 = \sqrt{2}$$

$$x_1 = 1$$

$$x_2 = 0$$

$$x_3 = -1$$

$$x_4 = 0$$

$$x_5 = -1$$

$$\vdots$$

which begins to cycle after the second iteration.

In contrast, if we choose the seed $x_0 = 0.5$, the orbit tends to the cycle of period 2 since

$$x_0 = 0.5$$
$$x_1 = (0.5)^2 - 1 = -0.75$$
$$x_2 = -0.4375$$
$$x_3 = -0.8086\ldots$$
$$\vdots$$
$$x_{20} = 0.00000\ldots$$
$$x_{21} = -1.00000\ldots$$
$$x_{22} = 0.00000\ldots$$
$$\vdots$$

It is important to realize that this orbit never actually reaches the cycle at 0 and -1. Rather, the orbit comes arbitrarily close to these two numbers, but because of round-off error, the calculator or computer eventually displays the numbers 0 and -1 in succession. Thus the orbit is not eventually periodic, it merely tends to the cycle of period 2. This behavior is technically different from that of an orbit that is eventually periodic, although in practice the fate of these orbits is essentially the same.

Discrete dynamical systems may have orbits that cycle with any period. For example, the difference equation

$$x_{n+1} = -\tfrac{3}{2}x_n^2 + \tfrac{5}{2}x_n + 1$$

admits a cycle of period 3, since for the seed $x_0 = 0$ we calculate

$$x_0 = 0$$
$$x_1 = 1$$
$$x_2 = 2$$
$$x_3 = 0.$$

Thus the orbit is the repeating sequence $0, 1, 2, 0, 1, 2, \ldots$. In general, the orbit of x_0 lies on a cycle of period n if $x_n = x_0$ and n is the smallest positive integer for which this happens. A fixed point would not be regarded as a cycle of period 10, even though $x_{10} = x_0$, since 10 is not the smallest integer for which the orbit repeats. Similarly, if x_0 has period 5, we also have $x_{10} = x_0$, but we would not call this a period 10 orbit either.

Since the orbit of a fixed point is a constant sequence $x_0, x_0, x_0, \ldots$, fixed points for discrete dynamical systems can be thought of as the analogs of equilibrium points for differential equations. Both represent constant solutions in which the given system is at rest. Similarly, cycles for a discrete system are analogous to periodic solutions of differential equations, as both return to their original position after some time.

Time Series and Histograms

One convenient method to describe orbits geometrically is via a **time series**. In these diagrams we plot the iteration count on the horizontal axis and the numerical values of the orbit on the vertical axis. It often helps to draw straight lines connecting successive points in the time series. In Figure 8.1 we plot the time series corresponding to the orbit of -0.5 under $G(x) = x^2 - 1$. Note that this orbit tends to a cycle of period 2. In Figure 8.2 we plot the time series corresponding to the orbit of 0 under $H(x) = x^2 - 1.3$. Here we see that this orbit tends to a cycle of period 4. These time series are the analogs of the $x(t)$- and $y(t)$-graphs we often plot for differential equations.

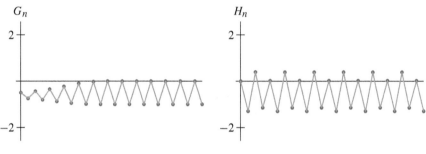

Figure 8.1
Time series for the orbit of -0.5 under
$G(x) = x^2 - 1$.

Figure 8.2
Time series for the orbit of 0 under
$H(x) = x^2 - 1.3$.

Figure 8.3 displays the time series for the orbit of 0.5 under iteration of the logistic function $L_{3.9}(x) = 3.9x(1 - x)$. This is neither fixed nor periodic behavior; it is difficult to see any sort of coherent pattern. We study this sort of time series in Section 8.5.

Another visual way of displaying orbits is a **histogram**. These images subdivide an interval that contains the points of an orbit into many small subintervals of equal length. Each time the orbit enters one of these subintervals, the histogram is incremented by one unit over that subinterval. Figures 8.4 and 8.5 display the histograms corresponding to the orbit of 0.123 under iteration of the logistic functions

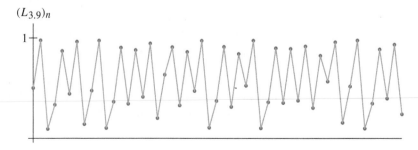

Figure 8.3
Time series for the orbit of 0.5 under $L_{3.9}(x) = 3.9x(1 - x)$.

Figure 8.4
Histogram for the orbit of 0.123 under
$L_{3.83}(x) = 3.83x(1 - x)$.

$L_{3.83}(x) = 3.83x(1 - x)$ and $L_{3.9}(x) = 3.9x(1 - x)$. In the first case the orbit quickly settles down on a period 3 cycle. In the second case we see a different view of Figure 8.3. The orbit apparently never settles down, instead visiting all of the subintervals, some more than others.

Figure 8.5
Two histograms for the orbit of 0.123 under $L_{3.9}(x) = 3.9x(1 - x)$. In the first figure we subdivide the unit interval into ten subintervals of length 0.1, while in the second figure there are 100 subintervals of length 0.01.

Finding Fixed Points

As we see from the examples above, there are many different types of orbits for a discrete dynamical system. Fixed points are the simplest orbits, and these are often the most important types of orbits in such a system. As in the case of differential equations, we use three different methods to find these special orbits: analytic, qualitative, and numerical techniques.

The analytic method involves solving equations. Given a function F, to find the fixed points for F, we need only solve the equation $F(x) = x$. For example, if $F(x) = x^2 - 2$, then we find that the fixed points are the solutions of the equation

$$x^2 - 2 = x,$$

which can be written as

$$x^2 - x - 2 = 0.$$

Factoring turns this into

$$(x + 1)(x - 2) = 0,$$

and the fixed points for F are -1 and 2.

Geometrically we can view fixed points by superimposing the graph of the diagonal line $y = x$ on the graph of F. Note that the diagonal $y = x$ meets the graph of $F(x) = x^2 - 2$ directly over -1 and 2, as shown in Figure 8.6. As another example, $K(x) = x^3$ has 3 fixed points, which occur at the roots of $x^3 - x = 0$, or at $0, -1$, and $+1$ (see Figure 8.7).

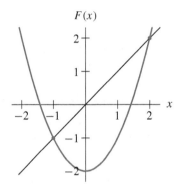

Figure 8.6
The fixed points of $F(x) = x^2 - 2$ are $x = -1, 2$.

Figure 8.7
The fixed points of $K(x) = x^3$ are $x = 0, -1, +1$.

For the logistic equation $L_k(x) = kx(1 - x)$, we find the fixed points by solving

$$kx(1 - x) = x,$$

which yields solutions $x = 0$ and $x = (k - 1)/k$. Recall that for our population model we assume that $0 \leq x \leq 1$, so this second fixed point is positive only if $k > 1$. (When $k < 1$, this fixed point is negative and so is not biologically interesting.) Also, $(k - 1)/k < 1$ for all $k > 1$. The fixed point at $x = 0$ represents population extinction, and the nonzero fixed point represents a population that never changes from generation to generation. Figure 8.8 shows the graph for $k = 4$, with fixed points at 0 and 0.75.

Given a function F, finding the fixed points by solving $F(x) = x$ algebraically may be difficult or impossible. If we need a more accurate value for the fixed point than we can get from graphical methods, then we can turn to numerical methods. We can use Newton's method to find values of x where $F(x) - x = 0$. For example, from Figure 8.9 we see that $C(x) = \cos x$ has a single fixed point, which we can determine numerically to be $0.73908\ldots$ (see Exercise 6).

$L_4(x)$

cos x

Figure 8.8
The fixed points for the logistic model $L_4(x) = 4x(1-x)$ are $x = 0, 3/4$.

Figure 8.9
The fixed point for cos x occurs at $x = 0.73908\ldots$.

Notation for Iteration

To simplify notation, we introduce the expression F^n to denote the nth iterate of F. For example, if $F(x) = x^4$, then

$$F^2(x) = F(F(x)) = F(x^4) = (x^4)^4 = x^{16},$$

and similarly $F^3(x) = x^{64}$. It is important to realize that $F^n(x)$ *does not* mean $F(x)$ raised to the nth power; rather, $F^n(x)$ means to first iterate F exactly n times, then evaluate this new function at x.

Finding Cycles

To find cycles for a discrete dynamical system, we proceed in essentially the same manner as for fixed points. For example, if $H(x) = -x^3$, the only fixed point of H is 0, since the equation for the fixed points is $-x^3 = x$ or $x(1 + x^2) = 0$. For the cycles of period 2, we must compute

$$H(H(x)) = H^2(x) = -(-x^3)^3 = x^9$$

and then solve $x^9 = x$. The solutions here are the fixed point $x = 0$ and two new solutions, $x = 1$ and $x = -1$. These latter two points form a cycle of period 2, since $H(1) = -1$ and $H(-1) = 1$.

Finding cycles is usually more difficult than finding fixed points. For example, to find cycles of period 2 for $F(x) = x^2 - 2$, we must find values of x for which $F(F(x)) = x$. We have

$$F^2(x) = (x^2 - 2)^2 - 2 = x^4 - 4x^2 + 2,$$

so that we must solve

$$x^4 - 4x^2 + 2 = x.$$

That is, we must find the roots of the fourth-degree equation

$$x^4 - 4x^2 - x + 2 = 0.$$

Luckily we know two solutions of this equation already, since we saw above that -1 and 2 are fixed points, and fixed points have orbits that repeat every two iterations as well as every iteration. So we can divide this expression by $(x + 1)(x - 2)$ to find

$$\frac{x^4 - 4x^2 - x + 2}{(x + 1)(x - 2)} = \frac{x^4 - 4x^2 - x + 2}{x^2 - x - 2} = x^2 + x - 1 = 0.$$

This quadratic equation has roots $(-1 \pm \sqrt{5})/2$. These points lie on a cycle of period 2 since we have

$$\left(\frac{-1 + \sqrt{5}}{2}\right)^2 - 2 = \frac{-1 - \sqrt{5}}{2}$$

and

$$\left(\frac{-1 - \sqrt{5}}{2}\right)^2 - 2 = \frac{-1 + \sqrt{5}}{2}.$$

In general, to find cycles of period n, we must first iterate the function F a total of n times. For example, if $L_3(x) = 3x(1 - x)$, then we compute

$$L_3^2(x) = 3[3x(1 - x)][1 - 3x(1 - x)] = -27x^4 + 54x^3 - 36x^2 + 9x$$

and

$$L_3^3(x) = L_3(L_3^2(x))$$
$$= 3\left[-27x^4 + 54x^3 - 36x^2 + 9x\right]\left[1 - (-27x^4 + 54x^3 - 36x^2 + 9x)\right],$$

which we will not bother to simplify. You should note that, when multiplied out, $L_3^3(x)$ has terms that involve x^8, so L_3^3 is an eighth-degree polynomial. Finding the solutions of $L_3^3(x) = x$ is therefore not a pleasant task.

Finding cycles geometrically

In general, finding cycles of period n involves solving the equation $F^n(x) = x$. Even for logistic functions of the form $L_k(x) = kx(1 - x)$, $L_k^n(x)$ is a polynomial of degree 2^n, and so we have little chance of finding explicit solutions. However, geometric information is relatively easy to come by. We can usually discover some information about the number of cycles of period n by sketching the graph of F^n and looking for places where this graph crosses the diagonal line $y = x$.

For example, if $H(x) = -x^3$, the graph of H shows that H has a single fixed point that occurs at 0. But $H^2(x) = x^9$ has a graph that meets the diagonal three times, at the fixed point and at a cycle of period 2, which as we saw above is given by ± 1 (see Figures 8.10 and 8.11).

$H(x)$

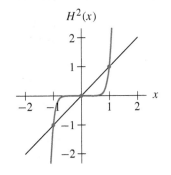

$H^2(x)$

Figure 8.10
The graph of $H(x) = -x^3$.

Figure 8.11
The graph of $H^2(x) = x^9$.

In Figure 8.12 we sketch the graph of $F(x) = x^2 - 2$ as well as F^2 and F^3. Note that we see two fixed points for F and four fixed points for F^2. Two of these are the fixed points for F, and the other two lie on a cycle of period 2. For F^3 there are eight points where the graph of $y = F^3(x)$ crosses $y = x$. Two of these points are again the fixed points of F, but the other six must be periodic with period 3. The cycle of period 2 does not appear in the graph of F^3 since these orbits do not repeat after 3 iterations.

$F(x)$

$F^2(x)$

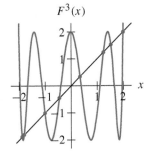

$F^3(x)$

Figure 8.12
The graphs of $F(x) = x^2 - 2$, $F^2(x)$, and $F^3(x)$.

EXERCISES FOR SECTION 8.1

In Exercises 1–8, compute the orbit of $x_0 = 0$ for each of the given difference equations. Determine whether this orbit is fixed, cycles with some period, is eventually periodic, tends to infinity, or is none of the above. (Use a calculator as needed.)

1. $x_{n+1} = x_n^2 - 2$

2. $x_{n+1} = \sin(x_n)$

3. $x_{n+1} = e^{x_n}$

4. $x_{n+1} = x_n^2 + 1$

5. $x_{n+1} = -x_n^2 + x_n + 2$

6. $x_{n+1} = \cos(x_n)$

7. $x_{n+1} = 4x_n(1 - x_n) + 1$

8. $x_{n+1} = -\frac{1}{2}x_n^2 + 1$

In Exercises 9–21, find all fixed points and periodic points of period 2 for each of the given functions. If you cannot determine these values explicitly, use the graph of F or F^2 to determine how many fixed points and periodic points of period 2 F has.

9. $F(x) = -x + 2$ **10.** $F(x) = x^4$

11. $F(x) = x^2 + 1$ **12.** $F(x) = x^2 - 3$

13. $F(x) = \sin x$ **14.** $F(x) = 1/x$

15. $F(x) = -2x - x^2$ **16.** $F(x) = e^x$

17. $F(x) = -e^x$ **18.** $F(x) = x^3$

19. $F(x) = -x$ **20.** $F(x) = -2x + 1$

21. $F(x) = 2$

22. Describe the fate of the orbit of any seed under iteration of $F(x) = x^3$.

23. Describe the fate of the orbit of any seed under iteration of $F(x) = -x + 4$.

In Exercises 24–35, describe the fate of the orbit of each of the following seeds under iteration of the function

$$T(x) = \begin{cases} 2x, & \text{if } x < 1/2; \\ 2 - 2x, & \text{if } x \geq 1/2. \end{cases}$$

24. 2/3	**25.** 1/6	**26.** 2/5	**27.** 2/7
28. 3/14	**29.** 1/8	**30.** 1/9	**31.** 6/11
32. 2/9	**33.** 0	**34.** 1/4	**35.** 1/2

36. Discuss the fate of the orbit of 0 under iteration of $F(x) = x^2 + c$ for each of the following c values: $c = 0.4, 0.3, 0.2, 0.1$. Use the graph and technology as necessary to explain your findings. Is there a change in the fate of the orbit of 0 as c changes? Explain in a sentence or two.

37. Consider the family of functions $F_c(x) = x^2 + c$ that depends on a parameter c. For which values of c does this function have real fixed points? How many?

38. Discuss the fate of the orbit of 0 under iteration of $F(x) = x^2 + c$ for each of the following c values: $c = -0.6, -0.7, -0.8, -0.9$. Use the graph of F and F^2 and technology as necessary to explain your findings. Is there a change in the fate of the orbit of 0 as c changes? Explain in a sentence or two.

39. Consider the family of functions $F_c(x) = x^2 + c$ that depends on a parameter c. For which values of c does this function have periodic points of period 2? How many?

8.2 FIXED POINTS AND PERIODIC POINTS

As was the case with equilibrium points for differential equations, fixed points and cycles for difference equations come in several distinct varieties, depending on what happens to nearby orbits. In this section we discuss analytic and qualitative methods for distinguishing different types of fixed points and cycles.

Graphical Iteration

As we saw in the previous section, the graph of a function and the diagonal line $y = x$ give us a geometric method for finding fixed points and cycles. In this section we see that the graph of F together with the diagonal line $y = x$ also provides a convenient way to sketch other orbits geometrically.

Given the function F and a seed x_0, we view the orbit of x_0 as follows: We begin at the point (x_0, x_0) on the diagonal and draw a vertical line from (x_0, x_0) to the graph of F; we reach the graph at (x_0, x_1), where $x_1 = F(x_0)$. Now we draw a horizontal line from this point back to the diagonal. Along this line, the y-values remain constant at $y = x_1$, so we reach the diagonal at (x_1, x_1). We interpret this point as the second point on the orbit.

Now we continue in the same fashion. First we draw a vertical line from (x_1, x_1) to the graph of F, reaching the graph at (x_1, x_2), where $x_2 = F(x_1)$. Then we draw a horizontal line back to the diagonal, reaching this line at (x_2, x_2). This is the second point on the orbit.

Continuing in this manner, the orbit of x_0 is displayed as a sequence of points along the diagonal (see Figure 8.13). The resulting picture, which sometimes resembles a staircase and sometimes a cobweb, is often called a **web diagram**. The process of drawing vertical lines to the graph and then horizontal lines to the diagonal is called **graphical analysis**, or **graphical iteration**. We sometimes add arrows to the web diagram to indicate the direction of the iteration.

Figure 8.14 depicts the web diagram for $G(x) = x^2 - 0.7$ using the seed $x_0 = 1.4$.

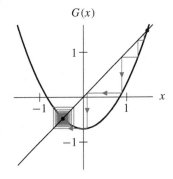

Figure 8.13
Graphical analysis showing the orbit of x_0.

Figure 8.14
Web diagram of $G(x) = x^2 - 0.7$ using the seed $x_0 = 1.4$.

We clearly see two fixed points for G, given by the points where the graph of G crosses the diagonal. Note that the orbit of $x_0 = 1.4$ tends away from one of these fixed points and toward the other fixed point.

Another important type of orbit for discrete systems is the cycle, or periodic orbit. Recall that for a function F, the seed x_0 lies on a cycle or periodic orbit of period n if $F^n(x_0) = x_0$. (Remember that F^n indicates the nth iterate of F, not the nth power.) Equivalently, x_0 lies on a cycle of period n if the orbit of x_0 is a repeating sequence of length n of the form

$$x_0, x_1, \ldots, x_{n-1}, x_0, x_1, \ldots, x_{n-1}, \ldots .$$

Graphical iteration allows us to visualize cycles without having to compute the graph of F^n. For example, 0 and -1 lie on a cycle of period 2 for $F(x) = x^2 - 1$, since $F(0) = -1$ and $F(-1) = 0$. In Figure 8.15 we see that this cycle is represented geometrically by a square in the web diagram, and we also see an orbit that tends to this cycle.

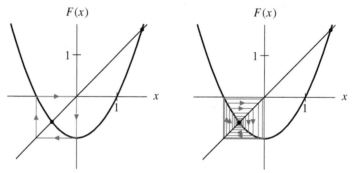

Figure 8.15
A 2-cycle for $F(x) = x^2 - 1$ and an orbit tending to the 2-cycle.

As we saw in the previous section, the logistic difference equation admits cycles of many periods. Figures 8.16 and 8.17 display some of these cycles using graphical iteration.

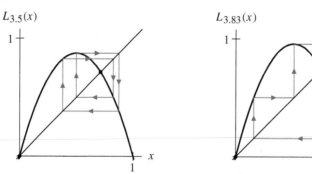

Figure 8.16
A 4-cycle for $L_{3.5}(x) = 3.5x(1 - x)$.

Figure 8.17
A 3-cycle for $L_{3.83}(x) = 3.83x(1 - x)$.

Attracting and Repelling Fixed Points

Figures 8.18 and 8.19 display the graphical analysis of two additional logistic functions, $L_{2.8}(x) = 2.8x(1-x)$ and $L_{3.2}(x) = 3.2x(1-x)$. In the first case observe that the displayed orbit tends to a fixed point, but when the parameter is changed to 3.2, this is no longer the case. The displayed orbit comes close to the fixed point but then tends toward a 2-cycle. This is an example of a bifurcation, a topic we discuss later. For now note that there is a dramatic difference between the behavior of orbits near the nonzero fixed point in these two cases. In the first case the orbit is attracted to the nonzero fixed point, whereas in the second the orbit seems to be repelled away. Just as with equilibrium points for a differential equation, there are different types of fixed points and cycles.

DEFINITION   A fixed point x_0 of a function F is called **attracting** if there is an interval around x_0 having the property that every seed in this interval has an orbit that remains in the interval and tends to x_0 under iteration of F. The fixed point is called **repelling** if there is an interval around x_0 having the property that every seed in this interval (except x_0) has an orbit that leaves the interval under iteration of F. A fixed point that is neither attracting nor repelling is called **neutral**.   ■

Attracting fixed points are the analogs of sinks for one-dimensional differential equations, whereas repelling fixed points correspond to sources.

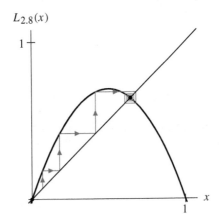

Figure 8.18
An attracting fixed point for
$L_{2.8}(x) = 2.8x(1-x)$.

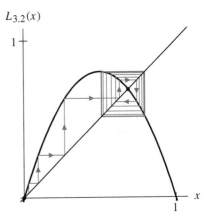

Figure 8.19
A repelling fixed point for
$L_{3.2}(x) = 3.2x(1-x)$.

Fixed points of linear functions

To see what makes certain fixed points attracting and others repelling, we first consider linear functions. For $F(x) = mx$, $x_0 = 0$ is always a fixed point. Graphical analysis shows that as long as $|m| < 1$, x_0 attracts all orbits (see Figure 8.20). Note that if $-1 < m < 0$, orbits hop back and forth about 0 as they tend to the fixed point.

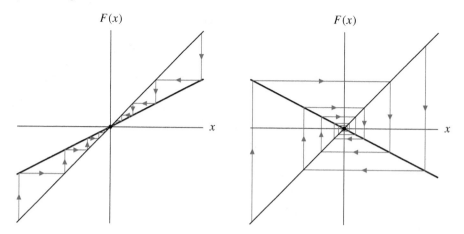

Figure 8.20
The function $F(x) = mx$ has an attracting fixed point at 0 if $|m| < 1$.

When $|m| > 1$, the situation is quite different. Now all orbits tend away from 0; the origin is a repelling fixed point (see Figure 8.21). Again orbits hop from side to side as they leave the origin when $m < -1$.

The intermediate cases $m = \pm 1$ are special. When $m = 1$, all points are fixed, whereas when $m = -1$, all nonzero points are periodic with period 2.

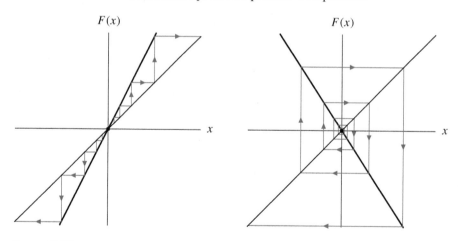

Figure 8.21
The function $F(x) = mx$ has a repelling fixed point at 0 if $|m| > 1$.

Classification of Fixed Points

When F is a nonlinear function, a similar story holds true, at least near fixed points. If F has a fixed point at x_0 and $|F'(x_0)| < 1$, then the graph of F is tangent to a straight line whose slope is less than 1 in absolute value. As in the case of linear functions,

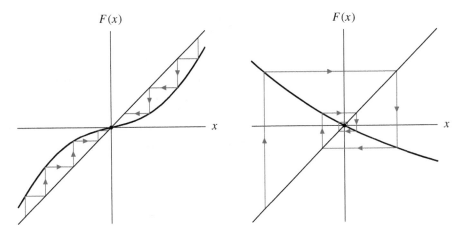

Figure 8.22
If $|F'(x_0)| < 1$, then x_0 is an attracting fixed point.

graphical analysis shows that nearby orbits must be attracted to x_0. Thus a fixed point x_0 for F for which $|F'(x_0)| < 1$ is attracting (see Figure 8.22). Note that this result is true only close to x_0; far away, orbits may behave very differently.

Similarly, if $|F'(x_0)| > 1$, nearby orbits are repelled away from x_0 just as in the linear case. Therefore if $|F'(x_0)| > 1$ at the fixed point, then x_0 is a repelling fixed point (see Figure 8.23).

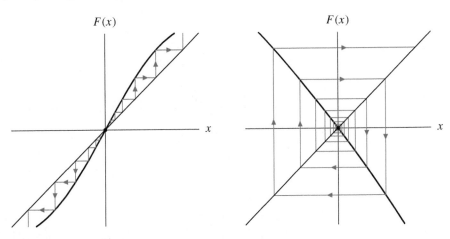

Figure 8.23
If $|F'(x_0)| > 1$, then x_0 is a repelling fixed point.

Fixed points for the logistic function

Consider the logistic function $L_{2.8}(x) = 2.8x(1 - x)$. The fixed points for $L_{2.8}$ are given by solving

$$2.8x(1 - x) = x.$$

Some algebra shows that these fixed points are $x = 0$ and $x = 1.8/2.8 \approx 0.64$. Now

$$L'_{2.8}(x) = 2.8(1 - 2x),$$

so $L'_{2.8}(0) = 2.8$ and $L'_{2.8}(0.64\ldots) = 2.8(-0.28\ldots) \approx -0.78$. Thus 0 is a repelling fixed point and $0.64\ldots$ is an attracting fixed point. We saw this qualitatively in Figure 8.18.

As a second example, suppose we consider instead the logistic function $L_{3.2}(x) = 3.2x(1 - x)$. Solving for the fixed points as above, we find fixed points at $x = 0$ and $x = 2.2/3.2 = 0.6875$. Now we have

$$L'_{3.2}(x) = 3.2(1 - 2x).$$

Thus $L'_{3.2}(0) = 3.2$ and $L'_{3.2}(0.6875) = -1.2$. In this case both fixed points are repelling. We saw this also in Figure 8.18.

These two examples are special cases of the general logistic equation $L_k(x) = kx(1 - x)$. Recall from the previous section that the fixed points of F occurred at 0 and at $(k - 1)/k$, provided $k > 1$. We compute

$$L'_k(x) = k - 2kx.$$

So $L'_k(0) = k$. Therefore 0 is an attracting fixed point when $0 \le k < 1$ and a repelling fixed point when $k > 1$. Also,

$$L'_k\left(\frac{k - 1}{k}\right) = k - 2(k - 1) = -k + 2.$$

So $(k - 1)/k$ is an attracting fixed point when $-1 < -k + 2 < 1$, that is, for $1 < k < 3$. When $k > 3$, this fixed point is repelling (see Figures 8.24 and 8.25).

$L_{0.5}(x)$

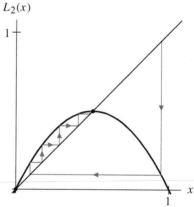

$L_2(x)$

Figure 8.24
Fixed points for $L_k(x) = kx(1 - x)$ when $k = 0.5$.

Figure 8.25
Fixed points for $L_k(x) = kx(1 - x)$ when $k = 2$.

Examples of neutral fixed points

Neutral fixed points can occur only if $F'(x_0) = \pm 1$. Orbits near neutral fixed points may behave in a variety of ways. For example, consider $F(x) = -x + 4$. We have $F(2) = 2$ and $F'(2) = -1$. All other seeds have orbits that lie on cycles of period 2, since $F^2(x) = -(-x + 4) + 4 = x$.

As another example, $G(x) = x + x^2$ has a fixed point at $x = 0$. Note that $G'(0) = 1$. The graph of G shows that 0 attracts from the left but repels from the right. Thus 0 is a neutral fixed point (see Figure 8.26). If we consider instead $H(x) = x + x^3$, then again 0 is a fixed point and $H'(0) = 1$. This time, however, $x = 0$ is a repelling fixed point, as we see from graphical iteration (see Figure 8.27).

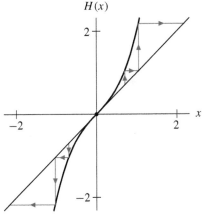

Figure 8.26
$G(x) = x + x^2$ has a neutral fixed point at $x = 0$ and $G'(0) = 1$.

Figure 8.27
$H(x) = x + x^3$ has a repelling fixed point at $x = 0$, despite the fact that $H'(0) = 1$.

Classification of Periodic Points

Periodic points can also be classified as attracting, repelling, and neutral. If x_0 lies on an n-cycle of a given function F, then the graph of F^n meets the diagonal at (x_0, x_0). That is, F^n has a fixed point at x_0. So it is natural to call the cycle attracting, repelling, or neutral depending on whether the fixed point for F^n has this property.

For example, the function $F(x) = x^2 - 1$ has a 2-cycle at 0 and -1. Since $F^2(x) = (x^2 - 1)^2 - 1 = x^4 - 2x^2$, we have

$$(F^2)'(x) = 4x^3 - 4x.$$

Therefore $(F^2)'(0) = 0$, so 0 lies on an attracting 2-cycle, as we have seen in Figure 8.15. Note also that $(F^2)'(-1) = 0$.

The fact that $(F^2)'(x_0)$ is the same at both points on the cycle in this example is no accident. The Chain Rule tells us why. Suppose x_0 and x_1 lie on a 2-cycle for F.

So $F(x_0) = x_1$ and $F(x_1) = x_0$. Then we have

$$(F^2)'(x_0) = F'(F(x_0)) \cdot F'(x_0)$$
$$= F'(x_1) \cdot F'(x_0).$$

That is, the derivative of F^2 at x_0 is just the product of the derivatives along the orbit of x_0. The same is true for x_1. More generally, if $x_0 \ldots, x_{n-1}$ lies on a cycle of period n for F, then

$$(F^n)'(x_0) = F'(F^{n-1}(x_0)) \cdot F'(F^{n-2}(x_0)) \cdots F'(x_0)$$
$$= F'(x_{n-1}) \cdot F'(x_{n-2}) \cdots F'(x_0),$$

which is again the product of the derivatives along the cycle.

As a check, for $F(x) = x^2 - 1$ we have $F'(0) = 0$ and $F'(-1) = -2$. Consequently,

$$(F^2)'(0) = F'(0) \cdot F'(-1) = 0(-2) = 0$$

as before.

As a final example, consider the function

$$T(x) = \begin{cases} 2x, & \text{if } x < 1/2; \\ \\ 2 - 2x, & \text{if } x \geq 1/2. \end{cases}$$

The seed $x_0 = 2/7$ leads to a cycle of period 3 for T with orbit $2/7, 4/7, 6/7, 2/7 \ldots$. We compute

$$T'(2/7) = 2$$
$$T'(4/7) = -2$$
$$T'(6/7) = -2.$$

Therefore we have

$$(T^3)'(2/7) = T'(2/7) \cdot T'(4/7) \cdot T'(6/7) = 8,$$

and so this cycle is repelling. Note that any cycle for this function is repelling, since the derivative will always be a product of 2's and -2's.

EXERCISES FOR SECTION 8.2

In Exercises 1–11, for each of the given functions, find all fixed points and determine whether they are attracting, repelling, or neutral.

1. $F(x) = x^2 - 2x$ 2. $F(x) = x^5$ 3. $F(x) = \sin x$

4. $F(x) = x^3 - x$ 5. $F(x) = \arctan x$ 6. $F(x) = 3x(1 - x)$

7. $F(x) = (\pi/2) \sin x$ 8. $F(x) = x^2 - 3$ 9. $F(x) = 1/x$

10. $F(x) = 1/x^2$ 11. $F(x) = e^x$

In Exercises 12–17, for each of the given functions, 0 lies on a periodic orbit. First determine the period of this orbit; then decide if this cycle is attracting, repelling, or neutral.

12. $F(x) = -x^5 + 1$

13. $F(x) = 1 - x^2$

14. $F(x) = (\pi/2) \cos x$

15. $F(x) = \begin{cases} x + 1, & \text{if } x < 3.5; \\ 2x - 8, & \text{if } x \geq 3.5. \end{cases}$

16. $F(x) = 1 - x^3$

17. $F(x) = |x - 2| - 1$

In Exercises 18–25, each of the given functions has at least one fixed point with derivative ± 1. First find these fixed points. Then, using graphical analysis with an accurate graph or other means, determine if this fixed point is attracting, repelling, or neutral.

18. $F(x) = x - x^2$

19. $F(x) = 1/x$

20. $F(x) = \sin x$

21. $F(x) = \tan x$

22. $F(x) = -x + x^3$

23. $F(x) = -x - x^3$

24. $F(x) = e^{x-1}$ (fixed point is 1)

25. $F(x) = -e \cdot e^x$ (fixed point is -1)

26. Find all fixed points for $F_c(x) = x^2 + c$ for all values of c. Determine for which values of c each fixed point is attracting, repelling, and neutral.

27. How many fixed points does $F(x) = \tan x$ have? Are they attracting, repelling, or neutral? Why?

28. What can you say about fixed points for $F_c(x) = ce^x$ with $c > 0$? What does the graph of F_c tell you about these fixed points? Note that when $c = 1/e$, $F_c(1) = 1$.

29. Consider the function

$$T(x) = \begin{cases} 4x, & x < 1/2; \\ 4 - 4x, & x \geq 1/2. \end{cases}$$

Does T have any attracting cycles? Why or why not?

30. Recall from calculus that Newton's method is an iterative procedure to find the roots of a given function P. Indeed, the Newton iteration is just

$$x_{n+1} = x_n - \frac{P(x_n)}{P'(x_n)}.$$

(a) Show that a root x of P for which $P'(x) \neq 0$ is a fixed point for the Newton iteration.

(b) Suppose $P(x) = x^3 - x$. Determine whether the fixed points for the corresponding Newton iteration are attracting or repelling.

8.3   BIFURCATIONS

Just as with differential equations, discrete dynamical systems may undergo changes in their orbit structure as parameters vary. These changes may include the birth or death of fixed points and cycles or changes in the type of these orbits. As before, such changes are known as bifurcations.

Tangent Bifurcation

One of the simplest types of bifurcation occurs when fixed or periodic points suddenly appear or disappear. As an example, consider the family of quadratic functions $F_c(x) = x^2 + c$. These functions have fixed points when

$$x^2 + c = x.$$

Using the quadratic formula, we see that there are two fixed points for F_c given by

$$p_\pm = \frac{1 \pm \sqrt{1 - 4c}}{2}.$$

These fixed points are real if $1 - 4c \geq 0$, so F_c has fixed points only if $c \leq 1/4$. If $c > 1/4$, F_c has no real fixed points. If $c = 1/4$, the two fixed points are the same, since $p_+ = p_- = 1/2$. For $c < 1/4$, this fixed point splits apart into distinct fixed points p_+ and p_-. Thus a bifurcation occurs at $c = 1/4$ since a pair of fixed points is born as c decreases through $1/4$.

Qualitatively, the graph of F_c lies above the diagonal $y = x$ when $c > 1/4$, so F_c has no fixed points. When $c = 1/4$, the graph is tangent to the diagonal at one point. As we see in Figure 8.28, this is a neutral fixed point. When $c < 1/4$, the graph meets the diagonal at two distinct fixed points $p_\pm$.

To determine the character of these fixed points, we compute $F_c'(x) = 2x$, so that

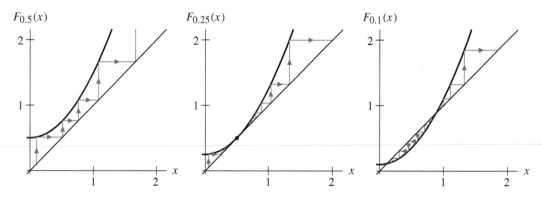

Figure 8.28
Graphical analysis of $F_c(x) = x^2 + c$ for $c = 0.5, 0.25$, and 0.1.

$F_c'(p_+) = 1 + \sqrt{1 - 4c}$. Now $\sqrt{1 - 4c} > 0$ if $c < 1/4$, so it follows that $F_c'(p_+) > 1$ for these c values. Therefore p_+ is a repelling fixed point for all $c < 1/4$. For p_-, the situation is more complicated. We have $F_c'(p_-) = 1 - \sqrt{1 - 4c}$, so $F_c'(p_-) < 1$ for all values of $c < 1/4$. To guarantee that p_- is attracting, we must also have $F_c'(p_-) > -1$. This happens provided

$$-1 < 1 - \sqrt{1 - 4c} = F_c'(p_-)$$
$$2 > \sqrt{1 - 4c}$$
$$-3/4 < c.$$

Therefore the fixed point p_- is attracting for $-3/4 < c < 1/4$ and repelling for $c < -3/4$.

Graphical analysis provides an alternative view of this bifurcation. Figure 8.28 shows some orbits before, at, and after the bifurcation when $c = 0.5, 0.25$, and 0.1. Of course, there really is no "before" and "after" the bifurcation. We could just as easily view this change as c increases, in which case this bifurcation consists of two fixed points that coalesce when $c = 1/4$ and then disappear. Note that this bifurcation is quite similar to some of the bifurcations of equilibrium points that we encountered for differential equations.

As with differential equations, a picture is a useful way to display the changes that occur in a bifurcation. Figure 8.29 shows the bifurcation diagram corresponding to the bifurcation above. In this picture the fixed points $p_\pm$ are plotted as functions of c. The parameter c is plotted on the horizontal axis, and the x-values of the fixed points are plotted on the vertical axis. As c decreases through $1/4$, we see the bifurcation: A neutral fixed point is born at the point $c = 1/4, x = 1/2$. This point immediately splits apart into a pair of fixed points, one attracting and one repelling. Any bifurcation of fixed points or cycles that proceeds in this manner is called a **tangent bifurcation** or **saddle node bifurcation**.

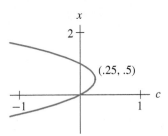

Figure 8.29
The tangent bifurcation diagram.

Pitchfork Bifurcation

The fixed-point structure of a discrete dynamical system may change in other ways as well. For example, consider the family of functions $G_\alpha(x) = x^3 - \alpha x$ depending on the parameter α. The fixed points of G_α are the solutions of $x^3 - \alpha x = x$, which are $x = 0, \pm\sqrt{1+\alpha}$. For $\alpha \leq -1$, G_α has only one fixed point, at $x = 0$, but when $\alpha > -1$, there are three fixed points; hence a bifurcation occurs at $\alpha = -1$. Note that $G'_\alpha(0) = -\alpha$, so the fixed point at 0 is attracting if $|\alpha| < 1$ and repelling if $|\alpha| > 1$. Also, $G'_\alpha(\pm\sqrt{1+\alpha}) = 3+2\alpha$. Since $\alpha > -1$, it follows that both of these fixed points are repelling, and the bifurcation is as shown in Figure 8.30. Because of the shape of this diagram, this bifurcation is called a **pitchfork bifurcation**.

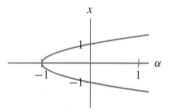

Figure 8.30
The pitchfork bifurcation diagram.

Note that two things happen at the bifurcation point. As α increases through -1, the fixed point at $x = 0$ suddenly changes from repelling to attracting. Meanwhile, a pair of new repelling fixed points is born. It is also possible for a fixed point to change from attracting to repelling and simultaneously give birth to a pair of new attracting fixed points. This type of bifurcation is also called a pitchfork bifurcation.

Period-Doubling Bifurcation

One of the most important types of bifurcations for discrete dynamical systems is the *period-doubling bifurcation*. This bifurcation produces a new cycle having twice the period of the original cycle. For example, consider again the family $F_c(x) = x^2 + c$. Recall that this function has two fixed points given by

$$p_\pm = \frac{1 \pm \sqrt{1 - 4c}}{2}.$$

The fixed point p_- is attracting if $-3/4 < c < 1/4$ and repelling if $c < -3/4$. The fact that $F'_c(p_-) = -1$ at $c = -3/4$ is our signal that a bifurcation may occur. Graphical analysis gives us a hint about what happens. In Figure 8.31, we display the web diagrams for F_c for the nearby c-values $c = -0.6$ and $c = -0.9$. Note that orbits tend to the attracting fixed point when $c = -0.6$, but they tend to an attracting cycle of period 2 when $c = -0.9$.

To find cycles of period 2 for F_c, we must solve the equation $F_c^2(x) = x$. Writing this equation out, we find

$$(x^2 + c)^2 + c = x,$$

or in simplified form,

$$x^4 + 2cx^2 - x + c^2 + c = 0.$$

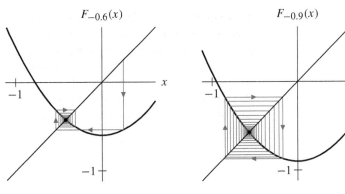

Figure 8.31
Graphical analysis of $F_c(x) = x^2 + c$ when $c = -0.6$ and $c = -0.9$.

This equation looks tough to solve. However, recall that any fixed point for F_c must also solve this equation, for if $F_c(x_0) = x_0$, then certainly $F_c^2(x_0) = x_0$ as well. But we know the fixed points already; they are $p_\pm$. Hence $(x - p_+)(x - p_-) = x^2 - x + c$ is a factor of $F^2(x) - x$.

After some long division, we find

$$\frac{x^4 + 2cx^2 - x + c^2 + c}{x^2 - x + c} = x^2 + x + c + 1.$$

Therefore to find the cycles of period 2, we need only solve the quadratic equation

$$x^2 + x + c + 1 = 0.$$

The roots of this equation are

$$q_\pm = \frac{-1 \pm \sqrt{-3 - 4c}}{2},$$

as given by the quadratic formula.

Note that $q_\pm$ are real numbers only if $c < -3/4$. Thus a cycle of period 2 appears precisely when c decreases through $c = -3/4$. We leave it to the reader to check that this cycle is attracting for c in the interval $-5/4 < c < -3/4$ and repelling for $c < -5/4$ (see the exercises). Thus we may augment the bifurcation diagram in Figure 8.29 as in Figure 8.32.

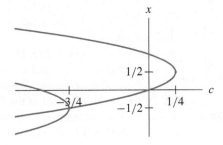

Figure 8.32
The bifurcation diagram for the function

$$F_c(x) = x^2 + c.$$

It shows fixed points and 2-cycles.

Any bifurcation in which a given cycle changes from attracting to repelling (or vice versa) with derivative passing through -1 and that is accompanied by the birth of a new cycle having twice the original period is called a **period-doubling bifurcation**.

A qualitative view of this period-doubling bifurcation is provided by the graphs of the second iterate of F_c, namely $F_c^2(x) = x^4 + 2cx^2 + c^2 + c$. Figure 8.33 presents these graphs for $c = -0.6$, $c = -0.75$ and, $c = -0.9$. Note how the graph of F_c^2 twists through the diagonal at $c = -0.75$ to produce a pair of new fixed points for F_c^2 when $c < -0.75$.

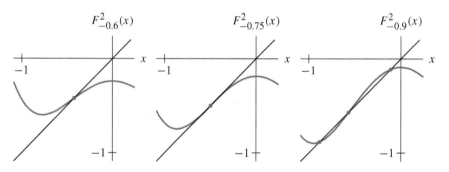

Figure 8.33
The graphs of $F_c^2(x)$ for $c = -0.6$, $c = -0.75$, and $c = -0.9$.

The Logistic Equation

The types of bifurcations discussed in this section are just the beginning of a long story of how discrete dynamical systems evolve as parameters are varied. To illustrate this in a concrete setting, we return to the logistic family of functions given by $L_k(x) = kx(1 - x)$. Recall that x represents the fraction of the maximum possible population of a species, so we only consider x-values in the interval $0 \le x \le 1$.

We will use a combination of qualitative, analytic, and numerical methods to analyze this family of functions. However, the fate of orbits for many parameter values in this system is still an unsolved problem, so we are not be able to describe completely what happens for every parameter value and seed.

The first step in analyzing this family of functions is to find the fixed points. As we have seen, L_k has fixed points at 0 and at the point $p_k = (k - 1)/k$. Note that p_k lies in the interval $0 \le x \le 1$ only if $k \ge 1$; otherwise, p_k is negative and so is not of interest in terms of a population model.

Differentiating L_k, we find $L_k'(x) = k(1 - 2x)$. Hence $L_k'(0) = k$. We conclude that 0 is an attracting fixed point for $0 \le k < 1$ and a repelling fixed point if $k > 1$. Graphical iteration gives us much more information. We see from Figure 8.34 that, as long as $k < 1$, the graph of L_k meets the diagonal at the fixed point 0 and lies below the diagonal for all other x values. This means that all orbits tend to the fixed point. In terms of our population, we conclude that if $k < 1$, the population eventually dies out.

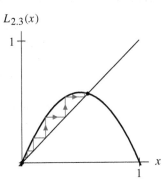

Figure 8.34
Populations die out when $k = 0.8$ but level off when $k = 2.3$.

At the other fixed point, we compute

$$L'_k(p_k) = k\left(1 - 2\left(\frac{k-1}{k}\right)\right) = -k + 2.$$

Therefore p_k is attracting for $-1 < -k + 2 < 1$ or $1 < k < 3$. Again using graphical iteration, we see from Figure 8.34 that all seeds in the interval $0 < x < 1$ have orbits that tend to this attracting fixed point. In terms of populations, we conclude that when $1 < k < 3$, the population of our species eventually levels off at the constant value $(k-1)/k$.

If $k = 3$, we have $L'_k(p_k) = -1$, and if $k > 3$, we find $L'_k(p_k) < -1$. Thus p_k changes from an attracting to a repelling fixed point as k increases through 3, and we expect a period-doubling bifurcation at this point. In Exercise 10 we ask you to verify this fact analytically. As usual, we can use graphical iteration to see this qualitatively. Figure 8.35 displays orbits of L_k when $k = 2.8$ and when $k = 3.2$. Note that these orbits tend to the attracting fixed point when $k = 2.8$ but to a 2-cycle when $k = 3.2$. In terms of populations, when k is slightly larger than 3, we expect that the populations will eventually cycle in a biennial cycle: One year the population will be high; the next year it will be low.

    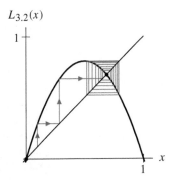

Figure 8.35
The period-doubling bifurcation in the logistic family when $k = 2.8$ and when $k = 3.2$.

The period-doubling route to chaos

It is impossible to use analytic methods to describe much of what occurs for the logistic family as k continues to increase, so we resort to qualitative and numerical methods. For the remainder of this section we give only a glimpse of the very complicated mathematics that arises. Figure 8.36 provides a "movie" of the graph of L_k for 6 k-values in the range $1 \leq k \leq 4$. As k increases, note how the slope of the tangent line to the graph of L_k at p_k decreases, eventually passing through the period-doubling point at which the slope is -1. Dynamically we see p_k change from attracting to repelling at this bifurcation point.

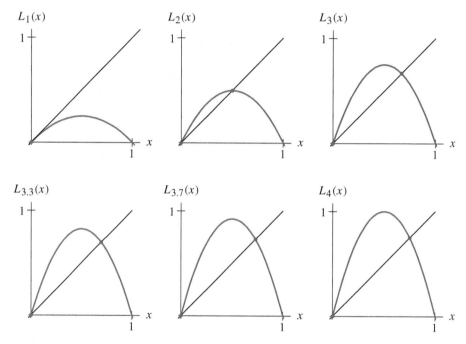

Figure 8.36
Graphs of $L_k(x)$ for $k = 1, 2, 3, 3.3, 3.7,$ and 4.

Figure 8.37 provides another movie, this time of the graphs of L_k^2 for 6 other k-values. The small boxes superimposed on each of these graphs are not web diagrams; rather, they are intended to draw your attention to the piece of the graph of L_k^2 inside each box. Note that these graphs look exactly like the graphs of L_k in Figure 8.36, only smaller and upside down. As we watch k change, we see these small pieces of the graph of L_k^2 behave just as the graphs of L_k do. Hence we expect L_k^2 to undergo the exact same bifurcations that L_k did in the previous pictures.

To make this precise, note that there is a fixed point for L_k^2 at the right-hand edge of the small boxes. This is actually a fixed point for L_k, not a periodic point of period 2. As k increases in these pictures, we see the birth of a new fixed point for L_k^2

(this point lies on a cycle of period 2 for L_k). Call this point q_k. The point q_k is at first attracting, then repelling, as the derivative of L_k^2 at q_k passes through -1. Thus we expect that L_k^2 will undergo a period-doubling bifurcation as k increases. That is, for some intermediate k-value, the period 2 cycle containing q_k will cease to be attracting and become repelling. Meanwhile, a new attracting cycle of twice the period, a 4-cycle, will be born.

Another way to view this would be to draw the graphs of L_k^4 over the intervals of x-values contained in each of the small boxes in Figure 8.37. If we do this, we see that the graphs of L_k^4 resemble the corresponding graphs of L_k^2 in Figure 8.37, only much smaller and turned upside down. So if we magnify these pictures and turn them around, we see that the pieces of the graphs of L_k^2 and L_k^4 look exactly like those of L_k and L_k^2 in Figures 8.36 and 8.37, respectively.

This process of iterating and zooming in to small portions of a graph is called **renormalization**. Borrowed from physics, this process allows us to recognize certain patterns that occur over and over. For example, we could now renormalize the graph of L_k^4. We would see that, over a smaller interval of k-values, the 4-cycle discussed above would itself undergo a period-doubling bifurcation. That is, the attracting 4-cycle would become repelling and a new attracting 8-cycle would be born. This scenario continues with the successive births of attracting cycles of periods 16, 32, 64, and so forth. This phenomenon has been called the "period-doubling route to chaos."

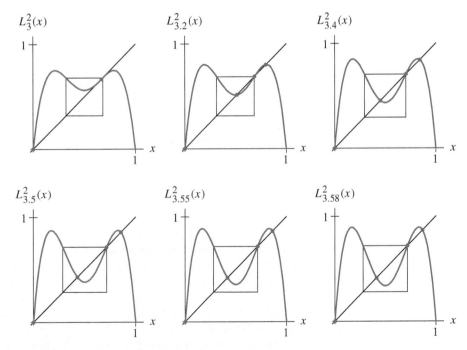

Figure 8.37
Graphs of $L_k^2(x)$ for $k = 3, 3.2, 3.4, 3.5, 3.55,$ and 3.58.

Chaos

As the parameter k continues to increase, we eventually leave the period-doubling inter-val and encounter a new type of behavior—chaos (see Lab 1 at the end of this chapter). For a large number of k-values in the interval $3.57 < k \leq 4$, the logistic family exhibits the type of behavior shown by the case $k = 3.9$ displayed in Figures 8.3 and 8.5. While we cannot fully explain what is happening to orbits for values in this regime, we can describe some simplified models of this chaotic behavior. This is the topic of the next section.

EXERCISES FOR SECTION 8.3

In Exercises 1–7, each function undergoes a bifurcation of fixed points at the given pa-rameter value. In each case use analytic or qualitative methods to identify this bifurca-tion as a tangent, pitchfork, or period-doubling bifurcation or as none of these. Discuss the behavior of orbits near the fixed point in question at, before, and after the bifurca-tion.

1. $F_\alpha(x) = x + x^2 + \alpha, \quad \alpha = 0$ **2.** $F_\alpha(x) = \alpha x, \quad \alpha = -1$

3. $F_\alpha(x) = \alpha \sin x, \quad \alpha = 1$ **4.** $F_\alpha(x) = \alpha \sin x, \quad \alpha = -1$

5. $F_\alpha(x) = \alpha - x^2, \quad \alpha = -1/4$ **6.** $F_\alpha(x) = \alpha \arctan x, \quad \alpha = 1$

7. $F_\alpha(x) = \alpha(x + x^2), \quad \alpha = 0$

8. Consider the family of functions $F_\beta(x) = x^3 + \beta$. Find all β-values for which this family undergoes a bifurcation of fixed points. Determine the type of bifurcation that occurs at each bifurcation point, and then sketch the bifurcation diagram for this family.

9. Consider the family of functions given by

$$T_\mu(x) = \begin{cases} \mu x, & \text{if } x < 1/2; \\ \mu(1 - x), & \text{if } x \geq 1/2. \end{cases}$$

Discuss in detail the bifurcation of fixed points that occurs at $\mu = 1$ for this family. Sketch the fixed-point bifurcation diagram.

10. For the logistic family $F_k(x) = kx(1 - x)$, show explicitly that there exists a cycle of period 2 for $k > 3$.

11. Consider the family of functions $F_c(x) = x^2 + c$. Sketch the graphs of F_c for $c = 0.25, 0, -0.75, -1, -1.5$, and -2 for $-2 \leq x \leq 2$. Then sketch the graphs of F_c^2 for $c = -0.75, -1, -1.25, -1.3, -1.35$, and -1.4. Is there any similarity be-tween these two families of graphs? Discuss in a brief essay, complete with pictures.

8.4 CHAOS

In this section we begin the study of chaotic behavior. Until a few years ago most scientists and mathematicians thought that typical differential and difference equations did not exhibit the kind of behavior we now call chaos—that this kind of behavior was extremely unlikely in simple mathematical models. Now we know that this is not the case. There are many models whose behavior is quite erratic, and this behavior persists when parameters are varied. Moreover, mathematicians have begun to develop techniques to predict the onset of chaos and to determine the scope or limits of this behavior. In this chapter we use a number of qualitative and numerical techniques to explain and illustrate the concept of chaos.

An Example without Chaos

Roughly speaking, chaos means unpredictability. Thus far, most of the differential and difference equations we have encountered have been completely predictable. For example, consider the simple discrete logistic model

$$P_{n+1} = 2P_n(1 - P_n).$$

As we have seen, this system has two fixed points, at 0 and at 1/2. Graphical analysis shows that the orbit of any seed x_0 in the interval $0 < x_0 < 1$ tends to 1/2 (see Figure 8.38). The only other point, $x_0 = 1$, has an orbit that is eventually fixed. So we know the fate of all orbits.

Now in real life we rarely know the seed for our orbit with complete accuracy. For example, we most likely do not know the precise value of the population at time 0; we probably miss a few individuals in our initial count. If our model is the simple logistic model above, however, then this inaccuracy does not really matter. A small fluctuation in our initial seed in the interval $0 < x_0 < 1$ does not alter our predictions at all.

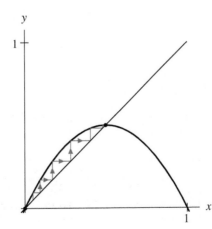

Figure 8.38
Graphical analysis of the function

$$F(x) = 2x(1 - x).$$

This system has two fixed points, one at 0 and one at 1/2. The orbit of any seed x_0 in the interval $0 < x_0 < 1$ tends to 1/2. The only other point, $x_0 = 1$, has an orbit that is eventually fixed.

A Chaotic Logistic Function

Now let's contrast the orbits of the logistic model above with those of

$$P_{n+1} = 4P_n(1 - P_n),$$

where we have simply changed the growth parameter from 2 to 4. As always, studying this system is the same as iterating the function $L_4(x) = 4x(1 - x)$. This function has fixed points at 0 and 3/4, and eventually fixed points at 1/2 and 1. But consider the fate of virtually any other orbit. In Figure 8.39 we compute the web diagram for the first 200 points on the orbit of 0.123 for this function. It is hard to distinguish any pattern in this picture. A histogram and a time series for this orbit further confirm this erratic behavior (see Figures 8.40 and 8.41). In the histogram we see that the orbit seems to visit every small subinterval in $0 \leq x \leq 1$ many times.

If you repeat this experiment for a variety of seeds in the interval $0 < x < 1$, you will see essentially the same results in each case. Of course, the fixed points and eventually fixed points yield quite different diagrams. However, virtually any other seed yields figures that resemble those in Figures 8.40 and 8.41 in their complexity.

$L_4(x)$

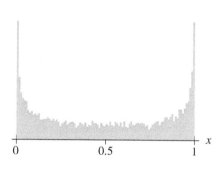

Figure 8.39
Web diagram of the orbit of 0.123
under $L_4(x) = 4x(1 - x)$.

Figure 8.40
Histogram of the orbit in Figure 8.39.

$(L_4)_n$

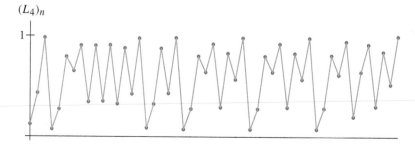

Figure 8.41
Time series for the orbit of 0.123 under $L_4(x) = 4x(1 - x)$.

$L_4^2(x)$

$L_4^3(x)$

$L_4^4(x)$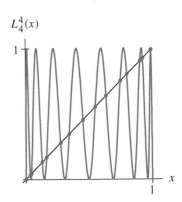

Figure 8.42
Graphs of L_4^2, L_4^3, and L_4^4.

These computer experiments suggest that most orbits of $L_4(x) = 4x(1 - x)$ behave erratically. However, there are many cycles for L_4 as well. In Figure 8.42 we plot the graphs of L_4^2, L_4^3, and L_4^4. Note that the graph of L_4^n appears to cross the diagonal $y = x$ at 2^n points, yielding infinitely many cycles of arbitrarily high period in the interval $0 \leq x \leq 1$.

Another important element in our discussion of chaotic behavior is **sensitive dependence on initial conditions**. This is illustrated in Figure 8.43, where we plot the time series for the orbits of 0.123 and 0.124. Note that these orbits initially are very close, but after very few iterations, the orbits separate and thereafter bear little resemblance to one another. Just a small change in the third decimal place of our seed has resulted in a large change in the ensuing orbit.

$(L_4)_n$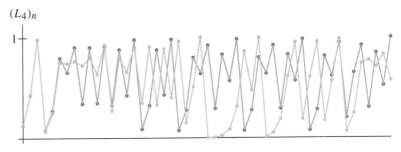

Figure 8.43
Time series for the orbit of 0.123 in dark blue; for 0.124 in light blue.

A Model Chaotic System

To characterize the chaotic behavior we observed for the logistic function $L_4(x) = 4x(1 - x)$, we turn to a discrete dynamical system that is much easier to understand. Consider the function T defined for $0 \leq x < 1$ by

$$T(x) = 10x \mod 1.$$

This notation means that to compute $T(x)$ we first multiply x by 10 and then drop the integer part, keeping only the fractional part of the result. For example

$$T\left(\tfrac{1}{3}\right) = \tfrac{10}{3} - 3 = \tfrac{1}{3},$$

$$T(0.232323\ldots) = 2.32323\ldots - 2 = 0.32323\ldots,$$

and

$$T(0.1111\ldots) = 1.111\ldots - 1 = 0.111\ldots.$$

More generally, if $0 \leq x < 1$, we can write the decimal expansion of x in the form

$$x = .a_1 a_2 a_3 \ldots,$$

where the a_i are digits between 0 and 9. Then

$$10x = a_1.a_2 a_3 a_4 \ldots$$

so that

$$T(x) = 10x - a_1 = 0.a_2 a_3 a_4 \ldots;$$

that is, T simply drops or chops off the first digit in the decimal expansion of x. For this reason, T is called the **chopping function**. For example,

$$T(0.\overline{123}) = 0.\overline{231}$$

and

$$T^2(.25) = T(.5) = 0.$$

The notation $0.\overline{123}$ denotes the repeating sequence $0.123123123\ldots$.

Density of periodic points

Because of the special definition of the chopping function, periodic points of T are easy to find. Any x whose decimal expansion is repeating lies on a cycle for T. For example, $2/9$ is a fixed point, since

$$\frac{2}{9} = 0.222\ldots$$

and $T(0.222\ldots) = 0.222\ldots$. Similarly, $0.\overline{37}$ lies on a 2-cycle and $0.\overline{1234}$ lies on a 4-cycle. Clearly T has infinitely many periodic points in $[0, 1)$, and the periods of these points may be arbitrarily large. Moreover, unlike the logistic functions, we can write down all of the periodic points for T explicitly.

There is more to this story, however. Given any x in the interval $[0, 1)$, we can find a periodic point of T arbitrarily close to x. This follows since, if x has the decimal expansion

$$x = 0.a_1 a_2 a_3 \ldots,$$

then the point

$$y = 0.\overline{a_1 a_2 \ldots a_n}$$

is close to x (within $1/10^n$ units) and moreover is periodic with period n.

For example, suppose we are given the number

$$x = \frac{1}{\sqrt{2}} = 0.707106781\ldots$$

and are asked to find a periodic point within 0.001 units of x. We could choose $x_1 = 0.\overline{707}$, since

$$|0.707106781\ldots - 0.\overline{707}| = 0.00060093\ldots$$

and x_1 has period 3. But there are many other choices that work as well, including $x_2 = 0.\overline{707106}$, which is much closer to x than x_1 and has period 6. Alternatively, we could choose $0.\overline{7071999222555}$, which is also within 0.001 units of x and is periodic with period 13.

Extrapolating from this example, we see that no matter how small a subinterval we choose in $[0, 1)$, we can always find periodic points inside this subinterval. In fact we can always find infinitely many distinct periodic points inside the subinterval. A subset of $[0, 1)$, which contains points in every subinterval of $[0, 1)$, no matter how small, is said to be a **dense subset** of the interval. For example, the set of all rational numbers between 0 and 1 is a dense subset of $0 \leq x < 1$. So too are the irrationals in this interval. The above argument shows that the set of periodic points for T is also a dense subset of $[0, 1)$.

Periodic orbits are by no means the only types of orbits for the chopping function; there are many eventually periodic points as well. Any point whose decimal expansion eventually begins to repeat is eventually periodic under T. For example,

$$x = 0.123\overline{4}\ldots$$

is a point whose orbit becomes fixed after 3 iterations of T. Similarly, the orbit of $0.1234\overline{567}$ is eventually periodic with period 3. Now recall that any rational number has decimal expansion that either repeats, eventually repeats, or terminates (which means it ends with a sequence of all zeroes). As a consequence, we see that any rational number in $[0, 1)$ has an orbit under T that is either periodic or eventually periodic. On the other hand, any irrational number in $[0, 1)$ has an orbit that never cycles, since the decimal expansions of irrational numbers never repeat. For example, the orbit of $1/\sqrt{2}$ is not periodic under T since $\sqrt{2}$ is an irrational number.

A dense orbit

Some of these irrational numbers have very interesting orbits. For example, consider the number x_0 whose decimal expansion begins

$$x_0 = .0123456789.$$

Suppose the next 200 terms of this expansion consist of all possible 2-blocks of digits, that is,

$$x_0 = 0.\underbrace{0123\ldots 9}_{\text{all 1-blocks}}\ \underbrace{00\ 01\ 02\ldots 10\ 11\ 12\ldots 20\ 21\ldots 99}_{\text{all 2-blocks}}$$

followed by all possible 3-blocks, 4-blocks, and so on, so we have

$$x_0 = 0. \underbrace{0\ldots9}_{\text{all 1-blocks}} \quad \underbrace{00\ldots99}_{\text{all 2-blocks}} \quad \underbrace{000\ldots999}_{\text{all 3-blocks}} \ldots .$$

What can we say about the orbit of x_0? Recall that each iteration of T simply drops the leading digit from this sequence. Thus given any x in $[0, 1)$, we see that there is a point on the orbit of x_0 that comes arbitrarily close to x. Indeed, we need only iterate T enough times so that the first n digits of the decimal expansion of x appear as the leading terms of this point on the orbit of x_0. Then these two points are within $1/10^n$ units of each other. For example, suppose we are given the number $1/3$ and are asked to find a point on the orbit of x_0 that lies within 0.001 of $1/3$. To do this, we simply iterate T until the 3-block that begins with 333 appears as the first entry of the sequence. Now it takes 10 iterations to remove the entries $0.012\ldots9$, and then another 200 iterations to remove $00\ 01\ 02\ldots\ 98\ 99$, and then finally $3 \cdot 333$ more iterations to bring 333 to the head of the list. Thus, $T^{1209}(x_0)$ lies within the required distance of $1/3$.

Of course, there are many other points on the orbit of x_0 that lie within 0.001 units of $1/3$. All we need do is iterate T enough times until any k-block of the form $33\ldots3$ is first in line and this point is close enough to $1/3$. As a consequence, the set of points that make up the orbit of x_0 is also a dense subset of $[0, 1)$.

Sensitive dependence on initial conditions

The final observation we wish to make about T is that this dynamical system exhibits **sensitive dependence on initial conditions**. By this we mean the following: Consider any x_0 in $[0, 1)$. Then there are points y_0 arbitrarily close to x_0 so that the orbit of y_0 is eventually "far" from that of x. For example, if we consider the fixed point $x_0 = 1/9 = 0.\overline{1}$ of T, then there are nearby points whose orbits behave very differently. For example, the seed $y_0 = 0.11\ldots1\overline{8}$ eventually lands on the fixed point $0.\overline{8} = 8/9$. How close y_0 is to $1/9$ depends on how many initial 1's are present in the sequence. Although these two seeds are close together initially, their orbits eventually are far apart in the interval $[0, 1)$ (see Figure 8.44). As another example, we can consider a seed of the form $0.11\ldots1\overline{35}$, which is eventually periodic with period 2, or $0.11\ldots1\overline{23456}$, which eventually has period 5. Both of these seeds are also close to x_0, but their orbits behave quite differently. Indeed, arbitrarily close to the given seed $x_0 = 1/9$, we can find another seed whose orbit eventually has any behavior whatsoever. All we need to do is precede a given orbit by a sufficiently long string of 1's. Again, this orbit is initially close to $1/9$, but it eventually diverges from this fixed point.

Figure 8.44
Time series for the orbit of $0.\overline{1}$ (in dark blue) and for $0.11\ldots1\overline{8}$ (in light blue).

The general situation is similar. Given any seed x_0 whose decimal expansion is

$$x_0 = 0.a_1a_2a_3\ldots,$$

we consider the nearby seed

$$y_0 = 0.a_1a_2\ldots a_nb_1b_2b_3\ldots,$$

where $b_1b_2b_3\ldots$ are arbitrary digits. Then the seed y_0 is close to x_0 (how close depends on n, the number of decimal places to which x_0 and y_0 agree), but the fate of the two orbits can be radically different. Thus T exhibits sensitive dependence throughout the interval $[0, 1)$.

The existence of sensitive dependence in dynamical systems has profound implications for scientists and mathematicians who use difference or differential equations as mathematical models. If a given system exhibits sensitive dependence on initial conditions, then numerical predictions about the fate of orbits or solutions are to be totally distrusted. For we can never know the exact seed or initial condition for our orbit or solution because we cannot make physical measurements with infinite precision. Even if we had exact measurements, we could never carry out the necessary computations. The small numerical errors that are always introduced in such numerical procedures throw us off our original orbit and onto another whose ultimate behavior may be radically different.

Unlike T, for which we have sensitive dependence at all points in its domain, many functions exhibit sensitivity only at isolated points. Consider for example the logistic function $L_{3.2}(x) = 3.2x(1 - x)$. As shown in Figure 8.45, most orbits of this function tend to an attracting cycle of period 2. However, not all orbits share this fate. As we see in Figure 8.45, there are two repelling fixed points for $L_{3.2}$, and there are infinitely many points whose orbits eventually land on the nonzero fixed point for $L_{3.2}$. The function $L_{3.2}$ exhibits sensitive dependence at each of these seeds, since we can change the seed only slightly and find an orbit that is neither fixed nor eventually fixed,

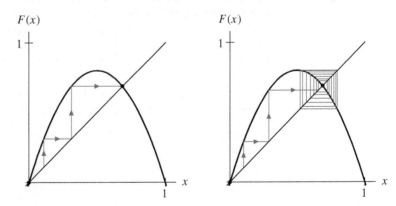

Figure 8.45
Two nearby orbits for $F(x) = 3.2x(1 - x)$; one is eventually fixed, the other tends to the 2-cycle.

but rather tends to the attracting cycle of period 2. This behavior is much different from what we see for T; it is commonly called *unstable behavior*. Chaos occurs when we have unstable behavior everywhere, or at least on a large set.

Summary

We have shown that our model system T possesses an amazing amount of complexity in the interval $[0, 1)$. The periodic points of T form a dense subset of this interval, and there is also a single orbit that forms another dense subset of $[0, 1)$. Moreover, this function exhibits sensitive dependence on initial conditions, which renders precise calculations of orbits impossible. The combination of these three properties makes this system quite unpredictable, or **chaotic**.

It may seem that this type of complexity is special to our model function T because T has constant slope 10. However, over the past few years mathematicians have verified that a number of dynamical systems—both discrete systems and systems of differential equations—exhibit this type of behavior. For example, there is a large set of k-values for which the logistic function $L_k(x) = kx(1 - x)$ is chaotic. The histograms and time series we displayed earlier give qualitative evidence for this behavior.

EXERCISES FOR SECTION 8.4

1. For the logistic function $L_4(x) = 4x(1 - x)$, compute a histogram of the first 10,000 points on the orbit of $x_0 = 0.3$ by dividing the interval $[0, 1]$ into 100 subintervals, each of length 0.01, and then calculating the successive subintervals into which the orbit of x_0 falls. Then repeat this calculation using the seed $y_0 = 0.3001$. Do you see any difference? Now compute the time series for the first 100 points on the orbits of x_0 and y_0 and compare the results. Do you see any differences? Discuss your findings in a brief essay. Do not attempt this exercise without access to computing and graphing technology.

Exercises 2–11 deal with the doubling function given by

$$T(x) = \begin{cases} 2x, & \text{if } 0 \le x < 1/2; \\ 2x - 1, & \text{if } 1/2 \le x < 1. \end{cases}$$

2. Compute the orbit of the following seeds under T. Which seeds are periodic and which are eventually periodic?

(a) $x_0 = 1/5$ (b) $x_0 = 2/7$ (c) $x_0 = 3/11$ (d) $x_0 = 1/10$
(e) $x_0 = 1/6$ (f) $x_0 = 4/14$ (g) $x_0 = 4/15$

3. What can you say about the orbit of the seed $x_0 = p/2^n$ under iteration of T?

4. Sketch the graphs of T^2, T^3, and T^4. What do you expect the graph of T^n to look like? How many fixed points should T^n have?

5. Using the graphs drawn in Exercise 4, discuss the question of density of the subset of periodic points of T in the interval $0 \le x < 1$.

6. Suppose x_0 has binary representation $.a_1a_2a_3\ldots$. That is,

$$x_0 = \frac{a_1}{2} + \frac{a_2}{2^2} + \frac{a_3}{2^3} + \cdots.$$

What is the binary expansion of $T(x_0)$? Of $T^n(x_0)$?

7. Using the results of Exercise 6, find all points in the interval $[0, 1)$ that are eventually fixed at 0 by T^n.

8. Using the results of Exercise 6, find all points in the interval $[0, 1)$ that are periodic with period 2, 3, or 4 under T.

9. How many points in $[0, 1)$ are fixed by T^n for each n?

10. Does T exhibit sensitive dependence on initial conditions?

11. Write down an expression for a seed x_0 in $[0, 1)$ whose orbit under T forms a dense subset of this interval.

Exercises 12–14 deal with the doubling function given by

$$T(x) = \begin{cases} 2x, & \text{if } 0 \le x < 1/2; \\ 2 - 2x, & \text{if } 1/2 \le x \le 1. \end{cases}$$

12. Sketch the graphs of T, T^2, and T^3 over the interval $[0, 1]$. How many fixed points does T^n have for each n?

13. Discuss the question of density of the set of periodic points of T.

14. Does T exhibit sensitive dependence on initial conditions?

8.5 CHAOS IN THE LORENZ SYSTEM

In this section we complete the discussion of the Lorenz system of differential equations started in Section 2.8 and continued in Section 5.5. Recall that this system is a nonlinear system of three differential equations given by

$$\frac{dx}{dt} = 10(y - x)$$
$$\frac{dy}{dt} = 28x - y - xz$$
$$\frac{dz}{dt} = -\frac{8}{3}z + xy.$$

We will first show how to reduce this problem to a two-dimensional model, and then we will further simplify the system to a one-dimensional iteration. Using techniques that we developed in the last section, we will then show how this simplified Lorenz system exhibits chaotic behavior.

The Lorenz Attractor

As we saw in Section 2.8, virtually every solution curve of the Lorenz system eventually has similar behavior: After initially meandering around three-dimensional space, the solution curve eventually winds in complicated fashion around a certain region in three-dimensional space. Our goal is to create a model of this behavior that will enable us to see more clearly how the solutions behave.

For example, Figure 8.46 shows the solution curve starting at the initial condition $(0, 1, 0)$. After a short period of time, the solution settles down on a region where it winds around two "holes" in seemingly random manner.

The object toward which solutions of the Lorenz equation tend is called a **strange attractor**. It is an "attractor" because all solutions tend to it. It is "strange" since it is very different from the simpler attractors we have encountered thus far: equilibrium points (sinks) and certain periodic orbits.

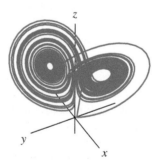

Figure 8.46
The solution of the Lorenz
equation with initial condition
$(0, 1, 0)$.

Projections

To understand this attractor, we will try to view it from many different vantage points. The easiest way to do this is to project the solution curve onto a coordinate plane. Geometrically this means that we simply forget one of the coordinates of the solution curve. For example, to project the solution to the xy-plane, we merely plot the curve $(x(t), y(t))$ in the plane, forgetting about the third coordinate of the solution, $z(t)$. In Figure 8.47, we sketch the xy-, yz-, and xz-projections of the solution tending to the Lorenz attractor. Note that in each of these projections the solution curve apparently crosses itself. By the Uniqueness Theorem, this does not happen in the full xyz-space. When the solution crosses itself in the projection, the missing coordinate of the solution

Figure 8.47
The xy-, yz-, and xz-projections of the solution in Figure 8.46.

assumes different values, so the real solution curve passes "above" or "below" itself in three dimensions.

The advantage of projections is that they can be viewed on computer screens easily, whereas three-dimensional images require some geometric intuition and perspective when viewed on a screen. The trick is to use various projections to piece together the behavior of solutions.

The Lorenz Template

The projections of the Lorenz attractor indicate that solution curves tend toward a region in the three-dimensional space that resembles the surface drawn in Figure 8.48. This image is called the **Lorenz template**, although it is sometimes called the Lorenz mask, for obvious reasons. Note that the two lobes of the template are joined along a straight line, with one lobe bending backward and the other forward of the line.

The attractor for the Lorenz system is actually a much more complicated object. It consists of infinitely many sheets packed tightly together in a complicated fashion. However, this template allows us to get a good idea of the mechanism that produces the chaotic behavior in the Lorenz system. What follows is not an exact description of the solutions of the equation; rather, it should be viewed as a simplification or model of the full system.

Figure 8.48
The Lorenz template.

To understand the full phase portrait of the Lorenz system, recall from Section 5.5 that there are three equilibrium points for the Lorenz system. There is a saddle with one positive and two negative eigenvalues at the origin. There is also a pair of equilibrium points that are spiraling saddles. Each of these points have two complex eigenvalues with positive real parts and a single negative eigenvalue. Figure 8.49 indicates the positions of these equilibrium points relative to the template. Note that at each of the equilibrium points, there is a stable eigenvector pointing toward the template. This forces nearby solutions to tend toward the attractor.

The Poincaré return function

Now we add a typical solution curve to the template. As we have seen, solutions wind about the attractor, sometimes circling about the left lobe, at other times about the right lobe. We think of this solution as actually lying on the template, wandering back and forth between lobes (see Figure 8.50). The solution depicted in this figure is really fictitious, intended only as a representation of the behavior of a real solution of the equations.

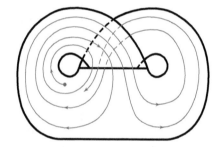

Figure 8.49
The equilibrium points for the Lorenz system and the template.

Figure 8.50
A fictitious solution lying on the Lorenz template.

Note that the typical solution on the template repeatedly crosses the line of intersection of the two lobes. We call this line L. To understand the qualitative behavior of solutions near the template, we will try to understand how this solution returns to L. Note that there is a point p on L through which the solution curve never returns to L. This is a solution that lies in both the template and the stable separatrix for the equilibrium point at $(0, 0, 0)$ (see Figure 8.51). Using the linearization near this equilibrium point, we see that solutions on either side of this solution through p evolve in different directions about the lobes (see Figure 8.52). Thus, this "disappearing" solution through p forms the boundary between those curves on the template that circle around the left lobe and those that circle around the right lobe.

Let L_ℓ denote the portion of L to the left of p (in Figure 8.51), and L_r the portion to the right of p. We wish to understand how solution curves that begin in L_ℓ or L_r return to L. To that end we will construct a function that assigns to any point in L_ℓ or L_r its next point of intersection with L. This function is called the Poincaré function,

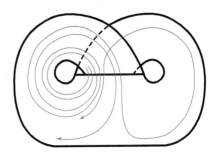

Figure 8.51
A disappearing solution on the Lorenz template.

Figure 8.52
Several solutions on the Lorenz template.

or the (first) return function on L (see Section 5.6). We denote it by ϕ. To understand how solution curves repeatedly intersect L, we need only iterate the function ϕ. It is impossible to give an analytic formula for ϕ. After all, the template itself is really fictitious. However, by observing the behavior of solutions on the xz- or yz-projections, we can give a good qualitative description of ϕ.

In Figure 8.53 we juxtapose the yz-projection and a model for ϕ. Note that ϕ is undefined at p. As indicated in the projection, ϕ takes the interval L_ℓ to an interval that covers L_r and a portion of L_ℓ. Also, ϕ takes L_r to an interval that covers L_ℓ and a portion of L_r. Therefore we can construct an approximation of the graph of ϕ as shown in Figure 8.53.

Note that the images of both L_ℓ and L_r when they return to L are longer than L_ℓ or L_r. That is, ϕ expands these intervals. This is the reason that we have drawn the graph of ϕ with slope larger than 1.

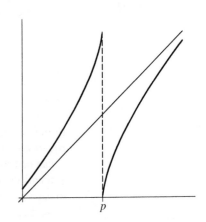

Figure 8.53
The yz-projection and a model for ϕ.

Iteration of the Poincaré Return Function

To understand the evolution of solutions as they continually reintersect L, we must iterate ϕ. As in the previous section, we see that ϕ exhibits sensitive dependence on initial conditions, since $\phi' > 1$ on L. Graphical iteration shows that there are a number of periodic points for ϕ as well (see Figure 8.54).

We need to be careful when we interpret the meaning of the observed behavior of ϕ. For example, consider the periodic orbit and the corresponding cycle for ϕ displayed in Figure 8.54. Presumably there is a real orbit for the Lorenz system that exhibits this behavior. However, such an orbit is impossible to find using numerical methods such as Euler's method, mainly because of the expansion in the direction along L. Moreover, since the Lorenz attractor is really much thicker than the template, there will be other orbits in the attractor that tend to this periodic solution. For example, note how far away solutions may eventually match up with the given periodic solution in the template. These correspond to eventually periodic points for ϕ.

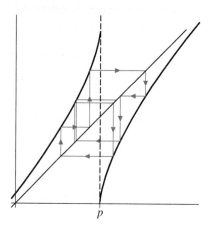

Figure 8.54
Graphical iteration of ϕ.

Summary

In this section we have described an important technique in the study of differential equations. When we observed complicated behavior of solutions in a higher dimensional setting, we tried to describe this behavior by reducing the dimension of the system and viewing it as a discrete dynamical system. Although it is not always possible to carry out such a procedure, this type of process does work in a number of important systems of differential equations.

REVIEW EXERCISES FOR CHAPTER 8

Short answer exercises: Exercises 1–10 focus on the basic ideas, definitions, and vocabulary of this chapter. Their answers are short (a single sentence or drawing), and you should be able to do them with little or no computation. However, they vary in difficulty, so think carefully before you answer.

1. Find all fixed points for the function $F(x) = x^3 - x/2$ and determine their type.

2. Find all fixed points for the logistic function $L_{3.5}(x) = 3.5x(1 - x)$ and determine their type.

3. Let

$$T(x) = \begin{cases} 2x, & \text{if } x \leq 1/2; \\ 2x - 1, & \text{if } x > 1/2. \end{cases}$$

Sketch the graph of the second iterate of T and find all cycles of period two for T.

4. Find all attracting cycles for the function T in Exercise 3.

5. Describe the bifurcation that occurs in the family $F_c(x) = cx - x^3$ at $c = 1$.

6. Is it possible for a function to have exactly two fixed points, both of which are attracting?

7. Using graphical iteration, describe the ultimate behavior of all orbits under iteration of $F(x) = \sin x$.

8. How many fixed points does the function $F(x) = \cos x$ have? Classify them as attracting, repelling, or neutral.

9. Using graphical iteration, determine the fate of all orbits under iteration of $F(x) = \cos x$.

10. How many fixed points does $T(x) = \tan x$ have?

True-false: For Exercises 11–16, determine if the statement is true or false. If it is true, explain why. If it is false, provide a counterexample or an explanation.

11. The function $F(x) = x^2 + c$ has at least one fixed point for each value of c.

12. The point $x = \sqrt{2}$ is a fixed point for $F(x) = x^2 - 2$.

13. The point $x = 1$ lies on an attracting cycle of period two for $F(x) = -x^3$.

14. The point $x = 1/2^{100}$ lies on a cycle of period 100 for

$$T(x) = \begin{cases} 2x, & \text{if } x < 1/2; \\ 2x - 1, & \text{if } x \geq 1/2. \end{cases}$$

15. The family $F_a(x) = a \sin x$ has a tangent bifurcation at $x = 0$ and $a = 1$.

16. The logistic function $L_k(x) = kx(1 - x)$ has a tangent bifurcation at $k = 4$.

17. List all points in the interval $0 \leq x < 1$ that lie on cycles of period two for the chopping function.

18. Consider the function

$$F_{a,b}(x) = ax + b,$$

where a and b are parameters. Sketch the regions in the ab-plane where this function has different behaviors (for example, an attracting fixed point, no fixed points, more than one fixed point, ...).

19. Using the graph of the function $F(x) = x^2 - 2$, determine the total number of points that lie on cycles.

20. The function

$$F_a(x) = \begin{cases} ax, & \text{if } 0 \leq x \leq 1/2; \\ a(1-x), & \text{if } 1/2 \leq x \leq 1 \end{cases}$$

is called the "tent function" if $a > 0$.

   (a) Why is it called the tent function?

   (b) Describe the bifurcation that occurs at $a = 1$.

   (c) Using the graph of F_2^n, determine the number of fixed points of F_2^n for all n.

LAB 8.1 Newton's Method as a Difference Equation

The **roots** of a function $f(x)$ are the values of x for which $f(x) = 0$. Given a function f, we can find approximate values for its roots using Newton's method. To use Newton's method, we first make an initial guess for the value of the root, say x_0. If this is not the root (that is, $f(x_0)$ is not 0), we can (hopefully) improve the guess by first computing the tangent line to the graph of f at x_0, then compute the point x_1, where this tangent line intersects the x-axis. The relationship between x_0 and x_1 (a good calculus review) is given by

$$x_1 = x_0 - \frac{f(x_0)}{f'(x_0)}.$$

If $f(x_1)$ is not sufficiently close to zero, then we improve the guess again by computing

$$x_2 = x_1 - \frac{f(x_1)}{f'(x_1)},$$

and so forth. What we are doing is creating the orbit of the seed x_0 under the Newton's method function

$$N(x) = x - \frac{f(x)}{f'(x)}.$$

The initial guess is the seed x_0, the first improvement is $x_1 = N(x_0)$, the second improvement is $x_2 = N(x_1)$, and so on. The hope is that the improved guesses really are improvements and that the sequence $x_0, x_1, \ldots$ approaches a root of the function f. In this lab we consider how the choice of seed affects the long-term behavior of the orbits of the Newton's method function by studying a particular example. Let

$$f(x) = x^3 - 3x^2 + 2x = x(x - 1)(x - 2).$$

This function has roots at $x = 0$, $x = 1$, and $x = 2$.

In your report, consider the following items:

1. For the function $f(x)$ chosen above, compute the corresponding Newton's method function $N(x) = x - f(x)/f'(x)$. Verify that this function has fixed points at the roots $x = 0$, $x = 1$ and $x = 2$. Verify that these are attracting fixed points. What does this imply for orbits of $N(x)$ for seeds x_0 near 0, 1, or 2?

2. Compute the first 20 points of the orbits under $N(x)$ for the initial seeds

$$x_0 = 0, 0.05, 0.10, 0.15, 0.20, \ldots, 0.90, 0.95, 1.$$

Sketch the segment $0 \leq x_0 \leq 1$. For each choice of x_0, if the 20th point x_{20} in its orbit under N is within 0.01 of 0, color the point red; if x_{20} is within 0.01 of 1, color it blue; if x_{20} is within 0.01 of 2, color it green; and if none of these conditions holds, color it black. Is this figure consistent with your conclusions from Part 1? What implications does this picture have concerning the choice of the seed x_0 for Newton's method?

3. Repeat Part 2 for the points

$$x_0 = 0.30, 0.32, 0.34, 0.36, 0.38, \ldots, 0.46, 0.48, 0.50.$$

Your report: In your report address each of the items above. Parts 2 and 3 require a computer or calculator. Do *not* submit lists of numbers. What is important is the color-coded sketches of the x-axis and your interpretation of these sketches.

LAB 8.2   The Delayed Logistic and Two-Dimensional Iteration

In Section 8.1 the logistic equation

$$L_k(y) = ky(1 - y)$$

was used to model populations that reproduce on a discrete time scale. In particular, if y_n represents the population (or population density) of some species in the nth generation, then the population in the $(n + 1)$st generation is given by

$$y_{n+1} = L_k(y_n) = ky_n(1 - y_n).$$

We have seen that the behavior of orbits can depend in a dramatic way on the value of the growth rate parameter k and on the seed y_0.

One hidden assumption in the discrete logistic model is that the population in the $(n + 1)$st generation depends only on the population in the nth. This may not always be the case for species that have a long maturation time or that migrate to nesting or breeding areas. For example, suppose a species returns to the same breeding area each year. The amount of food available to the nth generation in that area might depend on how much was eaten in the previous year. In this case the population of the $(n + 1)$st generation depends on the population in the nth and the $(n - 1)$st generations. This motivates the following model of population growth called the **delayed logistic** population model,

$$y_{n+1} = ky_n(1 - y_{n-1}).$$

As with the logistic model, y_n represents the population density in the nth generation. We restrict attention to the case where $0 \leq y_n \leq 1$. For the delayed logistic model, the population in the $(n + 1)$st generation is proportional to the population in the nth generation, as well as to how close the population in the $(n - 1)$st generation was to the maximum possible population density.

To study the delayed logistic model, we perform an operation very similar to the conversion of second-order differential equations into first-order systems. We introduce a new variable x_n, letting

$$x_n = y_{n-1}.$$

That is, x_n is the population density in the $(n - 1)$st generation. We then rewrite the delayed logistic equation as

$$x_{n+1} = y_n$$

$$y_{n+1} = ky_n(1 - x_n).$$

As with the conversion of second-order equations to first-order systems, the system is simpler in one way because the values x_{n+1} and y_{n+1} depend only on the values of x_n and y_n. However, it is more complicated because we have had to introduce a new variable.

We can interpret this model as an iteration by letting

$$F_k(x, y) = (y, ky(1 - x)).$$

Then

$$(x_{n+1}, y_{n+1}) = F_k(x_n, y_n);$$

that is, orbits of an initial seed (x_0, y_0) are made up of a sequence of points in the plane. Hence

$$(x_1, y_1) = F_k(x_0, y_0) = (y_0, ky_0(1 - x_0)),$$

$$(x_2, y_2) = F_k(x_1, y_1) = (y_1, ky_1(1 - x_1)),$$

and so on. In this lab we study the behavior of orbits for this two-dimensional interation. We can graph orbits in two different ways. We can form "time-series" graphs with n on the horizontal axis and x_n or y_n on the vertical axis. Alternately, we can make "phase plane" graphs, plotting the sequence of points (x_n, y_n) in the xy-plane. You should make both types of graphs for each orbit you compute.

Consider the following items:

1. For several different values of k between 1.5 and 2.5, compute the first 200 iterates of the orbit with seed $(x_0, y_0) = (0.1, 0.2)$. For each of the values of k that you choose, describe the long-term behavior of the orbit. What are the biological implications of your conclusions? (For some values of k, you may need more than 200 iterates.)

2. How do your results from Part 1 change if you change the initial seed? What are the biological implications of your conclusions? (Remember to keep the initial seed in the "physically meaningful" range $0 \le x_0 \le 1, 0 \le y_0 \le 1$.)

Your report: In your report describe your discoveries based on these numerical experiments. You may include a *limited* number of graphs and/or phase plane pictures of orbits to illustrate your description. You should *not* include a catalog of orbits for different initial conditions and values of k. The goal of the lab is to interpret the results of your experiments.

LAB 8.3 The Bifurcation Diagram

When we encountered bifurcations for first-order differential equations, we saw that the bifurcation diagram (see Figures 1.83 and 1.85) provided a neat way to summarize the changes that occur in the solutions of the system. Here we introduce an analogous picture that captures the many more complicated bifurcations that occur in the logistic family $L_k(x) = kx(1 - x)$.

For a variety of k-values in the interval $1 \le k \le 4$, you should compute the orbit of the seed $x_0 = 0.5$. Then you will display the **asymptotic orbit** for each of these chosen k-values in the bifurcation diagram. By asymptotic orbit we mean the "tail" of the orbit. To be specific, for each chosen k-value, you should compute the first 100 points on the orbit, but you will display only the last 75 iterations. The first 25 points on the orbit should be disregarded so that you will see only the fate of the orbit.

In the bifurcation diagram you should plot the k-axis horizontally ($1 \le k \le 4$) and the x-axis ($0 \le x \le 1$) vertically. For each chosen k-value you should record all 75 points on the vertical line over the chosen k. In general, there are many fewer points than 75 to record. For example, if $k = 2.8$, we show that the orbit tends quickly to an attracting fixed point that is located at approximately $x = 0.64$. So, neglecting the first 25 numbers on the orbit, you would plot the point $(2.8, 0.64)$ to indicate the presence of this fixed point.

Similarly, if $k = 3.2$, the orbit tends to a 2-cycle located at $x = 0.513$ and $x = 0.799$, so you would plot $(3.2, 0.513)$ and $(3.2, 0.799)$. Finally, if $k = 3.9$, the histogram in Figure 8.5 indicates that the orbit is distributed throughout a large subinterval of $0 \le x \le 1$. Over $k = 3.9$ you might sketch an interval to indicate this chaotic behavior (see Figure 8.55). The bifurcation diagram thus gives a record of the fate of the orbit of 0.5 for a collection of k-values.

In your report you should first collect and display the fate of orbits for at least 50 different k-values chosen as follows:

1. Choose 5 values in the interval $1 \le k \le 3$.

2. Choose 10 values in the interval $3 < k \le 3.44$.

3. Choose 5 values in the interval $3.44 < k \le 3.55$.

4. Choose 5 values in the interval $3.55 < k \le 3.56$.

5. Choose a number of other values just above $k = 3.56$.

6. Choose 5 values in the interval $3.57 < k \le 3.83$.

Figure 8.55
The beginning of the bifurcation diagram for
$L_k(x) = kx(1 - x)$.

7. Choose 10 values in the interval $3.83 < k \leq 3.86$.

8. Choose your remaining k-values in the regime $3.86 \leq k \leq 4$.

Your Report: In your report discuss the qualitative behavior of the fate of orbits of the logistic family. Can you "fill in" the diagram for other values of k with $1 \leq k \leq 3$? How about $3 \leq k \leq 3.4$? Describe a magnification of the bifurcation diagram in the interval $3.83 \leq k \leq 3.86$. You may need to choose additional k-values to see the structure here. Why is this interval called the "period 3 window"?

Further Projects:

1. Compute the bifurcation diagram for the quadratic family $F_c(x) = x^2 + c$. You should choose the parameter c in the interval $-2 \leq c \leq 0.25$ and use the seed $x_0 = 0$ in this case. The orbit should be plotted in the interval $-2 \leq x \leq 2$. Describe the similarity with the logistic bifurcation diagram. Can you find a period 3 window here?

2. Repeat the previous investigation using the sine family $S_\lambda(x) = \lambda \sin x$. Choose λ in the interval $1 \leq \lambda \leq \pi$ and use the seed $x_0 = \pi/2 \approx 1.57$. Use the interval $0 \leq x \leq \pi$ to plot your orbits. Now compare all three bifurcation diagrams.

APPENDICES

APPENDIX A
CHANGING VARIABLES

APPENDIX B
POWER SERIES: THE ULTIMATE GUESS

APPENDIX C
COMPLEX NUMBERS AND EULER'S FORMULA

A CHANGING VARIABLES

In Chapter 1 we described three types of techniques for studying first-order differential equations: analytic, qualitative, and numerical techniques. Qualitative and numerical techniques apply to many different differential equations while each analytic technique starts by identifying a particular form of differential equation to which it applies.

Given a differential equation, it is unlikely that it is already in a form that is appropriate for any particular technique. Indeed we are lucky when we know an analytic technique that applies to the equation at hand. To increase our luck, we need to have many different analytic techniques available from which to choose.

Many analytic techniques are basically some sort of change of variables. In this appendix, we begin with a review of the technique of changing variables as it appears in calculus. Then we give examples of this same technique as it applies to differential equations.

u-Substitution

The idea behind changing variables is not new. We use it in calculus when we do "u-substitutions" to compute antiderivatives. For example, given the integral

$$\int t \sin t^2 \, dt,$$

we define a new variable u by $u = t^2$ and rewrite the integral in terms of u instead of t. Because $du/dt = 2t$ (sometimes informally written as $du = 2t \, dt$), we obtain

$$\int t \sin t^2 \, dt = \int \frac{1}{2} \sin u \, du.$$

This integral is the same one expressed in terms of the new variable u. The variable u was chosen so that the new integral is easy to evaluate. We have

$$\frac{1}{2} \int \sin u \, du = -\frac{\cos u}{2} + c.$$

We now find the solution to the original problem by replacing u with t^2, and we obtain

$$\int t \sin t^2 \, dt = -\frac{\cos t^2}{2} + c.$$

Changing variables from t to u makes a difficult computation easy. After computing the integral in terms of the new variable, we can recover the desired integral by replacing the new variable with the original one.

Changing Variables and Differential Equations

The technique of changing variables is not a "magic bullet" with which all differential equations can be solved. Finding a change of variables that makes the differential equation manageable depends both on the form of the equation and on the goal of the analysis.

An example

We can use the method of changing variables in concert with qualitative and numerical methods. For example, consider the very complicated equation

$$\frac{dy}{dt} = y^2 - 4ty + 4t^2 - 4y + 8t - 3.$$

This equation is neither linear nor separable, so we might try to look for a change of variables that simplifies the equation algebraically. To make an intelligent guess of a new dependent variable, we first rewrite the equation using some algebra. After staring at the right-hand side for a while, we see that we can make it look a little simpler by collecting terms and factoring, that is,

$$\frac{dy}{dt} = y^2 - 4ty + 4t^2 - 4y + 8t - 3$$

$$= (y - 2t)^2 - 4(y - 2t) - 3.$$

This new form of the right-hand side of the equation suggests a possible choice of a new dependent variable. Let

$$u = y - 2t.$$

This new dependent variable u is a combination of the old dependent variable y and the independent variable t. To replace all the occurrences of the old variable y with the new dependent variable u, we compute dy/dt in terms of u by differentiating $u = y - 2t$. We get

$$\frac{du}{dt} = \frac{dy}{dt} - 2.$$

Thus the equation

$$\frac{dy}{dt} = (y - 2t)^2 - 4(y - 2t) - 3$$

becomes

$$\frac{du}{dt} + 2 = u^2 - 4u - 3,$$

which simplifies to

$$\frac{du}{dt} = u^2 - 4u - 5.$$

The equation

$$\frac{du}{dt} = u^2 - 4u - 5$$

is autonomous, so we can study it by drawing its phase line. Because $u^2 - 4u - 5 = (u-5)(u+1)$, the equilibrium points are $u = 5$ and $u = -1$. The phase line and graphs of several solutions are shown in Figure A.1. The equilibrium point $u = -1$ is a sink, and the equilibrium point $u = 5$ is a source.

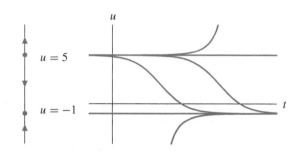

Figure A.1
Phase line and graphs of solutions for
$du/dt = u^2 - 4u - 5$.

What does this tell us about the original differential equation? Because the new equation is separable, we can find explicit solutions by integration. However, we can also obtain information about the solutions $y(t)$ even more directly. We know that $u_1(t) = -1$ is an equilibrium solution for the new equation, and any solution of the new equation corresponds to a solution of the original equation. Hence,

$$y_1(t) = u_1(t) + 2t = -1 + 2t$$

is a solution of the original equation.

Similarly, because $u_2(t) = 5$ is a solution of the new equation,

$$y_2(t) = u_2(t) + 2t = 5 + 2t$$

is a solution of the original equation. Every solution $u(t)$ of the new equation corresponds to a solution of the original equation by the change of variables $y(t) = u(t) + 2t$. Thus the graphs of solutions of the new equation in the tu-plane and the solutions of the original equation in the ty-plane are closely related. The ty-plane can be obtained from the tu-plane by adding $2t$ to every solution. Essentially we take the tu-plane and shear it upward with slope 2. The equilibrium solutions in u correspond to solutions whose graphs are lines with slope 2 in the ty-plane. Also, because $u = -1$ is a sink, solutions tend toward the solution $y_1(t) = -1 + 2t$ as t increases. Similarly, solutions tend away from the solution $y_2(t) = 5 + 2t$ because $u = 5$ is a source. So we may graph solutions in the ty-plane just from the corresponding graphs in the tu-plane (see Figure A.2).

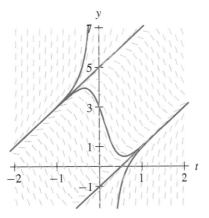

Figure A.2

The slope field and graphs of solutions for the equation

$$\frac{dy}{dt} = y^2 - 4ty + 4t^2 - 4y + 8t - 3.$$

Solutions tend toward the solution $y_1(t) = -1 + 2t$ as t increases. Similarly, solutions tend away from the solution $y_2(t) = 5 + 2t$.

A Linearization Problem

As we have noted, changing variables is not just a method for finding analytic solutions. It can also help to determine the qualitative behavior of solutions. For example, consider the logistic population model

$$\frac{dP}{dt} = 0.06P\left(1 - \frac{P}{500}\right),$$

where $P(t)$ is a population at time t measured in hundreds or thousands of individuals. Note that the growth-rate parameter is 0.06 and the carrying capacity is 500.

The equilibrium points are a source at $P = 0$ and a sink at $P = 500$. All solutions with positive initial conditions approach the sink $P = 500$ as $t \to \infty$ (see Figure A.3). We know that if P is small, the term $(1 - P/500)$ is close to 1 and solutions of this logistic equation are very close to solutions of the exponential growth model

$$\frac{dP}{dt} = 0.06P,$$

which have the form $P(t) = ke^{0.06t}$. So, for small P, the population grows at an exponential rate with a growth-rate parameter of 0.06.

After a long time, we expect that the population will be near $P = 500$. Suppose that some unexpected event (not included in the model) pushes the population away from $P = 500$. Such an event could be a period of uncharacteristically harsh weather, unlawful poaching, or the unexpected immigration of a small number of individuals. Any of these events will give a population near, but not at, $P = 500$. We know that, if conditions return to those of the model, then the population will again approach the carrying capacity of $P = 500$. A natural question is: How long will this recovery take? In other words, for $P(0)$ near 500, how will the solution behave?

The sort of answer we want is similar to the description of the behavior of small populations. We change variables so that the point $P = 500$ is moved to the origin. In the new variables, the behavior of solutions near the equilibrium point are easier to discover.

Figure A.3

Phase line for the familiar equation

$$\frac{dP}{dt} = 0.06P\left(1 - \frac{P}{500}\right).$$

Let $u = P - 500$. Then the new differential equation in terms of u is given by

$$\frac{du}{dt} = \frac{dP}{dt}$$

$$= 0.06P\left(1 - \frac{P}{500}\right)$$

$$= 0.06(u + 500)\left(1 - \frac{u + 500}{500}\right)$$

$$= 0.06(u + 500)\left(-\frac{u}{500}\right).$$

Collecting terms, this equation becomes

$$\frac{du}{dt} = -0.06u - 0.06\frac{u^2}{500}.$$

The equilibrium point $P = 500$ corresponds to the equilibrium point $u = 0$ in the new variable. For u very close to 0, the term containing u^2 is the square of a very small number. Hence, for u near 0, the behavior of solutions of this equation will be very close to the behavior of solutions of the equation where we neglect the u^2 term, that is,

$$\frac{du}{dt} = -0.06u.$$

Solutions of this equation are of the form $u(t) = ke^{-0.06t}$, so they decay toward $u = 0$ at an exponential rate with an exponent of $-0.06t$ (see Figure A.4). Because the new dependent variable u is just a translation of the original variable P, solutions for the logistic equation that have initial condition near $P = 500$ approach $P = 500$ at an exponential rate with an exponent of $-0.06t$ (see Figure A.5).

Figure A.4
Solutions of the equation
$du/dt = -0.06u$.

Figure A.5
Solutions of the equation
$dP/dt = 0.06P(1 - P/500)$ with initial
condition near $P = 500$.

Linearization

In general, to find the behavior of a solution near an equilibrium point, we first change variables moving that equilibrium point to the origin. Next we neglect all the "nonlinear" terms in the equation. Since we are only concerned with values of the new variable near 0, any higher power of that variable will be so close to 0 that it can be safely neglected, at least on first approximation. The equation that results is linear and hence easy to analyze.

This method involves an approximation. Thus, it is valid only for values of the variable where that approximation is reasonable. Once the new variable becomes large, the nonlinear terms in the equation become significant and cannot be ignored. Exactly how small is small and how large is large depends on the particular equation considered (see Exercises 17–20).

This technique is called **linearization**. The idea is to approximate a complicated equation with a simpler, linear equation. Hopefully, the linear equation will be simple enough so that we understand the behavior of solutions. Linearization is an important tool in our study of systems of differential equations, and we discuss it at length in Section 5.1.

The Bernoulli and Riccati Equations

Differential equations is an old subject. Over the centuries many differential equations have been solved analytically by clever changes of variables. Each of these special changes of variables can be a lifesaver, if your life ever depends on finding the general solution of the differential equation to which it applies. Since you never know what sorts of odd turns life might take, it is good to know lots of changes of variables.

For example, suppose you need to find the general solution of

$$\frac{dy}{dt} = r(t)y + a(t)y^n$$

for some functions $r(t)$ and $a(t)$ and some integer n. An equation of this form is called a **Bernoulli equation**. If $n = 1$, the equation is linear and separable, so you already know methods to solve the equation. If $n \geq 2$, the equation is nonlinear and you need a miracle.

The miracle comes in the form of a clever change of variables. Let

$$z = y^{1-n}.$$

Then

$$\frac{dz}{dt} = (1-n)y^{-n}\left(\frac{dy}{dt}\right)$$

$$= (1-n)y^{-n}\left(r(t)y + a(t)y^n\right)$$

$$= (1-n)\left(r(t)y^{1-n} + a(t)\right)$$

$$= (1-n)\left(r(t)z + a(t)\right).$$

The differential equation with respect to the new dependent variable z is

$$\frac{dz}{dt} = (1-n)\left(r(t)z + a(t)\right).$$

This equation is linear, so we have a chance of finding the general solution by the techniques of Sections 1.8 and 1.9.

For example, consider the Bernoulli equation

$$\frac{dy}{dt} = y + e^{-2t}y^2.$$

First, we note that the equilibrium solution $y(t) = 0$ for all t is a solution to this equation. So we assume that $y \neq 0$ when we change variables.

For this equation, $n = 2$, so we make the substitution $z = y^{-1}$. We obtain

$$\frac{dz}{dt} = -z - e^{-2t}.$$

Using the Extended Linearity Principle and the guessing technique described in Section 1.8, we calculate the general solution

$$z(t) = ke^{-t} + e^{-2t},$$

where k is an arbitrary constant.

Reversing the change of variables, we see that the general solution of the original Bernoulli equation $dy/dt = y + e^{-2t}y^2$ is

$$y(t) = (z(t))^{-1} = \frac{1}{ke^{-t} + e^{-2t}}$$

along with the equilibrium solution $y(t) = 0$ for all t.

Where do Bernoulli equations come from?

It is natural to ask where a Bernoulli equation might arise. Given the most general first-order differential equation

$$\frac{dy}{dt} = f(t, y),$$

we have no reason to hope that the function $f(t, y)$ has one of the special forms that lets us find a formula for the general solution. We can approximate solutions with qualitative and numerical techniques. In some applications it is useful to approximate the solutions using analytic methods as well.

If we cannot solve the given equation, we settle for the next best thing—solving an approximation to the equation. The type of approximation we use depends on the problem at hand. For example, if we are most concerned with the solution near $y = 0$, then it is natural to approximate the function $f(t, y)$ with its Taylor series in y centered at $y = 0$ (its Maclaurin series). This Taylor expansion has the form

$$f(t, y) \approx f(t, 0) + \left(\frac{\partial f}{\partial y}(t, 0)\right)y + \frac{1}{2!}\left(\frac{\partial^2 f}{\partial y^2}(t, 0)\right)y^2 + \frac{1}{3!}\left(\frac{\partial^3 f}{\partial y^3}(t, 0)\right)y^3 + \dots.$$

The coefficients of the powers of y are functions of t.

With this expansion, the first-order approximation of $dy/dt = f(t, y)$ has the form

$$\frac{dy}{dt} = r(t) + a(t)y$$

where

$$r(t) = f(t, 0) \quad \text{and} \quad a(t) = \frac{\partial f}{\partial y}(t, 0).$$

This first-order approximation is a linear equation, and we use the techniques of Sections 1.8 and 1.9 to find the general solution.

If more accuracy is required, we can try the second-order approximation

$$\frac{dy}{dt} = r(t) + a(t)y + b(t)y^2,$$

where

$$r(t) = f(t, 0), \quad a(t) = \frac{\partial f}{\partial y}(t, 0), \quad \text{and} \quad b(t) = \frac{1}{2!}\left(\frac{\partial^2 f}{\partial y^2}(t, 0)\right).$$

Equations of this form are called **Riccati equations**.

There is no general method for solving a Riccati equation. However, if we happen to know one particular solution, then it is possible to find the general solution. To see why, suppose $y_1(t)$ is a solution of

$$\frac{dy}{dt} = r(t) + a(t)y + b(t)y^2,$$

and define a new dependent variable w by $w = y - y_1$. We have

$$\frac{dw}{dt} = \frac{dy}{dt} - \frac{dy_1}{dt}$$

$$= \left(r(t) + a(t)y + b(t)y^2\right) - \left(r(t) + a(t)y_1 + b(t)y_1^2\right)$$

$$= a(t)(y - y_1) + b(t)(y^2 - y_1^2).$$

Now we need to express the right-hand side of this equation in terms of w rather than y. Using the fact that $y^2 - y_1^2 = (y - y_1)(y + y_1)$, we have

$$\frac{dw}{dt} = a(t)(y - y_1) + b(t)(y - y_1)(y + y_1)$$

$$= a(t)(y - y_1) + b(t)(y - y_1)(y - y_1 + 2y_1)$$

$$= a(t)(w) + b(t)(w)(w + 2y_1).$$

If we group terms by powers of w and recall that y_1 is really one given solution $y_1(t)$, we have the new differential equation

$$\frac{dw}{dt} = (a(t) + 2b(t)y_1(t))\, w + b(t)w^2.$$

But wait, another miracle has occurred. This equation is a Bernoulli equation, and we already know how to try to solve Bernoulli equations.

Solving a Riccati equation with this approach involves five steps. First, find a particular solution somehow. Second, use that particular solution to change to a new dependent variable, obtaining a Bernoulli equation in the new variable. Third, change to another new dependent variable to obtain a linear equation. Fourth, solve the linear equation if possible. Fifth, reverse the two changes of variables to get the general solution of the original Riccati equation in the original variables.

Avoiding depression

As mentioned earlier, there are many special purpose changes of variables that lead to the general solution of differential equations of special forms. Studying these special cases can be depressing. While it is possible to follow the steps, you may be haunted by the question "How could I ever have found this myself?"

Sometimes there are hints in the differential equation itself. For example, in a Bernoulli equation

$$\frac{dy}{dt} = r(t)y + a(t)y^n,$$

the annoying part of the equation is the y^n term. You could try defining a new dependent variable by $z = y^n$, but that just moves the exponent of n someplace else. However, if you try $z = y^\alpha$ and look for the choice of α that simplifies the equation the most, you end up with $\alpha = 1 - n$ (see Exercise 30). Still, it is difficult to imagine making the correct guess $z = y^\alpha$ on your first attempt.

It is important to remember that differential equations have been studied by a great many people over the last 300 years, and during most of that time there were no TVs, no cell phones, and no Internet. Given 300 years with nothing else to do, you might get quite adept at changing variables.

EXERCISES FOR APPENDIX A

In Exercises 1–4, change the dependent variable from y to u using the change of variables indicated. Describe the equation in the new variable (separable, linear, ...).

1. $\dfrac{dy}{dt} = y - 4t + y^2 - 8yt + 16t^2 + 4,$ let $u = y - 4t$

2. $\dfrac{dy}{dt} = \dfrac{y^2 + ty}{y^2 + 3t^2},$ let $u = \dfrac{y}{t}$

3. $\dfrac{dy}{dt} = t(y + ty^2) + \cos ty,$ let $u = ty$

4. $\dfrac{dy}{dt} = e^y + \dfrac{t^2}{e^y},$ let $u = e^y$

In Exercises 5–7, find a change of variables that transforms the equation into an autonomous equation. Sketch the phase line for the equation that you obtain and use it to sketch the graphs of solutions for the original equation.

5. $\dfrac{dy}{dt} = (y - t)^2 - (y - t) - 1$ **6.** $\dfrac{dy}{dt} = \dfrac{y^2}{t} + 2y - 4t + \dfrac{y}{t}$

7. $\dfrac{dy}{dt} = y \cos ty - \dfrac{y}{t}$

In Exercises 8–10, change variables as indicated and then find the general solution of the resulting equation. Use this information to give the general solution of the original equation.

8. $\dfrac{dy}{dt} = \dfrac{ty}{2} + \dfrac{e^{t^2/2}}{2y}$,   let $y = \sqrt{u}$

9. $\dfrac{dy}{dt} = \dfrac{y}{1+t} - \dfrac{y}{t} + t^2(1 + t)$,   let $u = \dfrac{y}{1+t}$

10. $\dfrac{dy}{dt} = y^2 - 2yt + t^2 + y - t + 1$,   let $u = y - t$

In Exercises 11–13, create your own change of variables problems. Starting with the simple equation $du/dt = (1 - u)u$, change variables as indicated. Sketch the graphs of solutions of the new equations using what you know about the solutions of $du/dt = (1 - u)u$.

11. $y = u + t$ **12.** $y = \sqrt{u}$ **13.** $y = u^2$

14. Consider a 20-gallon vat that at time $t = 0$ contains 5 gallons of clean water. Suppose water enters the vat from two pipes. From the first pipe, salt water containing 2 pounds of salt per gallon enters the vat at a rate of 3 gallons per minute. From the second pipe, salt water containing 0.5 pounds of salt per gallon enters the vat at a rate of 4 gallons per minute. Suppose the water is kept well mixed and salt water is removed from the vat at a rate of 2 gallons per minute.

   (a) Derive a differential equation for the rate of change of the total amount of salt $S(t)$ in the vat at time t.

   (b) Convert this equation to a differential equation for the concentration $C(t)$ of salt in the vat at time t.

   (c) Find the concentration of salt in the vat at the instant when the vat first starts to overflow.

15. Consider a very large vat that initially contains 10 gallons of clean water. Suppose water enters the vat from two pipes. From the first pipe, salt water containing 2 pounds of salt per gallon enters the vat at a rate of 1 gallon per minute. From the second pipe, salt water containing 0.2 pounds of salt per gallon enters the vat at a rate of 5 gallons per minute. Suppose the liquid is kept well mixed and salt water is removed from the vat at a rate of 3 gallons per minute.

(a) Derive a differential equation for the total amount of salt $S(t)$ in the vat at time t.

(b) Convert this equation to a differential equation for the concentration $C(t)$ of salt in the vat at time t.

(c) What will the concentration of salt in the vat be after a very long time?

(d) Find the concentration of salt in the vat at the time $t = 5$.

16. Given a differential equation of the form

$$\frac{dy}{dt} = g\left(\frac{y}{t}\right),$$

show that the change of variables $u = y/t$ gives a separable equation for du/dt. (First-order differential equations of this type are sometimes called homogeneous. This use of the term homogeneous is not the same as its use with linear equations as in Section 1.8.)

In Exercises 17–20:

(a) Identify the equilibrium points.

(b) For each equilibrium point, give the change of variables that moves the equilibrium point to the origin and give the linear approximation of the new system near the origin. Identify the equilibrium point as a source, a sink, or a node.

(c) For each equilibrium point, describe quantitatively the behavior of solutions with initial values that are very close to the equilibrium point. (For example, how quickly does a solution that starts close to the equilibrium point approach or recede from the equilibrium point as t increases?) In particular, approximate the amount of time it takes a point to halve or double its distance from the equilibrium point as t increases.

17. $\dfrac{dy}{dt} = 10y^3 - 1$

18. $\dfrac{dy}{dt} = (y + 1)(y - 3)$

19. $\dfrac{dy}{dt} = (y + 1)(3 - y)$

20. $\dfrac{dy}{dt} = y^3 - 3y^2 + y$

21. Consider the differential equation $dy/dt = f(y)$, where $f(y)$ is a smooth function (it can be differentiated as many times as we like). Suppose $y = y_0$ is an equilibrium point for this equation.

(a) Let $u = y - y_0$ and write the differential equation in terms of the new dependent variable u.

(b) Show that the linear approximation of the differential equation for du/dt near the equilibrium point at $u = 0$ is given by

$$\frac{du}{dt} = f'(y_0)u.$$

22. So far we have changed only the dependent variable. It is also possible to change the independent variable to obtain a new equation. This is analogous to changing the way we measure time. Consider the differential equation

$$\frac{dy}{dt} = 3t^2 y + 3t^5.$$

This differential equation could be solved using an integrating factor, but we will use a change of variables. Suppose we define a new time variable (a new independent variable) using the formula $s = t^3$. Thus $t = \sqrt[3]{s}$.

   (a) Show that $dy/ds = y + s$.

   (b) Find the general solution of the differential equation with the variable s.

   (c) What is the general solution of the original differential equation?

In Exercises 23–26, find the general solution of the differential equation specified.

23. $\dfrac{dy}{dt} = y + y^3$ **24.** $\dfrac{dy}{dt} = y + ty^3$

25. $\dfrac{dy}{dt} = \dfrac{1}{t}y - y^4$ **26.** $\dfrac{dy}{dt} = y + y^{50}$

In Exercises 27–29, a differential equation and a function $y_1(t)$ are specified.

   (a) Check that $y_1(t)$ is a solution of the equation.

   (b) Using $y_1(t)$, find the general solution of the equation.

27. $\dfrac{dy}{dt} = (2t + 1/t)y - y^2 - t^2, \quad y_1(t) = t$

28. $\dfrac{dy}{dt} = (t^4 - t^2 + 2t) + (1 - 2t^2)y + y^2, \quad y_1(t) = t^2$

29. $\dfrac{dy}{dt} = 2\sin t + \cos t + t^2 \sin^2 t - 2(1 + t^2 \sin t)y + t^2 y^2, \quad y_1(t) = \sin t$

30. Suppose we must solve a Bernoulli equation

$$\frac{dy}{dt} = r(t)y + a(t)y^n,$$

but we have forgotten the change of variables formula that converts it to a linear equation. Suppose all we remember is that the change of variables has the form

$$z = y^\alpha,$$

where α is a constant that depends on n. Use this change of variables to show that $\alpha = 1 - n$ is the only choice that gives a linear equation for dz/dt.

31. In the discussion of the Riccati equation $dy/dt = r(t) + a(t)y + b(t)y^2$ given in this appendix, we did one change of variables to convert to a Bernoulli equation and then another change of variables to obtain a linear equation. It is possible to combine the two into a single change of variables. If $y_1(t)$ is a particular solution of the Riccati equation above, show that changing to the new dependent variable

$$u(t) = \frac{1}{y(t) - y_1(t)}$$

yields a linear equation for du/dt.

32. Construct your own Bernoulli equation exercise of the form $dy/dt = a(t)y + b(t)y^5$ from the linear equation $dz/dt = z + e^{2t}$.

33. Construct your own Riccati equation exercise from the function $y_1(t) = t^2 + 2$ and the Bernoulli equation $dw/dt = w + t^2w^2$.

B POWER SERIES: THE ULTIMATE GUESS

One of our most commonly used methods for finding formulas for solutions of differential equations has been picking a reasonable guess for a solution which is then substituted into the equation to see if it is indeed a solution. We improve our chances of success by including unknown constants whose values are determined as part of the process.

The only difficulty with this method is deciding what to guess. It would be sweet if there was a guess that always, or almost always, works. In this appendix we present such a guess.

A First-Order Example

Consider the equation

$$\frac{dy}{dt} = ty + 1.$$

This is a linear equation, and we could try to solve it using an integrating factor (see Section 1.9). However, the integrals involved are problematic, so we would rather guess. But what should we guess? No function $y(t)$ jumps to mind.

Since we do not have any ideas regarding what guess to make, we make a guess that does not require any ideas. We know that almost all nice functions encountered in calculus have Taylor series, so we guess

$$y(t) = a_0 + a_1t + a_2t^2 + \ldots = \sum_{n=0}^{\infty} a_nt^n.$$

That is, we guess the Taylor series for $y(t)$ centered at $t = 0$ (a Maclaurin series).

As you remember from calculus, an infinite sum is really a limit and the first question with any limit is "Does it exist?". In this appendix we answer all convergence questions with "No comment." Ignoring questions of convergence is often done in this kind of analysis. In fact, there is even a name for it. It is called *formal analysis*. We essentially treat power series as polynomials with a really large degree. This kind of analysis is like skipping dinner and eating double dessert. It tastes good, but it isn't very healthy. There are theorems that justify this formal analysis for most reasonable differential equations including the examples that follow. However, there is also some danger, so if your life (or someone else's) depends on a power series calculation, you should consult an advanced calculus or real analysis textbook for the appropriate convergence result.

Given that disclaimer, we proceed with this power series as our guess. First, we compute the derivative of the power series by differentiating term by term. We get

$$\frac{dy}{dt} = a_1 + 2a_2t + 3a_3t^2 + \ldots = \sum_{n=1}^{\infty} na_n t^n.$$

So the differential equation $dy/dt = ty + 1$ becomes

$$a_1 + 2a_2t + 3a_3t^2 + 4a_4t^3 + 5a_5t^4 + \ldots$$
$$= t(a_0 + a_1t + a_2t^2 + a_3t^3 + a_4t^4 + \ldots) + 1,$$

and simplifying the right-hand side, we obtain

$$a_1 + 2a_2t + 3a_3t^2 + 4a_4t^3 + 5a_5t^4 + \ldots$$
$$= 1 + a_0t + a_1t^2 + a_2t^3 + a_3t^4 + a_4t^5 + \ldots.$$

These two series are equal if and only if the corresponding coefficients are equal (see Exercise 16). Hence, we can think of the equality of these two power series as the infinite list of equations

$$a_1 = 1$$
$$2a_2 = a_0$$
$$3a_3 = a_1$$
$$4a_4 = a_2$$
$$5a_5 = a_3$$
$$\vdots$$

with infinitely many unknowns.

Has guessing a power series helped? Basically we have replaced the differential equation for the function $y(t)$ with infinitely many algebraic equations for the infinitely many unknown coefficients of the power series. At first glance, we seem to have replaced one problem with another. However, looking closely at the list of equations for the coefficients, we notice something nice. We do not have to solve the entire infinite system of equations simultaneously. Rather, we can start at the top and work down, solving each successive equation in terms of the previous solutions. That is,

$$a_1 = 1;$$

$$2a_2 = a_0, \quad \text{which implies that} \quad a_2 = \frac{a_0}{2};$$

$$3a_3 = a_1, \quad \text{which implies that} \quad a_3 = \frac{1}{3};$$

$$4a_4 = a_2 = \frac{a_0}{2}, \quad \text{which implies that} \quad a_4 = \frac{a_0}{2 \cdot 4};$$

$$5a_5 = a_3 = \frac{1}{3}, \quad \text{which implies that} \quad a_5 = \frac{1}{3 \cdot 5};$$

$$6a_6 = a_4 = \frac{a_0}{2 \cdot 4}, \quad \text{which implies that} \quad a_6 = \frac{a_0}{2 \cdot 4 \cdot 6};$$

$$\vdots$$

Therefore, the solution $y(t)$ has the form

$$y(t) = a_0 + t + \frac{a_0}{2} t^2 + \frac{1}{3} t^3 + \frac{a_0}{2 \cdot 4} t^4 + \frac{1}{3 \cdot 5} t^5 + \frac{a_0}{2 \cdot 4 \cdot 6} t^6 + \dots$$

$$= a_0 \left(1 + \frac{1}{2} t^2 + \frac{1}{2 \cdot 4} t^4 + \frac{1}{2 \cdot 4 \cdot 6} t^6 + \dots \right) + \left(t + \frac{1}{3} t^3 + \frac{1}{3 \cdot 5} t^5 + \dots \right).$$

Since $y(0) = a_0$, we see that this power series is the general solution of the differential equation. The "..." represent higher degree terms that we could compute if we had the patience. We can think of this power series representation of the solution as giving a sequence of polynomials

$$y(t) \approx y_0(t) = a_0$$

$$y(t) \approx y_1(t) = a_0 + t$$

$$y(t) \approx y_2(t) = a_0 + t + \frac{a_0}{2} t^2$$

$$y(t) \approx y_3(t) = a_0 + t + \frac{a_0}{2} t^2 + \frac{1}{3} t^3$$

$$y(t) \approx y_4(t) = a_0 + t + \frac{a_0}{2} t^2 + \frac{1}{3} t^3 + \frac{a_0}{2 \cdot 4} t^4$$

$$\vdots$$

which form better and better approximations of the actual solution (see Figure B.1). This convergence is just like saying $\sqrt{2} \approx 1.4$, $\sqrt{2} \approx 1.41$, $\sqrt{2} \approx 1.414$, ... Each additional term makes the approximation a bit more accurate near $t = 0$. We graph these

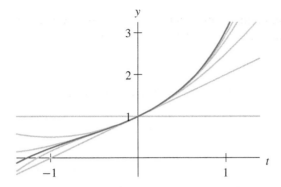

Figure B.1
The graph of the solution to the initial-value problem

$$\frac{dy}{dt} = ty + 1, \quad y(0) = 1,$$

is shown in dark blue. The other graphs are the graphs of the Taylor polynomial approximations to this solution of degrees zero through four.

Taylor approximations for $y(0) = a_0 = 1$ along with the actual solution in Figure B.1. Note that, like most Taylor polynomials, the approximations are accurate near $t = 0$ but not even close far from this point.

A Second-Order Example: Hermite's Equation

The power series technique is particularly useful on certain second-order equations that arise in physics. These equations are linear but do not have constant coefficients.

For example, consider the second-order equation

$$\frac{d^2y}{dt^2} - 2t\frac{dy}{dt} + 2py = 0,$$

where p is a parameter. This equation is called **Hermite's equation**. The coefficient $-2t$ of dy/dt is not a constant, so our usual guess-and-test technique for constant-coefficient equations does not work (try it and see).

Since no better guesses come to mind, we try the guess of last resort, that is, we guess the power series

$$y(t) = a_0 + a_1 t + a_2 t^2 + a_3 t^3 + \ldots = \sum_{n=0}^{\infty} a_n t^n.$$

To try this quess, we need

$$\frac{dy}{dt} = a_1 + 2a_2 t + 3a_3 t^2 + 4a_4 t^3 + 5a_5 t^4 + \ldots = \sum_{n=1}^{\infty} n a_n t^{n-1}$$

and

$$\frac{d^2y}{dt^2} = 2a_2 + (3 \cdot 2)a_3 t + (4 \cdot 3)a_4 t^2 + (5 \cdot 4)a_5 t^3 + \ldots = \sum_{n=2}^{\infty} n(n-1)a_n t^{n-2}.$$

Note that $y(0) = a_0$ and $y'(0) = a_1$, so we can think of the first two coefficients as the initial condition for an initial-value problem.

Substituting these series into Hermite's equation

$$\frac{d^2y}{dt^2} - 2t\frac{dy}{dt} + 2py = 0,$$

we get

$$\left(2a_2 + (3\cdot2)a_3t + (4\cdot3)a_4t^2 + (5\cdot4)a_5t^3 + \ldots\right)$$
$$-2t\left(a_1 + 2a_2t + 3a_3t^2 + 4a_4t^3 + \ldots\right)$$
$$+2p\left(a_0 + a_1t + a_2t^2 + a_3t^3 + \ldots\right) = 0.$$

Collecting terms on the left-hand side yields the power series

$$(2pa_0 + 2a_2)$$
$$+ (2pa_1 - 2a_1 + 6a_3)\,t$$
$$+ (2pa_2 - 4a_2 + 12a_4)\,t^2$$
$$+ (2pa_3 - 6a_3 + 20a_5)\,t^3$$
$$+ \ldots = 0$$

In order for this power series to be identically zero, all of the coefficients must be zero (see Exercise 16). Consequently, solving the differential equation turns into the problem of solving the infinite family of algebraic equations with infinitely many unknowns

$$2pa_0 + 2a_2 = 0$$
$$2(p-1)a_1 + 6a_3 = 0$$
$$2(p-2)a_2 + 12a_4 = 0$$
$$2(p-3)a_3 + 20a_5 = 0$$
$$\vdots$$

Because $a_0 = y(0)$ and $a_1 = y'(0)$ are determined by the initial condition, we can solve this list of equations starting from the top to obtain

$$a_2 = -p\,a_0$$
$$a_3 = -\frac{p-1}{3}\,a_1$$
$$a_4 = -\frac{p-2}{6}\,a_2 = \frac{(p-2)p}{6}\,a_0$$
$$a_5 = -\frac{p-3}{10}\,a_3 = \frac{(p-3)(p-1)}{30}\,a_1$$
$$\vdots$$

Note that the coefficients with even subscripts are multiples of a_0 while those with odd subscripts are multiples of a_1. The resulting general solution is

$$y(t) = a_0 + a_1 t - p a_0 t^2 - \frac{p-1}{3} a_1 t^3 + \frac{(p-2)p}{6} a_0 t^4 + \frac{(p-3)(p-1)}{30} a_1 t^5 - \cdots$$

$$= a_0 \left(1 - p t^2 + \frac{(p-2)p}{6} t^4 \mp \cdots \right)$$

$$+ a_1 \left(t - \frac{p-1}{3} t^3 + \frac{(p-3)(p-1)}{30} t^5 \mp \cdots \right).$$

With this series representation of the general solution of Hermite's equation, we are tempted to think that our discussion of this example is complete. However there are some important special cases that we should consider.

The parameter p is a nonnegative integer

Suppose the parameter p is a nonnegative integer. For example, consider $p = 0$. Then all of the coefficients having p as a factor are zero, and we have

$$y(t) = a_0 \left(1 - 0 \cdot t^2 + 0 \cdot t^4 - \cdots \right) + a_1 \left(t - \frac{0-1}{3} t^3 + \frac{(0-3)(0-1)}{30} t^5 - \cdots \right).$$

Moreover, if we specify the initial condition $y(0) = a_0 = 1$ and $y'(0) = a_1 = 0$, we get that the constant function

$$y(t) = 1$$

is the solution to Hermite's equation with $p = 0$ and $(y(0), y'(0)) = (1, 0)$.

Next we consider $p = 1$. Then all of the terms having $p - 1$ as a factor are zero, and we get

$$y(t) = a_0 \left(1 - t^2 + \frac{(1-2)p}{6} t^4 - \cdots \right) + a_1 \left(t - 0 t^3 + 0 t^5 - \cdots \right),$$

Taking $y(0) = a_0 = 0$ and $y'(0) = a_1 = 1$, we get that the function

$$y(t) = t$$

is the solution of Hermite's equation with $p = 1$ and $(y(0), y'(0)) = (0, 1)$.

Taking $p = 2$, we see that all of the coefficients having $p - 2$ as a factor equal zero. Again taking $y(0) = a_0 = 1$ and $y'(0) = a_1 = 0$, we get that the function

$$y(t) = 1 - 2t^2$$

is the solution of Hermite's equation with $p = 2$ and $(y(0), y'(0)) = (1, 0)$.

This pattern continues. If p is a positive even integer and $(y(0), y'(0)) = (1, 0)$, then the resulting Taylor series has only finitely many nonzero terms. The same holds if p is a positive odd integer and $(y(0), y'(0)) = (0, 1)$. These solutions are called the **Hermite polynomials** $H_p(t)$. We have shown that the first three Hermite polynomials are $H_0(t) = 1$, $H_1(t) = t$, and $H_2(t) = 1 - 2t^2$, and we can easily generate many more.

Our analysis of the Hermite's equation can be compared to our study of constant-coefficient, second-order equations such as

$$\frac{d^2y}{dt^2} + n^2 y = 0.$$

The solutions of this constant-coefficient equation are linear combinations of $\sin nt$ and $\cos nt$. The Hermite polynomials share many properties with the functions $\sin nt$ and $\cos nt$, and both families of functions appear frequently in applications. Legendre's equation, another equation with similar properties, is studied in Exercise 15.

EXERCISES FOR APPENDIX B

In Exercises 1–4, use the guess-and-test method to find the power series expansion centered at $t = 0$ for the general solution up to degree four, that is, up to and including the t^4 term. (You may find the general solution using other methods and then find the Taylor series centered at $t = 0$ to check your computation if you like.)

1. $\dfrac{dy}{dt} = y$

2. $\dfrac{dy}{dt} = -y + 1$

3. $\dfrac{dy}{dt} = -2ty$

4. $\dfrac{dy}{dt} = t^2 y + 1$

In Exercise 5–8, find the power series expansion for the general solution up to degree four, that is, up to and including the t^4 term.

5. $\dfrac{dy}{dt} = -y + e^{2t}$

6. $\dfrac{dy}{dt} = 2y + \sin t$

7. $\dfrac{d^2y}{dt^2} + 2y = \cos t$

8. $\dfrac{d^2y}{dt^2} + 5\dfrac{dy}{dt} + y = \sin 2t$

9. Verify that $y(t) = \tan t$ is a solution of

$$\frac{dy}{dt} = y^2 + 1,$$

and compute a power series solution to find the terms up to degree six (up to and including the t^6 term) of the Taylor series centered at $t = 0$ of $\tan t$.

In Exercises 10–13, find the general solution up to degree six, that is, up to and including the t^6 term.

10. $\dfrac{d^2y}{dt^2} + 2y = 0$

11. $\dfrac{d^2y}{dt^2} + \dfrac{dy}{dt} + y = 0$

12. $\dfrac{d^2y}{dt^2} + \dfrac{dy}{dt} + t^2 y = \cos t$

13. $\dfrac{d^2y}{dt^2} + t\dfrac{dy}{dt} + y = e^{-2t}$

14. Compute the Hermite polynomials $H_3(t)$, $H_4(t)$, and $H_5(t)$. Check that they are solutions of the appropriate Hermite's equation.

15. Legendre's equation is the second-order differential equation

$$(1 - t^2)\frac{d^2 y}{dt^2} - 2t\frac{dy}{dt} + \nu(\nu + 1)y = 0,$$

where ν is a constant. Guess the Taylor series centered at $t = 0$ as a solution $y(t)$, that is,

$$y(t) = a_0 + a_1 t + a_2 t^2 + a_3 t^3 + a_4 t^4 + \ldots.$$

(a) Compute the coefficients a_2, a_3, and a_4 in terms of the initial conditions $a_0 = y(0)$ and $a_1 = y'(0)$.

(b) By choosing ν to be a positive integer and using the initial conditions $(y(0), y'(0)) = (1, 0)$ or $(y(0), y'(0)) = (0, 1)$, show that there are polynomial solutions $P_\nu(t)$ of Legendre's equation.

(c) Show that the first three of these polynomials are $P_0(t) = 1$, $P_1(t) = t$, and $P_2(t) = 1 - 3t^2$.

(d) Compute $P_\nu(t)$ for $\nu = 3, 4, 5$, and 6.

(e) Verify that $k P_\nu(t)$ is a solution of the same Legendre's equation as $P_\nu(t)$ if k is a constant. (Consequently, $P_\nu(t)$ is only determined up to a constant. For example, $P_2(t)$ is sometimes given as $P_2(t) = (3t^2 - 1)/2$.)

16. In this appendix we used the fact that two power series in t are equal if and only if all of their coefficients are equal. To justify this statement, let

$$f(t) = a_0 + a_1 t + a_2 t^2 + a_3 t^3 + \ldots$$

and

$$g(t) = b_0 + b_1 t + b_2 t^2 + b_3 t^3 + \ldots$$

and suppose that $f(t) = g(t)$ for all t for which both sides are defined.

(a) Verify that $a_0 = b_0$.

(b) Compute $f'(t)$ and $g'(t)$. Since $f(t) = g(t)$, we must have $f'(t) = g'(t)$ for all t. Verify that $a_1 = b_1$.

(c) Explain why $a_n = b_n$ for all n.

17. Consider the linear differential equation

$$\frac{dy}{dt} = -y + e^{-t}.$$

The natural first guess for a particular solution of this equation is $y(t) = ae^{-t}$, but this guess does not yield a solution because it is a solution of the associated homogeneous equation.

Guess a power series solution $y(t) = a_0 + a_1 t + a_2 t^2 + a_3 t^3 + \ldots$ and use the initial condition $y(0) = a_0 = 0$ to find a particular solution. Check that this series is the power series for te^{-t}.

18. Consider the linear differential equation

$$\frac{d^2y}{dt^2} + 4y = \cos 2t.$$

The natural first guess for a particular solution of this equation is

$$y_p(t) = a \sin 2t + b \cos 2t,$$

but this guess does not yield a solution because it is already a solution of the associated homogeneous equation.

Guess a power series $y(t) = a_0 + a_1 t + a_2 t^2 + a_3 t^3 + a_4 t^4 + a_5 t^5 + \ldots$ and use this guess to find the solution with $y(0) = a_0 = 0$ and $y'(0) = a_1 = 1$. Identify the series you obtain as the Taylor series centered at $t = 0$ of a familiar function.

C COMPLEX NUMBERS AND EULER'S FORMULA

This appendix is a summary of some of the basic properties of complex numbers that we use in this text. Complex numbers and functions arise quite often when we attempt to solve differential equations, even if the equations involve only real numbers. For instance, complex numbers may appear as eigenvalues for linear systems or as roots of the characteristic equation of a second-order equation. Complex functions also appear as solutions to differential equations when we use Euler's formula.

Complex Numbers

A complex number is a number of the form $a + bi$, where a and b are real numbers and i is the "imaginary" number $\sqrt{-1}$. Equivalently, i is the number whose square is -1, that is, $i^2 = -1$. For example, $2 + 3i$, $-\pi - i$, and $\sqrt{7}\,i$ are all complex numbers. We may also regard the real number 7 as a complex number by writing it in the form $7 + 0i$.

For a complex number $z = a + bi$, the real number a is called the **real part** of z, and real number b is called the **imaginary part** of z. Note that we do not include the i in the imaginary part of a complex number. Thus, *the imaginary part of a complex number is a real number.*

The usual rules of arithmetic apply to complex numbers. For example, we may add two complex numbers by simply adding the corresponding real and imaginary parts. Therefore

$$(1 + 2i) + (3 + 4i) = (1 + 3) + (2 + 4)i = 4 + 6i.$$

Multiplication also obeys the usual rules of arithmetic. For example,

$$(7 + 2i)(3 + 4i) = 7 \cdot 3 + 7 \cdot 4i + 2i \cdot 3 + 2i \cdot 4i$$
$$= (7 \cdot 3 - 2 \cdot 4) + (7 \cdot 4 + 2 \cdot 3)\,i$$

because $i^2 = -1$, and the result is $13 + 34i$. An easier product to compute is

$$7i(1 + 2i) = -14 + 7i.$$

Division of complex numbers is a little trickier. For example, to compute the quotient

$$\frac{2+3i}{3+4i},$$

we first multiply numerator and denominator by $3-4i$, obtaining

$$\frac{2+3i}{3+4i} = \left(\frac{2+3i}{3+4i}\right)\left(\frac{3-4i}{3-4i}\right).$$

The denominator becomes

$$(3+4i)(3-4i) = 3^2 + 4^2 = 25,$$

and the numerator is

$$(2+3i)(3-4i) = 18 + i.$$

Therefore

$$\frac{2+3i}{3+4i} = \frac{18+i}{25} = \frac{18}{25} + \frac{1}{25}i.$$

In general, when computing $(a+bi)/(c+di)$, we multiply numerator and denominator by $c-di$. Then the denominator becomes a real number, since

$$(c+di)(c-di) = c^2 + d^2 + 0i.$$

The complex number $c-di$ is called the **complex conjugate** of $c+di$. Another example of division is

$$\frac{1+i}{1-i} = \left(\frac{1+i}{1-i}\right)\left(\frac{1+i}{1+i}\right) = \frac{2i}{2} = i.$$

Geometry of Complex Numbers

Complex numbers do not lie on the real line. Instead, they reside naturally in a plane. We think of the x-axis as the **real axis** in the plane and the y-axis as the **imaginary axis**. We then plot the complex number $a+bi$ at the point (a, b). For example, the imaginary number i is located at the point $(0, 1)$ in the plane, and $1+i$ is located at $(1, 1)$.

The **magnitude** of a complex number z is the distance from z to the origin in the complex plane. We denote the magnitude of z by $|z|$, so we have

$$|z| = \sqrt{a^2 + b^2}.$$

Figure C.1
The magnitude $|z|$ and polar angle θ of a complex number z.

The angle θ formed by the positive x-axis and a straight line from the origin to z (measured in the counterclockwise direction) is called the **polar angle** of z. For example, the number i has polar angle $\pi/2$ and magnitude 1. The complex number $-3i$ has polar angle $3\pi/2$ and magnitude 3, and $1+i$ has polar angle $\pi/4$ and magnitude $\sqrt{2}$. Note that to specify any complex number, all we have to do is give its real and imaginary parts. Alternatively, we can specify a complex number by giving its magnitude and polar angle (see Figure C.1).

The polar angle of a complex number is not unique. In fact, if θ is one polar angle, the angles $\theta \pm 2\pi$, $\theta \pm 4\pi$, $\theta \pm 6\pi$, ... are also polar angles for that number. However, there is a unique polar angle θ with $0 \le \theta < 2\pi$ for any complex number.

Euler's Formula

One of the most amazing formulas in all of mathematics is Euler's formula. This formula gives a surprising relationship between the exponential and the trigonometric functions. Euler's formula is

$$e^{i\theta} = \cos\theta + i\sin\theta$$

for any real number θ. We will see where this formula comes from in a moment, but before that, note that a special case of this result is

$$e^{i\pi} = \cos\pi + i\sin\pi = -1.$$

That is, when you combine three of the most important numbers in mathematics, namely e, i, and π, as above, the result is -1. Some people enjoy expressing this relationship as

$$e^{i\pi} + 1 = 0.$$

Euler's formula lets us compute the exponential of a complex number. Using the usual rules for exponentiation, we have

$$e^{a+bi} = e^a\, e^{ib} = e^a\,(\cos b + i\sin b) = e^a\cos b + ie^a\sin b.$$

For example,

$$e^{2+3i} = e^2\cos 3 + i\,e^2\sin 3,$$

and

$$e^{(2+3i)t} = e^{2t}\cos 3t + i\,e^{2t}\sin 3t$$

is a complex-valued function of t.

Euler's formula is most easy to verify using the power series for the functions involved. Recall that

$$e^x = 1 + x + \frac{x^2}{2!} + \frac{x^3}{3!} + \frac{x^4}{4!} + \cdots,$$

$$\cos x = 1 - \frac{x^2}{2!} + \frac{x^4}{4!} - \frac{x^6}{6!}\cdots, \quad \text{and} \quad \sin x = x - \frac{x^3}{3!} + \frac{x^5}{5!} - \frac{x^7}{7!} + \cdots.$$

If we substitute $i\theta$ for x in the power series for e^x, we obtain

$$e^{i\theta} = 1 + i\theta + \frac{i^2\theta^2}{2!} + \frac{i^3\theta^3}{3!} + \frac{i^4\theta^4}{4!} + \cdots$$

$$= 1 + i\theta - \frac{\theta^2}{2!} - i\frac{\theta^3}{3!} + \frac{\theta^4}{4!} + \cdots.$$

Collecting terms that comprise the real and imaginary parts, we find

$$e^{i\theta} = \left(1 - \frac{\theta^2}{2!} + \frac{\theta^4}{4!} - \frac{\theta^6}{6!} + \cdots\right) + i\left(\theta - \frac{\theta^3}{3!} + \frac{\theta^5}{5!} - \frac{\theta^7}{7!}\cdots\right)$$

$$= \cos\theta + i\sin\theta.$$

Geometry of Complex Multiplication

Using Euler's formula, we can give a geometric interpretation of multiplication of complex numbers. Suppose $z = a + bi$ has polar angle θ. Then z is located at the point in the plane whose coordinates are (a, b). Using trigonometry, we have $a = |z| \cos \theta$ and $b = |z| \sin \theta$, so we may write this complex number in **polar form** as

$$z = |z| \cos \theta + i|z| \sin \theta,$$

which with the aid of Euler's formula becomes

$$z = |z| \, e^{i\theta}.$$

Suppose we have two complex numbers z_1 and z_2 with polar angles θ_1 and θ_2, respectively. Then we may write these complex numbers in polar form as

$$z_1 = |z_1| \, e^{i\theta_1} \quad \text{and} \quad z_2 = |z_2| \, e^{i\theta_2}.$$

If we multiply z_1 and z_2, we have

$$z_1 z_2 = \left(|z_1| \, e^{i\theta_1} \right) \left(|z_2| \, e^{i\theta_2} \right) = |z_1||z_2| \, e^{i(\theta_1 + \theta_2)}.$$

Thus, $z_1 z_2$ is the complex number whose magnitude is $|z_1||z_2|$ and whose polar angle is $\theta_1 + \theta_2$. Therefore, the product of two complex numbers is the complex number whose magnitude is the product of the magnitudes of the factors and whose polar angle is the sum of the polar angles of the factors.

An important special case of this geometric interpretation of complex multiplication involves $w = 1/z$. First, we note that $|1| = 1$ and the polar angle of 1 is 0. Then, since $zw = 1$, we see that

$$|w| = \frac{1}{|z|}$$

and the polar angle of w is the negative of the polar angle of z. In other words, if

$$z = |z| \, e^{i\theta}, \quad \text{then} \quad \frac{1}{z} = \frac{1}{|z|} e^{-i\theta}.$$

HINTS AND ANSWERS

Hints and Answers for Section 1.1

1. The function that is constantly -3, that is, $y(t) = -3$ for all t, is the only equilibrium solution.

3. (a) $P = 0$ and $P = 230$

(b) $0 < P < 230$

(c) $P > 230$ and the nonphysical values $P < 0$

5. (a) $y = -3$, $y = 0$, and $y = 4$

(b) $-3 < y < 0$ and $y > 4$

(c) $y < -3$ and $0 < y < 4$

7. (a) $\lambda = (\ln 2)/5230 \approx 0.000132533$

(b) $\lambda = (\ln 2)/8 \approx 0.0866434$

(c) $1/\text{year}$ for C-14, $1/\text{day}$ for I-131

(d) Yes

9. (a) $\tau = 1/\lambda$

(b) If h is the half-life, then $\tau = h/(\ln 2)$.

(c) The time constant for Carbon 14 is approximately 7545 years, and the time constant for Iodine 131 is approximately 11.5 days.

(d) *Hint*: What is the derivative of $r(t)/r_0$ at $t = 0$?

(e) *Hint*: Is the exponential function ever zero? What is the value of $r(\tau)/r_0$?

11. $k \approx 0.58$ and $R(10) \approx 1{,}321{,}198$

13. $L = 0$

15. (a) Beth

(b) Aly

(c) They have the same rate.

17. (a) $dP/dt = k(1 - P/N)P - 100$.

(b) $dP/dt = k(1 - P/N)P - P/3$.

(c) $dP/dt = k(1 - P/N)P - a\sqrt{P}$, where a is a positive parameter.

19. For example, one model is $dR/dt = kR(R/M - 1)$, where M is a parameter that corresponds to the threshold at which the population is too small to sustain itself over the long term. There are other reasonable models based on the stated assumptions.

21. (a) x is predator; y is prey; no limits on prey; no other food source for predator.

(b) x is prey; y is predator; other limits on prey exist; other food sources available for predator.

23. (a) The species cooperate.

(b) The species compete.

Hints and Answers for Section 1.2

1. (a) Bob and Glen

(b) The function that is constantly -1, that is, the equilibrium solution $y(t) = -1$ for all t.

3. $\dfrac{dy}{dt} = 3t^2 y$

5. $y(t) = -3/(t^3 + k)$, where k is any real number, along with the equilibrium solution $y(t) = 0$ for all t.

7. $y(t) = ke^{2t} - 1/2$, where k is any real number

9. $y(t) = \ln(t + c)$, where c is any real number

11. $y(t) = -1/(t^2 + 3t + k)$, where k is any real number, along with the equilibrium solution $y(t) = 0$ for all t.

13. $y(t) = \pm\sqrt{\ln\left(k(t^2 + 1)\right)}$, where k is any positive real number. The choice of sign is determined by the initial condition.

15. $y(t) = \left(-1 \pm \sqrt{4t + c}\right)/2$, where c is any real number. The plus or minus sign is determined by the initial condition.

17. $y(t) = ke^t/(ke^t + 1)$, where k is any real number, along with the equilibrium solution $y(t) = 1$ for all t. Note that this is a special case of the logistic equation with growth-rate parameter 1 and carrying capacity 1.

19. $v(t) = -1 + ke^{-2t+t^3/3}$, where k is any real number

21. $y^2/2 + \ln|y| = e^t + c$, where c is any real number, along with the equilibrium solution $y(t) = 0$ for all t.

23. $w(t) = kt$ for $t \neq 0$, where k is any real number. (The differential equation is not defined at $t = 0$.)

25. $x(t) = e^{-t^2/2}/\sqrt{\pi}$

27. $y(t) = 1/(t + 2)$

29. $y(t) = 0$

31. $y(t) = 7e^{2t}/2 - 1/2$

33. $x(t) = -\sqrt{4 + (2/3)\ln(t^3 + 1)}$

35. $y(t) = \tan(t^2/2 + \pi/4)$

37. $y(t) = -1/(t^2 + t^3 - 1)$

39. **(a)** Amount of salt ≈ 0.238 lbs
(b) Amount of salt ≈ 1.58 lbs
(c) Amount of salt ≈ 2.49 lbs
(d) Amount of salt ≈ 2.50 lbs
(e) Amount of salt ≈ 2.50 lbs

41. **(a)** The initial-value problem is

$$\frac{dT}{dt} = -0.2(T - 70), \quad T(0) = 170.$$

(b) $t = (\ln 0.4)/ - 0.2 \approx 4.6$

43. **(a)** $v(t) = \sqrt{\frac{mg}{k}}\left(\frac{Ce^{2\sqrt{(kg/m)}\,t} - 1}{Ce^{2\sqrt{(kg/m)}\,t} + 1}\right)$

(b) $\sqrt{\dfrac{mg}{k}}$

Hints and Answers for Section 1.3

1.

3.

5.

13.

7. (a)

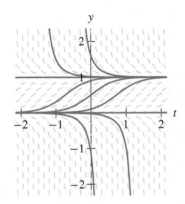

(b) The solution approaches the equilibrium value
$y = 1$ from below.

9. (a)

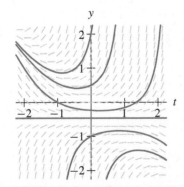

(b) $y(t) \to \infty$ as t increases and as t decreases

11. (a) On the line $y = 3$ in the ty-plane, the slope
marks all have slope -1.

(b) No. Solutions with $y(0) < 3$ satisfy the
inequality $y(t) < 3$ for all $t > 0$.

15.

17. (a) We can sketch the slope field on the entire
ty-plane.

(b) Translate the graph of the given solution up so
that it goes through the point $(0, 2)$.

19. (a) $I_1 = -0.1$

$I_2 = 0.0$

$I_3 = 0.1$

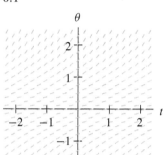

(b) For $I_1 = -0.1$, the equilibrium values are $\theta = 2\pi n \pm 0.613$ radians, where n is any integer.
For $I_2 = 0.0$, the equilibrium values are $\theta = 0 + 2\pi n$ radians, where n is any integer.
For $I_3 = 0.1$, there are no equilibrium values.

(c) If $I_1 = -0.1$, $\theta \to 2\pi n - 0.613$ as $t \to \infty$ for solutions between the successive equilibrium values $2\pi n - 0.613$ and $2\pi n + 0.613$. As $t \to -\infty$, $\theta \to 2\pi n + 0.613$. The reverse behavior occurs for solutions between $2\pi(n-1) + 0.613$ and $2\pi n - 0.613$.
If $I_2 = 0.0$, $\theta \to 2\pi(n+1)$ as $t \to \infty$ for solutions between the successive equilibrium values $2\pi n$ and $2\pi(n+1)$. As $t \to -\infty$, $\theta \to 2\pi n$.
If $I_3 = 0.1$, $\theta \to \infty$ as $t \to \infty$.

21. $v_c(t) = K + (v_0 - K)e^{-t/RC}$ where $v_0 = v_c(0)$

Hints and Answers for Section 1.4

1. Selected approximate values: $y_1 = 6.5$, $y_2 = 13.5$, and $y_4 = 55.5$.

3. Selected approximate values: $y_1 = 0.56$, $y_2 = 0.39$, and $y_8 = -2.69$.

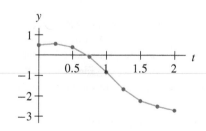

5. Selected approximate values: $w_1 = -1$, $w_2 = -1$, and $w_5 = -1$.

7. Selected approximate values: $y_1 = 3.36$, $y_2 = 4.27$, and $y_4 = 5.81$.

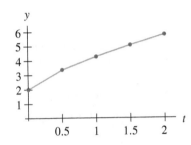

9. Selected approximate values: $y_1 = 0.203$, $y_2 = 0.206$, and $y_{100} = 0.991$.

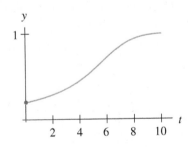

11. *Hint*: What are the constant solutions? From qualitative analysis, solutions with initial condition $w(0) = 0$ should increase until reaching the equilibrium solution at $w = 3$. However, the numerical solution indicates that the solution oscillates about $w = 3$.

13. The answer of Exercise 8 is a translate to the right of the answer of Exercise 7. Why?

15. The true solution is above all three approximations.

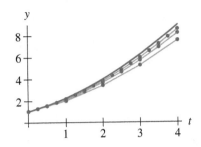

17. The neuron spikes at $t \approx 2.15$.

19.

Graph of approximate solution obtained using Euler's method with $\Delta t = 0.1$.

21.

Graph of approximate solution obtained using Euler's method with $\Delta t = 0.1$.

Hints and Answers for Section 1.5

1. $y(t) < 3$ for all t in the domain of $y(t)$

3. $-t^2 < y(t) < t + 2$ for all t

5. $y(t) > 3$ for all t in the domain of $y(t)$, $y(t) \to 3$ as t decreases, and $y(t)$ increases without bound as t increases.

7. $1 < y(t) < 3$ for all t, $y(t) \to 1$ as $t \to \infty$, and $y(t) \to 3$ as $t \to -\infty$.

9. (a) Substitute each solution into the differential equation and compute.

(b) Use the Existence and Uniqueness Theorem.

11. (a) If $y_1(t) = 0$, then $dy_1/dt = 0 = y_1/t^2$.

(b) For any real number c, let

$$y_c(t) = \begin{cases} 0, & \text{for } t \le 0; \\ ce^{-1/t}, & \text{for } t > 0. \end{cases}$$

The function $y_c(t)$ satisfies the equation for all $t \ne 0$. It is 0 for $t < 0$ and nonzero for $t > 0$.

(c) The differential equation is not defined at $t = 0$.

13. (a) $y(t) = 1/\sqrt{1 - 2t}$

(b) $t < 1/2$

(c) $y(t) \to \infty$ as $t \to 1/2^-$

15. (a) $y(t) = (3t + 27)^{1/3} - 2$

(b) $t > -9$ (note that the differential equation is not defined at $y = -2$).

(c) $y(t) \to \infty$ as $t \to \infty$; $y(t) \to -2$ as $t \to -9^+$.

17. (a) *Hint*: Differentiate $y_1(t)$ at t_0 and remember that $y_1(t)$ is a solution.

(b) *Hint*: Remember that the equation is autonomous.

(c) *Hint*: Look at the slope field, but check by substituting $y_2(t)$ into both sides of the differential equation.

(d) *Hint*: Uniqueness Theorem

(e) *Hint*: Do the same four steps again.

Hints and Answers for Section 1.6

1. The point $y = 0$ is a sink, and the point $y = 2$ is a source.

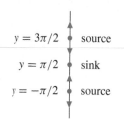

3. The sinks are $\ldots -3\pi/2$, $\pi/2$, $5\pi/2$, $\ldots$, and the sources are $\ldots -\pi/2$, $3\pi/2$, $7\pi/2$, $\ldots$.

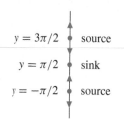

5. The sources are $w = 0$, -2π, -4π $\ldots$ and π, 3π, 5π $\ldots$. The sinks are $w = 1$, $w = -\pi$, -3π, $\ldots$ and $w = 2\pi$, 4π, 6π, $\ldots$.

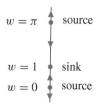

7. There are no equilibrium points.

9. The equilibria are $y = n\pi$, for $n = \pm 1, \pm 3, \ldots$, and they are all nodes.

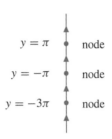

$y = \pi$ • node

$y = -\pi$ • node

$y = -3\pi$ • node

11. The equation is not defined at $y = 0$, but near $y = 0$ the solutions act as if they are going toward a sink at $y = 0$.

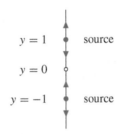

$y = 1$ • source

$y = 0$ ○

$y = -1$ • source

13.

15.

17.

19.

21.

23. Since the initial condition, $y(0) = 2$, is between the roots of $y^2 - 4y + 2$, the solution $y(t)$ to the initial-value problem is always between $2 - \sqrt{2}$ and $2 + \sqrt{2}$. The limit of $y(t)$ as $t \to \infty$ is $2 - \sqrt{2}$ and the limit of $y(t)$ as $t \to -\infty$ is $2 + \sqrt{2}$.

25. The solution remains below the equilibrium point $2 - \sqrt{2}$ and is increasing for all t for which it is defined. The limit of $y(t)$ as $t \to \infty$ is $2 - \sqrt{2}$.

27. Since the initial condition, $y(3) = 1$, is between the roots of $y^2 - 4y + 2$, the solution $y(t)$ to the initial-value problem is always between $2 - \sqrt{2}$ and $2 + \sqrt{2}$. The limit of $y(t)$ as $t \to \infty$ is $2 - \sqrt{2}$ and the limit of $y(t)$ as $t \to -\infty$ is $2 + \sqrt{2}$.

29.   **31.**

(b)

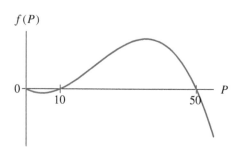

33. The following graph is one of many possible answers.

$f(P)$

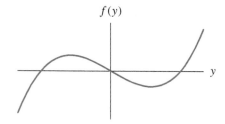

(c) The functions $f(P) = P(P - 10)(50 - P)$ and $f(P) = P(P - 10)^2(50 - P)$ are two possibilities.

35. The following graph is one of many possible answers.

41. *Hint*: The equilibrium points occur at solutions of $y^2 + a = 0$, so you should first study the solutions of this equation for different values of a.

43. (a) Source

(b) Sink

(c) Node

37. (a) viii.   **(b)** vi.   **(c)** i.   **(d)** vii.

45. The term βx is the effect of the passengers, and $-\alpha$ is the term that gives the rate of decrease of the time between trains when no passengers are present. Both α and β should be positive.

39. (a)

47. The only equilibrium point, $x = \alpha/\beta$, is a source. If the initial gap between trains is too large, then x will increase without bound. If it is too small, x decreases toward zero. It is very unlikely that the time between trains will remain constant for long.

Hints and Answers for Section 1.7

1.

3.

5.

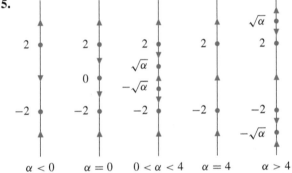

7. *Hint*: There is one equilibrium for $\alpha > 0$ and three equilibria for $\alpha < 0$.

9. The bifurcation values are $\alpha = \pm 1$. There are infinitely many nodes at these bifurcation values. For $\alpha > 1$ and $\alpha < -1$, there are no equilibria. For $-1 \leq \alpha \leq 1$, there are infinitely many equilibria.

11. *Hint*: Sketch the graph of $f(y) + \alpha$ for a variety of α values, both positive and negative. You should be able to give approximate bifurcation values of α from these graphs.

13. (a) iii.   **(b)** v.   **(c)** iv.   **(d)** i.

15.

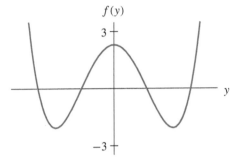

17. *Hint*: The graph of $f(y)$ can only cross the y-axis once, and it must cross from above to below.

19. (a) $L = 16$

  (b) If the population is above 40 when the 16 licenses are issued, the population heads toward 60. If the population is below 40 when fishing begins, the fish will become extinct.

  (c) As long as the fish population is well over 40 when 16 licenses are issued, unexpected perturbations in the system will not drastically affect the fish and the population will tend toward the equilibrium point near 60. (What happens if $16\frac{2}{3}$ licenses are issued?)

21. *Hint*: Once P is close to zero, C must be essentially zero so that the smaller of the two equilibria is less than the given value of P.

23. (a)

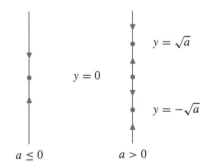

  (b) $a = 0$

(c) There is one bifurcation value a_0, and $a_0 > 0$. If $a < a_0$, the phase line has one equilibrium point, and it is a positive sink. If $a > a_0$, there are two more equilibria. Both are negative, and one is sink and the other is a source.

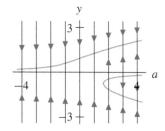

(d) There is one bifurcation value a_0, and $a_0 > 0$. If $a < a_0$, the phase line has one equilibrium point, and it is a negative sink. If $a > a_0$, there are two more equilibria. Both are positive, and one is sink and the other is a source.

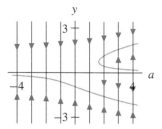

Hints and Answers for Section 1.8

1. $y(t) = ke^{-4t} + 3e^{-t}$

3. $y(t) = ke^{-3t} + \frac{12}{13}\cos 2t + \frac{8}{13}\sin 2t$

5. $y(t) = ke^{3t} - 4te^{3t}$

7. $y(t) = ke^{-2t} + \frac{3}{7}e^{t/3}$; take $k = 4/7$ for the initial-value problem.

9. $y(t) = ke^{-t} + \frac{1}{5}\cos 2t + \frac{2}{5}\sin 2t$; take $k = 24/5$ for the initial-value problem.

11. $y(t) = ke^{2t} + 7te^{2t}$; take $k = 3$ for the initial-value problem.

13. *Hint*: The guess $y_p(t) = \alpha \cos 3t$ leads to terms involving both $\cos 3t$ and $\sin 3t$ on the left.

15.

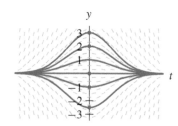

17. (a) *Hint*: Substitute $y_1(t)$ into the differential equation and note that the equation holds.

(b) *Hint*: Substitute $y_2(t)$ into the differential equation and note that the equation does not hold.

(c) The equation is not linear.

19. *Hint*: Substitute $y_h(t)$, $y_1(t)$, and $y_2(t)$ into their respective equations and add.

21. $y(t) = ke^{-2t} + \frac{1}{2}t^2 + \frac{1}{2}t + \frac{1}{4} + \frac{1}{6}e^{4t}$; take $k = -5/12$ for the initial-value problem.

23. $y(t) = ke^{3t} - \frac{2}{3}t - \frac{2}{9} - e^{4t}$; take $k = 11/9$ for the initial-value problem.

25. Tends to a solution that satisfies $-1/2 \le y(t) \le 1$.

27. $y(t) \to 3$ as $t \to \infty$

29. (a) $dP/dt = .011P + 1040$
(b) about \$6,402.20

31. Approximately 76.58 years (919 months)

33. *Hint*: For part (a), calculate $d(y_h + y_p)/dt$ and use the assumptions about $y_h(t)$ and $y_p(t)$. For part (b), calculate $d(y_p - y_q)/dt$ and use the assumptions about $y_p(t)$ and $y_q(t)$.

Hints and Answers for Section 1.9

1. $y(t) = t + c/t$

3. $y(t) = \dfrac{3t^4 + 4t^3 + 12c}{12(t+1)}$

5. $y(t) = (1 + t^2)(3\arctan(t) + c)$

7. $y(t) = \dfrac{t^2 + 2t + 3}{1 + t}$

9. $y(t) = t + \dfrac{2}{t}$

11. $y(t) = 2t^3 + 5t^2$

13. $y(t) = 4e^{-\cos t} \displaystyle\int e^{\cos t}\, dt$

15. $y(t) = 4e^{-1/t} \displaystyle\int e^{1/t} \cos t\, dt$

17. $y(t) = \left(e^{-\int e^{-t^2}\, dt}\right) \displaystyle\int \left(e^{\int e^{-t^2}\, dt}\right) \cos t\, dt$

19. $a = -2$

21. Either way you compute a great deal and get
$$v(t) = ke^{-0.4t} + \tfrac{15}{52}\cos 2t + \tfrac{75}{52}\sin 2t.$$

23. *Hint*: Use the integrating factor $\mu(t) = e^{2t}$ to calculate the general solution of the equation.

25. 4.25 parts per billion

27. **(a)** $dy/dt = 1/2 - y/(V_0 + t)$ with $y(0) = 0$
(b) Note that if $V_0 = 0$, then the differential equation is undefined at $t = 0$. The amount of salt in the tank at time t is $t/4$.

Hints and Answers for Review Exercises in Chapter 1

1. $dy/dt = 2$

3. No equilibrium solutions

5. $y(t) = -1$ for all t

7. $dy/dt = y$

9. The graph of $f(y)$ must cross the y-axis from negative to positive at $y = 0$.

11. True   **13.** True   **15.** False

17. True   **19.** False

21. **(a)** autonomous, separable, linear and nonhomogeneous
(b) $y(t) = \tfrac{3}{2} + ke^{-2t}$

23. **(a)** linear and nonhomogeneous
(b) $y(t) = ke^{3t} + \tfrac{1}{4}e^{7t}$

25. **(a)** linear and nonhomogeneous
(b) $y(t) = ke^{-5t} - \tfrac{3}{34}\cos 3t + \tfrac{5}{34}\sin 3t$

27. **(a)** autonomous, separable
(b) $y(t) = \sqrt{3}\tan\left(\sqrt{3}\,t + k\right)$

29. **(a)** linear and nonhomogeneous
(b) $y(t) = ke^{-3t} + e^{-2t} + \tfrac{1}{3}t^2 - \tfrac{2}{9}t + \tfrac{2}{27}$

31. **(a)** linear and nonhomogeneous
(b) $y(t) = \tfrac{11}{10}e^{2t} - \tfrac{1}{10}\cos 4t + \tfrac{1}{5}\sin 4t$

33. **(a)** separable
(b) $y(t) = -1/\sqrt{4 - 2t - (2/3)t^3}$

35. **(a)** linear and nonhomogeneous
(b) $y(t) = \left(\tfrac{3}{2}t^2 + 1\right)e^{t^2}$

37. **(a)** separable
(b) $y(t) = 1/(1 - t^2 - t^3)$

39. **(a)** separable
(b) $y(t) = -\sqrt{4 + (2/3)\ln(t^3 + 1)}$

41. **(a)**

(b)

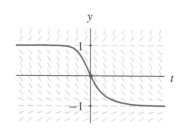

43. *Hint*: The behavior as t increases depends on whether $y(0) > 2$ or $y(0) < 2$. The behavior as t decreases depends on whether $y(0) > y_0$ or $y(0) < y_0$ where $y_0 \approx -0.1471$.

45. (a) $y(t) = -1 + ke^{t+t^3/3}$

   (b) $y_h(t) = ke^{t+t^3/3}$

   (c) $y(t) = -1$ for all t

47. (a) $y_0 = 0$, $y_1 = 1.5$, $y_2 = 1.875$, $y_3 = 1.617$, and $y_4 = 1.810$ (rounded to three decimal places).

(b)

(c) The graph of the actual solution is below and asymptotic to the line $y = \sqrt{3}$ as $t \to \infty$. The oscillations obtained using Euler's method come from numerical error.

49. (a) iv.   **(b)** vii.   **(c)** viii.   **(d)** vi.

51. (a)

   (b) $y(t) \to b$ as $t \to \infty$

   (c) The equation is separable. It is also a nonhomogeneous linear equation.

   (d) $y(t) = b + ke^{-t/a}$

   (e) See part (d).

   (f) Since $a > 0$, $e^{-t/a} \to 0$ as $t \to \infty$.

53. (a) $dC/dt = 1 - C/10$, $C(0) = 30$

   (b)

   (c) $t \approx 6.93$ hours

Hints and Answers for Section 2.1

1. System (i) corresponds to large predators and small prey. System (ii) corresponds to small predators and large prey.

3. *Hint*: Compute dy/dt for $y = 0$.

5. *Hint*: Compute dx/dt for $x = 0$.

7. The populations oscillate with decreasing amplitude about an equilibrium point with both populations nonzero.

9. Change only the dR/dt equation:

(i) $dR/dt = 2R - 1.2RF - \alpha$
(ii) $dR/dt = R(2 - R) - 1.2RF - \alpha$

11. In both systems, let $dF/dt = kF + 0.9RF$, where $k > 0$ is the growth-rate parameter for the predator population (ignoring any effects of overcrowding).

13. In both systems, make

$$\frac{dF}{dt} = -F + 0.9FR + k(R - 5F),$$

where k is an immigration-rate parameter.

15. System (i) corresponds to boa constrictors and system (ii) to small cats. *Hint*: Suppose there was one predator ($y = 1$). How many units of prey are required to keep $dy/dt = 0$?

17. (a) *Hint*: Follow the behavior of a solution with initial condition near $(0, 0)$ (after application of the pesticide kills almost all of both predator and prey).

(b) After applying the pest control, you may see an explosion of the pest population due to the absence of the predator.

19. (a) $\dfrac{d^2(\sin t)}{dt^2} + \sin t = -\sin t + \sin t = 0$

(b)

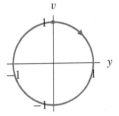

(c) They are the same curve in the yv-plane.
(d) They are parameterized differently.

21. 4 lbs. per in. or 48 lbs. per ft.

23. Large

25. $da/dt = -\alpha ab$
$db/dt = -\alpha ab$,
where α is the reaction-rate parameter

27. $da/dt = k_1 - \alpha ab$
$db/dt = k_2 - \alpha ab$,
where k_1 and k_2 are the rates at which A and B are added, respectively

29. $da/dt = k_1 - \alpha ab + \gamma b^2/2$
$db/dt = k_2 - \alpha ab - \gamma b^2$,
where γ is the reaction-rate parameter for the reaction that turns two B's into an A

Hints and Answers for Section 2.2

1. (a) $\mathbf{V}(x, y) = (1, 0)$
(b) See part (c).

(c)

5. (a) $\mathbf{V}(x, y) = (x, -y)$

  (b) See part (c).

  (c)

  (d)

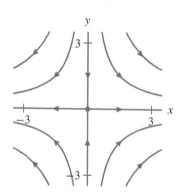

  (e) As t increases, solutions move toward the x-axis in the y-direction and away from the y-axis in the x-direction.

7. (a) $\mathbf{V}(y, v) = (v, y)$

  (b) See part (c).

  (c)

(d)

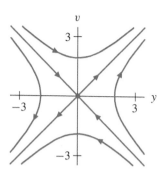

  (e) As t increases, solutions in the 2nd and 4th quadrants move toward the origin and away from the line $y = -v$. Solutions in the 1st and 3rd quadrants move away from the origin and toward the line $y = v$.

9. (a)

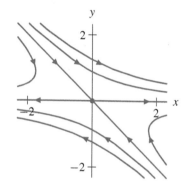

  (b) The solution tends to the origin along the line $y = -x$ in the xy-phase plane. Therefore both $x(t)$ and $y(t)$ tend to zero as $t \to \infty$.

11. (a) vii   **(b)** viii   **(c)** v   **(d)** vi

13. (a) Equilibrium is $(-1/9, 2/9)$.

  (b)

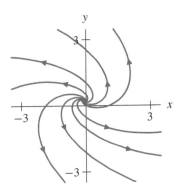

(c) As t increases, typical solutions spiral away from the origin in the counter-clockwise direction.

15. (a) Equilibria: $(\pi/2 + k\pi, \pi/2 + k\pi)$, where k is any integer.

(b)

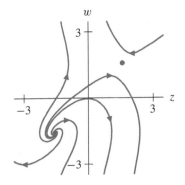

(c) As t increases, typical solutions move away from the line $z = w$, which contains the equilibrium points. The value of w is either increasing or decreasing without bound depending on the initial condition.

17. (a) Equilibria: $(\pi/2 + k\pi, 0)$, where k is any integer.

(b)

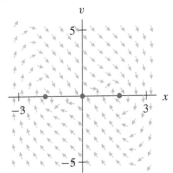

(c) As t increases, typical solutions spiral toward one of the equilibria on the x-axis. Which equilibrium point the solution approaches depends on the initial condition.

19. (a) $\mathbf{V}(x, v) = (v, 3x - x^3 - 2v)$
(b) $(0, 0), (\pm\sqrt{3}, 0)$
(c)

(d)

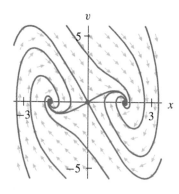

(e) As t increases, most solutions spiral toward one of the two equilibria $(\pm\sqrt{3}, 0)$. There is a curve of initial conditions for which the corresponding solutions tend to the equilibrium point at $(0, 0)$.

21. (a)

(b)

(c)

(d)

23.

25.

27.

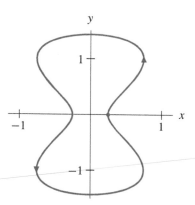

Hints and Answers for Section 2.3

1. (a) See part (c).

    (b) $y_1(t) = e^{-2t}$; $y_2(t) = e^{-5t}$

    (c)

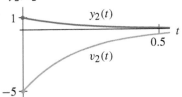

3. (a) See part (c).

    (b) $y_1(t) = e^{(-2+\sqrt{3})t}$; $y_2(t) = e^{(-2-\sqrt{3})t}$

    (c)

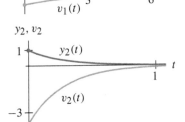

5. (a) See part (c).

    (b) $y_1(t) = e^{2t}$; $y_2(t) = e^{-5t}$

    (c)

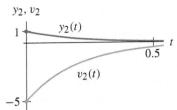

7. (a) *Hint:* Let $y(t)$ be a solution of the damped harmonic oscillator and k be any constant. Substitute the function $ky(t)$ into the equation.

(b) *Hint:* Substitute ke^{-t} and ke^{-2t} into the equation.

(c) An infinite number of solutions.

9. $m\dfrac{d^2x}{dt^2} + b\dfrac{dx}{dt} + (k_1 + k_2)x = k_1L_1 - k_2L_2 + k_2,$

where $x = 0$ is the left-hand wall.

Hints and Answers for Section 2.4

1. $(x(t), y(t))$ is a solution.

3. $(x(t), y(t))$ is not a solution.

5. $y(t) = e^{-2t}$ is not a solution.

7. $x(t) = k_2e^{2t} - \frac{k_1}{3}e^{-t}$, $y(t) = k_1e^{-t}$

9. (a) $x(t) = e^{2t}$, $y(t) = 0$

(b)

(c)

11. (a) $x(t) = \frac{1}{3}e^{2t} - \frac{1}{3}e^{-t}$, $y(t) = e^{-t}$

(b)

(c)

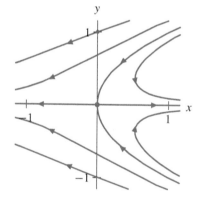

13. (a) $(x(t), y(t)) = (k_1e^{2t} + k_2^2e^{-6t}, k_2e^{-3t})$

(b) $(x, y) = (0, 0)$

(c) $(x(t), y(t)) = (e^{-6t} - e^{2t}, e^{-3t})$

(d)

Hints and Answers for Section 2.5

1. (a) It is a solution.

(b) At $t = 10$ the Euler approximation is approximately $(-9.21, 1.41)$ and the actual solution is approximately $(-0.84, -0.54)$.

(c) At $t = 10$ the Euler approximation is approximately $(-1.41, -0.85)$ and the actual solution

is approximately $(-0.84, -0.54)$.

(d) The approximation will always spiral away from the origin.

3. (a) $(x_5, y_5) \approx (0.65, -0.59)$

(b)

(c)

(c)

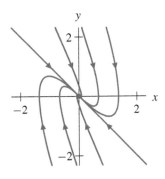

7. For $\Delta t = 0.25$, if $(y_0, v_0) = (2, 0)$, then we obtain $(y_5, v_5) \approx (-0.06, -2.81)$, $(y_{10}, v_{10}) \approx (-1.98, 1.15)$, $(y_{15}, v_{15}) \approx (0.87, 2.34)$, ...
For $\Delta t = 0.1$, if $(y_0, v_0) = (2, 0)$, then we obtain $(y_5, v_5) \approx (1.62, -1.73)$, $(y_{10}, v_{10}) \approx (0.57, -2.44)$, $(y_{15}, v_{15}) \approx (-0.60, -1.94)$, ...

5. (a) $(x_5, y_5) \approx (1.94, -0.72)$

(b)

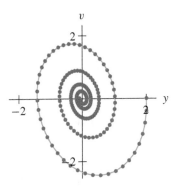

Hints and Answers for Section 2.6

1. (a) If $y = 0$, the system is

$$\frac{dx}{dt} = -x$$

$$\frac{dy}{dt} = 0.$$

Therefore, any solution that lies on the x-axis tends toward the origin. Solutions on negative half of the x-axis approach the origin from the left, and solutions on the positive half of the x-axis approach from the right. The third solution curve is the equilibrium point at the origin.

(b)

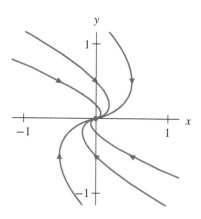

The solutions curves do not intersect the x-axis, nor do they touch the origin. (Note that the system is partially decoupled.)

3. *Hint:* Substitute $\mathbf{Y}_1(t) = (e^{-t} \sin(3t), e^{-t} \cos(3t))$ into the system.

5.

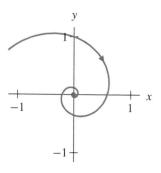

They both sweep out the same spiral in the phase plane. This fact does not contradict the Uniqueness Theorem because the two solutions are never in the same place at the same time.

7. Assuming the vector field satisfies the hypotheses of the Uniqueness Theorem, they will not collide.

9. *Hint:* See Exercise 8.

11. *Hint:* Note that $y(t) \geq 1$ for all positive t in the domain of $y(t)$. Use this observation to note that $dx/dt \geq x^2 + 1$. What are the solutions to the differential equation $dx/dt = x^2 + 1$?

Hints and Answers for Section 2.7

1. *Hint:* Show that $dS/dt + dI/dt + dR/dt = 0$ for all t.

3. (a) $0 \leq S(0) \leq \beta/\alpha$

(b) If $S(0) < \beta/\alpha$, then $dI/dt < 0$. In that case, any initial influx of infecteds will decrease toward zero, and the epidemic will die out. The fraction vaccinated must be at least $1 - \beta/\alpha$.

5. (a)

(b)

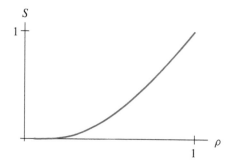

(c) As ρ increases, the fraction of the population that contracts the disease approaches zero.

7. (a) The only equilibrium points lie on the line $I = 0$.

(b) $I < \left(\dfrac{\alpha}{\beta}\right)^2 S^2$.

(c)

The model predicts that the entire population will become infected. Thus, $R(t) \to 1$ as $t \to \infty$.

9. (a) $\beta = 0.44$.

(b) As $t \to \infty$, $S(t) \approx 19$. Therefore, the total number of infected students is $763 - 19 = 744$ students.

(c) The small value of β (with respect to α) indicates that it will take a long time for the infected students to recover.

Hints and Answers for Section 2.8

1. (a) *Hint*: Substitute $(0, 0, 0)$ and $(\pm 6\sqrt{2}, \pm 6\sqrt{2}, 27)$ to see if all derivatives are zero.

(b) *Hint*: Set $dx/dt = dy/dt = dz/dt = 0$ and solve three simultaneous equations.

3. (a) Note that dx/dt and dy/dt vanish if $x = y = 0$.

(b) The solution is $x(t) = 0$, $y(t) = 0$, and $z(t) = e^{-8t/3}$.

(c) The solution is $x(t) = 0$, $y(t) = 0$, and $z(t) = z_0 e^{-8t/3}$.

5. (a)

(b)

(c)

Hints and Answers for Review Exercises in Chapter 2

1. The equilibrium solution $(x(t), y(t)) = (0, 0)$ for all t

3. $dy/dt = v$, $dv/dt = 1$

5. Every point on the x-axis

7. Yes

9. $(x(t), y(t)) = (k_1 e^{2t}, k_2 e^{-3t})$

11. There are many examples. One is $dx/dt = (x^2 - 1)(x^2 - 4)(x^2 - 9)(x^2 - 16)(x^2 - 25)$ and $dy/dt = y$.

13.

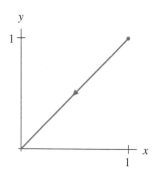

15. True   **17.** False

19. False   **21.** True

23. False   **25.** False

27. False

29. (a) $(x, y) =$
$\ldots, (-3\pi/2, -3\pi/4), (-\pi/2, -\pi/4),$
$(\pi/2, \pi/4), (3\pi/2, 3\pi/4), (5\pi/2, 5\pi/4), \ldots$

(b)

(c) Most solutions become unbounded in y as t increases. However, there appears to be a "curve" of solutions that tend toward the equilibria $\ldots, (-\pi/2, -\pi/4), (3\pi/2, 3\pi/4), \ldots$ as t increases.

31.

33.

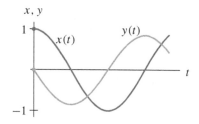

35. (a) $(x(t), y(t)) = (k_1 e^t + k_2 e^{3t} - 1, k_2 e^{3t})$

(b) $(x, y) = (-1, 0)$

(c) $(x(t), y(t)) = (-3e^t + 3e^{3t} - 1, 3e^{3t})$

(d)

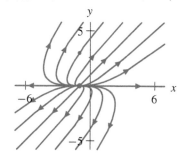

37. (a) For each D, there is one equilibrium point for which $-\pi < \theta < \pi$. It is $(\theta, s) = (-\arctan D, (1 + D^2)^{-1/4})$.

(b) The equilibrium point represents motion along a line at a given angle from the horizon with a constant speed.

Hints and Answers for Section 3.1

1. In this case, $dx/dt = dy/dt$, so both x and y change by identical amounts. Paul's profitability has a positive effect on both x and y while Bob's profitability has a negative effect. If Paul's profits are greater than Bob's, then the profits of both stores will increase.

3. Since $a = 1$ and $b = 0$, Paul's profits have a positive effect on his future profits, but Bob's profits have no effect. Since $c = 2$ and $d = 1$, profits from both stores positively affect Bob's profits. However, the effect of Paul's profits is twice that of Bob's profits.

5. $\mathbf{Y} = \begin{pmatrix} x \\ y \end{pmatrix}, \quad \dfrac{d\mathbf{Y}}{dt} = \begin{pmatrix} 2 & 1 \\ 1 & 1 \end{pmatrix} \mathbf{Y}$

7. $\mathbf{Y} = \begin{pmatrix} p \\ q \\ r \end{pmatrix}, \quad \dfrac{d\mathbf{Y}}{dt} = \begin{pmatrix} 3 & -2 & -7 \\ -2 & 0 & 6 \\ 0 & 7.3 & 2 \end{pmatrix} \mathbf{Y}$

9. $\dfrac{dx}{dt} = \beta y$ and $\dfrac{dy}{dt} = \gamma x - y$

11. (a)

(b)

(c)

13. (a)

(b)

(c)

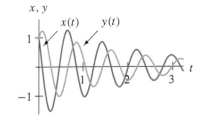

15. If $a \neq 0$, then the set of all equilibrium points (x_0, y_0) is exactly those points that lie on the line $x_0 = (-b/a)y_0$. If $a = 0$ and $b \neq 0$, then the set of all equilibrium points is the x-axis. Finally, if $a = b = 0$, then the set of all equilibrium points consists of the line defined by the equation $cx_0 + dy_0 = 0$.

17. (a) Equilibrium line $v = 0$.
(b) Equilibrium line $v = 0$.

19. *Hint:* Let $v = dy/dt$ and $w = d^2y/dt^2$.

21. $\gamma > 0$

23. $\delta < 0$

25. (a) Substitute $\mathbf{Y}(t)$ into the differential equation and check that the left-hand side equals the right-hand side.
(b) The solution with this initial condition is $(-2te^{2t}, (2t + 2)e^{2t})$.

27. (a) They are solutions.
(b) They are linearly independent.
(c) $\mathbf{Y}(t) = (e^{-3t} + e^{-4t}, e^{-3t} + 2e^{-4t})$

29. (a) $\mathbf{Y}_1(t)$ is a solution, but $\mathbf{Y}_2(t)$ is not.

31. We want to show that one of the vectors is a multiple of the other.

(a) $(x_1, y_1) = (0, 0)$ and any (x_2, y_2) are on the same line through the origin.
(b) (x_1, y_1) is a multiple of (x_2, y_2), which implies that they are on the same line through the origin.
(c) If $x_1 \neq 0$, we have $(x_2, y_2) = (x_2/x_1)(x_1, y_1)$. Then we can use part (b). If $x_1 = 0$ and $y_1 \neq 0$, then proceed as before. If both $x_1 = 0$ and $y_1 = 0$, then we have the situation of part (a).

33. We can give explicit solutions for (a), (c) and (d). For example, the explicit solution to part (a) is $(-2e^{-t}, 2e^{-t})$. What are the explicit solutions for parts (c) and (d)?

35. (a) $dW/dt = y_2(dx_1/dt) + x_1(dy_2/dt) - y_1(dx_2/dt) - x_2(dy_1/dt)$
(b) Replace dx_1/dt by $ax_1 + by_1$ (and so on) in the above formula.
(c) $W(t) = Ce^{(a+d)t}$, where C is a constant.
(d) Using the result of part (c), if $C = W(0) = 0$, then $W(t) = 0$ for all t. If $W(0) \neq 0$, then $W(t) \neq 0$ for all t. If $\mathbf{Y}_1(0)$ and $\mathbf{Y}_2(0)$ are linearly independent, then $W(0) \neq 0$ (see Exercise 32). But then $W(t) \neq 0$ for all t. Hence, $\mathbf{Y}_1(t)$ and $\mathbf{Y}_2(t)$ are linearly independent for all t.

Hints and Answers for Section 3.2

1. (a) Eigenvalues: $\lambda_1 = -2$, $\lambda_2 = 3$
(b) Eigenvectors: $\mathbf{V}_1 = (x_1, y_1)$ where $5x_1 = -2y_1$ and $\mathbf{V}_2 = (x_2, 0)$
(c)

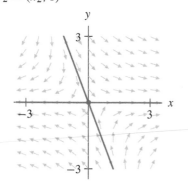

(d) $\mathbf{Y}_1(t) = e^{-2t}\begin{pmatrix} -2 \\ 5 \end{pmatrix}$, $\mathbf{Y}_2(t) = e^{3t}\begin{pmatrix} 1 \\ 0 \end{pmatrix}$

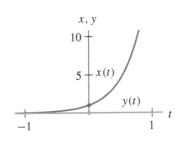

(e) $\mathbf{Y}(t) = k_1 e^{-2t} \begin{pmatrix} -2 \\ 5 \end{pmatrix} + k_2 e^{3t} \begin{pmatrix} 1 \\ 0 \end{pmatrix}$

3. (a) Eigenvalues: $\lambda_1 = -3$, $\lambda_2 = -6$

(b) Eigenvectors: $\mathbf{V}_1 = (x_1, y_1)$ where $y_1 = -x_1$
and $\mathbf{V}_2 = (x_2, y_2)$ where $2y_2 = x_2$

(c)

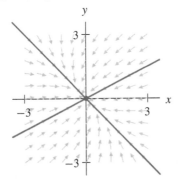

(d) $\mathbf{Y}_1(t) = e^{-3t} \begin{pmatrix} 1 \\ -1 \end{pmatrix}$, $\mathbf{Y}_2(t) = e^{-6t} \begin{pmatrix} 2 \\ 1 \end{pmatrix}$

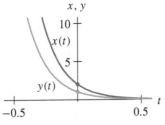

(e) $\mathbf{Y}(t) = k_1 e^{-3t} \begin{pmatrix} 1 \\ -1 \end{pmatrix} + k_2 e^{-6t} \begin{pmatrix} 2 \\ 1 \end{pmatrix}$

5. (a) Eigenvalue: $\lambda_1 = -1/2$

(b) Eigenvectors: $\mathbf{V}_1 = (0, y_1)$

(c)

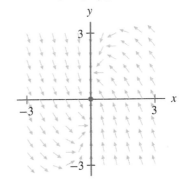

(d) $\mathbf{Y}_1(t) = e^{-t/2} \begin{pmatrix} 0 \\ 1 \end{pmatrix}$

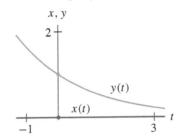

(e) We cannot form the general solution since we
do not have two distinct eigenvalues.

7. (a) Eigenvalues: $\lambda_1 = -1$, $\lambda_2 = 4$

(b) Eigenvectors: $\mathbf{V}_1 = (x_1, y_1)$ where $y_1 = -x_1$
and $\mathbf{V}_2 = (x_2, y_2)$ where $x_2 = 4y_2$

(c)

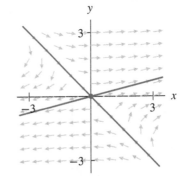

(d) $\mathbf{Y}_1(t) = e^{-t} \begin{pmatrix} 1 \\ -1 \end{pmatrix}$, $\mathbf{Y}_2(t) = e^{4t} \begin{pmatrix} 4 \\ 1 \end{pmatrix}$

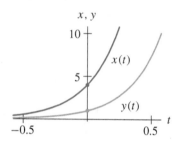

(e) $\mathbf{Y}(t) = k_1 e^{-t} \begin{pmatrix} 1 \\ -1 \end{pmatrix} + k_2 e^{4t} \begin{pmatrix} 4 \\ 1 \end{pmatrix}$

9. (a) Eigenvalues: $\lambda_1 = (3 + \sqrt{5})/2$, $\lambda_2 = (3 - \sqrt{5})/2$

(b) Eigenvectors: $\mathbf{V}_1 = (x_1, y_1)$ where $(-1 + \sqrt{5})x_1 = 2y_1$ and $\mathbf{V}_2 = (x_2, y_2)$ where $(1 + \sqrt{5})x_2 = -2y_2$

(c)

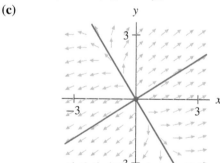

(d) $\mathbf{Y}_1(t) = e^{(3+\sqrt{5})t/2} \begin{pmatrix} 2 \\ -1 + \sqrt{5} \end{pmatrix}$,

$\mathbf{Y}_2(t) = e^{(3-\sqrt{5})t/2} \begin{pmatrix} -2 \\ 1 + \sqrt{5} \end{pmatrix}$

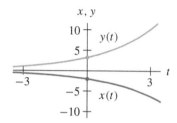

(e) $\mathbf{Y}(t) = k_1 e^{(3+\sqrt{5})t/2} \begin{pmatrix} 2 \\ -1 + \sqrt{5} \end{pmatrix} + k_2 e^{(3-\sqrt{5})t/2} \begin{pmatrix} -2 \\ 1 + \sqrt{5} \end{pmatrix}$

11. (a) $\mathbf{Y}(t) = \frac{1}{5}e^{2t} \begin{pmatrix} 1 \\ -2 \end{pmatrix} + \frac{2}{5}e^{-3t} \begin{pmatrix} 2 \\ 1 \end{pmatrix}$

(b) $\mathbf{Y}(t) = -\frac{2}{5}e^{2t} \begin{pmatrix} 1 \\ -2 \end{pmatrix} + \frac{1}{5}e^{-3t} \begin{pmatrix} 2 \\ 1 \end{pmatrix}$

(c) $\mathbf{Y}(t) = e^{2t} \begin{pmatrix} 1 \\ -2 \end{pmatrix}$

13. (a) $\mathbf{Y}(t) = \frac{2}{3}e^{-5t} \begin{pmatrix} 1 \\ -1 \end{pmatrix} + \frac{1}{3}e^{-2t} \begin{pmatrix} 1 \\ 2 \end{pmatrix}$

(b) $\mathbf{Y}(t) = e^{-5t} \begin{pmatrix} 1 \\ -1 \end{pmatrix} + e^{-2t} \begin{pmatrix} 1 \\ 2 \end{pmatrix}$

(c) $\mathbf{Y}(t) = -e^{-2t} \begin{pmatrix} 1 \\ 2 \end{pmatrix}$

15. Show that for any vector $\mathbf{Y}_0$, $\mathbf{AY}_0 = a\mathbf{Y}_0$. Therefore, every vector is an eigenvector associated with the eigenvalue a.

17. The characteristic polynomial is $\lambda^2 - (a+d)\lambda + ad - b^2$. This quadratic has the discriminant $D = (a-d)^2 + 4b^2$. Since $D \geq 0$, the characteristic polynomial always has real roots. In other words, the matrix always has real eigenvalues. If $b \neq 0$, then $D > 0$. In this case, the characteristic polynomial always has distinct real roots.

19. (a) $\dfrac{d\mathbf{Y}}{dt} = \begin{pmatrix} 0 & 1 \\ -q & -p \end{pmatrix} \mathbf{Y}$

(b) $\lambda^2 + p\lambda + q$

(c) $\left(-p \pm \sqrt{p^2 - 4q} \right)/2$

(d) The roots are distinct real numbers if $p^2 > 4q$.

(e) Note that $p^2 - 4q < p^2$.

21. (a) $dy/dt = v$ and $dv/dt = -10y - 7v$

(b) Eigenvalue $\lambda_1 = -5$ with eigenvectors $\mathbf{V}_1 = (y_1, v_1)$ where $v_1 = -5y_1$ and eigenvalue $\lambda_2 = -2$ with eigenvectors $\mathbf{V}_2 = (y_2, v_2)$ where $v_2 = -2y_2$

(c) $\mathbf{Y}_1(t) = e^{-5t} \begin{pmatrix} 1 \\ -5 \end{pmatrix}$, $\mathbf{Y}_2(t) = e^{-2t} \begin{pmatrix} 1 \\ -2 \end{pmatrix}$

23. (a) $dy/dt = v$ and $dv/dt = -y - 4v$

(b) Eigenvalue $\lambda_1 = -2 + \sqrt{3}$ with eigenvectors $\mathbf{V}_1 = (y_1, v_1)$ where $v_1 = (-2 + \sqrt{3})y_1$ and eigenvalue $\lambda_2 = -2 - \sqrt{3}$ with eigenvectors $\mathbf{V}_2 = (y_2, v_2)$ where $v_2 = (-2 - \sqrt{3})y_2$

(c) $\mathbf{Y}_1(t) = e^{(-2+\sqrt{3})t} \begin{pmatrix} 1 \\ -2 + \sqrt{3} \end{pmatrix}$,

$\mathbf{Y}_2(t) = e^{(-2-\sqrt{3})t} \begin{pmatrix} 1 \\ -2 - \sqrt{3} \end{pmatrix}$

25. The characteristic polynomial is $\lambda^2 + \lambda + 4$, and its roots are the complex numbers $(-1 \pm \sqrt{15}\,i)/2$. Therefore there are no straight-line solutions (see the direction field).

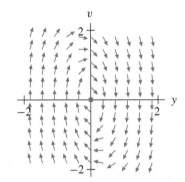

Hints and Answers for Section 3.3

1.

3.

5.

7.

9.

11.

13.

15.

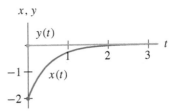

17. For any initial condition, Bob's profits y are eventually asymptotic to 0. For an initial condition on the line $y = -3x$, Paul's profits x also are eventually asymptotic to 0. For an initial condition to the left of the line $y = -3x$, Paul eventually goes broke. For an initial condition to the right of the line $y = -3x$, Paul eventually makes a fortune.

19. (a) sink

    (b) all solutions of the form $e^{-2t}(x, 0)$ and $e^{-t}(x, 2x)$

    (c)

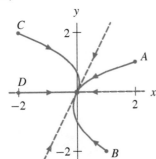

21. (a) $\dfrac{d^2y}{dt^2} + 7\dfrac{dy}{dt} + 6y = 0$;

    As a system: $dy/dt = v, \, dv/dt = -6y - 7v$

    (b) $\lambda^2 + 7\lambda + 6 = 0$

    (c) $\lambda_1 = -6, \lambda_2 = -1$

    (d) The solution never crosses the rest position. It tends to the rest position like e^{-t}.

23. (a) The presence of the new species has a negative effect on the native population, but the native fish do not affect the population of the new fish.

(b) The model agrees with the premises of the system.

(c)

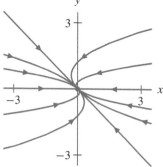

The straight-line solutions lie on the x-axis and the line $y = -x$ in the phase plane. Solutions along these lines tend toward the origin.

(d) If a small number of new fish are added to the lake, the native population drops from equilibrium. The new fish die out as the native fish return to equilibrium.

25. (a) The native fish have no effect on the population of the new fish. The introduction of new fish increases the population of the native fish.

(b) The model agrees with the system described in the problem.

(c)

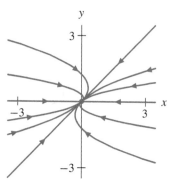

The lines containing the straight-line solutions are the x-axis and the line $y = x$. The solutions along these lines tend toward the origin. The origin is a sink.

(d) The population of the native fish initially increases and then returns toward its equilibrium value. The new fish will die out.

27. (a) Calculate the eigenvalues.

(b) Eigenvalues: $\lambda_1 = 2, \lambda_2 = -2$
Lines of eigenvectors: $y = 4x, y = 0$

(c)

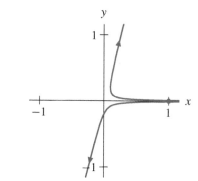

(d) $t \approx 1.96$

Hints and Answers for Section 3.4

1. $k_1 e^t \begin{pmatrix} 2\cos 3t - \sin 3t \\ \cos 3t \end{pmatrix} +$

$k_2 e^t \begin{pmatrix} \cos 3t + 2\sin 3t \\ \sin 3t \end{pmatrix}$

3. (a) $\lambda_1 = 2i, \lambda_2 = -2i$
   (b) Center
   (c) Natural period π; natural frequency $1/\pi$
   (d) Clockwise
   (e)

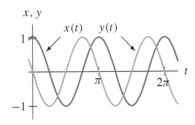

5. (a) $-1 \pm i\sqrt{11}$
   (b) Spiral sink
   (c) Natural period $= 2\pi/\sqrt{11}$, natural frequency
       $= \sqrt{11}/(2\pi)$
   (d) Counterclockwise

(e)

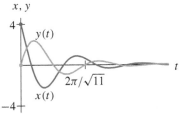

7. (a) $(3 \pm \sqrt{47}\,i)/2$
   (b) Spiral source
   (c) Natural period: $\frac{4\pi}{\sqrt{47}}$; natural frequency: $\frac{\sqrt{47}}{4\pi}$.
   (d) Counterclockwise
   (e)

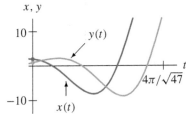

9. (a) $\mathbf{Y}(t) = k_1 \begin{pmatrix} \cos 2t \\ -\sin 2t \end{pmatrix} + k_2 \begin{pmatrix} \sin 2t \\ \cos 2t \end{pmatrix}$

(b) $\mathbf{Y}(t) = \begin{pmatrix} \cos 2t \\ -\sin 2t \end{pmatrix}$

(c)

11. (a) $\mathbf{Y}(t) =$

$k_1 e^{-t} \begin{pmatrix} -2\cos\sqrt{11}\,t - \sqrt{11}\sin\sqrt{11}\,t \\ 3\cos\sqrt{11}\,t \end{pmatrix} +$

$k_2 e^{-t} \begin{pmatrix} \sqrt{11}\cos\sqrt{11}\,t - 2\sin\sqrt{11}\,t \\ 3\sin\sqrt{11}\,t \end{pmatrix}$

obtained using eigenvalue $\lambda = -1 + i\sqrt{11}$ and eigenvector $\left(-2 + i\sqrt{11}, 3\right)$

(b) $\mathbf{Y}(t) = e^{-t} \begin{pmatrix} 4\cos\sqrt{11}\,t - \dfrac{8}{\sqrt{11}}\sin\sqrt{11}\,t \\ \dfrac{12}{\sqrt{11}}\sin\sqrt{11}\,t \end{pmatrix}$

(c)

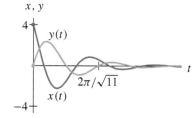

13. (a) $k_1 e^{3t/2} \begin{pmatrix} 12\cos\left(\frac{\sqrt{47}}{2}t\right) \\ \cos\left(\frac{\sqrt{47}}{2}t\right) + \sqrt{47}\sin\left(\frac{\sqrt{47}}{2}t\right) \end{pmatrix} +$

$k_2 e^{3t/2} \begin{pmatrix} 12\sin\left(\frac{\sqrt{47}}{2}t\right) \\ -\sqrt{47}\cos\left(\frac{\sqrt{47}}{2}t\right) + \sin\left(\frac{\sqrt{47}}{2}t\right) \end{pmatrix}$

(b) $e^{3t/2} \begin{pmatrix} 2\cos\left(\frac{\sqrt{47}}{2}t\right) - \frac{10}{\sqrt{47}}\sin\left(\frac{\sqrt{47}}{2}t\right) \\ \cos\left(\frac{\sqrt{47}}{2}t\right) + \frac{7}{\sqrt{47}}\sin\left(\frac{\sqrt{47}}{2}t\right) \end{pmatrix}$

(c)

15. (a) iv. and v.

(b) Graph (iv) has a natural period of approximately 1.5, and the origin is a sink. Graph (v) has a natural period of approximately 1.25, and the origin is a source.

(c) (i) The time between successive zeros is not constant. (ii) Oscillation stops at some t. (iii) The amplitude is not monotonically increasing nor decreasing. (vi) Oscillation starts at some t. There was no prior oscillation.

17. If $\lambda_1 = \alpha + i\beta$ satisfies the equation, then $\lambda_2 = \alpha - i\beta$ does as well. A quadratic equation has at most two solutions, so they must be λ_1 and λ_2.

19. *Hint*: Suppose $\mathbf{Y}_2 = k\mathbf{Y}_1$ for some constant k. Then, $\mathbf{Y}_0 = (1 + ik)\mathbf{Y}_1$. Derive a contradiction using the fact that $\mathbf{A}\mathbf{Y}_0 = \lambda\mathbf{Y}_0$. (Note that $\mathbf{A}$ only has real entries.)

21. (a) π/β

(b) π/β

(c) $2\pi/\beta$

(d) $(\arctan(\beta/\alpha))/\beta$

23. (a) $dy/dt = v,\ dv/dt = -qy - pv$

(b) $p^2 < 4q$

(c) Spiral sink: $q > 0$ and $0 < p < 2\sqrt{q}$; center: $q > 0$ and $p = 0$; spiral source: $q > 0$ and $-2\sqrt{q} < p < 0$

(d) $q > 0$

25. *Hint*: Draw the phase portrait and remember the Linearity Principle.

Hints and Answers for Section 3.5

1. (a) Eigenvalue: -3

(b) Eigenvectors: $x = 0$

(c)

(d)

(e)

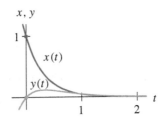

3. (a) Eigenvalue: -3

(b) Eigenvectors: $y = x$

(c)

(d)

(e)

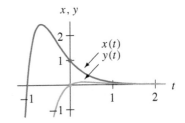

5. (a) $\mathbf{Y}(t) = e^{-3t} \begin{pmatrix} x_0 \\ y_0 \end{pmatrix} + t e^{-3t} \begin{pmatrix} 0 \\ x_0 \end{pmatrix}$

(b) $\mathbf{Y}(t) = e^{-3t} \begin{pmatrix} 1 \\ 0 \end{pmatrix} + t e^{-3t} \begin{pmatrix} 0 \\ 1 \end{pmatrix}$

(c)

7. (a) $\mathbf{Y}(t) = e^{-3t} \begin{pmatrix} x_0 \\ y_0 \end{pmatrix} + t e^{-3t} \begin{pmatrix} x_0 - y_0 \\ x_0 - y_0 \end{pmatrix}$

(b) $\mathbf{Y}(t) = e^{-3t} \begin{pmatrix} 1 \\ 0 \end{pmatrix} + t e^{-3t} \begin{pmatrix} 1 \\ 1 \end{pmatrix}$

(c)

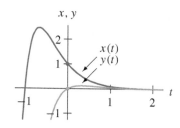

9. (a) $\alpha^2 = 4\beta$
  (b) $\beta = 0$

11. (a) $p^2 - 4q > 0$
  (b) $p^2 - 4q < 0$
  (c) $p^2 - 4q = 0$

13. A $\begin{pmatrix} 1 \\ 0 \end{pmatrix} = \begin{pmatrix} a \\ c \end{pmatrix} = \begin{pmatrix} \lambda \\ 0 \end{pmatrix}$,

  so we have $a = \lambda$ and $c = 0$.

  Likewise, **A** $\begin{pmatrix} 0 \\ 1 \end{pmatrix} = \begin{pmatrix} b \\ d \end{pmatrix} = \begin{pmatrix} 0 \\ \lambda \end{pmatrix}$,

  so we have $d = \lambda$ and $b = 0$.

15. *Hint:* Evaluate both functions at $t = 0$ and then at any other value of t.

17. (a) Eigenvalues: $0, -1$
  (b) Eigenvectors: $y_1 = 0, 2y_2 = -x_2$
  (c)

(d)

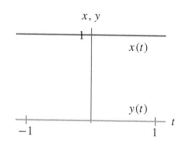

(e) $\mathbf{Y}(t) = k_1 \begin{pmatrix} 1 \\ 0 \end{pmatrix} + k_2 e^{-t} \begin{pmatrix} 2 \\ -1 \end{pmatrix}$

(f) $\mathbf{Y}(t) = (1, 0)$

19. (a) Eigenvalues: $0, 5$
  (b) Eigenvectors: $y_1 = -2x_1, x_2 = 2y_2$
  (c)

(d)

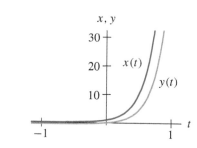

(e) $\mathbf{Y}(t) = k_1 \begin{pmatrix} 1 \\ -2 \end{pmatrix} + k_2 e^{5t} \begin{pmatrix} 2 \\ 1 \end{pmatrix}$

(f) $\mathbf{Y}(t) = \begin{pmatrix} \frac{1}{5} + \frac{4}{5}e^{5t} \\ -\frac{2}{5} + \frac{2}{5}e^{5t} \end{pmatrix}$

21. (a) $\lambda = 0$ is the only eigenvalue.

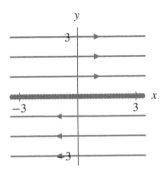

(b) $\lambda = 0$ is the only eigenvalue.

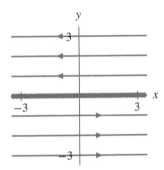

23. (a) Eigenvalues: a, d
(b) Eigenvectors: x- and y-axes respectively if $a \neq d$

(c) General solution: $\mathbf{Y}(t) = e^{at}\mathbf{Y}_0$, where $\mathbf{Y}_0$ is the initial condition

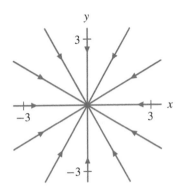

(d) General solution: same as part (c)

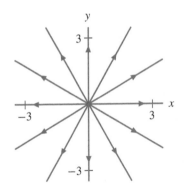

Hints and Answers for Section 3.6

1. $y(t) = k_1 e^{-t} + k_2 e^{7t}$

3. $y(t) = k_1 e^{-3t} + k_2 t e^{-3t}$

5. $y(t) = k_1 e^{-4t} \cos 3t + k_2 e^{-4t} \sin 3t$

7. $y(t) = 4e^t + 2e^{-3t}$

9. $y(t) = e^{2t} \cos 3t - 2e^{2t} \sin 3t$

11. $y(t) = 3e^{4t} - te^{4t}$

13. (a) $d^2y/dt^2 + 8dy/dt + 7y = 0$;
system: $dy/dt = v$, $dv/dt = -7y - 8v$
(b) $\lambda_1 = -1$ with eigenvectors $\mathbf{V}_1 = (y_1, v_1)$
where $v_1 = -y_1$; $\lambda_2 = -7$ with eigenvectors
$\mathbf{V}_2 = (y_2, v_2)$ where $v_2 = -7y_2$
(c) Overdamped

(d)

(e)

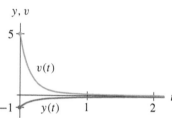

15. (a) $d^2y/dt^2 + 4dy/dt + 5y = 0$;
system: $dy/dt = v, dv/dt = -5y - 4v$

(b) $\lambda_1 = -2 + i$ with eigenvectors $\mathbf{V}_1 = (y_1, v_1)$
where $v_1 = (-2+i)y_1$; $\lambda_2 = -2 - i$ with eigenvectors $\mathbf{V}_2 = (y_2, v_2)$ where $v_2 = (-2 - i)y_2$

(c) Underdamped. Period is 2π.

(d)

(e)

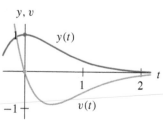

17. (a) $2d^2y/dt^2 + 3dy/dt + y = 0$;
system: $dy/dt = v, dv/dt = -y/2 - 3v/2$

(b) $\lambda_1 = -1$ with eigenvectors $\mathbf{V}_1 = (y_1, v_1)$
where $v_1 = -y_1$; $\lambda_2 = -1/2$ with eigenvectors
$\mathbf{V}_2 = (y_2, v_2)$ where $v_2 = -y_2/2$

(c) Overdamped

(d)

(e)

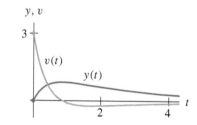

19. (a) $2d^2y/dt^2 + 3y = 0$;
system: $dy/dt = v, dv/dt = -3y/2$

(b) $\lambda = \pm i\sqrt{3/2}$ with eigenvectors $\mathbf{V} = (y, v)$ that
satisfy $v = \pm i\sqrt{3/2}\, y$

(c) Undamped with natural period $4\pi/\sqrt{6}$

(d)

(e)

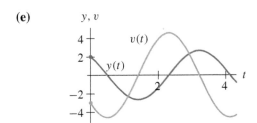

21. (a) $y(t) = k_1 e^{-7t} + k_2 e^{-t}$
  (b) $y(t) = -\frac{2}{3} e^{-7t} - \frac{1}{3} e^{-t}$
  (c) See the $y(t)$- and $v(t)$-graphs for Exercise 13.

23. (a) $y(t) = k_1 e^{-2t} \cos t + k_2 e^{-2t} \sin t$
  (b) $y(t) = e^{-2t} \cos t + 2e^{-2t} \sin t$
  (c) See the $y(t)$- and $v(t)$-graphs for Exercise 15.

25. (a) $y(t) = k_1 e^{-t} + k_2 e^{-t/2}$
  (b) $y(t) = -6e^{-t} + 6e^{-t/2}$
  (c) See the $y(t)$- and $v(t)$-graphs for Exercise 17.

27. (a) $y(t) = k_1 \cos\left(\sqrt{3/2}\, t\right) + k_2 \sin\left(\sqrt{3/2}\, t\right)$
  (b) $y(t) = 2\cos\left(\sqrt{3/2}\, t\right) - \sqrt{6}\sin\left(\sqrt{3/2}\, t\right)$
  (c) See the $y(t)$- and $v(t)$-graphs for Exercise 19.

29. *Hint:* The table will have four rows—one each for

the undamped, underdamped, critically damped, and overdamped cases.

31. *Hint:* Use the fact that a complex number is zero if and only if both its real and imaginary parts are zero.

33. (a) $\begin{pmatrix} 0 & 1 \\ -\lambda_0^2 & 2\lambda_0 \end{pmatrix}$

  (b) $\mathbf{Y}(t) = e^{\lambda_0 t} \begin{pmatrix} y_0 \\ v_0 \end{pmatrix} + te^{\lambda_0 t} \begin{pmatrix} -\lambda_0 y_0 + v_0 \\ -\lambda_0^2 y_0 + \lambda_0 v_0 \end{pmatrix}$

  (c) $y(t) = y_0 e^{\lambda_0 t} + (-\lambda_0 y_0 + v_0)te^{\lambda_0 t}$
  (d) Let $k_1 = y_0$ and $k_2 = -\lambda_0 y_0 + v_0$. Why are all k_1 and k_2 possible?

35. The solution will approach the equilibrium no faster than $e^{-\alpha t}$, where $\alpha = b/2$ if $b^2 < 12$ and $\alpha = (b + \sqrt{b^2 - 12})/2$ if $b^2 > 12$.

37. (a) $md^2 y/dt^2 - b_{mf}\, dy/dt + ky = 0$
  (b) $dy/dt = v, \ dv/dt = -(k/m)y + (b_{mf}/m)v$
  (c) *Hint:* Modify the case-by-case analysis of this section to reflect the change in sign of the dy/dt term.

39. $m = 1/(2\pi^2)$

Hints and Answers for Section 3.7

1. *Hint:* There are three types in Sections 3.2 and 3.3, and three types in Section 3.4. These six are the most common. The remaining types are discussed in Section 3.5.

3. (a)

  (b) Real sink: $a < -1$; repeated negative eigenvalue: $a = -1$; spiral sink: $-1 < a < 0$; repeated 0 eigenvalues: $a = 0$; saddle: $a > 0$

  (c) Bifurcations at $a = 0$ and $a = -1$

5. (a)

  (b) One zero eigenvalue: $a = \pm 1$; saddle: $-1 < a < 1$
  (c) Bifurcations at $a = \pm 1$

7. (a)

(b) Real source: $a > 1$; zero eigenvalue: $a = 1$; saddle: $0 < a < 1$; repeated zero eigenvalues: $a = 0$; spiral sink: $a < 0$

(c) Bifurcations at $a = 0$ and $a = 1$

9. Eigenvalues are $a \pm |b|$. Zero eigenvalue: $a = \pm b$. Repeated eigenvalue: $b = 0$, $a \neq 0$. Repeated zero eigenvalue: $a = b = 0$. Real source: $a > |b|$. Real sink: $-a > |b|$. Saddle otherwise.

11. (a) $dy/dt = v$, $dv/dt = -3y - bv$

(b)

(c) Overdamped: $b > 2\sqrt{3}$; critically damped: $b = 2\sqrt{3}$; underdamped: $0 < b < 2\sqrt{3}$; undamped: $b = 0$

13. (a) $dy/dt = v$, $dv/dt = (-2y - v)/m$

(b)

(c) Overdamped: $0 < m < 1/8$; critically damped: $m = 1/8$; underdamped: $m > 1/8$

Hints and Answers for Section 3.8

1. Differentiate $\mathbf{Y}_2(t)$ and $\mathbf{Y}_3(t)$ coordinate by coordinate to see if the system of differential equations is satisfied.

3. (a) not linearly independent
   (b) linearly independent
   (c) linearly independent
   (d) linearly independent

5. (a) $\pm 1, -5$

   (b) The system decouples into a two-dimensional system in the xy-plane and a one-dimensional system in z.

   (c) There is a sink on the z-phase line and a saddle in the xy-phase plane.

(d)

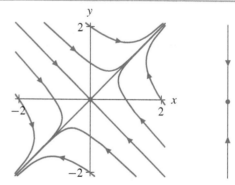

7. (a) 1, 3

(b) The system decouples into a two-dimensional system in the yz-plane and a one-dimensional system in x.

(c) Source on x-axis, source in yz-plane

(d)

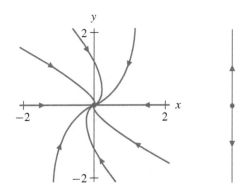

9. Using the fact that $p(a+ib) = 0$, show that the real and imaginary parts of $p(a-ib)$ are each zero.

11. (a) -2, 1

(b) source on the z-axis, repeated eigenvalue sink in the xy-plane

(c)

13. (a) 0, -5

(b) The z-axis is all equilibrium points; in the xy-plane there is a line of equilibrium points, and all solutions approach this line.

(c)

15. (a) $\lambda = 0$

(b) the x-axis

(c) $z(t)$ is a constant function, and the value of that constant is determined by the initial condition. For $z_0 = 0$, $y(t)$ is constant, and the solution curves lie on straight lines parallel to x-axis. For $y_0 > 0$, $x(t)$ is increasing, and for $y_0 < 0$, $x(t)$ is decreasing. For $z_0 \neq 0$, $x(t)$ is quadratic in y. Therefore, solution curves that satisfy the

initial condition $z(0) = z_0$ stay on the plane
$z = z_0$ and lie on a parabola.

17. (a) -1

(b) $-2 \pm i$

(c) Spiral sink

(d) *Hint*: Find the plane containing the real and
imaginary parts of an eigenvector for the
complex eigenvalue. This is the plane where
solutions spiral. Also, on the line $x = y$ and
$z = 0$, solutions tend directly toward the origin.

19. (a) Glen's profitability helps both Paul and Bob be
profitable (and they need all the help they can
get).

(b) The status of either Paul's or Bob's profits have
no effect on Glen's profit.

21. (a) *Hint*: If $z = 0$, $dz/dt = 0$, so this is a
two-dimensional problem.

(b)

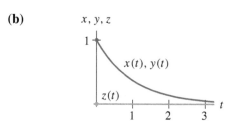

The graphs of $x(t)$ and $y(t)$ are
identical, and $z(t) = 0$ for all t.

(c) Glen continues to break even while Paul's and
Bob's profits tend toward the break-even point
(always staying equal).

Hints and Answers for Review Exercises in Chapter 3

1. $\lambda = 1, 2$

3. $\mathbf{Y}(t) = k_1 e^{3t} \begin{pmatrix} 1 \\ 0 \end{pmatrix} + k_2 e^{-2t} \begin{pmatrix} 0 \\ 1 \end{pmatrix}$

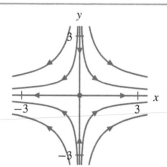

5. Overdamped: $b > \sqrt{20}$; critically damped: $b = \sqrt{20}$; underdamped: $0 < b < \sqrt{20}$; undamped: $b = 0$

7. the equilibrium solution $\mathbf{Y}(t) = (0, 0)$ for all t

9. $\dfrac{d\mathbf{Y}}{dt} = \begin{pmatrix} 0 & -6 \\ 2/3 & 0 \end{pmatrix} \mathbf{Y}$

11. False   **13.** True   **15.** False   **17.** True

19. (a) iv.   **(b)** iii.   **(c)** ii.   **(d)** vii.

21. (a) iii.   **(b)** vi.   **(c)** v.   **(d)** viii.

23. $y(t) = 2e^{-2t} - 2e^{-3t}$

25. $y(t) = e^{-t} + 2te^{-t}$

27. (a) $\mathbf{Y}(t) = k_1 e^{2t} \begin{pmatrix} 3 \\ 1 \end{pmatrix} + k_2 e^{-2t} \begin{pmatrix} 1 \\ -1 \end{pmatrix}$

(b)

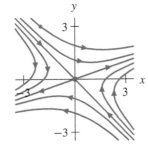

(c) $\mathbf{Y}(t) = \frac{1}{4} e^{2t} \begin{pmatrix} 3 \\ 1 \end{pmatrix} - \frac{11}{4} e^{-2t} \begin{pmatrix} 1 \\ -1 \end{pmatrix}$

(d)

29. (a) $\mathbf{Y}(t) = k_1 \begin{pmatrix} 3\cos\sqrt{2}\,t \\ 2\cos\sqrt{2}\,t - \sqrt{2}\sin\sqrt{2}\,t \end{pmatrix} + k_2 \begin{pmatrix} 3\sin\sqrt{2}\,t \\ \sqrt{2}\cos\sqrt{2}\,t + 2\sin\sqrt{2}\,t \end{pmatrix}$

(b)

(c) $\mathbf{Y}(t) = \begin{pmatrix} -2\cos\sqrt{2}\,t + 5\sqrt{2}\sin\sqrt{2}\,t \\ 2\cos\sqrt{2}\,t + 4\sqrt{2}\sin\sqrt{2}\,t \end{pmatrix}$

(d)

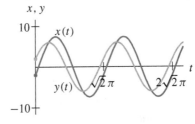

31. (a) $\mathbf{Y}(t) = e^{-2t} \begin{pmatrix} x_0 \\ y_0 \end{pmatrix} + te^{-2t} \begin{pmatrix} y_0 - x_0 \\ y_0 - x_0 \end{pmatrix}$

(b)

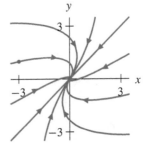

(c) $\mathbf{Y}(t) = e^{-2t} \begin{pmatrix} 4t - 3 \\ 4t + 1 \end{pmatrix}$

(d)

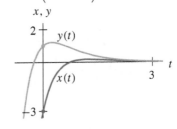

Hints and Answers for Section 4.1

1. $y(t) = k_1 e^{-2t} + k_2 e^{3t} + \frac{1}{6} e^{4t}$

3. $y(t) = k_1 e^{-t} + k_2 e^{2t} + \frac{5}{4} e^{3t}$

5. $y(t) = k_1 e^{-2t} \cos 3t + k_2 e^{-2t} \sin 3t - \frac{1}{3} e^{-2t}$

7. $y(t) = k_1 e^t + k_2 e^{4t} + \frac{1}{3} t e^{4t}$

9. $y(t) = -\frac{1}{2} e^{-2t} + \frac{1}{6} e^{-4t} + \frac{1}{3} e^{-t}$

11. $y(t) = \frac{1}{3} e^{-2t} \cos 3t - \frac{1}{3} e^{-2t}$

13. (a) $y(t) = k_1 e^{-t} + k_2 e^{-3t} + \frac{4}{5} e^{-t/2}$

   (b) $y(t) = -e^{-t} + \frac{1}{5} e^{-3t} + \frac{4}{5} e^{-t/2}$

   (c) All solutions are approximately equal to $\frac{4}{5} e^{-t/2}$ for t large (so they all tend to zero).

15. (a) $y(t) = k_1 e^{-t} + k_2 e^{-3t} + \frac{1}{3} e^{-4t}$

   (b) $y(t) = \frac{1}{6} e^{-t} - \frac{1}{2} e^{-3t} + \frac{1}{3} e^{-4t}$

   (c) All solutions tend to zero; most tend to zero like e^{-t}.

17. (a) $y(t) = k_1 e^{-2t} \cos 4t + k_2 e^{-2t} \sin 4t + \frac{1}{16} e^{-2t}$

   (b) $y(t) = -\frac{1}{16} e^{-2t} \cos 4t + \frac{1}{16} e^{-2t}$

   (c) All solutions tend to zero like e^{-2t}, and all but one oscillates with frequency $2/\pi$.

19. $y(t) = k_1 e^{-t} + k_2 t e^{-t} + t^2 e^{-t}/2$

21. (a) $y(t) = k_1 e^t + k_2 e^{4t} + \frac{5}{4}$

   (b) $y(t) = \frac{5}{4} - \frac{5}{3} e^t + \frac{5}{12} e^{4t}$

23. (a) $y(t) = k_1 e^{-t} \cos 3t + k_2 e^{-t} \sin 3t + 1$

   (b) $y(t) = -e^{-t} \cos 3t - \frac{1}{3} e^{-t} \sin 3t + 1$

25. (a) $y(t) = k_1 \cos 3t + k_2 \sin 3t + \frac{1}{10} e^{-t}$

   (b) $y(t) = -\frac{1}{10} \cos 3t + \frac{1}{30} \sin 3t + \frac{1}{10} e^{-t}$

   (c)

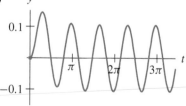

27. (a) $y(t) = k_1 \cos \sqrt{2}\, t + k_2 \sin \sqrt{2}\, t - \frac{3}{2}$

   (b) $y(t) = \frac{3}{2} \cos \sqrt{2}\, t - \frac{3}{2}$

   (c)

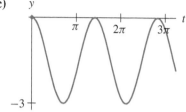

29. (a) $y(t) = k_1 \cos 3t + k_2 \sin 3t + \frac{2}{3}$

   (b) $y(t) = -\frac{2}{3} \cos 3t + \frac{2}{3}$

   (c)

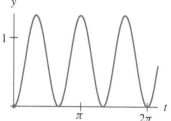

31. (a) $y(t) = k_1 \cos 2t + k_2 \sin 2t - \frac{3}{4} t^2 + \frac{1}{2} t + \frac{9}{8}$

   (b) $y(t) = \frac{7}{8} \cos 2t - \frac{1}{4} \sin 2t - \frac{3}{4} t^2 + \frac{1}{2} t + \frac{9}{8}$

33. (a) $y(t) = k_1 \cos 2t + k_2 \sin 2t + \frac{3}{4} t + \frac{1}{2}$

   (b) $y(t) = -\frac{1}{2} \cos 2t - \frac{3}{8} \sin 2t + \frac{3}{4} t + \frac{1}{2}$

   (c) The solution tends to ∞.

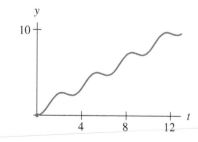

35. (a) $y(t) = k_1 \cos 2t + k_2 \sin 2t - \frac{1}{80}t^2 + \frac{1}{4}t + \frac{1}{160}$

(b) $y(t) = -\frac{1}{160} \cos 2t - \frac{1}{8} \sin 2t - \frac{1}{80}t^2 + \frac{1}{4}t + \frac{1}{160}$

(c) The solution tends to $-\infty$.

37. (a) $y(t) = k_1 e^{-3t} + k_2 e^{-2t} + \frac{1}{2}e^{-t} + \frac{2}{3}$

(b) $y(t) = \frac{11}{6}e^{-3t} - 3e^{-2t} + \frac{1}{2}e^{-t} + \frac{2}{3}$

(c) The solution tends to $2/3$.

39. (a) $y(t) = k_1 e^{-4t} + k_2 e^{-2t} - \frac{3}{16} + \frac{1}{4}t + \frac{1}{3}e^{-t}$

(b) $y(t) = \frac{5}{48}e^{-4t} - \frac{1}{4}e^{-2t} - \frac{3}{16} + \frac{1}{4}t + \frac{1}{3}e^{-t}$

(c) The solution tends to ∞.

41. (a) $y(t) = k_1 \cos 2t + k_2 \sin 2t + \frac{1}{4}t + \frac{1}{5}e^{-t}$

(b) $y(t) = -\frac{1}{5}\cos 2t - \frac{1}{40}\sin 2t + \frac{1}{4}t + \frac{1}{5}e^{-t}$

(c) The solution tends to ∞.

Hints and Answers for Section 4.2

1. $y(t) = k_1 e^{-2t} + k_2 e^{-t} + \frac{1}{10}\cos t + \frac{3}{10}\sin t$

3. $y(t) = k_1 e^{-2t} + k_2 e^{-t} - \frac{3}{10}\cos t + \frac{1}{10}\sin t$

5. $y(t) = k_1 e^{-4t} + k_2 e^{-2t} + \frac{7}{85}\cos t + \frac{6}{85}\sin t$

7. $y(t) = k_1 e^{-2t}\cos 3t + k_2 e^{-2t}\sin 3t + \frac{27}{145}\cos 2t + \frac{24}{145}\sin 2t$

9. $y(t) = k_1 e^{-2t}\cos 4t + k_2 e^{-2t}\sin 4t - \frac{3}{20}\sin 2t + \frac{3}{40}\cos 2t$

11. $y(t) = \frac{2}{17}e^{-4t} - \frac{1}{5}e^{-2t} + \frac{7}{85}\cos t + \frac{6}{85}\sin t$

13. $y(t) = -\frac{3}{40}e^{-2t}\cos 4t + \frac{3}{80}e^{-2t}\sin 4t - \frac{3}{20}\sin 2t + \frac{3}{40}\cos 2t$

15. (a) $a = -8/145$ and $b = 9/145$

(b) $A = -1/\sqrt{145}$ and $\phi = \arctan(9/8)$

17. (a) v   **(b)** ii   **(c)** i   **(d)** vi

19. (a) $y(t) = k_1 e^{-2t}\cos 4t + k_2 e^{-2t}\sin 4t + \frac{4}{65}e^{-t}\cos t + \frac{1}{130}e^{-t}\sin t$

(b) Solutions tend to zero.

21. ϕ is the polar angle of the complex number $z = a - bi$ and $k = |z| = \sqrt{a^2 + b^2}$ (see Appendix C).

23. *Hint:* Simplify $e^{-i\phi}e^{i\beta t}$ using the Laws of Exponents.

Hints and Answers for Section 4.3

1. $y(t) = k_1 \cos 3t + k_2 \sin 3t + \frac{1}{8}\cos t$

3. $y(t) = k_1 \cos 2t + k_2 \sin 2t - \frac{4}{15}\cos \frac{1}{2}t$

5. $y(t) = k_1 \cos 3t + k_2 \sin 3t + \frac{1}{3}t \sin 3t$

7. $y(t) = k_1 \cos \sqrt{3}\,t + k_2 \sin \sqrt{3}\,t - \frac{1}{6}\cos 3t$

9. $y(t) = -\frac{1}{8}\cos 3t + \frac{1}{8}\cos t$

11. $y(t) = -3\cos \sqrt{5}\,t + 3\cos 2t$

13. $y(t) = 2\cos 3t - 3\sin 3t + \frac{1}{3}t \sin 3t$

15. (a) $1/(16\pi)$

(b) $17/(16\pi)$

(c)

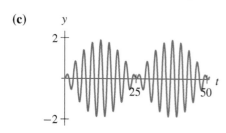

17. (a) $(\sqrt{5} - 2)/(4\pi)$
   (b) $(\sqrt{5} + 2)/(4\pi)$
   (c)

19. (a) $y(t) = k_1 \cos\sqrt{12}\,t + k_2 \sin\sqrt{12}\,t - \frac{3}{4}\cos 4t + \frac{2}{11}\sin t$

   (b) $y(t) = \frac{3}{4}\cos\sqrt{12}\,t - \frac{2}{11\sqrt{12}}\sin\sqrt{12}\,t - \frac{3}{4}\cos 4t + \frac{2}{11}\sin t$

   (c)

   (d) *Hint*: Which of the two forcing functions is more important?

21. (a) v   **(b)** iii   **(c)** i   **(d)** iv

23. *Hint*: Largest amplitude response occurs for $T = 3/2$.

Hints and Answers for Section 4.4

1. Rubbing a finger around the edge of the glass starts the glass vibrating. The finger then skips along the glass, giving a forcing term which has the same frequency as the natural frequency of vibration of the glass. Pressing harder changes the size of the forcing. Since the frequency is determined by the motion of the glass itself, pushing harder does not alter the frequency. (You can try this by rubbing a slightly damp finger *gently* around the rim of a good-quality glass.)

3. *Hint*: Use the Chain Rule applied to $y_p(t + \theta)$.

5. (a) $\left(\sqrt{(\omega_1^2 - \omega_2^2)^2 + p^2\omega_2^2}\right)/(p\omega_1)$

   (b) The ratio tends to infinity as $p \to 0$ and to ω_2/ω_1 as $p \to \infty$.

7. (a)

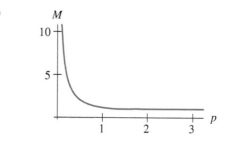

   (b) *Hint*: Use the formula in part (b) of Exercise 6. Recall that we are considering the case where $q > 0$ is fixed but $p \to 0^+$.

9. *Hint*: Guess $y(t) = \alpha \sin\omega t + \beta \cos\omega t$ and solve the resulting system of equations for α and β.

11. The amplitude of the oscillations will be very small.

Hints and Answers for Section 4.5

1. (a) β increases.

(b) The system becomes more like a linear system.

3. (a) γ increases.

(b) *Hint*: For values of y where $c(y) \neq 0$, $|c(y)|$ is increased.

5. *Hint*: $m\, d^2y/dt^2 + \epsilon\, dy/dt + pa^2\, y = mg$

What do y, ϵ, p, a, m, and g represent?

7. $m\, d^2y/dt^2 + \epsilon\, dy/dt = mg + b(y)$ where $b(y) = -pa^2 y$ if $y > 0$ and $b(y) = 0$ if $y < 0$.
What do ϵ, p, a, m, and g represent?

Hints and Answers for Review Exercises in Chapter 4

1. the constant function $y(t) = 1/k$ for all t

3. $1/\pi$

5. Yes, $\dfrac{\lambda}{\lambda^2 + \omega^2}\cos \omega t + \dfrac{\omega}{\lambda^2 + \omega^2}\sin \omega t$

(see page 118).

7. $T \approx \ln 100 \approx 4.6$

9. You can find the ratio of the spring constant and the mass.

11. True

13. False

15. $y(t) = k_1 e^{-2t} + k_2 e^{-4t} + \frac{1}{3}e^{-t}$

17. $y(t) = k_1 e^{-t} + k_2 e^{3t} + \frac{1}{4}t e^{3t}$

19. $y(t) = k_1 e^{-2t} + k_2 e^{-4t} + \frac{5}{8}$

21. $y(t) = k_1 e^{2t}\cos 3t + k_2 e^{2t}\sin 3t - \frac{3}{53}\cos 4t - \frac{16}{53}\sin 4t$

23. (a) vii   **(b)** ii   **(c)** v   **(d)** viii

25. (a) $y(t) = k_1 e^{-t}\cos \sqrt{2}\,t + k_2 e^{-t}\sin \sqrt{2}\,t - \frac{1}{17}\cos 2t + \frac{4}{17}\sin 2t$

(b) Amplitude $A = 1/\sqrt{17} \approx 0.24$; phase angle $\phi \approx -104°$

27. (a) $y(t) = k_1 e^{-3t} + k_2 e^{-t} - \frac{8}{13}\cos 2t - \frac{1}{13}\sin 2t$

(b) Amplitude $A = \sqrt{65}/13 \approx 0.62$; phase angle $\phi \approx -187°$. (The phase angle is not between $-180°$ and $0°$ due to the fact that the forcing is in terms of sine rather than cosine.)

Hints and Answers for Section 5.1

1. Systems (i) and (iii) have the same "local picture" near $(0, 0)$.

3. (a) The linearized system at the origin is
$$dx/dt = -2x + y, \, dy/dt = -y.$$

(b) Sink

(c)

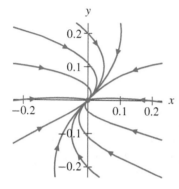

(d) The linearized system at $(2, 4)$ is
$$dx/dt = -2x + y, \, dy/dt = 4x - y.$$ Hence, $(2, 4)$ is a saddle.

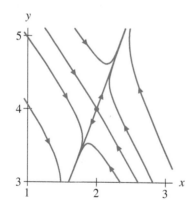

5. (a) $x(t) = x_0 e^{-t}$

(b) $y(t) = x_0^3 e^{-3t} + (y_0 - x_0^3)e^t$

(c) $x(t) = x_0 e^{-t}$ and $y(t) = x_0^3 e^{-3t} + (y_0 - x_0^3)e^t$

(d) y_0 must equal x_0^3

(e) x_0 must be 0

(f)

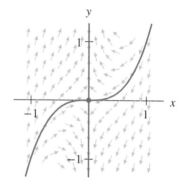

(g) For the nonlinear system, solutions tend to the origin along the curve $y = x^3$. For the linearized system, solutions tend to the origin along the x-axis.

7. (a) Equilibria: $(0, 0)$, $(0, 100)$, $(150, 0)$, $(30, 40)$. They are source, sink, sink, and saddle respectively.

(b)

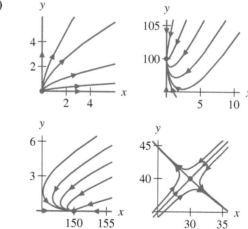

9. (a) Equilibria: $(0, 0)$, $(0, 25)$, $(100, 0)$, $(75, 12.5)$. They are source, saddle, saddle, and sink respectively.

(b)

11. (a) Equilibria: $(0, 0)$, $(0, 50)$, $(40, 0)$. They are source, sink, and saddle respectively.

(b)

13. (a) Equilibria: $(0, 0)$, $(0, 50)$, $(60, 0)$, $(30, 40)$, $(234/5, 88/5)$. They are source, saddle, sink, sink, saddle respectively.

(b)

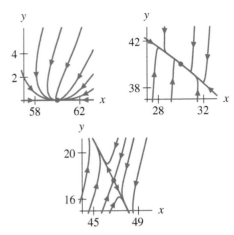

15. (a) Equilibria: $(0, 0)$, $(1, 1)$, $(2, 0)$. The point $(0, 0)$ has zero as an eigenvalue (see phase portrait to determine local behavior). $(1, 1)$ is a saddle, and $(2, 0)$ is a sink.

(b)

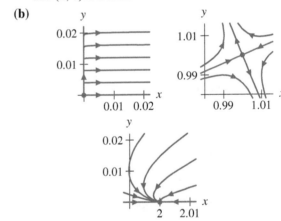

17. (a) $dx/dt = 0$, $dy/dt = -y$

(b) Eigenvalues 0 and -1 with eigenvectors $(1, 0)$ and $(0, 1)$ respectively.

(c) $dx/dt = 0, dy/dt = y$

(d) Eigenvalues 0 and 1 with eigenvectors $(1, 0)$ and $(0, 1)$ respectively.

(e)

(f) The equation for dx/dt contains only "higher-order" terms.

19. (a)

(b)

(c)

21. (a) If $a \neq 0$, there are no equilibrium points. If $a = 0$, the curve $y = x^2$ consists entirely of equilibrium points.

(b) $a = 0$

(c) If $a < 0$ all solutions decrease in the y-direction since $dy/dt < 0$. If $a > 0$ all solutions increase in the y-direction since $dy/dt > 0$. If $a = 0$ there is a curve of equilibrium points located along $y = x^2$.

Phase portrait for $a < 0$

Phase portrait for $a = 0$

Phase portrait for $a > 0$

23. (a) $(0, 0), (\pm 1/\sqrt{a}, \pm 1/\sqrt{a})$

(b) $a = 0$

(c) If $a < 0$, there are is only one equilibrium point at the origin, and this equilibrium point is a spiral source. If $a > 0$, the system has two additional equilibrium points, at $(\pm 1/\sqrt{a}, \pm 1/\sqrt{a})$. These equilibrium points come from infinity as a increases through 0.

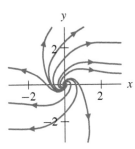

Phase portrait for $a < 0$

Phase portrait for $a = 0$

Phase portrait for $a > 0$

25. (a) $(\pm \sqrt{a/2}, -a/2)$

(b) $a = 0$

(c) If $a < 0$, there are no equilibrium points. If $a = 0$, an equilibrium point appears at the origin. If $a > 0$, the system has two equilibrium points, at $(\pm \sqrt{a/2}, -a/2)$.

Phase portrait for $a < 0$

Phase portrait for $a = 0$

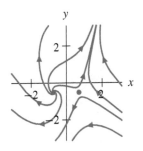

Phase portrait for $a > 0$

27. (a) If X and Y are not present on the island, then neither emigrate to the island.

(b) They are positive and large.

(c) They are negative.

(d) Source or saddle

(e)

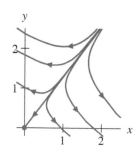

29. (a) At $(0, 0)$, $\partial f/\partial x$ is large, and $\partial g/\partial y$ is small.

(b) At $(0, 0)$, $\partial f/\partial y$ is negative with a large absolute value and $\partial g/\partial x = 0$.

(c) Source

(d)

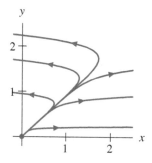

Hints and Answers for Section 5.2

1. Equilibria: $(1, 1)$ and $(-2, 4)$. All three solutions will go down and to the right without bound.

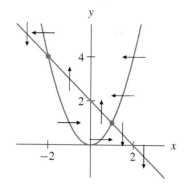

3. Equilibria: $(0, 0)$ and $(1, 1)$. Solutions with initial conditions (a) and (b) move toward the equilibrium

point $(0, 0)$, while the solution with initial condition (c) travels down the line $x = 1$ toward the equilibrium point at $(1, 1)$.

5. (a)

(b)

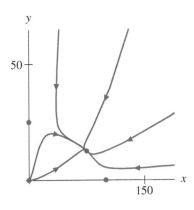

(c) *Hint*: See the phase portrait in part (b).

(b)

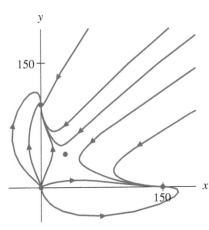

(c) *Hint*: See the phase portrait in part (b).

9. (a)

7. (a)

(b)

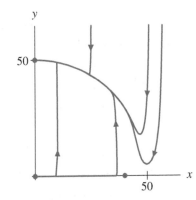

(c) *Hint*: See the phase portrait in part (b).

11. (a)

(b)

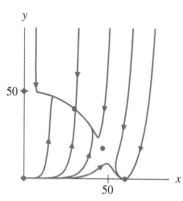

(c) *Hint*: See the phase portrait in part (b).

13. (a)

(b)

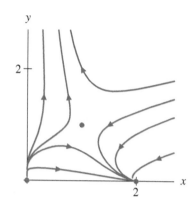

(c) *Hint*: See the phase portrait in part (b).

15. (a) Change the signs in front of both B and D.

(b) The nullclines are lines, and the points $(0, 0)$, $(0, F/E)$, and $(C/A, 0)$ are equilibrium points. In order for there to be an equilibrium point with both x and y positive, $AE - BD > 0$.

17. (a)

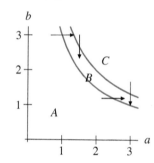

(b) In region A, the vector field points up and to the right. In region B, the vector field points down and to the right. In region C, the vector field points down and to the left.

(c) Once in region B, a solution cannot leave. As time increases, these solutions are asymptotic to the positive x-axis from above.

19. (a)

21.

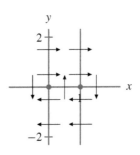

(b) In region A, the vector field points up and to the right. In region B, the vector field points down and to the right. In region C, the vector field points down and to the left.

(c) Solutions that enter region B with $b < 2$ stay in B and $a(t) \to \infty$ and $b(t) \to 0$ as t increases.

23. The linearized system at $(0, 0)$ is $dx/dt = y$ and $dy/dt = x$. The linearized system at $(1, 0)$ is $dx/dt = y$ and $dy/dt = -x$. We do not know if the equilibrium point at $(1, 0)$ is a nonlinear center. See Section 5.3.

Hints and Answers for Section 5.3

1. (a) *Hint*: Compute $\partial H/\partial x$ and $\partial H/\partial y$. Then compare.

(b) $(0, 0)$ is a local minimum; $(\pm 1, 0)$ are saddles.

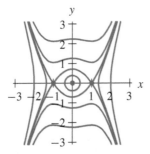

(c) $(0, 0)$ is a center; $(\pm 1, 0)$ are saddles that are connected by separatrix orbits.

3. (a) *Hint*: Compute $\partial H/\partial x$ and $\partial H/\partial y$. Then compare.

(b)

(c) The equilibrium points occur at points of the form $((1 - 4n)\pi, (2n - \frac{1}{2})\pi)$ and $((1 + 4n)\pi, (2n + \frac{1}{2})\pi)$ where n is an integer.

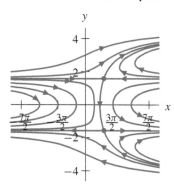

5. No, longer.

7. (a) Multiply by $\sqrt{2}$.

    (b) $\pi/\sqrt{gl}$

9. Phase portraits (a) and (d) might correspond to Hamiltonian systems, but portraits (b) and (c) do not.

11. Yes, $H(x, y) = xy - y^3$

13. Not Hamiltonian

15. (a) It is not Hamiltonian.

    (b)

    (c) Yes.

    (d) $H(x, y) = -y + 2 \arctan y - x^2/2$

    (e) Changing the length of the vectors doesn't change the direction field, so the phase portraits of the two systems are the same.

17. If $y > 2$, $H(x, y) = -x^2/2 + 2y + y^2/2 + 3\ln(y-2)$; if $y < 2$, $H(x, y) = -x^2/2 + 2y + y^2/2 + 3\ln(2-y)$.

19. *Hint:* For $a \neq 0$ the system has two equilibrium points, but for $a = 0$, it has only one.

Hints and Answers for Section 5.4

1. (a) Let $(x(t), y(t))$ be a solution and differentiate $L(x(t), y(t))$ with respect to t.

    (b) The level sets are circles.

    (c) All solutions tend toward the origin as t increases.

3. (a) The eigenvalues are $-0.05 \pm i\sqrt{15.99}/2$.

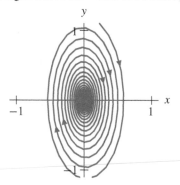

    (b) Let $(x(t), y(t))$ be the solution with $x(0) = -1$,

$y(0) = 1$ and compute the derivative with respect to t of $L(x(t), y(t))$ at $t = 0$.

    (c) Let $(x(t), y(t))$ be any solution and compute the derivative with respect to t of $K(x(t), y(t))$.

5. It runs fast.

7. $m > b\sqrt{l}/(2\sqrt{g})$

9. *Hint:* The amplitude of the solutions near $(0, 0)$ decreases like $e^{-tb/(2m)}$.

11. (a) $(0, 0)$ is a spiral source; $(\pi, 0)$ is a saddle.

    (b)

    (c) Oscillate with increasing amplitude

13. (a) $dx/dt = 2x$, $dy/dt = -2y$

(b) Saddle

(c)

(d)

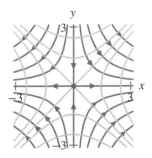

15. (a) The eigenvalues are 1 and -1.

(b) The eigenvalues are -2 and -1.

(c) *Hint*: $(0, 1)$ is an eigenvector for eigenvalue -1.

(d) The Jacobian matrices are the same.

17. (a)
$$\frac{dx}{dt} = 2x - x^3 - 6xy^2$$
$$\frac{dy}{dt} = 2y - y^3 - 6x^2y$$

(b)

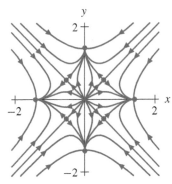

(c) There are four dead fish located at the sinks.

(d)

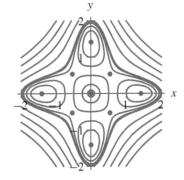

(e) S is negative for large $|x|$ or $|y|$.

19. (a) *Hint*: Mixed partials are equal.

(b) $\partial f/\partial y \neq \partial g/\partial y$.

21. Let $(y(t), v(t))$ be a solution of the second system and compute $d(H(y(t), v(t)))/dt$.

23. *Hint*: Differentiate $H(x_1(t), x_2(t), p_1(t), p_2(t))$ with respect to t using the Chain Rule and the fact that $(x_1(t), x_2(t), p_1(t), p_2(t))$ is a solution.

Hints and Answers for Section 5.5

1. (a) $\begin{pmatrix} -1 & -1 & 0 \\ 0 & 2 & 0 \\ 0 & 0 & 1 \end{pmatrix}$

(b) Eigenvalues 1, 2, −1 and eigenvectors (0, 0, 1), (1, −3, 0) and (1, 0, 0) respectively

(c) Saddle

(d)

(e) Solutions with $y = z = 0$ move toward the equilibrium point. All others move away.

3. (a) $\begin{pmatrix} 1 & 0 & 0 \\ 0 & 0 & 0 \\ 0 & 1 & -1 \end{pmatrix}$

(b) The eigenvalues are 1, 0 and −1 and eigenvectors are (1, 0, 0), (0, 1, 1) and (0, 0, 1), respectively.

(c) The linearized system has a line of equilibrium points, so it does not have a name.

(d)

(e) Solutions with $x = 0$ move toward an equilibrium point; solutions with $x \neq 0$ move in the direction of increasing x.

5. (a) *Hint*: Set $y = 0$ and look at the system of equations, then try $x = 0$ and $z = 0$.

(b) *Hint*: What is the connection between wolves and trees if moose are extinct?

7. (a) (0, 0, 0), (1, 0, 0), (0, 1, 0), (0, 0, 1), (1, 0, 1), and
$$\left(\frac{\zeta + 1}{2\zeta + 1}, \frac{\zeta}{2\zeta + 1}, \frac{2\zeta + 2}{2\zeta + 1} \right).$$

(b) As ζ decreases, x_0 increases, y_0 decreases, and z_0 increases.

(c) *Hint*: Compare how ζ comes into the equation for dz/dt with the system in the section.

Hints and Answers for Section 5.6

1. Graph (ii)

3. Graph (i)

5. Graph (iii)

7. Graph (iv)

9. (a) $y(t) = -\frac{\sqrt{3}}{30} \sin \sqrt{3}\, t + \cos \sqrt{3}\, t + \sin \frac{1}{10} t$

(b) Only the terms corresponding to the homogeneous solution affect the Poincaré map.

(c) The same, only larger

(d) *Hint*: There are no saddles for the (unforced) harmonic oscillator.

Hints and Answers for Review Exercises in Chapter 5

1. $dx/dt = x$ and $dy/dt = -2y$

3. $dx/dt = 3x$ and $dy/dt = 2y$

5.

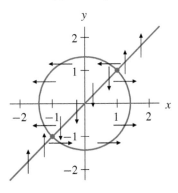

7. No

9. The solution could tend to infinity or to a periodic solution.

11. True   **13.** False

15. (a) Equilibria: $(12, 2)$ is a spiral sink, and $(3, -1)$ is a saddle.

(b)

(c)

17. (a) Equilibria: $(0, 0)$, $(4, 0)$, $(0, 3)$, $(2, 2)$. They are source, saddle, saddle, and sink respectively.

(b)

(c)

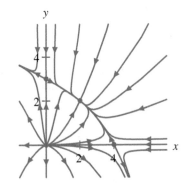

19. Equilibria: $(\sqrt{a}, \sqrt{b})$, $(-\sqrt{a}, \sqrt{b})$, $(\sqrt{a}, -\sqrt{b})$, $(-\sqrt{a}, -\sqrt{b})$. They are source, saddle, saddle, and sink respectively.

21. (a) Equilibria: $(0, \sqrt{b})$, $(0, -\sqrt{b})$. The point $(0, \sqrt{b})$ is a "saddle" for $x < 0$ and a "source" for $x > 0$. The point $(0, -\sqrt{b})$ is a "sink" for $x < 0$ and a "saddle" for $x > 0$.

(b)

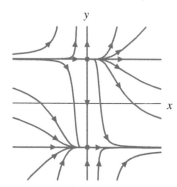

(c) As $a \to 0$ from above, the x-nullclines merge. The pair of equilibrium points $(\sqrt{a}, \sqrt{b})$ and $(-\sqrt{a}, \sqrt{b})$ coalesce at $(0, \sqrt{b})$, and the pair of equilibrium points $(\sqrt{a}, -\sqrt{b})$ and $(-\sqrt{a}, -\sqrt{b})$ coalesce at $(0, -\sqrt{b})$.

23. (a) The origin is the only equilibrium point. It is a "source" in the first quadrant, a "saddle" in the second and fourth quadrants, and a "sink" in the third quadrant.

(b)

25. (a) Equilibrium points: $(0, \pm 1)$. Both are saddles.

(b)

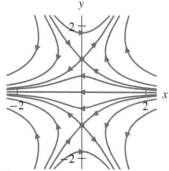

(c) This system is a Hamiltonian system with Hamiltonian $H(x, y) = (y^3/3) - (x^2 + 1)y$.

(d) This system is the gradient of $G(x, y) = x(y^2 - 1) - x^3/3$.

27. *Hint*: Use the method for construction Hamiltonian functions described on pages 499–502.

29. (a) There is one equilibruim point subject to
$$-\pi < \theta < \pi. \text{ It is } (\theta, s) = \left(-\pi/4, 1/\sqrt[4]{2}\right).$$

(b) Spiral sink

Hints and Answers for Section 6.1

1. $\mathcal{L}[f] = 3/s$

3. $\mathcal{L}[h] = -10/s^3$

5. *Hint*: Use the formula $\mathcal{L}[dy/dt] = s\mathcal{L}[y] - y(0)$.

7. e^{3t}

9. $(2e^{-5t/3})/3$

11. $\frac{4}{3}\left(1 - e^{-3t}\right)$

13. $5e^{2t} - 3e^t$

15. (a) $\mathcal{L}[dy/dt] = s\mathcal{L}[y] - y(0)$ and
$\mathcal{L}[-y + e^{-2t}] = -\mathcal{L}[y] + 1/(s+2)$
(b) $\mathcal{L}[y] = (2s+5)/((s+1)(s+2))$
(c) $y(t) = 3e^{-t} - e^{-2t}$

17. (a) $\mathcal{L}[dy/dt + 7y] = s\mathcal{L}[y] - y(0) + 7\mathcal{L}[y]$
$\mathcal{L}[1] = 1/s$
(b) $\mathcal{L}[y] = (3s+1)/(s(s+7))$

(c) $y(t) = 20e^{-7t}/7 + 1/7$

19. (a) $\mathcal{L}[dy/dt + 9y] = s\mathcal{L}[y] - y(0) + 9\mathcal{L}[y]$
$\mathcal{L}[2] = 2/s$
(b) $\mathcal{L}[y] = (-2s+2)/(s(s+9))$
(c) $y(t) = -20e^{-9t}/9 + 2/9$

21. (a) $\mathcal{L}[dy/dt + y] = s\mathcal{L}[y] - y(0) + \mathcal{L}[y]$
$\mathcal{L}[e^{-2t}] = 1/(s+2)$
(b) $\mathcal{L}[y] = (s+3)/((s+1)(s+2))$
(c) $y(t) = 2e^{-t} - e^{-2t}$

23. (a) $\mathcal{L}[dy/dt] = s\mathcal{L}[y] - y(0)$ and
$\mathcal{L}[-y + t^2] = -\mathcal{L}[y] + 2/s^3$
(b) $\mathcal{L}[y] = (2 + s^3)/(s^3(s+1))$
(c) $y(t) = t^2 - 2t + 2 - e^{-t}$

25. $-2e^{-3t}/5 + ce^{2t}$ where $c = y(0) + 2/5$

27. *Hint*: What do you need to do with $\mathcal{L}[y^2]$?

Hints and Answers for Section 6.2

1. (a) $g_a(t) = 1 - u_a(t)$
(b) $\mathcal{L}[g_a] = (1 - e^{-as})/s$

3. $(1 - e^{-as})/as^2$

5. $u_3(t)\left(e^{2(t-3)} - e^{(t-3)}\right)$

7. $u_1(t)\left(e^{4(t-1)} - e^{-2(t-1)/3}\right)$

9. $y(t) = -2e^{-9t} + \frac{1}{9}u_5(t)\left(1 - e^{-9(t-5)}\right)$

11. $y(t) = e^{-t} + u_2(t)\left(e^{-(t-2)} - e^{-2(t-2)}\right)$

13. $y(t) = u_1(t)\left((t-2) + e^{-(t-1)}\right) + 2e^{-t}$

15. $y(t) = u_a(t)\left(1 - e^{-(t-a)}\right) + y(0)e^{-t}$

17. $\mathcal{L}[w] = (1 - e^{-s})/(s(1 + e^{-s}))$

19. (a) $\mathcal{L}[y] = (1 - e^{-s})/(s(s+1)(1 + e^{-s}))$
(b) The function $w(t)$ is alternatively 1 and -1.
While $w(t) = 1$, the solution decays exponentially toward $y = 1$. When $w(t)$ changes to -1,
the solution then decays toward $y = -1$.

Hints and Answers for Section 6.3

1. $\mathcal{L}[\sin \omega t] = \omega/(s^2 + \omega^2)$

3. $\mathcal{L}[e^{at} \sin \omega t] = \omega/((s-a)^2 + \omega^2)$

5. *Hint*: Take the Laplace transform of both sides of
the equation and solve for $\mathcal{L}[y]$.

7. $\mathcal{L}[f] = (s^2 - \omega^2)/(s^2 + \omega^2)^2$

9. $\mathcal{L}[f] = 2/(s-a)^3$

11. $(s+1)^2 + 3^2$

13. $(s+1/2)^2 + (\sqrt{3}/2)^2$

15. $(1/3)e^{-t}\sin 3t$

17. $2e^{-t/2}\cos(\sqrt{3}t/2) + (4/\sqrt{3})e^{-t/2}\sin(\sqrt{3}t/2)$

19. Same as for real exponentials.

21. *Hint*: Recall that $e^{at}\cos\omega t$ is the real part of $e^{(a+ib)t}$.

23. $(1/3)e^{-t}\sin 3t$

25. $2e^{-t/2}\cos(\sqrt{3}\,t/2) + (4/\sqrt{3}\,)e^{-t/2}\sin(\sqrt{3}\,t/2)$

27. (a) $\mathcal{L}[d^2y/dt^2 + 4y] = (s^2 + 4)\mathcal{L}[y] - 11s - 5$
$\mathcal{L}[8] = 8/s$

(b) $\mathcal{L}[y] = 2/s + (9s + 5)/(s^2 + 4)$

(c) $y(t) = 2 + 9\cos 2t + (5/2)\sin 2t$

29. (a) $\mathcal{L}[d^2y/dt^2 - 4dy/dt + 5y] =$
$(s^2 - 4s + 5)\mathcal{L}[y] - 3s + 11$
$\mathcal{L}[2e^t] = 2/(s - 1)$

(b) $\mathcal{L}[y] = 1/(s - 1) + (2s - 8)/(s^2 - 4s + 5)$

(c) $y(t) = e^t + 2e^{2t}\cos t - 4e^{2t}\sin t$

31. (a) $\mathcal{L}[d^2y/dt^2 + 4y] = (s^2 + 4)\mathcal{L}[y] + 2s$
$\mathcal{L}[\cos 2t] = s/(s^2 + 4)$

(b) $\mathcal{L}[y] = -2s/(s^2 + 4) + s/(s^2 + 4)^2$

(c) $y(t) = -2\cos 2t + (t/4)\sin 2t$

33. (a) $\mathcal{L}[d^2y/dt^2 + 4dy/dt + 9y] =$
$(s^2 + 4s + 9)\mathcal{L}[y] - s - 6$
$\mathcal{L}[20u_2(t)\sin(t - 2)] = 20e^{-2s}/(s^2 + 1)$

(b) $\mathcal{L}[y] = (s + 6)/(s^2 + 4s + 9) +$
$20e^{-2s}/((s^2 + 1)(s^2 + 4s + 9))$

(c) $y(t) = e^{-2t}\cos\sqrt{5}\,t + (4/\sqrt{5})e^{-2t}\sin\sqrt{5}\,t +$
$u_2(t)\left[e^{-2(t-2)}\cos(\sqrt{5}\,(t - 2)) -\right.$
$\left.\cos(t - 2) + 2\sin(t - 2)\right]$

35. *Hint*: Differentiate the formula for $\mathcal{L}[f]$ with respect to s.

Hints and Answers for Section 6.4

1. *Hint*: Remember L'Hôpital's Rule: Differentiate numerator and denominator with respect to Δt.

3. $y(t) = e^{-t}\cos 2t + e^{-t}\sin 2t +$
$(1/2)u_3(t)e^{-(t-3)}\sin(2(t - 3))$

5. $y(t) = (1/\sqrt{2})u_1(t)e^{-(t-1)}\sin(\sqrt{2}(t - 1)) -$
$(3/\sqrt{2})u_4(t)e^{-(t-4)}\sin(\sqrt{2}(t - 4))$

7. (a) $\mathcal{L}[\delta_a] = e^{-as}$
$s\mathcal{L}[u_a] - u_a(0) = s(e^{-as}/s) - 0 = e^{-as}$

(b) This relationship suggests that $du_a/dt = \delta_a$.

(c) Approximate $u_a(t)$ with piecewise linear continuous functions and take their derivatives.

9. (a) $\mathcal{L}[y] = (1/(s^2 + 2))\sum_{n=1}^{\infty} e^{-ns}$

(b) $y(t) = (1/\sqrt{2})\sum_{n=1}^{\infty} u_n(t)\sin(\sqrt{2}(t - n))$

(c) *Hint*: It is not resonant forcing.

Hints and Answers for Section 6.5

1. $1 - e^{-t}$

3. $u_2(t)\sin(t - 2)$

5. $\cos t - \cos 2t$

7. $p = 4, q = 5$

9. (a) $a(s + p)\mathcal{L}[\zeta]$

(b) $b\mathcal{L}[\zeta]$

(c) $(\mathcal{L}[f] + a(s + p) + b)\mathcal{L}[\zeta]$

11. (a) $\mathcal{L}[y_1] = \mathcal{L}[f_1]/(s^2 + ps + q)$

(b) *Hint*: Calculate both and compare.

(c) *Hint*: Solve for $\mathcal{L}[y_2]$.

Hints and Answers for Section 6.6

1. (a) $\mathcal{L}[y] = (2s + 2)/(s^2 + 2s + 2) +$
$4/((s^2 + 4s + 20)(s^2 + 2s + 2))$

(b) Poles are at $s = -2 \pm 4i$ and at $s = -1 \pm i$.

(c) Since the poles have negative real part, solutions decrease to zero (at a rate of e^{-t}). Since the poles are complex, the solutions oscillate (the oscillations have periods 2π and $\pi/2$).

3. (a) $\mathcal{L}[y] = \dfrac{\cos(4)s + \sin(4) - se^{-4s}}{(s^2 + s + 8)(s^2 + 1)}$

(b) The poles are $s = \pm i$ and $s = -1/2 \pm i\sqrt{31}/2$.

(c) *Hint*: The forcing "turns off" at time $t = 4$.

5. (a) $\mathcal{L}[y] = (s + 1)/(s^2 + 16)$

(b) $s = \pm 4i$

(c) The solution is periodic and does not decay.

7. (a) $\mathcal{L}[y] = (s + 4)/(s^2 + 2s + 1)$

(b) Poles are at $s = -1$.

(c) The behavior of a second-order homogeneous linear equation with a double pole on the real axis corresponds to a first-order system with double real eigenvalues.

9. (a) $\mathcal{L}[y] = (s + 20)/(s^2 + 20s + 200) +$
$(1 - e^{-s})/(s(1 + e^{-s})(s^2 + 20s + 200))$

(b) The poles are $s = -10 \pm 10i$, $s = 0$, and $s = (2n + 1)i\pi$ for $n = 0, \pm 1, \pm 2, \ldots$

(c) *Hint*: The natural response dies out quickly, so the solution oscillates between the forced response for forcing $+1$ and forcing -1.

Hints and Answers for Review Exercises in Chapter 6

1. $2(e^t - e^{-t})$

3. $s/(s^2 + 5)$

5. $(1 - e^{-6s})/s$

7. e^{-4}

9. $y(t) = u_1(t) + u_2(t) + u_3(t) + \ldots$

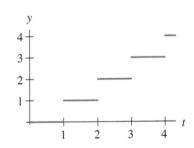

11. (a) iii. (b) xi. (c) i. (d) vi. (e) vii. (f) ix.

13. True

15. True

17. False

19. $3e^{2t} - 2e^{-4t}$

21. $u_3(t)\left(2e^{2(t-3)} + 3e^{3(t-3)}\right)$

23. te^{2t}

25. $y(t) = \left(1 + 3t + \tfrac{1}{2}t^2\right)e^{2t}$

27. $y(t) = 5e^{4t} +$
$u_2(t)\left(6e^{4(t-2)} - 6\cos(3(t - 2)) - 8\sin(3(t - 2)))\right)$

29. $y(t) = 2e^{-t}\cos\sqrt{2}\,t + (3/\sqrt{2})e^{-t}\sin\sqrt{2}\,t +$
$(u_4(t)/\sqrt{2})e^{-(t-4)}\sin\sqrt{2}\,(t - 4)$

Hints and Answers for Section 7.1

1. (a) $y(t) = 1/(1 + t^2)$

   (b) $y_{20} \approx 0.19334189$

   (c) $e_{20} \approx 0.0066581009$

   (d) $y_{1000} \approx 0.19987415, \ldots, y_{6000} \approx 0.19997904$

   (e) $e_{100} \approx 0.0012709461$, $e_{200} \approx 0.00063198471$, $e_{300} \approx 0.00042055770, \ldots, e_{6000} \approx .000020955836$

   (f) $K \approx 0.126731$

3. (a) $y(t) = e^{t^2/2}$

   (b) $y_{20} \approx 2.51066$

   (c) $e_{20} \approx 0.20762$

   (d) $y_{1000} \approx 2.71376, \ldots, y_{6000} \approx 2.71753$

   (e) $e_{100} \approx 0.0444901$, $e_{200} \approx 0.0224467, \ldots,$ $e_{6000} \approx 0.000754848$

   (f) $K \approx 4.47023$

5. (a) $y(t) = (11/9)e^{3t} + t/3 - 2/9$

   (b) $y_{20} \approx 20.1147$

   (c) $e_{20} \approx 4.54544$

   (d) $y_{1000} \approx 24.5501, \ldots, y_{6000} \approx 24.6417$

   (e) $e_{100} \approx 1.05955$, $e_{200} \approx 0.540843, \ldots,$ $e_{6000} \approx 0.0183987$

   (f) $K \approx 107.125$

7. (a) $M_1 = 8t^2 y^3 - 2y^2$, $M_2 = -4ty$

   (b) The estimated error is 0.00039984000 while the real error is approximately 0.00039984006.

   (c) The estimated error is 0.00079744230 while the real error is approximately 0.00079744409.

   (d) The third and fourth estimated errors are 0.0011896358 and 0.0015733200. The third

and fourth real errors are 0.0011896455 and 0.0015733534.

   (e) Compare your plots with Figures 7.5 and 7.6 in Section 7.1.

9. We cannot compare our estimate to the true error since we do not know how to calculate the true error for this differential equation. (We cannot find a closed-form solution.)

   (a) $M_1 = (y + t \sin ty) \cos ty$, $M_2 = t \cos ty$

   (b) The estimated error in the first step is 0.00134894.

   (c) The estimated error in the second step is 0.00269115.

   (d)

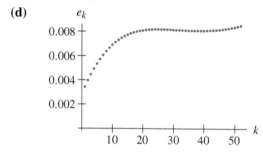

11. *Hint*: For parts (a)–(c), refer to Figures 7.3 and 7.4. For parts (d) and (e), use the inequality in part (c). For part (f), recall geometric sums from calculus. Parts (g)–(i) are just algebra. The expression $M_1(e^{M_2(t_n - t_0)} - 1)/(2M_2)$ is a constant in part (i) once the function $f(t, y)$ and the interval of time are given. In other words, $C = M_1/(2M_2)(e^{M_2(t_n - t_0)} - 1)$. The bounds that contribute to C do not depend on n.

Hints and Answers for Section 7.2

1.

Results of improved Euler's method.

k	t_k	y_k
0	0.0	3.0000
1	0.5	8.2500
2	1.0	21.3750
3	1.5	54.1875
4	2.0	136.2187

3.

Results of improved Euler's method

k	t_k	y_k
0	0.00	0.5
1	0.25	0.445801
2	0.50	0.103176
3	0.75	−0.501073
4	1.00	−1.16818
5	1.25	−1.70535
6	1.50	−2.09616
7	1.75	−2.39212
8	2.00	−2.63277

5.

Results of improved Euler's method

k	t_k	y_k
0	0.0	0.
1	0.5	1.6875
2	1.0	2.06728
3	1.5	2.22283
4	2.0	2.31738
5	2.5	2.38333
6	3.0	2.43292
7	3.5	2.47205
8	4.0	2.50398
9	4.5	2.53071
10	5.0	2.55351

7.

Results of improved Euler's method

k	t_k	y_k
0	1.0	2.
1	1.5	3.13301
2	2.0	4.01452
3	2.5	4.80396
4	3.0	5.54124

9. (a) The analytic solution is $y(t) = 1 - e^{-t}$.
  (b) $y_4 \approx 0.627471$, $e_4 \approx .00464959$
  (c) Use $n = 28$ steps. Then $e_{28} \approx 0.00008$.

11. (a) The analytic solution is $y(t) = 1/(t + 2)$.
  (b) $y_4 \approx 0.252281$, $e_4 \approx 0.002281$
  (c) Use $n = 20$ steps. Then $e_{20} \approx 0.00008$.

13. (a) 53 steps
  (b) $y_{53} = 0.200095$
  (c) $e_{53} = 0.000095$

15.

17.

Hints and Answers for Section 7.3

1.

Results of Runge-Kutta.

t_k	y_k
0.0	3.000
0.5	8.979
1.0	25.173
1.5	69.030
2.0	187.811

3.

Results of Runge-Kutta.

t_k	y_k	t_k	y_k
0.0	0.00000	3.0	2.99645
0.5	1.82290	3.5	2.99882
1.0	2.70058	4.0	2.99961
1.5	2.90368	4.5	2.99987
2.0	2.96803	5.0	2.99996
2.5	2.98935		

5.

Results of Runge-Kutta.

t_k	y_k
1.0	2.00000
1.5	3.10456
2.0	3.98546
2.5	4.77554
3.0	5.51352

7. (a)

Results of Runge-Kutta for the Predator-Prey System.

t_k	R_k	F_k
0	1.000000	1.00000
1	1.504120	1.91806
2	0.641301	2.48192
3	0.416812	1.62154
4	0.636774	1.08434

(b)

(c)

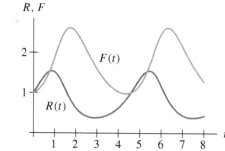

Hints and Answers for Section 7.4

1. Using single precision, we obtain $y(1) \approx 0.941274$. To make sure you are on the right track: $y_2 \approx 0.941885$, $y_4 \approx 0.924849$, $y_8 \approx 0.94101$

3. Using single precision, we obtain $y(2) \approx 1.25938$. To make sure you are on the right track: $y_2 \approx 1.331857$, $y_4 \approx 1.256846$, $y_8 \approx 1.259110$

Hints and Answers for Review Exercises in Chapter 7

1. True   **3.** False   **5.** False

7. (a)

(b)

(c)

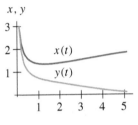

Hints and Answers for Section 8.1

1. The orbit is eventually fixed.

3. The orbit tends to infinity.

5. The orbit is periodic of period 2.

7. The orbit is eventually fixed.

9. Fixed points: $x = 1$
Period 2 points: all real numbers except $x = 1$

11. There are no fixed points or period 2 points.

13. Fixed points: $x = 0$
Period 2 points: none

15. For fixed points, $x = 0, -3$. For periodic points of period 2, $x = (-1 \pm \sqrt{5})/2$.

17. There is one point which is fixed and there are no periodic points of period 2.

19. There is one fixed point $x = 0$ and every other point is a periodic point of period two.

21. There is one fixed point $x = 2$ and no other periodic points.

23. The orbit of any real number is a periodic orbit of period 2, except $x = 2$, which is fixed.

25. It is eventually fixed.

27. It is periodic with period 3.

29. It is eventually fixed.

31. The orbit is periodic with period 5.

33. It is a fixed point.

35. It is eventually fixed.

37. For $c = 1/4$, there is one fixed point. For $c < 1/4$, there are two fixed points. For $c > 1/4$, there are no fixed points.

39. For $c \geq -3/4$, there are no periodic points of period 2, and for $c < -3/4$, there are two periodic points of period 2.

Hints and Answers for Section 8.2

1. The fixed points $x = 0, 3$ are both repelling.

3. $x = 0$ is attracting.

5. $x = 0$ is attracting.

7. $x = 0$ is repelling, and $x = \pm\pi/2$ are attracting.

9. $x = \pm 1$ are neutral.

11. There are no fixed points.

13. The period is 2 and the orbit is attracting.

15. The period is 5 and the cycle is repelling.

17. The period is 2 and the cycle is neutral.

19. The fixed points are $x = \pm 1$ and they are neutral.

21. The fixed point is $x = 0$ and it is repelling.

23. The fixed point is $x = 0$ and it is repelling.

25. The fixed point is attracting.

27. There are infinitely many fixed points and they are all repelling.

29. All cycles are repelling. Note that $|(T^n)'(x_0)| = 4^n$.

Hints and Answers for Section 8.3

1. The bifurcation is a tangent bifurcation. For small enough $-\alpha$, the fixed point $x = \sqrt{-\alpha}$ is repelling and the other fixed point $x = -\sqrt{-\alpha}$ is attracting. For $\alpha = 0$, $x = 0$ is neutral.

3. This is a pitchfork bifurcation. For α slightly smaller than or equal to 1, the fixed point is attracting. For $\alpha > 1$, two more fixed points appear and they are attracting. The origin becomes a repelling fixed point.

5. This is a tangent bifurcation. There is one neutral fixed point for $\alpha = -1/4$, and there are two fixed points for α slightly larger than $-1/4$. One is repelling and the other is attracting.

7. The bifurcation is none of the above. If $\alpha = 0$, everything maps to zero, which is a fixed point. If $0 < |\alpha| < 1$, then the origin is an attracting fixed point and $x = (1 - \alpha)/\alpha$ is a repelling fixed point.

9. If $\mu < 1$, the origin is the only fixed point. At $\mu = 1$, every point satisfying $x \leq 1/2$ is a fixed point. If $\mu > 1$, there are two fixed points.

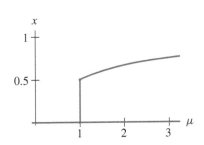

11. *Hint*: The function $F_c(x)$ is a parabola with minimum at $x = 0$ and $F_c(0) = c$. The function $F_c^2(x)$ is a quartic with local maximum at $x = 0$. Try looking at $F_c^2(x)$ upside down near $x = 0$.

Hints and Answers for Section 8.4

1.

Histogram for seed .3

Time series for seed .3

Histogram for seed .3001

$(L_4)_n$

Time series for seed .3001

3. $T^n(p/2^n) = 0$

5. Each of the intervals $[i/2^n, (i + 1)/2^n)$ for $i = 0, 1, 2, \ldots, 2^n - 2$ contains a fixed point of T^n.

7. The eventually fixed points have the form $p/2^n$ for any integer p between zero and $2^n - 1$.

9. T^n has $2^n - 1$ fixed points.

11. List all possible finite strings of zeros and ones, then concatenate them to form the binary expansion of x_0.

13. Note that T^n has a fixed point in each interval of the form $[i/2^n, (i + 1)/2^n]$, so there are 2^n fixed points for T^n.

Hints and Answers for Review Exercises in Chapter 8

1. The point $x = 0$ is attracting, and the points $x = \pm\sqrt{3/2}$ are both repelling.

3. There is one cycle of period two, $x = 1/3, 2/3$.

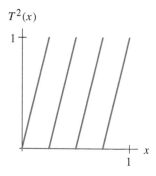

5. Pitchfork bifurcation

7. All orbits are attracted to the fixed point at 0.

9. Every orbit tends to the attracting fixed point at $0.73908\ldots$.

11. False

13. False

15. False

17. All points of the form $x = .abababab\ldots$ where $a \neq b$ are digits from 0 to 9 are periodic points of period two.

19. *Hint*: Compare to the logistic function $L_4(x)$

Hints and Answers for Appendix A

1. $du/dt = u^2 + u$

3. $du/dt = u/t + t(u + u^2 + \cos u)$

5. If $u = y - t$, the equation becomes
$du/dt = u^2 - u - 2$.

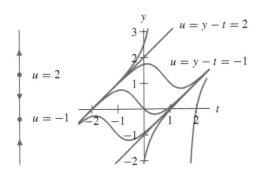

7. If $u = ty$, the equation becomes $du/dt = u \cos u$.

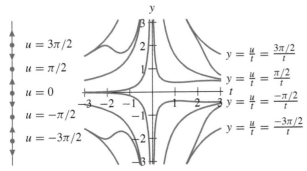

9. $y(t) = t^3(1 + t)/4 + c/t + c$

11. $dy/dt = -y^2 + 2ty + y - t^2 - t + 1$

13. $dy/dt = 2y(1 - \sqrt{y})$

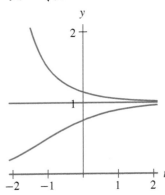

15. (a) $dS/dt = 3 - (3S)/(10 + 3t)$
(b) $dC/dt = (3 - 6C)/(10 + 3t)$
(c) $C = 1/2$
(d) $C(5) = 0.42$

17. (a) $y = \sqrt[-3]{10}$
(b) With $u = y - 10^{-1/3}$, the linear approximation is $du/dy = 3\sqrt[3]{10}\,u$. The origin is a source.
(c) The solution grows like $e^{3\sqrt[3]{10}\,t}$. The time necessary to double the distance from the equilibrium point is $t = (\ln 2)/(3\sqrt[3]{10})$.

19. (a) $y = -1$ and $y = 3$
(b) At $y = -1$, the linearization is $du/dt = 4u$. At $y = 3$, the linearization is $dv/dt = -4v$.
(c) In both cases, $t = (\ln 2)/4$.

21. (a) $du/dt = f(u + y_0)$
(b) Using the Taylor series about $u = 0$, we obtain $f(y) = f(y_0) + f'(y_0)\,u + \dots$. The linearization is $du/dt = f'(y_0)\,u$.

23. $y(t) = \dfrac{\pm 1}{\sqrt{ke^{-2t} - 1}}$ where $k > 0$ along with the equilibrium solution $y(t) = 0$ for all t

25. $y(t) = \left(\tfrac{3}{4}t + ct^{-3}\right)^{-1/3}$ along with the equilibrium solution $y(t) = 0$ for $t \neq 0$

27. $y(t) = \dfrac{(2+k)t + t^3}{t^2 + k}$

29. $y(t) = \dfrac{4}{2t^2 + 2t + 1 + ke^{2t}} + \sin t$ as well as

$\quad y(t) = \sin t$

31. *Hint*: Use

$$\frac{du}{dt} = -\frac{1}{(y - y_1)^2}\left(\frac{dy}{dt} - \frac{dy_1}{dt}\right),$$

substitute for dy/dt and dy_1/dt, and then simplify.

33. $\dfrac{dy}{dt} = (t^6 + 4t^4 + 3t^2 + 2t - 2) +$

$$(1 - 4t^2 - 2t^4)y + t^2 y^2$$

Hints and Answers for Appendix B

1. $y(t) = a_0 \left(1 + t + t^2/2! + t^3/3! + t^4/4! + \dots\right).$

This series is the power series for $a_0 e^t$.

3. $y(t) = a_0 \left(1 - t^2 + t^4/2 \mp \dots\right).$

This series is the power series for $a_0 e^{-t^2}$.

5. $y(t) = a_0 + (-a_0 + 1)t + ((a_0 + 1)/2!)t^2 +$
$\quad ((-a_0 + 3)/3!)t^3 + ((a_0 + 5)/4!)t^4 + \dots.$

7. $y(t) = a_0 + a_1 t + (1/2 - a_0)t^2$
$\quad - (a_1/3)t^3 + (a_0/6 - 1/8)t^4 + \dots.$

9. $\tan t = t + \frac{1}{3}t^3 + \frac{2}{15}t^5 + \dots$

11. $y(t) = a_0 + a_1 t - \frac{1}{2}(a_0 + a_1)t^2 + \frac{1}{6}a_0 t^3 +$
$\quad \frac{1}{24}a_1 t^4 - \frac{1}{120}(a_0 + a_1)t^5 + \frac{1}{720}a_0 t^6 + \dots$

13. $y(t) = a_0 + a_1 t + \frac{1}{2}(1 - a_0)t^2 - \frac{1}{3}(a_1 + 1)t^3 +$
$\quad \frac{1}{24}(3a_0 + 1)t^4 + \frac{1}{15}a_1 t^5 + \frac{1}{720}(11 - 15a_0)t^6 + \dots$

15. (a) $a_2 = \dfrac{-v(v+1)}{2}a_0,$

$\quad a_3 = \dfrac{2 - v(v+1)}{2}a_1,$

$\quad a_4 = \left(\dfrac{6 - v(v+1)}{12}\right)\left(\dfrac{-v(v+1)}{2}\right)a_0$

(b) *Hint*: Note that a_{2n} has a_0 as a factor and a_{2n+1} has a_1 as a factor. Also note that if $v = n$ is a positive integer, then $a_{n+2} = 0$.

(c) *Hint*: Use the formulas from part (a).

(d) $P_3(t) = t - \frac{5}{3}t^3,$

$\quad P_4(t) = 1 - 10t^2 + \frac{35}{3}t^4,$

$\quad P_5(t) = t - \frac{14}{3}t^3 + \frac{21}{5}t^5,$

$\quad P_6(t) = 1 - 21t^2 + 63t^4 - \frac{231}{5}t^6$

(e) *Hint*: Use linearity.

17. $y(t) = t - t^2 + t^3/2 - t^4/6 \pm \dots$

INDEX

Air resistance, 4
Alligood, Kathleen, 353
Aly, 18
Analytic methods, 14, 61
    systems, 161, 189
Angular frequency, 305
Annihilation parameter, 677
Apollo 13, 672
Asymptotic stability, 84
Attractor
    Lorenz, 716
    strange, 716
Autonomous equations, 25, 40, 74, 94
Autonomous systems, 177

Baby bottle, 385
Beating, 417
Bernoulli equation, 735
Beth, 18
Bifurcation, 94–105
    creating equilibrium points, 96, 99
    definition, 95
    discrete dynamical systems, 698
    linear systems, 354, 381
    period-doubling, 700
    pitchfork, 699
    tangent, 698
    value, 98, 101
Bifurcation diagram, 98
Biography
    Alligood, Kathleen, 353
    Curry, James H., 199

Kopell, Nancy, 38
Lorenz, Edward N., 219
McKenna, Joseph, 446
Smale, Stephen, 83
Xia, Zhihong, 501
Zeeman, Mary Lou, 105
Boat, v
Bob, 492, *See* Boat.
    I Am Legend, 215
Boston, 93
    Braves, 1914
Braun, M., 150, 439
Bridges
    Jeff, 825
    Millennium, 423
    Tacoma Narrows, 439, 454

Carbon dating, 16
Carrying capacity, 9–11
Cell phones, 19, 738
Center, 305, 307
Changing variables, 461, 730
    u-substitution, 730
Chaos, 222, 538, 549, 706, 707
    sensitive dependence on initial conditions,
    709, 712
Characteristic polynomial, 271
    three-dimensional matrix, 364, 367
Checking solutions, 22, 189
Chemical reactions, 166, 489
Chopping function, 710
Circuit theory, *See* RC and RLC circuits.

Cobweb, 689
Coefficient matrix, 244
    three-dimensional, 360
Competing species system, 175, 230, 459, 478
Complex numbers, 297, 750
Compound interest, 29, 123
Computers, 634, 648, 659
Conservation of energy, 498
Conserved quantity, 492, 494
Constant-coefficient equations, 331
Convolution, 611, 631
Cool picture, 160
Cooling model, 35, 139
Cooperative species system, 230, 489
Critical loci, 353
Critically damped harmonic oscillator, 337
Curry, James H., 199
Cycles, 680, 685, 689–696
    classification, 695

D'Ancona, U., 13
Damped harmonic oscillator, 183
Damping, 184
Damping coefficient, 184
Decoupled systems, 191
Delayed logistic equation, 724
Dempster, J. P., 150
Dense subset, 711
Dependent variables, 3–5, 21
    systems, 170
Determinant, 248, 348
    three-dimensional matrix, 363
    Wronskian, 264
Dick, See Glen.
Difference equation, See Discrete dynamical system.
Dirac delta function, 603
Direction field, 169
Discontinuous forcing, 578
Discrete dynamical system, 640, 676
    bifurcation, 698–702
        logistic difference equation, 702
        period-doubling, 700
        pitchfork, 699
        tangent, 698
    cycles, 680, 685, 689–696

        attracting, 695
        classification, 695
        neutral, 695
        repelling, 695
    eventually fixed points, 679
    eventually periodic points, 680
    fixed points, 679, 689–696
        attracting, 691
        classification, 692
        finding, 683
        neutral, 691, 695
        repelling, 691
    graphical iteration, 689
    histograms, 682
    Newton's method, 723
    orbits, 679
    period-doubling route to chaos, 704
    periodic orbits, 680
    renormalization, 705
    seed, 679
    time series, 682
    web diagram, 689
Discrete logistic equation, See Logistic difference equation.
Dissipative systems, 508–524
Dottie number, 684
Duffing equation
    periodically forced, 538
Dynamical system
    discrete time, 676

Eigenvalue, 269, 270
    complex, 296
    real distinct, 274
    repeated, 315, 316
    three-dimensional matrix, 363
    zero, 325
Eigenvector, 269, 272
    three-dimensional matrix, 363
Electric circuit theory, See RC and RLC circuits.
Energy, 495, 498
    dissipation, 511
Equilibrium points, 5, 68, 76
    bifurcation, 100–102, 354
    center, 305, 307

classification, 84, 290, 468
creation, 99
Hamiltonian systems, 502–503
linear systems, 246–248
linearization, 85, 735
node, 84
saddle, 280, 368, 469
sink, 84, 285, 368, 469
source, 84, 289, 368, 469
spiral saddle, 366
spiral sink, 305, 469
spiral source, 305, 469
stable, 84, 291
systems, 154, 173
unstable, 291
Equilibrium solutions, 5, 11, 68
systems, 150, 173
Eric, *See* Initial value.
Euler's formula, 299, 752
Euler's method, 52–59, 647
errors, 60, 70, 634
first step, 636
second step, 639
total, 641
formula, 54
improved, 647–653
order, 643
step size, 52, 635
systems, 196–201
Eventually fixed points, 679
Eventually periodic points, 680
Existence and Uniqueness Theorem
applications, 68–71
first-order equations, 66
systems, 205
Existence Theorem, 64
Explicit solutions, 13, 24
Exponential growth, *See* Population models., 568, 676
Extendability of solutions, 64, 80
Extended Linearity Principle, 114, 390
External forcing, 389
Extinction, 102
Extra firm mattress, 165

Fancy word, 241

Finite arithmetic, 665–668
First-order differential equation, 5, 21
autonomous, 25
Bernoulli equation, 735
bifurcation, 94
changing variables, 730
Euler's method, 52
existence, 66
linear, 110
one-parameter family, 95
power series, 742
Riccati equation, 737
separable, 21
slope fields, 36
uniqueness, 66
First-order numerical method, 643
Fixed points, 679, 689–696
attracting, 691
classification, 692
finding, 683
neutral, 691, 695
Poincaré return map, 542
repelling, 691
Floating cube, 447
Food chain model, 531
Forced equation, 389
Forced harmonic oscillator, 388
beating, 417
discontinuous forcing, 592
glass harmonica, 436
impulse forcing, 601, 604
phase angle, 427
phase portrait, 407
resonance, 421
sinusoidal forcing, 403, 419, 427, 595
undamped, 415
Forced response, 393
Fourier transform, 566
Frequency
angular, 305
natural, 305
WEEI, 305

General solution, 7, 24
systems, 191

Gib, 208
Gift, 302, 331
Glass armonica, *See* glass harmonica.
Glass harmonica, 436
Glen, *See* Dick.
Glider, 562
Gradient systems, 519
    definition, 522
    lobster motion, 519
    Lyapunov functions, 519
Graphical iteration, 689
Gravitational force, 495
Growth-rate coefficient, 5
Guess-and-test method, 185, 394
    ultimate guess, 742

Half-life, 15
Hamilton, William R., 494
Hamiltonian function, 494
    construction, 499
Hamiltonian systems, 490–503
    conserved quantity, 492
    definition, 494
    energy, 495
    equilibrium points, 502
    Hamiltonian function, 494
    ideal pendulum, 496
    phase portraits, 494
    undamped harmonic oscillator, 494
Hammer strike, 427, 601
Harmonic oscillator, 156, 158, 240, 256, 275, 321,
        383, 385
    as Hamiltonian system, 494
    classification, 336
    clock, 346
    critically damped, 337, 340–341
    damped, 183, 513
    damping coefficient, 184
    forced, *See* Forced harmonic oscillator.
    hard spring, 558
    modified damping, 232
    natural frequency, 336
    overdamped, 337, 339–340
    resonance, 421

    sinusoidal forcing, 403, 427
    soft spring, 558
    spring constant, 158
    trace-determinant plane, 352
    undamped, 307, 336
    underdamped, 337–338
Harry, 208
Harvesting, 144
Heaviside function, 579
    derivative of, 608
Heaviside, Oliver, 579
Hermite equation, 745
Hermite polynomials, 745
Hey, Jude, 426
Histograms, 682
Homogeneous equations, 112, 331, 389, 740
Hooke's law of springs, 157

Ideal pendulum, 496
Identity matrix, 270
Implicit solutions, 28
Improved Euler's method, 647–653
    cost, 652
    definition, 650
    errors, 650
    order, 652
Impulse forcing, 601
Independent variable, 3–5, 21
Initial condition, 6, 23, 152
    discrete dynamical system, *See* seed.
Initial point, 153
Initial-value problem, 7, 23–24
    system, 159
Integral equation, 648
Integral transform, 566
Integrating factor, 126
Internet, 738
Inverse Laplace transform, 572
Iteration, 676, 679
    graphical, 689
    notation, 685

Jacobian matrix, 466

Kopell, Nancy, 38

Laplace transform, 566
    convolution, 611, 631
    definition, 567
    Dirac delta function, 603
    discontinuous functions, 578
    exponential functions, 568
    first-order equations, 571
    Heaviside function, 579
    inverse, 572
    linearity, 570
    periodic functions, 586
    poles, 620, 630
    product rule, 609
    qualitative theory, 618
    ramp function, 585
    rules, 626
    sawtooth function, 587
    second-order equations, 587
    square wave, 587
    table, 626
    trigonometric functions, 588
Lazer, A. C., 439
Learning model, 17, 142
Legendre equation, 749
Legendre polynomials, 749
Levy, Matthys, 439
Linear approximation, *See* Linearization.
Linear combinations, 249
Linear equations, 110
    integrating factor, 126
    Linearity Principle, 112
      Extended, 114
    nonhomogeneous, 112, 389
    second-order, 331
Linear independence, 255
    in three dimensions, 361
Linear systems
    angular frequency, 305
    bifurcation, 354, 381
    center, 305
    characteristic polynomial, 271
    classification, 290
      three-dimensional, 367
    coefficient matrix, 244
    complex eigenvalues, 296

    constant coefficients, 243
    critical loci, 353
    determinant, 248
    dimension, 246
    eigenvalues, 269
    eigenvectors, 269
    equilibrium points, 246–248
    general solution, 256, 274, 298, 319
    harmonic oscillator, *See* Harmonic oscillator.
    Linearity Principle, 249
      Extended, 390
    linearly independent solutions, 256
    matrix notation, 243
    natural frequency, 305, 336
    natural period, 305
    Principle of Superposition, *See* Linearity Principle.
    real distinct eigenvalues, 274
    repeated eigenvalues, 315, 324
    saddle, 280, 368
    sink, 285, 368
    source, 289, 368
    spiral sink, 305
    spiral source, 305
    straight-line solutions, 273
    three dimensions, 360–371
    trace-determinant plane, 348
    Wronskian, 264
    zero eigenvalue, 325
Linearity Principle, 112, 249
    Extended, 114, 390
    three-dimensional systems, 361
Linearization, 85, 459, 464, 733
    failure, 470–471
Linearly independent solutions, 256
Logistic difference equation, 677
    annihilation parameter, 677
    bifurcation, 702
    chaotic example, 708
    delayed logistic, 724
    fixed points, 693
    without chaos, 707
Logistic equation, 9–11, 146, 733
    harvesting, 102
    modified version, 87

    predator-prey system, 155
    qualitative analysis, 10
Long-term behavior, 11, 14
Lorenz equations, 218, 374, 534, 715–720
    attractor, 716
    Poincaré return map, 718
    template, 717
Lorenz, Edward N., 219
Lori, *See* Gib and Harry.
Lost in space, 672
Lucky guess method, 116, 185, 331, 335
Lyapunov function, 513
    damped harmonic oscillator, 513
    nonlinear pendulum, 511
Lyapunov, Aleksandr M., 513

Magic fingers, 346
Magnetorheological fluid, *See* MR fluid.
Maple, 110
Mass-spring, *See* Harmonic oscillator.
Mathematica, 110
Matrices
    characteristic polynomial, 271
    degenerate, 248
    determinant, 348
    diagonal, 364
    eigenvalue, 269
    eigenvector, 269
    identity, 270
    Jacobian, 466
    singular, 248
    symmetric, 279
    trace, 348
    upper triangular, 279
Matrix notation, 243
McKenna, Joseph, 439, 446
Memorization model, 142
Metaphor of the parking lot, 172
Metaphor of the rope, 75
Millennium Bridge, 423
Miracle, 735
Miracle, another, 738
Missing solutions, 27
Mixing problems, 31–33, 43, 129
Modeling, 2–14

Models
    baby bottle, 385
    bridges
        Millennium, 423
        Tacoma Narrows, 439, 454
    carbon dating, 16
    chemical reactions, 166, 489
    cholesterol, 35
    circuit, *See* RC and RLC circuits.
    clock, 346
    coffee and tea cafés, 241
    competing species, 175, 230, 459, 478
    compound interest, 29, 123
    cooperative species, 230, 489
    delayed logistic, 724
    electric circuits, *See* RC and RLC circuits.
    exponential growth, 676
    extinction, 102
    extra firm mattress, 165
    food chain, 531
    forced harmonic oscillator, 388
    glass harmonica, 436
    glider, 562
    hammer strike, 427, 601
    harmonic oscillator, *See* Harmonic oscillator.
    harvesting, 102, 144
    housing market, 260
    ideal pendulum, 496
    learning, 17, 142
    lobster motion, 519
    logistic population model, *See* Logistic equation
        and Logistic difference equation.
    measuring mass in space, 383
    memorization, 142
    Millennium Bridge, 423
    mixing problems, 31–33, 43, 129
    modified logistic, 87
    mold growth, 143
    MR fluid, 236, 432
    Newton's law of cooling, 35, 139
    oil production, 147
    opera singer, 388, 426, 436
    pendulum, 495, 508
    population, *See* Population models.
    predator-prey, *See* Predator-prey system.

radioactive decay, 15
RC circuit, 44, 59–60, 574, 583
RLC circuit, 381, 397, 453
rubber band, 234, 454
saving, 29
shock absorbers, 345
    active, 236
state populations, 146
suspension bridge, 423, 439, 454
suspension system, 345
Tacoma Narrows Bridge, 439, 454
trolley cars, 93
truck seats, 236
tuned-mass damper, 515
Volterra-Lotka, 488
washing machines, 432
weather, 218
Moldy bread, 143
Mozart
    The Marriage of Figaro, 493
MR fluid, 236, 432

Natural frequency, 305, 336
Natural period, 305
Natural response, 393
Newton's law of cooling, 35, 139
Newton's method, 723
Newton's second law of motion, 157, 388, 496
Node
    first-order equation, 84
Nonhomogeneous equations, 112, 389
Nonlinear equation, 10
Nullcline, 478, 482–485
Numerical integration, 648
    Riemann sums, 648
    Simpson's rule, 656
    trapezoid rule, 648, 656
Numerical methods, 14, 61
    errors, 60, 70, 634, 650, 659
    Euler's method, 52–59, 196, 634
    finite arithmetic, 665–668
    improved Euler's method, 650
    in practice, 668
    order, 643, 647, 652
    round-off error, *See* Finite arithmetic.

Runge-Kutta, 655–664
    Mathematica program, 661
    C program, 662
    order, 659
    TI calculator program, 660
systems, 161, 196, 663

Oil Production, 147
Old joke, 2
Opera singer, 388, 426, 436
Orbit, 679
Order
    differential equation
        first-order, 5, 21
        second-order, 158
    numerical method, 643, 647
        Euler's method, 643
        Improved Euler's method, 652
        Runge-Kutta, 659
Ordinary differential equation, 5
Overdamped harmonic oscillator, 337

Palin, Sarah, 323
Parameterized family of differential equations, 94
Parameters, 3–5, 21
    bifurcation value, 98
    family of equations, 95
Parking lot, 172
Particular solution
    initial-value problem, 7
    nonhomogeneous equations, 390
Pendulum, 495
    damped, 463, 508
    higher order approximation of, 559
    ideal, 496
    periodically forced, 546
        Poincaré return map, 546
Period, 153
    forcing, 403
    natural, 305
Period-doubling bifurcation, 700
Period-doubling route to chaos, 704
Periodic solution, 152
Phase angle, 427

Phase line, 75
    equilibrium points, 76, 82
    metaphor of the rope, 75
    sketching solutions, 79
Phase plane, 154
    metaphor of the parking lot, 172
Phase portrait, 154
Phase shift, 407, 431
Phase space
    three-dimensional, 362
Phasor, 414
Pitchfork bifurcation, 699
Poincaré return map, 539
    fixed point, 542
    Lorenz equations, 718
Poincaré, Henri, 218, 540, 549
Poles of Laplace transform, 620, 630
Population models
    carrying capacity, 9–11
    competing species, 175, 230, 459, 478
    cooperative species, 230, 489
    delayed logistic, 724
    exponential growth, 4, 146, 676
    growth-rate coefficient, 5
    harvesting, 102, 144
    logistic equation, *See* Logistic equation and
        Logistic difference equation.
    modified logistic, 87
    predator-prey, *See* Predator-prey system.
      modified, 155
    state populations, 146
    three species, 531
    U.S. population, 8
    Volterra-Lotka, 488
Power series methods, 742
Predator-prey system, 12–13, 150–156
    modified, 155
    pesticides, 164
    Runge-Kutta method, 664
    three species, 531

Qualitative methods, 6, 14, 61
    systems, 161, 177

Radioactive decay, 15

Ramp function, 585
RC circuit, 44, 59–60, 574, 583
    constant input voltage, 45
    flashing light, 46
    on-off input voltage, 46
    zero input voltage, 44
Red Sox, 305, 1903, 1912, 1915, 1916, 1918,
    2004, 2007
Relative growth rate, 19
Renormalization, 705
Resonance, 421
    Millennium Bridge, 423
Restoring force, 157
Return map, *See* Poincaré return map.
Riccati equation, 737
Riemann sums, 648
RLC circuit, 381, 397, 453
Road kill, 4
Robo-lobster, 519
Rocky and Bullwinkle, 4
Rope, 75
Round-off error, *See* Finite arithmetic.
Rubber band, 234, 454
Runge-Kutta method, 655–664
    Mathematica program, 661
    C program, 662
    definition, 656
    error, 659
    finite arithmetic, 665
    order, 659
    step size, 659
    systems, 663
    TI calculator program, 660

Saddle, 280, 283, 368
    connection, 493
    nonlinear, 469
    separatrix, 293, 469
Saddle node bifurcation, *See* Tangent.
Salvadori, Mario, 439
Saving model, 29
Sawtooth function, 587
Second-order differential equation, 158
    forced, 389
    Hermite Equation, 745

homogeneous, 389
Legendre equation, 749
linear, 331
nonhomogeneous, 389
power series, 745
Second-order numerical methods, 652
Seed, 679
Sensitive dependence on initial conditions, 709, 712
Separable equations, 24–29
   missing solutions, 27–28
Separation of variables, 21–33
Separatrix, 293, 469
   stable, 469
   unstable, 469
Shock absorbers, *See* washing machines., 345
   active, 236
Simple harmonic motion, 337
Simpson's rule, 656
Singular matrix, 248
Sink
   first-order equation, 84
   linear systems, 285, 368
   nonlinear, 469
   spiral, 305, 469
Sinusoidal forcing, 403, 419, 427, 595
SIR model, 210
   nullcline, 212
   threshold value, 212
Skyscraper
   tuned-mass damper, 515
Slope fields, 36–43
   sketching, 37
   special cases
      autonomous equations, 40
      $dy/dt = f(t)$, 39–40
Smale, Stephen, 83
Sock absorbers, *See* washing machines.
Solution, 6–7, 21
   asymptotic, 80
   checking, 22, 189
   domain of definition, 64, 80
   existence and uniqueness, 63
   general, 7, 24, 191
   implicit, 28
   initial-value problem, 7, 23–24, 159, 190

linearly independent, 361
missing, 27–28
periodic, 152, 305
separatrix, 293
system, 13, 152
Solution curve, 153
Source
   first-order equation, 84
   linear systems, 289, 368
   nonlinear, 469
   spiral, 305, 469
Speed bumps, 427
Spiral saddle, 366
Spiral sink, 305, 469
Spiral source, 305, 469
Spring constant, 158
Springs, 157, *See* Harmonic oscillator.
   hard, 558
   soft, 558
Square wave, 587
Stable equilibrium point, 84, 291
Stable separatrix, 469
Steady-state response, 393
Steady-state solution, 118
Step function, *See* Heaviside function.
Step size, 52
Straight-line solutions, 273
Superposition, Principle of, *See* Linearity Principle.
Suspension
   automobile, 183
Suspension bridge, 423, 439, 454
   Lazer-McKenna light flexible long span, 447
Suspension system model, 345
Sustainability, 102
Systems, 150
   autonomous, 177
   checking solutions, 189
   coupled, 13
   decoupled, 191
   dimension, 246
   direction field, 169
   Euler's method, 196
   Existence and Uniqueness Theorem, 205
   first-order, 13
   general solution, 191

linear, 243
Metaphor of the parking lot, 172
nullcline, 478
phase plane, 154
phase portrait, 154
Runge-Kutta method, 663
three-dimensional, 360–371
vector field, 168

Tacoma Narrows Bridge, 439–446, 454
Tangent bifurcation, 698
Taylor polynomial, 465
Taylor's Theorem, 463, 637, 647
Template
    Lorenz, 717
Tent function, 722
Time constant, 16, 45
Time-series, 682
Trace, 348
Trace-determinant plane, 348
    critical loci, 353
Trapezoid rule, 648, 656
Trolley cars, 93
Truck seats, 236
Tuned-mass damper, 515
TV, 738

u-substitution, 27, 730
Undamped harmonic oscillator, 336
Underdamped harmonic oscillator, 337
Undetermined coefficients, method of, 116, 394
Unforced equation, 389
Uniqueness Theorem
    first-order equations
        applications, 68–71
    systems, 205
Unstable equilibrium point, 291, 305
Unstable separatrix, 469

Van der Pol equation, 198, 458
Vector field, 168, 170
Volterra, V., 13
Volterra-Lotka system, 488

Washing machines, 432
Weather, 218, 545
Web diagram, 689
WEEI, 305
Wronskian, 264

Xia, Zhihong, 501

Zeeman, Mary Lou, 105
Zombie, 215